Edited by
Dirk M. Guldi and
Nazario Martín

Carbon Nanotubes and
Related Structures

Further Reading

Martín, N., Giacalone, F. (eds.)

Fullerene Polymers

Synthesis, Properties and Applications

2009

ISBN: 978-3-527-32282-4

Hierold, C. (ed.)

Carbon Nanotube Devices

Properties, Modeling, Integration and Applications

2008

ISBN: 978-3-527-31720-2

Yudin, A.K. (ed.)

Catalyzed Carbon-Heteroatom Bond Formation

2010

ISBN: 978-3-527-32428-6

Kim, D.M.

Introductory Quantum Mechanics for Semiconductor Nanotechnology

2010

ISBN: 978-3-527-40975-4

Krüger, A.

Carbon Materials and Nanotechnology

2010

ISBN: 978-3-527-31803-2

Jorio, A.

Raman Spectroscopy in Nanoscience and Nanometrology

Carbon Nanotubes, Nanographite and Graphene

ISBN: 978-3-527-40811-5

Mittal, V. (ed.)

Optimization of Polymer Nanocomposite Properties

2010

ISBN: 978-3-527-32521-4

Monthioux, M.

Meta-Nanotubes - Modification of Carbon Nanotubes

2010

ISBN: 978-0-470-51282-1

Kumar, C. S. S. R. (ed.)

Nanomaterials for the Life Sciences

10 Volume Set

ISBN: 978-3-527-32261-9

Kumar, C. S. S. R. (ed.)

Nanotechnologies for the Life Sciences

10 Volume Set

ISBN: 978-3-527-31301-3

Edited by
Dirk M. Guldi and Nazario Martín

Carbon Nanotubes and Related Structures

Synthesis, Characterization, Functionalization,
and Applications

WILEY-VCH Verlag GmbH & Co. KGaA

The Editors

Prof. Dr. Dirk M. Guldi
Friedrich-Alexander-Universität
Department of Chemistry & Pharmacy
Egerlandstr. 3
91058 Erlangen
Germany

Prof. Dr. Nazario Martín
Universidad Complutense
Dept. de Quimica Organica
28040 Madrid
Spain

All books published by Wiley-VCH are carefully produced. Nevertheless, authors, editors, and publisher do not warrant the information contained in these books, including this book, to be free of errors. Readers are advised to keep in mind that statements, data, illustrations, procedural details or other items may inadvertently be inaccurate.

Library of Congress Card No.: applied for

British Library Cataloguing-in-Publication Data
A catalogue record for this book is available from the British Library.

Bibliographic information published by the Deutsche Nationalbibliothek
The Deutsche Nationalbibliothek lists this publication in the Deutsche Nationalbibliografie; detailed bibliographic data are available on the Internet at http://dnb.d-nb.de.

© 2010 WILEY-VCH Verlag GmbH & Co. KGaA, Weinheim

Composition   Thomson Digital, Noida, India
Printing & Bookbinding   Strauss GmbH, Mörlenbach
Cover Design   Formgeber, Eppelheim

Printed in the Federal Republic of Germany
Printed on acid-free paper

ISBN: 978-3-527-32406-4

Contents

Carbon Nanotubes and Related Structures. Edited by Dirk M. Guldi and Nazario Martín
Copyright © 2010 WILEY-VCH Verlag GmbH & Co. KGaA, Weinheim
ISBN: 978-3-527-32406-4

Preface

Following Sumio Iijima's groundbreaking discovery of multiwall carbon nanotubes (MWCNTs) in 1991, carbon nanostructures – and in particular carbon nanotubes (CNTs) – have been at the forefront of scientific research in physics, chemistry, materials science, and so on. The discovery of single-wall carbon nanotubes (SWCNTs) in 1993 set yet another milestone in an exponentially growing field. Conceptually, these new nanoforms of carbon allotropes with cylindrical geometry belong to the versatile family of fullerenes. In contrast to fullerenes with their esthetically pleasing spherical shape, CNTs have a length to diameter ratio of up to 28,000 : 1. Some CNTs reach lengths of up to several millimeters. From the very beginning, scientists realized that CNTs are inhomogeneous materials giving rise to different lengths, different diameters, and, most importantly, different structures. The structural features depend, to a large extent, on the means by which individual graphene sheets are wrapped around a chiral vector to form seamless cylinders. According to the specific indices of the chiral vector, *armchair, zigzag,* and *chiral* nanotubes are formed. However, they all exist together in a single sample without being controllable (!). To this end, separation and selective formation of homogeneous CNT samples constitute important challenges in contemporary CNT research. Unless significant breakthroughs are achieved in the foreseeable future, potential applications of these carbon materials will continue to be hampered by the lack of pure, homogeneous, and reproducible samples.

The initial excitement – associated with the pioneering discovery of CNTs – was based on their remarkable properties bearing enormous potential for applications in nanotechnology, materials science, and biomedicine: the outstanding tensile strength of CNTs or the fact that they reveal semiconducting and/or conducting electrical properties – to name a few of the exceptional properties of CNTs. Furthermore, CNTs have been found to efficiently interact with living cells. As a matter of fact, they function as transporters of molecules through cellular membranes and, in turn, have opened new avenues for biomedicine.

Considering the sheer endless interest in these unique compounds, a large number of authoritative reviews and comprehensive books have dealt with these carbon allotropes. It is an understatement to say that research on carbon

Carbon Nanotubes and Related Structures. Edited by Dirk M. Guldi and Nazario Martín
Copyright © 2010 WILEY-VCH Verlag GmbH & Co. KGaA, Weinheim
ISBN: 978-3-527-32406-4

nanostructures is a very active field with an enormous number of original studies published every year by groups from all over the world. This, in fact, constitutes the major thrust of this book, that is, gathering an elite number of world leading scientists from very different disciplines to highlight the recent advances in the area of CNTs and other related carbon nanostructures. Needless to emphasize the interdisciplinary nature of the different contributions – ranging from theory and production to practical applications in the field of solar energy conversion and biomedicine.

This book contains 16 contemporary chapters sharing one goal in common – carbon nanostructures. To this end, aspects associated with the production and formation of CNTs (M.H. Rümmeli, P. Ayala, and T. Pichler) are discussed in the opening chapter of the first part of this book followed by a chapter dedicated to the theory of electronic and optical properties (S.V. Rotkin and S.E. Snyder). Equally important are the perspectives that arose around the electrochemistry (M. Iurlo, M. Marcaccio, and F. Paolucci) and the photophysics (T. Hertel) of CNTs as compelling complements to the former topics. Breakthroughs in the chemistry of CNTs are covered in two matching contributions – one focusing exclusively on the covalent chemistry of CNTs (F. Hauke and A. Hirsch) and the other surveying exhaustively the field of noncovalent chemistry of CNTs (M.A. Herranz and N. Martín).

The objectives of the second part of this book are quite different. Here, properties and potential applications in different areas stand at the forefront. Biological applications and potential toxicity (P. Singh, T. Da Ros, K. Kostarelos, M. Prato, and A. Bianco) are certainly some of them, which have been brought together in one specific chapter. This is rounded out by several critical reviews with regard to the use of CNTs in emerging fields in materials science. Emphasis is placed, for example, on ground- and excited-state charge transfer (V. Sgobba and D.M. Guldi), photovoltaics in general (E. Kymakis), and layer-by-layer assembly (B.S. Shim and N. Kotov). Noteworthy are the additional two sections that describe fundamental issues related to the use of CNTs for catalytic applications (E. Castillejos and P. Serp) and the fascinating function of CNTs as containers (T.W. Chamberlain, M.d.C. Gimenez-Lopez, and A.N. Khlobystov).

As reflected in the multidisciplinary title of this book, other carbon nanostructures –besides just CNTs – that have received similar attention throughout the scientific communities, namely, nanohorns (M. Yudasaka and S. Iijima), nanographenes (W. Pisula, X. Feng, and K. Müllen), and endohedral fullerenes (L. Feng, T. Akasaka, and S. Nagase), are at the focus of the third part of this book. The closing statement belongs, however, to a chapter that deals with the calculations on energetics, thermodynamics, and general stability (Z. Slanina, F. Uhlik, S.-L. Lee, and T. Akasaka).

We – as editors of this book – share the view that all chapters should prove very useful to both students and researchers in different disciplines of carbon research. Simply, the fact that all chapters are solid and offer comprehensive discussions on the properties of carbon nanoforms assists in serving as a constructive tool for specialists in various fields and an updated reference for the broad readership of nonspecialists. In this regard, it is essential that each chapter begins with an excellent introduction to the topic at hand and then turns to more details for the expert in the area.

Finally, we would like to express our sincere gratitude to a unique set of internationally leading authors who accepted our invitation to join this venture and committed their valuable time and efforts to guarantee the success of this book in terms of allowing a better understanding of these carbon-based systems. Moreover, we would be very pleased if this book would turn into a source of inspiration for further adventures on so far nonimagined/nonexplored carbon nanoforms. Likewise, we would like to say thanks to the dedicated Wiley-VCH staff for their continuous support and enthusiasm, especially when informed at an early stage about our keen interest in editing this book.

Dirk M. Guldi
Nazario Martín

List of Contributors

Takeshi Akasaka
University of Tsukuba
Center for Tsukuba Advanced Research
Alliance
1-1-1 Tennodai, Tsukuba
Ibaraki 305-8577
Japan

Paola Ayala
University of Vienna
Faculty of Physics
Strudlhofgasse 4
01090 Wien
Austria

Alberto Bianco
CNRS
Institut de Biologie Moléculaire et
Cellulaire
Laboratoire d'Immunologie et
Chimie Thérapeutiques
67000 Strasbourg
France

Eva Castillejos
Toulouse University
Laboratoire de Chimie de Coordination
UPR CNRS 8241 composante
ENSIACET
4 allée Emile Monso
31432 Toulouse Cedex 4
France

Thomas W. Chamberlain
University of Nottingham
School of Chemistry
UK

Tatiana Da Ros
Università di Trieste
Dipartimento di Scienze Farmaceutiche
34127 Trieste
Italy

Lai Feng
University of Tsukuba
Center for Tsukuba Advanced Research
Alliance
Tsukuba 305-8577
Japan

Xinliang Feng
Max Planck Institute for
Polymer Research
Ackermannweg 10
55128 Mainz
Germany

Maria del Carmen Gimenez-Lopez
University of Nottingham
School of Chemistry
UK

Carbon Nanotubes and Related Structures. Edited by Dirk M. Guldi and Nazario Martín
Copyright © 2010 WILEY-VCH Verlag GmbH & Co. KGaA, Weinheim
ISBN: 978-3-527-32406-4

Dirk M. Guldi
Friedrich-Alexander-Universität
Erlangen-Nürnberg
Department of Chemistry and
Pharmacy & Interdisciplinary Center for
Molecular Materials (ICMM)
91058 Erlangen
Germany

Frank Hauke
University of Erlangen-Nürnberg
Department of Chemistry and
Pharmacy and Institute of Advanced
Materials and Processes (ZMP)
Henkestr. 42
91054 Erlangen
Germany

Ma Ángeles Herranz
Universidad Complutense de Madrid
Departamento de Química Orgánica
Facultad de Ciencias Químicas
28040 Madrid
Spain

Tobias Hertel
University of Würzburg
Institute of Physical and Theoretical
Chemistry
Am Hubland
97074 Würzburg
Germany

Andreas Hirsch
University of Erlangen-Nürnberg
Department of Chemistry and
Pharmacy and Institute of Advanced
Materials and Processes (ZMP)
Henkestr. 42
91054 Erlangen
Germany

Sumio Iijima
National Institute of Advanced
Industrial Science and Technology
Central 5-2, 1-1-1 Higashi
Tsukuba
Ibaraki 305-8565
Japan

Matteo Iurlo
Università di Bologna
Dipartimento di Chimica
Unit of Bologna
INSTM
Via Selmi 2
40126 Bologna
Italy

Andrei N. Khlobystov
University of Nottingham
School of Chemistry
UK

Kostas Kostarelos
University of London
The School of Pharmacy
Centre for Drug Delivery Research
Nanomedicine Lab
29-39 Brunswick Square
London WC1N 1AX
UK

Nicholas A. Kotov
University of Michigan
Departments of Chemical Engineering
Ann Arbor, MI 48109
USA
University of Michigan
Departments of Materials Science and
Engineering
Ann Arbor, MI 48109
USA
University of Michigan
Department of Biomedical Engineering
Ann Arbor, MI 48109
USA

Emmanuel Kymakis
Technological Educational Institute
(TEI) of Crete
Electrical Engineering Department
P.B 1939, Heraklion
71 004 Crete
Greece

Shyi-Long Lee
National Chung-Cheng University
Department of Chemistry and
Biochemistry
Chia-Yi 62117
Taiwan

Massimo Marcaccio
Università di Bologna
Dipartimento di Chimica
Unit of Bologna
INSTM
Via Selmi 2
40126 Bologna
Italy

Nazario Martín
Universidad Complutense de Madrid
Departamento de Química Orgánica
Facultad de Ciencias Químicas
28040 Madrid
Spain

Klaus Müllen
Max Planck Institute for
Polymer Research
Ackermannweg 10
55128 Mainz
Germany

Shigeru Nagase
Institute for Molecular Science
Department of Theoretical Molecular
Science
Myodaiji
Okazaki 444-8585
Japan

Francesco Paolucci
Università di Bologna
Dipartimento di Chimica
Unit of Bologna
INSTM
Via Selmi 2
40126 Bologna
Italy

Thomas Pichler
University of Vienna
Faculty of Physics
Strudlhofgasse 4
01090 Wien
Austria

Wojciech Pisula
Max Planck Institute for
Polymer Research
Ackermannweg 10
55128 Mainz
Germany

Maurizio Prato
Università di Trieste
Dipartimento di Scienze Farmaceutiche
34127 Trieste
Italy

Slava V. Rotkin
Lehigh University
Physics Department
16 Memorial Dr. E.
Bethlehem, PA 18015

and

Lehigh University
Center for Advanced Materials and
Nanotechnology
5 E. Packer Ave.
Bethlehem, PA 18015
USA

Mark H. Rümmeli
IFW-Dresden
01069 Dresden
Germany

Philippe Serp
Toulouse University
Laboratoire de Chimie de Coordination
UPR CNRS 8241 composante
ENSIACET
4 allée Emile Monso
31432 Toulouse Cedex 4
France

Vito Sgobba
Friedrich-Alexander-Universität
Erlangen-Nürnberg
Department of Chemistry and
Pharmacy & Interdisciplinary Center
for Molecular Materials (ICMM)
91058 Erlangen
Germany

Bong Sup Shim
University of Michigan
Departments of Chemical Engineering
Ann Arbor, MI 48109
USA

Prabhpreet Singh
CNRS
Institut de Biologie Moléculaire et
Cellulaire
Laboratoire d'Immunologie et
Chimie Thérapeutiques
67000 Strasbourg
France

Zdeněk Slanina
University of Tsukuba
Center for Tsukuba Advanced Research
Alliance
1-1-1 Tennodai, Tsukuba
Ibaraki 305-8577
Japan

Stacy E. Snyder
Lehigh University
Physics Department
16 Memorial Dr. E.
Bethlehem, PA 18015
USA

Filip Uhlík
Charles University of Prague
Faculty of Science
Department of Physical and
Macromolecular Chemistry
Hlavova 2030/8
128 43 Praha 2
Czech Republic

Masako Yudasaka
National Institute of Advanced
Industrial Science and Technology
Central 5-2, 1-1-1 Higashi
Tsukuba
Ibaraki 305-8565
Japan

1
Carbon Nanotubes and Related Structures: Production and Formation

Mark H. Rümmeli, Paola Ayala, and Thomas Pichler

1.1
Introduction

In 1996, Harry Kroto, Robert Curl, and Richard Smalley were awarded the Nobel prize in Chemistry for the discovery of a spherical molecule composed entirely of carbon atoms (Figure 1.1) [34]. This nanometer-scale structure was named "fullerene" due to its resemblance to the highly symmetric architectonic geodesic domes designed by the architect Richard Buckminster Fuller. In the 1980s and early 1990s extensive research on fullerene theory, synthesis, and its characterization was carried out. In 1991 Iijima [24], presented transmission electron microscopy observations of elongated and concentric layered microtubules made of carbon atoms, which, until then, had mostly been considered as filamentous carbon. This propelled the research related to one of the most actively investigated structures of the last century: nowadays called the carbon nanotubes (CNTs). Despite the ongoing polemic of who should be given credit for the discovery of CNTs [46], it is widely recognized that Iijima's work catapulted carbon nanotubes onto the global scientific stage.

Figure 1.1 shows the ideal molecular models of a C_{60} fullerene and a single-walled carbon nanotube. To understand the structure of a carbon nanotube it can be first imagined as a rolled up sheet of graphene (see Figure 1.2), which is a planar-hexagonal arrangement of carbon atoms distributed in a honeycomb lattice. From a fundamental point of view and for future applications the most noticeable features of single-walled CNTs are their electronic (semiconducting or metallic), mechanical (Young modulus $\sim 1\,TPa$), optical and chemical characteristics. On the other hand, multiwalled nanotubes are a collection of concentric single-walled nanotubes with different diameters, and their properties are very different from their single-walled counterparts. The promise afforded by carbon nanotubes is varied and they often utilizes their unique electrical properties, extraordinary strength, and heat conduction efficiency. Some applications seek to exploit their structural properties. Examples include high tensile strength fibers and fire resistant materials.

Carbon Nanotubes and Related Structures. Edited by Dirk M. Guldi and Nazario Martín
Copyright © 2010 WILEY-VCH Verlag GmbH & Co. KGaA, Weinheim
ISBN: 978-3-527-32406-4

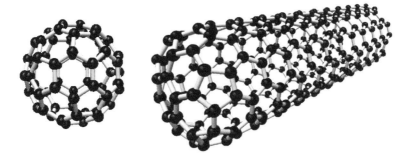

Figure 1.1 Molecular models representing a C$_{60}$ fullerene (left) and a single-walled carbon nanotube caped on one end (right). Fullerenes are closed-cage clusters relatively stable in gas phase, whereas nanotubes are elongated structures that can nowadays reach several micrometers in length but still have very small diameters (in the few nanometers order). The end cap of the nanotube can be thought as an incomplete fullerene.

Other applications utilize their electromagnetic properties, for example, artificial muscles [1]. Loudspeakers created from sheets of parallel carbon nanotubes by researchers at the Tsinghua-Foxconn Nanotechnology Research Center in Beijing exemplify their electroacoustic potential. Chemical applications include air and water filtration. Carbon nanotubes are the fastest known oscillators (>50 GHz) and this highlights their incredible mechanical properties. In addition, carbon nanotubes are highly suited as building blocks for molecular electronics. Single-walled CNT, in particular, are attractive due to their electronic structure (semiconducting or metallic)

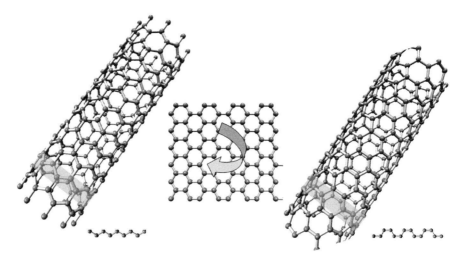

Figure 1.2 A single-walled carbon nanotube can be pictured as a rolled graphene sheet. Nanotubes are classified according to the rolling direction as zigzag, armchair, and chiral. Multiwalled nanotubes can be pictured as a set of concentric single-walled nanotubes differing from them in their properties.

being fully described by the tubes diameter and chirality. Field emission transistors (FETs) require semiconducting CNT and high-speed (ballistic) electrical connectors require metallic CNT. However, inherent difficulties in CNT handling pose serious difficulties in their development. For example, the control required to reliably synthesize SWCNT of a given diameter and chirality is still lacking. One of the more promising routes to overcome this difficulty is to functionalize CNT that then allows one to modify their electronic properties in a homogeneous manner. Further, functionalization extends their properties and consequently their application potential. Hence understanding their production and functionalization is crucial for fully exploiting their potential.

There are several techniques for producing single- and multiwalled nanotubes and all of them have advantages as well as disadvantages. In this direction we briefly introduce as examples two of the most established high temperature techniques, namely arc discharge and laser ablation, as well as chemical vapor deposition (CVD) with its most common variants. The literature regarding the synthesis of multiwalled carbon nanotubes is abundant, and in case of the single-walled nanotubes case also it is abundant, but still far from being fully comprehended. For this reason we encourage the reader interested in methods and the specific parameters involved in synthesis processes to deepen in the related review features, chapters in collected editions, books, and journal papers available in the literature [25, 44, 60]. In the following sections, we will rather focus on the contemporary understanding of the growth mechanisms, actual trends, and challenges in carbon nanotube manufacture.

1.2
Carbon Nanotube Production

The state-of-the-art carbon nanotube production encompasses numerous methods and new routes are continuously being developed. The following sections are devoted to give an overall description of the most established nanotube production routes but more detailed information can be found in several review articles.

1.2.1
Arc Discharge

The carbon arc-discharge method is a high temperature process that can be used for the production of nanotubes as well as fullerenes. In fact, mass production of fullerenes was first achieved using arc discharge with the Krätschmer–Huffman method [33]. The derived product and the yields are mainly dependent on the atmosphere and catalysts utilized. This method is probably one of the simplest methods for synthesizing nanotubes on a large scale. However, the simultaneous production of a multimorphology soot demands several purification steps. In the carbon arc-discharge method, an arc is ignited between two graphite electrodes in a gaseous background (usually argon/hydrogen) [11, 23, 24, 64] (Figure 1.3).

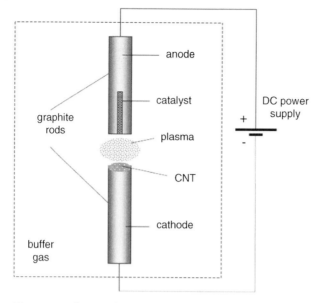

Figure 1.3 Schematic diagram of an arc-discharge system.

The arcing evaporates the carbon and meanwhile it cools and condenses that some of the product forms as filamentous carbon on the cathode. Iijima's work [24] exposed that this filamentous product comprised of multiwalled carbon nanotubes. The optimization of metals being included in the anode led to the growth of single-walled carbon nanotubes [27]. This, in turn, stimulated the successful use of metal catalysts in other techniques for both single-walled and multiwalled carbon nanotube synthesis.

1.2.2
Laser Ablation

Historically the fullerenes synthesized in the pioneering work of R.F. Curl, H.W. Kroto, and R.E. Smalley were produced using a laser evaporation technique [34]. They directed an intense pulse of laser light on a carbon surface in a stream of helium gas. The evaporated material condenses to yield fullerenes. However it was later noticed that the incorporation of a metal catalyst in the carbon target leads to the formation of SWNT with a narrow diameter distribution and high yields [68]. Further, the SWNT yield and diameter distribution can be varied by controlling the process parameters in the reaction [3, 26, 56]. This technique is not suitable for mass production; however the quality and controlled diameters and diameter distributions offered by laser ablation make them choice SWNT samples for fundamental studies (Figure 1.4).

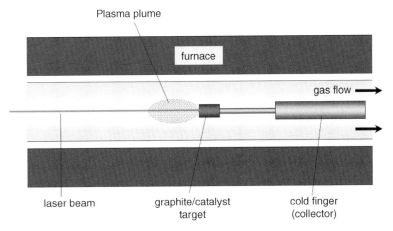

Plasma plume

furnace

gas flow ➡

laser beam   graphite/catalyst   cold finger
                         target   (collector)

Figure 1.4   Schematics of a laser ablation setup

1.2.3
Chemical Vapor Deposition

In principle, chemical vapor deposition can be understood as a chemical process in which volatile precursors are used to provide a carbon feed source to a catalyst particle or pore at elevated temperatures (350–1000 °C). The literature available with chemical vapor deposition methods is really extensive, which also shows how this multivariable process can been adjusted in several manners to render various CVD-based methods such as thermochemical CVD (traditional), plasma enhanced CVD, aerosol assisted (AA-CVD), aerogel supported, high pressure CO disproportionation (HiPCO), alcohol catalytic CVD (AACCVD) in ambient or vacuum [5] base pressure, the CoMoCat process, and even a hybrid laser assisted thermal CVD (LCVD) among others. These various forms of CVD in wide use make it the most common route by which CNTs are formed. The relative ease with which one can set up a CVD system also makes of it the most promising route for the mass production of CNTs (Figure 1.5).

In a CVD reaction the catalyst particles can reside in free space, the so-called floating catalysts, or they can reside on a substrate (supported catalysts). In such process, carbon saturates on the catalyst particles and precipitates in the form of a carbon nanotube and is discussed in much detail later (see Figure 1.9). The catalyst particles can also be formed by the decomposition of a volatile precursor as in the HiPCO method that produces relatively larger batches of high quality single-walled carbon nanotubes. This was developed by Nikolaev *et al.* [48]. Catalysts are formed by the decomposition of a metal containing compound, for example, ferrocene or iron pentacarbonyl in a heated flow of carbon monoxide at pressure between 1 and 10 atm and temperatures between 800 and 1200 °C. Modification of the process parameters provides control of the SWNT diameter distribution and yield.

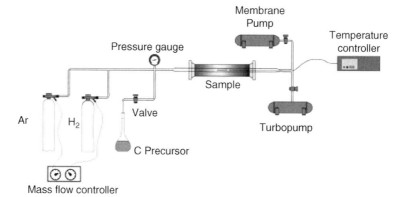

Figure 1.5 Schematic diagram of a chemical vapor deposition system adapted to work in high vacuum conditions, which is only one example of the various CVD-based methods. (Modified from Ref. [5].)

1.2.4
Miscellaneous Synthesis Methods

Apart from the above-mentioned primary synthesis routes there are other variations to these methods and in addition many other synthesis techniques. For example, CNT can be synthesized by dipping hot graphite in cold water [28], ball milling [50], flame synthesis [45], SiC decomposition [35, 37], and graphene scrolling [69] (Figure 1.6).

1.3
Catalysts

The success of catalyst particles in the synthesis of carbon nanotubes is immense. However, it is worth mentioning that carbon nanotubes can be formed without using a catalyst particle [24]. In the case where catalyst particles are used the catalyst particles are considered to play various roles. It is argued the catalyst both nucleates a carbon nanotube and sustains growth. In addition, in CVD, the catalyst is believed to catalytically decompose the carbon feedstock (hydrocarbon).

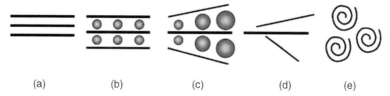

(a)   (b)   (c)   (d)   (e)

Figure 1.6 Schematic of the chemical intercalation/exfoliation process. First graphite (a) is intercalated with potassium metal (b) and then exfoliated using ethanol (c) to form a dispersion of carbon sheets (d). A sonication process produces carbon nanoscrolls (e).

1.3.1
Metallic Catalysts

Historically, the most successful catalysts were transition metals. However, in recent times the successful implementation of nontransition metals and even nonmetals in the synthesis of carbon nanotubes are challenging contemporary concepts in carbon nanotube synthesis.

As a generalization the most commonly used catalysts are Ni, Co, and Fe. These are used independently or mixed with another metal. Ternary catalysts are used often since they usually lead to higher CNT yields. Some examples are, the so-called CoMo Cat CVD synthesis [22], in arc discharge (Rh-Pt) [58] and laser ablation (NiCo) [41]. Rather interestingly though, Fe does not work as a catalyst in laser ablation [38] unless activated with oxygen or hydrogen [11, 38, 53]. In arc discharge it is shown that hydrogen activates Fe [11]. In CVD, Fe is highly effective, however, it should be borne in mind that hydrogen is inherently always present due to the use of decomposing hydrocarbons and often the addition of pure H_2 is also implemented. The controversy with iron extends to its oxidation state. This and the roles of hydrogen are discussed later in the enhanced growth section. In terms of using a single metal catalyst for carbon nanotube production only a select few transition metals (Ni, Co, Pt, and Rh) are shown to be active in laser ablation [26] As laser ablation conditions are so pure (only pure carbon and the pure metal) it is arguable that only Ni, Co, Pt, and Rh are true catalysts that can catalyze nanotube formation without the assistance of an add-catalyst or enhancer.

Developments over the last few years have demonstrate that many atypical metals, including poor metals can be used for the catalytic growth of carbon nanotubes. Various laser ablation studies showed that atypical metals such as Cu [54] and poor metals, for example, Pb and In [53] can be used when activated in the presence of oxygen. In CVD, Takagi *et al.* developed an activation procedure in which samples preannealed in air yielded SWNT [66]. The technique was used to produce SWNT from Au, Ag, Pd, and Cu. CVD synthesized SWNT from gold were also synthesized from particles prepared by a block copolymer technique [10]. Gold is generally considered catalytically inactive; however at reduced dimensions it is catalytically active and may have melting points as low as 300 °C. Zang *et al.* showed that Pb, as found in laser ablation studies, could be used as a catalyst in CVD [74]. Further, they used Pb as a catalyst for horizontally aligned arrays of SWNT. Yuan *et al.* synthesized horizontally aligned SWNT from a variety of atypical metal catalysts, including, Mg, Al, Cr, Mo, Sn, and Mn [72].

1.3.2
Ceramic Catalysts

The use of nonmetallic catalysts is of interest in that purification can be easier than for conventional metal catalysts and potentially offers an attractive approach integrating carbon nanotubes with conventional silicon-based technology. The development of nonmetallic catalysts opens up new insights into the growth mechanisms of CNT.

Most nonmetallic catalysts are ceramics, for example, SiC and Al_2O_3. The first nonmetallic synthesis of carbon nanotubes from a ceramic was shown from the decomposition of SiC in 1997 [35]. However, the success of transition metal-based catalysts and the shear number of investigators working with CVD left this work largely ignored. However, with recent developments in novel catalysts, including ceramics, and also the ever rising interest in graphene, interest in CNT (and graphene) formation from SiC and other ceramics is rising. Another ceramic of interest is Al_2O_3. Porous alumina is often employed as a template for highly ordered CNT arrays. Usually the catalyst resides at the base of a pore and once subjected to CVD, CNT grow within a pore. However, more recent studies have shown that carbon nanotubes grow from porous alumina without a catalyst particle [6, 59]. This is in agreement with other studies that show many oxides, often used as supports for catalyst or as porous templates for carbon growth, are able to graphitize carbon. The graphene layers apparently stem from step sites [55]. This graphitization potential from alumina is extended to supported catalysis where it is argued that nucleation of a carbon nanotube occurs via the catalysts (Fe) but growth stems from the oxide support rather than the catalyst [57]. Takagi *et al.* [65] showed the ceramic catalyst concept in its full glory in a study in which they prepared nanosized Ge and Si particles on SiC and nanosized SiC in Si. In all cases they successfully synthesized SWNT. More recently Huang *et al.* showed that one can actually grow SWNT from SiO_2, Al_2O_3, and TiO_2 on supports confirming the work by Rümmeli *et al.* Often these ceramic techniques are referred to as catalyst free, however, other techniques in which CNT are formed from the vapor phase without a catalytic particle or catalytic support are more aptly described as "catalyst-free" synthesis routes.

1.3.3
Catalyst Free

The most well known, yet ironically often forgotten, example of catalyst-free synthesis of carbon nanotubes is the work by Iijima [24] that played a key role in the development of carbon nanotube synthesis. In this work MWNT were formed in an arc discharge similar to that used for fullerene synthesis. Another example of "catalyst-free" CNT production was established by using a modified flame route. In this study no catalysts were used and a counter oxygen flow was directed into a methane flame. MWCNT were formed with an oxygen enrichment of 50% [45]. More recently, MWNT were obtained on a porous carbon surface, at 800 °C by the decomposition of diluted ethylene [39]. Another catalyst-free route is that in which a hot graphite rod is dipped in cooled water. A sediment is formed that consists of high-purity MWNT [28].

1.4
Growth Enhancement

The use of additives in the synthesis of carbon nanotubes can be used to enhance growth and/or yield. Various additives have been developed with the most notable being the use of nominal amounts of H_2O in supported catalyst CVD. The route Hata developed is commonly known as supergrowth since it yields very long nanotubes with very high purity [18]. The role of H_2O is argued to be due to the preferential etching of amorphous carbon species [12], water removing undesired surface carbon layers on the catalyst, and are hence, reactivated and the catalytic activity is also dramatically increased [70]. Zhang *et al.* explored the roles of oxygen and hydrogen [73]. They argue hydrogen has a negative impact that as H radicals also etch the nanotubes themselves. They suggest that the presence of O reduces H radical species via the formation of OH. Alcohol CVD studies by Maruyama *et al.* suggest OH removes amorphous carbon [43]. Laser evaporation studies with nominal amounts of water included in the reaction support the notion of amorphous carbon reduction and also suggest oxygen and hydrogen enhance growth and activate the catalysts [12]. Indeed oxygen activation can allow atypical metals to be used as catalysts [53] and hydrogen activates Fe [11, 38]. Sulfur can also be used to enhance CNT formation [3, 9, 38]. The exact nature of the growth enhancement and catalyst activation through additives still remains elusive. However, it is likely that they also play a role in the actual graphitization of carbon species. For example, oxygen-containing gases have been shown to graphitize carbon [49]. In addition, metal oxides are well known to graphitize carbon [55]. However, the oxide state of catalyst particles is not clear. Early work by Baker *et al.* [7] on filamentous carbon formed in CVD, pointed to the highest activity occurring from the unstable FeO phase. They argue that this may be due to the defect structure of FeO. More recent *in situ* XPS studies show conflicting view points; some argue that oxides do not lead to CNT formation (e.g., [19]) while others argue that oxides do play a role (e.g., [42]). The matter is further complicated by *in situ* TEM data pointing to CNT growth clearly occurring from iron carbide [71]. The issue of growth enhancement and growth are intimately linked and both remain poorly understood. Even so, some important advances in growth have been made and are now briefly discussed.

1.5
Growth Mechanisms

The ever increasing amount of information on CNT growth prevents an exhaustive review in a book chapter like this. Instead the aim here is to focus on major elements in the current understanding of growth, highlight points of controversy, and present new results. In terms of the underlying science carbon nanotube formation may be viewed comprising of two major processes: initial cap formation (nucleation) and circumferential carbon addition (growth).

The formation of an initial cap was first put forward by Dai *et al.* [14] because the formation of a cap reduces the total surface energy since the basal plane of graphite has a low surface energy. This hemispherical cap formation is referred to as the *Yarmulke* mechanism. The edges at the base of the cap, it is argued, chemisorb to the metal, hence the existence of energetically costly dangling bonds are avoided. This is supported by calculations by Fan *et al.* [16] in which they used a series of total energy calculations using density functional theory to show that nucleation

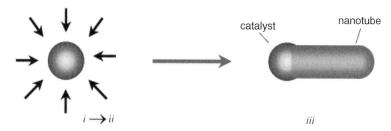

Figure 1.7 VLS schematic: (i, vapor) A semiconductor vapor source adsorbs (black arrows) into the catalyst particle causing it to form a liquid eutectic (ii, liquid). Upon supersaturation a solid precipitates (iii, solid).

of a closed cap or a capped single-walled carbon nanotube is overwhelmingly favored compared to any structure with dangling bond or to a fullerene. Once the hemispherical cap forms, further addition of carbon around the circumference in a cylindrical fashion constitutes the emergence of a tubular structure and is called growth. There are various models proposed for the manner in which the cap formation and addition of carbon to the root of a growing tube occur. However, the vast majority are based around the so-called vapor–solid–liquid (VLS) mechanism [8] to greater or lesser degrees. Hence we briefly introduce the reader to the VLS process before looking at some of the different models proposed for the different catalyst systems. In the VLS process the metal clusters are heated above their metal-carbon eutectic temperature in the presence C. The catalyst particles adsorb carbon, forming a liquid metal-carbide, which eventually consumes the whole catalyst particle. Eventually supersaturation is reached, which leads to the nucleation of solid carbon as the system aims to reach equilibrium. The process is illustrated in Figure 1.7. From this figure it is clear that the role of the catalyst is not only provide a medium through which the nanotube can nucleate and then grow, but also to play an important role in determining the diameter of the tube. Variations to the central VLS model and other suggested models are now briefly discussed according to the manner in which the catalyst particle exists in the reaction; residing in free space (floating catalyst), residing on a substrate (supported catalyst), or no catalyst particle is used (catalyst free).

1.5.1
Floating Catalyst Methods

Synthesis routes that use floating catalysts include arc discharge, laser evaporation, HiPCO, and thermal CVD methods. Common to all these methods is the formation of the catalyst clusters from the vapor phase. In the arc discharge and laser evaporation routes the catalyst material and the carbon are evaporated due to the arc or laser, respectively. In CVD, the decomposition of a metal-based compound (e.g., ferrocene) is the most common way to provide metal catalyst species. The carbon is provided in a similar manner via the decomposition of a hydrocarbon.

Once the catalyst and carbon species have been vaporized they begin to form clusters as they cool.

As they cool further they coalesce yielding catalyst particles saturated with carbon. As the system cools further the catalyst particles precipitate carbon, forming a nucleation cap and as more carbon is added, a carbon nanotube is extruded. With floating catalyst systems, usually SWNT are obtained. While there is a general agreement that a cap is required for nucleation of a tube, there are disparate ideas as to how this occurs. For example, Gorbunov *et al.* [17] suggest that precipitating carbon at the particle surface forms graphitic islands which have an initial curvature due to the particle morphology. They argue that the continued addition of carbon will include defects introducing buckling and hence allow curvature until at some stage, a hemispherical cap is arrived at. Further addition of carbon at the cap edges will continue in a cylindrical fashion (tube growth). This mechanism suggests that there is no dependence on the tube diameter and that of the catalyst particle at the point of nucleation. Nasibulin *et al.* propose that at nucleation the graphitic structure that forms on the surface of the catalyst exceeds that of a Yarmulke-like cap. Here defect forming heptagons provide the required rearrangement for cap lift of (nucleation) and that growth occurs through the constant rearrangement at the root via heptagon formation [47]. This model was developed to explain the larger diameters of the catalyst particles as compared to the obtained nanotubes as observed in postsynthesis TEM studies (Figure 1.8).

However, as pointed out by Dai *et al.* [14], such studies are complicated because they do not show what the catalyst particle is like at the point of nucleation, for example, ripening process would lead to enlargement of the catalyst particle after cap formation. Dai's findings, on the other hand, suggest there is a direct relationship between the catalyst particle diameter and that of the SWNT. Rümmeli *et al.* also proposed that the Yarmulke-like nucleation cap is actually dictated by the catalyst particle diameter in the so-called catalyst volume to surface area [56]. They suggested that at the point at which a cooling particle precipitates its carbon, the formation of a Yarmulke cap is not guaranteed. If the particle is too small, insufficient carbon would be available for a stable cap to form and if the particle was too big, the particle would have an excess of carbon that would encapsulate the particle, as also pointed out by Dai *et al.* [14]. However if the catalyst particle's volume to surface area is just right, a graphitic hemisphere would form providing the nucleation cap (see Figure 1.9). Hence a nucleation window exists. Further carbon addition to the ends would then lead to tube growth. Most theoretical studies tend to point to a strong correlation between the catalyst particle size and the resultant carbon nanotube. Furthermore, theoretical investigations also suggest deviations for very small catalyst particles and larger catalyst particles. For example, molecular dynamics studies show, in general, a strong correlation between the catalyst particle size and the tube diameter at the point of nucleation; however, the studies indicate that for very small catalyst particles (less than 0.5 nm), the diameter of the resultant tubes is slightly larger (0.6–0.7 nm) [15]. Shibuta and Maruyama also conducted molecular dynamics studies to elucidate carbon nanotube nucleation in floating catalyst CVD. They also found a strong correlation between the catalyst

(a)

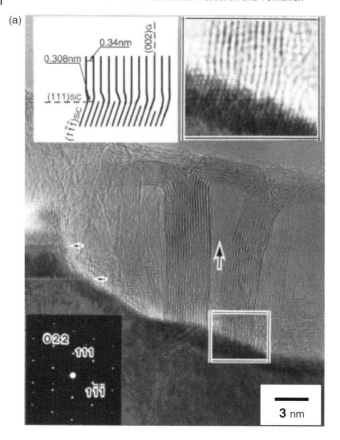

(b)

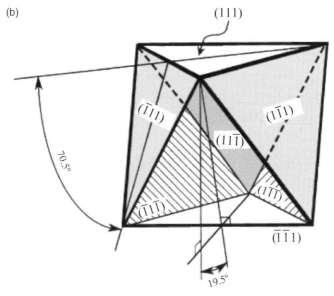

Figure 1.8   (caption see on P. 13)

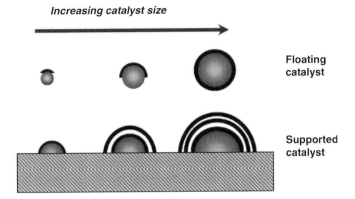

Increasing catalyst size

Floating
catalyst

Supported
catalyst

Figure 1.9 Illustration of nucleation variants with floating catalysts (upper) and supported catalysts (lower) under the volume to surface area model. At the point of nucleation, carbon precipitates out of the catalyst particle and in the case of floating catalysts, if insufficient carbon is available then no stable cap is formed, and if too much carbon is available the catalyst particle is encapsulated. Only when the catalyst volume to surface area provides just the right amount of carbon for a stable cap to form, nucleation occur. With supported catalysts since particle encapsulation is prevented by the particle/support interaction, as the amount of carbon increases (increasing particle size) the number of caps forming also increases.

particle and the nucleation cap (tube diameter). Their work also showed that for larger particles, hump formation on the catalyst particle could lead to cap formations around the hump. In addition their work suggests that very small particle sizes do not lead to tube growth since small caps cannot be lifted as the curvature energy is too high [61]. The various processes are illustrated in Figure 1.10. Hence, molecular dynamics studies point to a nucleation window for SWNT formation with floating catalysts, in agreement with experiment.

1.5.2
Supported Catalyst Routes

In supported catalysis of carbon nanotubes the effect of the substrate on the catalytic process is still unknown since there are many chemical reactions between the substrate and the catalyst metals [22]. However, two modes of growth mechanisms in supported catalysts have been identified: the base growth mode, in which the catalyst particle resides on the support throughout the process, and the tip growth mode, in which the catalyst particle detaches from the support [8]. Indeed, recent advances in transmission electron microscopy have allowed *in situ* observations of

◄

Figure 1.8 (a) HREM micrograph of the interface between the graphite constructing a carbon nanotube and beta-SiC on the surface of (111) beta-SiC. (b) Schematic of the orientation relationship between one (111) SiC plane, on which carbon nanotubes are standing perpendicularly, and the other [111] SiC planes (Reprinted with permission from Ref. [36]. Copyright 2009, The American Institute of Physics).

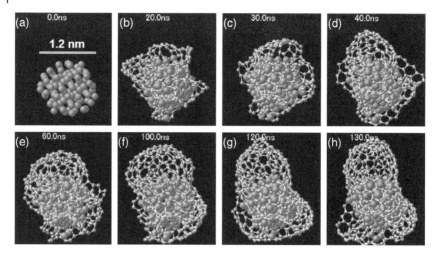

Figure 1.10 Snapshots illustrating the metal-catalyzed nucleation (cap formation) and growth process: a molecular dynamics calculation at 2500 K for Ni108 from 0 to 130 ns (images a–h). Blue spheres represent nickel atoms, and green spheres are carbon atoms. (Images courtesy of Shibuta and Maruyama [63].)

both tip growth [40] and base growth [71]. TEM observations show that for nanotube walls rooted in the catalysts particles, the graphitic walls reside at step sites [21, 52]. They also show that active catalyst particles during growth exhibit liquid-like properties while still retaining solid properties (lattice fringes are clearly observed). This liquid-like behavior does not answer a controversial area for both floating catalysts and supported catalysts, namely, does carbon delivery to a growing graphitic wall occur through bulk diffusion or surface diffusion. Experimental data point to both bulk diffusion [52] and surface diffusion [21] being possible. Likewise, some theoretical investigations suggest that surface diffusion is possible [20]. Other theoretical studies point to bulk diffusion [75]. In terms of specific simulations for catalysts residing on a support, the calculations are more complicated. Shibuta and Maruyama have made preliminary molecular dynamics studies focusing on the effect of the substrate on the metal (catalyst) atoms and have neglected other possible contributions from the substrate in the reaction. Their study show that cap nucleation can occur from catalyst deformation (humps) or from the catalyst particle overall surface similar to their floating catalyst results [62]. A variety of experimental data show a strong correlation between the resultant carbon nanotubes diameter and catalyst particle. In addition, the number of walls increases as the catalyst particles size increases. This is in keeping with the catalyst volume to surface area model described for floating catalysts, but modified to account for the catalyst particle interaction with the support. In this scenario encapsulation of a catalyst particle is prevented by the catalyst/support interaction and results in multiple cap formation as illustrated in Figure 1.9. Regards the root of a carbon nanotube, while most studies assume or show growth to stem from the catalyst particle recent studies by Rümmeli et al. suggest growth may also occur from the support itself after nucleation from the catalyst particle has occurred [57]. Such growth is similar to ceramic catalyzed/templated growth. This will be further discussed in the next section.

1.5.3
Catalyst-Free Routes

In this section *catalyst free* refers to routes in which no catalyst particle is used. However, in some instances it may be that the catalytic activity of the catalyst particle is replaced by some other means. An example is the template synthesis of carbon nanotubes in porous alumina with no catalysts particle present in the pores. In this instance the porous alumina is subjected to CVD conditions similar to catalytic CVD and graphitic walls are found to form in the pores forming MWNT [6, 59]. Similar studies using a variety of oxide nanoparticles found several layers of graphene to form on the surface [55]. Further they appear to root from step sites on the oxide surface. Step sites are well known as catalytic sites. Another route to form graphitic layers or carbon nanotubes is via the decomposition of SiC. This is achieved by annealing SiC in high vacuum at temperatures between 1400 and 1700 °C. As the SiC decomposes, graphitic structures like carbon nanotubes emerge. The process is only found to work with trace amounts of oxygen and local EELS studies show oxygen species near and at the surface of the SiC [35, 37]. TEM studies show the graphitic layers align with the lattice fringes of the SiC, thus rooting directly into the crystal. This is illustrated in Figure 1.8. The diameter of the CNT and the number of walls formed by the surface decomposition of the SiC are directly proportional [36]. In addition, the diameter of the nanotubes are determined by the initial cap diameter. The decomposition of the SiC is argued to occur as

$$SiC + CO \rightarrow SiO + 2C \tag{1.1}$$

The growth rate of the CNTs is argued to be dependent on the diffusion of SiO and/or CO gas through the CNT film. Catalyst-free carbon nanotube formation can also occur without the presence of a ceramic, the most famous being the synthesis of MWNT in the arc discharge without the use of catalysts as first presented by Iijima [24]. A possible argument for their formation is the scroll mechanism in which graphene or graphitic layers roll up. The idea was first proposed in 1960 by Bacon [51] who was studying the filamentous deposit found while arcing two graphite electrodes, which was later shown by Iijima to be MWNT. Theoretical studies [2] show that scrolling is possible. Nucleation is also attributed to the initial formation of a fullerene cap or dome. Recent experiments show such nanoscrolls can also be formed chemically by first exfoliating graphite and then sonicating the sample [69] (see Figure 1.9). The scrolling mechanism is also argued to explain the formation of CNT when hot graphitic rods are dipped in cold water [28]. MWNT have also been synthesized in flame synthesis when nominal amounts of oxygen are included [45]. This route is similar to that used for soot formation, and it is likely that polyaromatic hydrocarbons, PAHs, are involved in their formation.

1.6
Functionalization

The outer wall of pristine carbon nanotubes is, in principle, conceived as chemically inert. However, this is not always desirable for applications and for this reason, and to further modify the properties of carbon nanotubes in a controlled manner, several functionalization routes have been thought to make them chemically active in the last years. It is worth mentioning that in many cases the word functionalization could be too general to describe the type of changes desired. It can, in principle, mean in-lattice doping, intercalation, molecule or particle adsorption, encapsulation, or even other nonexplored modifications. The several functionalization routes will be discussed in detailed in the following contributions. However, for completeness of this chapter some lines are devoted to the functionalization during synthesis (endohedral doping and sidewall functionalization by heteroatoms).

One typical example for the functionalization of fullerenes was the chemical route for producing $C_{59}NH$ [29]. Smith *et al.* pioneered the endohedral functionalization with molecular structures with the introduction of C_{60} fullerenes in opened nanotubes, forming the well-known peapods simultaneously [63]. Moreover, endohedral doping of fullerenes was also achieved by the synthesis of different metallofullerenes in the Krätschmer–Huffman furnace [32], and later on, these metallofullerenes were introduced into opened SWCNT by vacuum sublimation in a similar way as Smith *et al.* had proposed. In this case it is important to notice that this method has also been reported as a unique template reaction in which nanotubes and fullerenes allow the production of metallic nanowires inside SWCNT [31] and MWCNT (Fe filled SWCNT and MWCNT). A second functionalization route can be applied for the formation of heterofullerenes and heteronanotubes by the substitution of carbon by heteroatoms. Boron and nitrogen doped (CBx and CNx) nanotubes have been successfully incorporated in multiwalled carbon nanotubes by several methods [67]. Although the synthesis of their single-walled counterparts has only recently been reported with successful identification of low doping levels [4, 30].

1.7
Purification

A remaining challenge is the purification of the nanotube material obtained from the different processes. Depending on the synthesis method employed, the yield of undesirable byproducts in the soot varies and these impurities interfere with the properties of the nanotubes. Determining the purity degree of nanotube material is still controversial as no unified standard has been established yet. This is also due to the perspective uses of the synthesized materials. Methods that aim at single-walled nanotube purification usually employ various oxidation steps, sonication, thermal treatments, chemical acid treatments, among others. However, none of the patented methods is universal. For instance, in a sample free of metal residuals after purification, there still could exist a mixture of nanotubes of different numbers of walls. Another clear example could be the case of separated nanotubes. Nowadays it is possible to perform separation of nanotubes with electrophoresis-based as well as ultracentrifugation-based techniques. However, the concept of purity in such a sample is still not fully understood regarding the number of parameters that could be considered. This topic will be discussed more in detail in the later chapters.

1.8
Future Perspectives

The above discussion is intended to summarize the actual research status in the production of carbon nanotubes and to show the line of reasoning regarding growth mechanisms theories. There are several techniques to produce single and multi-walled nanotubes and of course all of them have advantages and disadvantages. The growth mechanisms occurring within these synthesis variants still need to be fully understood and further research is expected in this respect. A notable aspect in this sense is the emergence on nonmetal catalysts and these may well replace metallic catalyst systems due to their compatibility with silicon technology and potential to yield high-purity samples. The literature regarding the synthesis of multiwalled carbon nanotubes is countless and in the single-walled case it is also abundant but still far from being fully comprehended. Great advances have been made in the last few years regarding separation of single-walled carbon nanotubes according to metallicity. However, the purity of those samples and the structure conservation upon the processing methods still needs to be analyzed.

Positive developments have been made in controlling and optimizing the nano-tube synthesis, as well as in purification and separation of nanotubes by metallicity and chirality. However, in particular the length separation and the control of the defect concentration and the chirality separation of thick nanotubes still needs to be addressed in order to achieve the final goal of monodisperse samples with full length, chirality, and diameter control.

References

1 Aliev, A.E., Oh, J., Kozlov, M.E., Kuznetsov, A.A., Fang, S., Fonseca, A.F., Ovalle, R., Lima, M.D., Haque, M.H., Gartstein, Y.N., Zhang, M., Zakhidov, A.A., and Baughman, R.H. (2009) Giant-stroke, superelastic carbon nanotube aerogel muscles. *Science*, **323**, 1575.

2 Amelinckx, S., Bernaerts, D., Zhang, X.B., Van Tendeloo, G., and Van Landuyt, J. (1995) A structure model and growth mechanism for multishell carbon nanotubes. *Science*, **267**, 1334.

3 Arepalli, S. (2004) Laser ablation process for single-walled carbon nanotube production. *J. Nanosci. Nanotechnol.*, **4**, 317.

4 Ayala, P., Grueneis, A., Gemming, T., Grimm, D., Kramberger, C., Rümmeli, M.H., Freire, F.L., Kuzmany, H., Pfeiffer, R., Barreiro, A., Buechner, B., and Pichler, T. (2007) Tailoring n-doped single and double wall carbon nanotubes from a ondiluted carbon/nitrogen feedstock. *J. Phys. Chem. C*, **111**, 2879.

5 Ayala, P., Grüneis, A., Kramberger, C., Grimm, D., Engelhard, R., Rümmeli, M.H., Schumann, J., Kaltofen, R., Büchner, B., Schaman, C., Kuzmany, H., Gemming, T., Barreiro, A., and Pichler, T. (2008) Cyclohexane triggers staged growth of pure and vertically aligned single wall carbon nanotubes. *Chem. Phys. Lett.*, **454**, 332.

6 Bae, E.J., Choi, W.B., Jeong, K.S., Chu, J.U., Park, G.S., Song, S., and Yoo, Y.U. (2001) Selective growth of carbon nanotubes on pre-patterned porous anodic aluminum oxide. *Adv. Mater.*, **14**, 277.

7 Baker, R.T.K., Alonzo, J.R., Dumesic, J.A., and Yates, D.J.C. (1982) Effect of the surface state of iron on filamentous carbon formation. *J. Catal.*, **77**, 74.

8 Baker, T. (1982) Formation of filamentous carbon. *Chem. Ind. (London)*, **18**, 698.

9 Barreiro, A., Kramberger, C., Rümmeli, M.H., Grüneis, A., Grimm, D., Hampel, S., Gemming, T., Büchner, B., Bachtold, A., and Pichler, T. (2007) Control of the single-wall carbon nanotube mean diameter in sulphur promoted aerosol-assisted chemical vapour deposition. *Carbon*, **45**, 55.

10 Bhaviripudi, S., Mile, E., Steiner, S.A., III, Zare, A.T., Dresselhaus, M.S., Belcher, A.M., and Kong, J. (2007) CVD synthesis of single-walled carbon nanotubes from gold nanoparticle catalysts. *J. Am. Chem. Soc.*, **129**, 1516.

11 Bystrzejewski, M., Rümmeli, M.H., Lange, H., Huczko, A., Baranowski, P., Gemming, T., and Pichler, T. (2008) Single-walled carbon nanotubes synthesis: a direct comparison of laser ablation and carbon arc routes. *J. Nanosci. Nanotechnol.*, **8**, 1.

12 Bystrzejewski, M., Schönfelder, R., Cuniberti, G., Lange, H., Huczko, A., Gemming, T., Pichler, T., Büchner, B., and Rümmeli, M.H. (2008) Exposing multiple roles of H_2O in high-temperature enhanced carbon nanotube synthesis. *Chem. Mater.*, **20**, 6586.

13 Ci, L., Rao, Z., Zhou, Z., Tang, D., Yan, X., Liang, Y., Liu, D., Yuan, H., Zhou, W., Wang, G., Liu, W., and Xie, S. (2002) Double wall carbon nanotubes promoted by sulfur in a floating iron catalyst CVD system. *Chem. Phys. Lett.*, **359**, 63.

14 Dai, H., Rinzler, A., Nikolaev, P., Thess, A., Colbert, D.T., and Smalley, R.E. (1996) Single-wall nanotubes produced by metal-catalyzed disproportionation of carbon monoxide. *Chem. Phys. Lett.*, **260**, 471.

15 Ding, F., Rose, A., and Bolton, K. (2004) Molecular dynamics study of the catalyst particle size dependence on carbon nanotube growth. *J. Chem. Phys.*, **121**, 2775.

16 Fan, X., Buczko, R., Puretzky, A.A., Geohegan, D.B., Howe, J.Y., Pantelides, S.T., and Pennycook, S.J. (2003) Nucleation of single-walled carbon nanotubes. *Phys. Rev. Lett.*, **90**, 145501/1.

17 Gorbunov, A., Jost, O., Pompe, W., and Graff, A. (2002) Solid–liquid–solid growth mechanism of single-wall carbon nanotubes. *Carbon*, **40**, 113.

18 Hata, K., Futaba, D.N., Mizuno, K., Namai, T., Yumura, M., and Iijima, S. (2004) Water-assisted highly efficient synthesis of impurity-free single-walled carbon nanotubes. *Science*, **306**, 1362.

19 Hofmann, S., Blume, R., Wirth, C.T., Cantoro, M., Sharma, R., Ducati, C., Hävecker, M., Zafeiratos, S., Schnoerch, P., Oestereich, A., Teschner, D., Albrecht, M., Knop-Gericke, A., Schlögl, R., and Robertson, J. (2009) State of transition metal catalysts during carbon nanotube growth. *J. Phys. Chem. C*, **113**, 1648.

20 Hofmann, S., Csanyi, G., Ferrari, A.C., Payne, M.C., and Robertson, J. (2005) Surface diffusion: the low activation energy path for nanotube growth. *Phys. Rev. Lett.*, **95**, 036101.

21 Hofmann, S., Sharma, R., Ducati, C., Du, G., Mattevi, C., Cepek, C., Cantoro, M., Pisana, S., Parvez, A., Cervantes-Sodi, F., Ferrari, A.C., Dunin-Borkowski, R., Lizzit, S., Petaccia, L., Goldoni, A., and Robertson, J. (2007) *In situ* observations of catalyst dynamics during surface-bound carbon nanotube nucleation. *Nano Lett.*, **7**, 602.

22 Hu, M., Murakami, Y., Ogura, M., Maruyama, S., and Okubo, T. (2004) Morphology and chemical state of Co–Mo catalysts for growth of single–walled

carbon nanotubes vertically aligned on quartz substrates. *J. Catal.*, **225**, 230.

23 Huczko, A., Lange, H., Bystrzejewski, M., Baranowski, P., Ando, Y., Zhao, X., and Inoue, S. (2006) Formation of SWCNTs in arc plasma: effect of graphitization on Fe-doped anode and optical emission studies. *J. Nanosci. Nanotechnol.*, **6**, 1.

24 Iijima, S. (1991) Helical microtubules of graphitic carbon. *Nature*, **354**, 56.

25 Jorio, A., Dresselhaus, G., Dresselhaus, M.S. (eds) (2008) *Carbon Nanotubes: Advanced Topics in the Synthesis, Structure, Properties and Applications, Topics in Applied Physics*, vol. **111**, 1st edn, Springer.

26 Jost, O., Gorbunov, A., Liu, X., Pompe, W., and Fink, J. (2004) Single-walled carbon nanotube diameter. *J. Nanosci. Nanotechnol.*, **4**, 433.

27 Journet, C., Maser, W.K., Bernier, P., Lamy de la Chapelle, M., Loiseau, A., Lefrant, S., Deniard, P., Lee, R., and Fischer, J.E. (1977) Large-scale production of single-walled carbon nanotubes by the electric-arc technique. *Nature*, **388**, 756.

28 Kang, Z., Wang, E., Gao, L., Lian, S., Jiang, M., Hu, C., and Xu, L. (2003) One-step water-assisted synthesis of high-quality carbon nanotubes directly from graphite. *J. Am. Chem. Soc.*, **125**, 13652.

29 Keshavarz, M., Gonzalez, K.R., Hicks, R.G., Srdanov, G., Srdanov, V.I., Collins, T.G., Hummelen, J.C., Bellavia-Lund, C., Pavlovich, J., Wudl, F., and Holczer, K. (1996) Synthesis of hydroazafullerene $C_{59}HN$, the parent hydroheterofullerene. *Nature*, **383**, 147.

30 Kim, S.Y., Lee, J., Na, C.W., Park, J., Seo, K., and Kim, B. (2005) N-doped double-walled carbon nanotubes synthesized by chemical vapor deposition. *Chem. Phys. Lett.*, **413**, 300.

31 Kitaura, R., Imazu, N., Kobayashi, K., and Shinohara, H. (2008) Fabrication of metal nanowires in carbon nanotubes via versatile nano-template reaction. *Nano Lett.*, **8**, 693.

32 Kitaura, R. and Shinohara, H. (2007) Endohedral metallofullerenes and nano-peapods. *Jpn. J. Appl. Phys.*, **46**, 881.

33 Krätschmer, W., Lamb, L.D., Fostiropoulos, K., and Huffman (1990)

D.R. C_{60}: a new form of carbon. *Nature*, **347**, 354.

34 Kroto, H.W., Heath, J.R., O'Brien, S.C., Curl, R.F., and Smalley, R.E. (1985) C_{60}: buckminsterfullerene. *Nature*, **318**, 14.

35 Kusunoki, M., Rokkaku, M., and Suzuki, T. (1997) Epitaxial carbon nanotube film self-organized by sublimation decomposition of silicon carbide. *Appl. Phys. Lett.*, **71**, 2620.

36 Kusunoki, M., Suzuki, T., Honjo, C., Usami, H., and Kato, H. (2007) Closed-packed and well-aligned carbon nanotube films on sic. *J. Phys. D*, **40**, 6278.

37 Kusunoki, M., Suzuki, T., Kaneko, K., and Ito, M. (1999) Formation of self-aligned carbon nanotube by surface decomposition of silicon carbide. *Phil. Mag. Lett.*, **79**, 153.

38 Lebedkin, S., Schweiss, P., Renker, B., Malik, S., Hennrich, F., Neumaier, M., Stoermer, C., and Kappes, M.M. (2002) Single-wall carbon nanotubes with diameters approaching 6 nm obtained by laser vaporization. *Carbon*, **40**, 417.

39 Lin, J.-H., Chen, C.-S., Ma, H.-L., Chang, C.-W., Hsu, C.-Y., and Chen, H.-W. (2008) Self-assembling of multi-walled carbon nanotubes on a porous carbon surface by catalyst-free chemical vapor deposition. *Carbon*, **46**, 1611.

40 Lin, M., Ying Tan, J.P., Boothroyd, C., Loh, K.P., Tok, E.S., and Foo, Y.-L. (2007) Dynamical observation of bamboo-like carbon nanotube growth. *Nano Lett.*, **7**, 2234.

41 Lin, X., Rümmeli, M.H., Gemming, T., Pichler, T., Valentin, D., Ruani, G., and Taliani, C. (2007) Single-wall carbon nanotubes prepared with different kinds of NiCo catalysts: Raman and optical spectrum analysis. *Carbon*, **45**, 196.

42 De los Arcos, T., Oelhafen, P., Thommen, V., and Mathys, D. (2007) The influence of catalyst's oxidation degree on carbon nanotube growth as a substrate-independent parameter. *J. Phys. Chem. C*, **111**, 16392.

43 Maruyama, S., Kojima, R., Miyauchi, Y., Chiashi, S., and Kohno, M. (2002) Low-temperature synthesis of high-purity single-walled carbon nanotubes from alcohol. *Chem. Phys. Lett.*, **360**, 229.

44 Maser, W., Benito, A.M., Munoz, E., and Teresa Martinez, M. (2008) *Carbon Nanotubes: From Fundamental Nanoscale Objects Towards Functional Nanocomposites and Applications*, p. 101 Springer, Netherlands.

45 Merchan-Merchan, W., Saveliev, A., Kennedy, L.A., and Fridman, A. (2002) Formation of carbon nanotubes in counter-flow, oxy-methane diffusion flames without catalysts. *Chem. Phys. Lett.*, **354**, 20.

46 Monthioux, M. and Kuznetsov, Vladimir L. (2006) Who should be given the credit for the discovery of carbon nanotubes? *Carbon*, **44**, 1621.

47 Nasibulin, A.G., Pikhitsa, P.V., Jiang, H., and Kauppinen, E.I. (2005) Correlation between catalyst particle and single-walled carbon nanotube diameters. *Carbon*, **43**, 2251.

48 Nikolaev, P., Bronikowski, M.J., Bradley, R.K., Rohmund, F., Colbert, D.T., Smith, K.A., and Smalley, R.E. (1999) Gas-phase catalytic growth of single-walled carbon nanotubes from carbon monoxide. *Chem. Phys. Lett.*, **313**, 91.

49 Noda, T. and Inagaki, M. (1964) Effect of gas phase on graphitization of carbon. *Carbon*, **2**, 127.

50 Pierard, N., Fonseca, A., Konya, Z., Willems, I., Van Tendeloo, G., and Nagy, J.B. (2001) Production of short carbon nanotubes with open tips by ball milling. *Chem. Phys. Lett.*, **335**, 1.

51 Bacon, R. (1960) Growth, structure, and properties of graphite whiskers. *J. Appl. Phys.*, **31**, 283.

52 Rodriguez-Manzo, J.A., Terrones, M., Terrones, H., Kroto, H.W., Sun, L., and Banhart, F. (2007) *In situ* nucleation of carbon nanotubes by the injection of carbon atoms into metal particles. *Nat. Nanotechnol.*, **2**, 307.

53 Rümmeli, M.H., Borowiak-Palen, E., Gemming, T., Pichler, T., Knupfer, M., Kalbac, M., Dunsch, L., Jost, O., Silva, S.R.P., Pompe, W., and Büchner, B. (2005) Novel catalysts, room temperature, and the importance of oxygen for the synthesis of single-walled carbon nanotubes. *Nano Lett.*, **5**, 1209.

54 Rümmeli, M.H., Güneis, A., Löffler, M., Jost, O., Schönfelder, R., Kramberger, C., Grimm, D., Gemming, T., Barreiro, A., Borowiak-Palen, E., Kalbac, M., Ayala, P., Hübers, H.-W., Büchner, B., and Pichler, T. (2006) Novel catalysts for low temperature synthesis of single wall carbon nanotubes. *Phys. Stat. Sol. (B)*, **243**, 3050.

55 Rümmeli, M.H., Kramberger, C., Grüneis, A., Ayala, A., Gemming, T., Büchner, B., and Pichler, T. (2007) On the graphitization nature of oxides for the formation of carbon nanostructures. *Chem. Mater.*, **19**, 4105.

56 Rümmeli, M.H., Kramberger, C., Loeffler, M., Jost, O., Bystrzejewski, M., Grueneis, A., Gemming, T., Pompe, W., Buechner, B., and Pichler, T. (2007) Catalyst volume to surface area constraints for nucleating carbon nanotubes. *J. Phys. Chem. B*, **111**, 8234.

57 Rümmeli, M.H., Schäffel, F., Kramberger, C., Gemming, T., Bachmatiuk, A., Kalenczuk, R.J., Rellinghaus, B., Büchner, B., and Pichler, T. (2007) Oxide-driven carbon nanotube growth in supported catalyst CVD. *J. Am. Chem. Soc.*, **129**, 15772.

58 Saito, Y., Tani, Y., Miyagawa, N., Mitsushima, K., Kasuya, A., and Nishina, Y. (1998) High yield of single wall carbon nanotubes by arc discharge using Rh-Pt mixed catalysts. *Chem. Phys. Lett.*, **294**, 593.

59 Schneider, J.J., Maksimova, N.I., Engstler, J., Joshi, R., Schierholz, R., and Feile, R. (2008) Catalyst free growth of a carbon nanotube–alumina composite structure. *Inorg. Chim. Acta*, **361**, 1770.

60 Sgobba, V. and Guldi, D.M. (2009) Carbon nanotubes-electronic/electrochemical properties and application for nanoelectronics and photonics. *Chem. Soc. Rev.*, **38**, 165.

61 Shibuta, Y. and Maruyama, S. (2003) Molecular dynamics simulation of formation process of single-walled carbon nanotubes by CCVD method. *Chem. Phys. Lett.*, **382**, 381.

62 Shibuta, Y. and Maruyama, S. (2007) A molecular dynamics study of the effect of a substrate on catalytic metal clusters in nucleation process of single-walled carbon nanotubes. *Chem. Phys. Lett.*, **437**, 218.

63 Smith, B., Monthioux, M., and Luzzi, D. Encapsulated C_{60} in carbon nanotubes. (1998) *Nature*, **396**, 323.

64 Sun, X., Bao, W., Lv, Y., Deng, J., and Wang, X. (2007) Synthesis of high quality single-walled carbon nanotubes by arc discharge method in large scale. *Mater. Lett.*, **61**, 3956.

65 Takagi, D., Hibino, H., Suzuki, S., Kobayashi, Y., and Homma, Y. (2007) Carbon nanotube growth from semiconductor nanoparticles. *Nano Lett.*, **7**, 2272.

66 Takagi, D., Homma, Y., Hibino, H., Suzuki, S., and Kobayashi, Y. (2006) Single-walled carbon nanotube growth from highly activated metal nanoparticles. *Nano Lett.*, **6**, 2642.

67 Terrones, M., Jorio, A., Endo, M., Rao, A.M., Kim, Y.A., Hayashi, T., Terrones, H., Charlier, J.-C., Dresselhaus, G., and Dresselhaus, M.S. (2004) New direction in nanotube science. *Mater. Today*, 10, 30–45.

68 Thess, A., Lee, R., Nikolaev, P., Dai, H., Petit, P., Robert, J., Xu, C., Lee, Y.H., Kim, S.G., Rinzler, A.G., Colbert, D.T., Scuseria, G.E., Tománek, D., Fischer, J.E., and Smalley, R.E. (1996) *Science*, **273**, 483.

69 Viculis, L.M., Mack, J.J., and Kaner, R.B. (2009) A chemical route to carbon nanoscrolls. *Science*, **299**, 1361.

70 Yamada, T., Maigne, A., Yudasaka, M., Mizuno, K., Futaba, D.N., Yumura, M., Iijima, S., and Hata, K. (2008) Revealing the secret of water-assisted carbon nanotube synthesis by microscopic observation of the interaction of water on the catalysts. *Nano Lett.*, **8**, 4288.

71 Yoshida, H., Takeda, S., Uchiyama, T., Kohno, H., and Homma, Y. (2008) Atomic-scale *in situ* observation of carbon nanotube growth from solid state iron carbide nanoparticles. *Nano Lett.*, **8**, 2082.

72 Yuan, D., Ding, L., Chu, H., Feng, Y., McNicholas, T.P., and Liu, J. (2008) Horizontally aligned single-walled carbon nanotube on quartz from a large variety of metal catalysts. *Nano Lett.*, **8**, 2576.

73 Zhang, G., Mann, D., Zhang, L., Javey, A., Li, Y., Yenilmez, E., Wang, Q., McVittie, J.P., Nishi, Y., Gibbons, J., and Dai, H. (2005) Ultra-high-yield growth of vertical single-walled carbon nanotubes: Hidden roles of hydrogen and oxygen. *Proc. Natl. Acad. Sci. USA*, **102**, 16141.

74 Zhang, Y., Zhou, W., Jin, Z., Ding, L., Zhang, Z., Liang, X., and Li, Y. (2008) Direct growth of single-walled carbon nanotubes without metallic residues by using lead as a catalyst. *Chem. Mater.*, **20**, 7521.

75 Zhu, Y.-A., Dai, Y.-C., Chen, D., and Yuan, W.-K. (2007) First-principles study of carbon diffusion in bulk nickel during the growth of fishbone-type carbon nanofibers. *Carbon*, **45**, 21.

2
Theory of Electronic and Optical Properties of DNA–SWNT Hybrids

Slava V. Rotkin and Stacy E. Snyder

2.1
Introduction

Single-walled nanotubes (SWNTs) are being studied for use as components in the next generation of electronic and optical devices, including transistors, photodetectors, and biological sensors [1], but to realize large scale production of these and other potential devices, methods must be developed to gain control over the material. This task is multifaceted. First, as-synthesized carbon nanotubes vary in physical structures. Differences in diameter and chiral angle lead to differences in electronic structure, resulting in wide-ranging electronic and optical properties [2, 3]. Thus, a sorting technique must be developed. Next nanotubes tend to clump together in bundles due to van der Waals attraction, further complicating efforts to sort them by type. In order to take advantage of the unique properties of a *specific* SWNT electronic structure, starting with currently available raw materials, solution-based processing can be used to disperse as-grown SWNTs [4], differentiate them by structure [5], and then selectively place the desired type of nanotube on a substrate. This processing requires a robust dispersion method that must overcome nanotube insolubility and the strong attraction between tubes. It was shown [6, 7] that the dispersion of hydrophobic SWNTs in solution can be facilitated by noncovalent functionalization with helically wrapped single-stranded DNA (ssDNA). A SWNT forms a stable hybrid structure with ssDNA, thus allowing further processing. Since a DNA–SWNT hybrid is not readily dismantled and a DNA wrapped nanotube material may be used for electronic or optical device applications, it is important to determine the extent to which the DNA changes the relevant properties of SWNTs. The range and scope of these effects are presented in this study.

The chapter is organized as follows: we start with a brief overview of the existing experimental and theoretical studies on DNA–SWNT hybrids. Then our own model is discussed. Next, we present results on the modulation of electronic/optical properties such as optical bandgap of SWNTs upon hybridization with ssDNA. Different geometries are studied systematically to reveal scaling relations with charge

Carbon Nanotubes and Related Structures. Edited by Dirk M. Guldi and Nazario Martín
Copyright © 2010 WILEY-VCH Verlag GmbH & Co. KGaA, Weinheim
ISBN: 978-3-527-32406-4

and chirality. We assess the component of cohesion energy of the hybrid due to polarization of nanotube in response to the charged helical wrap. Last, we briefly discuss how the optical properties of DNA–SWNT hybrids, resulting from helical symmetry breaking, can be used to detect and identify the wrapping.

2.2
Physical Structure and Bonding in Nanotube–DNA Hybrids: A Short Review

A variety of polymers including ssDNA have been found to disperse SWNTs. It has been suggested that polyvinyl pyrrolidone (PVP) may wrap helically around nano-tubes [4] and that wrapping of SWNTs by water-soluble polymers is a general phenomenon resulting from the thermodynamic penalty associated with the hydro-phobic interface of the SWNTs and aqueous solvent. Although many polymers have been used to disperse SWNTs, ssDNA seems to be more efficient [8].

Different DNA sequences were used to find those most efficient at dispersion [9], including genomic sequences [10, 11]. Zheng *et al.* [7] successfully used poly(T) type DNA in 30-mer length, which is the most studied oligomer in this chapter. Ion exchange chromatography has been used with success to sort poly(GT) hybrids by SWNT structure and electronic properties [6, 7, 12]. With the addition of size exclusion chromatography to sort by length, separation resolution was improved to obtain the first substantial enrichment of a sample with a SWNT of a single chirality [13] and finally to provide separation by chirality [14].

At present, we do not have exact knowledge of the geometric structure of the hybrids, though various experimental groups have performed nanocharacterization of ssDNA–SWNT hybrids. Recent AFM images [15] and modeling suggest that the strands arrange approximately end-to-end on the SWNT surface and have fixed pitch, almost independent of the sequence. High-resolution transmission electron micros-copy (TEM) has been used recently to corroborate optical studies [16] of the DNA-wrapped SWNT material. Figure 2.1 from Ref. [17] shows clear images of SWNTs with helical ssDNA wrapping. Kawamoto *et al.* [18] found evidence that the DNA wrap persists even when DNA-wrapped nanotubes are dried. When dried, DNA–SWNT hybrids rebundle, but this process is reversible upon hydration.

Modeling indicates that DNA can wrap around a nanotube in a helix and can bind via π-stacking interactions with the sidewall of the nanotube [19–23]. DNA is generally oriented with bases stacked parallel to the walls [24]. DNA wrapping is presumed to be noncovalent and to leave the essential electronic characteristics of the nanotubes intact. For low ionic strength, adhesive interactions between the SWNT and DNA and electrostatic repulsion of the charges on the DNA are the main structure-determining factors and these allow a helical geometry [19]. Using Molecular Dynamics simulation, Johnson *et al.* [20] found a variety of possible geometries including right- and left-handed helical wrapping of DNA, along with more disordered structures. For lower salt concentrations, the helical pitch was less than 10 nm, and for higher concentrations, the helical conformation was not favored. Martin *et al.* also found a variety of distinct geometries that were dependent on the

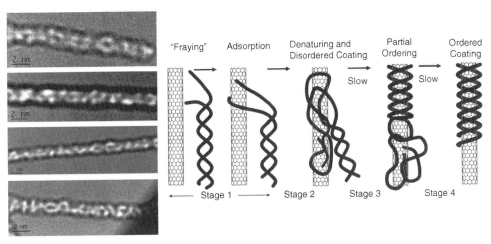

Figure 2.1 (Left) High-resolution TEM images of SWNTs wrapped with single-stranded DNA and (right) the scheme of the hybrid formation. (Reproduced from Ref. [17] with permission of the ACS.)

sequence and SWNT diameter [21]. Kinetics of wrapping [25] was studied with Monte Carlo methods [26] and penetration of DNA inside SWNT received attention in [27]. The electronic component of the cohesion energy in the hybrid has been studied in Refs [22, 28, 29].

While multiple theories have been developed on the physical structure of DNA–SWNT hybrids, their electronic structure did not received much attention until recently. Before it was known that ssDNA forms noncovalent bonds with the nanotube, earlier studies examined the effects of covalent functionalization [30, 31] of SWNT walls with a single DNA base on the nanotube band structure [32, 33]. More realistic approaches that take into account the helical geometry of the wrap and the absence of chemical bonding include a few heuristic models [34–37] where the perturbation of the DNA was limited to a single line of atoms on the SWNT surface or to a stripe of finite width. Finally in our works [38–40], an accurate theory of the electronic structure of the hybrid under helical symmetry breaking was developed. Double-stranded DNA hybrids were also modeled [41, 42].

In a standard experimental setup, with the polarization of both the incident light and the collected fluorescence signal along the SWNT axis, the peak shift upon the wrapping is small [43–48] but detectable, even at nanometer scale [49]. Such small changes in the optical bandgap, in fact, are consistent with the available theoretical results for parallel polarization [50]. Below, we consider optical absorption of light with perpendicular polarization, which has been measured for SWNTs dispersed using a surfactant [51–53]. In this polarization, the effect of helical symmetry breaking is evident and can be used to characterize the wrap. One useful feature for identifying the wrapping is the appearance of circular dichroism in nonchiral SWNTs. We stress that the contribution of the ssDNA to this circular dichroism can be discounted because the DNA absorption is higher in energy [54, 55] than the relevant features seen in our simulations, and it also develops upon *unwrapping* of the DNA from the SWNT [56, 57]. We note that other macromolecules can bind to the nanotube sidewalls and change the electronic structure substantially due to the breaking of charge neutrality and/or the symmetry of the system [58], which is beyond the scope of this chapter.

2.3
Quantum Mechanical Modeling of the Hybrid Structure: Tight Binding Band Structure Calculations

Among various theories developed for pristine nanotubes, several semiempirical approaches have been shown to reproduce band structures with high accuracy at low computational cost [3, 59–63]. We choose a tight binding model here to calculate the band structures of DNA–SNWT hybrids. Additionally, our method allows to determine a self-consistent charge density on the surface of the nanotube. Our theory is one-electron theory and as such takes into account higher-order many-body terms only approximately as discussed below.

The one-electron Schrödinger equation can be derived using the mean-field approximation in the spirit of Hartree theory [64]. The Hamiltonian

$$H = H_{\mathrm{TB}} + V_{\mathrm{DNA}} + W_{\mathrm{C}} \qquad (2.1)$$

consists of three main terms: H_{TB} is the bare electronic Hamiltonian of the SWNT taken in the tight binding approximation; V_{DNA} is the perturbation of the SWNT term due to the potential from the DNA wrap; W_{C} is the Coulomb term that we treated by using a self-consistent mean-field procedure. The DNA perturbation depends parametrically on the geometry of the wrap, which is considered to be fixed and known in our approach. Refinement of the DNA geometry is possible in a multiscale algorithm, which is beyond the scope of this publication. Next we consider all terms of the Hamiltonian in detail.

The simplest implementation of tight binding theory applied to carbon nanotubes uses an orthogonal basis [3, 59, 62], for example, it does not correct the electronic structure for nonzero curvature of the tube surface as compared to flat graphene and considers only the nearest neighbor interactions. Extended versions of tight binding that use a nonorthogonal basis (nonzero overlap between atomic orbits) and go beyond the nearest neighbor interactions [60, 63, 65] can be easily implemented in a similar way (see, e.g., how the nonorthogonal method changes the SWNT symmetry breaking in the external electric field in Ref. [66]). For the sake of clarity we present only an orthogonal basis tight binding theory below.

Bandstructure of a bare tube is derived using the Hamiltonian

$$H_{\mathrm{TB}} = \sum_{\langle ij \rangle} \gamma a_i^\dagger a_j, \qquad (2.2)$$

where a_i is a standard tight binding electron annihilation operator at the atomic site i, $\gamma \simeq 2.9$ eV is the hopping integral (characteristic energy scale of the model), and the sum is taken over the nearest neighbor carbon atoms in the SWNT. We note that all the many-body effects that persist in the one-electron approximation of the bare nanotube (such as Hartree–Fock terms) are already included in the empirical tight binding parameter γ and must not be included again. New terms, however, can appear with the perturbation. These will be included explicitly in the last term of the Hamiltonian (2.1).

This bare Hamiltonian matrix H_{TB} includes connectivity information for the carbon atoms of the whole SWNT. Using a standard Bloch theory [64], it can be reduced to those that are within an elementary translational period of our system. Then the matrix of dimensions $B \times B$ must be diagonalized, where B, in the general case, is the number of nanotube carbon atoms in the supercell. The high symmetry of the unit cell of the bare tube [62, 67, 68] allows substantial reduction in the dimension of the matrix problem for an unwrapped SWNT. However, this is not possible for the hybrid, whose symmetry is lowered due to the DNA wrapping. As the matrix for realistic supercells is large, special computational algorithms are needed to speed up the eigenproblem solution.

The main cause of helical symmetry distortion of the hybrid is a combination of the Coulomb potentials of the partial charges of the atoms of ssDNA and the non-compensated charges of the electrons on the SWNT itself that move in the field of the DNA. Many atomic charges of the DNA are small and partially cancel the effects of nearby charges. Thus, we proposed a model that approximates the effect of the partial atomic charges of the nucleotides with point charges having the same net charge as the whole ionized nucleotide (base + sugar + phosphate group). Such a quantity can be computed from CHARMM (Chemistry at HARvard Macromolecular Mechanics) parameterization [69]. Since much of the significant net charges are on the phosphate group, our model places the point charges along a regular, right circular, infinite helix, corresponding to the idealized positions of the phosphate groups.

We verified that the model gives a qualitatively correct picture of the DNA potential as well as quantitatively correct scaling of the perturbation with charge density on the DNA backbone (quadratic as it must be in case of the centrosymmetric graphene lattice). Figure 2.2 shows a segment of poly(T) single-stranded DNA. It consists of a backbone of alternating phosphate groups and deoxyribose sugar attached to a nitrogenous base (thymine for this case). Figure 2.3b shows partial atomic charges for a DNA nucleotide with a thymine base. Figure 2.3a shows the same nucleotide with an additional phosphate group, so that the backbone can be seen clearly. Here the labeled charges correspond to the strongest of the partial charges, which are highlighted in Figure 2.3b.

For our tight binding simulations, a periodic boundary condition is applied. It is a nontrivial problem to make a choice of the unit cell of the complex; a commensurate one-dimensional periodic structure does not exist for every (helically symmetric) DNA potential and the given SWNT symmetry. The SWNT lattice itself may possess spiral symmetry incommensurate with the DNA, which would prohibit the use of the periodic boundary condition and the Bloch theory. Here, we limit ourselves to commensurate structures where the angle of the DNA wrap is chosen such that

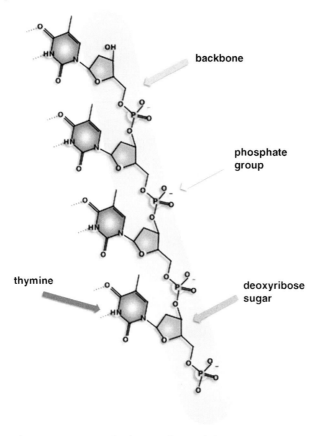

Figure 2.2 Segment of poly(T) single-stranded DNA.

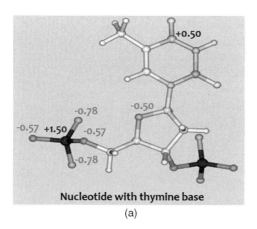

Nucleotide with thymine base

(a)

Figure 2.3 Structure of single-stranded DNA and partial atomic charges. (a) Nucleotide of DNA with a thymine base (additional phosphate group shown to clearly see backbone). (b) Partial atomic charges for a DNA nucleotide with a thymine DNA base from CHARMM [69].

```
RESI THY              -1.00
ATOM P       P         1.50
ATOM O1P     ON3      -0.78
ATOM O2P     ON3      -0.78
ATOM O5'     ON2      -0.57
ATOM C5'     CN8B     -0.08
ATOM H5'     HN8       0.09
ATOM H5''    HN8       0.09
GROUP
ATOM C4'     CN7       0.16
ATOM H4'     HN7       0.09
ATOM O4'     ON6B     -0.50
ATOM C1'     CN7B      0.16
ATOM H1'     HN7       0.09
GROUP
ATOM N1      NN2B     -0.34
ATOM C6      CN3       0.17
ATOM H6      HN3       0.17
ATOM C2      CN1T      0.51
ATOM O2      ON1      -0.41
ATOM N3      NN2U     -0.46
ATOM H3      HN2       0.36
ATOM C4      CN1       0.50
ATOM O4      ON1      -0.45
ATOM C5      CN3T     -0.15
ATOM C5M     CN9      -0.11
ATOM H51     HN9       0.07
ATOM H52     HN9       0.07
ATOM H53     HN9       0.07
GROUP
ATOM C2'     CN8      -0.18
ATOM H2'     HN8       0.09
ATOM H2''    HN8       0.09
GROUP
ATOM C3'     CN7       0.01
ATOM H3'     HN7       0.09
ATOM O3'     ON2      -0.57
```

(b)

Figure 2.3 *(Continued)*

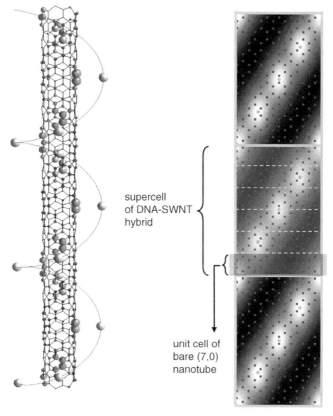

Figure 2.4 (Left) Schematic of a DNA–SWNT hybrid. (Right) Unscrolled nanotube surface showing three supercells for the (7,0) nanotube with one charged helical wrap over 6 unit cells of the bare nanotube.

one wind (or more winds) covers an integer number of nanotube lattice constants. For noncommensurate wrappings, it is always possible to find a commensurate wrapping with a wrapping angle that is infinitesimally close to the noncommensurate one, though the translational period of such a structure may be very large. An example of a supercell is shown in Figure 2.4. This case has four charges per helical wrap, which is an adjustable parameter in the model. The DNA is schematically represented by a backbone helix (all DNA atoms are removed from the drawing) with gray spheres indicating the positions of the backbone charges of the phosphate groups.

In Figure 2.5, we compare our simplified point charge model with the actual case of partial charges on all atoms of the DNA. We calculated the unscreened potential on the surface of a nanotube for both scenarios and showed that they are close, though not exactly the same. In solution, the charges are additionally screened by water molecules and counterions. We included this effect phenomenologically.

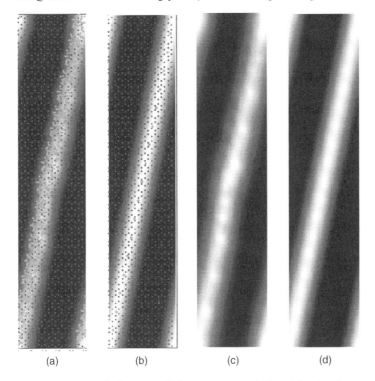

(a)   (b)   (c)   (d)

Figure 2.5 Supercell of an unscrolled DNA-wrapped (10,0) nanotube. (a) All DNA atoms with partial charges from CHARMM [14] are shown projected on the nanotube with the resulting potential. Atomic positions are from MD simulation. (b) Idealized helical perturbation with point charges evenly spaced along a circular helix a fixed distance from the nanotube surface. (c) and (d) show only the electric potential on the surface of the nanotube due the charges given in (a) and (b), respectively.

The electric potential of the carbon atom at position r_i due to q_j, the charges of the helical DNA wrap, is

$$V(i) = \sum_j \frac{q_j}{|r_i - r_j|} e^{-|r_i - r_j|/\lambda}, \tag{2.3}$$

where $|r_i - r_j|$ is the distance from the atomic site to the charge, λ is a cutoff parameter, and Gaussian units are used. The potential of charged wraps of the neighboring supercells must be included in the summation. Due to the cutoff factor, which we set equal to 10 radii in our calculation, the helical charges far away from the observation point are irrelevant. The physical reason for the cutoff is the charge screening in solution, where the potential of the ionized DNA always decays at large distances.

The Coulomb perturbation of the DNA backbone in our model is represented by the diagonal term in the Hamiltonian

$$V_{DNA} = \sum_i V(i) a_i^\dagger a_i = \sum_i e\varphi^{xt}(i) a_i^\dagger a_i, \tag{2.4}$$

where the sum over i runs over all atoms of the supercell. The potential φ^{xt} is different from the acting electrostatic potential at the lattice site, φ^{act}, as it is seen by the nanotube electrons. The acting potential also includes the polarization of the SWNT in the field of the DNA, that is, its induced potential. This extra term appears from the Coulomb interactions between the SWNT electrons. Like Hartree theory in solids, we account for it with

$$W_C = \sum_{ijnm} W_{ijnm} a_i^\dagger a_j^\dagger a_n a_m \simeq \sum_{in} W_{inni} a_i^\dagger a_i \langle a_n^\dagger a_n \rangle, \qquad (2.5)$$

where W_{ijnm} is the bare Coulomb matrix element between the tight binding electronic states, and we include the first nonvanishing correction through the mean-field charge density induced by the DNA perturbation $\delta\varrho \propto \langle a_n^\dagger a_n \rangle$. We emphasize once again that the similar effects in the bare tube are already included in the bare tight binding Hamiltonian, H_{TB}, and taking it into account here would result in a double-counting.

The charge density $\delta\varrho$ is determined by states below the Fermi level (at zero temperature; the Fermi–Dirac distribution function must be integrated to obtain $\delta\varrho$ at $T > 0$)

$$\delta\varrho(\vec{x}) = \langle e a^\dagger a \rangle = e \sum_{\lambda \leq \lambda_F} \left(|\psi_\lambda(\vec{x})|_\varphi^2 - |\psi_\lambda(\vec{x})|_0^2 \right), \qquad (2.6)$$

where λ is the full set of quantum numbers and λ_F corresponds to the Fermi level. The Fermi level is always in the middle of the bandgap in all the cases presented here to ensure charge neutrality of the SWNT. This means that the DNA wrapping happens in an aqueous solution where the electron transport to and from the nanotube is restricted. $|\psi_\lambda|_\varphi^2$ is the probability density at a given value of the electric potential $\varphi \neq 0$, as obtained from the Schrödinger equation with the Hamiltonian (2.1)

$$\hat{H}\psi_{\lambda,\varphi} = E_\lambda[\varphi]\psi_{\lambda,\varphi}, \qquad (2.7)$$

while the same quantity at zero potential (for the bare SWNT) is $|\psi_\lambda|_0^2$ and derived from $\hat{H}_{TB}\psi_{\lambda,0} = E_\lambda[0]\psi_{\lambda,0}$.

Both the DNA perturbation term and the Coulomb term can be considered as potential energy terms, diagonal in the real space representation of the tight binding technique and additive. However, the difference is in their physics: the term V_{DNA} is an external potential, independent of the polarization properties of the nanotube, while the term (2.5) is due to the induced charge density and as such depends on both the V_{DNA} and the induced charge density itself.

To dismantle the problem we represent the total (or acting) potential by the sum of two components

$$V = e\varphi^{act} = e(\varphi^{xt} + \varphi^{ind}) = e \left(\varphi^{xt} + \int \frac{\delta\varrho(\vec{\xi})}{|\vec{x} - \vec{\xi}|} d\vec{\xi} \right), \qquad (2.8)$$

where φ^{xt} is the unscreened potential due to DNA charges, cf. Eq. (2.4), and the induced (screening) potential φ^{ind} is calculated as the Poisson integral, the last term of Eq. (2.8).

These equations are solved iteratively to yield the self-consistent polarization charge density and resulting self-consistent induced potential.

2.4
Self-Consistent Computation Scheme: Acting Potential

The nominal linear charge density of the ionized ssDNA in vacuum is e/l, where $l \simeq 7$ Å (size of the DNA base), and $e = 1.6 \times 10^{-19}$ C is the elementary charge. In solution, however, the charge density decreases due to the surrounding counterion contribution and polarization of the solvent. This effect may be addressed phenomenologically by changing the elementary charge to a smaller effective charge, $e \rightarrow e^* \leq e$. A more accurate approach would require a microscopic description of the molecules and ions in solution around the hybrid. Significant modeling volume of the representative unit cell, computational difficulties of weakly converging Coulomb potentials of the particles, and even the absence of a well-established physical model prohibit such an approach at the present.

Other papers treated the DNA perturbation as a linear charge with given charge density [35, 36]. Our study shows that such approximation results in electrostatic potentials of slightly different shapes (than for the point charges), though it properly captures the symmetry breaking of the bandstructure. Such a model is convenient for analytical calculations [36] but not for a numerical study as presented in this chapter. Thus, we use the following model potential to describe the DNA perturbation

$$\varphi(i) = \frac{e^* \exp[-\sqrt{(z_i-Z)^2 + r^2 + R^2 - 2rR\cos(\theta_i-\Theta)}/\lambda]}{\sqrt{(z_i-Z)^2 + r^2 + R^2 - 2rR\cos(\theta_i-\Theta)}} \sum_j \delta(Z-Z_j)\delta(\Theta-\Theta_j),$$

(2.9)

where i and j label the atomic positions, where we determine the potential and the positions of the DNA phosphate groups with the cylindrical coordinates (r, θ_i, z_i) and (R, Θ_j, Z_j), respectively. Here r and R are the SWNT radius and the average radius of the DNA wrapping, and λ is the cutoff parameter.

The second, induced, component of the perturbation requires solving the Schrödinger equation (2.7) whose solutions are the eigenvectors of the Hamiltonian matrix (2.1) with the perturbation. Eigenvectors of that matrix give the amplitudes of the tight binding wave function at atomic site positions, $\psi_\lambda(i)$.

The trace of the density matrix $|\psi_\lambda(i)\rangle\langle\psi_\lambda(j)|$ calculated from the Hamiltonian with the perturbation proportional to φ^{act}, minus the trace of the density matrix determined from the Hamiltonian with zero perturbation, H_{TB}, gives $\delta\varrho$. Knowing the charge density (Eq. (2.6)), we can calculate the induced (screening) potential via the Poisson equation

$$\varphi^{ind}(\vec{x}) = \int \frac{\delta\varrho(\vec{\xi})}{|\vec{x}-\vec{\xi}|} \, d\vec{\xi} = \sum_i \int \frac{\delta(\vec{\xi}-\vec{r}_i)}{|\vec{x}-\vec{\xi}|} \, d\vec{\xi} \sum_{\lambda \leq \lambda_F} |\delta\psi_\lambda(i)|^2.$$

(2.10)

This expression formally does not converge for the values of $\vec{x}$ at the atomic site positions because $\delta\varrho(\vec{\xi})$ is zero anywhere except for those atomic site positions $\vec{r}_i$ (as a solution of the tight binding matrix problem), and the induced potential, as it should be used in the Hamiltonian, diverges as $1/|\vec{x}-\vec{\xi}|$. This divergence is rather non-physical and due to the fact that the tight binding eigenvector is not a smooth wave function of the coordinates as it should be. Thus, a simple regularization technique is applied: we replace the point-like source $\delta(\vec{\xi}-\vec{r}_i)/|\vec{x}-\vec{\xi}|$ in the integral (2.10) with a potential of a Gaussian distribution of charge, located at the same site i: $G(|\vec{\xi}-\vec{r}_i|) \propto \exp[-|\vec{\xi}-\vec{r}_i|^2/b^2]$, where b is a characteristic length of the order of the interatomic distance in the SWNT (whose exact value does not influence the results). We ensure that the normalization condition for the wave function is satisfied and also use an adaptive mesh to numerically evaluate the induced potential (2.10). This regularization of the tight binding charge density function gives us $\delta\varrho^R(\vec{\xi}) = e\sum_i\sum_{\lambda\leq\lambda_F}|\delta\psi_\lambda(i)|^2 G(|\vec{\xi}-\vec{r}_i|)$.

We note that for illustrative purposes it is still convenient to plot the nonregularized eigensolution of the matrix problem because the square of the absolute value of the wave function $|\delta\psi_\lambda(i)|^2$ also has a simple physical interpretation as partial charges on the carbon atoms of the SWNT induced upon the hybridization. The numerical computation, though, uses only the regularized charge density function $\delta\varrho^R(\vec{x})$.

We apply an iterative procedure to determine the self-consistent charge density. We start with the polarization charge density of nth step as determined from the Schrödinger equation (2.7) with the perturbation proportional to the induced potential determined at the step $(n-1)$ (for the first step we use the Hamiltonian with the external potential V_{DNA} only, which corresponds to the sole φ^{xt} term in Eq. (2.8)). We then iterate perturbation potential $\varphi^{act}[\delta\varrho_{(n-1)}]$ and the polarization charge density $\delta\varrho[H_{\varphi n}]$. Figure 2.6 shows polarization charges on atomic sites as a sequence of 11 steps of this iterative process. Clear oscillations of the charge density are seen; special numerical convergence procedures can be applied for damping such oscillations and speeding up the convergence.

2.5
Screening Factor and the Dielectric Permittivity

The SWNT responds to the external potential induced by the helical chain of point charges by redistributing the electronic charge density to have maximum screening of the external potential. As a result the total (acting) potential is substantially lower

Figure 2.6 Polarization charges for the hybrid discussed above are shown for several iterations of the self-consistent procedure.

than the external (bare) potential. In an analogy with solid state, we can try to define the self-consistent value of the acting electric potential via the value of the external potential of the DNA and a constant that describes the screening due the polarization of the SWNT: $\varphi^{act} = \varphi^{xt}/\varepsilon$. This constant ε is similar to the macroscopic dielectric permittivity of an insulator or a semiconductor. Since the system lacks total three-dimensional translation symmetry, such a dielectric constant cannot be defined equally everywhere in the space. Moreover, the electron charge density is localized on the surface of the nanotube, but the DNA charge is above this surface. Since these two charge densities have different helical symmetries (in the general case), the ratio $\varphi^{xt}/\varphi^{act}$ is not necessarily the same at different carbon lattice sites. Figures 2.6 and 2.7 show that the distribution of the polarization charge follows the external potential and is nonuniform along the circumference and along the axis of the nanotube.

We emphasize the difference in description of the depolarization of the external potential of the helical wrap of the DNA and the classical formalism of the dielectric function of the medium: the "dielectric constant" of the hybrid depends on the SWNT type, on the symmetry of the DNA wrap, and even on the properties of the surrounding medium, because the polarization charge of the nanotube is sensitive to all of these parameters. However, for a complex of given symmetry, in vacuum, we can accurately calculate the polarization charge at each of the atoms and compute the (position dependent) screening factor.

Very surprisingly we found (within the model) quite a weak variation in the screening factor. With high accuracy, the ratio $\varphi^{xt}/\varphi^{act} = \varepsilon$ is nearly constant and is 2.2 ± 0.5 over the atomic sites on the SWNT surface within the unit cell, independent of the NT type and wrap geometry, which means that nanotube electrons provide a coherent screening response to the DNA potential.

The result of the calculation for a typical case of (7,0) SWNT wrapped by fully ionized DNA with the charge of a phosphate group $e^* = e$ is shown in Figure 2.7. In this given complex, the DNA has four bases per supercell. The positions of the phosphate groups are easily distinguished in Figure 2.7 as the maxima of the positive charge density.

2.6
Polarization Component of Cohesion Energy of the SWNT–ssDNA Hybrid

A thorough understanding of the binding between DNA and a carbon nanotube is critical to both optimizing SWNT dispersion and allowing the development of accurate biological and chemical sensors based on these hybrids. Within the framework of our quantum mechanical model, we have calculated the polarization component of the energy of cohesion of an ionized DNA molecule to the surface of a nanotube.

The polarization interaction appears in complexes of DNA with highly polarizable substrates because the backbone of the DNA molecule possesses substantial charge density in solution [69]. The ionization of phosphate groups between bases of ssDNA results in the DNA having a linear charge density of $\sim e/(7\,\text{Å})$, neglecting compensation due to counterions in solution. In response to the net negative charge of the phosphate groups of the DNA, electrons (and holes) rearrange themselves, as shown in Figure 2.7.

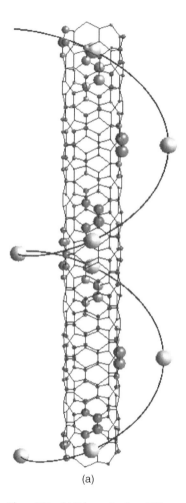

(a)

Figure 2.7 (a) Schematic of a ssDNA-wrapped carbon nanotube. Here a (7,0) nanotube is helically wrapped by DNA, with periodic charges corresponding to the charged phosphate groups equally spaced along an idealized helix. This particular case shows one helical wrap per six unit cells of the bare nanotube. The DNA charges cause the π-electrons of the nanotube to redistribute on the nanotube surface. Holes are attracted to the areas near the phosphate charges, and electrons migrate to the areas furthest from the phosphate groups, resulting in a polarization of surface charge. (b) Surface charge for a supercell of the hybrid structure of (a) unscrolled. The grayscale indicates relative electric potential at the surface of the nanotube due to the charged DNA wrap.

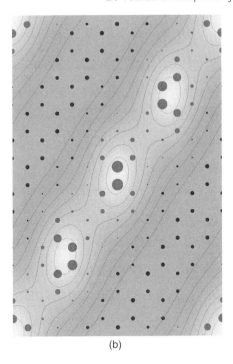

(b)

Figure 2.7 *(Continued)*

Calculation of the polarization interaction is a complex problem because in principle it must self-consistently take into account evolution of the geometry of the hybrid and electronic structure upon wrapping. In our method, DNA geometry is fixed and obtained from molecular dynamics with force constants parameterized without specifics of the polarization interaction. Thus, the Hamiltonian of the system uses "frozen" coordinates of all the atoms. Moreover, we assume that the polarization of the electronic subsystem of the DNA molecule is much weaker than the polarization of the electronic subsystem of the SWNT and neglect the former compared to the latter. This approximation is valid because the valence electrons of the single-walled nanotube form a highly polarizable electron shell with a density of states in the order of $4/\pi\hbar V_F$, where $V_F \simeq 10^8$ cm/s is the Fermi velocity in the nanotube.

Knowing the electron density induced at the SWNT surface by the DNA charges, we calculated the component of the cohesion energy due to the polarization interaction. The reduced cohesion energy (per single base of the DNA) equals

$$\delta\varepsilon = \frac{l}{\mathcal{L}}\sum_{\text{DNA}}\int \frac{e^*_{\text{DNA}}\delta\varrho(\vec{\xi})}{\left|\vec{x}_{\text{DNA}} - \vec{\xi}\right|}\,d\vec{\xi}, \tag{2.11}$$

where $\mathcal{L}$ is the length of the DNA per period of a single unit cell, and l is the distance between bases (that is, their ratio is the number of charges of the DNA per unit cell,

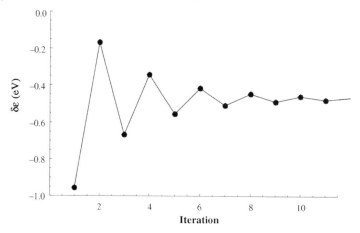

Figure 2.8 The convergence of the polarization component of the cohesion energy for the DNA–SWNT complex described in the work. Eleven steps of iteration are shown, including the nonself-consistent value for the first step.

which is four in the case presented in Figure 2.7). We obtain the value $\delta\varepsilon \simeq 0.47$ eV for the complex shown above. The effective screening coefficient of the nanotube, the effective SWNT permittivity discussed above, is approximately equal to two in this case. Correspondingly, the polarization component of the cohesion energy decreases by half as compared to the nonself-consistent value $\simeq 0.96$ eV at the first iteration (Figure 2.8).

If we compare the energy of the polarization component of cohesion and the classical attraction energy of an elementary charge to a metal surface

$$\delta\varepsilon_{cl} \sim \frac{e^2}{2(R-r)}, \tag{2.12}$$

where the distance $(R-r) \simeq 4.35\text{Å}$ corresponds to the average radial distance between a phosphate group and the nanotube surface, we obtain a quantity almost four times the correct value. Thus, the estimation of the polarization interaction via the interaction with the metal cylinder gives an essentially overestimated value. On the other hand, the transverse permittivity of the nanotube has a typical value $\varepsilon \simeq 5$, and still for the interaction of an elementary charge with the charge of the image in the insulator $e\frac{\varepsilon-1}{\varepsilon+1}$, we obtain a quantity more than two times the actual one. (We note that this modeling has been performed for the hybrid structure in "vacuum" as no external screening medium has been accounted for.)

Figure 2.7 presents a sample hybrid geometry. We varied the angle of the DNA helix, effectively varying the pitch of the wrap, as shown in Figure 2.9. An attempt was made to keep the linear charge density of the hybrids as constant as possible, and this determined the number of charges placed along the helical wrap. For these structures, the average polarization cohesion energy per DNA base (or per phosphate group/charge) varied approximately linearly from -0.24 eV/base for the smallest

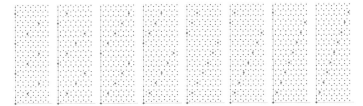

Figure 2.9 Hybrid geometries for (7,0) SWNT with varied wrapping angles. The first structure shows one wrap over three unit cells of the bare tube, and the wrapping angle increases incrementally to the final structure shown, where one wrap covers 10 unit cells of the bare nanotube. SWNT surfaces are shown unscrolled.

pitch to −0.91 eV/base for the largest pitch. Cohesion energy is larger for the larger pitch, which might be expected due to the decrease in the electrostatic repulsion of charges. The total polarization component of cohesion energy of the supercell was proportional to the square of the number of charges along the wrap, or said another way, the average polarization cohesion energy per DNA base is proportional to the number of DNA bases (or number of charges) per supercell (see Figure 2.10). The polarization cohesion energy per base (or per length of the nanotube) varied as the fourth power of the angle of the wrap with respect to the circumference of the nanotube, as shown in Figure 2.10b. The effective screening coefficient of the nanotube for this series of geometries varies from 1.83 to 2.09, in reasonable agreement with our previous result.

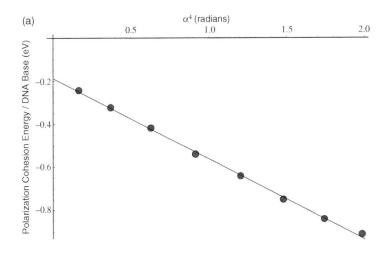

Figure 2.10 (a) Polarization cohesion energy per DNA base is plotted versus α^4, where α is the wrapping angle of the DNA with respect to the circumference of the nanotube. (b) Polarization cohesion energy per DNA base is plotted versus the number of DNA bases (or the number of charges) in the supercell.

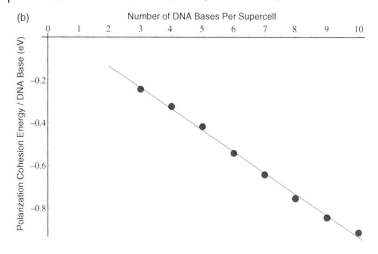

Figure 2.10 (*Continued*)

2.7
Optical Absorption of SWNT–DNA Hybrids

The main influence of the DNA wrap on the electronic structure of the SWNT in the helical hybrid is symmetry breaking; for a nonchiral tube, the helical perturbation generates "natural" optical activity in the DNA–nanotube complex. We demonstrated that the one-electron absorption spectrum for light polarized *perpendicular* to the nanotube axis is sensitive to the wrapping while the *parallel* light polarization shows features that are similar to bare SWNT. This may seem counterintuitive considering the large linear charge density of the ionized DNA and the large strength of the perturbation. We explain it by noting that for the parallel polarization absorption (conventions for polarization directions are shown in Figure 2.11), the effect of the DNA wrap is averaged over the SWNT circumference that smears out the induced charge and, in case of a charge neutral SWNT, gives almost zero net effect. On the contrary, the perpendicular polarization absorption is proportional to the first nonvanishing angular term, breaking the cylindrical symmetry, and therefore having a significant perturbation strength.

Lifting of optical selection rules results in new optical transitions and circular dichroism of the complex.

As the optical properties of a SWNT are dependent on the radius and chirality of the nanotube, spectroscopy can be used to characterize bulk samples of the nanotubes. The large aspect ratio and quasi-one-dimensional nature of SWNTs results in sharp peaks in their one-electron density of states, the so-called van Hove singularities, as shown in Figure 2.11b. This leads to distinct signatures for individual nanotube structures when emission intensity is plotted versus parallel excitation and emission energies in an absorption/luminescence plot [70].

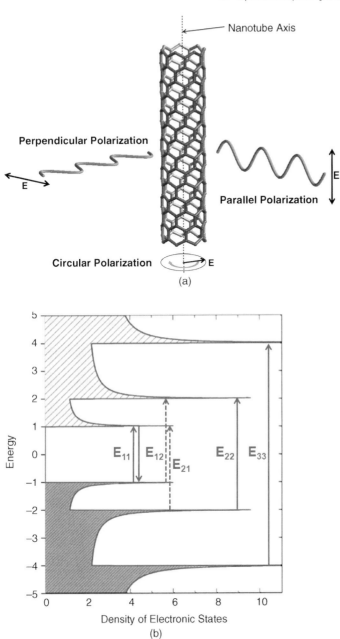

Figure 2.11 (a) Schematic of polarization directions given by the relative orientation of the electric field vector of the absorbed or emitted light and the nanotube axis. (b) Schematic of density of states for a semiconducting single-walled carbon nanotube.

For a pristine semiconducting SWNT, an E_{11} transition denotes absorption across the bandgap between the first van Hove peak below the Fermi level and the first peak above the Fermi level. In this case, the peaks correspond to states having the same angular momentum (the solid arrows in Figure 2.11b). In the many-body picture, this corresponds to the first excitonic transition. An E_{12} transition occurs between states with angular momenta, which differ by ± 1 (the dashed arrows in Figure 2.11b). The selection rule for the angular momentum quantum number of $\Delta m = 0$ for absorption of light polarized parallel to the nanotube axis, and $\Delta m = \pm 1$ for light polarized perpendicularly to the nanotube axis, is dictated by the odd symmetry of the momentum operator and the even symmetry of the product of wave functions with the same angular momentum for the electron and the hole.

Our simulations predict new peaks in the frequency range near the E_{11} transitions, prohibited for bare, pristine nanotubes, in the cross-polarized absorption spectra of DNA–SWNTs. In Figure 2.12, we have plotted the calculated optical strength versus simulated emission along the tube axis and simulated absorption of light polarized across the tube axis for the (7,0) SWNT, with and without a DNA wrap. The figure shows a dramatic difference in the absorption/luminescence map in the region of the first van Hove singularity, E_{11}. The optical absorption coefficients for light polarized across the axis of the semiconducting zigzag (7,0) DNA–SWNT hybrids drastically differ from the bare tube absorption in the same polarization. The first absorption peak in cross-polarization for the bare SWNT corresponds to E_{12} and E_{21} transitions. This peak is also present for the DNA–SWNT hybrid, although it shifts to higher frequency. In addition, a shoulder at lower frequency near the bare E_{11} transitions appears as a consequence of the lifting of selection rules [38].

We note that the excitonic effects can be easily included in the picture. The presence of excitons decreases the bandgap and, thus, decreases the energy of emission. For E_{11} transitions, analytical exciton theory shows that exciton binding energy is a fraction of the bare optical bandgap [48], which allows simple rescaling.

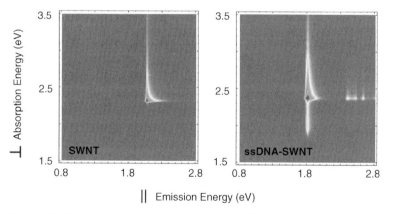

Figure 2.12 Calculated optical strength versus energy of the absorbed photon with perpendicular polarization and energy of the emitted photon with parallel polarization (absorption/luminescence map) for the bare (7,0) nanotube (left) and the DNA-wrapped (7,0) nanotube (right).

On the other hand, the exciton correction is expected to be essentially weaker for E_{12} transitions due to decay of the direct Coulomb matrix element with nonzero angular momentum transfer $\Delta m \neq 0$ [71], and thus it is neglected in this study.

In contrast to the bare tube, the subbands of the wrapped SWNT do not have definite angular momentum since the electron (and hole) wave functions are helically polarized by the Coulomb potential of the DNA. Thus, cross-polarized optical transitions are allowed at a lower frequency near that of the prohibited transition at E_{11}. The physical interpretation of this effect is that the polarization of the electron (hole) under the perturbation of the transverse electric field of a nearby DNA phosphate group creates across the tube a permanent dipole, which may be excited by the perpendicular electric field of incident light. Overall, (negative) electron density shifts to the opposite side of the tube and (positive) hole density shifts toward the DNA backbone [29].

To further prove the helical symmetry of the perturbed electronic structure of a DNA–SWNT hybrid, we plot in Figure 2.13 some representative wave functions for electronic states near the Γ-point in conduction and valence bands in a bare nanotube and a hybrid. The left panel shows the hole wave function for the highest occupied molecular orbital (HOMO), the highest valence subband, and the electron wave function for the lowermost unoccupied molecular orbital (LUMO), lowermost conduction subband, of the bare (5,0) nanotube.

Panels (b) and (e) show wave functions for the same subbands of the DNA-wrapped nanotube. For this particular wrap, the two lowest optical transitions in perpendicular polarization occur between HOMO-2 (shown in panel c) and LUMO and between HOMO and LUMO + 2 (shown in panel f), as indicated red arrows. It can be seen that corresponding wave functions of the hybrid have helical symmetry replicating the DNA wrap geometry.

To observe such an effect experimentally, one must tune to a singularity in the optical density of states. In the bare tube this happens only at the edges of the Brillouin zone. Additional singularities arise with the wrapping due to subband flattening.

For a DNA–SWNT hybrid, the supercell length (translational period of the complex) contains at least one wrap and typically includes many unit cells of the bare tube (Figure 2.4b, inset). Thus, the Brillouin zone of the complex is folded, and electron energy dispersion becomes flat at the new zone boundary due to the reflection of electrons at half the reciprocal lattice vector (when $\Delta k = n\pi/a$, where n is an integer and a is the supercell length), that is, near the new Brillouin zone boundary.

The partial absorption coefficient is as follows:

$$
\begin{aligned}
\alpha_{\pm}(\hbar\omega, k) \propto \sum_i \sum_f &\frac{e^2 \left| \langle \psi(k, m_f, \lambda_f) | \vec{p} \cdot \vec{E}_{\pm} | \psi(k, m_i, \lambda_i) \rangle \right|^2}{m_0 \omega} \\
&\times \frac{f(E_i(k, m_i, \lambda_i))[1 - f(E_f(k, m_f, \lambda_f))]\Gamma}{[(E_f(k, m_f, \lambda_f) - E_i(k, m_i, \lambda_i)) - \hbar\omega]^2 + \Gamma^2}.
\end{aligned}
\tag{2.13}
$$

Here $f(E)$ is the Fermi–Dirac function, Γ is the level broadening (set at 0.008 eV for these calculations),

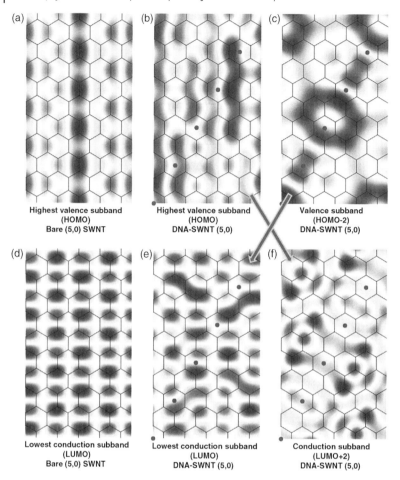

Figure 2.13 Selected wave functions for zero wave vector as projected onto the unscrolled cylindrical nanotube surface. Projected positions of charges along an idealized helix are shown as gray dots. (a) and (d) Closest conduction and valence subbands of the bare (5,0) nanotube. (b–f) and (c–e) Selected subbands of the ssDNA–SWNT hybrid that contribute to the lowest strong absorption peak in cross-polarization. Arrows indicate the optical transitions.

$$|p_\pm| = er |\langle \psi(k, m_f, \lambda_f)| \, e^{\pm i\theta} |\psi(k, m_i, \lambda_i)\rangle| \tag{2.14}$$

is the matrix element for a transition from an initial state in the valence band $|k, m_i, \lambda_i\rangle$ to a final state in the conduction band $|k, m_f, \lambda_f\rangle$ for the case of circularly polarized incident light, $\vec{E} \propto e^{\pm i\theta}$, with the selection rule for angular momentum, $\Delta m = m_f - m_i = \pm 1$. For a wrapped tube absorption is different for left- and right-polarized light, thus light with linear polarization (which is a sum of left- and right-polarized components) will change its linear polarization direction while propagating through the DNA–SWNT hybrid medium.

Such "natural" optical activity of the material is directly related to the helicity of the electron states. The total absorption spectrum is obtained by integrating the partial absorption coefficient (Eq. (2.13)) over the wave vector inside the first Brillouin zone, $\alpha_\pm(\hbar\omega) = \int_{BZ}\alpha_\pm(\hbar\omega, k)\,dk$, and is plotted in Figure 2.14 for the left and right circular polarizations $\Delta m = \pm 1$ (dashed and dotted curve, respectively) for an (8,0) DNA–SWNT hybrid. Absorption of circularly polarized light drastically differs for the hybrid and the bare tube (solid curve in Figure 2.14).

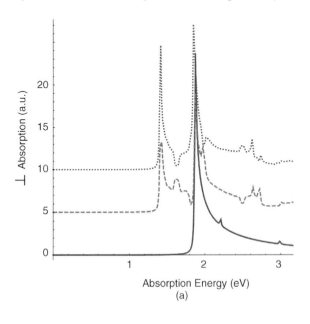

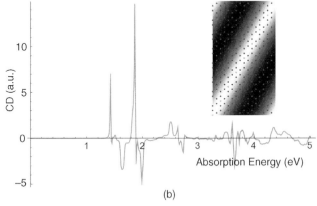

Figure 2.14 (a) $\perp$ absorption of the bare (8,0) SWNT (solid curve) compared to left and right polarized light absorption in the DNA-wrapped tube (dashed and dotted curves, respectively). Curves are offset for clarity. (b) Circular dichroism for an (8,0) ssDNA–SWNT hybrid. Inset shows geometry of the hybrid with charges projected onto the unscrolled surface of the nanotube.

The difference in absorption for two polarizations gives the circular dichroism spectrum of the DNA–SWNT hybrid: $CD = \alpha_+ - \alpha_-$. We calculated the optical circular dichroism spectra for a variety of SWNT–DNA hybrids and found strong dichroism for originally achiral as well as chiral tubes. Figure 2.14b shows the CD spectrum of the (8,0) DNA–SWNT hybrid. We stress that this circular dichroism is unrelated to the possible chirality of the DNA. First, the single-stranded DNA does not make a clear helix on its own, without wrapping around the SWNT. Second, DNA absorption is not included in this work at all, although a recent study [48] showed an interesting DNA hypochromicity effect in the hybrids. One can exclude DNA from the spectral analysis because the DNA absorption edge [48, 54] is at approximately 4.1 eV, which is well above the main IR spectral features of the SWNT itself. Hence, the optical activity must be fully attributed to the nanotube itself and to its helical

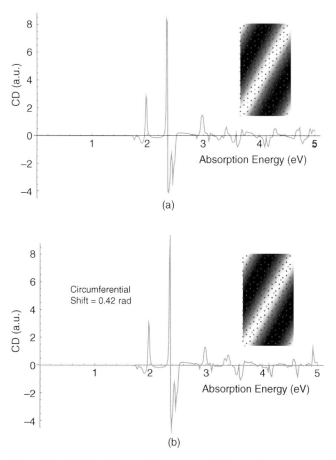

Figure 2.15 Calculated circular dichroism for a (7,0) nanotube with (a) wrap along zigzag direction with charges centered above carbon hexagons and (b) wrap with same wrapping angle, but shifted about nanotube circumference.

symmetry breaking. This conclusion is further supported by the experimental data in Ref. [45].

We found that the strength of the perturbation is proportional to the total charge of DNA in the supercell as well to the SWNT radius. Thus the typical splitting of the subbands scales as $\propto r^4$ and becomes stronger for tubes of larger radius. We found that chiral symmetry breaking is also present in wraps with different wrapping angles, with circular dichroism and absorption peak splitting clearly observed, though with very different absorption spectra. We observed that the symmetry breaking is most sensitive to the symmetry matching between the SWNT lattice and the DNA backbone helical angle.

For various angles and for various tubes we calculated different absorption spectra, though a general qualitative feature of the helical symmetry breaking, the appearance of new van Hove singularities in the optical data, was present. At the same time, when the axial or circumferential shift of the wrap was different, the CD spectrum had the same major features, though varied slightly, as shown in Figure 2.15. This suggests that exact placement of the ssDNA on the SWNT surface may not be important, given that the wrapping angle is fixed due to the microscopic interaction between the two. In such a case, cross-polarized optical spectroscopy can be a nondestructive optical characterization technique to determine wrapping efficiency by comparing intensities of the spectral lines corresponding to the bare and wrapped tubes.

2.8
Summary

In this chapter we consider the recently discovered hybrid structures of single-stranded DNA and single-walled carbon nanotubes and the influence of the charged DNA wrap on the electronic and optical properties of the SWNT.

We describe the tight binding model developed to calculate electronic band structure of DNA–SWNT hybrids and the self-consistent procedure for calculating electric potential on the surface of the nanotube. These methods have been applied to a variety of hybrids to study the dependence of the main parameters of the wrap, such as the cochirality of the wrap and the underlying lattice structure of the NT, the axial position of the wrap, the angular position of the wrap, the diameter of the NT, and the charge density of the wrap (distance between subsequent backbone charges as well as the charge of a single base), on the electronic structure. We verified that the model properly describes the broken symmetry and shows correct scaling of the perturbation with the charge density on the DNA backbone. We found that neither the axial, nor the circumferential position of the wrap plays a significant role in the perturbation, which is explained based on the large range of the Coulomb potential that results in self-averaging of the local fields and local atomic effects. However, there is a significant and nonmonotonic dependence on the chiral angle of the wrap (same as the pitch of the wrap for a fixed chirality and diameter of the NT), which is explained by a global symmetry breaking. A similar strong dependence was found on the diameter of the SWNT (for a fixed chirality).

As examples of our approach, the polarization component of the DNA–SWNT hybrid cohesion energy, cross-polarized absorption coefficient, and circular dichroism were calculated. We also estimated the screening effect of the SWNT polarization for various hybrids. We found a general feature of the broken symmetry of the nanotube electronic structure to be circular dichroism for originally nonhelical nanotubes as well as the appearance of new spectral lines in the optical absorption. These spectral lines should be experimentally sought for the case of polarization of light perpendicular to the nanotube axis. DNA wrapping should result in an extra peak at a frequency below the allowed E_{12} optical transitions in the bare tube, which may be used for nondestructive identification of the wrapping.

References

1 Avouris, P. (2007) Electronics with carbon nanotubes. *Phys. World*, **20**, 40–45.

2 Jorio, A., Dresselhaus, M.S., and Dresselhaus, G. (2008) *Carbon Nanotubes: Advanced Topics in the Synthesis, Structure, Properties and Applications*, Springer, ISBN: 0303-4216.

3 Reich, S., Thomsen, C., and Maultzsch, J. (2004) *Carbon Nanotubes: Basic Concepts and Physical Properties*, Wiley-VCH Verlag GmbH, ISBN: 3-527-40386-8.

4 O'Connell, M.J., Boul, P., Ericson, L.M., Huffman, C., Wang, Y., Haroz, E., Kuper, C., Tour, J., Ausman, K.D., and Smalley, R.E. (2001) Reversible water-solubilization of single-walled carbon nanotubes by polymer wrapping. *Chem. Phys. Lett.*, **342**, 265–271.

5 Arnold, M.S., Green, A.A., Hulvat, J.F., Stupp, S.I., and Hersam, M.C. (2006) Sorting carbon nanotubes by electronic structure using density differentiation. *Nat. Nanotechnol.*, **1**, 60–65.

6 Zheng, M., Jagota, A., Strano, M.S., Santos, A.P., Barone, P., Chou, S.G., Diner, B.A., Dresselhaus, M.S., Mclean, R.S., Onoa, G.B., Samsonidze, G.G., Semke, E.D., Usrey, M., and Walls, D.J. (2003) Structure-based carbon nanotube sorting by sequence-dependent DNA assembly. *Science*, **302**, 1545–1548.

7 Zheng, M., Jagota, A., Semke, E.D., Diner, B.A., Mclean, R.S., Lustig, S.R., Richardson, R.E., and Tassi, N.G. (2003) DNA-assisted dispersion and separation of carbon nanotubes. *Nat. Mater.*, **2**, 338–342.

8 Jagota, A., Diner, B.A., Boussaad, S., and Zheng, M. (2005) Carbon nanotube–biomolecule interactions: applications in carbon nanotube separation and biosensing, in *Applied Physics of Carbon Nanotubes, NanoScience and Technology* (eds S.V. Rotkin and S. Subramoney), Springer, Berlin Heidelberg, ISBN: 3-540-23110-2.

9 Vogel, S.R., Müller, K., Plutowski, U., Kappes, M.M., and Richert, C. (2007) DNA–carbon nanotube interactions and nanostructuring based on DNA. *Phys. Stat. Sol. B*, **244** (11), 4026.

10 Gigliotti, B., Sakizzie, B., Bethune, D.S., Shelby, R.M., and Cha, J.N. (2006) Sequence-independent helical wrapping of single-walled carbon nanotubes by long genomic DNA. *Nano Lett.*, **6** (2), 159–164.

11 Kim, S.N., Kuang, Z., Grote, J.G., Farmer, B.L., and Naik, R.R. (2008) Enrichment of (6,5) single wall carbon nanotubes using genomic DNA. *Nano Lett.*, **8** (12), 4415.

12 Lustig, S.R., Jagota, A., Khripin, C., and Zheng, M. (2005) Theory of structure-based carbon nanotube separations by ion-exchange chromatography of DNA/CNT hybrids. *J. Phys. Chem. B*, **109** (7), 2559–2566.

13 Zheng, M. and Semke, E.D. (2007) Enrichment of single chirality carbon nanotubes. *J Am. Chem. Soc.*, **129** (19), 6084–6085.

14 Tu, X. and Zheng, M. (2008) A DNA-based approach to the carbon nanotube sorting problem. *Nano Res.*, **1**, 185–194.

15 Campbell, J.F., Tessmer, I., Thorp, H.H., and Erie, D.A. (2008) Atomic force microscopy studies of DNA-wrapped carbon nanotube structure and binding to quantum dots. *J. Am. Chem. Soc.*, **130** (32), 10648–10655.

16 Malik, S., Vogel, S., Rösner, H., Arnold, K., Hennrich, F., Köhler, A.-K., Richert, C., and Kappes, M.M. (2007) Physical chemical characterization of DNA-SWNT suspensions and associated composites. *Compos. Sci. Technol.*, **67** (5), 916.

17 Cathcart, H., Nicolosi, V., Hughes, J.M., Blau, W.J., Kelly, J.M., Quinn, S.J., and Coleman, J.N. (2008) Ordered DNA wrapping switches on luminescence in single-walled nanotube dispersions. *J. Am. Chem. Soc.*, **130** (38), 12734–12744.

18 Kawamoto, H., Uchida, T., Kojima, K., and Tachibana, M. (2006) The feature of the Breit–Wigner–Fano Raman line in DNA-wrapped single-wall carbon nanotubes. *J. Appl. Phys.*, **99**, 094309.

19 Manohar, S., Tang, T., and Jagota, A. (2007) Structure of homopolymer DNA-CNT hybrids. *J. Phys. Chem. C*, **111** (48), 17835–17845.

20 Johnson, R.R., Johnson, A.T.C., and Klein, M.L. (2008) Probing the structure of DNA-carbon nanotube hybrids with molecular dynamics. *Nano Lett.*, **8** (1), 69–75.

21 Martin, W., Zhu, W., and Krilov, G. (2008) Simulation study of noncovalent hybridization of carbon nanotubes by single-stranded DNA in water. *J. Phys. Chem. B*, **112** (50), 16076–16089.

22 Enyashin, A.N., Gemming, S., and Seifert, G. (2007) DNA-wrapped carbon nanotubes. *Nanotechnology*, **18** (24), 245702.

23 Gowtham, S., Scheicher, R.H., Pandey, R., Karna, S.P., and Ahuja, R. (2008) First-principles study of physisorption of nucleic acid bases on small-diameter carbon nanotubes. *Nanotechnology*, **19**, 125701.

24 Meng, S., Wang, W.L., Maragakis, P., and Kaxiras, E. (2007) Determination of DNA-base orientation on carbon nanotubes through directional optical absorbance. *Nano Lett.*, **7** (8), 2312.

25 Esther, J.D.N., Jeng, S., Barone, P.W., and Strano, M.S. (2007) Hybridization kinetics and thermodynamics of DNA adsorbed to individually dispersed single-walled carbon nanotubes. *Small*, **3** (9), 1602.

26 Gurevitch, I. and Srebnika, S. (2008) Conformational behavior of polymers adsorbed on nanotubes. *J. Chem. Phys.*, **128**, 144901.

27 Gao, H. and Kong, Y. (2004) Simulation of DNA-nanotube interactions. *Annu. Rev. Mater. Res.*, **34**, 123–150.

28 Ranjan, N., Seifert, G., Mertig, M., and Heine, T. (2005) Wrapping carbon nanotubes with DNA: a theoretical study. AIP Conference Proceedings, Electronic Properties of Novel Nanostructures, CP786, pp. 448–451.

29 Snyder, S.E. and Rotkin, S.V. (2006) Polarization component of cohesion energy in single-wall carbon nanotube-DNA complexes. *JETP Lett.*, **84** (6), 348–351.

30 Katz, E. and Willner, I. (2004) Biomolecule-functionalized carbon nanotubes: applications in nanobio-electronics. *Chem. Phys. Chem.*, **5**, 1084–1104.

31 Czerw, R., Guo, Z., Ajayan, P.M., Sun, Y.-P., and Carroll, D.L. (2001) Organization of polymers onto carbon nanotubes: a route to nanoscale assembly. *Nano Lett.*, **1** (8), 423.

32 Song, C., Xia, Y., Zhao, M., Liu, X., Li, F., and Huang, B. (2005) *Ab initio* study of base-functionalized single walled carbon nanotubes. *Chem. Phys. Lett.*, **415**, 183–187.

33 Song, C., Xia, Y., Zhao, M., Liu, X., Huang, B., Li, F., and Ji, Y. (2005) Self-assembly of base-functionalized carbon nanotubes. *Phys. Rev. B*, **72**, 165430.

34 Wall, A. and Ferreira, M.S. (2006) Electronic contribution to the energetics of helically wrapped nanotubes. *Phys. Rev. B*, **74** (23), 233401.

35 Wall, A. and Ferreira, M.S. (2007) Electronic density of states of helically-wrapped carbon nanotubes. *J. Phys.: Condens. Mater.*, **19** (40), 406227.

36 Puller, V.I. and Rotkin, S.V. (2007) Helicity and broken symmetry of DNA-nanotube hybrids. *Europhys. Lett.*, **77** (2), 27006.

37 Michalski, P.J. and Mele, E.J. (2008) Carbon nanotubes in helically modulated potentials. *Phys. Rev. B*, **77**, 085429.

38 Snyder, S.E. and Rotkin, S.V. (2008) Optical identification of a DNA-wrapped carbon nanotube: signs of helically broken symmetry. *Small*, **4** (9), 1284–1286.

39 Rotkin, S.V. and Snyder, S.E. (2008) Cross-polarized spectroscopy of DNA-wrapped nanotubes: unraveling the nature of optical response. *Proc. SPIE, NanoSci. Eng.*, **7037**, 703–706.

40 Puller, V.I. and Rotkin, S.V. (2006) Characterization of DNA-carbon nanotube hybrids by their van hove singularities. *Proc. SPIE*, **6328**, 63280.

41 Lu, G., Maragakis, P., and Kaxiras, E. (2005) Carbon nanotube interaction with DNA. *Nano Lett.*, **5** (5), 897–900.

42 Zhao, X. and Johnson, J.K. (2007) Simulation of adsorption of DNA on carbon nanotubes. *J. Am. Chem. Soc.*, **129**, 10438–10445.

43 Chou, S.G., Ribeiro, H.B., Barros, E.B., Santos, A.P., Nezich, D., Samsonidze, G.G., Fantini, C., Pimenta, M.A., Jorio, A., Filho, F.P., Dresselhaus, M.S., Dresselhaus, G., Saito, R., Zheng, M., Onoa, G.B., Semke, E.D., Swan, A.K., Uenlue, M.S., and Goldberg, B.B. (2004) Optical characterization of DNA-wrapped carbon nanotube hybrids. *Chem. Phys. Lett.*, **397**, 296–301.

44 Jeng, E.S., Moll, A.E., Roy, A.C., Gastala, J.B., and Strano, M.S. (2006) Detection of DNA hybridization using the near-infrared band-gap fluorescence of single-walled carbon nanotubes. *Nano Lett.*, **6** (3), 371–375.

45 Dukovic, G., Balaz, M., Doak, P., Berova, N.D., Zheng, M., Mclean, R.S., and Brus, L.E. (2006) Racemic single-walled carbon nanotubes exhibit circular dichroism when wrapped with DNA. *J. Am. Chem. Soc.*, **128** (28), 9004–9005.

46 Fagan, J.A., Landi, B.J., Mandelbaum, I., Simpson, J.R., Bajpai, V., Bauer, B.J., Migler, K., Hight Walker, A.R., Raffaelle, R., and Hobbie, E.K. (2006) Comparative measures of single-wall carbon nanotube dispersion. *J. Phys. Chem. B*, **110** (47), 23801–23805.

47 Heller, D.A., Jeng, E.S., Yeung, T.-K., Martinez, B.M., Moll, A.E., Gastala, J.B., and Strano, M.S. (2006) Optical detection of DNA conformational polymorphism on single-walled carbon nanotubes. *Science*, **311** (5760), 508–511.

48 Hughes, M.E., Brandin, E., and Golovchenko, J.A. (2007) Optical absorption of DNA-carbon nanotube structures. *Nano Lett.*, **7** (5), 1191–1194.

49 Qian, H., Araujo, P.T., Georgi, C., Gokus, T., Hartmann, N., Green, A.A., Jorio, M.C.H.A., Novotny, L., and Hartschuh, A. (2008) Visualizing the local optical response of semiconducting carbon nanotubes to DNA-wrapping. *Nano Lett.*, **8** (9), 2706.

50 Rotkin, S.V. (2004) Formalism of dielectric function and depolarization in SWNT: application to nano-optical switches and probes. *Proc. SPIE*, **5509**, 145–159.

51 Miyauchi, Y., Oba, M., and Maruyama, S. (2006) Cross-polarized optical absorption of single-walled nanotubes by polarized photoluminescence excitation spectroscopy. *Phys. Rev. B*, **74** (20), 205440.

52 Islam, M.F., Milkie, D.E., Kane, C.L., Yodh, A., and Kikkawa, J. (2004) Direct measurement of the polarized optical absorption cross section of single-wall carbon nanotubes. *Phys. Rev. Lett.*, **93** (3), 037404.

53 Lefebvre, J. and Finnie, P. (2007) Polarized photoluminescence excitation spectroscopy of single-walled carbon nanotubes. *Phys. Rev. Lett.*, **98** (16), 167406.

54 Okada, T., Kaneko, T., Hatakeyama, R., and Tohji, K. (2006) Electrically triggered insertion of single-stranded DNA into single-walled carbon nanotubes. *Chem. Phys. Lett.*, **417**, 288–292.

55 Ostojic, G.N., Ireland, J.R., and Hersam, M.C. (2008) Noncovalent functionalization of DNA-wrapped single-walled carbon nanotubes with platinum-based DNA cross-linkers. *Langmuir*, **24** (17), 9784.

56 Gao, X., Xing, G., Yang, Y., Shi, X., Liu, R., Chu, W., Jing, L., Zhao, F., Ye, C., Yuan, H., Fang, X., Wang, C., and Zhao, Y. (2008) Detection of trace Hg^{2+} via induced circular dichroism of DNA wrapped

around single-walled carbon nanotubes. *J. Am. Chem. Soc.*, **130** (29), 9190.

57 Jin, H., Jeng, E.S., Heller, D.A., Jena, P.V., Kirmse, R., Langowski, J., and Strano, M.S. (2007) Divalent ion and thermally induced DNA conformational polymorphism on single-walled carbon nanotubes. *Macromolecules*, **40** (18), 6731–6739.

58 Johnson, A.T.C., Staii, C., Chen, R.J.M.L.K.M., Khamis, S., and Gelperin, A. (2006) DNA-decorated carbon nanotubes for chemical sensing. *Semicond. Sci. Technol.*, **21**, S17–S21.

59 Saito, R., Dresselhaus, G., and Dresselhaus, M.S. (1998) *Physical Properties of Carbon Nanotubes*, Imperial College Press.

60 Reich, S. and Thomsen, C. (2000) Chirality dependence of the density-of-states singularities in carbon nanotubes. *Phys. Rev. B*, **62** (7), 4273–4276.

61 Reich, S., Maultzsch, J., Thomsen, C., and Ordejon, P. (2002) Tight-binding description of graphene. *Phys. Rev. B*, **66** (3), 035412.

62 Milosevic, I., Vukovic, T., Dmitrovic, S., and Damnjanovic, M. (2006) Polarized optical absorption in carbon nanotubes: a symmetry-based approach. *Phys. Rev. B*, **73** (3), 035415.

63 Popov, V.N. and Henrard, L. (2004) Comparative study of the optical properties of single-walled carbon nanotubes within orthogonal and nonorthogonal tight-binding models. *Phys. Rev. B*, **70** (11), 115407.

64 Ziman, J.M. (1972) *Principles of the Theory of Solids*, 2nd edn, Cambridge University Press, ISBN: 0 52 29733 8.

65 Popov, V.N. (2004) Curvature effects on the structural, electronic and optical properties of isolated single-walled carbon nanotubes within a symmetry-adapted non-orthogonal tight-binding model. *New J. Phys.*, **6**, 17.

66 Li, Y., Ravaioli, U., and Rotkin, S.V. (2006) Metal-semiconductor transition and Fermi velocity renormalization in metallic carbon nanotubes. *Phys. Rev. B*, **73** (3), 035415.

67 Damnjanovic, M., Milosevic, I., Vukovic, T., and Sredanovic, R. (1999) Full symmetry, optical activity, and potentials of single-wall and multiwall nanotubes. *Phys. Rev. B*, **60** (4), 2728–2739.

68 Damnjanovic, M., Vukovic, T., and Milosevic, I. (2000) Fermi level quantum numbers and secondary gap of conducting carbon nanotubes. *Solid State Commun.*, **116**, 265–267.

69 Brooks, B.R., Bruccoleri, R.E., Olafson, B.D., States, D.J., Swaminathan, S., and Karplus, M. (1983) Charmm: A program for macromolecular energy, minimization, and dynamics calculations. *J. Comp. Chem.*, **4** (2), 187–217.

70 Weisman, R.B. and Subramoney, S. (2006) Carbon nanotubes. *The Electrochem. Soc. Interf.*, (Summer), 42–46.

71 Bulashevich, K.A. and Rotkin, S.V. (2002) Nanotube devices: microscopic model. *JETP Lett.*, **75** (4), 205–209.

.

3
Electrochemistry

Matteo Iurlo, Massimo Marcaccio, and Francesco Paolucci

3.1
Introduction

Ever since their appearance in the field of new molecular materials, electrochemistry has played a special role in investigating the electronic properties of carbon nanotubes (CNTs). In view of their structural and electronic uniqueness, carbon nanotubes have been proposed either as nanostructured bulk electrode materials for sensing and biosensing [1–8] and in advanced electrochemical devices [9–11] or as molecular-sized electrodes for investigating very fast electrode kinetics [12]. Alternatively, electron transfer has been used to probe the electronic properties of carbon nanotubes either by direct voltammetric inspection [13] or by coupling it with spectroscopic techniques [14], thus allowing, in the case of true solutions of individual uncut single-walled carbon nanotubes (SWNTs), to single out their redox potentials as a function of diameter [15]. Finally, for their optical and redox properties, as borne by these studies, SWNTs constitute new nanomaterials with enormous potential for a wide range of *in vivo* and *in vitro* bioapplications, ranging from drug delivery to highly sensitive sensors for the electronic and electrochemical detection of biomolecules [16], for energy-related applications and applications in catalysis [17, 18], redox-driven biomachines and/or compact bioreactors for the production of chemicals, industrial materials, and drugs [19], and, finally, for photofunctional nanosystems to be used in efficient light energy conversion devices [20].

Advances and applications in catalysis and sensing and other technologically relevant fields [21, 22], as well as the fundamental chemical, photochemical, optical, and optoelectronic features of carbon nanotubes, have been thoroughly surveyed in a series of excellent recent reviews [20, 23, 24] to which the readers are directed.

3.2
Electronic Properties of SWNTs

The electronic properties of a SWNT are univocally determined by its diameter and helicity, although some additional quantum confinement was recently shown in

Carbon Nanotubes and Related Structures. Edited by Dirk M. Guldi and Nazario Martín
Copyright © 2010 WILEY-VCH Verlag GmbH & Co. KGaA, Weinheim
ISBN: 978-3-527-32406-4

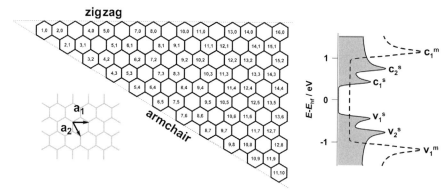

Scheme 3.1

ultrashort SWNTs [25]. A single-walled carbon nanotube is often described as the result of seamless wrapping of a graphene sheet into a cylinder along the (n, m) roll-up vector. The (n, m) indices fully define the SWNT radius and chirality and determine univocally its electronic structure (Scheme 3.1) [26, 27]. If $n - m = 3q$, where q is an integer, the SWNT is metallic, whereas for $n - m \neq 3q$, it is semiconducting with a bandgap in the density of states (DOS) whose size is inversely proportional to the diameter. As a consequence of the size-dependent quantization of electronic wave functions around the circumference of the SWNT, the DOS shows typical singularities, the so-called van Hove singularities, consisting of a singular increase in the DOS at energies ε_{vH} followed by a $(\varepsilon - \varepsilon_{vH})^{-1/2}$ decrease (Scheme 3.1).

About one-third of all the SWNTs are metallic and always have wider energy gaps between the first van Hove spikes than the semiconducting ones with similar diameters. The presence of the van Hove singularities dominates the spectral and redox features of these species [28].

3.3
Electrode Potentials Versus Work Functions

Materials with redox properties have many important technological applications, and understanding the redox properties of these materials is an important issue. The electroactive properties of materials are often characterized by cyclic voltammetry, in which the oxidation ratio of the material is changed by applying a potential. In a cyclic voltammogram, the current, which reflects the oxidation/reduction process of the material, depends nonlinearly on the applied potential. Redox couple peaks in the cyclic voltammogram, which are typical for reduction and oxidation processes of redox materials, are analyzed to obtain information on the energetics and kinetics of the redox process itself and of any possible chemical reaction associated with it. Finally, the standard potential for the redox process is obtained from the above analysis that is ultimately related to the work function of the material. The work function Φ of electrons of a solid phase material is defined as the difference between the Fermi energy, E_F, and the Volta energy, $-e\Psi$, which is the energy required to move an electron in vacuum from a distance of 100 Å from the solid surface of the material to infinity:

$$E_F = -\Phi - e\Psi. \tag{3.1}$$

When an electroactive material is deposited on a metal electrode, the Fermi levels of the metal and of the electroactive material are equal. This leads to a difference in the work function of electrons in the metal and in the material that is equal to the difference in the Volta energies:

$$^{Met}\Delta^M\Phi = \Phi_{Met} - \Phi_M = -e(\Psi_{Met} - \Psi_M). \tag{3.2}$$

If the electroactive material takes part in a redox process that is realized either by using a chemical oxidant/reductant or by applying an external current or potential,

$$O^z + e \leftrightarrow R^{(z-1)}, \tag{3.3}$$

where O and R are the oxidized and reduced partners of the characteristic redox couple, a change in the work function of the material takes place due to a change in the oxidation ratio of the material. In fact, the work function of the material is given by [29]

$$\Phi_M = \mu_e + zF\chi, \tag{3.4}$$

where χ, z, and F are the surface potential, the valence of the ions involved in the redox process, and the Faraday constant, respectively, and μ_e is the chemical potential of the electrons in the electroactive material, given by

$$\mu_e = \mu_R^0 - \mu_O^0 + RT \ln\left(\frac{c_R}{c_O}\right), \tag{3.5}$$

where μ_R^0 and μ_O^0 are the standard chemical potentials of the reduced and oxidized forms of the redox material, respectively. Thus, Eq. (3.4), through the use of Eq. (3.5), states that the work function of the electroactive material is a function of the oxidation ratio in the material and, conversely, that the redox reaction of the electroactive material can be studied by investigating changes in its work function, by using the conventional Kelvin probe technique or the electrolyte metal oxide semiconductor field effect transistor (MOSFET) device.

In an electrochemical experiment, the measured quantity is the electrode potential E relative to a reference system

$$E = \phi_{Met} - \phi_{sol} - \frac{\mu_{e,Met}}{e} \tag{3.6}$$

that combines the inner (Galvani) potential drop across the electrochemical interface with the chemical potential of the electrons in the metal. The Nernst equation finally relates the electrode potential to the standard potential of the relevant redox couple, E^0, and to its oxidation ratio:

$$E = E^0 + \frac{RT}{F} \ln\left(\frac{c_O}{c_R}\right). \tag{3.7}$$

While the electrode potential is necessarily measured against a reference system, for example, the *normal hydrogen electrode* (NHE) or the *standard calomel electrode* (SCE), for comparison with data obtained by different approaches, it is of interest to have an estimate of the *absolute* or *single electrode potential* (i.e., measured versus the potential of a free electron in vacuum). The absolute potential of the NHE can be estimated as $4.5 \pm 0.1\,V$, based on certain extrathermodynamic assumptions, such as about the energy involved in moving a proton from the gas phase into an aqueous solution [29, 30].

3.4
Electrochemistry at SWNTs Versus Electrochemistry of SWNTs

The large interest in using SWNTs as electrodes for electrochemistry mostly originates from their unique structural and electronic properties. Individual SWNTs may, in fact, be used as carbon nanoelectrodes making very fast electrode kinetics accessible or assembled in large surface area carbon electrodes for electrochemical sensor applications.

When a potential is applied between an SWNT supported onto an electrode and an electrolyte (Figure 3.1), a net charge may build up in the SWNT that is screened by charge in the electrical double layer in solution.

The electrostatic capacitance of this interface is given by the capacitance of the electrical double layer, C_{dl}. However SWNTs, analogous to a semiconductor electrode, have a rather low density of electronic states per unit energy around the Fermi level: bands are therefore pinned at the surface and a significant part of the interfacial potential thus takes the form of a change of the chemical potential, V_{ch}, of the SWNT instead of an electrostatic potential drop over the double layer. Such a behavior was recently modeled by Heller *et al.* [31] who represented the interfacial capacitance between SWNT and electrolyte as the electrostatic double-layer capacitance C_{dl} in series with the chemical quantum capacitance C_q (Figure 3.1). In the initial situation, the Fermi level is positioned in the valence band of the SWNT. In the classical case $C_q \gg C_{dl}$, application of a positive voltage $V_{appl} > 0$ will shift the band structure by eV_{dl} with respect to its initial position, while the position of the Fermi level with respect to the band edges remains unchanged. In the extreme opposite case, $C_q \ll C_{dl}$, the position of the bandgap remains unchanged with respect to the reference energy, while the Fermi level shifts by eV_{ch} with respect to the bands. Electrochemical gating and electrochemical charge transfer theories were then combined, according to the Gerischer–Marcus model of heterogeneous electron transfer kinetics, and showed that the interfacial capacitances have a large impact on the behavior of electron transfer at SWNT electrodes. Among the most important predictions derived from such an approach are the following: (i) the semiconducting or metallic SWNT band structure and its distinct van Hove singularities should be

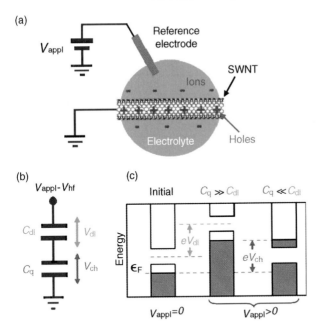

Figure 3.1 (a) Schematic representation of a single SWNT/electrolyte interface comprising an electrochemical circuit. The potential V_{appl} is applied between the SWNT and a reference electrode in solution, and an electrical double layer develops at the SWNT/electrolyte interface. (b) The interfacial capacitance between SWNT and electrolyte comprises the electrostatic double-layer capacitance C_{dl} in series with the chemical quantum capacitance C_q. (c) Energy diagram showing the influence of the quantum capacitance on the alignment and occupation of electronic states in the SWNT. The reference energy in this diagram is set by the reference electrode. The Fermi level is initially located in the valence band of the SWNT; in the classical case $C_q \gg C_{dl}$, application of a positive voltage $V_{appl} > 0$ will shift the band structure with respect to its initial position, leaving the position of the Fermi level with respect to the band edges unchanged. In the opposite case, $C_q \ll C_{dl}$, the position of the bandgap remains unchanged with respect to the reference energy, while the Fermi level shifts with respect to the bands. Reprinted with permission from Ref. [31]. Copyright 2006 The American Chemical Society.

resolved in voltammetry, in a manner analogous to scanning tunneling spectroscopy and (ii) a non-zero interfacial transfer rate is expected even when the Fermi level of a semiconducting SWNT resides in the bandgap. In such a case, in fact, the model predicts that the rate of reaction remains *constant* over the range of V_{appl}.

In Figure 3.2b, the rate of reduction k_{red} of a redox couple having formal potential $V^{0\prime} = -0.25$ V and reorganization energy $\lambda = 1$ eV, calculated according to the above model, at two different kinds of SWNTs, that is, a (10, 10) metallic SWNT or a (10, 11) semiconducting one, or at a graphene layer, is reported. In this example, V_{hf}, the potential at which the Fermi level in the SWNT equals the half-filling energy, was set to 0.2 V with respect to a Ag/AgCl reference potential. As illustrated in Figure 3.2a, the distribution of oxidized states is positioned close to the half-filling energy. Figure 3.2b shows that while a metallic SWNT yields an approximately exponential

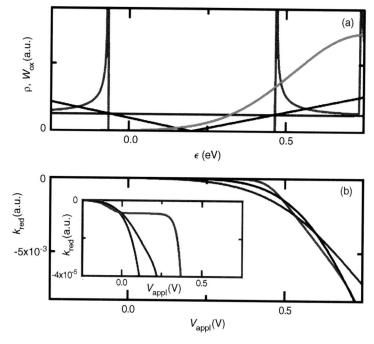

Figure 3.2 Calculations for a *reduction* reaction at a graphene sheet (black), a (10, 10) metallic SWNT (blue), and a (10, 11) semiconducting SWNT (red). (a) Density of states as a function of energy. The green line shows the distribution of the oxidized solution states. (b) Rate of reduction k_{red} as a function of applied potential. The inset shows the rate of reduction near V_{hf}. Reprinted with permission from Ref. [31]. Copyright 2006 The American Chemical Society.

$k_{red}-V_{appl}$ curve, a semiconducting SWNT would reveal a plateau in the curve as the Fermi level crosses the bandgap.

In most cases, experimental data concerning heterogeneous electron transfer kinetics at SWNTs were obtained using solid films of bundled SWNTs [32], that is, immobilized systems where the SWNTs were either spin-coated or drop-cast onto conductive supports or otherwise narrowly packed in self-standing bucky paper. From the point of view of fundamental investigation of their electrochemical and spectroscopic properties, such systems present two different levels of complexity: on the one hand, the films are typically composed of complex mixtures of tubes with chiralities and diameters ranging over many different indices (n, m); on the other hand, each individual SWNT experiences strong interactions either with the other tubes in the bundle or with the substrate with expected strong and unknown effects on its electronic properties. As a matter of fact, electrochemical and spectroelectrochemical experiments carried out on such systems did not allow, in general, to resolve the contribution from each SWNT to the overall experimental response [33]. Furthermore, charge injection to SWNTs may not occur homogeneously and be strongly limited by the low mobility of the counterions within the bundles [34].

3.5
Carbon Nanotubes for Electrochemical Energy Storage Devices

The large empty spaces inside SWNTs and MWNTs, coupled with their large geometric surface areas, ignited an intense investigation of the energy storage characteristics of carbon nanotube materials. The hypothesis that the hollow cores of nanotubes could be excellent sites for the storage of small molecules, such as hydrogen, was thoroughly investigated and, at the same time, studies aimed to develop novel storage materials with high capacity, low mass, and high stability based on CNTs have been carried out [35–37].

Investigating the electrochemical hydrogen storage characteristics of CNTs, capacities of up to 170 mAh/g for unpurified arc-discharge produced samples, containing only a small percentage of SWNTs, were found [38] that increased to 800 mAh/g in the case of purified SWNT samples [39]. Hydrogen storage capacities of around 200 mAh/g for both MWNTs and SWNTs were reported [40–42]. However, all these CNT materials were synthesized using transition metals and well-known hydride-forming materials as catalysts. More recently, a maximum discharge capacity of 130 mAh/g was reported for an as-produced SWNT material. Measurements indicate that the electrochemical activity of the matrix material, used to manufacture composite electrodes, should also not be ignored. Highly pure CNTs yield a substantially lower response: the measured response of CNT materials is related to a number of processes, including the irreversible oxidation of carbonaceous material, the reversible oxidation/reduction of residual metal catalyst or carbonaceous impurities, and an electrostatic charging component [43].

Very recently, a composite structure consisting of well-dispersed manganese oxide nanoflowers on a vertically aligned carbon nanotube array was reported [44]. The CNT array was found to be excellent for electrodepositing transition metal oxides because of its regular pore structure, high surface area, homogeneous property, and excellent conductivity. The composite was reported to exhibit superior rate performance (50.8% capacitance retention at 77 A/g), thus making it very promising for high-rate electrochemical capacitive energy storage applications.

3.6
Carbon Nanotubes for Electrochemical Sensors and Biosensors

Carbon is commonly used as an electrode material for electrochemistry. sp^2 carbon, in particular, combines a wide useful potential window with electronic properties similar to metals [45]. In addition, chemical surface modification of carbon is an important viable approach to functional electrodes [46]. Covalent grafting of organic monolayers on carbon (graphite, HOPG, carbon fibers, etc.) can be obtained by the electrochemical *in situ* reduction of aryl diazonium salts or the oxidation of aromatic and/or aliphatic amines. Such electrochemical methods produce a monolayer of an organic species by forming a carbon–carbon (or carbon–nitrogen) bond between the carbon surface and the functionalizing compound [46]. Carbon nanotubes represent a new kind of carbon materials that are superior to other kinds of carbon materials such as glassy carbon, graphite, and diamond. For their special structural features and unique electronic properties, NTs have found applications in electrocatalysis, direct electrochemistry of proteins, and electroanalytical devices such as electrochemical sensors and biosensors. The electroanalytical applications of carbon nanotubes have been thoroughly described in a series of recent reviews [47].

Innovations in carbon nanotube-assisted biosensing technologies such as DNA hybridization, protein binding, antibody–antigen, and aptamers have recently been reviewed, particularly focusing on the two major schemes for electronic biodetection, namely, biotransistor- and electrochemistry-based sensors (Scheme 3.2) [48].

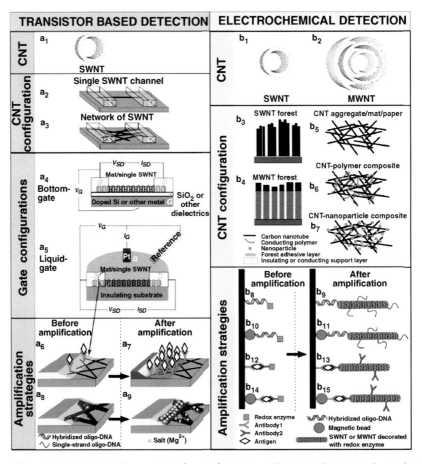

Scheme 3.2 Reprinted with permission from Ref. [48]. Copyright 2007 Wiley-VCH Verlag GmbH.

The redox properties of SWNTs have recently been exploited for realizing an efficient bioinspired reactor where a reductant (dithionite) supplies electrons to SWNTs that subsequently transfer and donate electrons to the acceptor molecules (cytochrome *c*) (Scheme 3.3) [49].

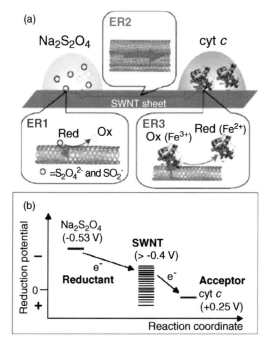

Scheme 3.3 Reprinted with permission from Ref. [49]. Copyright 2008 Wiley-VCH Verlag GmbH.

According to a novel, original approach, functionalized MWNT assemblies were assembled onto the surface of silicon electrodes and investigated with the scanning electrochemical microscopy (SECM) (Scheme 3.4) [50].

The study showed that the nanotubes allow an efficient electrical communication between the semiconductor surface and a redox probe in solution (Scheme 3.4), thus showing their potential application for electrically transducing a biomolecular recognition event.

3.7
Electrochemistry of Carbon Nanotubes

Electrochemical charging has been widely used to provide important information on the electronic structure of these materials. *In situ* Raman spectroelectrochemistry is a well-established method for investigating the change in physical properties of SWNTs during charging [14, 33, 51]. For instance, electrochemical charging has a strong impact on the development of the Raman spectra of SWNT bundles. In the presence of large bundles of SWNTs, however, the interpretation of the results is difficult because (i) the individual properties of specific tubes are averaged and difficult to single out and (ii) the properties of nanotubes in a bundle are different from those of individual nanotubes. Recently, Raman spectroelectrochemical experiments, carried out using small bundles and individual SWNTs supported

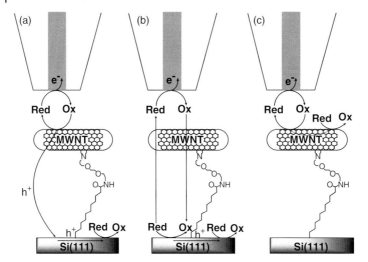

Scheme 3.4 SECM in feedback mode for studying charge transfer kinetics at a non-connected MWNT-modified p-type Si(111) surface. Three major pathways are described for regenerating the initial form of the mediator: (a) direct hole tunneling through the monolayer linker, (b) diffusion of the electroactive species through the pinholes of the monolayer followed by the injection of a hole into the valence band of silicon, and (c) charge transfer at the attached conducting MWNTs. Reproduced from Ref. [50] by permission of The Royal Society of Chemistry.

onto a conducting substrate [52], demonstrated the importance of making measurements at the single nanotube level in order to investigate their size-dependent electronic properties. The magnitude and direction (upshift or downshift) of the tangential mode band was analyzed as a function of the applied potential and was found to depend on the diameter of the semiconducting tubes (Figure 3.3).

For negative charging, the small-diameter tubes exhibit a downshift while the large-diameter tubes exhibit an upshift. This behavior was associated with a phonon renormalization effect and a C–C bond weakening during the charging process.

Optical Vis/NIR spectroelectrochemistry studies were performed on thin semi-transparent solid films of SWNTs (CoMoCat) enriched with (6, 5) tubes via density-gradient ultracentrifugation. Despite the partial bundling of the nanotubes, bands corresponding to optical transitions of tubes with distinct chiralities were detectable in the spectra and, from the effects of electrochemical redox doping, a partial quantification of changes in the electronic density of states of such SWNTs was possible [53].

However, when the electrochemistry investigation is carried out on SWNT films, the proximity of the sample to the conducting substrate may have significant effects on the electrochemical properties of the SWNTs [54]. A way to minimize such interactions and thus obtain valuable information on the redox (electronic) properties of non-interacting individual SWNTs, complementary to those given by spectroscopy, is to carry out electrochemical investigations in proper solutions of *individual* SWNTs. Among the reasons why *solution electrochemistry* of SWNTs has so far found very few examples vis-à-vis spectroscopic investigations are (i) the relatively lower sensitivity of electrochemical techniques with respect to spectroscopic ones, (ii) the interference in

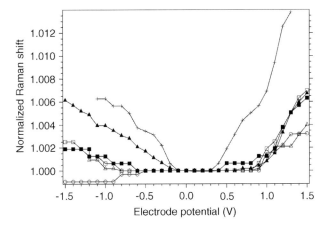

Figure 3.3 A plot of the relative frequency of the G$^+$ band for five different SWNTs in the range of electrode potential from $V_e = -1.5$ to 1.5 V. (The frequency is normalized to the frequency at 0 V.) The empty and filled squares correspond to tube B, where $E_{laser} = 1.70$ and 1.68 eV, respectively (RBM at 135 cm^{-1}). The empty circles, empty triangles, filled triangles, and crosses correspond to tubes with the RBM frequencies at 175, 145, 125, and 120 cm^{-1}, respectively. The spectra were excited by 2.03, 2.33, 1.62, and 1.72 eV lasers, respectively. Adapted with permission from Ref. [52]. Copyright 2008 The American Chemical Society.

the electrochemical measurements from most of the molecular systems (surfactants, polymer chains, DNA) used for making SWNT suspensions, such as their blocking effects on the electrical contact of SWNT with a metal electrode in an electrochemical cell, (iii) the high ionic strength typical of electrochemical experiments (due to the addition of a supporting electrolyte) that promotes flocculation of the SWNT suspension, and (iv) the limited electrochemical stability window of the aqueous medium.

3.8
Cyclic Voltammetric Investigations of Solutions of Individual SWNTs

Organic functionalization of SWNTs, either covalent or noncovalent, has been shown to be an effective way to make stable solutions of isolated SWNTs [55]. In 2003, we reported the first electrochemical investigation of individual SWNTs carried out by cyclic voltammetry and absorption electronic spectroelectrochemistry in (N-TEG)-pyrrolidine functionalized SWNT (*f*-NT) tetrahydrofuran solutions (Figure 3.4) [56]. Either CH$_3$- or amido-ferrocenyl-terminated *f*-NTs were investigated. The steady-state voltammetric curve obtained at an ultramicroelectrode (UME) displays a continuum of diffusion-controlled current, with onset, in the negative potential region, at ∼−0.5 V, which is attributed to the reduction of *f*-NTs. In the case of ferrocenyl-terminated *f*-NTs, the reversible oxidation of ferrocenyls was also detected. A combination of electrochemical measurements and quantum chemical calculations was made in order to address two major issues with *f*-NTs, namely, the nature of their bulk electronic properties and to what extent they retain those of pristine NT.

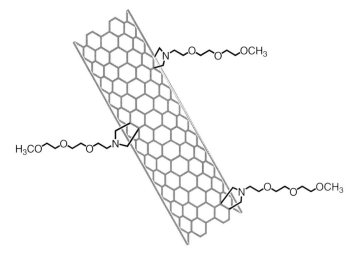

Figure 3.4 N-TEG pyrrolidine-SWNT.

In this case, experiments were carried out with a gold UME modified by a self-assembled monolayer of dodecanethiols according to a procedure proposed by Miller and coworkers [57] (Figure 3.5). The presence of the organic monolayer slows down the heterogeneous electron transfer process, thus avoiding competition from diffusion. The measured current thus becomes representative of the SWNT DOS and could therefore be compared with that calculated for pristine SWNTs from the electronic density of states, according to a simple equation (Eq. (3.8)) [58]:

$$i(E_f) \propto \int_0^{E_f} (DOS)\, dE. \tag{3.8}$$

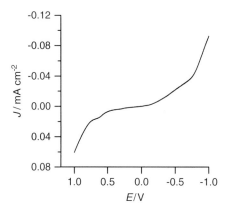

Figure 3.5 Steady-state CV curve of saturated f-NTs, 10 mM TBAH, THF solution. Scan rate = 0.5 V/s, T = 298 K at a Au electrode derivatized with a dodecanethiol self-assembled monolayer. Potentials measured versus silver quasireference electrode.

The DOS of pristine SWNTs was calculated according to a tight binding model [59] and averaged over 39 different NTs with diameters between 0.88 and 1.22 nm and relative abundances that correspond to the population of tubes contained in HiPCO samples probed in the voltammetric experiments [60].

The comparison showed that the overall DOS is not strongly affected by the chemical modification. However, the DOS of *f*-NT shows a smoothing of the peaks that appear in pristine CNT at about ±0.5 eV. Such peaks are related to the van Hove singularities between which electronic transitions take place, and in fact, experimentally, this class of *f*-NT shows a rather flat near-infrared spectrum.

Proper *solution* electrochemical experiments on pristine SWNTs were only possible after the recent discovery of an innovative way to form thermodynamically stable solutions of unmodified and uncut SWNTs [61]. Upon reduction with alkali metals, SWNTs produce polyelectrolyte salts (Figure 3.6) that are soluble in polar organic solvents without the use of sonication, surfactants, or functionalization. Polyelectrolyte SWNT salts were obtained by reacting arc-discharge samples (*a*-NT) or HiPCO samples (*h*-NT) with different alkali metals (Na, K) [61].

Voltammetric experiments were carried out in oxygen-free, ultra-dry DMSO where, to avoid flocculation of the SWNTs, only low concentrations of the supporting electrolyte were used [15]. Figure 3.7 compares the voltammetric curves of a saturated DMSO solution of Na[*a*-NT] and Na[*h*-NT], at potentials close to the open circuit potential (OCP), with that of *f*-NT. The curves show that (i) the current onset is anticipated for Na[*a*-NT] with respect to Na[*h*-NT] and to the *f*-NT (that were also produced from HiPCO SWNTs); (ii) the curves of *h*-NT and *a*-NT (for both K$^+$ and Na$^+$ salts) are richer in structure than that of *f*-NT. In particular, the broad waves at −0.81 and −1.05 V (*h*-NT) and −0.56 and −0.99 V (*a*-NT) are missing for *f*-NT.

The anticipated onset indicates a smaller energy gap, consistent with the larger average diameters of the arc-SWNTs with respect to the HiPCO ones. The richer

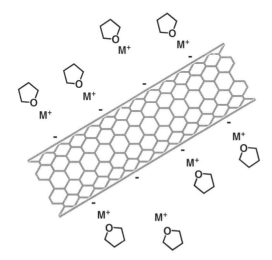

Figure 3.6 Schematic structure of the soluble SWNTs salts M$^+$(THF)C$_n$. M$^+$ = Na$^+$, K$^+$.

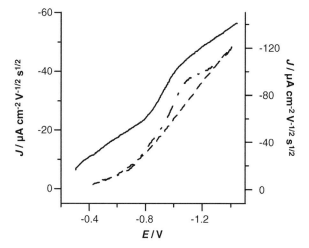

Figure 3.7 CV curves of DMSO solutions saturated with Na[a-NT] (black line), saturated Na[h-NT] (dash-dot line), or saturated f-NT (dash line). The black and blue curves are referred to the left axis, the red curve to the right axis. Experimental conditions: 2 mM TBAH/DMSO; working electrode: Pt disk ($r = 25 \mu$m); $T = 298$ K; scan rate = 1 V/s. Potentials are referenced to SCE.

curve morphology reflects, instead, the more complex electronic structure of the pristine materials, compared to f-NT, thus confirming that functionalization significantly affects the low-lying electronic states of the nanotubes.

3.9
Vis–NIR Spectroelectrochemical Investigation of True Solutions of Unfunctionalized SWNTs

The voltammetric response of individual SWNTs in solution reflects average collective properties rather than single SWNT ones: voltammetry is therefore not a suitable technique to determine the standard redox potentials of individual semiconducting SWNTs as a function of the tube structure. This was instead achieved via an extensive Vis–NIR spectroelectrochemical investigation of true polyelectrolyte SWNT solutions that was carried out in the ± 1500 mV range (versus SCE), where doping is expected to modify the population of the first and second van Hove singularities of the electronic bands of SWNTs. The electronic transitions affected by charging are therefore the semiconducting S_{11} ($v_s^1 \rightarrow c_s^1$) and S_{22} ($v_s^2 \rightarrow c_s^2$), and possibly the metallic M_{11} ($v_m^1 \rightarrow c_m^1$) [28].

As expected for heavily n-doped SWNTs, the starting solutions do not show significant absorption except for the plasmon band, typical of graphitic materials [62, 63]. Spectral changes were recorded as long as the applied potential was increased above -1.2 V, associated with the progressive undoping of SWNTs: the intensity progressively increased until the maximum was reached for the complete depletion of the conduction bands.

(a)

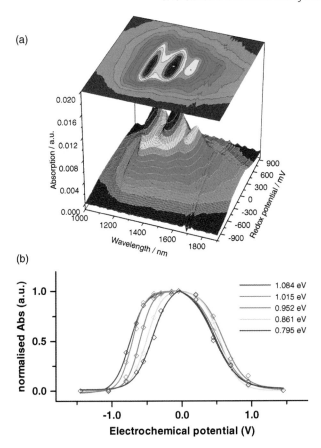

(b)

Figure 3.8 (a) Smoothed surface plot of optical absorption intensity as a function of wavelength and electrode potential in the S_{11} region for K[*h*-NT]. (b) Intensity of each of the five bands selected in the spectra above after normalization as a function of potential. The fitting curves (full lines) were calculated using Eq. (3.2), whence $E^0_{ox,i}$ and $E^0_{red,i}$ values were obtained (corresponding to the inflection points in each sigmoidal curve). In all plots, raw electrochemical data, that is, uncorrected for ohmic drop, are referenced to SCE.

The 3D plot of Figure 3.8a displays a summary of the spectroelectrochemical investigation of *h*-NT's S_{11}. On both reduction and oxidation sides, the absorption bands have different onset potentials, thus suggesting that different redox potentials are associated with different tube structures.

3.10
Standard Redox Potentials of Individual SWNTs in Solution

The knowledge of the oxidation and reduction potentials of SWNTs is instrumental to the rational design of electrochemical, photochemical, and electronic devices, such as sensors, photoactive dyads, light emitting diodes, or other electroluminescent devices, in which the energetics of electrochemical and electric generation of electrons and holes in the SWNTs play an important role. The analysis of the spectroelectrochemical results of *h*-NT and *a*-NT shown above allowed to obtain the standard redox potential as a function of the tube diameter of a large number of semiconducting SWNTs. In the analysis, we assumed that the absorbance is only due to neutral SWNTs (and that they do it according to the Lambert–Beer law) and that the Nernst equation describes the ratios of activity (concentration) of the neutral-to-reduced (or -oxidized) SWNTs. Therefore, the absorbance (*A*) of the samples is described as a function of electrochemical potential by two equations for the n-doped and p-doped SWNTs [64]:

$$A_{\text{n-dop},i} = \varepsilon_{N,i} bc_i^* \frac{e^{(F/RT)(E-E_{\text{red},i}^0)}}{1+e^{(F/RT)(E-E_{\text{red},i}^0)}}, \quad A_{\text{p-dop},i} = \varepsilon_{N,i} bc_i^* \frac{1}{1+e^{(F/RT)(E-E_{\text{ox},i}^0)}}, \quad (3.9)$$

where *i* is a single (*n*, *m*) nanotube that contributes to a specific transition band, ε_N is the extinction coefficient of the neutral (*n*, *m*) nanotube (for simplicity, assumed to be the same for all tubes), c^* is its bulk concentration, *E* is the electrochemical potential, *b* is the path of the spectroscopic cell, and $E_{\text{red},i}^0$ and $E_{\text{ox},i}^0$ are the standard potentials of its reduction (n-doping) and oxidation (p-doping). After normalization (Figure 3.8b), the intensity of each band in the spectra was analyzed according to Eq. (3.9), to give a set of standard potentials for the oxidations and reductions of *h*-NT and *a*-NT.

The assignment of the redox potentials, reported in Table 3.1, was obtained by adopting the empirical Kataura plot of Ref. 65 that gives the excitation energies of the van Hove optical transitions as a function of the structure for a wide range of semiconducting nanotubes. Finally, the E^0 values vary linearly with the excitation energy: to first order, the redox potential fits the equations

$$E_{\text{red}}^0 = -1.02 E_{\text{exc}} + 0.26 \quad (R_2 = 0.989) \tag{3.10}$$

$$E_{\text{ox}}^0 = 0.36 E_{\text{exc}} + 0.22 \quad (R^2 = 0.704) \tag{3.10'}$$

that can be employed in the design of devices mentioned above. The chirality map of Figure 3.9, where color codes are used to gather the structures that share the same values, shows that, in analogy with optical transition frequencies, a simple dependence of redox properties of SWNTs on diameter is not strictly followed.

As in most molecular systems, for example, fullerenes [66], the redox properties of SWNTs are affected by ion–solvent interaction effects [67]. This is evident, for instance, by the comparison of two sets of redox potentials obtained, respectively, in DMSO (Table 3.1) and water solutions [68]. The latter set of data was determined using a biohybrid approach: hydrogenase was integrated into the SWNT surface to mediate electron injection from H_2 into nanotubes having appropriately positioned lowest occupied molecular orbital levels when the H_2 partial pressure is varied (Figure 3.10).

Table 3.1 Electrochemical and optical data for C_{60} and unfunctionalized semiconducting SWNTs.[a]

E_{exc}[b] (eV)	E_{ox}[c] (eV)	E_{red}[c] (eV)	E_{gap} (eV)	E_{bind}[d] (eV)	Assignment[e]
—	1.71, 2.18, 2.58[f]	−0.62, −1.02, −1.45[f]	2.33[f]	—	C_{60}
1.084	0.575	−0.840	1.415	0.331	(9, 4) (8, 4) (7, 6)
1.015	0.550	−0.806	1.356	0.341	(8, 6) (12, 1) (11, 3)
0.953	0.606	−0.705	1.311	0.360	(9, 5) (10, 3) (10, 5) (8, 7) (11, 1)
0.861	0.587	−0.597	1.184	0.323	(12, 4) (11, 4) (12, 2) (13, 0) (9, 8)
0.795	0.541	−0.540	1.081	0.286	(13, 3) (12, 5) (14, 1)
0.760	0.485	−0.515	1.000	0.240	(12, 7) (17, 0) (10, 9) (16, 2)
0.743	0.483	−0.492	0.975	0.232	(11, 9) (15, 4)
0.726	0.470	−0.477	0.947	0.221	(14, 4) (13, 6) (14, 6)
0.712	0.466	−0.478	0.944	0.232	(12, 8)
0.701	0.458	−0.471	0.929	0.228	(13, 8) (18, 1) (17, 3)
0.679	0.432	−0.449	0.881	0.202	(16, 5) (17, 1) (16, 3)

a) Data collected for saturated solution of h-NT and a-NT in 2 mM TBAH/DMSO, $T = 298$ K.
b) Excitation energy.
c) Electrochemical standard potentials referred to SCE.
d) Exciton binding energy.
e) From Ref. [65], corrected by 35 meV (redshift) for taking the solvent effects into account.
f) Determined by cyclic voltammetry, in dichloromethane/tetrabutylammonium hexafluoroarsenate 50 mM solution, $T = 213$ K. W: Pt.

Using the Nernst equation, the LUMO energy (i.e., the conduction band edge) of various individual SWNTs was estimated from the partial H_2 pressure at which luminescence at each specific wavelength was quenched because of the electron transfer process. The difference between the two sets of data is inversely proportional to the nanotube diameter (Figure 3.11), in line with prevalent ion–solvent interaction effects (Born model).

It is, finally, worth comparing the redox properties of SWNTs with those of fullerenes and, in particular, of C_{60}. It is well known that much of the chemistry and the applications of C_{60} are based on its electron-accepting capacity. The dynamics of reduction of C_{60} and its derivatives has been thoroughly characterized in solution by electrochemical means [69]. On the other hand, the facile reduction of C_{60} contrasts with its difficult oxidation. In 1993, Echegoyen et al., employing scrupulously dried tetrachloroethane (TCE), tetrabutylammonium hexafluorophosphate (TBAPF$_6$) as a supporting electrolyte, and low temperatures, were the first to observe the electrochemical reversible one-electron oxidation of C_{60} at $E_{1/2} = 1.26$ V (versus Fc$^+$/Fc) [70]. By contrast, the further oxidation of C_{60}^+ was only recently obtained by adopting suitable experimental conditions that comprise ultra-dry solvents and TBAAsF$_6$ as a supporting electrolyte, whose anion is known for its high oxidation resistance and low nucleophilicity [71]. Comparison of the E^0 values for the first reduction and first oxidation of C_{60} with the corresponding values for SWNT

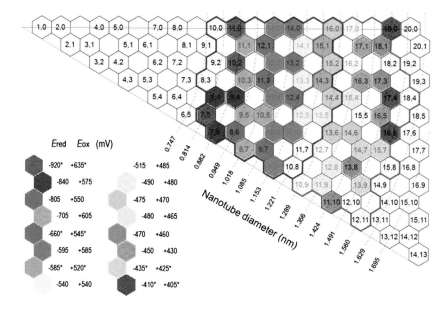

Figure 3.9 Chirality map displaying the average standard potentials associated with SWNT structures typically present in commercial samples. HiPCO SWNTs are located inside the red line, while arc-discharge SWNTs are inside the blue line.

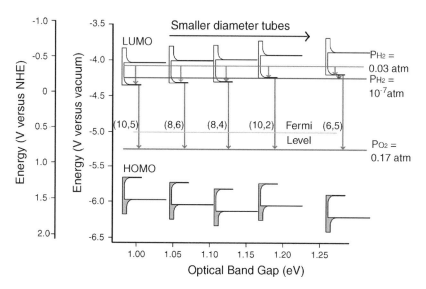

Figure 3.10 Diagram of nanotube LUMO and HOMO levels versus optical bandgap and decreasing nanotube diameter determined as described in the text. The electrochemical potential of the H^+/H_2 redox couple is shown for two P_{H_2} values, along with the potential of the O_2/H_2O redox couple for a P_{O_2} value. Reprinted with permission from Ref. [68]. Copyright 2007 The American Chemical Society.

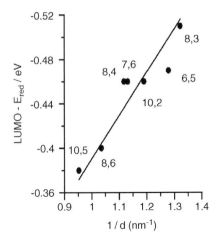

Figure 3.11 Comparison of the energies of the LUMO published in Ref. 68, relative to aqueous solutions, with E_{red} values calculated using the equation $E_{red} = -1.02 \times E_{exc} + 0.26$, relative to DMSO solutions: The difference between the two sets of data scales linearly with the reciprocal of nanotube diameter, as expected for prevailing solvent–ion interaction effects.

(Table 3.1) shows that SWNTs behave, in general, as both better electron acceptors and better electron donors than C_{60} but, while the electron-accepting properties are rather comparable, oxidation of C_{60} is energetically much more demanding than p-doping of SWNTs. Interestingly, more similar electron-donating properties were very recently observed in the case of a covalently linked C_{60} homopolymer, obtained electrochemically [72].

3.11
Fermi Level and Excitonic Binding Energy of the Nanotubes

The energy of the Fermi level in the neutral state (E_f) calculated from the standard redox potentials is almost constant for a-NTs, while for h-NTs it is linear with the energy of the optical gap [15]. The trend is in line with other reported experimental values and is also reproduced by *ab initio* calculations. Furthermore, standard redox potentials were finally used to establish the exciton binding energy for individual tubes in solution. For each SWNT, the difference of the redox potentials represents the gap between the van Hove singularities of the electronic band structure ($E_{gap} = E_{ox} - E_{red}$). E_{gap} is larger than the excitation energy because of the Coulomb attraction experienced by the excited electron with the positive hole left in the valence band, $E_{exc} < E_{gap}$ [73]. The excitonic binding energy calculated from the electrochemical data given by Eq. (3.10) is inversely proportional to the nanotube diameter and matched very well the excitonic binding energies obtained elsewhere.

3.12
Conclusions and Perspectives

Carbon is unique in nature. Besides its capability to form complex networks that are fundamental to organic chemistry, elemental carbon shows unrivalled complexity among the elements, forming many allotropes, from 3D diamond and graphite to 0D fullerenes, through 1D nanotubes and the most recently obtained 2D graphene [74]. Over the past two decades, we have contributed to the subsequent rise of the three different low-dimensional carbon allotropes. After being chemists' superstar in the 1990s, C_{60} is still at the center of considerable attention in many different fields of science [75]. Carbon nanotubes are perhaps the most notable representatives of the present nanoworld. However, they are very likely bound to soon share the stage with graphene, the "rapidly rising star on the horizon of materials science and condensed matter physics" [76]. Graphene consists of a two-dimensional hexagonal lattice of sp^2 carbon, through which electronic conduction can occur via the π-conjugated electron system, and is sometimes classified as a zero-gap semiconductor, since the density of states per unit area vanishes at the Fermi level.

Graphene may have a number of very peculiar electronic properties – from an anomalous quantum Hall effect to the absence of localization. As new procedures for the large-scale production of graphene are expected to be developed in the near future, most of such properties – including the electrochemical ones – will be soon experimentally demonstrated, thus permitting the development of the many important technological applications foreseen for this material.

Acknowledgments

The authors gratefully acknowledge the contributions of their collaborators and coworkers mentioned in the cited references. This work was performed with partial support from the University of Bologna, the Fondazione Cassa di Risparmio in Bologna, and the Italian Ministry of University and Research.

References

1 Kong, J., Franklin, N.R., Zhou, C., Chapline, M.G., Peng, S., Cho, K., and Dai, H. (2000) Science, 287, 622–625.

2 Wang, J. (2005) Electroanalysis, 17, 7–14.

3 Lin, Y., Lu, F., Tu, Y., and Ren, Z. (2004) Nano Lett., 4, 191–195.

4 Barone, P.W., Baik, S., Heller, D.A., and Strano, M.S. (2005) Nat. Mater., 4, 86–92.

5 Wohlstadter, J.N., Wilbur, J.L., Sigal, G.B., Biebuyck, H.A., Billadeau, M.A., Dong, L., Fischer, A.B., Gudibande, S.R., Jameison, S.H., Kenten, J.H., Leginus, J., Leland, J.K., Massey, R.J., and Wohlstadter, S.J. (2003) Adv. Mater., 15, 1184–1187.

6 Callegari, A., Cosnier, S., Marcaccio, M., Paolucci, D., Paolucci, F., Georgakilas, V., Tagmatarchis, N., Vázquez, E., and Prato, M. (2004) J. Mater. Chem., 14, 807–810.

7 Yan, Y.-M., Yehezkeli, O., and Willner, I. (2007) Chem. Eur. J., 13, 10168–10175.

8 Yuna, Y.-H., Dongb, Z., Shanovc, V., Heinemand, W.R., Halsalld, H.B., Bhattacharyae, A., Confortif, L., Narayang,

R.K., Ballh, W.S., and Schulz, M.J. (2007) *Nano Today*, **2** (6), 30–37.

9 Dillon, A.C., Jones, K.M., Bekkedahl, T.A., Kiang, C.H., Bethune, D.S., and Heben, M.J. (1997) *Nature*, **386**, 377–399.

10 Niu, C., Sichel, E.K., Hoch, R., Moy, D., and Tennent, H. (1997) *Appl. Phys. Lett.*, **70**, 1480–1482.

11 Portet, C., Taberna, P.L., Simon, P., and Flahaut, E. (2005) *J. Power Sources*, **139**, 371–378.

12 Nugent, J.M., Santhanam, K.S.V., Rubio, A., and Ajayan, P.M. (2001) *Nano Lett.*, **1**, 87–91.

13 Melle-Franco, M., Marcaccio, M., Paolucci, D., Paolucci, F., Georgakilas, V., Guldi, D.M., Prato, M., and Zerbetto, F. (2004) *J. Am. Chem. Soc.*, **126**, 1646–1647.

14 Kavan, L. and Dunsch, L. (2007) *ChemPhysChem*, **8**, 974–998.

15 Paolucci, D., Melle-Franco, M., Iurlo, M., Marcaccio, M., Prato, M., Zerbetto, F., Pénicaud, A., and Paolucci, F. (2008) *J. Am. Chem. Soc.*, **130**, 7393–7399.

16 Lu, F., Gu, L., Meziani, M.J., Wang, X., Luo, P.G., Veca, L.M., Cao, L., and Sun, Y.-P. (2009) *Adv. Mater.*, **21**, 139–152.

17 Vairavapandian, D., Vichchulada, P., and Lay, M.D. (2008) *Anal. Chim. Acta*, **626**, 119–129.

18 Emmenegger, C., Mauron, P., Sudan, P., Wenger, P., Hermann, V., Gallay, R., and Zuttel, A. (2003) *J. Power Sources*, **124**, 321–329.

19 Saito, T., Matsuura, K., Ohshima, S., Yumura, M., and Iijima, S. (2008) *Adv. Mater.*, **20**, 2475–2479.

20 Sgobba, V. and Guldi, D.M. (2009) *Chem. Soc. Rev.*, **38**, 165–184.

21 Tans, S.J., Devoret, M.H., Dai, H.J., Thess, A., Smalley, R.E., Geerligs, L.J., and Dekker, C. (1997) *Nature*, **386**, 474–477.

22 Bachtold, A., Hadley, P., Nakanishi, T., and Dekker, C. (2001) *Science*, **294**, 1317–1320.

23 Lu, X. and Chen, Z. (2005) *Chem. Rev.*, **105**, 3643–3696.

24 Gong, K., Yan, Y., Zhang, M., Su, L., Xiong, S., and Mao, L. (2005) *Anal. Sci.*, **21**, 1383–1393.

25 Sun, X., Zaric, S., Daranciang, D., Welsher, K., Lu, Y., Li, X., and Dai, H. (2008) *J. Am. Chem. Soc.*, **130**, 6551–6555.

26 Saito, R., Fujita, M., Dresselhaus, G., and Dresselhaus, M.S. (1992) *Appl. Phys. Lett.*, **60**, 2204–2206.

27 Dekker, C. (1999) *Phys. Today*, **52** (5), 22–28.

28 Saito, R., Fujita, M., Dresselhaus, G., and Dresselhaus, M.S. (2000) *Phys. Rev. B: Condens. Matter*, **61**, 2981–2990.

29 Trasatti, R. (1986) *Pure Appl. Chem.*, **58**, 955–966.

30 Parsons, R. (1985) Chapter 1, in *Standard Potentials in Aqueous Solution* (eds A.J. Bard, R. Parsons, and J. Jordan), Marcel Dekker, New York.

31 Heller, I., Kong, J., Williams, K.A., Dekker, C., and Lemay, S.G. (2006) *J. Am. Chem. Soc.*, **128**, 7353–7359.

32 Trojanowicz, M. (2006) *Trends Anal. Chem.*, **25**, 480–489.

33 Kavan, L. and Dunsch, L. (2008) *Carbon Nanotubes* (Topics in Applied Physics), vol. 111 (eds A. Jorio, G. Dresselhaus, and M.S. Dresselhaus), p. 567, Springer, Heidelberg.

34 Kavan, L., Frank, O., Kalbac, M., and Dunsch, L. (2009) *J. Phys. Chem. C*, **113**, 2611–2617.

35 Dresselhaus, M.S., Dresselhaus, G., and Avouris, P. (2001) *Carbon Nanotubes* (Topics in Applied Physics), vol. **80**, p. 391, Springer, Heidelberg.

36 Ding, R.G., Lu, G.Q., Yan, Z.F., and Wilson, M.A. (2001) *J. Nanosci. Nanotechnol.*, **1**, 7–29.

37 Frackowiak, E. and Béguin, F. (2001) *Carbon*, **39**, 937–950.

38 Nützenadel, C., Züttel, A., Chartouni, D., and Schlapbach, L. (1999) *Electrochem. Solid-State Lett.*, **2**, 30–32.

39 Rajalakshmi, N., Dhathatreyan, K.S., Govindaraj, A., and Satishkumar, B.C. (2000) *Electrochim. Acta*, **45**, 4511–4515.

40 Gao, X.P., Lan, Y., Pan, G.L., Wu, F., Qu, J.Q., Song, D.Y., and Shen, P.W. (2001) *Electrochem. Solid-State Lett.*, **4**, A173.

41 Züttel, A., Nützenadel, C., Sudan, P., Mauron, P., Emmenegger, C., Rentsch, S., Schlapbach, L., Weidenkaff, A., and Kiyobayashi, T. (2002) *J. Alloys Compd.*, **330–332**, 676–682.

42 Zhang, H., Fu, X., Chen, Y., Yi, S., Li, S., Zhu, Y., and Wang, L. (2004) *Physica B*, **352**, 66–72.

43 Niessen, R.A.H., de Jonge, J., and Notten, P.H.L. (2006) *J. Electrochem. Soc.*, **153**, A1484–A1491.

44 Zhang, H., Cao, G., Wang, Z., Yang, Y., Shi, Z., and Gu, Z. (2008) *Nano Lett.*, **8**, 2664–2668.

45 McCreery, R.L. (1999) *Electrochemical Properties of Carbon Surfaces*, Dekker, New York, p. 631.

46 (a) Downard, A.J. (2000) *Electroanalysis*, **12**, 1085–1096; (b) Pinson, J. and Podvorica, F. (2005) *Chem. Soc. Rev.*, **34**, 429–439.

47 Gong, K., Yan, Y., Zhang, M., Su, L., Xiong, S., and Mao, L. (2005) *Anal. Sci.*, **21**, 1383–1393.

48 Kim, S.N., Rusling, J.F., and Papadimitrakopoulos, F. (2007) *Adv. Mater.*, **19**, 3214–3228.

49 Saito, T., Matsuura, K., Ohshima, S., Yumura, M., and Iijima, S. (2008) *Adv. Mater.*, **20**, 2475–2479.

50 Hauquier, F., Pastorin, G., Hapiot, P., Prato, M., Bianco, A., and Fabre, B. (2006) *Chem. Commun.*, 4536–4538.

51 Kavan, L., Rapta, P., and Dunsch, L. (2000) *Chem. Phys. Lett.*, **328**, 363–368.

52 Kalbac, M., Farhat, H., Kavan, L., Kong, J., and Dresselhaus, M.S. (2008) *Nano Lett.*, **8**, 3532–3537.

53 Frank, O., Kavan, L., Green, A.A., Hersam, M.C., and Dunsch, L. (2008) *Phys. Stat. Sol. (b)*, **245**, 2239–2242.

54 (a) Kim, P., Odom, T.W., Huang, J.-L., and Lieber, C.M. (1999) *Phys. Rev. Lett.*, **82**, 1225–1228; (b) Ouyang, M., Huang, J.-L., and Lieber, C.M. (2002) *Annu. Rev. Phys. Chem.*, **53**, 201–220; (c) Kwon, Y.-K., Saito, S., and Tomanek, D. (1998) *Phys. Rev. B*, **58**, R13314–R13317.

55 Hirsch, A. (2002) *Angew. Chem., Int. Ed.*, **41**, 1853–1859.

56 Melle-Franco, M., Marcaccio, M., Paolucci, D., Paolucci, F., Georgakilas, V., Guldi, D.M., Prato, M., and Zerbetto, F. (2004) *J. Am. Chem. Soc.*, **126**, 1646–1647.

57 Becka, A.M. and Miller, C.J. (1992) *J. Phys. Chem.*, **96**, 2657–2668.

58 Ouyang, M., Huang, J.-L., and Lieber, C.M. (2002) *Acc. Chem. Res.*, **35**, 1018–1025.

59 Reich, S., Maultzsch, J., and Thomsen, C. (2002) *Phys. Rev. B*, **66**, 035412.

60 Kukovecz, A., Kramberger, C., Georgakilas, V., Prato, M., and Kuzmany, H. (2002) *Eur. Phys. J. B*, **28**, 223–230.

61 Pénicaud, A., Poulin, P., Derr, A., Anglaret, E., and Petit, P. (2005) *J. Am. Chem. Soc.*, **127**, 8–9; (b) Fischer, J., Petit, P., and Pénicaud, A. *Metananotubes* (ed. M. Monthioux), John Wiley & Sons, Inc., New York, in press.

62 Niyogi, S., Hamon, M.A., Hu, H., Zhao, B., Bhowmik, P., Sen, R., Itkis, M.E., and Haddon, R.C. (2002) *Acc. Chem. Res.*, **35**, 1105–1113.

63 (a) Rao, A.M., Eklund, P.C., Bandow, S., Thess, A., and Smalley, R.E. (1997) *Nature*, **388**, 257–259; (b) Kazaoui, S., Minami, N., Jacquemin, R., Kataura, H., and Achiba, Y. (1999) *Phys. Rev. B*, **60**, 13339–13342; (c) Petit, P., Mathis, C., Journet, C., and Bernier, P. (1999) *Chem. Phys. Lett.*, **305**, 370–374.

64 Bard, A.J. and Faulkner, L.R. (2001) Chapter 17, in *Electrochemical Methods: Fundamentals and Applications*, John Wiley & Sons, Inc., New York.

65 Weisman, R.B. and Bachilo, S.M. (2003) *Nano Lett.*, **3**, 1235–1238.

66 Iurlo, M., Paolucci, D., Marcaccio, M., and Paolucci, F. *Electrochemistry of Functional Supramolecular Systems* (eds M. Venturi, P. Ceroni, and A. Credi), John Wiley & Sons, Inc., New York, in press.

67 Bockris, J.O'M. and Reddy, A.K.N. (1998) Chapter 2, in *Modern Electrochemistry, Ionics*, vol. **1**, Plenum Press, New York.

68 McDonald, T.J., Svedruzic, D., Kim, Y.-H., Blackburn, J.L., Zhang, S.B., King, P.W., and Heben, M.J. (2007) *Nano Lett.*, **7**, 3528–3534.

69 (a) Xie, Q., Perez-Cordero, E., and Echegoyen, L. (1992) *J. Am. Chem. Soc.*, **114**, 3978–3980; (b) Oshawa, Y. and Saji, T. (1992) *J. Chem. Soc., Chem. Commun.*, 781–782.

70 Xie, Q., Arias, F., and Echegoyen, L. (1993) *J. Am. Chem. Soc.*, **115**, 9818–9819.

71 (a) Garcia, E., Kwak, J., and Bard, A.J. (1988) *Inorg. Chem.*, **27**, 4377–4382; (b) Bard, A.J., Garcia, E., Kukharenko, S., and Strelets, V.V. (1993) *Inorg. Chem.*, **32**, 3528–3531; (c) Ceroni, P., Paolucci, F., Paradisi, C., Juris, A., Roffia, S., Serroni,

S., Campagna, S., and Bard, A.J. (1998) *J. Am. Chem. Soc.*, **120**, 5480–5487.

72 Bruno, C., Marcaccio, M., Paolucci, D., Castellarin-Cudia, C., Goldoni, A., Streletskii, A.V., Drewello, T., Barison, S., Venturini, A., Zerbetto, F., and Paolucci, F. (2008) *J. Am. Chem. Soc.*, **130**, 3788–3796.

73 Scholes, G.D. and Rumbles, G. (2006) *Nat. Mater.*, **5**, 683–696.

74 Katsnelson, M.I. (2007) *Mater. Today*, **10** (1–2), 20–27.

75 Langa, F. and Nierengarten, J.-F. (2007) *Fullerenes. Principles and Applications*, RSC, Cambridge; (b) Guldi, D.M., Martin, N., and Prato, M. (2008) *J. Mater. Chem.* (Theme Issue: Carbon Nanostructures),

18, 1401–1607; (c) Yang, C., Kim, J.Y., Cho, S., Lee, J.K., Heeger, A.J., and Wudl, F. (2008) *J. Am. Chem. Soc.*, **130**, 6444–6450; (d) Nanjo, M., Cyr, P.W., Liu, K., Sargent, E.H., and Manners, I. (2008) *Adv. Funct. Mater.*, **18**, 470–477; (e) Bakry, R., Vallant, R.M., Najam-Ul-Haq, M., Rainer, M., Szabo, Z., Huck, C.W., and Bonn, G.K. (2007) *Int. J. Nanomed.*, **2**, 639–649.

76 (a) Geim, A.K. and Novoselov, K.S. (2007) *Nat. Mater.*, **6**, 183–191; (b) Vallés, C., Drummond, C., Saadaoui, H., Furtado, C.A., He, M., Roubeau, O., Ortolani, L., Monthioux, M., and Pénicaud, A. (2008) *J. Am. Chem. Soc.*, **130**, 15802–15804.

4
Photophysics

Tobias Hertel

4.1
Introduction

The photophysical exploration of single-walled carbon nanotubes (SWNTs) provides a window into the unique properties of these one-dimensional (1D) molecular nanoparticles. The realization that – depending on diameter and chirality – SWNTs could become semiconductors or metals, sparked interest in their electronic and optical properties [22, 41]. Long before their experimental verification, early theoretical investigations also predicted some of the key optical properties of SWNTs such as the existence of excitons [3]. The first experimental report on the optical properties of SWNT soot surfaced around 1999 [26] and, soon thereafter, absorption spectroscopy was established as a useful tool for characterizing nanotube sample composition. However, in 2002 the field of nanotube optics was propelled forward by the successful isolation and stabilization of individual SWNTs in colloidal suspensions that, for the first time, facilitated photoluminescence studies and stimulated most of the work described in this section [40].

In this article we focus on a discussion of phenomena that relate to continuous wave (CW) and time-resolved optical spectroscopy of carbon nanotubes (CNTs). Raman and electron spectroscopic techniques are discussed elsewhere. This chapter will start out with a brief review of the theoretical background of 1D photophysics, borrowing both from the well-established concepts of solid state semiconductor phenomena as well as from more recent developments specifically aimed at a better understanding of CNT photophysics. We will then discuss a selection of the key experiments and observations that provide us with a basis for further studies of SWNT photophysics. The last section of this chapter is dedicated to an introductory discussion of the dynamics of optically excited states as studied by different time-resolved optical probes.

4.2
Molecular Nanoparticles: Carbon Nanotubes Have it All

The boundaries between what constitutes a nanoparticle or a macromolecule sometimes appear fuzzy. However, a clear distinction can be made based on the

Carbon Nanotubes and Related Structures. Edited by Dirk M. Guldi and Nazario Martín
Copyright © 2010 WILEY-VCH Verlag GmbH & Co. KGaA, Weinheim
ISBN: 978-3-527-32406-4

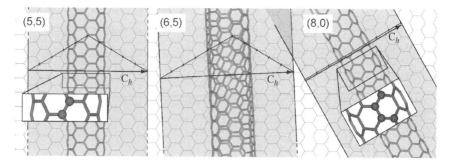

Figure 4.1 Illustration of how SWNTs are formed by wrapping from graphene. The 2- and 4-atom, circumferential and axial repeat units of the metallic $(5,5)$ armchair- and semiconducting $(8,0)$ zigzag tubes, respectively are magnified and emphasized in the insets. The unit cell of the chiral semiconducting $(6,5)$ tube extends beyond the graph. C_h designates the chiral or wrapping vector.

molecular identity of the material properties, rather than their sizes. For semiconductor nanocrystals, for example, one finds that important properties such as bandgap and luminescence wavelength are more or less continuously tunable with particle size [38]. Variations in size, surfaces, and defects naturally give rise to variations in particle properties and nanocrystals therefore generally lack a molecular identity of their properties [16]. On the other hand, molecular systems, even viruses, are characterized by a structure that is reproduced in each individual to the atomic level and, thus, exhibit a molecular identity in their photophysical and other properties. Similarly CNTs can be characterized by a set of indices (n, m) that uniquely specify their chiralities as well as diameters (see Figure 4.1), and consequently their electronic- and thereby optical properties [21, 47, 49].

Specific aspects of CNT structure and properties are, nevertheless, frequently characterized using terms and concepts of condensed matter or nanoparticle research, specifically when reference is made to doping, defects, or other structural heterogeneities such as length variations. As we will see later in this chapter, these may affect photophysical properties to some degree and thus endow CNTs with typical traits of solids. This ambivalence appears to justifiably qualify nanotubes as *molecular nanoparticles*, a material class in itself.

4.3
Understanding Optical Properties

Not surprisingly, a comprehensive description of the electronic and optical properties of CNTs is developed using a combination of language and concepts borrowed from both condensed matter- and organic molecular photophysics. Here we begin by a brief discussion of the electronic structures of two types of closely related 1D conjugated polymers, one with metallic and the other with semiconducting properties. An extension from these to the electronic structure of metallic and semiconducting nanotubes is straightforward and can be found in original articles and text books [21, 47, 49]. Later in the section we will extend these simple concepts to include many particle effects.

4.3.1
A Tight Binding Description

Perhaps, the simplest approach to explore the electronic structure of SWNTs is by following a tight binding (TB) description of 1D conjugated polymers. *trans*-polyacetylene (see Figure 4.2c) with its structural semblance of (n, n) tubes (see Figure 4.1) is used as a precursor of SWNTs with metallic properties [47]. Within the TB approximation, the expansion of molecular wave functions Ψ_j for the jth electron in terms of a linear combination of p_z atomic orbitals $\phi_{j'}$ is given by

$$\Psi_j = \sum_{j'} \hat{c}_{jj'} \phi_{j'}. \tag{4.1}$$

The eigenenergies of $\Psi_j(\vec{k}, \vec{r})$ are calculated in the usual manner using the sum of single-particle Hamiltonians $H = \sum h_j = \sum [(-\hbar^2/2m)\vec{\nabla}_j^2 + V(\vec{r}_j)]$. On minimizing the total energy, this yields the set of Hückel equations

$$\sum_{j=1}^{N} [H_{jj'} - ES_{jj'}]\hat{c}_{jj'} = 0, \tag{4.2}$$

where N is the number of carbon atom p_z orbitals in the system. The so-called transfer- and overlap integrals are given by $H_{jj'} = \langle j|H|j'\rangle$ and $S = \langle j|j'\rangle$, respectively. Solutions in terms of molecular orbital energies E_i are found by solving the determinant:

$$\det[H - ES] = 0. \tag{4.3}$$

For periodic structures, such as the ones considered here, the coefficients $\hat{c}_{jj'}$ have to be chosen such that they satisfy the constraints offered by the Bloch theorem $\Psi_j(\vec{r} + \vec{a}) = e^{i\vec{k}\vec{a}}\Psi_j(\vec{r})$, which accounts for translational invariance with respect to lattice vectors $\vec{a}$. Here, $\vec{k}$ is a so-called wave vector, which imposes a periodic modulation on the coefficients used to build TB crystal orbitals. Periodicity and its reflection in the Bloch theorem here simplifies matters in the sense that the summation in Eq. (4.1) only needs to cover n atomic orbitals within the unit cell of the periodic lattice. Thus, instead of N distinct molecular orbitals, one groups solutions into n energy bands composed of N/n states each. The N/n states within one such band are indexed by the continuous variable $\vec{k}$ (since N is typically very large). More details on the TB approach can be found in condensed matter text books [7].

The combination of two atomic carbon p_z orbitals from Figure 4.2a allows the formation of the bonding and antibonding π and π^* orbitals of ethylene shown in Figure 4.2b. Correspondingly, the combination of two p_z orbitals per unit cell, in

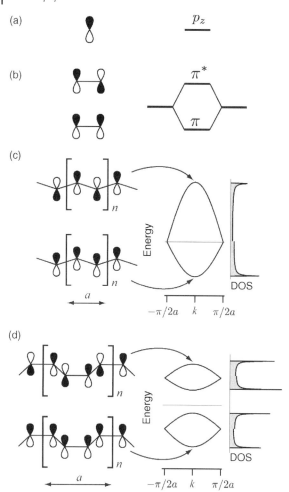

Figure 4.2 Orbitals and energy level diagrams of sp^2 bonded systems. (a) single p$_z$ orbital, (b) ethylene, and (c) *trans*-polyacetylene as well as (d) *cis*-polyacetylene. The band structure of the latter two is here sketched for the nondimerized molecules. The electronic structure of *trans*- and *cis*-polyacetylene can be considered to be closely related to that of metallic (n, n) and semiconducting $(n, 0)$ tubes (the latter with $n \neq 3, 6, \ldots$).

trans-polyacetylene (see Figure 4.2c) leads to the formation of dispersive π and π^* bands, each composed of a large number of states indexed by a continuous quantum number $\vec{k}$, also referred to as wave vector. The density of states (DOS) is shown on the right of the band dispersion. The splitting of these bands is governed by the transfer integral $\gamma_0 = \langle p_z^A | H | p_z^B \rangle$ between neighboring p$_z$ orbitals on atoms A and B and by the corresponding overlap integral S. Note that valence v and conduction c bands are joined at the boundary of the Brillouin zone. The energy of the polymer can be further reduced by lowering its symmetry in the form of alternating C–C bond lengths (dimerization). This Peierls distortion opens a gap between π and π^* bands that makes *trans*-polyacetylene semiconducting. In (n, n) CNTs (Figure 4.1) however, the repeat unit cell in Figure 4.2c remains undistorted and endows such CNTs with the metallic properties of the band structure in Figure 4.2c.

cis-polyacetylene shown in Figure 4.2d, on the other hand, has four atomic p_z orbitals per unit cell, which give rise to two bonding π and two antibonding π^* bands, which are again degenerate at the Brillouin zone boundaries. In contrast to nondimerized *trans*-polyacetylene, π and π^* bands are separated here by a bandgap, whose width scales with the transfer integral γ_0. The structural similarity of the repeat units of *cis*-polyacetylene with those of $(n, 0)$ CNTs (see Figure 4.1) again, allows to establish a relationship between this polymer and most of the semiconducting $(n, 0)$ tubes. In CNTs the different subbands are labeled using their subband index μ, which specifies the angular momentum projection onto the tube axis.

A more detailed analysis of TB band structure calculations reveals that all tubes with $(n-m)\mathrm{mod}3 = 0$ have no bandgap and are metallic, while those with $(n-m)\mathrm{mod}3 = \pm1$ have a bandgap and are semiconducting [21, 47, 49]. The bandgap E_g roughly scales with the inverse tube diameter $d_{n,m}$ according to $E_g \simeq \gamma_0 a_{\mathrm{C-C}}/d_{n,m}$, where $a_{\mathrm{C-C}}$ is the nearest neighbor carbon distance of 1.44 Å [47]. An important correction to transfer integrals is induced by diameter and chirality dependent tube curvature and the resulting realignment of p_z orbitals [15, 18]. This not only leads to a stronger deviation from the simple $d_{n,m}^{-1}$ scaling in semiconducting tubes but also to the opening of a small gap in metallic SWNTs except for (n, n) tubes, which remain truly metallic within the TB approximation.

Within the TB approximation, the optical properties at frequency ν are obtained from the absorption coefficient α_{free} for free e–h pair excitations. This is calculated using $\alpha_{\mathrm{free}} = \varepsilon_2(\nu)2\pi\nu c^{-1}$, where ε_2 is the imaginary part of the dielectric function and c is speed of light. Within the dipole approximation, the dielectric function is obtained by summation over the transition dipole moments $M_{\mathrm{vc}}(k)$ between all initial and finial states as well as over all momenta $\vec{k}$

$$\varepsilon_2(\nu) = \frac{e^2}{m^2\nu^2} \sum_{\mathrm{v,c}} \sum_{\vec{k}} |M_{\mathrm{vc}}(\vec{k})|^2 n_{\mathrm{v}}(1-n_{\mathrm{c}})\delta(E_{\mathrm{c}}(\vec{k})-E_{\mathrm{v}}(\vec{k})-h\nu), \tag{4.4}$$

where n_{v} and n_{c} refer to the Fermi distributions in valence and conduction bands, respectively, and $E_{\mathrm{v}}(\vec{k})$ and $E_{\mathrm{c}}(\vec{k})$ refer to the corresponding initial and final state energies. The projection of the orbital angular momentum along the tube axis cannot change if the incident light is polarized parallel to the CNT axis. Matrix elements for transitions from the valence to the conduction bands are thus found to be nonzero only for transitions with the same subband index μ, that is, $\Delta\mu = 0$. For light polarized perpendicular to the tube axis nonvanishing matrix elements have $\Delta\mu = \pm1$. However experimentally, transitions with $\Delta\mu = \pm1$ are found to be comparatively weak [3, 35].

In Figure 4.3 we plot Δ, the calculated energy splitting for transitions with $\Delta\mu = 0$ between the single-particle valence and conduction subbands of semiconducting S and small gap semiconducting and metallic tubes M. The gaps are given in units of

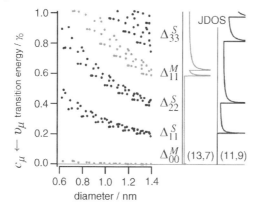

Figure 4.3 "Kataura" plot of transition energies for excitation of free carriers from valence to conduction bands, as obtained from curvature-corrected TB calculations for light polarized parallel to the tube axis [18]. On the right is the JDOS for a metallic and a semiconducting tube with about 1.4 nm diameter (light grey and black lines, respectively). Note that all metallic tubes except for those with $n = m$ have a small curvature-induced pseudogap at the Fermi level.

the transfer integral, which is frequently assumed to be on the order of 3 eV. However, recent photoluminescence experiments [28] suggest that $\gamma_0 \simeq 6\,\mathrm{eV}$ gives better agreement of optical bandgaps with calculated Δ_{11}^S TB gaps of free tubes, unless Coulomb interactions between valence and conduction band states are explicitly accounted for.

For a qualitative assessment of the optical spectra derived from the band structure it is convenient to neglect the energy dependence of matrix elements $(|M_{vc}(\vec{k})|^2 = \mathrm{const.})$ in which case the expression in Eq. (4.4) reduces to the so-called joint density of states (JDOS). The latter serves as a measure of the number of transitions at a specific energy that satisfy energy and momentum conservation. For parabolic valence and conduction bands of dimension d, the JDOS above the bandgap energy E_g takes on the familiar $(h\nu - E_g)^{(d/2-1)}$ dependence. For 1D valence and conduction bands, this leads to the characteristically diverging van Hove singularities at the bandgap (see Figure 4.3).

The following section will deal with many particle correlations and the influence of Coulomb interactions on the electronic structure, effects that are particularly important in 1D systems but that are not accounted for by the simple single-particle picture discussed up to here.

4.4
The Coulomb Interaction and Bound States

The Coulomb interaction has a profound influence on the optical spectra of semiconductors with its extent depending on the dimensionality of the system. This section will discuss the effects both on a qualitative and quantitative basis.

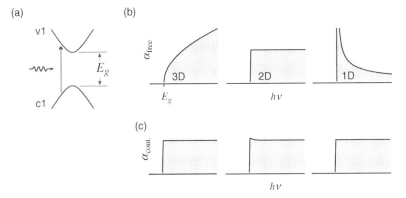

Figure 4.4 (a) Generic semiconductor band diagram. (b) Schematic illustration of the free carrier absorption coefficient α_{free} with no carrier–carrier interactions as obtained from the JDOS in three-, two-, and one-dimensional systems and (c) absorption coefficients $\alpha_{cont.}$ corrected by the corresponding continuum Sommerfeld factors.

Coulomb interactions not only modify the energetics of free electron and hole continua and the optical bandgaps, but they also give rise to the formation of new states, the so-called excitons that are identified with bound electron–hole pairs. Moreover, the strength, energetics, and the coupling of states are also strongly affected by the Coulomb interaction, that is, by screening and thus by the dimensionality of the systems under investigation [3,41].

The free carrier electron–hole pair continuum is modified by Coulomb interactions in the manner schematically illustrated in Figure 4.4b and c. The Sommerfeld correction term $C(h\nu)$ describes how the free carrier spectrum α_{free}, discussed in the previous subsection, is modified by absorption to ionized states to yield the continuum absorption coefficient $\alpha_{cont} = C(h\nu)\alpha_{free}$ [17, 41]. Specifically, one finds that the Sommerfeld factor increases continuum absorption at the band edge of 3D semiconductors, which makes the observation of weakly bound excitons challenging. In 1D systems the effect is equally dramatic due to the wiping out of the characteristic 1D van Hove singularity at the band edge.

The excitonic states can be described qualitatively within the effective medium approximation where screening of e–h pairs with effective mass μ is simply captured by the dielectric function ε. The resulting effective medium Hamiltonian is thus

$$H = -\frac{\hbar^2 \nabla^2}{2\mu} - \frac{e^2}{(4\pi\varepsilon_0)\varepsilon r}. \tag{4.5}$$

The character of solutions of $H\Psi = E\Psi$ in different dimensions, and, in particular, in 1D systems has been the subject of some debate [5, 29]. For fractal dimensionality α, this problem equation has been elegantly solved by He [22] and the energy of the nth eigenstate is found to scale with dimensionality according to

$$E_n = \frac{\mu}{m_e \varepsilon^2} \left(n + \frac{\alpha-3}{2}\right)^{-2} E_R, \tag{4.6}$$

where E_R is the Rydberg constant. Note that the 1s, that is, $E_{n=1}$ binding energy diverges when α approaches 1. This purely geometrical effect is compounded by changes in screening through ε in Eq. (4.6) which tends to be reduced in low-D systems owing to the general scarcity of mobile charges. Thus, in low-D systems and, in particular, in 1D the exciton binding is expected to be much larger than in the 3D systems. Exciton binding energies in SWNTs with 1 nm diameter have indeed been found to be in the order of 0.5–1 eV [28], while typical exciton binding energies in conventional bulk semiconductors are the order of a few to a few tens of meV.

In accordance with the conventions used in molecular photochemistry, exciton states associated with transitions across a specific free carrier gap $\Delta_{11}, \Delta_{22}, \Delta_{33}, \ldots$ will, here, be referred to as $S_1, S_2, S_3, \ldots$ singlet or $T_1, T_2, T_3, \ldots$ triplet states. We thus avoid the frequently used but somewhat ambiguous $E_{11}, E_{22}, E_{33}, \ldots$ designation, which suffers from over-usage as free carrier gap-, optical gap-, and exciton energy.

Last, the effect of Coulomb interactions increases f_{ab}, the oscillator strength of excitons, which is particularly pronounced in 1D [17]. According to the Thomas–Reiche–Kuhn sum rule the total oscillator strength of all the transitions from a specific initial state a to all final states b is conserved, that is, $\sum_b f_{ab} = Z$, where Z is the number of electrons participating in the transitions under consideration. An increase in the exciton oscillator strengths thus entails a decrease in the oscillator strength in the associated free electron–hole pair continua. The combined effects of Coulomb interactions on the spectra of semiconductors and the intriguing interplay with dimensionality are summarized schematically in Figure 4.5.

The first more quantitative description of excitonic effects in CNTs by Ando used a conventional effective mass, screened Hartree–Fock approximation within the $\vec{k} \cdot \vec{p}$

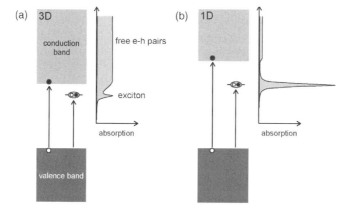

Figure 4.5 Schematic illustration of the effect of dimensionality on the optical properties of semiconductors. Particularly striking is a renormalization of the bandgap as well as an increase of the exciton binding energy due to a combination of reduced screening and geometrical factors in 1D systems. In addition, oscillator strength is transferred from the continuum band to the exciton. Spectra of ideal 1D semiconductors are thus expected to be dominated by excitonic transitions.

(a)
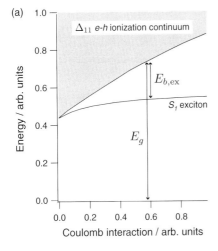

(b)
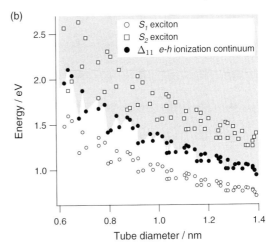

Figure 4.6 (a) Qualitative illustration of the effect of Coulomb interactions on the bandgap E_g for free e–h pair excitations as well as on the binding energy $E_{b,ex}$ of the first subband exciton (after [3]). (b) Experimental transition energies for excitation of S_1, the first and S_2, the second subband excitons in semiconducting nanotubes in aqueous SDS suspensions, open circles and open squares, respectively (after Ref. [59]). The position of the free carrier ionization continuum Δ_{11} (solid circles) is estimated from single tube PL and PLE experiments [28]. Due to the very large binding energies, some of the second subband excitons are within the first subbandgap.

scheme and could thereby account for many of the excitonic effects discussed later in this section [3]. Specifically, Ando predicted the formation of strongly bound exciton states whose binding energy and oscillator strength depend on the magnitude of the Coulomb interaction. Moreover, Ando noticed that the role of the Coulomb interaction in increasing the free carrier bandgap is expected to be more pronounced than its role for increasing the exciton binding energy such that the net effect of an increase in the strength of the Coulomb interaction is an increase in the exciton transition energy (see Figure 4.6a). This was later confirmed experimentally [28].

Last, we need to account for valence and conduction band degeneracies as well as for the spin of electrons and holes. The single-particle band structure for a typical semiconducting CNT with two degenerate valence and conduction bands in the CNT Brillouin zone is illustrated schematically in Figure 4.7a. The properly symmetrized combinations of these valence and conduction band states lead to the formation of four singlet- and four triplet excitons with different symmetry properties and energies [4, 43, 53]. The dispersions of the resulting four excitons are sketched in Figure 4.7b for the chiral (6,5) tube. Two of the four singlet excitons correspond to indirect bandgap transitions across the first Brillouin zone. The band minimum of the other two excitons is located at the Γ point. Only transitions from the ground state to excitons with symmetry A_2 are dipole allowed [9]. Transitions from the ground state to the first subband A_1 state are dipole forbidden and the state is thus often referred to as "dark state." A simplified energy level scheme is shown in Figure 4.7c,

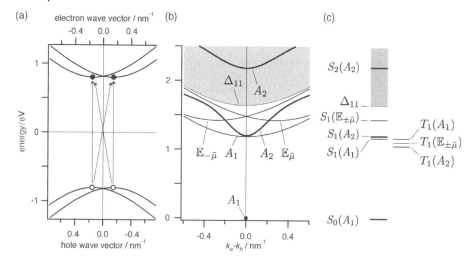

Figure 4.7 (a) Schematic illustration of the lowest single-particle valence- to conduction-band excitations in of a chiral (6,5) tube. (b) The corresponding two-particle spectrum contains so-called "bright" first- and second subband excitons with wave vector symmetry A_2 [9]. (c) Schematic energy level diagram with low points of different bands including triplet states on the right.

where S_n and T_n are used for a designation of the nth singlet- and triplet state manifolds.

The calculation of optical spectra from first principles is difficult and computationally costly [51]. Starting point of one approach is the calculation of ground state properties using density functional theory (DFT) within the local density approximation (LDA) [51]. DFT, however, is ill-equipped to deal with excited states and one has to resort to many body perturbation theory. One-particle Greens functions are thus used to calculate quasiparticle energies and amplitudes within the GW approximation. The quasiparticle states then correspond to a system with one added or removed electron and therefore need to be combined to calculate the neutral electron–hole pair spectrum. This requires solving the Bethe–Salpeter equation (BSE), which accounts for long-range Coulomb and short-range exchange interaction [11, 42, 52]. The agreement of the calculated with experimental transition energies is within 0.1–0.2 eV [51].

The binding energy of the bright excitons in colloidal (6,5) tubes, found to be on the order of 0.35 eV, is in good agreement with experimental observations [34, 57]. The optically silent A_1 exciton of the first singlet manifold on the other hand is found to be a few meV lower in energy than the bright exciton [36, 50]. The highest of the first subband triplet states is predicted to be only a few meV below the dark A_1 singlet state and the lowest triplet state is predicted to be about 20 meV below the lowest dark singlet state [43, 53]. See Figure 4.7c for an appropriately scaled graphic summary of these energies. Qualitatively similar results are also obtained from a TB approach for singlet and triplet excitons in conjugated polymers [1]. In these systems the lowest singlet state, however is bright and a similar degeneracy of singlet and triplet states as in CNTs occurs at energies well above the lowest singlet excitons [1].

A less costly approach approximates the quasiparticle energies using the parametrized TB Hamiltonian with a nearest neighbor transfer integral $\gamma_0 = 3.0$ eV [42]. Screening by the environment is accounted for by an effective dielectric constant, which for colloidal suspensions is assumed to be 3.3 [44]. The resulting bright exciton binding energies E_b with respect to the free carrier continuum are found to scale as $E_b \approx A_b R^{\alpha-2} \mu^{\alpha-1} \varepsilon^{-\alpha}$, where A_b and α are empirical parameters, μ is the effective mass and R is the tube radius [42]. Perebeinos *et al.* use $\alpha = 1.4$, which thereby gives a slightly weaker dependence of binding energies on the dielectric constant if compared, for example, with the hydrogen like system of the effective medium approximation where $E_b \propto \varepsilon^{-2}$ (see Eq. (4.6)).

The effective masses of all the optically silent excitons are predicted to be given by $\mu \approx 1.5(m_e + m_h)$, where m_e and m_h are the electron and hole masses, respectively, as obtained from curvature corrected TB calculations [19, 43]. The enhancement in the mass by 50% over the sum of the electron and hole masses is fairly independent of the tube diameter [43]. For the optically silent exitons of the (6,5) tube, this yields effective masses on the order of $0.28m_e$ except for the bright A_2 exciton whose dispersion attains a logarithmic correction $k_{eh}^2 \ln|k_{eh}|$ that reduces the effective mass by slightly over one order of magnitude [43]. The exciton size l_{ex}, which refers to the RMS displacement of electron and hole, is predicted to increase with the tube diameter d and with the dielectric constant while it decreases with the effective exciton mass. For a (6,5) tube this yields an e–h RMS of about 2 nm [53, 54], which is in good agreement with experiment [30].

4.5
Colloidal Chemistry Facilitates Detailed Study of Nanotube Optics

Advances in our understanding of nanotube electronic structure and optics have repeatedly been initiated by progress in sample preparation procedures. Some of the key developments in this respect are illustrated by the spectra in Figure 4.8. Early investigations of CNTs by absorption spectroscopy were confined to thin films of nanotube soot that had been spray coated onto glass substrates or to the spectroscopy of bucky paper films that were made by vacuum filtration [26]. Normalized absorption spectra from such films are shown in Figures 4.8 a and b for a sample obtained by pulsed-laser-vaporization and from the high pressure CO decomposition (HiPCO) process. These samples are each characterized by broad absorption bands in the near-infrared 850–1800 nm and in the 450–850 nm range of the spectrum, which can both be attributed to clusters of first (S_1) and second (S_2) subband transitions of the semiconducting nanotubes, respectively. Due to strong spectral congestion, however, the only quantitative information that can be obtained from these spectra is a rough estimate of the spread and magnitude of the mean diameter in these ensembles that is estimated to be 1.2 and 1.0 nm for the PLV and HiPCO samples, respectively. The spectra also indicate that the samples contain a significant fraction of metallic

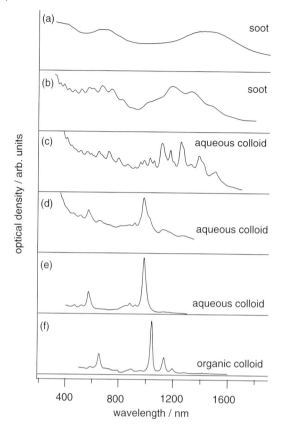

optical density / arb. units

(a) soot

(b) soot

(c) aqueous colloid

(d) aqueous colloid

(e) aqueous colloid

(f) organic colloid

400   800   1200   1600
wavelength / nm

Figure 4.8 Absorption spectra of different types of 1D samples. (a) and (b) are spectra obtained for thin solid films of laser oven- and HiPCO-synthesis samples. (c) is a colloidal sample of ultracentrifuged HiPCO soot. (d) Comocat raw material in a similar colloidal suspension. (e) Same as (d) but after density gradient ultracentrifugation. (f) Organic colloidal suspension of ultracentrifuged Comocat material.

nanotubes as evident from the weak M_{11} absorption band in Figure 4.8 a between 350 and 450 nm. It is generally assumed that the concentrations of metallic and semiconducting tubes in strongly polydisperse samples correspond to a randomized mixture of chiralities of $1:2$. The relatively small contribution of the metallic absorption feature thus indicates that the spectral weight of transitions in metallic tubes is distributed over a broader frequency range than the M_{11} feature might suggest, that is, a quantitative assessment of the concentration of metallic tubes from such spectra seems difficult.

The development of methods for the dispersion of soot in aqueous colloidal suspensions received widespread interest following the work of O'Connel *et al.*, which for the first time allowed to detect bandgap emission from semiconducting SWNTs [40]. Characteristic of these colloidal samples is that they are mostly composed of mutually isolated and well-separated tubes or small nanotube aggregates, which are encapsulated by surfactant micelles [14]. One of the most striking spectral signatures of surfactant dispersed nanotube soot is the rich structure found in absorption (see Figure 4.8c–f) and emission spectra. In contrast to this, soot or thin film samples are generally composed of nanotube aggregates, quasicrystalline hexagonal arrangements of bundled SWNTs with tens to hundreds of tubes per aggregate where the presence of the nearby tubes greatly reduces the effective PL quantum yields and tends to smear out exciton features in absorption spectra [14].

Further developments and higher degree of control over the diameter distribution as achieved by the Comocat process [27] have allowed further decongestion of optical spectra as seen in Figure 4.8d. Additional improvements of sample purity and structural selectivity were also obtained by density gradient ultracentrifugation (DGU), which – in combination with an optimization of surfactant coatings – allows to sort nanotube suspensions by aggregate size, diameter, chirality, and metallicity [6, 14]. The spectrum in Figure 4.8e, for example, was obtained by DGU with a $1:3$ Na-dodecylsulfate to Na-cholate cosurfactant mixture [6].

Recently there has also been considerable progress with the dispersion of CNTs with the help of organic polymers in organic solvents [39], see Figure 4.8f. These samples are typically characterized by reduced inhomogeneous broadening, very small background signals, and a seemingly stronger selectivity toward specific tube types in the dispersion process.

As illustrated in Figure 4.9a, a resonant excitation of micellar tubes in the S_2 energy range gives rise to pronounced emission at the corresponding S_1 energy. A measurement of both energies in fact provides a clear fingerprint of a specific (n, m) tube type, which may be used for structural identification. Typical absorption and emission spectra for a structurally sorted nanotube suspension are shown in Figure 4.9c along with a two-dimensional false color photoluminescence excitation (PLE) map in Figure 4.9b. PLE maps, such as the one in Figure 4.9b, generally have a rich fine structure and exhibit a variety of sidebands and weaker features that provide clues about exciton–phonon interactions, the role of defects for photoluminescence, as well as about internal conversion processes.

The small Stokes shift of the S_1 emission energy of about 4 nm with respect to the corresponding absorption feature is typical for colloidal SWNT samples. The absence of vibrational sidebands indicates that the excited state dynamics does not lead to significant vibrational relaxation in the excited state. This is taken as evidence for both, rapid internal conversion as well as for small forces acting on the nuclear wave packet in the excited state.

The absorption coefficient α can be used in the usual manner to estimate the concentration c of SWNTs in the suspension from $\alpha = \varepsilon_C \bar{n} c$, where ε_C is the molar carbon atom extinction coefficient and $\bar{n}$ is the average number of carbon atoms per tube. The extinction coefficient $\varepsilon_C = \sigma_C^{SWNT} N_A$ can be estimated from the carbon atom absorption cross section σ_C^{SWNT} at the S_1 or S_2 SWNT resonances of slightly less than 1×10^{-17} cm^2 [61]. For the spectrum in Figure 4.9c we then obtain a carbon atom concentration of 4.5×10^{-5} 1/(mol cm), which – for an average tube length on the order of 200 nm – roughly corresponds to a 2 nM concentration of (6,5) tubes or

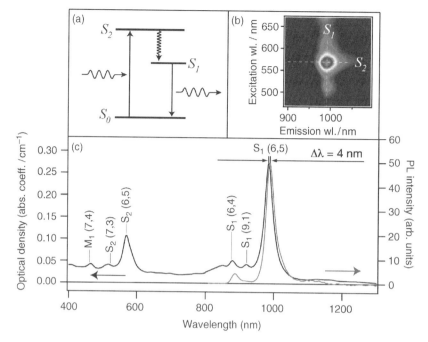

Figure 4.9 (a) Schematic illustration of absorption and emission of light from a specific tube type. (b) A two-dimensional photoluminescence excitation map of a structurally sorted nanotube sample as composed from photoluminescence spectra recorded at different excitation energies. The location of the main spectral feature in this image reveals the energies of S_2 and S_1 exciton resonances, here at 570 and 986 nm, respectively. (c) Comparison of the corresponding absorption and photoluminescence spectrum for resonant excitation at the energy of the S_2 resonance. The S_1 state is found to exhibit only a small Stokes shift of 4 nm.

about one tube per μm^3 of solution. The greatest uncertainty here is the value of the carbon atom absorption cross section, which is currently still under debate.

Photoluminescence excitation maps of the type shown in Figure 4.9b can also be used to characterize polydisperse nanotube samples, such as the HiPCO type, for which a PLE map has been reproduced in Figure 4.10. Such PLE maps were crucial for the structural assignment of S_1 and S_2 and in some cases of S_3 and S_4 subband features in absorption or emission spectra [8, 59]. Here, the key to structural assignment were patterns recognizable in the S_2 excitation and S_1 emission ranges. These patterns are indicated in Figure 4.10 by a connectivity grid of lines that can be thought of as connections between certain families of tubes whose (n, m) indices have common properties, that is, $(n-2m) = $ const., $(2n-m) = $ const., or $(n-m)\mathrm{mod}(3) = \pm 1$. In essence these connectivities are a reflection of the way in which the electronic structure of different SWNTs is derived from the electronic structure of their graphene parent by zone-folding.

Photoluminescence excitation spectroscopy has also been used in conjunction with two photon excitation to probe the excited states other than the bright

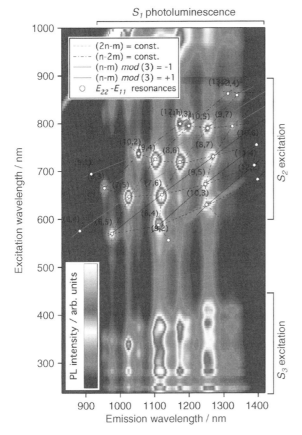

Figure 4.10 PLE spectrum of a polydisperse colloidal HiPCO sample. Tube families are indicated by the connectivity grids [24].

state [34, 57]. The transition from the ground state to the two higher lying degenerate states for example is two-photon allowed. The position of the higher lying exciton state can then be used in combination with the bright state energy to estimate the exciton binding energy. The results of Wang *et al.* as well as of Maultzsch *et al.* indicate that excitons in colloidal suspensions with diameters between 0.68 and 0.9 nm are bound by 0.3–0.4 eV [34, 57].

Lefebvre *et al.* used one-photon PLE of single vacuum-suspended SWNTs to map out bright excitons and higher Rydberg states [28]. The authors were able to confirm that scaling of the exciton binding energies with inverse tube diameter is nearly linear but values estimated for the binding energies were considerably higher than those obtained from the aforementioned two-photon work. Extrapolation of binding energies from Lefebvre's work to the diameter range explored by the two-photon PLE work suggests that energies in the small tube range are almost a factor of 2 higher in vacuum-suspended SWNTs, presumably due to screening by the surfactant- and hydration layers in suspensions. Using the $\varepsilon^{-1.4}$ scaling predicted by theory [42], this implies a 60% increase of screening in colloidally over bare vacuum-suspended SWNTs.

External factors like the surfactant type in colloidal suspensions, substrate type in single tube studies, pressure, magnetic field, and temperature have also been studied to better understand the electronic and optical properties of CNTs. Here we give a short account of some the results from the broad range of work in these areas.

The response of the electronic states of CNTs to external magnetic fields was first discussed by Ando [2]. The predictions of shifting valence and conduction band edges at high magnetic flux and the shift in the associated exciton transitions were later confirmed experimentally using pulsed magnetic fields in excess of 50 Tesla [50, 60]. The experiments show that a tube-threading, parallel magnetic field leads to a shift in the PL emission features, an effect that can be rationalized using a model of 1D magnetoexcitonic bands based on the Aharonov–Bohm effect [60]. Most notably, with increasing magnetic field, the lowest dark exciton state undergoes a redshift and brightening due to field-induced mixing with the bright exciton state, which in turn loses oscillator strength [50].

The effect of the environment on emission from bright excitons has been studied by various groups [12, 23, 28]. Comparison of emission from vacuum-suspended, adsorbate covered, and colloidally suspended tubes reveals a successive redshift of emission energies of up to 100 meV [12, 28]. With the transition energy $E_{ex} = E_g - E_{b,ex}$ depending on both, the free carrier gap E_g as well as on the exciton binding energy $E_{b,ex}$ this can be attributed to the aforementioned strong dependence of bandgap renormalization on the dielectric function. Considering the counterbalancing effect of weaker exciton binding energies with stronger screening, one can estimate that the free carrier bandgap increases by about 30% when the tube is removed from the influence of the colloidal suspension and is placed in vacuum.

4.6
Excited State Dynamics and Nonlinear Optics

One of the primary aims of time-resolved spectroscopy is the exploration of the coupling and interaction of electronic states with their environment, one another, and with static or dynamic lattice imperfections.

A generic energy diagram with some of the most relevant processes to be discussed in this context is shown in Figure 4.11. The rate of excitation from the ground to the excited states $k_{ab} = BI/c$ is determined by the intensity of light I and Einstein's B coefficient, which in turn can be obtained from the corresponding transition matrix elements. The rate for spontaneous emission or fluorescence $k_F = A = \tau_{rad}^{-1}$ on the other hand is given by Einstein's A coefficient, which is related to B by $A = \hbar\omega^3\pi^{-2}c^{-3}B$. Another useful expression relates the A coefficient to the oscillator strength f for the transition. In the case of nondegenerate initial and final states we have

$$\frac{1}{A} = \tau_{rad} = \frac{\varepsilon_0 m_e c \lambda^2}{2\pi e^2 f} = \frac{\lambda^2}{f} 1.499 \times 10^{-14} \, \text{nm}^{-2} \, \text{s}, \tag{4.7}$$

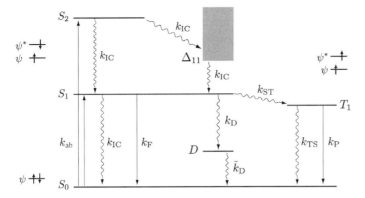

Figure 4.11 Generic state energy diagram for semiconducting SWNTs. Singlet states are labeled *S*, triplets *T*, free carrier continua Δ and defect or dark states are labeled *D*. Straight lines indicate transitions involving the emission or absorption of photons while wavy lines signify nonradiative transitions.

where e is the elementary charge, c is the speed of light, and λ is the wavelength of the emitted light.

Other decay processes are different internal conversion paths with rate constants k_{IC} and intersystem crossings from singlet to triplet states and or from the lowest triplet to the ground state k_{ST} and k_{TS}, respectively. Radiative return from the triplet- to the ground state is characterized by a phosphorescence rate k_P while transitions from the singlet to defect states D and their relaxation to the ground state are given by the rate constants k_D and $\tilde{k}_D$, respectively.

CW measurements of photoluminescence quantum yields Φ_{PL} can provide a simple means to gain information about excited state dynamics. The PL quantum yield is defined as the quotient of emitted N_{PL} and absorbed photons N_{ab}, which can also be rephrased in terms of the ratio of radiative and and the sum of all decay rates.

$$\Phi_{PL} = \frac{N_{PL}}{N_{ab}} = \frac{k_{rad}}{k_{rad} + \sum_i k_{i,nr}} = \frac{\tau_{PL}}{\tau_{rad}}, \tag{4.8}$$

where the sum over $k_{i,nr}$ represents the combination of different nonradiative decay processes from the state under consideration, as indicated by the wavy lines in Figure 4.11.

Early PL-quantum yield measurements of SWNT suspensions suggested discouragingly low values in the order of 10^{-4}–10^{-3} [40, 56], implying that nonradiative processes in such samples outweigh radiative decay. However, reported PL quantum yields have risen continuously and currently stand above 1% with some reports quoting 8% or even 20% [14, 25, 55]. One observation was that early sample preparation techniques produced not only structurally very heterogeneous but also aggregated samples with a majority of nanotubes still bound into small quasicrystalline aggregates, sometimes referred to as nanotube "ropes" [14]. Possible mechanisms that reduce the quantum yields in suspensions with aggregates are PL quenching by residual metallic tubes in the aggregates or a decrease if the exciton oscillator strength due to stronger screening in aggregates, which leads to an increase in the radiative lifetime in Eq. (4.7).

Additional PL quenching appears to be facilitated by tube ends, which act as efficient exciton sinks [48]. This was concluded by a study using length selected SWNT suspensions for which the PL quantum yield is found to increase up to a tube length of about 1 μm (see Figure 4.12a) [48]. The effect of quenching at tube ends with the rate k_{nr}^{end} was described using a 1D diffusion model that yields

$$\Phi_{PL} = \sum_{n=1}^{\infty} (2-2(-1)^n)^2 [n^2\pi^2(D\tau_{PL}^{\infty}(n\pi/L)^2 + 1)(1 + (n\pi/k_{nr}^{end})^2)]^{-1}, \qquad (4.9)$$

where D is the exciton diffusion constant, τ_{PL}^{∞} is the asymptotic photoluminescence lifetime for infinitely long tubes, and L is the tube length. A fit of this expression to the length dependence of PL quantum yields in Figure 4.12a an exciton diffusion length $l_{ex} = \sqrt{4D\tau_{PL}^{\infty}}$ of 130 nm and an asymptotic PL quantum yield of 3%. This is in good agreement with other reports of exciton diffusion lengths in the range of 100 nm [13, 46].

Measurements of the PL lifetime of nanotube suspensions have been numerous [20, 23, 31, 56]. The results of these experiments appear to depend on the type of sample and PL lifetimes range from ~10 ps to hundreds of ps. Some experiments appear to have been plagued by sample heterogeneities such as aggregate, length, and diameter dispersions. Here we reproduce recent time-correlated single-photon counting (TCSPC) measurements of a suspension with (6,5) tubes at low excitation densities, see Figure 4.12b. The PL decay is found to be nonexponential and the kinetics can either be analyzed in terms of a polyexponential decay with dominant components on the 20 ps (75%) and 300 ps (25%) scale or by using a stretched

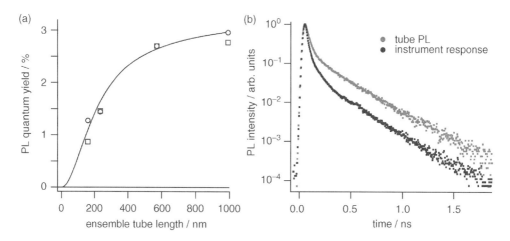

Figure 4.12 (a) Photoluminescence quantum yields for colloidal (6,5) enriched nanotube suspensions. The solid line is a fit to Eq. (4.9) and yields a diffusion length of 130 nm as well as an asymptotic PL quantum yield of 3%. (b) Photoluminescence decay of colloidally suspended (6,5) tubes.

exponential as expected for quenching kinetics with successively saturated traps [45]. Here, the average PL lifetime obtained form a multiexponential fit is about 90 ps. In combination with the PL quantum yield on the order of 1%, this suggests that the fluorescence decay rate k_F is on the order of 1–10 ns [56]. It should be noted that time resolved single tube PL experiments have previously reported mono as well as multiexponential kinetics [10, 20], with the substrate type or local environment possibly playing a role in determining the kinetics observed.

A mechanism that can account for the rapid nonradiative decay of the S_1 state in terms of phonon mediated internal conversion has been proposed by Perebeinos and Avouris [44]. This mechanism assumes that local doping opens new decay channels by coupling to a continuum of e–h pair excitations. Due to zone folding, this continuum of two particle excitations may also become resonant with bright exciton states (see Figure 4.13). Perebeinos and Avouris calculate the rate of phonon mediated decay by coupling the free carrier excitations with the S_1 state, and obtain nonradiative internal conversion k_{IC} rates in the range from 100 to $5\,\mathrm{ps}^{-1}$ in agreement with experimental observations [20, 23, 31, 56].

Complementary information on excited state dynamics is obtained from an investigation of optically induced spectral transients. This is done by pump-probe spectroscopy, which allows to study transient phenomena on a time-scale of 50 fs or less over the entire visible- or near-infrared spectral range. The technique probes changes in the absorbed light intensity $\Delta A/A$ of a laser impulse following short pulse excitation. Changes in absorption can be associated with changes in the initial and final state populations n_v and n_c of Eq. (4.4), respectively. The technique thereby

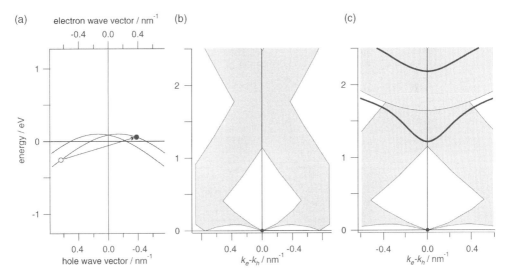

Figure 4.13 (a) Qualitative band diagram of a p-doped (6,5) tube. (b) The resulting free e–h pair excitation spectrum. (c) Superposition of the doping induced continuum of excitations with the spectrum of the undoped tube. Zone folding of the dopant induced free carrier continua leads to a direct overlap with the bright states of the undoped tube.

serves as a probe of the rate at which excited states are depopulated and at which the ground state is repopulated. For excitation of the S_1 state in Figure 4.11, for example, the measured rate of change of the optical transient reflects the combined effect of all processes leading to a depletion of S_1 and to replenishment of S_0.

A representative pump-probe spectrum for a (6,5) enriched nanotube suspension excited at 570 nm is shown in Figure 4.14. The transient at the S_1 energy has been reported to rise with a delay of about 40 fs [33]. This suggests that internal conversion between higher exciton subbands is extremely efficient. The mechanism for this rapid internal conversion was suggested to be zone-boundary phonon mediated interband scattering from the S_2 to the silent S_1 bands with a rate in excess of $0.01\ \text{fs}^{-1}$, followed by rapid intraband scattering and another internal conversion process to the bottom of the bright S_1 exciton band [24]. The same study also estimates that a minority of about 10% or less of the higher subband excitons undergo internal conversion to the free carrier Δ_{11} band.

The transient spectrum shown in the right part of Figure 4.14 exhibits a pronounced photobleach (PB) centered on the bright excitons as well as a smaller contribution from a photoabsorption signal (PA) at shorter wavelengths in the case of the S_1 state, and at lower as well as higher wavelengths in the case of the S_2 state. The

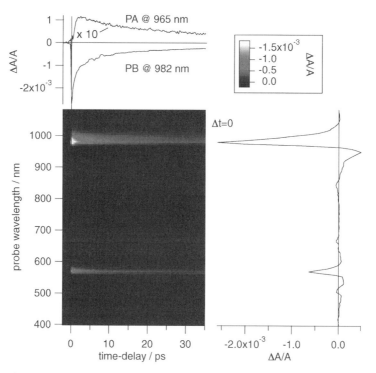

Figure 4.14 Pump-probe data from a (6,5) enriched nanotube suspension after 570 nm excitation. The upper and right sections of the Figure 4.4 show 1D cuts through the 2D dataset along time and spectral axes, respectively.

PB bands can be attributed to state filling of the higher exciton states and to a concomitant depletion of the ground state. The origin of the PA bands however remains somewhat less clear. Phenomenologically, the S_1 PA band and the spectral shape of the PA–PB region can be attributed to the overlap of the PB with a slightly broadened and similarly strong PA band that is blue shifted by 7 meV [62]. The formation of this state appears to be delayed by about 300 fs as evident from the rise time of the corresponding PA band.

The nonexponential recovery of the ground state is currently also difficult to attribute to any specific microscopic mechanism. From a semilogarithmic $\Delta A/A$ over $\ln t$ plot however, one finds that the transients closely follow a $t^{-1/2}$ power law behavior, which has tentatively been attributed to 1D diffusion limited ground state recovery [62].

The magnitude of optical transients in Figure 4.15 can be used in conjunction with the phase space filling model to directly determine the size of excitons [30]. In short, the phase space filling model assumes that the rate of absorption $k_{ab} = BI/c$ is reduced according to the decrease of phase space $\Delta p_z \, \Delta z$ occupied by each exciton. This implies that the rate of absorption is governed by $(1 - n_{ex}/n_{sat})$, the fraction of available exciton sites on the tube according to the simple saturation law $\tilde{k}_{ab} = (BI/c)(1 - n_{ex}/n_{sat})$ where n_{sat} is the saturation density of excitons. Experimentally available parameters are the rate of absorption as obtained from optical transients and the number of excitons on a tube as obtained from published absorption cross sections and the experimentally known pulse fluence. Here, the S_1 exciton size for the (6,5) tube is determined to be 2.0 ± 0.7 nm [30] in agreement with theoretical estimates [42, 52, 54].

However, the level of $\Delta A/A \approx -0.025$ at which the optical transients saturate in Figure 4.15, is small. The magnitude of this transient suggests that saturation of the

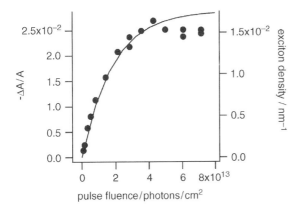

Figure 4.15 Saturation of the transient absorption signal at the S_1 exciton energy of the (6,5) tube for resonant excitation of the S_2 state by sub 50 fs laser impulses. Saturation occurs at low pulse fluences well below 10^{14} photons/cm^2 at an exciton density of roughly one exciton per 65 nm of tube length. This can be attributed to efficient exciton–exciton annihilation at higher excitation densities [37]. The solid line is a guide to the eye.

excited state population is reached at an density of about one exciton per 65 nm of tube length. This is easily an order of magnitude smaller than anticipated if phase space filling and the physical size of the excitons were to be the limiting factors. This is generally attributed to exciton–exciton annihilation at excitation densities where the probability for exciton–exciton encounters within their diffusion length of about 130 nm increases (see discussion above) [30, 32, 37, 58].

4.7
Outlook

The photophysical properties of SWNTs continue to fascinate and stimulate new research. Several photophysical properties and phenomena have not been discussed here but nevertheless present us with interesting questions regarding, for example, the role of defects, of impurities, or of lattice vibrations for optical spectra and excited state dynamics in SWNTs. Additional incentives for further studies can be obtained from the potential of CNTs for applications that directly benefit from their unique optical properties, for example, as contrast agents in the life sciences, as sensors, as transparent flexible conductors, and many more.

References

1 Abe, S., Yu, J., and Su, W.P. (1992) Singlet and triplet excitons in conjugated polymers. *Phys. Rev. B*, **45** (15), 8264–8271.

2 Ajiki, H. and Ando, T. (1994) Aharonov–Bohm effect in carbon nanotubes. *Physica B*, **201**, 349–352.

3 Ando, T. (1997) Excitons in carbon nanotubes. *J. Phys. Soc. Jpn.*, **66** (4), 1066–1073.

4 Ando, T. (2006) Effects of valley mixing and exchange on excitons in carbon nanotubes with Aharonov-Bohm flux. *J. Phys. Soc. Jpn.*, **75** (2)24707.

5 Andrews, M. (1966) Ground state of one-dimensional hydrogen atom. *Am. J. Phys.*, **34** (12), 1194–1195.

6 Arnold, M.S., Green, A.A., Hulvat, J.F., Stupp, S.I., and Hersam, M.C. (2006) Sorting carbon nanotubes by electronic structure using density differentiation. *Nat. Nanotechnol.*, **1** (1), 60–65.

7 Ashcroft, N.W. and Mermin, N.D. (1976) *Solid State Phys.*, Saunders College Publishing.

8 Bachilo, S.M., Strano, M.S., Kittrell, C., Hauge, R.H., Smalley, R.E., and Weisman, R.B. (2002) Structure-assigned optical spectra of single-walled carbon nanotubes. *Science*, **298** (5602), 2361–2366.

9 Barros, E.B., Capaz, R.B., Jorio, A., Samsonidze, G.G., Souza, A.G., Ismail-Beigi, S., Spataru, C.D., Louie, S.G., Dresselhaus, G., and Dresselhaus, M.S. (2006) Selection rules for one- and two-photon absorption by excitons in carbon nanotubes. *Phys. Rev. B*, **73** (24)241406.

10 Berciaud, S., Cognet, L., and Lounis, B. (2008) Luminescence decay and the absorption cross section of individual single-walled carbon nanotubes. *Phys. Rev. Lett.*, **101** (7)77402.

11 Chang, E., Bussi, G., Ruini, A., and Molinari, E. (2004) Excitons in carbon nanotubes: an *ab initio* symmetry-based approach. *Phys. Rev. Lett.*, **92** (19)196401.

12 Chiashi, S. Watanabe, S. Hanashima, T. Homma, Y. (2008) Influence of gas adsorption on optical transition energies

of single-walled carbon nanotubes. *Nano Lett.*, **8** (10), 3097–3101.

13 Cognet, L., Tsyboulski, D.A., Rocha, J.-D.R., Doyle, C.D., Tour, J.M., and Weisman, R.B. (2007) Stepwise quenching of exciton fluorescence in carbon nanotubes by single-molecule reactions. *Science*, **316** (5830), 1465–1468.

14 Crochet, J., Clemens, M., and Hertel, T. (2007) Quantum yield heterogeneities of aqueous single-wall carbon nanotube suspensions. *J. Am. Chem. Soc.*, **129** (26), 8058.

15 Ding, J.W., Yan, X.H., and Cao, J.X. (2002) Analytical relation of band gaps to both chirality and diameter of single-wall carbon nanotubes. *Phys. Rev. B.*, **66** (7) 73401.

16 Empedocles, S.A., Neuhauser, R., Shimizu, K., and Bawendi, M.G. (1999) Photoluminescence from single semiconductor nanostructures. *Adv. Mater.*, **11** (15), 1243–1256.

17 Koch, S.W. and Haug, H. (2004) *Quantum Theory of the Optical and Electronic Properties of Seimconductors*, World Scientific, Singapore.

18 Hagen, A. and Hertel, T. (2003) Quantitative analysis of optical spectra from individual single-wall carbon nanotubes. *Nano Lett.*, **3** (3), 383–388.

19 Hagen, A., Moos, G., Talalaev, V., and Hertel, T. (2004) Electronic structure and dynamics of optically excited single-wall carbon nanotubes. *Appl. Phys. A: Mater.*, **78** (8), 1137–1145.

20 Hagen, A., Steiner, M., Raschke, M.B., Lienau, C., Hertel, T., Qian, H.H., Meixner, A.J., and Hartschuh, A. (2005) Exponential decay lifetimes of excitons in individual single-walled carbon nanotubes. *Phys. Rev. Lett.*, **95** (19)197401.

21 Hamada, N., Sawada, S., and Oshiyama, A. (1992) New one-dimensional conductors – graphitic microtubules. *Phys. Rev. Lett.*, **68** (10), 1579–1581.

22 He, X.F. (1991) Excitons in anisotropic solids – the model of fractional-dimensional space. *Phys. Rev. B*, **43** (3), 2063–2069.

23 Hertel, T., Hagen, A., Talalaev, V., Arnold, K., Hennrich, F., Kappes, M., Rosenthal, S., McBride, J., Ulbricht, H., and Flahaut, E. (2005) Spectroscopy of single- and double-wall carbon nanotubes in different environments. *Nano Lett.*, **5** (3), 511–514.

24 Hertel, T., Perebeinos, V., Crochet, J., Arnold, K., Kappes, M., and Avouris, P. (2008) Intersubband decay of 1-D exciton resonances in carbon nanotubes. *Nano Lett.*, **8** (1), 87–91.

25 Ju, S.-Y., Kopcha, W.P., and Papadimitrakopoulos, F. (2009) Brightly fluorescent single-walled carbon nanotubes via an oxygen-excluding surfactant organization. *Science*, **323** (5919), 1319–1323.

26 Kataura, H., Kumazawa, Y., Maniwa, Y., Umezu, I., Suzuki, S., Ohtsuka, Y., and Achiba, Y. (1999) Optical properties of single-wall carbon nanotubes. *Synth. Met.*, **103** (1–3), 2555–2558.

27 Kitiyanan, B., Alvarez, W.E., Harwell, J.H., and Resasco, D.E. (2000) Controlled production of single-wall carbon nanotubes by catalytic decomposition of CO on bimetallic Co-Mo catalysts. *Chem. Phys. Lett.*, **317** (3–5), 497–503.

28 Lefebvre, J. and Finnie, P. (2008) Excited excitonic states in single-walled carbon nanotubes. *Nano Lett.*, **8** (7), 1890–1895.

29 Loudon, R. (1959) One-dimensional hydrogen atom. *Am. J. Phys.*, **27** (9), 649–655.

30 Lueer, L., Hoseinkhani, S., Polli, D., Crochet, J., Hertel, T., and Lanzani, G. (2009) Size and mobility of excitons in (6,5) carbon nanotubes. *Nat. Phys.*, **5** (1), 54–58.

31 Ma, Y.Z., Stenger, J., Zimmermann, J., Bachilo, S.M., Smalley, R.E., Weisman, R.B., and Fleming, G.R. (2004) Ultrafast carrier dynamics in single-walled carbon nanotubes probed by femtosecond spectroscopy. *J. Chem. Phys.*, **120** (7), 3368–3373.

32 Ma, Y.Z., Valkunas, L., Dexheimer, S.L., Bachilo, S.M., and Fleming, G.R. (2005) Femtosecond spectroscopy of optical excitations in single-walled carbon nanotubes: evidence for exciton–exciton annihilation. *Phys. Rev. Lett.*, **94** (15) 157402.

33 Manzoni, C., Gambetta, A., Menna, E., Meneghetti, M., Lanzani, G., and Cerullo, G. (2005) Intersubband exciton relaxation dynamics in single-walled carbon nanotubes. *Phys. Rev. Lett.*, **94** (20)207401.

34 Maultzsch, J., Pomraenke, R., Reich, S., Chang, E., Prezzi, D., Ruini, A., Molinari, E., Strano, M.S., Thomsen, C., and Lienau, C. (2005) Exciton binding energies in carbon nanotubes from two-photon photoluminescence. *Phys. Rev. B*, **72** (24) 241402.

35 Miyauchi, Y., Oba, M., and Maruyama, S. (2006) Cross-polarized optical absorption of single-walled nanotubes by polarized photoluminescence excitation spectroscopy. *Phys. Rev. B*, **74** (20)205440.

36 Mortimer, I.B. and Nicholas, R.J. (2007) Role of bright and dark excitons in the temperature-dependent photoluminescence of carbon nanotubes. *Phys. Rev. Lett.*, **98** (2)27404.

37 Murakami, Y. and Kono, J. (2009) Nonlinear photoluminescence excitation spectroscopy of carbon nanotubes: exploring the upper density limit of one-dimensional excitons. *Phys. Rev. Lett.*, **102** (3)37401.

38 Murray, C.B., Norris, D.J., and Bawendi, M.G. (1993) Synthesis and characterization of nearly monodisperse CDE (E = S, SE, TE) semiconductor nanocrystallites. *J. Am. Chem. Soc.*, **115** (19), 8706–8715.

39 Nish, A., Hwang, J.-Y., Doig, J., and Nicholas, R.J. (2007) Highly selective dispersion of single-walled carbon nanotubes using aromatic polymers. *Nat. Nanotechnol.*, **2** (10), 640–646.

40 O'Connell, M.J., Bachilo, S.M., Huffman, C.B., Moore, V.C., Strano, M.S., Haroz, E.H., Rialon, K.L., Boul, P.J., Noon, W.H., Kittrell, C., Ma, J.P., Hauge, R.H., Weisman, R.B., and Smalley, R.E. (2002) Band gap fluorescence from individual single-walled carbon nanotubes. *Science*, **297** (5581), 593–596.

41 Ogawa, T. and Takagahara, T. (1991) Interband absorption-spectra and Sommerfeld factors of a one-dimensional electron–hole system. *Phys. Rev. B*, **43** (17), 14325–14328.

42 Perebeinos, V., Tersoff, J., and Avouris, P. (2004) Scaling of excitons in carbon nanotubes. *Phys. Rev. Lett.*, **92** (25)257402.

43 Perebeinos, V., Tersoff, J., and Avouris, P. (2005) Radiative lifetime of excitons in carbon nanotubes. *Nano Lett.*, **5** (12), 2495–2499.

44 Perebeinos, V. and Avouris, P. (2008) Phonon and electronic nonradiative decay mechanisms of excitons in carbon nanotubes. *Phys. Rev. Lett.*, **101** (5)57401.

45 Phillips, J.C. (1996) Stretched exponential relaxation in molecular and electronic glasses. *Rep. Prog. Phys.*, **59** (9), 1133–1207.

46 Qian, H., Georgi, C., Anderson, N., Green, A.A., Hersam, M.C., Novotny, L., and Hartschuh, A. (2008) Exciton transfer and propagation in carbon nanotubes studied by near-field optical microscopy. *Phys. Stat. Sol. (B) – Basic Solid State Phys.*, **245** (10 Suppl. Issue SI), 2243–2246.

47 Dresselhaus, G., Saito, R., and Dresselhaus, M.S. (1998) *Physical Properties of Carbon Nanotubes*, Imperial College Press, London.

48 Rajan, A., Strano, M.S., Heller, D.A., Hertel, T., and Schulten, K. (2008) Length-dependent optical effects in single walled carbon nanotubes. *J. Phys. Chem. B*, **112** (19), 6211–6213.

49 Saito, R., Fujita, M., Dresselhaus, G., and Dresselhaus, M.S. (1992) Electronic-structure of chiral graphene tubules. *Appl. Phys. Lett.*, **60** (18), 2204–2206.

50 Jonah, S., Junichiro, K., Portugall, O., Krstic, V., Rikken, G.L.J.A., Miyauchi, Y., Maruyama, S., and Perebeinos, V. (2007) Magnetic brightening of carbon nanotube photoluminescence through symmetry breaking. *Nano Lett.*, **7** (7), 1851–1855.

51 Spataru, C.D., Ismail-Beigi, S., Rodrigo, B., Capaz, S., and Louie, G. (2008) Quasiparticle and excitonic effects in the optical response of nanotubes and nanoribbons, in *Carbon Nanotubes, Topics in Applied Physics*, Springer, Berlin vol. **111**, 195–227.

52 Spataru, C.D., Ismail-Beigi, S., Benedict, L.X., and Louie, S.G. (2004) Excitonic effects and optical spectra of single-walled carbon nanotubes. *Phys. Rev. Lett.*, **92** (7)77402.

53 Spataru, C.D., Ismail-Beigi, S., Capaz, R.B., and Louie, S.G. (2005) Theory and ab initio calculation of radiative lifetime of

excitons in semiconducting carbon nanotubes. *Phys. Rev. Lett.*, **95** (24)247402.

54 Tretiak, S. (2007) Triplet state absorption in carbon nanotubes: A TD-DFT study. *Nano Lett.*, **7** (8), 2201–2206.

55 Tsyboulski, D.A., Rocha, J.-D.R., Bachilo, S.M., Cognet, L., and Weisman, R.B. (2007) Structure-dependent fluorescence efficiencies of individual single-walled cardon nanotubes. *Nano Lett.*, **7** (10), 3080–3085.

56 Wang, F., Dukovic, G., Brus, L.E., and Heinz, T.F. (2004) Time-resolved fluorescence of carbon nanotubes and its implication for radiative lifetimes. *Phys. Rev. Lett.*, **92** (17)177401.

57 Wang, F., Dukovic, G., Brus, L.E., and Heinz, T.F. (2005) The optical resonances in carbon nanotubes arise from excitons. *Science*, **308** (5723), 838–841.

58 Wang, F., Wu, Y., Hybertsen, M.S., and Heinz, T.F. (2006) Auger recombination of excitons in one-dimensional systems. *Phys. Rev. B*, **73** (24)245424.

59 Weisman, R.B. and Bachilo, S.M. (2003) Dependence of optical transition energies on structure for single-walled carbon nanotubes in aqueous suspension: an empirical Kataura plot. *Nano Lett.*, **3** (9), 1235–1238.

60 Zaric, S., Ostojic, G.N., Shaver, J., Kono, J., Portugall, O., Frings, P.H., Rikken, G.L.J.A., Furis, M., Crooker, S.A., Wei, X., Moore, V.C., Hauge, R.H., and Smalley, R.E. (2006) Excitons in carbon nanotubes with broken time-reversal symmetry. *Phys. Rev. Lett.*, **96** (1)16406.

61 Zheng, M. and Diner, B.A. (2004) Solution redox chemistry of carbon nanotubes. *J. Am. Chem. Soc.*, **126**, 15490–15494.

62 Zhu, Z., Crochet, J., Arnold, M.S., Hersam, M.C., Ulbricht, H., Resasco, D., and Hertel, T. (2007) Pump-probe spectroscopy of exciton dynamics in (6,5) carbon nanotubes. *J. Phys. Chem. C*, **111** (10), 3831–3835.

5
Noncovalent Functionalization of Carbon Nanotubes

Mª Ángeles Herranz and Nazario Martín

5.1
Introduction

Carbon nanotubes (CNTs) have mechanical, optical and electrical properties that make them ideal nanoscale materials [1, 2]. However, all the proposed applications have so far been limited by their practical insolubility in aqueous and organic solvents [3, 4]. Due to their high polarizability and smooth surface, CNTs – in particular, single-walled carbon nanotubes (SWCNTs) – form bundles and ropes, where several nanotubes are aligned parallel to each other with a high van der Waals attraction (0.5 eV/μm). In addition, CNTs are obtained as mixtures that exhibit different chiralities, diameters and length, in which non-NT carbon and metal catalysts are also present in the final material.

Some of these limitations can be overcome by the controlled defect and sidewall functionalization of CNTs (see next chapter) [5–18]. The formation of covalent linkages can drastically enhance the solubility of CNTs in various solvents at the same time that guarantees the structural integrity of the nanotube skeleton, but it also alters the intrinsic physical properties of the CNTs because of a modification of the sp^2 carbon framework [18]. The most notable effect is that the inherent conductivity of the CNTs is basically destroyed.

An alternative strategy for preserving the intrinsic electronical and mechanical properties of CNTs consists in the noncovalent or supramolecular modification of CNTs [5, 14, 19–21]. Such interactions primarily involve hydrophobic, van der Waals, and electrostatic forces, and require the physical adsorption of suitable molecules onto the sidewalls of the CNTs. Noncovalent functionalization is achieved by polymer wrapping, adsorption of surfactants or small aromatic molecules, and interaction with porphyrins or biomolecules such as DNA and peptides. A special case is the endohedral functionalization of CNTs [19] – filling of the tubes with atoms or small molecules (peapods) – which is the objective of another chapter in this book.

Here we assess the field of exohedral noncovalent interactions between molecules and CNTs. The diversity of mechanisms of molecule–CNTs interactions and the

Carbon Nanotubes and Related Structures. Edited by Dirk M. Guldi and Nazario Martín
Copyright © 2010 WILEY-VCH Verlag GmbH & Co. KGaA, Weinheim
ISBN: 978-3-527-32406-4

range of experimental techniques employed for their characterization are analyzed by considering the different types of molecules and illustrated with examples from recent reports. In addition, some practical applications of systems considering noncovalent CNTs interactions with several molecular structures are also discussed.

5.2
Early Insights in the Noncovalent Interaction of CNTs with Solvents and Classical Macrocyclic Scaffolds

At early stages, most of the investigations on CNTs' reactivity focused on separating and dispersing them by chemical oxidation in acidic media [22], where the acid not only dissolves any remaining metal catalyst but also removes the CNT caps, leaving behind carboxylic acid ($-COOH$) residues. These oxidized CNTs are easily dispersible in various amide-type organic solvents under the influence of an ultrasonic force field [23]. In this sense, Haddon and coworkers [24] reported the direct reaction of acid-purified short SWCNTs with long-chain amines yielding very soluble zwitterionic materials. In a following work [25], treating multiwalled carbon nanotubes (MWCNTs) by sonication in water led to the implementation of O-containing functionalities (alcohol, carbonyl, and carboxylic acid) and to no appreciable damage of the basic CNTs structure. The production of functional groups was reflected in the disappearance of $-CH_n$ groups existing on the pristine CNTs and the existence of hydrogen bonding between the CNTs and the aqueous medium, which was established by spectroscopic analysis [25].

Wong and coworkers, prepared soluble oxidized SWCNTs by the supramolecular attachment of functionalized organic crown ethers, namely 2-aminomethyl-18-crown-6 [26]. The obtained adducts yielded concentrations of dissolved nanotubes in the range of 1 g/l in water, as well as in methanol, according to optical measurements. The structure of these adducts are likely to be a consequence of a noncovalent, zwitterionic chemical interaction between carboxylic groups (i.e., oxidized tubes) and amine moieties (i.e., attached to the side chain of the crown derivative) (Figure 5.1).

The interaction of SWCNTs with other classical supramolecular architectures, such as cyclodextrins (CDs), has also been investigated [27]. Purified SWCNTs and CDs mixed by a mechanochemical "high-speed vibration milling technique" (HSVM) are soluble in aqueous solutions because of the formation of SWCNT–CDs complexes and the debundling of SWCNTs. In particular, the SWCNT–γ-CD complex can be dissolved in concentrations as high as 1.15 mg/ml. The formation of multipoint hydrogen bonding between the hydroxyl groups of the CDs and the carboxylic acids of the SWCNTs was confirmed by the control experiments carried out with methylated CDs. In comparison with β-CD, which dissolves 0.73 mg/ml of SWCNTs, the solubilities of SWCNTs by heptakis-(2,6-di-O-methyl)-β-cyclodextrin (DM-β-CD) and heptakis-(2,3,6-tri-O-methyl)-β-cyclodextrin (TM-β-CD) were notably decreased as a function of hydroxyl groups of CDs (0.50 and 0.17 mg/ml, respectively) [27].

Figure 5.1 Optimized geometry for crown ether functionalized (5,5) SWCNTs. Adduct formation likely arises from a zwitterionic interaction between the carboxylic acid groups on the SWCNT and the amino functionality on the derivatized crown ether. Reprinted from Ref. [9] with permission from Wiley-VCH Verlag GmbH.

5.3
Noncovalent Interactions of CNTs with Small Aromatic Molecules

The interactions between CNTs and a series of structurally related molecules: cyclohexane, cyclohexene, cyclohexadiene, and benzene, were investigated in the gas phase by Sumanasekera *et al.*, and demonstrated that π–π interactions are fundamental for the adsorption on CNTs [28]. In particular, the highest adsorption energy was found for the molecule with the largest π-electronic system (benzene) and the lowest one for the molecule without π-electrons (cyclohexane). This result clearly indicates that the CNT–molecule interactions in this series are controlled by coupling of the π-electrons of the molecules with the electronic π-system of the CNT.

In a similar investigation, the interactions between benzene, cyclohexane, and 2,3-dichloro-5,6-dicyano-1,4-benzoquinone (DDQ) were theoretically studied using first principles calculations [29]. The coupling of π-electrons between CNTs and aromatic molecules was also observed, demonstrating that noncovalent functionalization of CNTs with aromatic molecules may be an efficient way to solubilize individual CNTs and to control the electronic properties of CNTs as discussed in the following sections.

5.3.1
Anthracene Derivatives

The adsorption of anthracene **1** and several derivatives, substituted by groups with different electronic properties and volume size (**2–5**), onto the sidewalls of cut

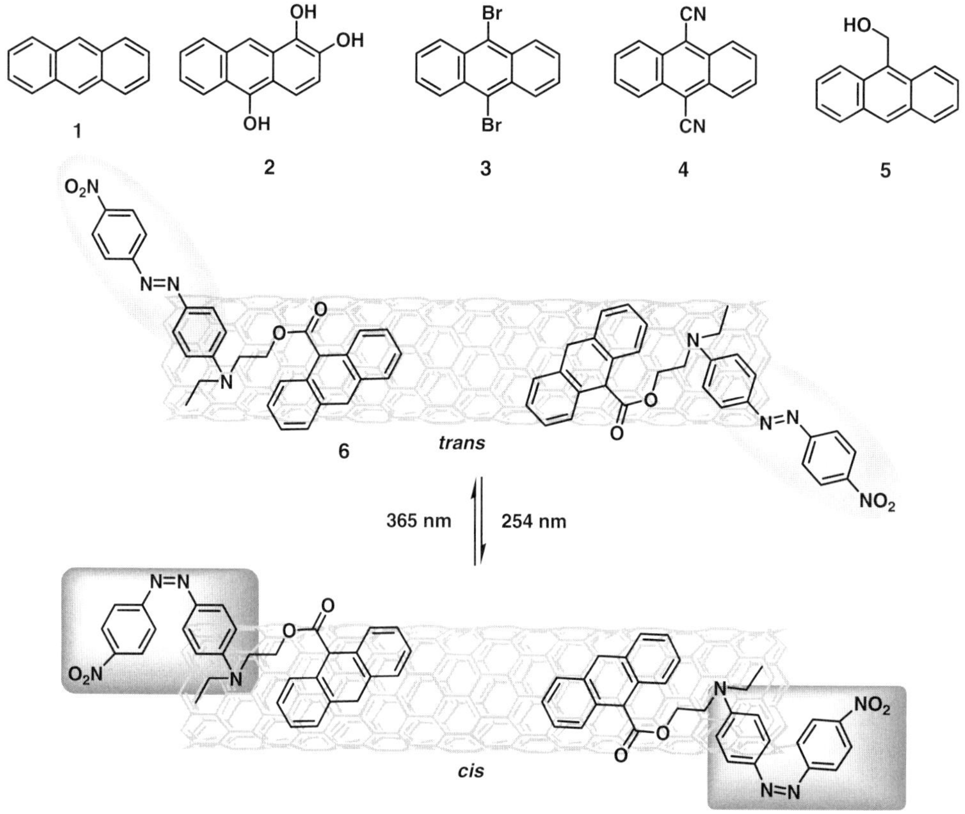

Figure 5.2 Different anthracene derivatives considered in the noncovalent functionalization of CNTs. Under UV light, DR1 **6** isomerizes from the equilibrium *trans* conformation to the metastable *cis* conformation. During this isomerization, the molecular dipole moment changes from 9 to 6 D.

SWCNTs (Figure 5.2) was investigated by considering the absorbance and fluorescence spectra of the aggregates formed. The absorption spectra of anthracenes **1–5** on SWCNTs seem to be unchanged as compared to free molecules in solution, whereas the fluorescence spectra of the same molecules on SWCNTs are redshifted, indicating a partial electron transfer from SWCNTs onto anthracenes. In fact, the adsorption coverage varied with the aromatic ring substitution, since the π–π bond interactions governing the adsorption process are accompanied by an electron donor–acceptor charge transfer interaction between the aromatic adsorbates and the SWCNT sidewall. The extent in which the strength of the charge transfer interaction controls the relative degree of adsorption coverage follows the order: **4 > 3 > 5,1 > 2** [30].

Simmons *et al.*, also considered an anthracene-based dye, disperse Red 1 (DR1) **6**, in the noncovalent functionalization of SWCNTs [31]. DR1 undergoes a highly reversible *trans–cis* molecular isomerization when exposed to light of different wavelengths. This photoisomerization involves the rearrangement of the dye's neighboring molecular groups around the double bond that cannot rotate itself. Ultraviolet light of 254 nm initiates the conversion from the *trans* to the *cis* form, whereas the blue light of 365 nm reverts the conversion from *cis* to *trans* (Figure 5.2). Interestingly, the reconfiguration of the dye molecule causes a significant change in the electrical dipole moment along its principal axis. This shift also modifies the local electrostatic potential in the CNTs and, in turn, modulates its conductance by shifting the threshold voltage at which current flows. For a ~1–2% coverage, a shift in the threshold voltage of up to 1.2 V is observed. In addition, the conductance change is reversible and repeatable over long periods of time, indicating that the chromophore-functionalized CNTs are useful for integrated nanophotodetectors.

5.3.2
Pyrene Derivatives

The pyrene group is an aromatic system that demonstrates a very high affinity for the CNT surface. Derivatives of pyrene can be efficiently deposited onto CNTs in all kind of solvents with a high surface coverage. In fact, the interaction of the aromatic system of pyrene with CNTs is so effective that functionalized pyrenes have been used for anchoring all kind of systems to CNTs with diverse applications.

Ammonium amphiphiles carrying a pyrenyl group act as excellent solubilizers of SWCNTs in water. Sonication of raw SWCNTs in an aqueous solution of **7** (Figure 5.3) gave a transparent solution of the CNTs, which was characterized by transmission electron microscopy (TEM), UV/Vis absorption, fluorescence, and ^{1}H NMR spectroscopies [32]. In addition, the aqueous solution of **7** has a tendency to dissolve, specifically, semiconducting SWCNTs with diameters in the range of 0.89–1.0 nm. This fine discrimination in the diameters of the SWCNTs is of particular interest in the design of CNT solubilizers that recognize a single nanotube chiral index, and could be very useful for the development of many potential applications [33].

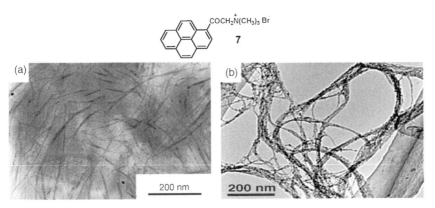

Figure 5.3 TEM images of aqueous solutions/dispersions of raw-SWCNTs/**7** (a) and insoluble raw-SWCNTs (b). Reprinted from Ref. [33] with permission from Wiley-VCH Verlag GmbH.

Functionalized pyrenes have also proved to be very useful in anchoring proteins or small biomolecules on CNTs. For example, the bifunctional *N*-succinimidyl-1-pyrenebutanoate **8** irreversibly adsorbs onto the hydrophobic surface of SWCNTs. In these conjugates, the succinimidyl group could nucleophilically be substituted by primary or secondary amino groups from proteins such as ferritin or streptavidin, and cause inmobilization of the biopolymers on the tubes [34]. In another example, pyrene-based glycodendrimers (**9**) demonstrated to function as homogeneous bioactive coatings for SWNTs that also mitigate their cytotoxicity [35]. Pyrene structures have also been combined with bioactive monosaccharides such as *N*-acetyl-D-glucosamine (GlcNAc) (**10**) and served to prepare glycosylated CNTs that are able to interface biocompatibly with living cells and detect the dynamic secretion of biomolecules from them [36] (Figure 5.4).

The development of reliable and reproducible methodologies to integrate CNTs into functional structures such as donor–acceptor hybrids, able to transform sunlight into electrical or chemical energy, has emerged as an area of intensive research

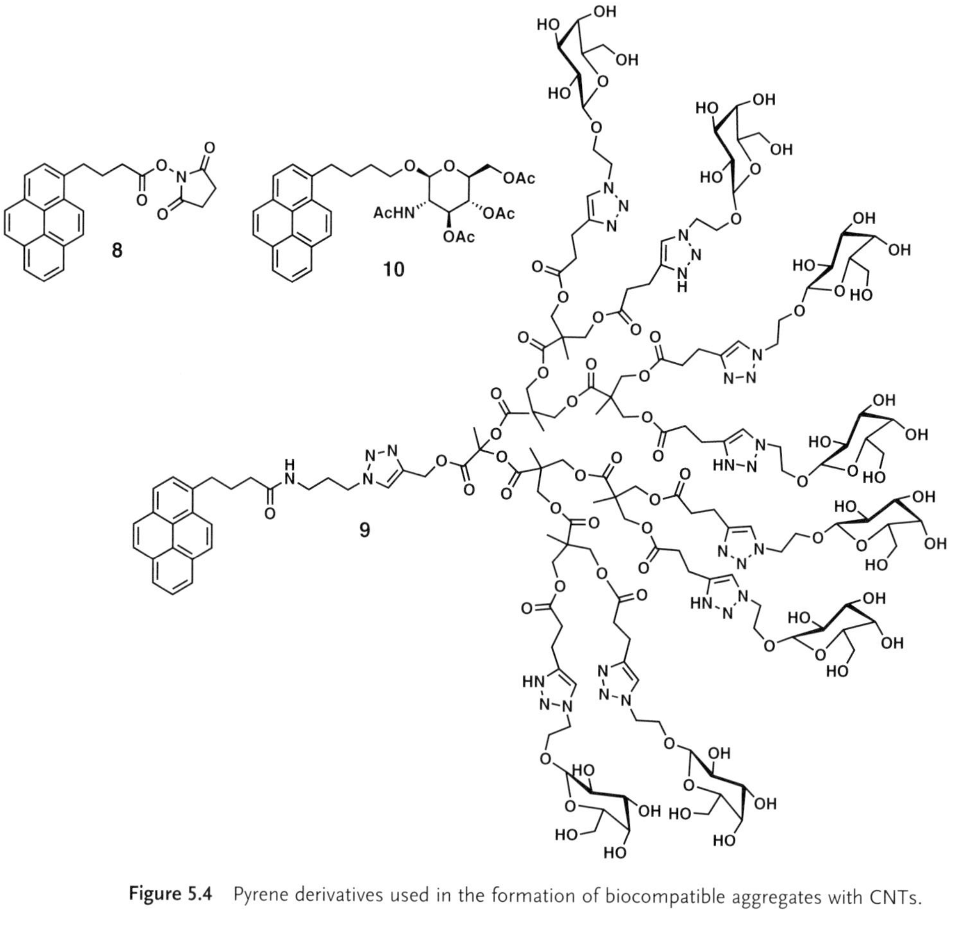

Figure 5.4 Pyrene derivatives used in the formation of biocompatible aggregates with CNTs.

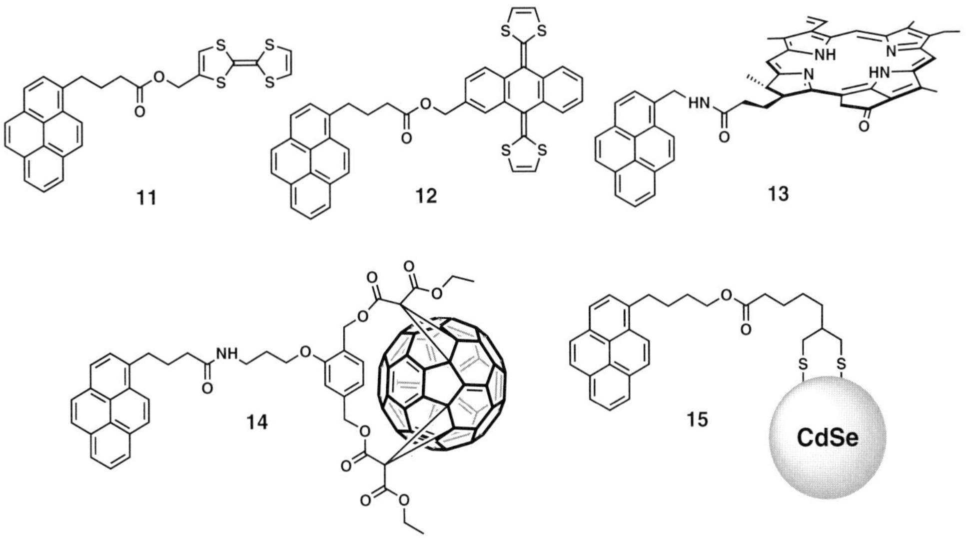

Figure 5.5 Different pyrene-bifunctional systems used for the immobilization of electroactive moieties onto CNT surfaces.

[16–18]. In this context, the use of pyrene derivatives has been particularly crucial in the construction of versatile nanosized electron donor–acceptor ensembles through directed π–π interactions. A considerable effort has been dedicated to combine CNTs with different donor or acceptor units by using different pyrene-bifunctional systems, for example, pyrene-tetrathiafulvalene (TTF) (**11**) [37], pyrene-π-extended-TTF (exTTF) (**12**) [38], pyrene-pyro-pheophorbide *a* (**13**) [39], pyrene-C_{60} (**14**) [40], or pyrene-CdSe nanoparticles (NPs) (**15**) [41] (Figure 5.5).

The synthesis of these molecules was based on the covalent linkage of both units through a flexible and medium length chain that favors a facile interaction with the CNTs surface, and where the pyrene fragment functions solely as a template that guarantees the immobilization of the electroactive unit on the CNT surface. The investigation of the photophysical properties of these supramolecular systems by different techniques (i.e., steady-state and time-resolved fluorescence, transient absorption spectroscopy) seems to indicate a very rapid intrahybrid electron transfer that yields a photogenerated radical ion pair with a lifetime in the nanosecond scale. For the supramolecular aggregates that pyrene-CdSe nanoparticles (**15**) form with SWCNTs, the magnitude of the electron transfer shows a strong dependency on the light intensity and wavelength, reaching a maximum of 2.2 electrons per pyrene-CdSe nanoparticle [41].

The noncovalent functionalization of CNTs using pyrenes bearing positive or negative charges [42–44], nitrogenated bases [45], or alkyl ammonium ions [46–48], through π–π interactions followed by the assembling of the electron/energy donor molecules by complementary electrostatics, axial coordination, or crown ether–alkyl ammonium ion interactions, respectively, has resulted in stable donor–acceptor systems with maximum preservation of the electronic and mechanical properties of CNTs (Figure 5.6).

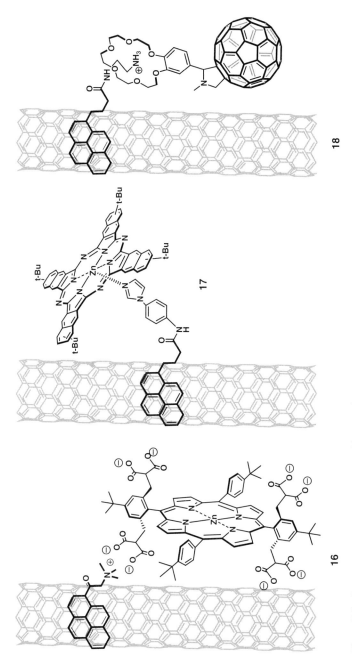

Figure 5.6 Donor–acceptor systems prepared by considering pyrene-π–π interactions followed by complementary electrostatics (**16**), axial coordination (**17**), or crown ether–alkyl ammonium ion interactions (**18**).

With the CNT surface covered with positively or negatively charged ionic head groups, van der Waals and electrostatic interactions were utilized to complex oppositely charged electron donors. Water soluble porphyrins (i.e., octapyridinium ZnP/H$_2$P salts or octacarboxylate ZnP/H$_2$P salts) have been used to form SWCNT-(π–π interaction)-pyrene$^+$-electrostatic-ZnP/H$_2$P, **16** or SWCNT-(π–π interaction)-pyrene$^-$-electrostatic-ZnP/H$_2$P electron donor–acceptor nanohybrids [42–44]. Photoexcitation of all the resulting nanohybrids with visible light revealed the formation of long-lived radical ion pairs, with lifetimes in the range of microseconds. The better delocalization of electrons in MWCNTs enhanced the stability of the radical ion pairs formed $(5.8 \pm 0.2 \, \mu s)$ when compared to the analogous SWCNT systems $(0.4 \pm 0.05 \, \mu s)$.

The imidazole ligand of soluble imidazol-pyrene-SWCNTs aggregates, such as zinc tetraphenylporphyrin (ZnP) and zinc naphthalocyanine (ZnNc) (**17**), served to anchor donor entities to the SWCNTS surface by axial coordination of the nitrogen to the metallic center of the macrocycle. Utilization of ZnNc as the electron donor in these supramolecular structures enabled to observe the donor cation radical (ZnNc$^{\bullet +}$), which acts as the direct evidence for the photoinduced electron transfer within these systems [45].

Among the noncovalent methodologies reported to date, self-assembly by using an ammonium ion–crown ether interaction is regarded as one of the most powerful methods as it offers a high degree of directionality with binding energies of up to 50–200 kJ/mol. The efficient selectivity of the 18-crown-6 moiety toward ammonium cations and the ease of formation of the size–fit complex even in polar solvents, enabled to build porphyrin and fullerene-based donor–acceptor supramolecular systems with CNTs [46–48]. In the SWCNTs-PyrNH$_3$$^+$-crown-C$_{60}$ nanohybrids **18**, free energy calculations suggested the possibility of electron transfer from the CNT to the singlet excited fullerene, resulting in the formation of a SWCNTs$^{\bullet +}$-PyrNH$_3$$^+$-crown-C$_{60}$$^{\bullet -}$ charge separated state [47]. Transient absorption spectroscopy confirmed the electron transfer as the quenching mechanism affording a lifetime for the radical ion pair over 100 ns, which suggests a further charge stabilization due to the supramolecular assembly.

Polymeric CNT aggregates bearing pyrene-terminal functionalities have demonstrated thermo- or chemically responsive abilities [49, 50] and, pyrene derivatives have, in addition, been used as initiators of norbornene ring-opening metathesis polymerization on the surface of CNTs. The adsorption of the organic precursor was followed by cross-metathesis with a ruthenium alkylidene, resulting in a homogeneous noncovalent poly(norbornene) coating [51].

In a very recent report, it has been described the fabrication of the first CNT–single molecule magnet (SMM) nanohybrids that have been prepared using a tailor-made tetrairon(III) SMM, [Fe$_4$(L)$_2$(dpm)$_6$] (**19**; Hdpm = dipivaloylmethane), which features a ligand (L) with a terminal pyrenyl group (Figure 5.7) [52]. The grafting of the SMMs was controlled to the single-molecule level and a single-SMM sensitivity was also demonstrated for the fabricated CNT-field effect transistors (FETs). These results pave the way for constructing "double-dot" molecular

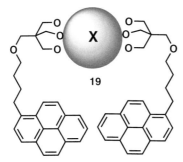

X = Fe$_4$(OMe)$_2$(dpm)$_6$, Hdpm = dipivaloylmethane

Figure 5.7 Pyrene derivative used in the preparation of single molecule magnet–CNT hybrids.

spintronic devices, where a controlled number of nanomagnets are coupled to an electronic nanodevice [53].

5.3.3
Other Polyaromatic Derivatives

As the physical and chemical properties of CNTs depend strongly on their structure, the separation of crude samples is one of the primary technological hurdles for their investigation and future applications. The ideal separation method should be of a high degree of selectivity and also should be nondestructive for the selected tubes. According to Hirsch and coworkers, a molecule containing a hydrophilic dendritic structure and simultaneously bearing an apolar polyaromatic subunit, ensuring a strong interaction with the CNTs, would be perfectly suited for CNT crude samples separation. Therefore, they have recently focused on the use of Newkome-dendronized perylenetetracarboxydiimides [54] as aromatic amphiphiles for the dispersion of SWCNTs in aqueous media [55]. The asymmetric perylene bisimide **20** (Figure 5.8) bears a polycyclic aromatic moiety, which adsorbs on the nanotube surface via π–π stacking, and a solvophilic moiety to aid dissolution in dipolar solvents such as water. SWCNTs were efficiently dispersed with a pronounced degree of individualization in buffered aqueous media by **20** even at concentrations as low as 0.01 wt%. In addition, the characteristic nanotube (van Hove singularities) and perylene features in the UV/Vis/NIR absorption spectrum appears redshifted compared to independent SWCNTs or perylene **20** samples. This, together with the observed quenching of the perylene fluorescence caused by the π–π stacking interaction with the sidewalls of the SWCNTs demonstrates that the adsorption of **20** is accompanied by pronounced electronic communication such as photoinduced energy or electron transfer [16–18].

Other small amphiphiles have been able to extract and solubilize in water-specific SWCNTs subsets with (*n*, *m*) selectivities based on the nanotube helicity angle [56]. For example, compound **21** is a quaterrylene with a hydrophilic side chain, which is negatively charged at basic pH. The π–π stacking interactions are particularly

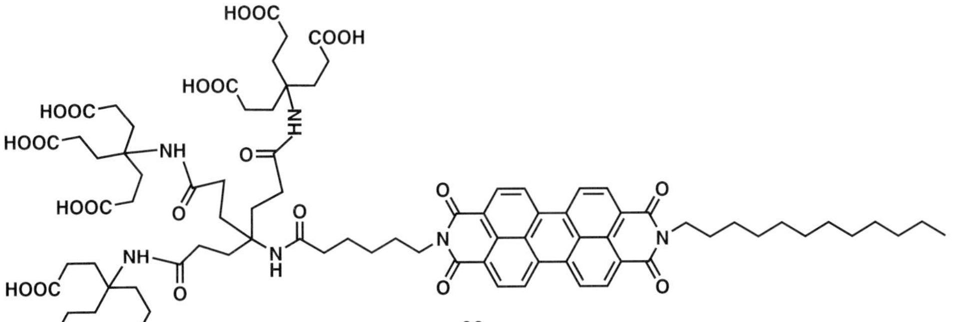

Figure 5.8 Different aromatic amphiphiles synthesized for the solubilization of CNTs.

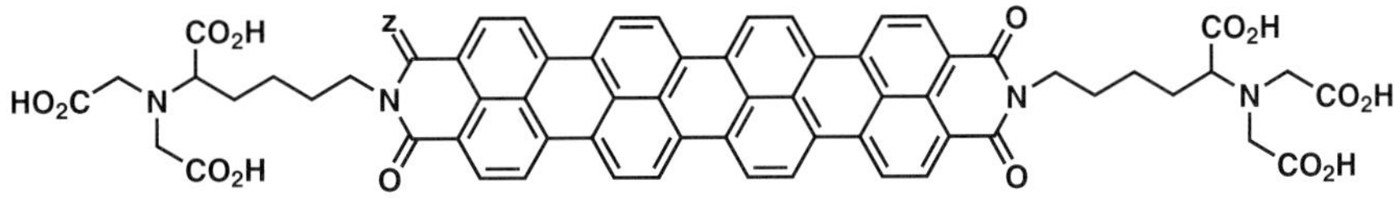

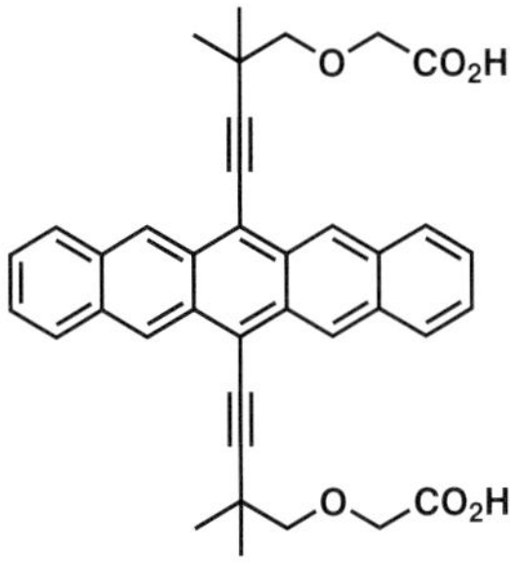

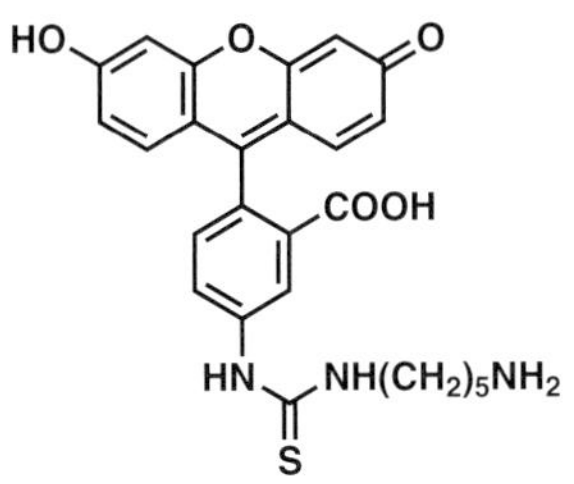

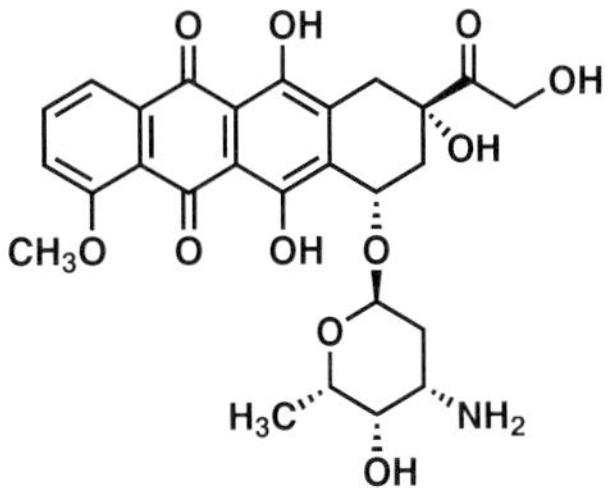

favorable for the adsorption of **21** to the zigzag SWCNTs sidewalls. In the case of compound **22**, which contains a pentacenic moiety and two hydrophilic side chains that are negatively charged in basic aqueous media, the adsorption happens preferentially on the armchair SWCNTs sidewalls in order to maximize the π–π stacking interactions by simultaneous optimization of the effective contact area and the atomic correlation with the CNT.

The π-stacking of other molecules of interest, such as the widely used fluorescent molecule, fluorescein **23**, or the cancer chemotherapy drug, doxorubicin **24**, with an ultrahigh loading capacity of a ~400% by weight, has also been effectively achieved on pre-PEGylated SWCNTs [57, 58]. Consequently, the π–π stacking of small aromatic molecules could represent a very simple and efficient method to solubilize CNTs and introduce new opportunities for applications in chemistry, material sciences, or medicine.

5.4
Noncovalent Interactions of CNTs with Heterocyclic Polyaromatic Systems

Diverse heterocyclic polyaromatic molecules can form effective interactions with CNTs. It is believed that the nature of the interaction is essentially the same van der Waals forces as in the case of the polyaromatic compounds mentioned in the previous section. However, in some cases, the incorporation of a metal ion in the systems seems to weaken their interactions with the CNT sidewalls [59, 60].

5.4.1
Porphyrins, Phthalocyanines, and Sapphyrins

The combination of CNTs with phthalocyanines [59] and porphyrins or oligomers and polymers that contain them [60–70], has proved to be particularly beneficial in the preparation of donor–acceptor nanohybrids bearing visible light harvesting chromophores. SWNTs were found to strongly interact with free-base porphyrins, Zn-porphyrins (ZnP) (**25**) [61–66], fused ZnP-trimers (**26**) [67], and ZnP in conjugated polymers (**27**) [68, 69] (Figure 5.9). Successful complexation with, for example, the ZnP-polymer **27** was manifested in a 127 nm bathochromic shift of the Q-band absorption of porphyrins. Additional evidence for interactions between **25**, **26**, or **27** and SWNTs was obtained from fluorescence spectra, where the ZnP fluorescence was significantly quenched. The fluorescence quenching has been ascribed to energy transfer between the photoexcited porphyrin and SWNTs.

In general, ZnPs showed a weaker binding ability to SWCNTs as compared to the metal-free molecules. The free base forms nucleate on the CNTs sidewall in the form of J- and H-type aggregates, which sometimes drives a further assembly in linear or wheel structures [63, 65, 66]. For the linear structures, when photocurrent measurements were performed using the SWCNT-free base porphyrin as a component of the photoanode, under the standard three-electrode conditions, a maximum incident photon-to-photocurrent efficiency (IPCE) value of 13% was obtained at a bias potential of 0.2 V versus SCE [66].

The formation of nanohybrids of SWCNTs with flexible porphyrinic polypeptides has been described by Fukuzumi *et al.* [70]. In this approach, the polypeptide bearing 16 porphyrin units $P(H_2P)_{16}$, **28**, was synthesized via oligomerization of a porphyrin functionalized enantiopure L-lysine derivative bearing a free-base chromophore (Figure 5.9). Supramolecular aggregation to SWCNTs occurred through the wrapping of the peptide backbone and π–π interaction with the porphyrinic units, which resulted in a stable nanohybrid that facilitated the extraction of the large-diameter CNTs (~1.3 nm). Laser photoexcitation of the $[P(H_2P)_{16}]$/SWCNTs aggregates in DMF resulted in a photoinduced electron transfer from the singlet and triplet excited states of $P(H_2P)_{16}$ to SWCNTs. The lifetime of the charge separated state was determined to be 0.37 ± 0.03 ms.

For use in photovoltaic devices, noncovalent functionalization of SWCNTs with porphyrin derivatives seems to be an attractive approach, although a precise understanding of the dynamics of the photoinduced electron/energy transfer (PET)

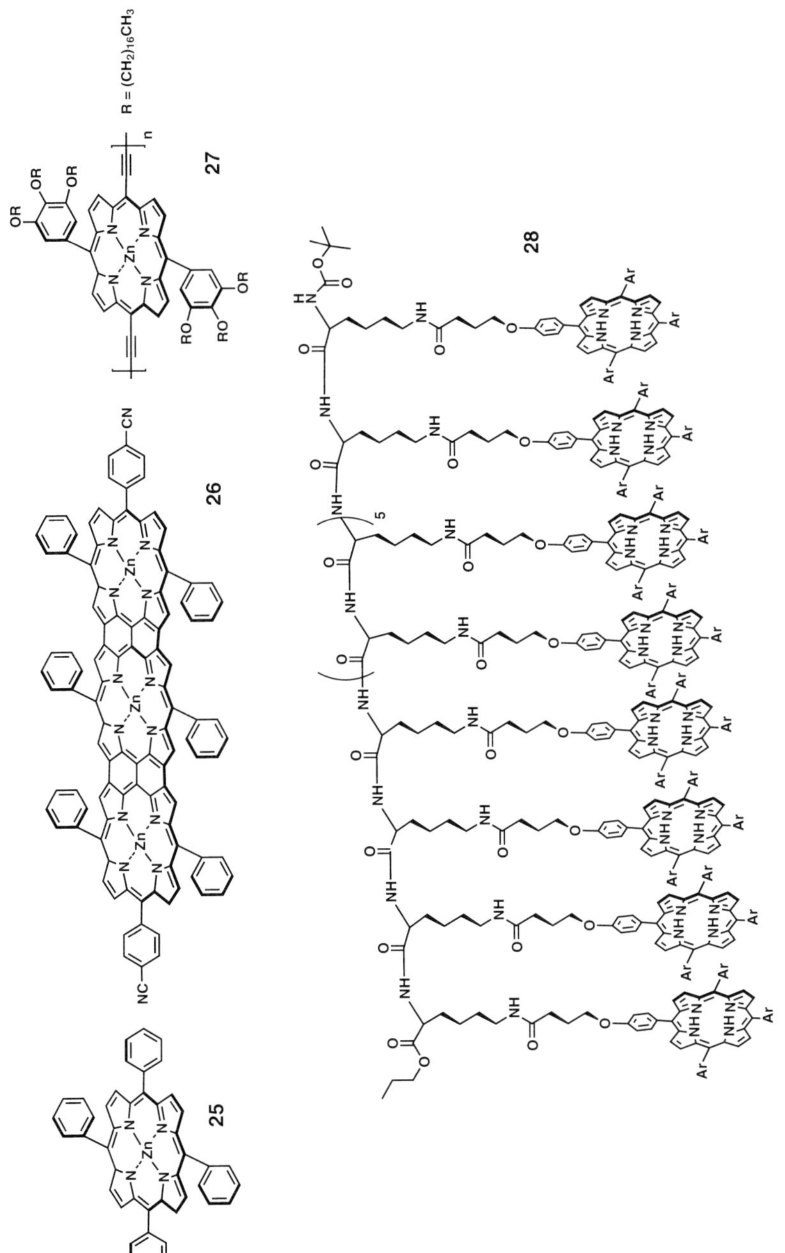

Figure 5.9 Leading examples of ZnP and H_2P used for the preparation of SWCNTs charge transfer nanohybrids.

processes occurring within the D/A composites is essential for its full realization. With this aim, the ZnP/SWCNTs system **29** was synthesized and integrated in a field-effect transistor that was irradiated with visible light [71]. Unlike the natural process, in **29** the absorption of light initiates the transfer of holes rather than electrons from the porphyrin. The rate and magnitude of the PET were found to depend both on the wavelength and on the intensity of the light used, reaching a maximum value of 0.37 electrons/porphyrin molecule.

Sapphyrins are well-characterized pentapyrrolic macrocycles that have proved particularly attractive for complexation to SWCNTs as they present larger π surfaces than typical tetrapyrrolic macrocycles [72]. They are also fluorescent through excitation at lower energy, being very striking as mimics of naturally occurring chlorophylls. The preparation of water suspendable SWCNTs/sapphyrin complexes **30** (Figure 5.10) has recently been reported [73]. The resulting modified CNTs undergo an intramolecular electron transfer upon photoexcitation. As in the case of **29**, pulse radiolysis experiments and femtosecond spectroscopic data confirmed that the SWCNTs act as the electron donor and the sapphyrin molecules as the electron acceptor in systems **30**.

5.4.2
Metallic Coordination

The interaction of porphyrins with pyridyl functionalized SWCNTs through an apical coordinate bond between the porphyrin transition metal atom and the pyridyl ligand has been object of several reports [74–76]. The formation of the complexes was firmly established by optical and electrochemical measurements. In particular, ruthenium porphyrin functionalized SWCNTs arrays have been prepared by coordinating the

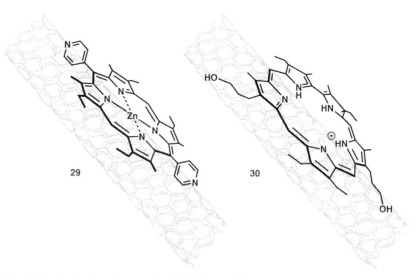

Figure 5.10 Porphyrin- and sapphyrin-SWCNTs assemblies.

axial position of the metal ion with 4-aminopyridine preassembled SWCNTs directly anchored to a silicon (100) surface. The surface concentration of the ruthenium porphyrin molecules (3.44×10^{-8} mol/cm^2) and the good immobilization suggest that these ruthenium porphyrin modified electrodes are excellent candidates for molecular memory devices and light harvesting antennae [76].

The use of planar metallic complexes is another alternative that has been considered in the solubilization and debundling of CNTs [77, 78]. Furthermore, the solubilization of CNTs can be controlled by considering the redox properties of the metallic complex [78]. In this sense, the Cu^{II} complex in which the ligand (**31**) is a 2,2-bipyridine derivative bearing two chloresterol groups (Figure 5.11) shows a reversible sol–gel phase transition on changing the redox state of the Cu^I/Cu^{II} complexes by

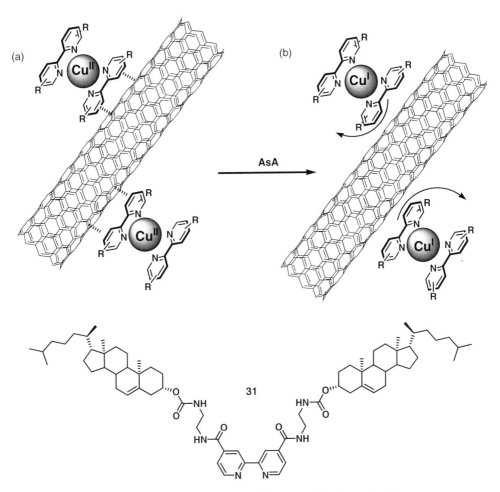

Figure 5.11 Schematic illustration of (a) the presence of the π–π stacking interaction between [Cu^{II}(**31**)$_2$] and the SWCNTs and (b) the absence of the π–π stacking interaction between [Cu^I(**31**)$_2$] and the SWCNTs. AsA is ascorbic acid.

chemical oxidation and reduction [79]. This property has also been explored in combination with SWCNTs, **31** alone has only a low ability to solublize SWCNTs, but they are highly soluble when combined with [Cu^{II}(**31**)$_2$]. Furthermore, the addition of a solution of ascorbic acid to the [Cu^{II}(**31**)$_2$]–SWCNTs mixture immediately resulted in the precipitation of the SWCNT from the solution, although the precipitate could be redissolved by bubbling air through it. This selective control on the solubility of CNTs using the redox properties of the solubilizing agents could be useful for the purification of metallic and semiconducting SWCNTs, which is still a real challenge in CNTs chemistry [80].

5.5
Noncovalent Interactions of CNTs with Surfactants and Ionic Liquids

A significant progress toward the solubilization of CNTs in water, which is important because of potential biomedical applications and biophysical processing schemes, has been achieved by using surfactants. In particular, the anionic surfactant sodium dodecyl sulfate (SDS, **32**) has been the most widely used [81–85] surfactant. TEM unambiguously demonstrated that SDS is adsorbed on the surface of CNTs and arranged into rolled-up half-cylinders with the alkyl group of each molecule pointed toward the nanotube. Depending on the symmetry and the diameter of the CNTs, ring, helices, or double helices were observed [83].

Surfactant molecules containing aromatic groups are capable of forming more specific and more directional π–π stacking interactions with the graphitic surface of CNTs. Interactions of SDS and the structurally related sodium dodecylbenzene sulfonate (SDBS, **33**) with SWCNTs have been compared to demonstrate the role of the aromatic rings [85]. SDS and SDBS have the same length of the alkyl chain but the latter has a phenyl ring attached between the alkyl chain and the hydrophilic group (Figure 5.12). The presence of the phenyl ring makes SDBS more effective for solubilization of CNTs than SDS, due to the aromatic stacking formed between the SWCNTs and the phenyl rings of the SDBS within the micelle. The diameter distribution of CNTs in the dispersion of SDBS, measured by atomic force micros-copy (AFM), showed that even at 20 mg/ml ∼63 ± 5% of the SWCNT bundles exfoliate in single tubes [85].

Bile salt detergents such as the sodium salts of deoxycholic acid (DOC, **34**) and its taurine analogue taurodeoxycholic acid (TDOC, **35**) are extremely efficient in solubilizing pristine individual SWCNTs, as evidenced by highly resolved optical absorption spectra, bright bandgap fluorescence, and the unprecedented resolution (∼2.5 cm^{-1}) of the radial breathing modes in Raman spectra. This is attributed to the formation of very regular and stable micelles around the CNTs providing an unusual homogeneous environment [86].

The hemimicelle self-assembly of amphiphilic molecules on CNTs provides an effective template for the homogeneous and dense deposition of noble metal nanoparticles on CNTs. In this regard, surfactants **36** and **37** (Figure 5.12) were designed with a long alkyl chain for interacting with the CNT surface and a polar head

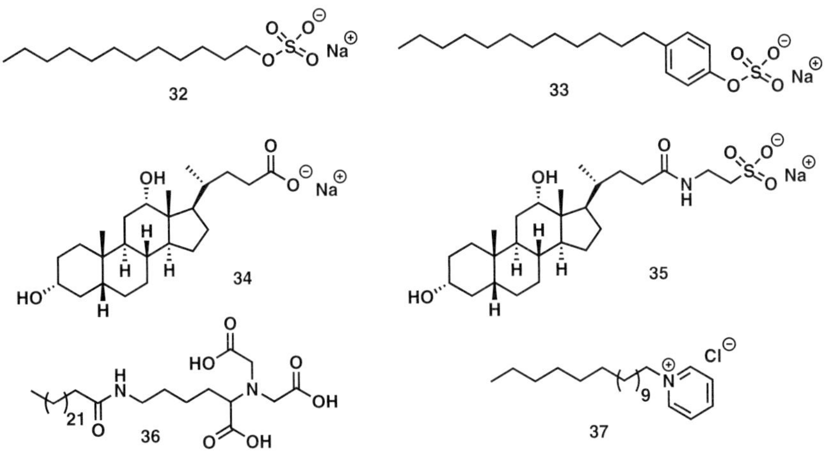

Figure 5.12 Surfactants used in the solubilization of CNTs.

was made of either nitrilotriacetic acid (NTA) or pyridinium, for the binding of metallic nanoparticles. The electrocatalytic applications of the resulting nanohybrids were evaluated, and a superior activity, in comparison to analogous systems, was observed in certain oxidation reactions [87].

Recently, the separation of SWCNTs by exploiting subtle differences in the buoyant density of surfactant encapsulated CNTs in density gradient ultracentrifugation (DGU) has been reported [88–92]. As outlined by Green and Hersam [93], the buoyant density of a dispersed CNT is both a function of the geometry of the CNT (diameter) and the surfactant coating. On the one hand, a completely uniform surfactant coating results in separation by diameter [88], while on the other hand, the use of a surfactant (or a combination of surfactants) that encapsulates CNTs in a manner dependent on the electronic structure, allows separation by electronic properties, as shown by cosurfactant DGU with 1 : 4 SDS/sodium cholate (SC) [88]. Therefore, the investigation of different surfactant systems and various cosurfactant experiments is crucial to exploit the tunability of the separation process by DGU.

Imidazolium-ion-based ionic liquids (IL), which are organic salts with melting points below 100 °C, have been studied extensively in recent years as CNTs solubilizers (Figure 5.13). They have some unique properties such as negligible vapor pressure, good thermal stability, wide liquid temperature range, designable properties, and an ample electrochemical window [94, 95].

When SWCNTs were mixed with an IL such as 1-butyl-3-methylimidazolium tetrafluoroborate ([bmim][BF$_4$], Figure 5.13), a gel was formed by means of cation–π and/or π–π interactions with the π-electronic surface of the SWNTs [96]. The gelation occurs in a variety of imidazolium-ion-based ionic liquids (Figure 5.13) upon grinding with 0.5–1 wt% (critical gel concentration) of SWNTs. When SWNTs are ground into IL in excess with respect to the critical gel concentration, the gel and IL phases are clearly separated from each other by centrifugation (Figure 5.14a),

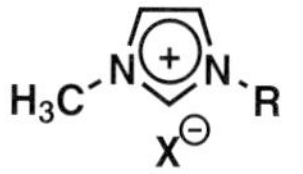

[emim][BF$_4$]: R = C$_2$H$_5$, X = BF$_4$
[emim][Tf$_2$N]: R = C$_2$H$_5$, X = (CF$_3$SO$_2$)$_2$N
[bmim][BF$_4$]: R = C$_4$H$_9$, X = BF$_4$
[bmim][PF$_6$]: R = C$_4$H$_9$, X = PF$_6$
[bmim][Tf$_2$N]: R = C$_4$H$_9$, X = (CF$_3$SO$_2$)$_2$N
[omim][PF$_6$]: R = C$_8$H$_{17}$, X = PF$_6$

H$_3$C-N(+)N-R
X(−)

Figure 5.13 Molecular formulas and schematic structures of typical imidazolium-ion-based ionic liquids.

indicating that the gel can trap a limited amount of the IL. Bucky gels thus obtained are easy to be processed into any shape. For example, through extrusion from a needle, one can fabricate a cable-like material that cannot be easily torn apart even when suspended (Figure 5.14b). Due to the negligible volatility of IL, bucky gels, in sharp contrast with ordinary organogels and hydrogels, are highly stable and can retain their physical properties even under reduced pressure. By means of TEM, bucky gels were found to contain highly exfoliated SWNT bundles (Figure 5.14c and d) that are not free from one another, but are glued together by a great number of weak physical cross-links, for which secondary ordering of ILs by means of interionic interactions may be responsible. This CNT network is preserved even after the IL components are polymerized, so that the resulting polymer composites (bucky plastics) are mechanically reinforced and show an excellent electrical conductivity [94].

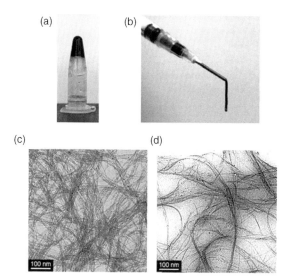

Figure 5.14 (a) Phase-separation behavior of a bucky gel formed from SWCNTs and [bmim][BF$_4$] (upper phase) and excess [bmim][BF$_4$] (lower phase). (b) Extrusion of the gel from a needle. (c) TEM micrographs of as-received SWCNTs. (d) TEM micrographs obtained by dropping a bucky gel of [bmim] [BF$_4$] into deionized water. Reprinted from Ref. [94] with permission from Wiley-VCH Verlag GmbH.

The combination of ILs with CNTs and the bucky gels formed are particularly interesting for the fabrication of CNT-modified electrodes, since the large specific surface area of the highly dispersed CNTs can facilitate electron transfer processes more efficiently. Another advantage is a broader window of ILs in the applicable electrochemical potential than those of traditional electrolyte solutions, which enables electrochemical doping of CNTs. Bucky-gel-modified electrodes can readily be fabricated from a variety of CNTs and ILs, and their integration in different electrochemical devices, such as sensors, capacitors, and actuators, could be very beneficial for the development of this field [94].

5.6
Noncovalent Interactions of CNTs with Polymers

Besides possibly improving the mechanical and electrical properties of polymers, the formation of polymer/NTs composites is considered to be a useful approach for incorporating NTs into polymer-based devices. For noncovalently modified CNTs with polymers, several strategies have been developed. Currently, these strategies involve physical mixing in solution, *in situ* polymerization of monomers in the presence of CNTs, surfactant-assisted processing of composites, and chemical functionalization of the incorporated tubes. Since the field of CNT-based polymers has been recently reviewed [15], no attempt will be made in this chapter to cover all the early results, and only a selective overview with a bias toward the interests of the authors is given in the following paragraphs.

5.6.1
Polymeric Amphiphiles

A good deal of research has been carried out with nonconducting polymers. Linear polymers that bear polar octyloxy alkyl side-chains, such as polyvinyl pyrrolidine (PVP, **38**) and polystyrene sulfonate (PSS, **39**) (Figure 5.15), are known to form stable composite materials with CNTs [97].

Various approaches have been developed to improve the solubility of CNTs in different polymer matrices. The addition of 1 wt% nonionic surfactants improves, for instance, the glass transition temperature. Moreover, the elastic modulus increased by more than 30% is relative to the scenario in the absence of a surfactant [98].

Water soluble poly(diallyldimethylammonium) chloride (PDDA, **40**)/MWCNTs aggregates have also been obtained by mild sonication [99]. In these composites, it has been suggested that the interaction between PDDA and MWCNTs is not electrostatic in nature, but the presence of unsaturated impurities in the PDDA chain drives π–π interactions. Thus, CNTs coated with PDDA function as positively charged polyelectrolytes, which exhibit electrostatic repulsion between CNTs/PDDA and likewise guarantees CNTs hydrophilicity.

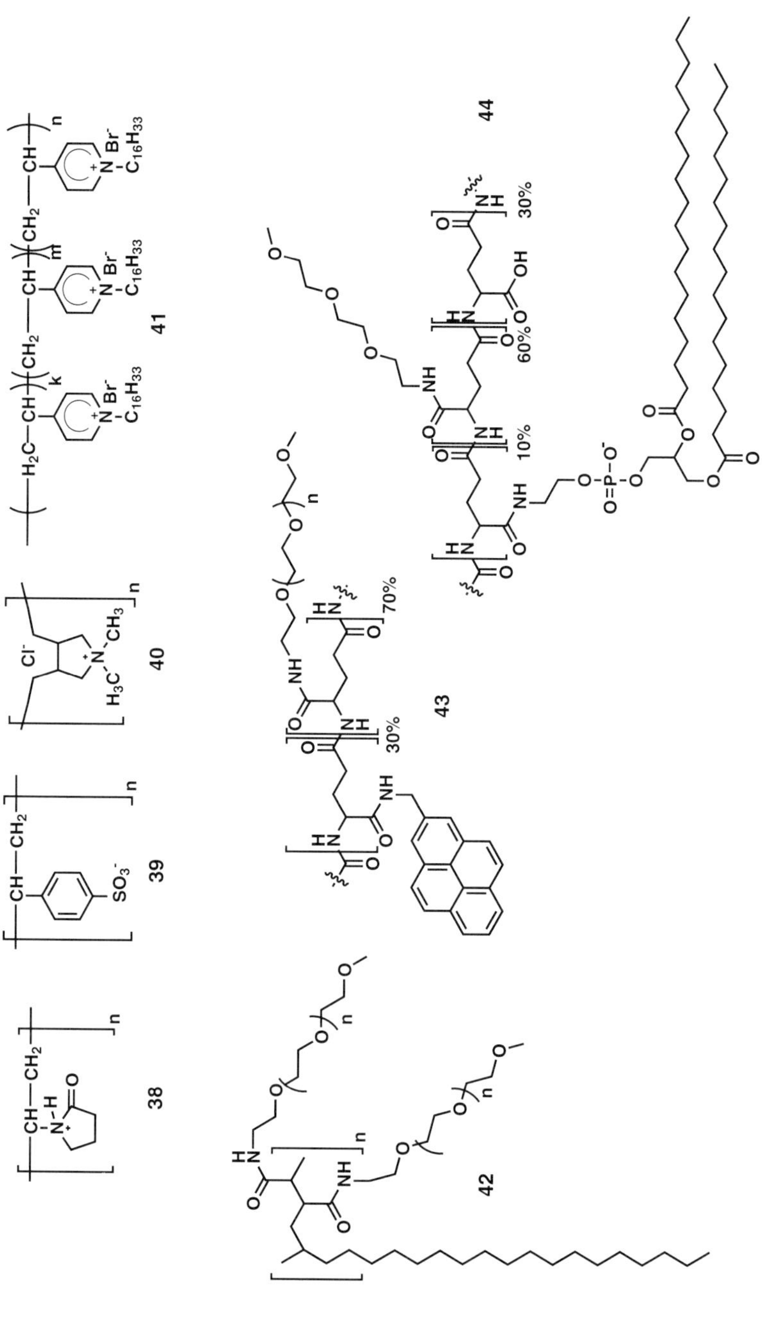

Figure 5.15 Polymeric amphiphiles that noncovalently functionalize CNTs.

The combination of PSS and PDDA has proved to be very useful for the systematic and molecularly controlled organization of SWNTs and electron donor molecules on electrodes in the so-called *layer-by-layer* approach [100]. The assembled films were integrated on ITO electrodes, and the photoelectrochemical measurements with a standard three-electrode system revealed maximum IPCE values of about a 8–13% [66, 101–103].

Polymeric surfactants also self-assembled on the CNTs surface [83]. In particular, poly(*N*-acetyl-4-vinylpyridinium bromide-*co*-*N*-ethyl-4-vinylpyridinium bromide-*co*-4-vinylpyridine, **41**) was found to form exceptionally stable CNT dispersions [104]. It is suggested that the efficiency of macromolecular dispersion agents for CNTs solubilization correlates with the topological and electronic similarity of polymer–CNT and CNT–CNT interactions in the nanotube bundles. Raman spectroscopy and atomic force and transmission electron microscopies data indicate that the polycations are wrapped around CNTs forming a uniform coating 1.0–1.5 nm thick [104].

As already mentioned several times along this chapter, to solubilize CNTs in aqueous media is essential to improve their biocompatibility and enable their environmental friendly characterization, separation, and self-assembly. To exploit the unique properties of CNTs into biological relevant systems, SWCNTs were suspended in water soluble amylose in the presence of iodine [105]. Very likely, iodine plays a key role in the initial preorganization of amylose in support of a helical conformation. Since this process is reversible at high temperatures, it opens the exciting opportunity to employ the strategy to separate SWCNTs from amorphous carbon.

Another common strategy to impart functionality, water solubility, and biocompatibility to CNTs is PEGylation. A series of polymeric amphiphiles based upon PEGylated poly(γ-glutamic acid) (γPGA) and poly(maleic anhydride-*alt*-1-octadecene) (PMHC$_{18}$) (**42–44**) have recently been used in the supramolecular modification of SWCNTs [106]. The formed aggregates exhibit high stability in aqueous solutions at different pH values, at elevated temperatures, and in serum. Moreover, the polymer-coated SWCNTs exhibit remarkably long blood circulation ($t_{1/2} = 22.1$ h) upon intravenous injection into mice, far exceeding the previous record of 5.4 h [106]. These characteristics make PEGylated/CNTs very powerful materials for *in vivo* applications, including drug delivery or imaging.

5.6.2
Conjugated Polymers

An interesting class of polymer composites that has attracted much attention is that of the conjugated polymers such as poly(phenylenevinylene) (PPV), that was one of the first to be investigated [107, 108]. The SWCNTs/poly(*m*-phenylenevinylene) (PmPV) complexes exhibit a conductivity eight times higher than that of the pure polymer, without any restriction on its luminescence properties. Similar conjugated luminescent polymers, such as poly(*m*-phenylenevinylene-*co*-2,5-dioctyloxy-*p*-phenylenevinylene) (PmPV-*co*-DOctOPV) (**45**) [109] and its derivatives, such as poly

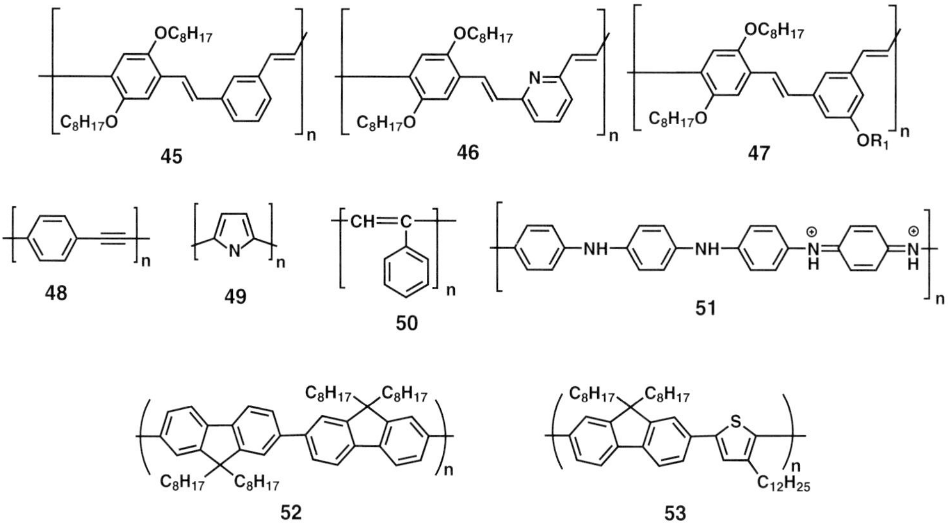

Figure 5.16 Conjugated polymers that wrap around CNTs.

(2,6-pyridinylenevinylene-*co*-2,5-dioctyloxy-*p*-phenylenevinylene) (PPyPV-*co*-DOc-tOPV) (**46**) [110] and poly(5-alkoxy-*m*-phenylenevinylene)-*co*-2,5-dioctyloxy-*p*-phenylenevinylene) (PAmPV-*co*-DOctOPV) (**47**) (Figure 5.16) [111] have formed supramolecular complexes with CNTs. Stoddart *et al.* [112] also synthesized the stilbenoid dendrimers, a hyperbranched variant of the PmPV polymer, which exhibits an appropriate degree of branching and it was found to be more efficient in breaking up nanotube bundles, provided it is employed at higher polymer-to-CNTs ratios than the parent PmPV polymer.

Homogeneous nanocomposites of poly(phenyleneethynylene) (PPE) (**48**)-SWCNTs/polystyrene (PS) and PPE-SWCNTs/polycarbonate were noncovalently functionalized and revealed dramatic improvements in the electric conductivity with very low percolation thresholds (i.e., 0.05–0.1 wt% SWNTs loading) [113].

Among the many conducting polymers, polypyrrole (PPY, **49**) stands out as a highly promising material in terms of commercial applications. Nanotube-polypyrrole composites have been engineered by *in situ* chemical or electrochemical polymerization [114–116]. These types of composites have been used as active electrode materials in the assembly of supercapacitors, for the selective detection of glucose [117] and the selective measurement of DNA hybridization [118]. The detection approach relied on the doping of glucose oxidase and nucleic acid fragments within electropolymerized polypyrrole onto the surface of CNTs. Composites of CNTs/PPY have also been studied as gas sensors for NO_2 [119].

Tang and Xu [120] reported the soluble MWCNTs-containing photoconductive poly(phenylacetylenes) (**50**) (CNTs/PPAs) preparation by *in situ* polymerizations of phenylacetylene catalyzed by WCl_6-Ph_4Sn and $[Rh(nbd)Cl]_2$ (nbd = 2,5-norbornardiene) in the presence of the CNTs. They demonstrated that the CNTs in the composite solutions can easily be aligned by mechanical shear, and an efficient optical limiting property was observed in these nanocomposite materials.

Molecular dynamics studies carried out on the interactions of PS, PPA, PmPV, and PPV (Figure 5.16) with SWCNTs in vacuum, indicate that the interaction between the CNTs and the polymer is strongly influenced by the specific monomer structure [121]. CNTs–polymer interactions are the strongest for conjugated polymers with aromatic rings on the polymer backbone, as these rings are able to align parallel to the CNT surface and thereby provide strong interfacial adhesion. In the presence of well-separated CNTs, different polymer chains get disentangled and align the CNTs to cover their surface. This is a general observation for all the investigated polymers, although the effect is most pronounced for PmPV, which combines certain flexibility in the backbone structure, flexible side chains, and strong interaction with the CNTs surface.

Another fascinating class of composites is the one that results from the combination of CNTs and aniline. Polyaniline (PANI, **51**) bears particularly great potential in synthesizing polymer/CNT composites due to its environmental stability, good processability, and reversible control of conductivity both by protonation and by charge transfer doping. Electrochemical polymerization of aniline on CNT electrodes for the deposition of conducting polymeric films has been reported by independent groups [122, 123]. Alternative strategies involve the chemical polymerization of aniline or solution mixing of CNTs and the conjugated polymer [124, 125]. The blends exhibited an increased order of magnitude in the electrical conductivity over the neat polymer [126]. Liu *et al.* have successfully assembled poly(aminobenzenesulfonic acid) (PABS) modified SWCNTs with polyaniline via the simple *layer-by-layer* approach. The obtained PANI/PABS-SWCNT multilayer films were very stable and showed a high electrocatalytic ability toward the oxidation of reduced β-nicotinamide adenine dinucleotide (NADH) at a much lower potential (about $+50\,mV$ versus Ag/AgCl) [127].

Although poly(9,9-dialkylfluorenes) exhibit the extended conjugation required for π-stacking to the CNTs surface, they have so far, attracted limited attention as supramolecular counterparts for SWNTs [128]. Only recently, poly(9,9-dialkylfluorene) (PF, **52**) and poly(9,9-dialkylfluorene-*co*-3-alkylthiophene) (PFT, **53**) were used to prepare discrete polymer–SWNTs complexes which showed excellent solubilities in organic solvents in the absence of excess free polymer [129]. Both the polymer structure and the solvent used strongly influence the dispersion of the CNTs, and can prepare SWCNT solutions with different solubilities as well as perform selective solubilization of SWCNTs with the ability to selectively tune the distribution of solubilized CNT species [130].

5.6.3
Biopolymers

Biological systems have evolved as complex architectures created through the self-assembly of biological macromolecules. Evolution has thus provided biological molecules with a wide variety of intricate structures tailored for controlled self-assembly that have somehow been adapted for the nanoscale self-assembly of nonbiological systems such as CNTs [15]. One structural motif used by proteins to promote self-assembly is the amphiphilic α-helix, which is a protein secondary structure in which the polypeptide backbone forms a folded spring-like conformation stabilized by hydrogen, with the amino acid side chain extending outward from the exterior surface of the helix. These amphiphilic helical peptides were found to fold around the graphitic surface of CNTs and to disperse them in aqueous solutions by noncovalent interactions. Most importantly, the size and morphology of the coated fibers can be controlled by peptide–peptide interactions, affording highly ordered structures [131].

Another example of assembly on the CNTs surface involves the synthetic single-chain lipids [83, 132, 133]. At the nanotube–water interface, permanent assemblies were produced from mixed micelles of SDS and different water-insoluble double-chain lipids after dialysis of the surfactant [83]. Moreover, the polar part of the lipids could participate in the selective immobilization of histidine-tagged protein through metal ion chelates. In a different approach, Artyukhin *et al.* [132] deposited alternating layers of cationic and anionic poly-electrolytes on templated CNTs. The authors demonstrated the occurrence of spontaneous self-assembly of common phospholipid bilayers around the hydro-philic polymer coating CNT. The lipid membrane was found to maintain its fluidity and the mobility of lipid molecules, and can still be described by a simple diffusion model [133].

Within biological polymers, DNA is especially effective in dispersing CNTs in water, with resulting CNT–DNA hybrid solutions being stable for months at room temperature [134–140]. The oligonucleotides are found to readily adsorb on the CNT surface, after which they undergo a slow structural rearrangement. Cluster analysis of bound DNA conformations as well as population distribution maps computed as a function of several local and global order parameters show that the hybrids exhibit a complex morphology with DNA strands assuming a number of distinct backbone geometries, which depend on both DNA sequence and CNT diameter. In contrast, the nucleotide bases are found to align parallel to the CNT surface with a high degree of orientational order. While the binding appears to be primarily driven by energetically favorable π-stacking of DNA bases onto the CNT surface, equilibrium distribution of hybrid conformations is modulated by a complex interplay of forces, including the DNA conformation strain and solvent interactions [141] (Figure 5.17).

In addition, DNA-dispersed SWCNTs can be separated on the basis of diameter using ion-exchange chromatography [138, 139] or by ultracentrifugation through an aqueous density gradient [142]. Additionally, oligonucleotides can be removed using small aromatic molecules, such as rhodamine 6G or the cDNA strand, once any necessary processing is complete [143]. A variety of different applications, such as fiber spinning [144], self-assembled nanotube field-effect transistors [145], stabili-zation of colloidal particles [146], chemical sensing [147], and applications in both medical diagnostic and biological fields [148–150], have been investigated for DNA-dispersed SWCNTs.

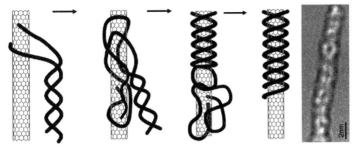

Figure 5.17 DNA wrapping of CNTs. On the right side, the TEM image shows a 32 days old sample. Reprinted with permission from Ref. [140]. Copyright 2008, The American Chemical Society.

5.7
Optically Active SWCNTS

The properties and technological applications of SWCNTs are largely determined by their structures [151]. As it is well known, samples of chiral SWCNTs contain the same amounts of right- and left-handed helical structures [152]. Amazingly, the separation of both the optical isomers has not been properly addressed and, only recently, optically active SWCNTs have been obtained from a commercial sample by means of favored supramolecular π–π interactions [153, 154]. Thus, molecular recognition of the helical isomers (left-handed, M)- or (right-handed, P)-SWCNTs by a chiral nanotweezer (**54**) formed by two chiral Zn-porphyrins covalently connected through a 1,3-phenylene bridge has allowed the structural discrimination of chiral CNTs for the first time (Figure 5.18).

According to theoretical calculations, the formed supramolecular complexes exhibit slightly different enthalpies of association for the right-hand (RH) and left-hand (LH) helicities of (6,5)-SWCNTs with one (*S*)-diporphyrin molecule (−56.09 *versus* −56.36 kcal/mol, respectively). Although a small energy difference was found between the two complexes, the high number of tweezers bound to the SWCNTs result in a significantly more stable complex, thus affording the RH:(*S*)-diporphyrin molecule formed preferentially. Similarly, using (*R*)-diporphyrin tweezer led to the corresponding LH:(*R*)-diporphyrin complex as the most stable one, thus reinforcing the calculations performed.

To complete the chiral separation of SWCNTs, the diporphyrins were efficiently liberated from the respective complexes by washing with pyridine several times, thus affording optically enriched SWCNTs. As expected, the solutions of SWCNTs extracted by (*R*)- and (*S*)-dipophyrin showed symmetrical circular dichroism with peaks at 341 and 562 nm, corresponding to E_{33} and E_{22} transactions of the (6,5)-SWCNTs [155].

5.8
Noncovalent Interactions of CNTs with Nanoparticles

CNTs and monodispersed nanoparticles are two of the most important building blocks proposed to create nanodevices. In particular, the combination of functional

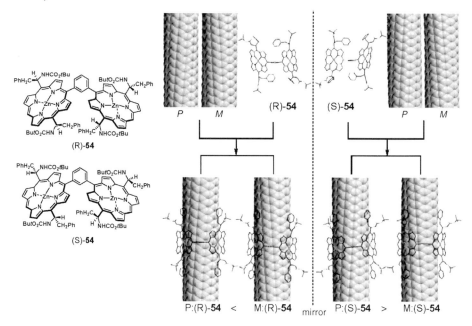

Figure 5.18 Side views of the computer-generated molecular models of the 1.1 complexes of (*R*)-**54** and (*S*)-**54** with (P) and (M)-(6,5)-SWCNTs. Reprinted with permission from Ref. [154]. Copyright 2007, The American Chemical Society.

semiconducting NPs and CNTs has important implications for the development of advanced materials and has been the objective of recent investigations [156–158]. The main noncovalent strategies for the construction of semiconducting NP/CNT aggregates consist of the adsorption of surface active molecules, which in turn may bind nanoparticles electrostatically [159, 160], or the self-assembly of CdSe and InP nanoparticles using the grooves in bundles of SWCNTs as a one-dimensional template [161]. As an alternative to these synthetic approaches, the hot injection method was recently used to obtain CdSe/CNT composites [162]. It was observed that, in the presence of CNTs, CdSe nanorods evolved into pyramidal-like NPs that connected to CNTs by the wurtzite (001) facets (Figure 5.19). In the mechanism of NP/CNT hybrids formation, the presence of water improves the CNT coverage while 1,2-dichloroethane, HCl or in general chlorine, is responsible for the shape transformation of the NPs and further attachment to the carbon lattice. The experiments also show that the mechanism taking place involves the right balance of several factors, namely, a low passivated NP surface, particles with well-defined crystallographic facets, and interactions with an organics-free sp^2 carbon lattice [163].

The noncovalent interactions between semiconducting NPs and CNTs are particularly advantageous for combining the outstanding electrical properties of CNTs with the unique possibility of band gap tuning of quantum dots, and because of these reasons, a large impact of these materials in optoelectronics and photovoltaics is expected in the near future.

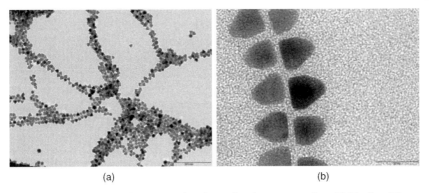

(a)   (b)

Figure 5.19 TEM images of CdSe particles obtained in the presence of MWCNTs after 48 h.
(a) Scale bar 200 nm. (b) Scale bar 20 nm.

5.9
Summary and Conclusions

In the past few decades a considerable research activity has been focused in developing the chemistry of new carbon allotropes. Although the chemistry of fullerenes has become a mature field and a large number of fullerene derivatives have been synthesized, the inherent difficulties in processing CNTs, specifically their poor solubility in most common solvents and the tendency of CNTs to form large bundles, still hamper their full exploitation in many applications. A variety of different noncovalent methods, such as specific π–π interactions with small aromatic and heterocyclic molecules, supramolecular interactions with ambiphilic molecules or ionic liquids, and wrapping with polymers of different nature, have been explored in the search for ideal separation methods that will allow the sorting of CNTs with the highest degree of selectivity, without modifying the properties of the tubes, and with the ultimate challenge of producing materials that will allow access to the properties of individual CNTs. Through the elegant combination of methodologies discussed in this chapter, design principles that constitute the foundation for a realistic application of CNTs in material sciences, biology, and medicine have been thoroughly discussed in detail.

Acknowledgments

The authors wish to express their gratitude to the MEC of Spain (projects CTQ2008-00795 and Consolider-Ingenio 2010 CSD2007-0010 Nanociencia Molecular) and the CAM (project P-PPQ-000225-0505) for the generous financial support.

References

1 Popov, V.N. and Lambin, P. (2006) *Carbon Nanotubes*, Springer, Dordrecht.

2 Reich, S., Thomsen, C., and Maultzsch, J. (2004) *Carbon Nanotubes: Basic Concepts*

and Physical Properties, VCH, Weinheim, Germany.

3 (2002) Special issue on carbon nanotubes. *Acc. Chem. Res.*, **35**, 997.

4 Harris, P.J.F. (2001) *Carbon Nanotubes and Related Structures: New Materials for the Twenty-First Century*, Cambridge University Press, Cambridge.

5 Hirsch, A. (2002) *Angew. Chem. Int. Ed.*, **41**, 1853–1859.

6 Barh, J.L. and Tour, J.M. (2002) *J. Mater. Chem.*, **12**, 1952–1958.

7 Nigoyi, S., Hamon, M.A., Hu, H., Zhao, B., Bhomwik, P., Sen, R., Itkis, M.E., and Haddon, R.C. (2002) *Acc. Chem. Res.*, **35**, 1105–1113.

8 Sun, Y.-P., Fu, K., Lin, Y., and Huang, W. (2002) *Acc. Chem. Res.*, **35**, 1096–1104.

9 Banerjee, S., Kahn, M.G.C., and Wong, S.S. (2003) *Chem. Eur. J.*, **9**, 1898–1908.

10 Tasis, D., Tagmatarchis, N., Georgakilas, V., and Prato, M. (2003) *Chem. Eur. J.*, **9**, 4001–4008.

11 Dyke, C.A. and Tour, J.M. (2004) *Chem. Eur. J.*, **10**, 812–817.

12 Banerjee, S., Hemraj-Benny, T., and Wong, S.S. (2005) *Adv. Mater.*, **17**, 17–29.

13 Guldi, D.M., Rahman, G.M.A., Zerbetto, F., and Prato, M. (2005) *Acc. Chem. Res.*, **38**, 871–878.

14 Hirsch, A. and Vostrowsky, O. (2005) *Top. Curr. Chem.*, **245**, 193–237.

15 Tasis, D., Tagmatarchis, N., Bianco, A., and Prato, M. (2006) *Chem. Rev.*, **106**, 1105–1136.

16 Guldi, D.M., Rahman, G.M.A., Sgobba, V., and Ehli, C. (2006) *Chem. Soc. Rev.*, **35**, 471–487.

17 Guldi, D.M. (2007) *Phys. Chem. Chem. Phys.*, **7**, 1400–1420.

18 Sgobba, V. and Guldi, D.M. (2009) *Chem. Soc. Rev.*, **38**, 165–184.

19 Britz, D.A. and Khlobystov, A.N. (2006) *Chem. Soc. Rev.*, **35**, 637–659.

20 Herranz, M.A., Giacalone, F., Sánchez, L., and Martín, N. (2007) Hydrogen bonding donor–acceptor carbon nanostructures, *Fullerenes: Principles and Applications* (eds F. Langa and J.F. Nierengarted), Royal Society of Chemistry, Cambridge, pp. 152–190.

21 Sánchez, L., Herranz, M.A., Martín, N., and Guldi, D.M. (2009) Carbon nanostructures built around hydrogen-bonding motifs, *Bottom-up Nanofabrication: Supramolecules, Self-Assemblies, and Organized Films* (ed. K. Ariga), American Scientific Publishers, pp. 407–427.

22 Liu, J., Rinzler, A.G., Dai, H., Hafner, J.H., Bradley, R.K., Boul, P.J., Lu, A., Iverson, T., Shelimov, K., Huffman, C.B., Rodriguez-Macias, F., Shon, Y.S., Lee, T.R., Colbert, D.T., and Smalley, R.E. (1998) *Science*, **280**, 1253–1256.

23 Asuman, K.D., Poner, R., Lourie, O., Rufo, R.S., and Korobov, M. (2000) *J. Phys. Chem. B*, **104**, 8911–8915.

24 Hamon, M.A., Chen, J., Hu, H., Chen, Y., Rao, A., Eklund, P.C., and Haddon, R.C. (1999) *Adv. Mater.*, **11**, 834–840.

25 Yang, D.-Q., Rochette, J.-F., and Sacher, E. (2005) *J. Phys. Chem. B*, **109**, 7788–7794.

26 Kahn, M.G.C., Banerjee, S., and Wong, S.S. (2002) *Nano Lett.*, **2**, 1215–1218.

27 Ikeda, A., Hayashi, K., Konishi, T., and Kikuchi, J.-I. (2004) *Chem. Commun.*, 1334–1335.

28 Sumanasekera, G.U., Pradhan, B.K., Romero, H.E., Adu, K.W., and Eklund, P.C. (2002) *Phys. Rev. Lett.*, **89** (4), 166801.

29 Zhao, J., Lu, J.P., Han, J., and Yang, C.K. (2003) *Appl. Phys. Lett.*, **82**, 3746–3748.

30 Zhang, J., Lee, J.-K., Wu, Y., and Murray, R.W. (2003) *Nano Lett.*, **3**, 403–407.

31 Simmons, J.M., In, I., Campbell, V.E., Mark, T.J., Léonard, F., Gopalan, P., and Eriksson, M.A. (2007) *Phys. Rev. Lett.*, **98** (4), 086802.

32 Nakashima, N., Tomonari, Y., and Murakami, H. (2002) *Chem. Lett.*, 638–639.

33 Tomonari, Y., Murakami, H., and Nakashima, N. (2006) *Chem. Eur. J.*, **12**, 4027–4034.

34 Chen, R.J., Zhang, Y., Wang, D., and Dai, H. (2001) *J. Am. Chem. Soc.*, **123**, 3838–3839.

35 Wu, P., Chen, X., Hu, N., Tam, U.C., Blixt, O., Zettl, A., and Bertozzi, C.R. (2008) *Angew. Chem. Int. Ed.*, **47**, 5100–5103.

36 Sudiya, H.G., Ma, J., Dong, X., Ng, S., Li, L.J., Lu, X.-W., and Chen, P. (2009) *Angew. Chem. Int. Ed.*, **48**, 2723–2726.

37 Ehli, C., Guldi, D.M., Herranz, M.A., Martín, N., Campidelli, S., and Prato, M. (2008) *J. Mater. Chem.*, **18**, 1498–1503.

38 Herranz, M.A., Ehli, C., Campidelli, S., Gutiérrez, M., Hug, G.L., Ohkubo, K., Fukuzumi, S., Prato, M., Martín, N., and Guldi, D.M. (2008) *J. Am. Chem. Soc.*, **130**, 66–73.

39 Kavakka, J.S., Heikkinen, S., Kilpeläinen, I., Mattila, M., Lipsanen, H., and Helaja, J. (2007) *Chem. Commun.*, 519–521.

40 Guldi, D.M., Rahman, G.M.A., Qin, S., Tchoul, M., Ford, W.T., Marcaccio, M., Paolucci, D., Paolucci, F., Campidelli, S., and Prato, M. (2006) *Chem. Eur. J.*, **12**, 2152–2161.

41 Hu, L., Zhao, Y.-L., Ryu, K., Zhou, C., Stoddart, J.F., and Grüner, G. (2008) *Adv. Mater.*, **20**, 939–946.

42 Guldi, D.M., Rahman, G.M.A., Jux, N., Balbinot, D., Tagmatarchis, N., and Prato, M. (2005) *Chem. Commun.*, 2038–2040.

43 Guldi, D.M., Rahman, G.M.A., Jux, N., Balbinot, D., Hartnagel, U., Tagmatarchis, N., and Prato, M. (2005) *J. Am. Chem. Soc.*, **127**, 9830–9838.

44 Ehli, C., Rahman, G.M.A., Jux, N., Balbinot, D., Guldi, D.M., Paolucci, F., Marcaccio, M., Paolucci, D., Melle-Franco, M., Zerbetto, F., Campidelli, S., and Prato, M. (2006) *J. Am. Chem. Soc.*, **128**, 11222–11231.

45 Chitta, R., Sandanayaka, A.S.D., Schumacher, A.L., D'Souza, L., Araki, Y., Ito, O., and D'Souza, F. (2007) *J. Phys. Chem. C.*, **111**, 6947–6955.

46 D'Souza, F., Chitta, R., Sandanayaka, A.S.D., Subbaiyan, N.K., D'Souza, L., Araki, Y., and Ito, O. (2007) *Chem. Eur. J.*, **13**, 8277–8284.

47 D'Souza, F., Chitta, R., Sandanayaka, A.S.D., Subbaiyan, N.K., D'Souza, L., Araki, Y., and Ito, O. (2007) *J. Am. Chem. Soc.*, **129**, 15865–15871.

48 Chitta, R. and D'Souza, F. (2008) *J. Mater. Chem.*, **18**, 1440–1471.

49 Chen, G., Wright, P.M., Geng, J., Mantovani, G., and Haddleton, D.M. (2008) *Chem. Commun.*, 1097–1099.

50 Ogoshi, T., Takashima, Y., Yamaguchi, H., and Harada, A. (2007) *J. Am. Chem. Soc.*, **129**, 4878–4879.

51 Gómez, F.J., Chen, R.J., Wang, D., Waymounth, R.M., and Dai, H. (2003) *Chem. Commun.*, 190–191.

52 Bogani, L., Danieli, C., Biavardi, E., Bendiab, N., Barra, A.-L., Dalcanale, E., Wernsdorfer, W., and Cornia, A. (2009) *Angew. Chem. Int. Ed.*, **48**, 746–750.

53 Bogani, L. and Wernsdorfer, W. (2008) *Nat. Mater.*, **7**, 179–186.

54 Schmidt, C.D., Böttcher, C., and Hirsch, A. (2007) *Eur. J. Org. Chem.*, 5497–5505.

55 Backes, C., Schmidt, C.D., Hauke, F., Böttcher, C., and Hirsch, A. (2009) *J. Am. Chem. Soc.*, **131**, 2172–2184.

56 Marquis, R., Greco, C., Sadokierska, I., Lebedkin, S., Kappes, M.M., Michel, T., Alvarez, L., Sauvajol, J.-L., Meunier, S., and Mioskowski, C. (2008) *Nano Lett.*, **8**, 1830–1835.

57 Nakayama-Ratchford, N., Bangsaruntip, S., Sun, X., Welsher, K., and Dai, H. (2007) *J. Am. Chem. Soc.*, **129**, 2448–2449.

58 Liu, Z., Sun, X., Nakayama-Ratchford, N., and Dai, H. (2007) *ACS Nano*, **1**, 50–56.

59 Wang, X.B., Liu, Y.Q., Qiu, W.F., and Zhu, D.B. (2002) *J. Mater. Chem.*, **12**, 1636–1639.

60 Li, H., Zhou, B., Lin, Y., Gu, L., Wang, W., Fernando, K.A.S., Kumar, S., Allard, L.F., and Sun, Y.-P. (2004) *J. Am. Chem. Soc.*, **126**, 1014–1015.

61 Wang, W., Fernando, K.A.S., Lin, Y., Meziani, M.J., Veca, L.M., Cao, L., Zhang, P., Kimani, M.M., and Sun, Y.-P. (2008) *J. Am. Chem. Soc.*, **130**, 1415–1419.

62 Rahman, G.M.A., Guldi, D.M., Campidelli, S., and Prato, M. (2006) *J. Mater. Chem.*, **16**, 62–65.

63 Chen, J. and Collier, C.P. (2005) *Phys. J. Phys. Chem. B*, **109**, 7605–7609.

64 Tu, W., Lei, J., and Ju, H. (2009) *Chem. Eur. J.*, **15**, 779–784.

65 Geng, J., Ko, Y.K., Youn, S.C., Kim, Y.-H., Kim, S.A., Jung, D.-H., and Jung, H.-T. (2008) *J. Phys. Chem. C*, **112**, 12264–12271.

66 Hasobe, T., Fukuzumi, S., and Kamat, P.V. (2005) *J. Am. Chem. Soc.*, **127**, 11884–11885.

67 Cheng, F., Zhang, S., Adronov, A., Echegoyen, L., and Diederich, F. (2006) *Chem. Eur. J.*, **12**, 6062–6070.

68 Cheng, F. and Adronov, A. (2006) *Chem. Eur. J.*, **12**, 5053–5059.

69 Satake, A., Miyajima, Y., and Kobuke, Y. (2005) *Chem. Mater.*, **17**, 716–724.

70 Saito, K., Troiani, V., Qiu, H., Solladié, N., Sakata, T., Mori, H., Ohama, M., and Fukuzumi, S. (2007) *J. Phys. Chem. C*, **111**, 1194–1199.

71 Hecht, D.S., Ramirez, R.J.A., Briman, M., Artukovic, E., Chichak, K.S., Stoddart, J.F., and Grüner, G. (2006) *Nano Lett.*, **6**, 2031–2036.

72 Sessler, J.L., Cyr, M.J., Lynch, V., McGhee, E., and Ibers, J.A. (1990) *J. Am. Chem. Soc.*, **112**, 2810–2813.

73 Boul, P.J., Cho, D.-G., Rahman, G.M.A., Marquez, M., Ou, Z., Kadish, K.M., Guldi, D.M., and Sessler, J.L. (2007) *J. Am. Chem. Soc.*, **129**, 5683–5687.

74 Guldi, D.M., Rahman, G.M.A., Qin, S., Tchoul, M., Ford, W.T., Marcaccio, M., Paolucci, D., Paolucci, F., Campidelli, S., and Prato, M. (2006) *Chem. Eur. J.*, **12**, 2152–2161.

75 Alvaro, M., Atienzar, P., de la Cruz, P., Delgado, J.L., Troiani, V., García, H., Langa, F., Palkar, A., and Echegoyen, L. (2006) *J. Am. Chem. Soc.*, **128**, 6626–6635.

76 Yu, J., Mathew, S., Flavel, B.S., Johnston, M.R., and Shapter, J.G. (2008) *J. Am. Chem. Soc.*, **130**, 8788–8796.

77 Wang, P., Moorefield, C.N., Li, S., Hwang, S.-H., Shreiner, C.D., and Newkome, G.R. (2006) *Chem. Commun.*, 1091–1093.

78 Nobusawa, K., Ikeda, A., Kikuchi, J., Kawano, S., Fujita, N., and Shinkai, S. (2008) *Angew. Chem. Int. Ed.*, **47**, 4577–4589.

79 Kawano, S., Fujita, N., and Shinkai, S. (2004) *J. Am. Chem. Soc.*, **126**, 8592–8593.

80 Kanungo, M., Lu, H., Malliaras, G.G., and Blanchet, G.B. (2009) *Science*, **323**, 234–237.

81 O'Connell, M.J., Bachilo, S.M., Huffman, C.B., Moore, V.C., Strano, M.S., Haroz, E.H., Rialon, K.L., Boul, P.J., Noon, W.H., Kittrell, C., Ma, J., Hauge, R.H., Weisman, R.B., and Smalley, R.E. (2002) *Science*, **297**, 593–596.

82 Moore, V.C., Strano, M.S., Haroz, E.H., Hauge, R.H., and Smalley, R.E. (2003) *Nano Lett.*, **3**, 1379–1382.

83 Richard, C., Balavoine, F., Schultz, P., Ebbesen, T.W., and Mioskowski, C. (2003) *Science*, **300**, 775–778.

84 Wiltshire, J.G., Li, L.J., Khlobystov, A.N., Padbury, C.J., Briggs, G.A.D., and Nicholas, R.J. (2005) *Carbon*, **43**, 1151–1155.

85 Islam, M.F., Rojas, E., Bergey, D.M., Johnson, A.T., and Yodh, A.G. (2002) *Nano Lett.*, **3**, 269–273.

86 Wenseleers, W., Vlasov, I.I., Goovaerts, E., Obraztsova, E.D., and Lobach, A.S. (2004) *Adv. Funct. Mater.*, **14**, 1105–1112.

87 Mackiewicz, N., Surendran, G., Remita, H., Keita, B., Zhang, G., Nadjo, L., Hagège, A., Doris, E., and Mioskowski, C. (2008) *J. Am. Chem. Soc.*, **130**, 8110–8111.

88 Arnold, M.S., Green, A.A., Hulvat, J.F., Stupp, S.I., and Hersam, M.C. (2006) *Nat. Nanotechnol.*, **1**, 60–65.

89 Crochet, J., Clemens, M., and Hertel, T. (2007) *J. Am. Chem. Soc.*, **129**, 8058–8059.

90 Miyata, Y., Yanagi, K., Maniwa, Y., and Kataura, H. (2008) *J. Phys. Chem. C*, **112**, 3591–3596.

91 Hersam, M.C. (2008) *Nat. Nanotechnol.*, **3**, 387–394.

92 Backes, C., Hauke, F., Schmidt, C.D., and Hirsch, A. (2009) *Chem. Commun.*, 2643–2645.

93 Green, A.A. and Hersam, M.C. (2007) *Mater. Today*, **10**, 59–60.

94 Fukushima, T. and Aida, T. (2007) *Chem. Eur. J.*, **13**, 5048–5058.

95 Zhou, X., Wu, T., Ding, K., Hu, B., Hou, M., and Han, B. (2009) *Chem. Commun.*, 1897–1899.

96 Fukushima, T., Kosaka, A., Ishimura, Y., Yamamoto, T., Takigawa, T., Ishii, N., and Aida, T. (2003) *Science*, **300**, 2072–2074.

97 O'Connel, R.J., Boul, P.J., Ericson, L.M., Huffman, C., Wang, Y., Haroz, E., Kuper, C., Tour, J., Ausman, K.D., and Smalley, R.E. (2001) *Chem. Phys. Lett.*, **342**, 265–271.

98 Gong, X., Liu, J., Baskaran, S., Voise, R.D., and Young, J.S. (2000) *Chem. Mater.*, **12**, 1049–1052.

99 Yang, D.-Q., Rochette, J.-F., and Sacher, E. (2005) *J. Phys. Chem. B*, **109**, 4481–4484.

100 Guldi, D.M. (2005) *J. Phys. Chem. B*, **109**, 11432–11441.

101 Sgobba, V., Rahman, G.M.A., Guldi, D.M., Jux, N., Campidelli, S., and Prato, M. (2006) *Adv. Mater.*, **18**, 2264–2269.

102 Guldi, D.M., Rahman, G.M.A., Prato, M., Jux, N., Qin, S., and Ford, W. (2005) *Angew. Chem. Int. Ed.*, **44**, 2015–2018.

103 Hasobe, T., Fukuzumi, S., and Kamat, P.V. (2006) *J. Phys. Chem. B*, **110**, 25477–25484.

104 Sinani, V.A., Gheith, M.K., Yaroslavov, A.A., Rakhnyanskaya, A.A., Sun, K., Mamedov, A.A., Wicksted, J.P., and Kotov, N.A. (2005) *J. Am. Chem. Soc.*, **127**, 3463–3472.

105 Star, A., Steuerman, D.W., Heath, J.R., and Stoddart, J.F. (2002) *Angew. Chem. Int. Ed.*, **41**, 2508–2512.

106 Prencipe, G., Tabakman, S.M., Welsher, K., Liu, Z., Goodwin, A.P., Zhang, L., Henry, J., and Dai, H. (2009) *J. Am. Chem. Soc.*, **131**, 4783–4787.

107 Kymakis, E. and Amaratunga, G.A. (2002) *J. Appl. Phys. Lett.*, **80**, 112–114.

108 Coleman, J.M., Dalton, A.B., Curran, S., Rubio, A., Davey, A.P., Drury, A., MacCarthy, B., Lahr, B., Ajayan, P.M., Roth, S., Barklie, R.C., and Blau, W.J. (2000) *Adv. Mater.*, **12**, 213–216.

109 Star, A., Stoddart, J.F., Steuerman, D., Diehl, M., Boukai, A., Wong, E.W., Yang, X., Chung, S.-W., Choi, H., and Heath, J.R. (2001) *Angew. Chem. Int. Ed.*, **40**, 1721–1725.

110 Steuerman, D.W., Star, A., Narizzano, R., Choi, H., Ries, R.S., Nicolini, C., Stoddart, J.F., and Heath, J.R. (2002) *J. Phys. Chem. B*, **106**, 3124–3130.

111 Star, A., Liu, Y., Grant, K., Ridvan, L., Stoddart, J.F., Steuerman, D.W., Diehl, M.R., Boukai, A., and Heath, J.R. (2003) *Macromolecules*, **36**, 553–560.

112 Star, A. and Stoddart, J.F. (2002) *Macromolecules*, **35**, 7516–7520.

113 Ramasubramanian, R., Chen, J., and Liu, J. (2003) *Appl. Phys. Lett.*, **83**, 2928–2930.

114 Zhang, X., Zhang, J., Wang, R., Zhu, T., and Liu, Z. (2004) *ChemPhysChem*, **5**, 998–1002.

115 Chen, G.Z., Shaffer, M.S.P., Coleby, D., Dixon, G., Zhou, W.Z., Fray, D.J., and Windle, A.H. (2000) *Adv. Mater.*, **12**, 522–526.

116 Gao, M., Huang, S.M., Dai, L.M., Wallace, G.C., Gao, R.P., and Wang, Z.L. (2000) *Angew. Chem., Int. Ed.*, **39**, 3664–3667.

117 Gao, M., Dai, L., and Wallace, G.C. (2003) *Electroanalysis*, **15**, 1089–1094.

118 Xu, Y., Jiang, Y., Cai, H., He, P.-G., and Fang, Y.-Z. (2004) *Anal. Chim. Acta*, **516**, 19–27.

119 An, K.H., Jeong, S.Y., Hwang, H.R., and Lee, Y.H. (2004) *Adv. Mater.*, **16**, 1005–1009.

120 Tang, B.Z. and Xu, H.Y. (1999) *Macromolecules*, **32**, 2569–2576.

121 Yang, M., Koutsos, V., and Zaiser, M. (2005) *J. Phys. Chem. B*, **109**, 10009–10014.

122 Downs, C., Nugent, J., Ajayan, P.M., Duquette, D.J., and Santhanam, K.S.V. (1999) *Adv. Mater.*, **11**, 1028–1031.

123 Ferrer-Anglada, N., Kaempgen, M., Skakalova, V., Dettlaf-Weglikowska, U., and Roth, S. (2004) *Diamond Relat. Mater.*, **13**, 256–260.

124 Li, X.H., Wu, B., Huang, J.E., Zhang, J., Liu, Z.F., and Li, H.I. (2003) *Carbon*, **41**, 1670–1673.

125 Bavastrello, V., Carrara, S., Kumar Ram, M., and Nicolini, C. (2004) *Langmuir*, **20**, 969–973.

126 Tchmutin, I.A., Ponomarenko, A.T., Krinichnaya, E.P., Kozub, G.I., and Efimov, O.N. (2003) *Carbon*, **41**, 1391–1395.

127 Liu, J., Tian, S., and Knoll, W. (2005) *Langmuir*, **21**, 5596–5599.

128 Chen, F.M., Wang, B., Chen, Y., and Li, L.J. (2007) *Nano Lett.*, **7**, 3013–3017.

129 Cheng, F., Imin, P., Maunders, C., Botton, G., and Adronov, A. (2008) *Macromolecules*, **41**, 2304–2308.

130 Hwang, J.-Y., Nish, A., Doig, J., Douven, S., Chen, C.-W., Chen, L.-C., and Nicholas, R.J. (2008) *J. Am. Chem. Soc.*, **130**, 3543–3553.

131 Dieckmann, G.R., Dalton, A.B., Johnson, P.A., Razal, J., Chen, J., Giordano, G.M., Muñoz, E., Musselman, I.H., Baughman, R.H., and Draper, R.K. (2003) *J. Am. Chem. Soc.*, **125**, 1770–1777.

132 Artyukhin, A.B., Shestakov, A., Harper, J., Bakajin, O., Stroeve, P., and Noy, A. (2005) *J. Am. Chem. Soc.*, **127**, 7538–7542.

133 Qiao, R. and Ke, P.C. (2006) *J. Am. Chem. Soc.*, **128**, 13656–13657.

134 Xin, H. and Woolley, A.T. (2003) *J. Am. Chem. Soc.*, **125**, 8710–8711.

135 Cathcart, H., Quinn, S., Nicolosi, V., Kelly, J.M., Blau, W.J., and Coleman, J.N. (2007) *J. Phys. Chem. C*, **111**, 66–74.

136 Gigliotti, B., Sakizzie, B., Bethune, D.S., Shelby, R.M., and Cha, J.N. (2006) *Nano Lett.*, **6**, 159–164.

137 Nakashima, N., Okuzono, S., Murakami, H., Nakai, T., and Yoshikawa, K. (2003) *Chem. Lett.*, **32**, 456–457.

138 Zheng, M., Jagota, A., Semke, E.D., Diner, B.A., McLean, R.S., Lustig, S.R., Richardson, R.E., and Tassi, N.G. (2003) *Nat. Mater.*, **2**, 338–342.

139 Zheng, M., Jagota, A., Strano, M.S., Santos, A.P., Barone, P., Chou, S.G., Diner, B.A., Dresselhaus, M.S., McLean, R.S., Onoa, G.B., Samsonidze, G.G., Semke, E.D., Usrey, M., and Walls, D.J. (2003) *Science*, **302**, 1545–1548.

140 Cathcarth, H., Nicolosi, V., Hughes, J.M., Blau, W.J., Kelly, J.M., Quinn, S.J., and Coleman, J.N. (2008) *J. Am. Chem. Soc.*, **130**, 12734–12744.

141 Martin, W., Zhu, W., and Krilov, G. (2008) *J. Phys. Chem. B*, **112**, 16076–16089.

142 Arnold, M.S., Stupp, S.I., and Hersam, M.C. (2005) *Nano Lett.*, **5**, 713–718.

143 Chen, R.J. and Zhang, Y. (2006) *J. Phys. Chem. B*, **110**, 54–57.

144 Barisci, J.N., Tahhan, M., Wallace, G.G., Badaire, S., Vaugien, T., Maugey, M., and Poulin, P. (2004) *Adv. Funct. Mater.*, **14**, 133–138.

145 Keren, K., Berman, R.S., Buchstab, E., Sivan, U., and Braun, E. (2003) *Science*, **302**, 1380–1382.

146 Hobbie, E.K., Bauer, B.J., Stephens, J., Becker, M.L., McGuiggan, P., Hudson, S.D., and Wang, H. (2005) *Langmuir*, **21**, 10284–10287.

147 Staii, C., Johnson, A.T., Chen, M., and Gelperin, A. (2005) *Nano Lett.*, **5**, 1774–1778.

148 Heller, D.A., Jeng, E.S., Tsun-Kwan, Y., Martinez, B.M., Moll, A.E., Gastala, J.B., and Strano, M.S. (2006) *Science*, **311**, 508–511.

149 Rao, K.S., Daniel, S., Rao, T.P., Rani, S.U., Naidu, G.R.K., Heayeon, L., and Kawai, T. (2007) *Sens. Actuators B*, **122**, 672–682.

150 Lee, C.K., Shin, S.R., Mun, J.Y., Han, S.-S., So, I., Jeon, J.-H., Kang, T.M., Kim, S.I., Whitten, P.G., Wallace, G.G., Spinks, G.M., and Kim, S.J. (2009) *Angew. Chem. Int. Ed.* doi: 10.1002/anie.200804788

151 Baughman, R.H., Zakhidov, A.A., and de Heer, W.A.A. (2002) *Science*, **297**, 787–792.

152 Dukovic, G., Balaz, M., Doak, P., Berova, N.D., Zheng, M., Mclean, R.S., and Brus, L.E. (2006) *J. Am. Chem. Soc.*, **128**, 9004–9005.

153 Peng, X., Komatsu, N., Bhattacharya, S., Shimawaki, T., Aonuma, S., Kimura, T., and Osuka, A. (2007) *Nat. Nanotechnol.*, **2**, 361–365.

154 Peng, X., Komatsu, N., Kimura, T., and Osuka, A. (2007) *J. Am. Chem. Soc.*, **129**, 15947–15953.

155 Weisman, R.B. and Bachilo, S.M. (2003) *Nano Lett.*, **3**, 1235–1238.

156 Daniel, M.-C. and Astruc, D. (2004) *Chem. Rev.*, **104**, 293–346.

157 Love, J.C., Estroff, L.A., Kriebel, J.K., Nuzzo, R.G., and Whitesides, G.M. (2005) *Chem. Rev.*, **105**, 1103–1169.

158 Kamat, P.V. (2007) *J. Phys. Chem. C*, **111**, 2834–2860.

159 Guldi, D.M., Rahman, G.M.A., Sgobba, V., Kotov, N.A., Bonifazi, D., and Prato, M. (2006) *J. Am. Chem. Soc.*, **128**, 2315–2323.

160 Chaudhary, S., Kim, J.H., Singh, K.V., and Ozkan, M. (2004) *Nano Lett.*, **4**, 2415–2419.

161 Engtrakul, C., Kim, Y.H., Nedeljkovic, J.M., Ahrenkiel, S.P., Gilbert, K.E.H., Alleman, J.L., Zhang, S.B., Micic, O.I., Nozik, A.J., and Heben, M.J. (2006) *J. Phys. Chem. B*, **110**, 25153–25157.

162 Juárez, B.H., Klinke, C., Kornowski, A., and Weller, H. (2007) *Nano Lett.*, **7**, 3564–3568.

163 Juárez, B.H., Meyns, M., Chanaewa, A., Cai, Y., Klinke, C., and Weller, H. (2008) *J. Am. Chem. Soc.*, **130**, 15282–15284.

6
Covalent Functionalization of Carbon Nanotubes

Frank Hauke and Andreas Hirsch

6.1
Introduction

Until the 1980s, the carbon universe was built on the well-known modifications, graphite and diamond. This perspective totally changed with the discovery of the molecular carbon allotropes – fullerenes, carbon nanotubes (CNT), carbon nanohorns, and carbon onions. Owing to their outstanding properties, scientists became immediately interested in the investigation and modification of these new substance classes. Soon after the most abundant fullerene, C_{60}, became available in macroscopic amounts, chemists started to explore the possibilities to alter the structural framework of this carbon-based building block. Within 10 years, a tremendous variety of chemical functionalization sequences were investigated and are now widely used for the construction of multifunctional architectures with C_{60} as integral building unit. The chemical functionalization and the combination of its properties with other substance classes, such as porphyrins, polymers, and so on have been extensively reviewed in the past [1, 2].

Six years after the discovery of the zero-dimensional "quantum dot" C_{60}, the existence of a one-dimensional carbon allotrope – named carbon nanotubes – was reported by Iijima [3]. Their unique one-dimensional structure and therefore their outstanding electronical, thermal, and mechanical properties even surpass the potential of their smaller relatives – the fullerenes – and are unrivalled by any other substance class. Within 6 years, the production procedures for carbon nanotubes had been refined to provide a sufficient amount of material to the scientific community. Therefore, a multitude of applications for this new supermaterial as fundamental part in novel electronic devices, as single-molecule sensors, as fuel storage materials, and even as gene vectors in biomedical field have been envisaged for the near future [4]. Now, 18 years after the first report on multiwalled carbon nanotubes (MWCNTs) was published, we are still trying to get a grip on this revolutionary nanomaterial, and the question why this is so difficult has to be addressed. One major

Carbon Nanotubes and Related Structures. Edited by Dirk M. Guldi and Nazario Martín
Copyright © 2010 WILEY-VCH Verlag GmbH & Co. KGaA, Weinheim
ISBN: 978-3-527-32406-4

drawback in the processing of carbon nanotubes is their intrinsic poor solubility in organic and aqueous solvents [5, 6]. It was soon recognized that chemical functionalization of carbon nanotubes is key to overcoming this obstacle and that a targeted framework alteration would also help solve other problems based on the polydispersity of CNTs [7]. Until now, all different production techniques for CNTs yield a mixture of carbon nanotubes in terms of length, diameter, chirality, and, directly related to that, a mixture of metallic and semiconducting tubes that impedes their entry in nanoworld devices. And exactly this polydispersity of CNTs in combination with their low reactivity hampers the chemical functionalization and the characterization of the corresponding reaction products. Nevertheless, the chemical functionalization and structural alteration of carbon nanotubes has blossomed in the last decade, which is nicely documented by a series of review articles on this topic [8–17]. We have now arrived at a point where solubility of CNTs can be improved, their physical properties could be modified and fine-tuned, and the combination of this extraordinary carbon allotrope with the properties of other substance classes is finally possible. Therefore, the door for the entry of carbon nanotubes into CNT-based technological applications is open.

6.2
Chemical Functionalization of Carbon Nanotubes

6.2.1
Derivatization Strategies

In general, CNT functionalization can be divided into different categories (Figure 6.1a–d). The approaches toward chemical functionalization of CNTs comprise both supramolecular and molecular approaches. Supramolecular functionalization is mainly based on the noncovalent interaction of organic and inorganic moieties with the carbon nanotube surface by using varying interaction forces such as van der Waals, charge transfer, and π–π interactions (Figure 6.1a). A special case of noncovalent functionalization is the endohedral filling of CNTs with atoms or small molecules (Figure 6.1b). The main area of noncovalent, supramolecular functionalization comprises the exohedral decoration of the carbon nanotubes. Herein, the wrapping of CNTs by polymers, polypeptides, and DNA has to be mentioned and is widely used for the integration of CNTs into different matrices of other substance classes. The π–π interaction of aromatic moieties, for example, pyrene derivatives, with the extended π surface of CNTs has also been extensively used for the construction of supramolecular CNT-based architectures [18, 19]. One major application of the noncovalent CNT functionalization is the use of surfactant molecules to stabilize individual CNTs in an aqueous medium [20]. Owing to their huge surface area, single-walled carbon nanotubes (SWCNTs) tend to aggregate into bundles very efficiently by van der Waals interactions. By using ultrasonication, these bundles can be exfoliated, and in combination with suitable surfactants, the rebundling of the individual CNTs can be inhibited. This individ-

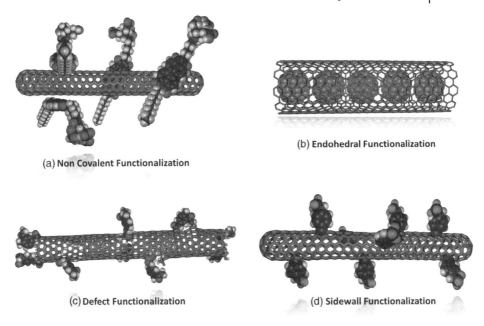

(a) **Non Covalent Functionalization**

(b) **Endohedral Functionalization**

(c) **Defect Functionalization**

(d) **Sidewall Functionalization**

Figure 6.1 Derivatization strategies for carbon nanotubes: (a) noncovalent exohedral functionalization with functional moieties based on specific CNT–molecule interactions; (b) endohedral functionalization (C_{60} incorporation); (c) defect-group functionalization; (d) direct covalent sidewall functionalization.

ualization opened the door for the photophysical characterization of SWCNTs by NIR emission spectroscopy [21]. The advantage of the supramolecular functionalization regime is reflected in the preservation of the intrinsic physical properties of the CNTs as the carbon π-network is not destroyed by the noncovalent interaction.

However, exactly this chemical alteration of the CNT framework can be used to tailor the interaction of the nanotubes with other entities. Covalently functionalized CNTs will have different mechanical and electrical properties due to the interference with the attached moieties and the alteration of the structural π-network.

Herein, we will exclusively deal with the covalent functionalization of the carbon nanotube framework, that is, the covalent attachment of functional entities onto the CNT scaffold. This structural alteration can take place at the termini of the tubes and/ or at the sidewalls. On the one hand, the direct sidewall functionalization (Figure 6.1d) is associated with a rehybridization of one or more sp^2 carbon atoms of the carbon network into a sp^3 configuration and a simultaneous loss of conjugation. On the other hand, the so-called defect functionalization sequence (Figure 6.1c) of CNTs is based on the chemical transformation of defect sites of the carbon nanotube material, intrinsically present or intentionally introduced by oxidation.

6.2.2
Topology and Reactivity of Carbon Nanotubes

The chemical modification and reactivity of CNTs needs to be discussed in the context of the reactivity of their closest carbon relatives – graphite and fullerenes. Graphite and graphene are chemically inert due to their planar, aromatic sp^2-bonded carbon framework. Only a few, very reactive compounds such as fluorine are able to derivatize the planar conjugated π-surface. Contrarily, as illustrated by the broad variety of derivatization sequences, the functionalization of the 3D curved π-surface of fullerenes can be carried out very easily [1]. This tremendous difference in reactivity of these two carbon allotropes is based on the introduction of strain by the bending of the planar graphene sheet into a three-dimensional carbon framework [22]. In nonplanar conjugated carbon structures, two basic factors are responsible for the chemical reactivity: (a) curvature-induced pyramidalization at the single carbon atoms (Figure 6.2a) [23, 24] and (b) π-orbital misalignment between adjacent pairs of carbon atoms (Figure 6.2b) [9, 25]. Fullerenes and carbon nanotubes both represent spherical carbon structures. Fullerenes are curved in two dimensions, whereas carbon nanotubes are curved in one dimension only. As a consequence, the reactivity of fullerenes is primarily driven by the release of pyramidalization strain energy. In contrast to sp^2 carbon atoms with a pyramidalization angle of $\Theta_p = 0°$, sp^3 carbon atoms exhibit a pyramidalization angle of $\Theta_p = 19.5°$. In the spherical C_{60}, the smallest stable fullerene, the pyramidalization angle of the 60 equivalent carbon atoms is $\Theta_p = 11.6°$. Therefore, an addition reaction toward the convex surface of C_{60} and the release of strain energy based on the rehybridization of a sp^2 carbon atom into

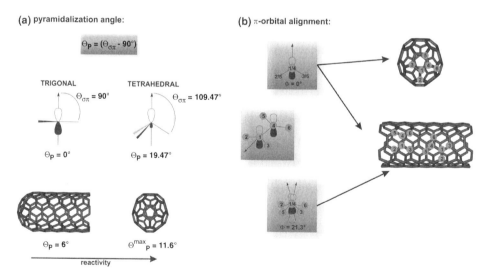

Figure 6.2 Representation of (a) the pyramidalization angle (Θ_p) and (b) the π-orbital misalignment angle (Φ) along the C1—C4 bond in a (5,5)-SWCNT and in C_{60}.

a sp^3 carbon atom is the foundation for the rich addition chemistry observed with fullerenes. A carbon nanotube with the same radius as C_{60} exhibits a less distorted π-framework due to its 1D curvature. As a consequence, CNTs are less reactive than fullerenes and the fullerene-like end caps of CNTs are more susceptible toward addition reactions than their sidewalls.

As mentioned above, the other factor that governs reactivity in conjugated nonplanar molecules is the π-orbital misalignment (Φ) and for carbon nanotubes this is the governing factor for their reactivity [9, 14]. The difference between the π-orbital alignment of a C_{60} fullerene and of an armchair (5,5) SWCNT is depicted in Figure 6.2b. The π-orbital alignment in fullerenes is almost perfect with a π-orbital misalignment angle of $\Phi = 0°$. In CNTs, all sidewall carbon atoms are equivalent. Nevertheless, two sets of different bonds can be found in CNTs: one type runs parallel to the circumference (or perpendicular to the nanotube axis) with a π-orbital misalignment angle of $\Phi = 0°$ and the other set of bonds exhibits an angle to the circumference with a π-orbital misalignment angle of $\Phi = 21.3°$. This π-orbital misalignment is the origin of torsional strain in nanotubes, and the relief of this strain energy controls the extent to which addition reactions occur with nanotubes.

Based on the two topological induced key factors for the reactivity of a conjugated carbon network, the chemical inertness of graphite and graphene can be easily understood. With a pyramidalization angle of $\Theta_p = 0°$ and a π-orbital misalignment angle of $\Phi = 0°$, no strain energy is stored within the carbon framework of graphite and graphene. Therefore, addition reactions will build up strain energy in contrast to the spherical carbon allotropes.

For carbon nanotubes, there should be a direct correlation between tube diameter and reactivity toward addition reactions [26], as both the pyramidalization angle and the π-orbital misalignment angle scale inversely with the diameter. As a consequence, smaller CNTs are expected to be more reactive than their larger counterparts [24, 27, 28]. MWCNTs with an outer tube diameter in the range of 5–50 nm generally exhibit a much lower tendency toward sidewall addition reactions than SWCNTs with diameter distributions in the range of 0.7–2 nm. Furthermore, a variety of investigations on the chemical oxidation of SWCNTs has confirmed this expected relationship, as smaller diameter carbon nanotubes are attacked faster by the oxidizing agent resulting in a pronounced destruction of the systems with higher internal strain [29–33]. As a consequence, the diameter selective separation of SWCNT samples by specific oxidation processes is basically possible.

Although Θ_p and Φ have been proven to be useful indices for describing the diameter-dependent reactivity, they are not sufficient if CNTs with the same diameter but different chiralities are taken into account. Therefore, a new reactivity criterion – the local curvature radius – was introduced by Li *et al.* [34], as the curvature of CNTs only expresses the curvature of every carbon atom on the tubular circumference and cannot be correlated with the strain energies of various C−C bonds, present in different tube chiralities. For a given tube radius, the strain in a C−C−bond perpendicular to the tube axis is more pronounced than the strain in C−C−bonds parallel to the tube axis. This refined concept can explain chirality selective functionalization reactions, which has been experimentally proven by the group of Kataura [35, 36]. A detailed investigation on the thermal oxidation of HiPCO and CoMoCat SWCNTs by air has shown that the oxidation rates do not depend solely on nanotube diameter and that the concept of local curvature of the different C—C bonds has to be taken into account in order to explain the observed chirality dependence of the air-driven degradation process. In addition to the observed reduced fluorescence emission of small diameter SWCNTs, the investigators have detected that the fluorescence signal of lower angle SWCNTs (near zigzag tubes) is also dramatically attenuated in comparison to the photoluminescence response of the SWCNT material with higher chiral angel (near armchair tubes). In the case of equal radius SWCNTs, the local curvature radius of zigzag SWCNTs is 4/3 times larger than that of an armchair SWCNT [37]. Therefore, a zigzag tube exhibits a more pronounced stability compared to an armchair tube and as a consequence, thinner and higher chiral angle SWCNTs are preferentially decomposed by the oxidation process.

The physical properties of carbon nanotubes can be described within the terminology of solid-state physics by using band structure representations in reciprocal space. For the prediction of the chemical properties, such as selectivity and reactivity, a molecular orbital theory description, most familiar to chemists, would be more desirable. Especially, understanding the interplay between the electronic structure and the chemical reactivity is of uttermost interest for finding suitable chemical reaction sequences in order to differently address metallic and semiconducting SWCNTs. As noted by Joselevich [38], pyramidalization and misalignment just deal with the localized electronic structure of carbon nanotubes without including the delocalized electronic band structure in the same molecular orbital picture. By merging the solid-state physics description of band structure with the chemical molecular orbital theory, the different reactivities of metallic and semiconducting SWCNTs can be interpreted and predicted by a classical HOMO–LUMO description [38]. Therefore, semiconducting nanotubes can be described as analogous to aromatic $[4n + 2]$ annulenes, whereas metallic nanotubes are analogous to antiaromatic $[4n]$ annulenes.

6.3
Defect Group Functionalization of Carbon Nanotubes

Defects in CNTs are of great importance for generating CNT-derivatives since they can serve as anchor groups for further functionalization. Defects are therefore a promising starting point for the development of the covalent chemistry of CNTs. Despite the partial destruction of the CNT backbone, defect functionalization of CNTs has led to the development of a widely used versatile tool that enables the combination of material properties of carbon nanotubes with other substance classes generating highly functional nanotube derivatives.

6.3.1
Defect Types and Defect Generation

An ideal carbon nanotube can be described as a perfect sp^2 carbon atom network with two hemifullerene units at the ends. In general, based on the production conditions a broad variety of defects and dislocated carbon atoms are generated in the CNT framework [39–41]. These areas of disorder are preferentially located at the ends of the nanotubes where the catalyst particles were attached and the growth process was started, leading to disordered fullerene caps and dangling bonds. Defects can also be located directly in the CNT sidewalls. Theory predicts that the degree of point defects (vacancies) should be higher in smaller CNT as the generation of defects leads to a relaxation of neighboring atoms and therefore the strain energy of the CNT framework can be reduced [42]. In addition, the amount and type of intrinsic defects also varies with the production and subsequent purification technique of the CNTs and encompasses the introduction of pentagon–heptagon defects, resulting in strain and bending of the CNTs; [43] Stone–Wales defects, obtained by a $90°$ rotation of a $C–C$ bond creating a 5-7-7-5 ring pattern; sp^3 defects saturated with hydrogen atoms or other groups; structural holes, the so-called vacancies, with dangling bonds [44, 45] and open tube ends terminated by carboxylic acid functionalities. Besides carboxy termini, the existence of other terminating groups as nitro, hydroxy, carbonyl functionalities or the saturation by protons has also been reported. It has been theoretically shown that these topological defects are the primary target locations for further functionalization reactions. For instance, Duan and coworkers have predicted that carbon atoms adjacent to vacancies in the nanotube framework should be the positions where the introduction of carboxylic acid functionalities takes place [46]. It has also been shown that Stone–Wales defects are more reactive than the pristine tube walls, exhibiting the most pronounced reactivity in the peripheral 5-6 and 6-7 ring fusions [47–49]. In general, present defects in the carbon nanotube sp^2 network represent locations of higher reactivity toward aggressive chemicals such as oxidizing agents and therefore the opening of the CNT framework at defect sides is the dominant pathway for oxidation-driven functionalization.

As all carbon nanotube production methods yield CNTs with a varying amount of impurities such as amorphous carbon, fullerenes, and metal catalyst particles, the purification of this inhomogeneous starting material is the fundamental first processing step [50]. Very efficient purification sequences are primarily based on liquid-phase or gas-phase oxidative treatment procedures. Widely used oxidants for carbon nanotubes are boiling concentrated nitric and sulfuric acids [51–57], mixtures of hydrogen peroxide and sulfuric acid (piranha solution) [31, 33, 58], gaseous oxygen [29, 30, 35, 37], ozone [52, 59–62], and potassium permanganate [63]. As a consequence, the removal of amorphous carbon and catalyst impurities is accompanied by the introduction of carboxylic acid functionalities and other oxygen bearing groups at the ends and defect sites of the carbon nanotube framework. Actually, it has turned out that it is not an easy task to develop purification strategies that are capable of removing the impurities only, without touching the structural integrity of the carbon nanotube framework [29, 64–66]. For instance, fluoromoderated purification of HiPCO tubes is expected to induce less sidewall damage to the nanotubes due to the interaction of fluorine with the exposed metal catalysts hindering the oxidative degradation of the nanotubes [29]. Another critical aspect is the complete removal of the carbon-containing oxidation debris, in order to verify that proceeding functionalization sequences take place on the carbon nanotube material and not on present carbonaceous fragments [53, 54, 67–69]. Green and coworkers have shown that the majority of the carboxylic acid functionalities created by nitric acid treatment are present on carboxylated carbonaceous fragments, adsorbed onto the SWCNT sidewalls [54]. Removing this debris by base treatment procedures is crucial for subsequent CNT functionalization.

When more drastic reaction conditions, such as concomitant ultrasonic treatment [70, 71], are applied, even the CNT material is shortened and the amount of carboxylic acid functionalities is increased. Shortened and carboxylated CNT material exhibits a pronounced solubility in polar organic media [55, 57, 72, 73] and therefore a multitude of cutting and purification techniques have been refined and adapted to the different types of carbon nanotubes during the past 5 years [33, 52, 58, 59, 61, 74–78]. In order to find an efficient cutting process, a high weight loss of the starting CNT material has to be avoided. Therefore, the pyrolysis of sidewall-fluorinated SWCNTs has become a preferred route that yields a material 50 nm in length [79]. Treating fluorinated HiPCO SWCNTs with Caro's acid leads to a breakage of sidewall $C-C-$bonds followed by etching at these damage sites creating short SWCNTs in very high yield [78].

The concentration of the carboxylic acid groups, which act as starting anchor groups in subsequent functionalization sequences, can be determined by spectroscopic [80–82] and analytical methods [83–85]. By a systematic study of the evolution of gaseous CO and CO_2 in heat-treated H_2SO_4/HNO_3 oxidized SWCNT samples, Mawhinney *et al.* have determined the degree of carboxylic acid functionalization in laser ablation SWCNTs to be in the range of about 5% [86]. Taking into account the high aspect ratio of carbon nanotubes, this value implies that in addition to an oxidative opening of the SWCNT end caps, defect introduction into the CNT-sidewall also take place. By using a simple acid–base titration method, the Haddon group has determined the total percentage of acidic sites in full-length purified SWCNT samples. The forward titration was carried out with sodium bicarbonate ($NaHCO_3$) and the back titration with HCl yielding carboxyl acid percentages in the nitric acid-treated samples in the order of 1–2% [84]. Using a detailed XPS analysis in combination with *ab initio* calculations, Hirsch and coworkers were able to give a quantitative picture of the different oxygen bearing functionalities (carboxyl, carbonyl, alcohol) generated through oxidative treatment of HiPCO nanotubes with nitric acid [82]. The calculation of the Mulliken charges for the carbon atoms in oxygen-containing functionalities ($-COOH$, $CONH_2$, $-CHO$, $-COH$) and the correlation to the chemical shifts in the X-ray photoelectron spectra allowed a C 1s core level fit of the experimentally obtained spectra and therefore a quantification of the present functional groups. After a reaction time of 3 h, a carboxylic acid content of 7.38% in a HiPCO sample was obtained. The XPS results further revealed that both the oxygen and the nitrogen content of the oxidized SWCNT material increased almost linearly with the oxidation time. An exact knowledge and control of the carboxylic acid functionality content is of utmost importance for defect group functionalization sequences, as these functionalities are the foundation for the attachment of other functional entities – the final functionalization degree is determined by the starting "defect" density.

6.3.2
Functional CNT-Derivatives by Defect Functionalizing Sequences

6.3.2.1 Soluble CNT-Derivatives
One primary objective in the introduction of CNT carboxylic acid functionalities is to overcome their intrinsic insolubility and to generate soluble carbon nanotube derivatives. The foundation for CNT-based defect functionalization reactions was laid by the seminal work of the Smalley group in 1998 (Scheme 6.1) [87]. By treating with thionyl chloride (SOCl$_2$), they were able to convert carboxylic acid functionalities in shortened SWCNTs (1) into the corresponding acid chlorides (2), which have subsequently been reacted with NH$_2$-(CH$_2$)$_{11}$-SH yielding a covalently coupled carbon nanotube amide derivative (3). The free thiol groups were attached to 10 nm diameter gold nanoparticles and investigated by AFM. Haddon and coworkers have expanded this defect functionalization concept by reacting the intermediately generated SWCNT acyl chloride functionalities with long-chain amines, for example, octadecylamine (ODA) [88]. This reaction protocol yields chemically modified SWCNT material that is soluble in organic media. The activation of the SWCNT carboxylic acid functions can also be carried out under mild, neutral conditions by using cyclohexylcarbodiimide (DCC) or other coupling reagents. In addition, by

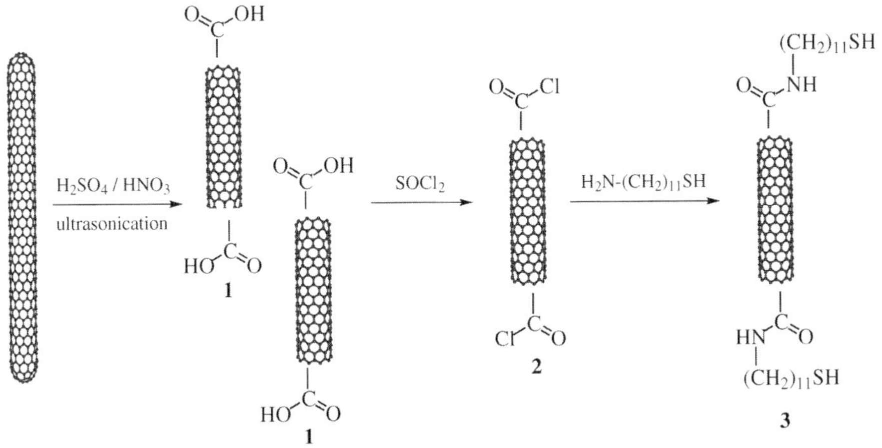

Scheme 6.1 Schematic representation of the oxidatively induced cutting and generation of carboxylic acid functionalities as well as their subsequent conversion into amide derivatives according to the protocol introduced by Smalley and coworkers [87].

microwave-assisted functionalization, the coupling of amines with carboxylated carbon nanotubes can be achieved without intermediate activated species on a timescale of 20–30 min in high yields [89].

Since the past decade, owing to its versatility, the carboxylic acid-based amidation and esterification sequence has become a widely used tool for the construction of soluble carbon nanotube derivatives. In an extensive study on the solubility of different types of MWCNT derivatives, Marsh *et al.* compared the effects of thermal oxidation, chemical functionalization, surfactant wrapping, and charge-doping on the solution behavior of MWCNTs [73]. Mere shortening of carbon nanotubes without introduction of functional groups or solubilizing agents substantially increases the stability of MWCNT dispersions in water. The modification of the exterior of shortened MWCNT by carboxylation, defect functionalization, and the use of surfactant molecules, or the reduction of the carbon nanotube derivatives further increases the stability of nanotubes in water. In another study, it has been shown that the solubility of MWCNT ester derivatives in chloroform depends on the alkyl chain length of the attached aliphatic moieties [90]. The highest solubility has been determined for alkyl chain lengths with 12 carbon atoms, whereas longer chains do not lead to a pronounced solubility. The possibility of solubilizing SWCNTs by a solvent-free amidation sequence has been demonstrated by Ford *et al.* [91]. Treating carboxylated SWCNTs with molten urea, which functions both as a solvent and as a reactant, leads to the attachment of ureido groups and the generation of excellent water-soluble material. Similarly, the coupling of another biocompatible molecule, the amino acid lysine, to acyl chloride-activated MWCNTs leads to a drastically improved water solubility of the carbon nanotube material [92]. Stable concentrations as high as 10 mg/ml can be obtained, which is nearly two orders of magnitude higher than that obtained for the carboxylated starting MWCNTs. The solubility of carbon nanotube derivatives in a wide range of solvents can be fine-tuned by attaching different types of amines. Coupling of aliphatic amines leads to an increased solubility in nonpolar solvents (ether, chloroform, etc.) whereas SWCNTs modified with perfluorinated amines are soluble in polar solvents (ethanol, DMD, etc.) but not in perfluorinated solvents [93]. The amidation of SWCNTs by aromatic amines yields derivatives with an intermediate behavior. In a comparative study using structurally related diamines, Shen *et al.* found a greatly reduced dispersibility of the amino functionalized MWCNTs compared to MWCNT terminated with carboxylic acid functionalities [94]. SEM images of the functionalized material point toward a pronounced cross-linking of carbon nanotubes through diamine connecting units leading to larger insoluble aggregates.

On the basis of a comprehensive microscopic study on alkyl chain functionalized SWCNTs, Bonifazi *et al.* demonstrated that the oxidative treatment of carbon nanotubes indeed generates carboxylic acid functions both on the tube tips and on the sidewalls [31]. STM images display the covalently attached aliphatic moieties (4) (Figure 6.3) as bright protuberances located near the SWCNT termini and as bright lumps that form on the SWCNT sidewalls. In another microscopic study, Green and coworkers were able to visualize organic moieties covalently attached to SWCNTs in atomic resolution by HRTEM [95]. As high-scattering heavy elements are clearly visible in HRTEM, they have coupled iodide tagged carbohydrates to SWCNTs by

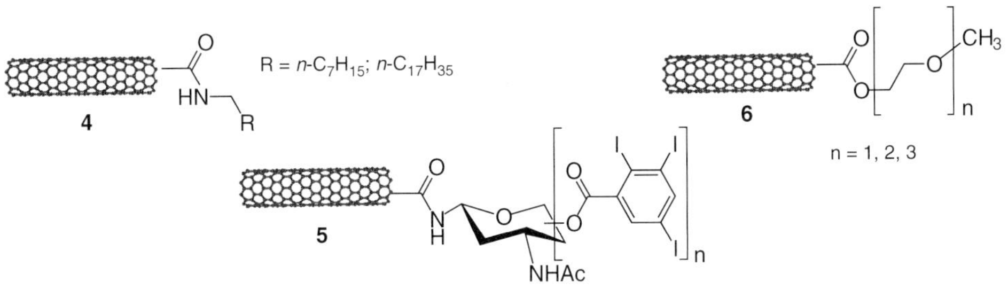

Figure 6.3 Examples of defect functionalized SWCNT derivatives for microscopic investigations and solubility studies.

amidation (**5**). The HRTEM images display iodides along the sidewall of the oxidatively purified SWCNTs. Bergeret *et al.* synthesized a variety of soluble ester derivatives by coupling polyether derivatives with different chain length to carboxylated HiPCO SWCNTs (**6**) (Figure 6.3) [96]. In a detailed analysis of the covalently functionalized SWCNTs, they were able to show that a loss of metallic character of the carbon nanotubes is initiated by the nitric acid treatment of the pristine SWCNT material and is maintained in the final SWCNT ester derivatives.

The overall functionalization sequence appears to be selective toward metallic species, with the nitric acid purification of pristine SWCNTs dramatically affecting the electronic properties of the finally derived functionalized SWCNTs.

6.3.2.2 Cofunctionalization of CNTs

By coupling pentanol to HNO_3-treated HiPCO SWCNTs, Alvaro *et al.* synthesized highly *o*-dichlorobenzene-soluble SWCNT derivatives, providing the foundation for supramolecular SWCNT donor/acceptor assemblies [97]. The sidewall functionalization of this pentyl ester derivative by a 1,3-dipolar cycloaddition of a nitrile oxide derivative leads to the formation of SWCNT architectures exhibiting pyridyl functionalities covalently attached to the sidewalls (**7**) (Figure 6.4). In the presence of a zinc porphyrin, a supramolecular SWCNT–porphyrin conjugate is formed through a coordination of the nitrogen atom of the pyridyl unit to the zinc metal center of the porphyrin. This doubly derivatized highly functional SWCNT assembly shows a deactivation pathway after photoexcitation by an energy transfer process from the porphyrin toward the CNT core. In a similar approach, the Tour group has used polyethylene glycol for water solubilization of the SWCNT material and attached in the second step functional aryl moieties by direct sidewall functionalization using diazonium salt chemistry (**8**) [98] (Figure 6.4).

The latter two examples express the possibility to use amide- and ester-derivatized CNT-systems for the generation of cofunctionalized carbon nanotubes, where the first moiety is used for CNT solubilization and the second addend exhibits functionality. In addition, the Menna group was able to show that it is possible to use the introduced carboxylic acid groups in SWCNTs for the simultaneous attachment of solubilizing poly(ethylene glycol)amines and amine derivatives of polycyclic aromatic fluoro-

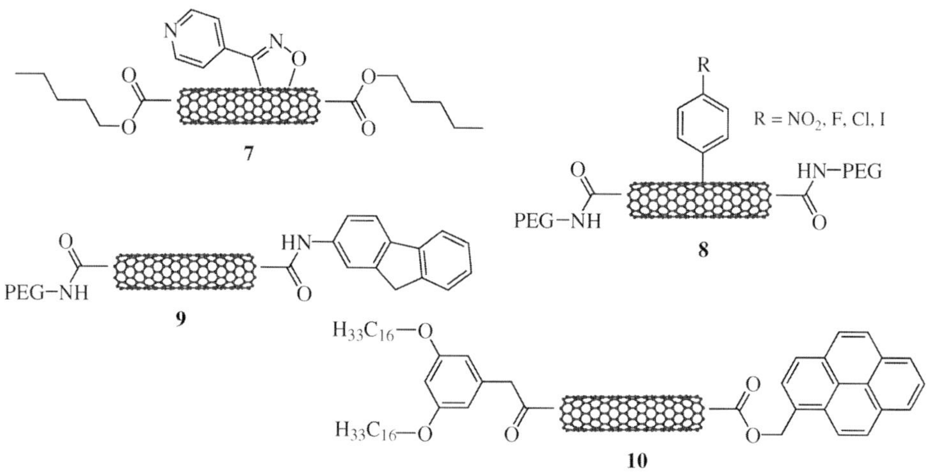

Figure 6.4 Examples of cofunctionalized CNT derivatives.

phores [99]. The spectroscopic characterization of the soluble fluorene (**9**), anthracene, and naphthalene derivatives exhibits a quenching of the dye fluorescence by intra-molecular interactions between the carbon nanotube and the attached aromatic moieties. In another study, SWCNTs were cofunctionalized with pyrenemethanol and 3,5-dihexadecanyloxybenzyl alcohol in different ratios with the purpose of altering the content of the functional pyrene moieties attached to the SWCNT surface [100]. Again, the absorption and emission properties of the soluble SWCNT derivatives (**10**) suggest that there are significant intramolecular excited state energy transfers from both pyrene monomer and excimer to the linked carbon nanotube. The excimer formation could be limited by reducing the pyrenemethanol fraction.

6.3.2.3 Asymmetric End-Functionalization of Carbon Nanotubes

Carbon nanotubes with two well-differentiated ends are attractive building blocks for a variety of applications [101]. The asymmetric end-functionalization should signif-icantly advance the self-assembling of carbon nanotubes into many new functional structures with a molecular-level control. One of the first steps toward these structures was performed by Chopra *et al.* [102]. A vertically aligned array of MWCNTs was embedded into a protecting polystyrene layer in order to prevent sidewall damage during the functionalization sequences. A water plasma oxidation opens the CNT tips and the oxidized CNT ends are exposed on both sides of the protecting polystyrene membrane. By floating the membrane assembly on top of a reactive solution of 2-aminoethanethiol with EDC coupling reagent, it is possible to selectively derivatize only one end of the carboxylated carbon nanotubes. The dissolution of the protecting layer in toluene results in a suspension of bifunctional MWCNTs, bearing carboxylic acid functionalities on the one side and aminoetha-nethiol chains on the other termini. Colloidal gold nanoparticles have been used to visualize the bifunctionality of the MWCNT derivatives by TEM, showing gold cluster decoration only at one CNT end.

6.3.2.4 **Nanoparticle and Quantum Dot: CNT Conjugates**

Hu *et al.* demonstrated that the attachment of functional entities on carboxylated carbon nanotubes does not always have to follow the pathway of amidation and esterification in the classical way [103]. By the treatment with NaOH, they transferred the carboxylic acid functionalities to the corresponding carboxylate ions. Their alkylation with 2-bromoethanethiol leads to the formation of MWCNT–SH derivatives. Now it is possible, either by direct attachment of preformed nanoparticles or by the *in situ* formation of gold nanoparticles, to generate nanoparticle–MWCNT conjugates. Metal nanoparticles are often used, like in the previous two examples, for the decoration of functional sites on carbon nanotubes and the visualization by microscopic methods. Furthermore, carbon nanotubes are ideal templates for the immobilization of nanoparticles allowing the construction of designed functional nanoarchitectures, which are extremely attractive as supports for heterogeneous catalysis and electronic and magnetic applications. A multitude of different strategies for the construction of these types of assemblies have been exploited during the past 5 years and have recently been reviewed [104, 105]. Of special interest is the combination of the unique electronic properties of carbon nanotubes with the properties of another quantum-confined material, the so-called quantum dots (QD). These semiconducting particles have formed the foundation for new photovoltaic cells and light-emitting diodes. The possibility of the covalent attachment of quantum dots on the surface of suitable derivatized MWCNTs has been demonstrated by Pan *et al.* [106]. MWCNT carboxylic acid functionalities on the ends and sidewalls were reacted with ethylenediamine leading to an amine-decorated CNT surface. CdSe quantum dots were decorated with mercaptoacetic acid, providing a carboxylic acid anchor group for attachment to the derivatized MWCNT systems. The resulting QD–MWCNT hybrid material shows an excellent solubility in aqueous solution. In a similar approach, *para*-phenylenediamine was used as anchor moiety to graft CdTe quantum dots onto MWCNT [107]. In the same study, a poly(2-dimethylaminoethyl) methacrylate coating was grown from the surface of the carboxylated MWCNTs by *in situ* atom transfer radical polymerization and subsequently, the amino functionalities were quaternized yielding a cationic polyamine coating for the attachment of CdTe quantum dots. Luminescence measurements have indicated that these MWCNT/CdTe complexes retain the luminescent properties of the quantum dots, whereas the photoluminescence of the pheneylenediamine-anchored composite is quenched by an electron transfer process. In the same and in a subsequent investigation, magnetic Fe_3O_4 nanoparticles were attached to the polymer-coated MWCNTs affording paramagnetic CNT-hybrids [108]. The Newkome group used a third-generation amino-polyester dendron connected to SWCNTs via amidation as a platform for the attachment of CdS nanoparticles [109]. The quantum dots were formed *in situ* within the dendron shell yielding quantum-confined particles exhibiting the characteristic photoluminescence. Therefore, SWCNT surface-anchored dendrons stabilize the CdS nanoparticles against coalescence into bulk CdS. Willner and coworkers attached a SWCNT/CdS conjugate to the surface of an electrode yielding a functional architecture that reveals efficient photoelectrochemical functions [110]. The gold electrode was functionalized with a primary monolayer of cysteamine to which the carboxylated CNTs were covalently coupled. The cysteamine-covered CdS nanoparticles were subsequently attached to the free CNT end by EDC-promoted amidation. Light excitation of the CdS quantum dots leads to the generation of a photo current by a photoinduced charge transfer from the nanoparticle along the SWCNT to the electrode.

6.3.2.5 Surface Attachment of CNTs

The attachment of carbon nanotubes onto surfaces in a well-defined fashion has been a major subject in research. Treating a glass substrate with aminosiloxanes leads to the formation of a self-assembled monolayer of aminosiloxanes exhibiting a terminal amino functionality for coupling shortened CNTs, resulting in a surface-bound array of SWCNTs (**11**) (Figure 6.5) [111]. Addition of a diamine linker and subsequent repeated condensation reactions lead to a high-density SWCNT multilayer. It has been shown that this surface-bound SWCNT assembly exhibits good electrical conductivity and can be used as a field emission device [112]. Yu *et al.* used self-assembled monolayers of ethyl undecanoate in a similar approach, which was subsequently reduced to the corresponding alcohol, for the aligned covalent attachment of shortened CNTs [113]. The cyclovoltammometric studies have shown that self-assembled monolayers with long alkyl chains inhibit an electronic communication between the substrate and the CNTs. Therefore, they have refined the process and succeeded in the covalent attachment of carboxylated CNTs to silicon surfaces without the use of the intermediate self-assembled monolayer of interconnecting molecules. Electrochemical investigations demonstrate an excellent conductivity to the substrate [114].

The covalent coupling of ferrocene methanol to the remaining SWCNT carboxylic acid functionalities leads to the formation of redoxactive electrodes, with a high

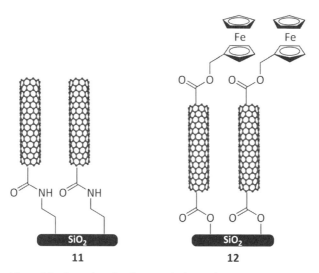

Figure 6.5 Examples of surface attached CNT derivatives.

surface density of ferrocene units (**12**) (Figure 6.5) [115]. The vertical orientation of the surface-anchored CNT derivatives has been confirmed by AFM investigations of the same group [116, 117]. Taking this concept further leads to ruthenium porphyrin-functionalized SWCNT arrays with the use of the axial coordination of the central ruthenium metal onto 4-aminopyridine preassembled SWCNTs directly anchored to a silicon surface. The photophysical investigations have indicated that the immobilization of ruthenium porphyrins enhances the light absorption of SWCNT-Si surfaces in the visible light region. Moreover, mixed assemblies of ferrocene/porphyrin donors covalently connected to the carbon nanotubes have been constructed [118].

6.3.2.6 Molecular Electronic Devices Based on Carbon Nanotubes

As outlined above, the special electronic properties of CNTs render this type of molecular carbon a promising candidate for future nanoscale electronics. For the construction of CNT-based nanoscale wires, the conductive interconnection of two end-capped opened CNTs is a fundamental prerequisite. Defect functionalization of the terminal carboxylic acid functionalities with bifunctional conjugated organic moieties leads to an electronic communication of the newly joined CNT wires [119]. The ability to functionalize the ends of cut nanotubes has led to a variety of functional nanoscale devices. Feldman *et al.* grew individual SWCNTs spanning metal electrodes, and by a spatially resolved oxidation, a section of the nanotube was removed, leaving two ends separated by a gap of 1–10 nm. A variety of different diamines have been used as gap bridging units. Depending on the molecular structure of the diamines, functionality is installed into the devices, yielding nanoscale pH sensors and detection units for metal ions and light [120, 121]. For instance, bridging the gap by oligoanilines leads to an electrical resistance that depends on the pH of the surrounding medium. In addition, by incorporating a biotin detection unit into the conjugated bridge, the binding of one corresponding protein can be monitored by the induced change in electrical resistance [122].

6.3.2.7 Carbon Nanotubes as Integrated Unit in Donor/Acceptor Assemblies

The former concept can be expanded by the covalent attachment of redox active moieties onto CNTs [123, 124]. The ability of nanotubes to respond to photochemical excitation is one of their most useful properties and may lead to efficient photovoltaic setups. The covalent coupling of 9,10-diphenylanthracene, a well-studied organic chromophore that exhibits electroluminescence, with carboxylic acid groups of short, soluble SWCNTs leads to the construction of a functional donor/acceptor assembly [125]. The comparison of the steady-state and time-resolved photoemission properties of the anthracene moiety with the transient optical spectra of the anthracene–SWCNT adduct proves an efficient quenching of the anthracene excited state by an energy transfer process of the attached donor moiety toward the carbon nanotube. By an EDC-assisted coupling of a flavin mononucleotide derivative (FMN) with the present defect functionalities of HiPCO SWCNTs, the group of Papadimitrakopoulos generated a cofactor–SWCNT hybrid system [126]. This assembly has been characterized by a multitude of spectroscopic methods and it has been shown that the strong π–π interaction of the aromatic flavin moiety causes a collapsed FMN configuration onto the sidewalls of the SWCNT. As a consequence, the fluorescence of the cofactor unit is totally quenched. The interaction of light with photosensitive moieties, directly attached to the carbon nanotube framework, can lead to a remarkable increase in the solubility of the donor/acceptor assembly upon illumination [127]. Herein, HiPCO SWCNTs are coupled to triarylpyrylium units, which act as a photosensitizer and exhibit strong electron acceptor capability. The illumination of this donor/acceptor assembly leads to a photoinduced electron transfer from the SWCNT to the pyrylium subunits, producing a positively charged CNT and negatively charged substituents. Coulombic repulsion between positive nanotube walls is the driving force responsible for the deaggregation and separation of the tubes into individual water-soluble molecules. In another study, Holleitner and coworkers demonstrated the possibility of coupling an entire functional protein complex – the photosystem I reaction center – to SWCNTs by means of a defect functionalization pathway [128]. These hybrid architectures can be used as building blocks for molecular optoelectronic circuits. For the construction of artificial photovoltaic cells, extended tetrathiafulvalenes (exTTF) are attractive candidates in correspondence with potent electron-accepting units. Guldi and coworkers have been able to investigate the photophysical properties of a covalent exTTF/SWCNT conjugate (**13**) [129] (Figure 6.6).

Hereby, time-resolved spectroscopy has enabled the identification of reduced SWCNT and oxidized exTTF units as metastable states in a series of donor–acceptor nanoconjugates with remarkable lifetimes in the range of hundreds of nanoseconds. By fine-tuning the donor/acceptor distance, the electron transfer rate can be controlled. Another potent electron acceptor class is manifested in viologenes and their derivatives. The covalent attachment of a substituted viologene unit with SWCNTs yields donor/acceptor assemblies where the intrinsic SWCNT fluorescence is quenched. Theses viologene-modified SWCNTs form charge transfer complexes with electron donor compounds such as 2,6-dimethoxynaphthalene leading to supramolecular donor/acceptor dyads [130]. Furthermore, the most frequently used

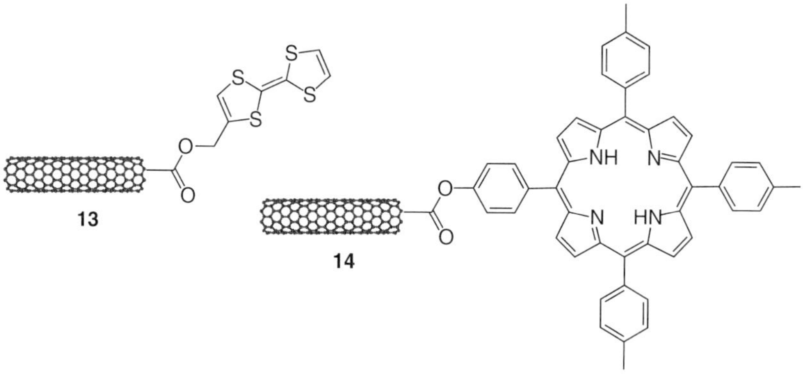

Figure 6.6 Examples of redox-active SWCNT donor/acceptor assemblies.

electron donor units in redox-active donor/acceptor assemblies are porphyrins. Therefore, the covalent integration of this substance class in CNT-based donor/acceptor systems is of fundamental interest. The reaction of a tetraphyenylporphyrin derivative, equipped with a hydroxyl functionality with carboxylated CNTs leads, to the generation of a redox-active assembly (14) (Figure 6.6) with porphyrin amounts in the range of 8–22% depending on the porphyrin concentration used for the coupling reaction [131]. Upon light excitation of the porphyrin moiety, a nearly quantitative fluorescence quenching of the chromophore unit is observed, based on an efficient light-induced electron transfer to the carbon nanotube framework. In another study, a similar kind of a multiporphyrin-linked SWCNT derivative has been electrophoretically deposited on nanostructured SnO_2 electrodes yielding a CNT-based photoelectrochemical device [132]. The film of porphyrin-linked SWCNTs exhibits photocurrent efficiencies as high as 4.9% under an applied potential of 0.08 V versus SCE. Besides energy conversion, the photoinduced electron or energy transfer in covalently linked porphyrin/SWCNT units can be used to enhance the optical limiting of the corresponding composite [133]. Furthermore, the photophysical interaction between photoactive units and the CNT core can be used for the construction of sensor units. Cho *et al.* demonstrated that the quenching of dye fluorescence can be utilized in a pH sensing setup [134]. The CNT-based pH sensor consists of a pH sensitive polymeric linker that contains a sulfadimethoxine (PSDM) group and a pyrene unit. As the PSDM moiety undergoes a conformational transition with pH variation, the distance between the fluorescent pyrene unit and the CNT changes with the pH value. When the pH value is lower than the pK_a of PSDM, the fluorescence intensity of pyrene is decreased, which indicates that the fluorescence quenching is effectively induced as a result of the conformational change in the linking unit.

6.3.2.8 Functional CNT Composite Architectures

The rational combination of two distinct materials leads to the generation of materials equipped with novel properties that are not found in the individual components [135]. Based on their unprecedented spectrum of properties, CNTs are ideal candidates as integral parts of functional architectures. The covalent combination of CNTs with ionic liquids leads to switchable systems, where the aggregate solubility can be reversibly shifted between water solubility and nonwater solubility by switching the electrolyte solution [136, 137]. The coupling of carboxylated MWCNTs with 1-hydroxethyl-3-hexyl imidazolium chloride yields a hybrid material (15), where the counterion can easily be exchanged with other target anions. Herein, the solubility depends on the anions coupled to the cationic imidazol ring (Figure 6.7). Counterions of the type Cl^-, Br^-, NO_3^-, and CH_3COO^- SO_4^{2-} exhibit water solubility, whereas lipophilic anions such as BF_4^-, PF_6^-, ClO_4^-, and $(CF_3SO_2)_2N^-$ mediate solubility in apolar solvents such as CH_2Cl_2. In addition, the imidazolium salt-functionalized MWCNTs show a high preferential solubility in ionic liquids per se.

These types of architecture can be used for preparing rhodium-based supported ionic liquid-phase catalysts [138]. Herein, the imidazolium-functionalized MWCNTs

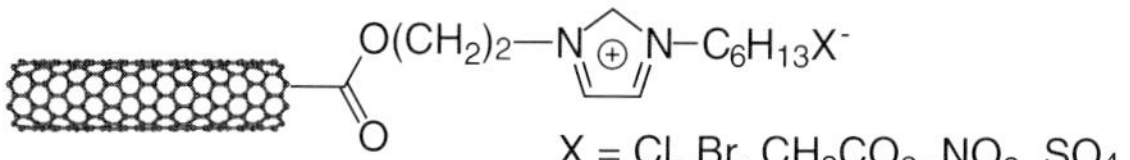

O(CH$_2$)$_2$—N$\overset{\oplus}{\smile}$N—C$_6$H$_{13}$X$^-$

15

X = Cl, Br, CH$_3$CO$_2$, NO$_3$, SO$_4$

Figure 6.7 Examples of a SWCNT with covalently attached imidazolium moiety.

have been solubilized in the ionic liquid and the active rhodium complex has been immobilized in an ionic liquid film. The resulting CNT-based catalyst composite exhibits a significantly higher activity than the corresponding systems with silica or activated carbon supports. A similar system, a palladium catalyst supported onto imidazolium-functionalized MWCNTs, exhibits high catalytic activity in Suzuki, Heck, and Sonogashira coupling reactions with high turnover numbers (about 500) [139]. Cyclodextrins are another highly functional substance class. It has been shown that cyclodextrins can act as good solubilizers for CNTs in water. Furthermore, cyclodextrins act as host units with a wide variety of molecules in supramolecular aggregates. Ritter and coworker have connected cyclodextrins to MWCNT via "click type" reactions [140]. Herein, propagylamine has been condensed with carboxylated MWCNTs, followed by a 1,3-dipolar cycloaddition of cyclodextrinazide. A stable MWCNT dispersion has been obtained by adding a water-soluble polymer bearing *tert*-butylphenyl guest moieties. The increased solution viscosity shows the formation of a supramolecular network between the polymer and the covalently attached cyclodextrin unit. Another type of a supramolecular CNT/cyclodextrin assembly has been prepared by the reaction of surface-bound carboxylic chloride groups of MWCNTs with aminoethyleneamino-deoxy-β-cyclodextrin, yielding highly water-soluble conjugates [141]. Upon addition of the cyclodextrin-decorated CNTs to an aqueous solution of tetrakis(4-carboxyphenyl)porphyrin, the fluorescence of the porphyrin is quenched as a result of a photoinduced electron transfer reaction between the cyclodextrin cavity-bound porphyrin donor system and the carbon nanotube. A fluorescence recovery can be achieved in the presence of 1-adamantane acetic acid, acting as a better guest, displacing the bound porphyrins from the cavity. In a similar fashion, direct interconnection of the two new carbon allotropes C$_{60}$ and CNTs yields multifunctional hybrid materials. The first SWCNT–C$_{60}$ conjugate (**16**) was synthesized by an amidation reaction between carboxylated SWCNTs and anilinie–fullerene derivatives [142] (Figure 6.8).

Small-angle neutron scattering experiments reveal that this type of assembly forms a complex aggregate network in solution [143]. In a similar way, an amino-terminated monoadduct was coupled to HiPCO SWCNTs by an EDC coupling sequence and the electronic properties of the resulting grapevine assembly were investigated [144]. The cyclovoltammetric response of the conjugate film remarkably resembles that of the pristine C$_{60}$ derivative in solution and exhibits reversible multiple step electro-chemical reactions, which reflect the covalent attachment of C$_{60}$ molecules to the surface of the SWCNT. The results of ESR spectroscopic investigations reveal a

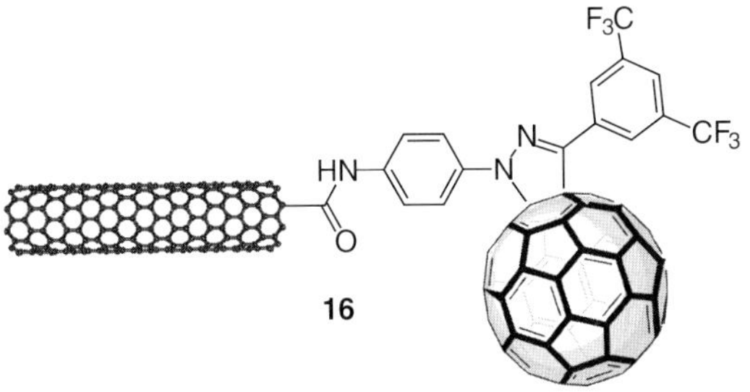

Figure 6.8 First example of a multifunctional C$_{60}$-SWCNT assembly.

ground state electron transfer trend from the SWCNT backbone to the attached fullerene moieties. With this concept, the construction of functional aggregates based on redox center (ferrocenes or porphyrins) coupled C$_{60}$ moieties is possible [145].

6.3.2.9 Carbon Nanotubes as Polymer Reinforcement Additives

Fiber-reinforced polymers comprise a soft matrix encapsulating load-bearing filler in the form of fibers or particles. For high-strength composites, the fibers should be stiff and exhibit a high aspect ratio. CNTs in polymer composites would serve to increase stiffness, strength, and toughness and provide additional properties. One fundamental prerequisite for an efficient load transfer from the matrix to the incorporated carbon nanotubes is their covalent attachment to the surrounding matrix. Current methods to covalently couple polymers onto carbon nanotubes include *grafting from* and *grafting to* approaches, mainly based on defect functionalization strategies. The field of CNT-based composite materials has rapidly developed during the past decade. Therefore, a detailed picture of the different strategic approaches would go far beyond the scope of this chapter and the reader may refer to the corresponding review articles [13, 146–148].

6.3.2.10 Functional CNT Derivatives in the Biological Context

The possibility of structural alteration by chemical modification of CNTs allowed the introduction of this carbon allotrope in the field of biology and medicine. By rational synthesis of biocompatible CNT derivatives through defect functionalization-based coupling reactions of DNA, enzymes, proteins, and other relevant biological systems, these highly functional assemblies can be used, for example, as sensors, for specific recognition and in drug delivery. As with polymer-based CNT composite materials, the field has blossomed in the last few years and a thorough account is given in literature [149–152].

6.3.3
Functional Group Interconversion

As outlined, carboxylic acid functionality-based functionalization of carbon nanotubes has become a versatile tool for the construction of carbon nanotube derivatives.

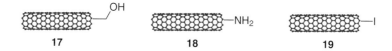

Figure 6.9 SWCNT derivatives through functional group interconversion starting from carboxylated SWCNTs.

The chemical alteration of the oxidatively introduced carboxylic acid group into other types of functional groups would expand the scope of possible chemical functionalization sequences. Ramanathan *et al.* have confirmed the possibility of functional group interconversion by reducing the carboxylic acid functionality to the corresponding hydroxymethyl group by lithium aluminum hydride [153]. In addition, these hydroxymethyl functionalities (**17**) (Figure 6.9) can be subsequently transformed into aminomethyl groups by phthalimide coupling followed by hydrolysis. In another study, Campbell and coworkers were able to produce SWCNTs with amino groups directly attached to the tube open ends (**18**) by Hofmann rearrangement of carboxylic acid amides and Curtius rearrangement of carboxylic acid azides [154]. Coleman *et al.* have synthesized iodinated SWCNTs (**19**) by a modified Hundsdiecker reaction with elemental iodine and iodosobenzene diacetate, starting from carboxylated CNTs [155] (Figure 6.9).

The electronic properties of the CNTs have largely been retained and the introduced iodine atoms have been visualized by transmission electron microscopy. On the basis of the different types of newly introduced functional groups, the scope of defect group-based CNTs derivatization will further advance in the years to come.

6.4
Direct Sidewall Functionalization of Carbon Nanotubes

One unique characteristic of carbon nanotubes is their high aspect ratio. As outlined, a defect functionalization-based chemical attachment of functional moieties on CNTs will primarily proceed only at the nanotube tips. As a consequence, the functionalization degree will be relatively low as the huge surface area of the CNT sidewall is hardly involved in the transformation sequence. Direct sidewall functionalization therefore offers the opportunity to install a great amount of functional entities on the nanotube framework and should lead in principle to higher functionalization degrees. However, the direct sidewall functionalization has two drawbacks. In contrast to defect functionalization sequences, sidewall derivatization and therefore the formation of a covalent bond between the attacking species and a carbon atom of the CNT framework is always associated with rehybridization of a sp^2 carbon atom into a sp^3 configuration and, based on that, a disturbance of the π-system of the CNTs. This fact is directly correlated with a negative influence on the electronic [28, 156, 157] and mechanical properties [158] of the carbon nanotubes. The second point is that the direct chemical alteration of the CNT sidewall is a challenging task due to the close relationship of the CNT sp^2 carbon network with planar graphite and its low reactivity. This directly leads to the question: which reaction sequences of a chemist's toolkit are in principle capable of functionalizing nanotube sidewalls? Furthermore, is there a selectivity of these reactions in terms of tube diameter or electronic type? The direct chemical functionalization of nanotube sidewalls by addition reactions has dramatically developed within the past 5 years and opened a new, intensively studied field in the science of carbon nanotube chemistry.

6.4.1
Fluorination and Nucleophilic Substitution Reactions of Fluorinated Carbon Nanotubes

Historically, based on the low reactivity of the CNT sidewall, direct sidewall addition of elemental fluorine was one of the first successful alterations of the sp^2 carbon framework. During the past decade, a variety of pathways leading to fluorinated carbon nanotubes, including plasma-based functionalization [159], have been developed. The different access routes to and the properties of fluorinated CNTs have recently been reviewed [160, 161]. Fluorinated nanotubes have become a widely used starting material for further chemical transformation steps, based on the nucleophilic substitution of fluorine atoms. Therefore, the treatment of fluorinated SWCNTs with *n*-butylamine leads to the formation of *n*-butylamine-functionalized derivatives [162]. By using bifunctional amines, such as alkoxysilanes [163–165] and polyethyleneimines [166], a facile introduction of functionality on the basis of sidewall derivatization is possible. Barron and coworkers have shown that the reaction of fluorinated HiPCO SWCNTs with ω-amino acids leads to water-soluble SWCNT derivatives [167]. In addition, the solubility in water is controlled by the length of the hydrocarbon chain of the amino acid and does strongly depend on the pH value of the solution. The nucleophilic substitution reaction of fluorine with urea, guanidine, and thiourea leads to a partial substitution of the fluorine atoms and the generation of the covalently attached amino moieties [168]. These bifunctional derivatives show improved dispersibility in DMF and water. Kelley and coworkers have prepared thiol and thiophene functionalized SWCNTs (**21**) via the reaction of bifunctional amines with fluoronanotubes (**20**) (Scheme 6.2) [169, 170].

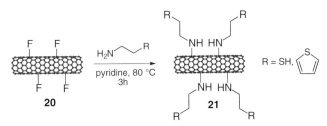

Scheme 6.2 Thiol and thiophene functionalized SWCNTs by nucleophilic substitution of fluorine atoms by bifunctional amines.

By coupling the free thiol group to gold nanoparticles and investigating in detail the corresponding conjugates by AFM and STM, Kelley and coworkers were able to show that both amine derivatives yield similar functionalization degrees, but the distribution of the functional groups is quite distinct. This leads to the conclusion that while AFM and the complexation of a "chemical marker" nanoparticle is an excellent method for determining the presence of functionalization on SWCNT sidewalls, STM provides excellent information on the substituents' location.

6.4.2
Hydrogenation of Carbon Nanotubes

Hydrogenation of SWCNTs has been the first reaction sequence that is based on the reduction of CNTs with alkali metals in liquid ammonia, the so-called Birch reduction. Pekker *et al.* were able to show that the protonation of Li reduced SWCNTs intermediates with methanol leads to the formation of hydrogenated SWCNTs [171, 172]. In a detailed spectroscopic study, Hirsch and coworkers investigated the electronic-type selectivity of the protonation of the intermediately charged SWCNTs derivatives with ethanol [173]. No pronounced preference for the reaction of metallic and small diameter tubes was observed during the hydrogenation process, which is in stark contrast to comparable alkylation reactions. In addition, the degree of functionalization is considerably lower than that observed for the corresponding alkylations. A novel access route to hydrogenated SWCNTs is provided by the use of high boiling polyamines as hydrogenation reagents. Tománek and coworkers have shown that the polyamine-based hydrogenation is both efficient and thermally reversible [174]. The functionalized tubes disperse well in common organic solvents such as methanol, ethanol, chloroform, and benzene. Mechanistically, they proposed an electron transfer from the polyamine to the carbon nanotube, followed by a proton transfer from the N-H group. In an accompanying theoretical investigation, they identified the preferential hydrogen adsorption geometries at different coverage densities.

6.4.3
Epoxidation of Carbon Nanotubes

Theoretical calculations have suggested that the direct epoxidation of SWCNTs may be accomplished by the reaction with highly reactive dioxiranes [175]. Billups and coworkers demonstrated that the epoxidation of HiPCO SWCNTs can be carried out by the reaction with either trifluorodimethyldioxirane or with 3-chloroperoxybenzoic acid (Scheme 6.3) [52, 176]. Their spectroscopic analysis of the oxidized materials using either acid peroxide or dioxirane is consistent with the formation of epoxides (**22**).

In addition, with $ReMeO_3/H_2O_2/4$-bromopyrazole, a catalytic oxygenation system, an epoxidation of SWCNTs is also possible [177]. Furthermore, the same rhenium catalyst in correspondence with PPh_3 as oxygen acceptor leads to a deoxygenation of sidewall SWCNT epoxy functionalities and therefore a quantification of the epoxide

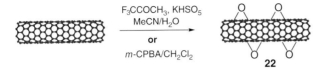

Scheme 6.3 Epoxidation of SWCNTs by reaction with triflourodimethyldioxirane or 3-chloroperoxybenzoic acid (*m*-CPBA).

content [176]. An earlier study by Liu and coworkers also showed that the treatment of purified SWCNTs with peroxytrifluoracetic acid under ultrasonication leads to the introduction of epoxy functionalities among other oxygen-based derivatives [178]. This type of SWCNT derivatization can be extended to other peroxy organic acids, including *m*-chloroperbenzoic acid and 2-bromo-2-methylperpropionic acid [179]. The initially introduced epoxides are subsequently cleaved through an acid-catalyzed pathway yielding SWCNTs bearing hydroxyl and ester functionalities as foundation for an ongoing atom transfer radical polymerization.

6.4.4
[2 + 1]-Cycloaddition Reactions

6.4.4.1 The Addition of Carbenes and Nitrenes
The first functionalization of the carbon nanotube sidewalls with a carbon-based substituent has been reported by the Haddon group in 1998 [180]. Dichlorocarbene generated *in situ* from chloroform with potassium hydroxide or from phenyl (bromodichloromethyl)mercury [88] has covalently been attached to soluble SWCNTs. The electrophilic carbene addition leads to a low functionalization degree of about 2%. This functionalization degree can be increased with soluble alkyl chain-functionalized HiPCO SWCNTs as starting material by up to 23% [181, 182]. The strength of the interband electronic van Hove transitions in this kind of adduct is suppressed as a result of the presence of sp³ hybridized carbon atoms within the sidewalls, based on a conversion of metallic SWCNTs into semiconducting species in the addition step [183]. In an early study, the Hirsch group was able to show that the SWCNT sidewalls could also be attacked by nucleophilic carbenes (Scheme 6.4) [184].

In contrast to the proposed cyclopropane structure in the case of dichlorocarbene, the addition of the nucleophilic dipyridyl imidazolidene leads to a zwitterionic reaction product due to an extraordinary stability of the resultant aromatic 14π perimeter of the imidazole addend. Nitrenes, a related class of subvalent neutral reactive intermediates, could also be successfully added to the CNT sidewalls. Nitrenes can be generated thermally *in situ* from alkoxycarbonyl azide precursors [185]. The driving force for the generation of the highly reactive nitrene

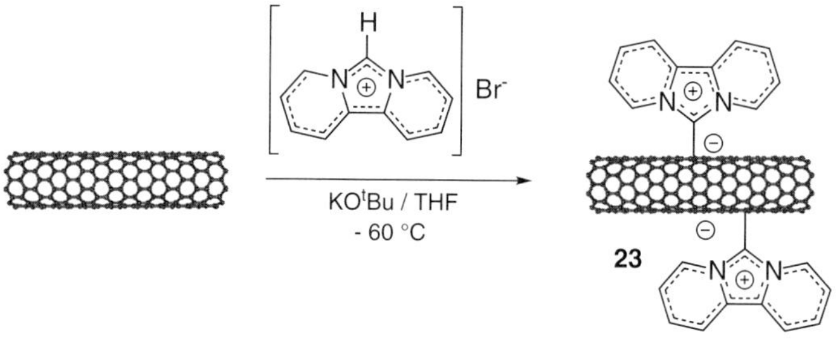

Scheme 6.4 Sidewall functionalization by the addition of nucleophilic carbenes.

intermediates is the thermally induced N_2 extrusion. As alkoxycarbonyl azide with different functional moieties are easily accessible, this SWCNT sidewall functionalization sequence is a versatile tool leading to a broad variety of SWCNT adducts with different functionalities, such as alkyl chains, aromatic units, crown ethers, and even dendritic structures. In a similar reaction, Moghaddam *et al.* used photochemically generated nitrene intermediates from a azidothymidine precursor to functionalize the sidewalls of solid support-bound MWCNTs [186]. Alkylazides can also be used as nitrene precursor compounds. In a photochemical-driven reaction sequence, one tip of aligned MWCNTs was functionalized through the reaction with 3'azido-3'deoxythymidine and a perfluorooctyl chain was attached to the opposite CNT end. The asymmetrically modified CNTs with the hydrophilic thymidine derivative on one tube end and the hydrophobic perfluorooctyl moiety on the other could effectively self-assemble at the hydrophilic/hydrophobic interface in a two-phase solvent system. In a similar fashion, Fréchet and coworkers modified the surface properties of a patterned array of a CNT forest by nitrene addition reactions with a variety of perfluoroarylazide precursor compounds (Scheme 6.5) [187]. By altering the append-

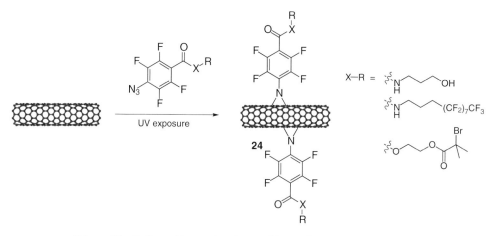

Scheme 6.5 Perfluroazides employed to modify CNT-forests.

ing alkyl chain functionalities (**24a–c**), the surface properties of the CNT forest could by changed from superhydrophobic to superhydrophilic and vice versa.

Yinghuai *et al.* attached azide-functionalized C_2B_{10} carborane cages to the sidewalls of SWCNTs by a thermal-driven nitrene addition reaction [188]. The decapitation of these C_2B_{10} carborane cages with the appended SWCNTs has been accomplished by the reaction with sodium hydroxide in refluxing ethanol. During base reflux, the three-membered ring formed by the nitrene and the SWCNT π-system was opened to produce water-soluble SWCNTs in which the sidewalls were functionalized by both substituted nido-C_2B_9-carborane units and ethoxide moieties. In a similar fashion, SWCNTs can be covalently functionalized by conjugated polyacetylenes through their cyclization reactions with poly(1-phenyl-1-alkynes) and poly-(diphenylacetylene) derivatives carrying azido functional groups at the ends of their alkyl pendants [189]. The resultant polyene nanotube aggregates are soluble in common solvents, emit intensive visible lights, and strongly attenuate the power of harsh laser pulses. The critical question whether carbene and nitrene addition reactions onto CNT sidewalls lead to closed cyclopropane and aziridine structures or if the carbene and nitrene insertion into a C−C bond is associated with an opening of the sidewalls has been addressed in a multitude of theoretical studies [190–197]. According to Bettinger [196], the topology of the cycloaddition adducts depends on the chirality of the CNT and the type of π bond where the addition takes place. In armchair tubes ranging from (3,3) to (12,12), the insertion into the circumferential bond to form a "methano-homonanotube" is always more exothermic than cyclopropanation of the axial bond. Circumferential C−C−bonds in zigzag SWCNTs also undergo insertion of carbenes, while the axial C−C bonds undergo cyclopropanation. The energetic preference of the reaction with circumferential bonds is less pronounced for metallic zigzag tubes than for semi-conducting ones. In addition, for a given diameter, the addition to the circumferential bond of armchair tubes should be more exothermic than to any other bond in zigzag tubes.

6.4.4.2 Nucleophilic Cyclopropanation: The Bingle Reaction

One of the most frequently used reactions for the functionalization of fullerenes is the so-called Bingel reaction, a nucleophilic cyclopropanation of one double bond of the fullerene framework by the reaction of a bromomalonate in the presence of a base [1]. Green and coworkers transferred this derivatization method to SWCNTs [198]. The reaction of diethyl bromomalonate with SWCNTs leads to the sidewall-functionalized material that was subsequently transformed into a thiolated derivative. By "tagging" the cyclopropane moiety with gold nanoparticles and visualizing the adduct by AFM, the authors were able to prove the covalent attachment of the malonate addend. To further confirm the derivatization of SWCNTs, a perfluorinated marker was introduced by transesterification making the adduct accessible to ^{19}F NMR and XPS. In a modified functionalization pathway, Ashcroft *et al.* used negatively charged ultrashort SWCNTs as starting material obtained by potassium reduction. Under these strongly reducing conditions, a malonate reacted in the presence of CBr_4 and DBU with the shortened CNTs, yielding Bingel-type sidewall functionalized material in 33% yield [199]. That the solubility of the CNT starting material is of utmost importance for a successful Bingel reaction has been proven by the Imahori group [200]. Shortened SWCNTs with alkyl chains at open ends and defect sites (25) have been functionalized in a microwave-assisted Bingel reaction (Scheme 6.6). The reaction time can be significantly reduced by microwave irradiation and it is worth noting that the electronic properties (distinct van Hove transitions in the UV/Vis region) are largely retained in the sidewall-functionalized material (26) exhibiting functionalization degrees of one diester unit per 75–300 carbon atoms of the SWCNT.

This is in marked contrast to the observation of conventional sidewall functionalization sequences of SWCNTs where a functionalization degree of one functional group per 10–100 carbon atoms on the SWCNT sidewall leads to disappearance of the van Hove transitions. No apparent selective reactivity for metallic and semiconducting SWCNTs could be observed by the nucleophilic cyclopropanation under these conditions.

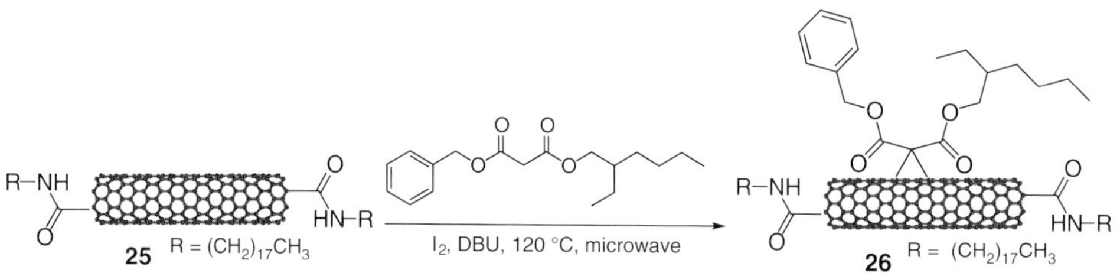

Scheme 6.6 Microwave-assisted Bingel reaction of soluble SWCNT derivatives.

6.4.4.3 Silylation of Carbon Nanotubes

The [2 + 1] cycloaddition of silylene has been theoretically investigated and it has been predicted that the sidewall addition of silylenes should be site selective and diameter specific [190, 201], favoring open structures [202]. Soon after, Wong and coworkers reported on the successful sidewall functionalization of SWCNTs by the reaction with trimethoxysilane and hexaphenyldisilane under UV irritation [203]. The Raman analysis demonstrates a selective reactivity of smaller diameter semiconducting nanotubes toward the addition of the photochemical-generated silylenes. It has been shown that the direct covalent sidewall functionalization is accompanied by an introduction of oxygen functionalities to the CNT ends and at present defect sites.

6.4.5
1,3-Cycloaddition Reactions

6.4.5.1 Cycloaddition of Zwitterionic Intermediates

Recently, a versatile and highly efficient cycloaddition approach both for the exohedral derivatization of C_{60} and for the sidewall functionalization of SWCNTs, based on

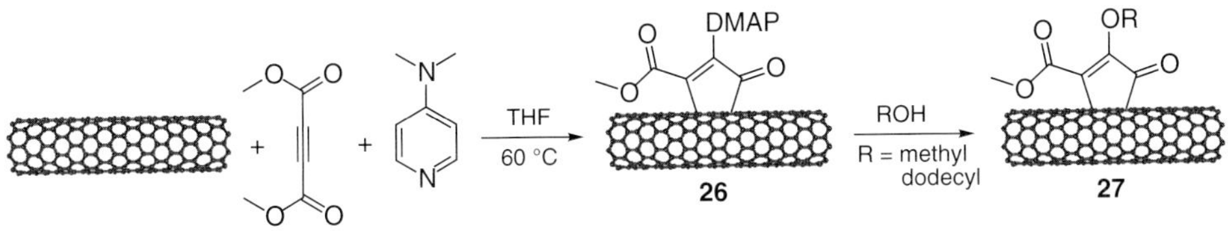

Scheme 6.7 Modular approach toward sidewall functionalized SWCNTs by 1,3-cycloaddition.

zwitterionic intermediates, has been reported (Scheme 6.7) [204]. In a modular reaction sequence, the zwitterionic intermediate resulting from the attack of 4-dimethylaminopyridine (DMAP), a very potent nucleophile, on a dimethyl acetylenedicarboxylate (DMAD) is formed in the first step. The addition to a double bond of the π-system is followed by an elimination of an alkoholate yielding a SWCNT derivative with a positively charged DMAP moiety (**26**).

In the last step, this DMAP moiety is replaced with other nucleophiles, methanol or 1-dodecanol, leading to the corresponding DMAD–SWCNT derivative (**27**). This reaction sequence yields very soluble, highly sidewall-functionalized material with one functional group per nine carbon atoms.

6.4.5.2 Azomethine Ylide Addition

One of the most commonly used sidewall derivatization reactions is the 1,3-dipolar cycloaddition of azomethine ylides, generated by the condensation of an α-amino acid and an aldehyde [205]. This reaction was originally developed for the exohedral modification of C_{60} and can be easily transferred to CNT sidewall derivatization. The thermal treatment of SWCNTs in DMF for 5 days in the presence of an aldehyde and α-amino acid derivative results in the formation of substituted pyrrolidine moieties on the SWCNT surface, well soluble in common organic solvents. In order to reduce the reaction time to the order of minutes, an alternative microwave-induced reaction sequence can be used (Scheme 6.8) [89].

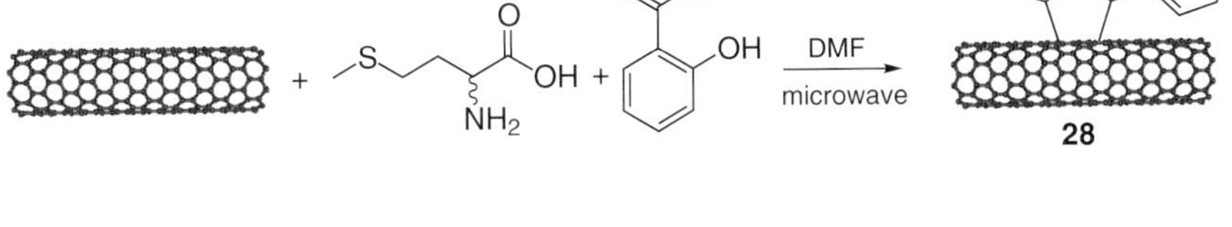

Scheme 6.8 Microwave-assisted 1,3-dipolar cycloaddition of azomethine ylides.

This procedure can also be extended to the functionalization of soluble, defect-functionalized MWCNTs. Starting from carboxylated MWCNTs equipped with C_{16} ester chains, the reaction of 4-octadecyloxybenzaldehyde with a suitable α-amino acid under microwave irradiation yields sidewall-functionalized material with the same functionalization degree as with the thermal reaction pathway but 50–60 times faster [206]. In addition, the amount of addends can be controlled by altering the reaction time and temperature, and the added groups are evenly distributed over the MWCNTs even in the case of lower degrees of functionalization. Aziridines are well-known precursors to azomethine ylides. In a solvent-free reaction sequence, aziridine can be reacted with HiPCO SWCNTs or MWCNTs under microwave irradiation yielding the corresponding functionalized material [207]. For MWCNTs, the 1,3-dipolar cycloaddition is a reversible process, as the thermal treatment of the functionalized MWCNT material in the presence of an excess of C_{60} in o-dichlorobenzene with $CuTf_2$ as catalyst leads to retro-cycloaddition, where the intermediate ylide is trapped by the C_{60}. Microwave-driven reaction sequences can also be used for the generation of multiple functionalized SWCNT derivatives, where the first substituent is introduced by 1,3-dipolar cycloaddition and the second one by arene diazonium chemistry [208]. The results show that the radical arene addition saturates more reactive sites than the cycloaddition. Therefore, for attaching two different functional groups to SWCNTs, the order of introduction is crucial.

The azomethine-based functionalization sequence has become a valuable tool for CNT sidewall functionalization due to its versatility. In principle, this method allows the attachment of any moiety to the carbon framework of the CNTs by a simple variation of the chemical structure of the adducts. Therefore, the use of CNT derivatives for applications in the field of medicinal chemistry [209–211], solar energy conversion [123, 211], and recognition of chemical species has been extensively investigated. In order to study the electronic communication between CNTs and redox-active chromophores, a broad variety of donor units have been covalently linked to the CNT sidewalls. Therefore, soluble carbon nanotube ensembles (**29**) (Figure 6.10) with ferrocene units covalently introduced by 1,3-dipolar cycloaddition have been studied for light-induced electron transfer interactions [212]. It has been found that the introduction of pyrrolidine moieties onto the π-system has a dramatic influence on the electronic properties of the resulting assembly. For each pyrrolidine moiety (and therefore for each ferrocene unit) that is connected to the SWCNT sidewall, two electrons are removed from the delocalized π-system. As a consequence, in an efficient D/A-aggregate, the degree of sidewall damage should be kept low whereas the amount of redox-active moieties should be as high as possible. This can be realized in an approach where polyamidoamine dendrimer units are attached to the SWCNT sidewall by azomethine ylide chemistry and subsequently a variety of tetraphenylprophyrin moieties are covalently attached to the periphery of the dendrimers, allowing the investigation of the photophysical properties of the resulting nanoconjugate (**30**) [213] (Figure 6.10).

As a result, an intraconjugate charge separation from one part of the photoexcited porphyrins to the SWCNT core can be observed, while the other part of the porphyrin units do not interact with the nanotube, thus exhibiting a fluorescence lifetime

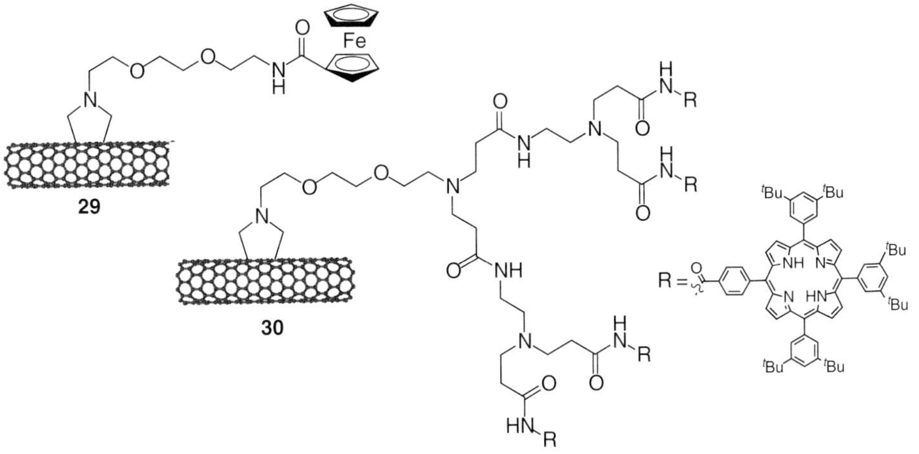

Figure 6.10 Examples of redox-active SWCNT derivatives through sidewall addition of azomethine ylides.

similar to that of the free porphyrins. Applying the same concept, a second-generation cyanophenyl-based dendrimer was coupled to MWCNTs and SWCNTs [214]. In a primary modification step, the CNTs were functionalized by a 1,3 dipolar cycloaddition reaction to install a carboxylic acid-bearing anchor unit. Subsequently, the dendrimer was attached by DCC coupling yielding a high density of dendritic anchor points with a minor damage to the conjugated π-system. Recently, phthalocyanines were coupled to HiPCO SWCNTs through a sidewall-bound pyrrolidine linker [215]. A significant quenching of the fluorescence of the phthalocyanine is observed due to the electronic interaction between the photoexcited redox chromophore and the SWCNT framework. In another study, phthalocyanines were coupled to HiPCO SWCNTs by two alternative pathways. In the stepwise approach, first an anchor unit was installed by 1,3-dipolar cycloaddition and the corresponding phthalocyanine derivative was coupled afterward (1 functional unit per 380 C atoms), whereas in the direct functionalization sequence, a formyl-bearing phthalocyanine system was coupled directly to a slightly lower degree of functionalization (1 functional unit per 580 C atoms) [216]. Again, an electron transfer from the photoexcited donor moiety to the accepting SWCNT unit can be observed. The physical properties and chemical functionality of the covalently modified CNT conjugates can be tuned by the type of substituent attached. Superhydrophobic MWCNTs can be derived in a three-step procedure that comprises oxidation, 1,3-cycloaddition, and silylation with a perfluoroalkylsilane [217]. Herein, 3,4-dihydroxybenzaldehyde, as starting material, yields the required silane anchor unit. The same anchor unit can be used for the combination of functionalized CNTs with polymers and layered aluminosilicate clay minerals yielding homogeneous, coherent, and transparent CNT thin films and gels [218]. The reaction of SWCNTs with formaldehyde and N-(4-hydroxy)phenyl glycine yields sidewall-functionalized material with covalently attached phenol groups [219]. These SWCNT derivatives exhibit higher zeta potential values showing their increased dispersant ability in water and acetone in comparison to the pristine material. Sidewall-derivatized, water-soluble SWCNTs with low functionalization degrees can be aligned in weak magnetic fields [220]. Herein, water solubility is based on free amine functionalities generated by deprotection of Boc-protected pyrrolidine derivatives. Recently, Prato and coworkers synthesized covalently functionalized MWCNT derivatives via 1,3-dipolar cycloaddition of azomethine ylides with orthogonally protected amino functions that can be selectively deprotected and subsequently modified with an anticancer drug and fluorescent probes [221]. Furthermore, the fluorescence properties of water-soluble, sidewall amino-terminated CNTs obtained by the azomethine ylide functionalization route can be used for studying aggregation phenomena with macromolecules, such as surfactants, polymers, and nucleic acids [222]. Recently, Ménard-Moyon *et al.* investigated an alternative access route to azomethine ylide intermediates and were able to show that this type of sidewall functionalization is selective in terms of the separation of semiconducting SWCNTs from metallic CNTs [223] (Scheme 6.9).

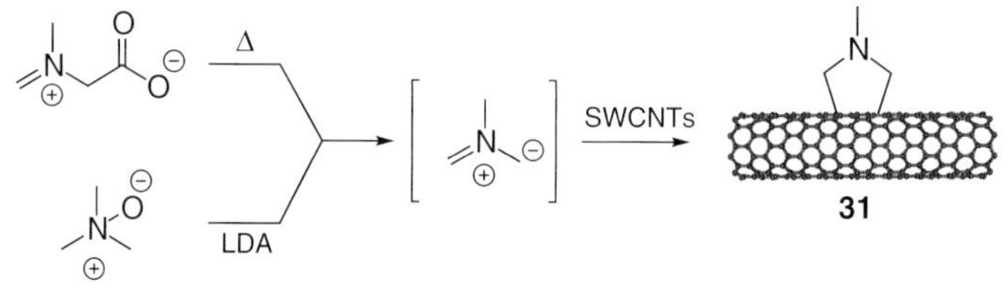

Scheme 6.9 1,3-Dipolar cycloaddition with azomethine ylides generated under mild conditions.

This mild reaction sequence is based on trialkylamine-*N*-oxides and leads to a preferred functionalization and therefore solubilization of semiconducting SWCNTs.

6.4.6
[4 + 2]-Cycloaddition Reactions: Diels–Alder Reaction

The Diels–Alder cycloadditions of *o*-quinodimethane to the sidewalls of SWCNTs has been theoretically predicted to be viable due to the aromatic stabilization of the transition state and reaction product [224, 225]. In this reaction sequence, the SWCNT double bonds should act as dienophiles. This has experimentally been verified by a microwave-assisted reaction of soluble SWCNTs and 4,5-benzo-1,2-oxathiin-2-oxide (*o*-quinodimethane precursor) [226]. Given that the rate of Diels–Alder addition reactions is enhanced by the use of electron withdrawing substituents on the monoene (the "dienophile"), fluorinated SWCNTs have been used as starting materials [227]. These SWCNTs with activated sidewalls have been thermally reacted with 2,3-dimethyl-1,3-butadiene, anthracene, and 2-trimethylsiloxyl-1,3-butadiene yielding highly functionalized SWCNT derivatives. XPS and IR investigations have shown that the addition of the butadiene unit is associated with a significant reduction in the fluorine content – detachment of about seven fluorine atoms per Diels–Alder addition substituent. For the Diels–Alder derivatization of pristine SWCNTs, without any covalent modification for the activation or solubilization of the nanotubes, transition metal catalysts in correspondence with high pressure can be used [228]. The reaction of HiPCO SWCNTs with $Cr(CO)_6$ in 1,4-dioxane leads to the formation of a chromium–SWCNT complex that subsequently is added to a solution of 2,3-dimethoxy-1,3-butadiene in THF. After a reaction time of 60 h under 1.3 GPa pressure, the reaction mixture was irradiated with a visible light to induce the decomplexation of chromium. This reaction sequence can be extended to other electron-rich dienes and yields sidewall-derivatized material bearing 1 functional group per 33 sidewall carbon atoms. In contrast, the reaction of 4-phenyl-1,2,4-triazoline-3,5-dione, a very strong electron acceptor and powerful dienophile, with SWCNTs under thermal reflux, yields no sidewall addition product, but the corresponding deaza dimer of the triazolinedione could be detected [229]. Presumably, the carbon nanotubes initiate the self-condensation of the triazolinedione by acting as an electron donation source. The Diels–Alder derivatization of CNTs can be extended to MWCNTs. The reaction of MWCNTs with maleic anhydride in the molten state and in solution results in the formation of well dispersible CNT derivatives [230]. MWCNTs can also be functionalized by the use of various substituted benzocyclobutenes yielding the corresponding MWCNT derivatives (**32**) (Scheme 6.10) [231]. Herein, the strained ring of the benzocyclobutene is susceptible to thermal ring opening to form intermediate *o*-quionodimethanes.

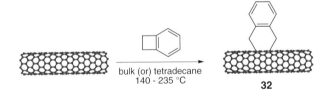

bulk (or) tetradecane
140 - 235 °C

32

Scheme 6.10 Diels–Alder cycloaddition by the use of benzocyclobutenes.

The extent of functionalization can be controlled by using an appropriate temperature and mode of addition of the reagent and also by controlling the reaction duration. The reaction of MWCNTs with 3,6-diaminotetrazine under heating leads to the formation of different reaction products depending on the reaction time [232]. In the first step, π–π interaction and cycloaddition take place. The second step is the junction formation, which occurs after 24–36 h. The final stage is a partial destruction of the MWCNTs, which is very pronounced after 48 h of heating as a result of excessive functionalization.

6.4.7
Alkali Metal-Based Reduction of Carbon Nanotubes with Subsequent Sidewall Functionalization

6.4.7.1 Naphthalenides as Electron Transfer Reagents

Owing to their unique electronic properties, CNTs are very potent electron acceptors. Therefore, SWCNT bundles can be readily exfoliated in standard organic solvents by reduction with an alkali metal, lithium, or sodium, mediated by lithium or sodium naphthalenide [233]. The alkali metal intercalation and the Coulomb repulsion of the anionic SWCNTs leads to a very efficient debundling of the CNTs and therefore to an individualization. Elemental analysis yields a chemical formula of $A(THF)C_{10}$ (A = Li, Na), hence, 1 negative charge per 10 carbon atoms. The fact that lithium or sodium salts of SWCNTs spontaneously dissolve, that is, with no supplied energy, and that the solutions are stable confirms that they can thermodynamically be considered as solutions. The reduced CNTs can be reversibly oxidized by suitable oxidizing agents. This oxidation step of the nanotube salt solutions has been monitored *in situ* by Raman spectroscopy upon exposure to air and by addition of benzoquinone as electron acceptor [234]. The results indicate a direct charge transfer reaction from the acceptor molecule to the SWCNT, leading to their gradual charge neutralization followed by subsequent precipitation of the CNT material.

6.4.7.2 Reductive Alkylation of CNTs

Negatively charged SWCNT anions are widely used for subsequent sidewall CNT derivatization [235]. The reaction of reduced SWCNTs (**33**) with diacyl peroxides at room temperature in THF yields sidewall-alkylated SWCNT derivatives (**34**) with a wide variety of functionalities (Scheme 6.11) [236]. The degree of functionalization can be increased and hence controlled by repeating the redox process by reducing the previously functionalized SWCNTs and resubjecting them to another reaction cycle. This repetitive process also affords the possibility of anchoring functional groups of different types, one type per cycle, to produce multifunctional SWCNT derivatives.

The addition of alkyl halides to the THF-dissolved SWCNT salts yields sidewall-alkylated material, which has been extensively investigated by TGA/MS [237].

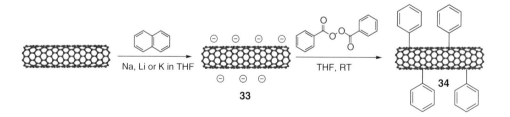

Scheme 6.11   Reductive alkylation of negatively charged SWCNT intermediates by reaction with diacyl peroxides.

Negative charged CNT intermediates can also be generated under classical Birch conditions. Herein, the reaction of SWCNTs with alkali metals such as lithium or sodium in liquid ammonia leads to the reduction of the carbon nanotubes and the generation of soluble CNT anions. In their seminal work, Billups and coworkers reacted a broad variety of alkylhalides, for example, dodecyl iodide, with reduced SWCNTs in liquid ammonia [238]. Again, a Raman spectroscopic investigation of the reduction process has shown that the charge transfer-induced lithiation of SWCNTs leads both to an intercalation of lithium into the SWCNT ropes and to the semi-covalent lithiation of the SWCNTs [239]. Afterward, the addition of the alkyl halides leads to the formation of radical anions that dissociate readily to yield halide and the corresponding alkyl radicals, which subsequently attack the SWCNT sidewall yielding well soluble, alkylated CNT derivatives [240]. The assumption of a radical pathway is supported by GC–MS analysis of the side products, where radical dimerization and disproportionation products were detected. This functionalization protocol can be easily expanded to aryl iodides yielding arylated SWCNT derivatives [172, 241]. A detailed thermal gravimetric analysis in combination with ^{13}C NMR spectroscopy of dodecylated SWCNTs has shown that the alkali metal used for the CNT reduction, for example, lithium, sodium, and potassium, has an influence on the structure of the alkylated CNT derivatives [242]. The observed differences in the thermal stability may be explained by different relative amounts of 1,2- and 1,4-addition patterns of the introduced alkyl groups. An alternative approach toward alkylated and arylated SWCNT derivatives is based on a high-speed vibrational mechanochemical milling technique. Herein, SWCNTs, alkyl halides, and potassium are vigorously shaken in a steal capsule for 40 min under nitrogen atmosphere at room temperature yielding the corresponding sidewall-functionalized SWCNT derivatives. MWCNTs are generally less reactive toward sidewall functionalization; nevertheless, reductive alkylation and arylation under Birch conditions yield highly functionalized MWCNT derivatives, which exhibit a pronounced solubility in organic solvents and even in water [243]. Water-soluble alkylated SWCNT derivatives are accessible by treating SWCNT with lithium in liquid ammonia with ϖ-bromocarboxylic acid and subsequent coupling of amine-terminated polyethyleneglycol chains to the introduced SWCNT sidewall carboxylic acid functionalities [244, 245]. This kind of sidewall carboxylated SWCNT derivative has also been used for preparing a nylon 6,10 nanocomposite material [246]. Therefore, the polycondensation between hexamethylenediamine and sebacoyl chloride has been carried out in the presence of the corresponding acyl chloride SWCNT derivative yielding a matrix-coupled SWCNT–polymer composite with improved tensile modulus, strength, and toughness. Similarly, the intermediately generated SWCNT anions can also serve as initiators for the *in situ* polymerization of methylmethacrylate [247]. The poly(methylmethacrylate)-grafted SWCNTs composite material displays a pronounced solubility in organic solvents and exhibits mostly individualized SWCNTs embedded in the polymer matrix. Recently, it has been shown that sulfur, nitrogen, and oxygen-centered radicals can also be attached to SWCNT sidewalls via the reductive pathway. The treatment of lithium-reduced SWCNTs in liquid ammonia with alkyl disulfide yields a transient radical anion by a single electron transfer process, which dissociates into a mercaptide and a sulfur-

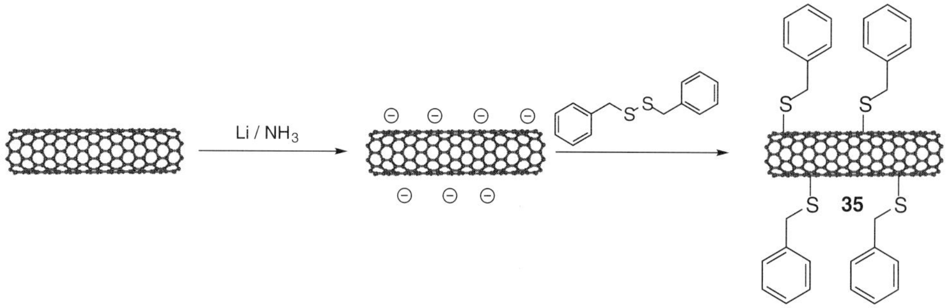

Scheme 6.12 Reductive alkylation of negatively charged SWCNT intermediates by disulfides.

centered radical that subsequently attacks the SWCNT sidewall (Scheme 6.12) [240]. When alkyl sulfides or aryl sulfides are used instead of disulfides, the intermediately generated radical anion is decomposed into a mercaptide and a carbon-centered radical yielding alkylated and arylated derivatives (**35**), respectively.

Similarly, *N*-halosuccinimides yield the corresponding succinimidyl radicals and halide anions and alkyl peroxides give rise to alkoxy radicals and alkoxides [248]. The free radicals add to the sidewalls of nanotube salts to form succinimidyl and alkoxy-derivatized SWCNTs. The functionalized SWCNTs can then be converted to amino and hydroxyl SWCNTs by hydrolysis with hydrazine or with fuming sulfuric acid, respectively. A selectivity investigation of the covalent sidewall functionalization of SWCNTs by reductive alkylation with butyl iodide reveals a pronounced SWCNT diameter dependence, with higher reactivity for smaller SWCNTs [173]. Moreover, this reaction sequence also favors the preferred functionalization of metallic over semiconducting SWCNTs.

6.4.7.3 Other Electron Transfer Mediators

As outline, lithium naphthalenides are commonly used electron transfer reagents for the reduction of CNTs in THF. Another access route to soluble CNT anions is provided via their reduction by Na/Hg amalgam in the presence of dibenzo-18-crown-6 [249]. The solubility of the negatively charged SWCNT intermediates is greatest in CH_2Cl_2 and DMD and the corresponding [Na(dibenzo-18-crown-6)]$_n$[SWCNT] complex has extensively been characterized by means of UV/Vis spectroscopy, Raman spectroscopy, AFM, and MALDI–MS. In an analogy to the well-established functionalization pathway, the reaction with alkyl halides yields alkylated SWCNT derivatives with an SWCNT:alkyl ratio of about 40:1, which is significantly lower than that observed by the analogous reaction of alkyl halides with reduced SWCNTs in liquid ammonia. In a recent study, it has been shown that lithium metal can be used to reduce SWCNTs to carbide-like species under the catalytic effect of di-*tert*-butyl-biphenyl (DTBP) [250]. The resulting nanotube poly-carbanions show a significantly increased dispersibility in THF and react *in situ* with trimethylchlorsilane, methylmethacrylate, and methyl-*N*-acetamidoacrylate to afford covalently functionalized silyl nanotubes and polymer-wrapped SWCNT derivatives. The density and average length of the polymer chains attached to the nanotube walls can be widely modified by varying the amount of monomer, lithium, and catalyst.

6.4.8
Sidewall Functionalization of Carbon Nanotubes Based on Radical Chemistry

6.4.8.1 Carbon-Centered Free Radicals
In general, reactive radical species can be generated from a variety of precursors either thermally or by photophysical routes. The experimental proof for a covalent sidewall functionalization of SWCNTs by a radical species was given 1 year after the theoretical prediction. The addition of perfluorinated alkyl radicals, obtained by photoinduced cleavage of the carbon–iodine bond from heptadecafluorooctyl iodide onto SWCNTs yielded soluble, perfluorooctyl-derivatized CNTs [184]. Alkyl- or aryl peroxides are commonly used radical initiators. It has been shown that they are suitable candidates for the sidewall functionalization of SWCNTs. The thermal cleavage of benzoyl and lauroyl peroxides leads to the generation of highly reactive phenyl and alkyl radicals that efficiently attack the SWCNTs, resulting in sidewall-derivatized materials that have been characterized by a broad variety of spectroscopic methods [251, 252]. The phenylated SWCNTs can be subsequently sulfonated in a scalable way yielding highly exfoliated water-soluble SWCNT derivatives [253]. The combination of the benzoyl peroxide as radical starter with alkyl iodides allows the covalent attachment of a large variety of different functionalities to single-walled carbon nanotube sidewalls [254]. This concept can also be used for grafting polymers onto the sidewall of MWCNTs. Treating poly(oxyalkylene)amines, equipped with a terminal maleic anhydride, with benzoyl peroxide leads to the intermediate formation of a free radical species that is grafted onto the sidewall of the dispersed MWCNTs [255]. The thermal treatment of dibenzoyl peroxide in the presence of methanol and CNTs leads to sidewall-derivatized material exhibiting hydroxyalkyl groups through a radical addition scheme [256, 257]. Herein, the peroxide-assisted hydrogen abstraction on solvent molecules produces various active species that covalently attach to the surface of the CNTs. By reacting SWCNTs with succinic or glutaric acid acyl peroxides, this concept of radical-based functionalization can be extended to the generation of highly functional SWCNT building blocks [258]. In addition, this procedure can be modified in order to open a fast and intrinsically scalable functionalization pathway of SWCNTs with carboxylic acid moieties [259]. Herein, by grinding SWCNTs and succinic acyl peroxide and heating the corresponding powder at 100 °C for about 4 min, sidewall carboxylic acid-functionalized SWCNT material with yield in excess of 90% is produced. The amount of functionalization can be easily adjusted by repeating this solvent-free derivatization procedure. In a detailed study, Engel *et al.* investigated the observed rate acceleration of peroxide thermolysis caused by SWCNT and attributed this phenomenon to an electron transfer-induced decomposition reaction with the SWCNT acting as the electron donation source [260]. SWCNTs reduce peroxides to radical anions, which immediately undergo O–O bond scission. Following decarboxylation, the radicals rapidly react with SWCNT$^+$ intermediates leading to functionalization. Compared to

organic peroxide radical sources, the azobis-type radical initiators contain a terminal carboxylic, hydroxylic, amideine, nitrile, or imidazoline group that can be quite general and extendable to react with CNTs. The thermal decomposition of azobisisobutyronitrile leads to the intermediate formation of a tertiary radical, which can undergo addition reactions to the sidewalls of SWCNTs [261, 262]. In a recent study, Su and coworkers demonstrated that this type of free radical-based sidewall addition can be extended to double-walled carbon nanotubes [263]. The thermal or ultrasonic decomposition of a broad variety of azobis compounds leads to sidewall-derivatized CNTs exhibiting heterofunctionalities such as carboxylic acid and hydroxyl moieties. These functional groups on the sidewalls of DWCNTs can serve as reactive sites to covalently couple DWCNTs with other materials, such as Pt/Ru nanoparticles. Liu *et al.* used 4-methoxyphenylhydrazine as a radical precursor for the free radical functionalization of HiPCO SWCNTs (Scheme 6.13) [264]. The hydrazine derivative is converted into a phenyl radical in the presence of oxygen either in a classical thermal reaction or in a microwave-assisted reaction yielding SWCNT sidewall-derivatized species (**36**), exhibiting a functionalization degree of 1 functional group per 27 carbon atoms [265].

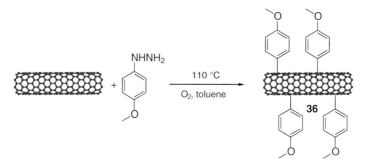

Scheme 6.13   Free radical functionalization of SWCNTs through 4-methoxyphenylhydrazine.

With a similar yield of functionalization, the microwave-assisted reaction is far more rapid with an optimal time of around 5 min in contrast to the thermal route that is often measured in days. Anionic carbon-based radicals are formed by the one-electron reduction of benzophenone in the presence of potassium. These radical anions readily add to the sidewalls of SWCNTs yielding diphenylcarbinol-derivatized material with a functionalization degree of 1 functional unit per 120 carbon atoms [266]. Taking the ketone-based functionalization concept further, the same research group has functionalized SWCNTs via photoreduction of aromatic ketones by alcohols [267]. The irradiation of benzophenone, benzhydrol, and SWCNTs in benzene leads to a covalent attachment of benzhydrol to the sidewalls of the SWCNTs. The intermediate radical is formed via hydrogen abstraction from a C—H bond by the photoexcited benzophenone in the triplet state. Since excited benzophenone is reactive to many types of C—H bonds such as alcohols, amines, ethers, hydrocarbons, or aliphatic sulfides, a broad variety of radicals can be generated using this photochemical method. Another nonthermal radical formation process with subsequent sidewall functionalization is ball milling CNTs in the presence of alkyl halides [268]. Herein, trifluoromethane, trichlormethane, tetrachloroethylene, and hexafluoropropene can be used for a large-scale production of functionalized short SWCNTs.

Radical-based functionalization sequences can be used to attach well-defined polymers onto carbon nanotubes leading to sidewall derivatized composite materials. Polystyrene-based homopolymers and copolymers can be grafted to SWCNTs through a radical coupling reaction involving polymer-centered radicals via the loss of stable free radical nitroxide capping agents [269]. The resulting polymer–SWCNT composites were found to be highly soluble in a variety of organic solvents. In another approach, well-defined poly(t-butyl acrylate) were prepared by atom transfer radical polymerization, where radicals are formed at the end of the polymer chain through an atom transfer [270]. The generated radicals at the chain ends added to MWCNTs to generate covalently functionalized polymer–MWCNTs composite materials.

6.4.8.2 Sulfur-Centered Free Radicals
The photolysis of cyclic disulfides in the presence of SWCNTs leads to the sidewall addition of the intermediately formed sulfur radical [271, 272]. The introduction of the sulfur-containing substituents has been confirmed by a broad variety of spectroscopic methods. It has been shown that the degree of functionalization is proportional to the irradiation time. Furthermore, the comparison of spectra between pristine and modified SWCNTs indicate that sidewall functionalization of SWCNTs with sulfur-containing substituents maintains the original electronic structure of the SWCNTs without heavy modifications. In addition, the free thiol group of the substituent can be attached to gold nanoparticles and the corresponding SWCNT-nanoparticle assembly has been characterized by microscopic methods.

6.4.8.3 Oxygen-Centered Free Radicals
Hydrogen peroxide can be used as a starting reagent for generating hydroxyl radicals. Li *et al.* treated MWCNTs with Fenton reagent (Fe^{2+}/H_2O_2) and systematically investigated the effect of hydroxyl radicals on the structure of MWCNTs with the aid of infrared and Raman spectroscopy [273]. The results revealed that oxygen-containing functional groups such as OH, COOH, and quinone groups could be introduced into the sidewall structure of MWCNTs. An investigation of the addition mechanism of the intermediately generated hydroxyl radicals has shown that the introduction of oxygen-containing groups can be regarded as the result of an attack of the hydroxyl radicals both on present defect sites and on double bonds of the CNT framework. In a fluorescence spectroscopic study by Iijima and coworkers, the spectral changes of SWCNTs through light-assisted oxidation with hydroperoxide has been investigated [32]. Herein, they observed that light irradiation accelerates the oxidation of SWCNTs through intermediately generated hydroxyl radicals. Furthermore, based on a wavelength dependence of the oxidation step, they have postulated that the absorption of the SWCNT fluorescence by H_2O_2 is the main factor that causes the photodissociation of H_2O_2 and the generation of hydroxyl radicals. The effect of

light-assisted oxidation depends on the SWCNT diameter, where larger diameter systems are consumed faster than SWCNTs with moderate diameters.

6.4.8.4 Amination of Carbon Nanotube Sidewalls

In a solvent-free approach, Basiuk *et al.* reported on a direct sidewall addition of alkylamines to closed cap MWCNTs [274, 275]. The characterization of the functionalized material and theoretical calculations lead to the assumption that the addition takes place only on five-member rings of the graphitic network of the nanotubes and present topological defects, whereas benzene rings are inert to the direct amination. Shortly thereafter, another group reported on the successful amino-functionalization of SWCNTs by the reaction with aromatic hydrazines [276]. The functionalized material exhibits a pronounced solubility in a variety of organic solvents, and a direct dependence of the functionalization degree with the type of substituent attached in *para* position to the hydrazine unit has been observed. Electron-donating substituents yield higher functionalization degrees than electron withdrawing entities. Recently, the reaction of 2-aminoethanol in DMF with SWCNTs has been used in order to install surface-bound hydroxyl groups for the further attachment of oligoquinolines and poly(3-octylthiophenes) [277, 278].

6.4.8.5 Diazonium-Based Functionalization Sequences

In 2001, Tour and coworkers discovered a convenient route to sidewall-functionalized carbon nanotubes by reacting aryl diazonium salts with SWCNTs in an electrochemical reaction, using a bucky paper as working electrode [279]. By varying the aryl diazonium salt, a broad variety of highly functionalized and well-soluble SWCNT adducts became accessible. The corresponding reactive aryl radicals are generated from the diazonium salt via one-electron reduction. Soon after, this concept was extended to a solvent-based thermally induced reaction with *in situ* generation of the diazonium compound by reacting aniline derivatives with isoamyl nitrite (Scheme 6.14) [280].

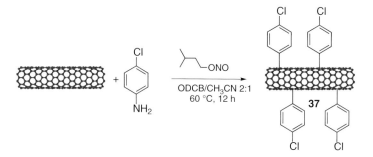

Scheme 6.14 Solvent-based sidewall arylation of SWCNTs.

These basic studies of diazonium-based SWCNT derivatization led shortly thereafter to a very efficient functionalization sequence of individualized carbon nanotubes [281]. Herein, using water-soluble diazonium compounds in combination with

SDS-coated individual SWCNTs yields highly functionalized nanotubes, with up to 1 in 9 carbons along their backbones bearing an organic moiety.

The diazonium-based functionalization of SWCNTs is a very versatile functionalization method, as it is even possible to conduct the reaction in the total absence of any solvent [282]. Another access route was elaborated soon after by Tour and coworkers [283]. SWCNTs have been exfoliated and functionalized predominantly as individuals by grinding them for minutes at room temperature with aryldiazonium salts in the presence of ionic liquids and potassium carbonate. This constitutes an extremely rapid and mild green chemical functionalization process. This environment friendly functionalization can even be extended to water as solvent [284]. Herein, SWCNTs are exfoliated and functionalized into small bundles and individuals by vigorous stirring "on water" in the presence of a substituted aniline and an oxidizing agent. The range of suitable solvents for this type of CNT functionalization reaction can even be extended to Caro's acid, a mixture of 96% sulfuric acid and ammonium persulfate [285]. This mixture can be used for the substitution of the solvent oleum [286, 287], which leads to a sulfonation of the appended aromatic SWCNT substituents in a side reaction. Recently, it has been shown that the *in situ* generation of aryl diazonium salts in the presence of ammonium persulfate provides a scalable and solvent-free functionalization route for MWCNTs leading to highly functionalized and dispersible MWCNT derivatives [288]. In addition, organic triazene compounds are stable diazonium salt precursors for the functionalization of SWCNTs in aqueous media. This method is particularly useful when a target molecule has functionality that will not tolerate the usual diazotation conditions, for example, biologically relevant biotin moieties [98, 289]. As with the addition of azomethine ylides, the reaction of CNT with diazonium precursors can be accelerated by microwave assistance [208, 290]. With *o*-dichlorobenzene as solvent, the reaction of SWCNTs with *p*-chloroaniline and isoamyl nitrite under microwave assistance leads to shorter reaction times, and sidewall functionalization degrees in the range of nearly 50% can be achieved [291]. Subsequent transformation of the chloro substituent into a thiol functionality leads to DMSO soluble material, enabling the acquisition of a COSY ^{1}H NMR spectrum of a covalently sidewall-functionalized SWCNT derivative for the first time.

Meanwhile, owing to its versatility, the diazonium-based functionalization has become the most frequently used reaction to access functional sidewall CNT derivatives, expanding the diversity of functionalized SWCNTs available to chemists and material scientists. In a diazonium salt-based strategy, a covalently attached multilayer of diphenyl disulfide has been introduced on the surface of SWCNTs and MWCNTs [292]. The interlayer S−S bonds are subsequently reductively degraded to essentially produce a single layer of thiophenols. The functionalization has been achieved in what is effectively a one-pot procedure since the involved transformations are performed without intermediate workup. This "formation-degradation" modification approach is unique in the sense that it allows the generation of a thin well-defined molecular layer in spite of involving higher reactive radicals. The direct implementation of heteroatom functionality on the CNT sidewalls is also possible by the addition of 1,4-benzenediamine in a single-step diazonium reaction yielding phenyl amine-attached SWCNT derivatives suitable for more chemical transformations [293]. Heteroatom-bearing moieties covalently linked to the CNT sidewalls by the diazonium route can further be used for the attachment of nanoparticles and therefore for the synthesis of functional CNT assemblies. The covalent attachment of *N-N*-diethylaninile groups, using the corresponding diazonium salt as coupling agent, renders the SWCNTs sensitive toward protonation [294]. Furthermore, gold particles with sizes of a few tens of nanometers deposited onto individual SWCNTs have been found to locally enhance the Raman signals of the underlying nanotube. In a related study, two types of nanoscale hybrid materials have been synthesized by conjugating CdSe–ZnS quantum dots and gold nanoparticles to sidewall-functionalized SWCNT templates and the photoluminescence properties of the hybrid materials and the corresponding precursors have been investigated [295]. Phenyl amine-derivatized SWCNTs with a very high functionalization degree have been synthesized in a multistep procedure and used for the coupling of two types of quantum dots. Diazonium-based anchor units can also be used for the directed self-assembly of polyhistidine-tagged proteins [296]. Therefore, CNTs have covalently been functionalized with 4-carboxybenze diazonium salts, rendering the SWCNTs water-soluble. The acid moieties on the CNTs have been subsequently reacted with Na,Na-bis(carboxymethyl)-ʟ-lysine hydrate. The nirilotriacetic acid moiety of the lysine linker has then been complexed with Ni^{2+} and used to specifically bind a polyhistidine-tagged photosynthetic reaction center. This approach allows both positional and orientation control over protein binding to carbon nanotubes.

Owing to their pronounced electron acceptor capacity, SWCNTs have been used in sidewall-attached donor/acceptor assemblies. Herein, a tetraphenyl porphyrin derivative, bearing an amino group, has been mounted on the sidewalls of SWCNTs by the diazonium pathway. In this soluble donor–acceptor nanohybrid, fluorescence of photoexcited porphyrin is effectively quenched through intramolecular electron transfer process [297]. More significantly, these novel covalently modified SWCNTs show superior optical limiting effects for nanosecond laser pulses. Sidewall-functionalized CNT derivatives have been used as starting materials in coupling photoactive anthracene moieties in Heck-type reactions [298], for the subsequent derivatization with functional entities by "click" coupling reactions [299–301] or by connecting porphyrins, fluorenes, and thiophenes with SWCNTs by Suszuki coupling reactions [302]. Furthermore, by the diazonium pathway, sidewall-functionalized SWCNT derivatives can also be used for the chemically assisted directed assembly of carbon nanotubes on surfaces. Tour and coworkers have covalently attached individualized SWCNTs to silicon surfaces via orthogonally functionalized oligo(phenylene ethynylene) aryldiazonium salts [303, 304]. This represents the first example for a support-bound SWCNT that does not require a CVD growth process. Thereafter, Kagan and coworkers showed the possibility to chemically functionalize SWCNTs through diazonium chemistry with hydroxamic acid end groups that both render the SWCNTs water-soluble and discriminately bind the SWCNT to basic metal oxide surfaces [305]. This method allows directed assembly into predefined positions of oriented SWCNTs by confining the SWCNT derivatives to trenches of patterned dielectric stacks.

The directed sidewall functionalization by diazonium chemistry has developed to a frequently applied tool for the construction of CNT polymer composite materials. Herein, the "grafting from" approach is widely used for the establishment of a direct covalent linkage between the CNT framework and the surrounding polymer matrix, resulting in enhanced mechanical properties of the composite material. A broad variety of functional moieties, equipped with heteroatom functionalities for a subsequent *in situ* polymerization, have been coupled to SWCNT and MWCNTs by the diazonium route in the last 4 years and the corresponding composite materials have been thoroughly investigated according to their novel physical properties [306–312].

In a groundbreaking investigation, Strano *et al.* demonstrated for the first time that the chemical sidewall functionalization of SWCNTs can be selective in terms of metallic and semiconducting CNTs [313]. By a very detailed UV/Vis and Raman study, they have been able to show that the diazonium-based SWCNT functionalization reactions lead to a higher functionalization degree for metallic carbon nanotubes compared to their semiconducting relatives. The mechanistical details of this electronic type specific addition reaction and the structure-dependent reactivity of semiconducting SWCNTs with benzenediazonium salts have been extensively investigated by various spectroscopic methods [314–319]. The reaction of SWCNTs with diazonium follows a two-step process [320]. An initial (n,m)-selective adsorption step is followed by a covalent reaction step. The adsorption step is mediated by an electron transfer from the nanotube to the diazonium molecule leading to an intermediate aryl radical, while the covalent bond results in the localization of electrons in its vicinity and the formation of an impurity state at the Fermi level. Therefore, the type specific addition of diazonium compounds has been used in separation approaches for semiconducting and metallic SWCNTs [321]. Strano and coworkers applied this concept of electronic type selective addition of *p*-hydroxy-benzene diazonium salts to enrich metallic and semiconducting fractions separately using the induced differences in electrophoretic mobilities by electrophoresis [322]. Nakashima and coworkers applied long-chain carrying benzenediazonium salts, which selectively react with metallic SWCNTs, generating soluble SWCNT derivatives [323]. By solvent extraction of the reacted metallic SWCNTs, they were able to separate the unreacted THF-insoluble SWCNTs, consisting mainly of semiconducting species. The type specific covalent attachment of amino-substituted phenyl diazonium compounds can even be used for the construction of pH sensor devices. The attached organic moieties function as pH-dependent carrier scattering sites, as a result of which the resistance of the modified metallic SWCNTs varies as a function of pH. Meanwhile, because of all these facts and its versatility the diazonium-based functionalization has become the most frequently used functionalization sequence for CNT sidewall derivatization.

6.4.9
Sidewall Functionalization Through Electrophilic Addition

The first report on the derivatization of SWCNTs by electrophilic addition of chloroform in the presence of $AlCl_3$ was published in 2002 [324]. The hydrolysis of the labile intermediate yields hydroxyl group-terminated SWCNTs. These functionalized SWCNTs can be used as starting material to graft polystyrene from the surface via atom transfer radical polymerization (ATRP) [325]. The ATRP initiators have been attached to the SWCNTs by esterification of 2-chloropropyl chloride with the surface-bound hydroxyl functionalities. Subsequent polymerization of styrene yields styrene-grafted SWCNT composite materials. The concept of the Friedel–Crafts-type electrophilic addition to SWCNT sidewalls has been generalized by Wang and coworkers who have developed a microwave-based functionalization sequence that allows the use of a broad variety of alkyl halides for the CNT derivatization [326]. By this sequence, SWCNT derivatives simultaneously bearing alkyl and hydroxyl substituents are accessible. The polyacylation of SWCNTs under Friedel–Crafts conditions is possible by reacting an acylchloride in the presence of sodium chloride and $AlCl_3$ at high temperatures (Scheme 6.15) [327].

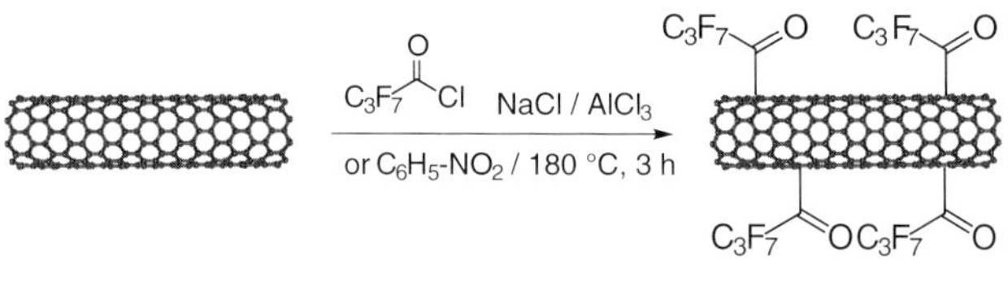

Scheme 6.15 Friedel–Crafts polyacylation of SWCNTs.

The acylated material (**38**) is highly soluble in organic solvents. The use of perfluorinated acyl chlorides leads to perfluoroalkyl decorated SWCNTs with interesting solubility characteristics in fluorocarbons. Polyphosphoric acid can also be used as Friedel–Crafts catalyst. Sidewall acylation of MWCNTs with various 4-substituted benzoic acid derivatives can be conducted by direct Friedel–Crafts acylation in mild polyphosphoric acid/phosphorous pentoxide [328]. The resultant functionalized MWCNTs form bundles with average diameters of 40–70 nm, depending on the polarity of the surface functionality introduced. The dispersibility and thermal properties are also greatly influenced by the nature of the groups attached. This reaction sequence can also be used for a one-pot purification and acylation of SWCNTs [65].

6.4.10
Sidewall Functionalization Through Nucleophilic Addition

6.4.10.1 Carbon-Based Nucleophiles
Stabilized carbanions such as of organolithium compounds readily react with SWCNTs through nucleophilic sidewall addition. The treatment of HiPCO SWCNTs with *sec*-butyllithium leads to the nucleophilic addition of the alkyl chain and the

intermediate generation of exfoliated, negatively charged SWCNTs anions [329]. The negatively charged nanotubes are separated from the bundles and stay in solution due to mutual electrostatic repulsion between individual tubes, which is confirmed by long-term solution homogeneity. These anionic charges can be used as initiators for the subsequent polymerization of styrene or acrylates yielding covalently polymer grafted SWCNT composite materials with homogeneous nanotube distribution in the polystyrene matrix [329, 330]. Similar composite materials are obtained by the initial treatment of MWCNTs with *n*-butyllithium followed by the reaction with chlorinated polypropylene and chlorinated polystyrene [331, 332]. The corresponding MWCNT–polymer composite material shows a substantial increase in mechanical properties due to the covalent sidewall attachment of the polymer chains.

In a detailed mechanistical study, Hirsch and coworkers were able to show that the SWCNT carbanions (**39**) generated by the nucleophilc addition of *t*-butyllithium can be easily reoxidized yielding alkylated SWCNT derivatives (**40**) (Scheme 6.16) [333]. XPS C 1s core-level spectra confirmed that the nucleophilic attack of the *t*-butyl carbanions on the sidewalls of the SWCNTs leads to the generation of negatively charged SWCNT derivatives.

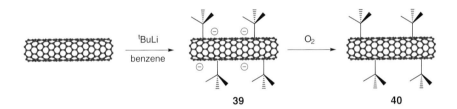

Scheme 6.16 Nucleophilic alkylation–reoxidation sequence for the direct sidewall functionalization of SWCNTs.

The C 1s core level of the functionalized sample is located at a binding energy, that is, by 2.6 eV higher than that of the starting material, which is explained by the transfer of negative charge to the SWCNT accompanied by the formation of the charged intermediates. Addition of oxygen yields SWCNT material with a C 1s core level position almost identical to that of the neutral starting material, proving that the negative charge is removed upon oxidation. This nucleophilic alkylation reoxidation cycle could be repeated, and the functionalization degree increases with each cycle. In addition, the covalently bound *t*-butyl group could be visualized by STM. At 4.7 K the rotation about the C–C bond between the addend and the sidewall of the tubes is frozen leading to a detectable threefold symmetry of the addend. According to the results obtained by Raman spectroscopy, metallic SWCNTs show a preferential reactivity toward the addition of carbon-based nucleophiles compared to semiconducting SWCNTs. This selectivity can be explained by the presence of electronic states close to the Fermi level characteristic for metallic but not for semiconducting CNTs. In addition, further Raman investigations on the selectivity of the nucleophilic sidewall addition of organolithium and organomagnesium compounds confirmed these observations and also revealed a clear diameter dependence for this kind of addition reaction, where smaller diameter SWCNTs exhibit higher reactivity [334–336]. Furthermore, the reactivity also depends on the steric demand of the addend as the binding of the bulky *t*-butyl addend is less favorable than the addition of primary alkyl groups [336]. The functionalization degree obtained by the nucleophilic addition of organolithium compounds is in the range of 1 addend per 31 carbon atoms and comparable to the data obtained by reductive alkylation. SWCNT anions generated by the nucleophilic addition of organolithium compounds can also be used as a versatile tool for the generation of higher functionalized SWCNT assemblies. The reaction of SWCNTs with *sec*-butyllithium and the subsequent treatment with carbon dioxide yields functionalized SWCNTs bearing simultaneously alkyl groups and carboxylic acid functionalities, which can be used for further derivatization [337]. In an alternative approach, hydroxyethylated SWCNT derivatives have been obtained by treating HiPCO SWCNTs with *sec*-butyllithium and subsequently with epoxy ethane [338]. The functionalization degree is in the range of 1 hydroxyl group per 31 carbon atoms according to TGA data, leading to water-dispersible SWCNT material. Pénicaud and coworkers intercepted the negatively charged SWCNTs and MWCNTs, generated by methyl lithium addition, by adding different alky and aryl halides [339]. This functionalization route combines the nucleophilc sidewall addition and the reductive alkylation pathway yielding multifunctional CNT derivatives.

6.4.10.2 Nitrogen-Based Nucleophiles

The concept of nucleophilic sidewall addition can be extended to other substance classes containing heteroatoms. Recently, it was shown that *in situ* generated *n*-propylamide can covalently be attached to the sidewalls of HiPCO SWCNTs by a direct nucleophilic addition reaction (Scheme 6.17) [340]. The reaction of *n*-propylamine with *n*-butyllithium yields the corresponding amide, which attacks the SWCNTs generating negatively charged SWCNT derivatives (**41**) that are subsequently reoxidized by air into the corresponding neutral amino-functionalized SWCNT derivatives (**42**).

The reaction product exhibits a drastically improved solubility in organic solvents and first concentration-dependent studies indicate that the functionalization degree is directly related to the concentration of amide used for the functionalization sequence.

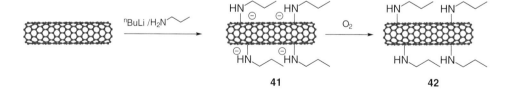

Scheme 6.17 Amidation sequence of SWNTs with lithium *n*-propylamide followed by reoxidation.

6.5
Conclusions

The expansion of knowledge in the field of the chemical functionalization of carbon nanotubes can be compared with the evolution of the universe. Soon after the big bang – the discovery of carbon nanotubes – the nascent universe passed through a phase of exponential expansion, the so-called cosmic inflation. For the past 5 years, we are in a phase of knowledge expansion on carbon nanotube reactivity, which is nicely reflected by an exponential increase in scientific articles published on the topic. Especially, the technique of carbon nanotube defect functionalization has meanwhile become a commonly used tool in the chemists' toolbox for the generation of functional CNT architectures. On the basis of the simple introduction of carboxylic acid functionalities and the subsequent coupling with amine or hydroxyl functionalized moieties, the unique electronic and mechanical properties of carbon nanotubes can be merged with other substance classes. In addition, novel composite materials can be generated exhibiting tailor-made properties for various applications in the field of nanotechnology. Keeping in mind that carbon nanotubes are close relatives of the unreactive sp^2 π-system graphene, the multitude of direct sidewall functionalization sequence explored in the past 5 years demonstrates that CNTs can be understood as classical molecules, only a little bit bigger in size. Nevertheless, the polydispersibility of this ultimate carbon molecule is still a big obstacle that hampers the entry of this new carbon allotrope in everyday life. Clearly, carbon nanotube chemistry will play a fundamental role in preparing the ground for a carbon-based nanotechnology future. So, new worlds out there are still waiting to be discovered and our journey through the carbon-based universe has just started.

References

1 Hirsch, A. and Brettreich, M. (2005) *Fullerenes: Chemistry and Reactions*, Wiley-VCH Verlag GmbH, Weinheim.

2 Langa, F. (2007) *Fulellerenes: Principles and Applications*, RSC Publishing, Cambridge.

3 Iijima, S. (1991) Helical microtubules of graphitic carbon. *Nature*, **354**, 56–58.

4 Baughman, R.H., Zakhidov, A.A., and de Heer, W.A. (2002) Carbon nanotubes: the route toward applications. *Science*, **297**, 787–792.

5 Ausman, K.D., Piner, R., Lourie, O., Ruoff, R.S., and Korobov, M. (2000) Organic solvent dispersions of single-walled carbon nanotubes: toward solutions of pristine nanotubes. *J. Phys. Chem. B*, **104**, 8911–8915.

6 Bahr, J.L., Mickelson, E.T., Bronikowski, M.J., Smalley, R.E., and Tour, J.M. (2001) Dissolution of small diameter single-wall carbon nanotubes in organic solvents? *Chem. Commun.*, 193–194.

7 Dyke, C.A. and Tour, J.M. (2004) Overcoming the insolubility of carbon nanotubes through high degrees of sidewall functionalization. *Chem. Eur. J.*, **10**, 812–817.

8 Sun, Y.-P., Fu, K., Lin, Y., and Huang, W. (2002) Functionalized carbon nanotubes: properties and applications. *Acc. Chem. Res.*, **35**, 1096–1104.

9 Niyogi, S., Hamon, M.A., Hu, H., Zhao, B., Bhowmik, P., Sen, R., Itkis, M.E., and Haddon, R.C. (2002) Chemistry of single-walled carbon nanotubes. *Acc. Chem. Res.*, **35**, 1105–1113.

10 Bahr, J.L. and Tour, J.M. (2002) Covalent chemistry of single-wall carbon nanotubes. *J. Mater. Chem.*, **12**, 1952–1958.

11 Lin, T., Bajpai, V., Ji, T., and Dai, L. (2003) Chemistry of carbon nanotubes. *Aust. J. Chem.*, **56**, 635–651.

12 Tasis, D., Tagmatarchis, N., Georgakilas, V., and Prato, M. (2003) Soluble carbon nanotubes. *Chem. Eur. J.*, **9**, 4000–4008.

13 Tasis, D., Tagmatarchis, N., Bianco, A., and Prato, M. (2006) Chemistry of carbon nanotubes. *Chem. Rev.*, **106**, 1105–1136.

14 Hirsch, A. and Vostrowsky, O. (2005) Functionalization of carbon nanotubes. *Top. Curr. Chem.*, **245**, 193–237.

15 Banerjee, S., Hemraj-Benny, T., and Wong, S.S. (2005) Covalent surface chemistry of single-walled carbon nanotubes. *Adv. Mater.*, **17**, 17–29.

16 Hirsch, A. (2002) Functionalization of single-walled carbon nanotubes. *Angew. Chem. Int. Ed.*, **41**, 1853–1859.

17 Peng, X. and Wong, S.S. (2009) Functional covalent chemistry of carbon nanotube surfaces. *Adv. Mater.*, **21**, 625–642.

18 Ehli, C., Guldi, D.M., Angeles Herranz, M., Martin, N., Campidelli, S., and Prato, M. (2008) Pyrene-tetrathiafulvalene supramolecular assembly with different types of carbon nanotubes. *J. Mater. Chem.*, **18**, 1498–1503.

19 Ehli, C., Rahman, G.M.A., Jux, N., Balbinot, D., Guldi, D.M., Paolucci, F., Marcaccio, M., Paolucci, D., Melle-Franco, M., Zerbetto, F., Campidelli, S., and Prato, M. (2006) Interactions in single wall carbon nanotubes/pyrene/porphyrin nanohybrids. *J. Am. Chem. Soc.*, **128**, 11222–11231.

20 Fujigaya, T. and Nakashima, N. (2008) Methodology for homogeneous dispersion of single-walled carbon nanotubes by physical modification. *Polym. J.*, **40**, 577–589.

21 O'Connell, M.J., Bachilo, S.M., Huffman, C.B., Moore, V.C., Strano, M.S., Haroz, E.H., Rialon, K.L., Boul, P.J., Noon, W.H., Kittrell, C., Ma, J., Hauge, R.H., Weisman, R.B., and Smalley, R.E. (2002) Band gap fluorescence from individual single-walled carbon nanotubes. *Science*, **297**, 593–596.

22 Lu, X. and Chen, Z. (2005) Curved pi-conjugation, aromaticity, and the related chemistry of small fullerenes (<C60) and single-walled carbon nanotubes. *Chem. Rev.*, **105**, 3643–3696.

23 Haddon, R.C. (1988) p-Electrons in three dimensional. *Acc. Chem. Res.*, **21**, 243–249.

24 Chen, Z., Thiel, W., and Hirsch, A. (2003) Reactivity of the convex and concave surfaces of single-walled carbon nanotubes (SWCNTs) towards addition reactions: dependence on the carbon-atom pyramidalization. *ChemPhysChem*, **4**, 93–97.

25 Hamon, M.A., Itkis, M.E., Niyogi, S., Alvaraez, T., Kuper, C., Menon, M., and Haddon, R.C. (2001) Effect of rehybridization on the electronic structure of single-walled carbon nanotubes. *J. Am. Chem. Soc.*, **123**, 11292–11293.

26 Zheng, G., Wang, Z., Irle, S., and Morokuma, K. (2006) Origin of the linear relationship between $CH_2/NH/O$-SWNT reaction energies and sidewall curvature: armchair nanotubes. *J. Am. Chem. Soc.*, **128**, 15117–15126.

27 Lu, X., Tian, F., Xu, X., Wang, N., and Zhang, Q. (2003) A theoretical exploration of the 1,3-dipolar cycloadditions onto the sidewalls of (n,n) armchair single-wall carbon nanotubes. *J. Am. Chem. Soc.*, **125**, 10459–10464.

28 Park, H., Zhao, J., and Lu, J.P. (2005) Distinct properties of single-wall carbon nanotubes with monovalent sidewall additions. *Nanotechnology*, **16**, 635–638.

29 Xu, Y.-Q., Peng, H., Hauge, R.H., and Smalley, R.E. (2005) Controlled multistep purification of single-walled carbon nanotubes. *Nano Lett.*, **5**, 163–168.

30 Arrais, A., Diana, E., Pezzini, D., Rossetti, R., and Boccaleri, E. (2006) A fast effective route to pH-dependent water-dispersion of oxidized single-walled carbon nanotubes. *Carbon*, **44**, 587–590.

31 Bonifazi, D., Nacci, C., Marega, R., Campidelli, S., Ceballos, G., Modesti, S., Meneghetti, M., and Prato, M. (2006) Microscopic and spectroscopic

characterization of paintbrush-like single-walled carbon nanotubes. *Nano Lett.*, **6**, 1408–1414.

32 Zhang, M., Yudasaka, M., Miyauchi, Y., Maruyama, S., and Iijima, S. (2006) Changes in the fluorescence spectrum of individual single-wall carbon nanotubes induced by light-assisted oxidation with hydroperoxide. *J. Phys. Chem. B*, **110**, 8935–8940.

33 Ziegler, K.J., Gu, Z., Peng, H., Flor, E.L., Hauge, R.H., and Smalley, R.E. (2005) Controlled oxidative cutting of single-walled carbon nanotubes. *J. Am. Chem. Soc.*, **127**, 1541–1547.

34 Li, J., Jia, G., Zhang, Y., and Chen, Y. (2006) Bond-curvature effect of sidewall [2 + 1] cycloadditions of single-walled carbon nanotubes: a new criterion to the adduct structures. *Chem. Mater.*, **18**, 3579–3584.

35 Miyata, Y., Kawai, T., Miyamoto, Y., Yanagi, K., Maniwa, Y., and Kataura, H. (2007) Bond-curvature effect on burning of single-wall carbon nanotubes. *Phys. Stat. Sol. (B)*, **244**, 4035–4039.

36 Miyata, Y., Kawai, T., Miyamoto, Y., Yanagi, K., Maniwa, Y., and Kataura, H. (2007) Chirality-dependent combustion of single-walled carbon nanotubes. *J. Phys. Chem. C*, **111**, 9671–9677.

37 Kawai, T. and Miyamoto, Y. (2008) Chirality-dependent C–C bond breaking of carbon nanotubes by cyclo-addition of oxygen molecule. *Chem. Phys. Lett.*, **453**, 256–261.

38 Joselevich, E. (2004) Electronic structure and chemical reactivity of carbon nanotubes: A chemist's view. *ChemPhysChem*, **5**, 619–624.

39 Ebbesen, T.W. and Takada, T. (1995) Topological and sp3 defect structures in nanotubes. *Carbon*, **33**, 973–978.

40 Louie, S.G. (2001) Electronic properties, junctions, and defects of carbon nanotubes. *Top. Appl. Phys.*, **80**, 113–145.

41 Charlier, J.C. (2002) Defects in carbon nanotubes. *Acc. Chem. Res.*, **35**, 1063–1069.

42 Carlsson, J.M. (2006) Curvature and chirality dependence of the properties of point defects in nanotubes. *Phys. Stat. Sol. (B)*, **243**, 3452–3457.

43 Suenaga, K., Wakabayashi, H., Koshino, M., Sato, Y., Urita, K., and Iijima, S. (2007) Imaging active topological defects in carbon nanotubes. *Nat. Nanotechnol.*, **2**, 358–360.

44 Lu, A.J. and Pan, B.C. (2004) Nature of single vacancy in achiral carbon nanotubes. *Phys. Rev. Lett.*, **92**, 105504/105501–105504/105504.

45 Urita, K., Suenaga, K., Sugai, T., Shinohara, H., and Iijima, S. (2005) In situ observation of thermal relaxation of interstitial-vacancy pair defects in a graphite gap. *Phys. Rev. Lett.*, **94**, 155502/155501–155502/155504.

46 Wang, C., Zhou, G., Wu, J., Gu, B.-L., and Duan, W. (2006) Effects of vacancy-carboxyl pair functionalization on electronic properties of carbon nanotubes. *Appl. Phys. Lett.*, **89**, 173130/173131–173130/173133.

47 Bettinger, H.F. (2005) The reactivity of defects at the sidewalls of single-walled carbon nanotubes: the Stone–Wales defect. *J. Phys. Chem. B*, **109**, 6922–6924.

48 Wang, C., Zhou, G., Liu, H., Wu, J., Qiu, Y., Gu, B.-L., and Duan, W. (2006) Chemical functionalization of carbon nanotubes by carboxyl groups on Stone–Wales defects: a density functional theory study. *J. Phys. Chem. B*, **110**, 10266–10271.

49 Lu, X., Chen, Z., and Schleyer, P.v.R. (2005) Are Stone–Wales defect sites always more reactive than perfect sites in the sidewalls of single-wall carbon nanotubes? *J. Am. Chem. Soc.*, **127**, 20–21.

50 Park, T.-J., Banerjee, S., Hemraj-Benny, T., and Wong, S.S. (2006) Purification strategies and purity visualization techniques for single-walled carbon nanotubes. *J. Mater. Chem.*, **16**, 141–154.

51 Osorio, A.G., Silveira, I.C.L., Bueno, V.L., and Bergmann, C.P. (2008) H_2SO_4/HNO_3/HCl – functionalization and its effect on dispersion of carbon nanotubes in aqueous media. *Appl. Surf. Sci.*, **255**, 2485–2489.

52 Chakraborty, S., Chattopadhyay, J., Peng, H., Chen, Z., Mukherjee, A., Arvidson, R.S., Hauge, R.H., and Billups, W.E. (2006) Surface area measurement of

functionalized single-walled carbon nanotubes. *J. Phys. Chem. B*, **110**, 24812–24815.

53 Yu, H., Jin, Y., Peng, F., Wang, H., and Yang, J. (2008) Kinetically controlled side-wall functionalization of carbon nanotubes by nitric acid oxidation. *J. Phys. Chem. C*, **112**, 6758–6763.

54 Salzmann, C.G., Llewellyn, S.A., Tobias, G., Ward, M.A.H., Huh, Y., and Green, M.L.H. (2007) The role of carboxylated carbonaceous fragments in the functionalization and spectroscopy of a single-walled carbon-nanotube material. *Adv. Mater.*, **19**, 883–887.

55 Wang, Y., Iqbal, Z., and Mitra, S. (2006) Rapidly functionalized, water-dispersed carbon nanotubes at high concentration. *J. Am. Chem. Soc.*, **128**, 95–99.

56 Barros, E.B., Souza Filho, A.G., Lemos, V., Mendes Filho, J., Fagan, S.B., Herbst, M.H., Rosolen, J.M., Luengo, C.A., and Huber, J.G. (2005) Charge transfer effects in acid treated single-wall carbon nanotubes. *Carbon*, **43**, 2495–2500.

57 Lee, G.-W. and Kumar, S. (2005) Dispersion of nitric acid-treated SWNTs in organic solvents and solvent mixtures. *J. Phys. Chem. B*, **109**, 17128–17133.

58 Wang, Y., Gao, L., Sun, J., Liu, Y., Zheng, S., Kajiura, H., Li, Y., and Noda, K. (2006) An integrated route for purification, cutting and dispersion of single-walled carbon nanotubes. *Chem. Phys. Lett.*, **432**, 205–208.

59 Li, M., Boggs, M., Beebe, T.P., and Huang, C.P. (2008) Oxidation of single-walled carbon nanotubes in dilute aqueous solutions by ozone as affected by ultrasound. *Carbon*, **46**, 466–475.

60 Byl, O., Liu, J., and Yates, J.T., Jr. (2005) Etching of carbon nanotubes by ozone: a surface area study. *Langmuir*, **21**, 4200–4204.

61 Chen, Z., Ziegler, K.J., Shaver, J., Hauge, R.H., and Smalley, R.E. (2006) Cutting of single-walled carbon nanotubes by ozonolysis. *J. Phys. Chem. B*, **110**, 11624–11627.

62 Simmons, J.M., Nichols, B.M., Baker, S.E., Marcus Matthew, S., Castellini, O.M., Lee, C.S., Hamers, R.J., and

Eriksson, M.A. (2006) Effect of ozone oxidation on single-walled carbon nanotubes. *J. Phys. Chem. B*, **110**, 7113–7118.

63 Zhang, J., Zou, H., Quan, Q., Yang, Y., Li, Q., Liu, Z., Guo, X., and Du, Z. (2003) Effect of chemical oxidation on the structure of single-walled carbon nanotubes. *J. Phys. Chem. B*, **107**, 3712–3718.

64 Wang, Y., Shan, H., Hauge, R.H., Pasquali, M., and Smalley, R.E. (2007) A highly selective, one-pot purification method for single-walled carbon nanotubes. *J. Phys. Chem. B*, **111**, 1249–1252.

65 Han, S.-W., Oh, S.-J., Tan, L.-S., and Baek, J.-B. (2008) One-pot purification and functionalization of single-walled carbon nanotubes in less-corrosive poly (phosphoric acid). *Carbon*, **46**, 1841–1849.

66 Tobias, G., Shao, L., Salzmann, C.G., Huh, Y., and Green, M.L.H. (2006) Purification and opening of carbon nanotubes using steam. *J. Phys. Chem. B*, **110**, 22318–22322.

67 Verdejo, R., Lamoriniere, S., Cottam, B., Bismarck, A., and Shaffer, M. (2007) Removal of oxidation debris from multi-walled carbon nanotubes. *Chem. Commun.*, 513–515.

68 Fogden, S., Verdejo, R., Cottam, B., and Shaffer, M. (2008) Purification of single walled carbon nanotubes: the problem with oxidation debris. *Chem. Phys. Lett.*, **460**, 162–167.

69 Shao, L., Tobias, G., Salzmann, C.G., Ballesteros, B., Hong, S.Y., Crossley, A., Davis, B.G., and Green, M.L.H. (2007) Removal of amorphous carbon for the efficient sidewall functionalization of single-walled carbon nanotubes. *Chem. Commun.*, 5090–5092.

70 Xing, Y., Li, L., Chusuei, C.C., and Hull, R.V. (2005) Sonochemical oxidation of multiwalled carbon nanotubes. *Langmuir*, **21**, 4185–4190.

71 Hennrich, F., Krupke, R., Arnold, K., Rojas Stuetz, J.A., Lebedkin, S., Koch, T., Schimmel, T., and Kappes, M.M. (2007) The mechanism of cavitation-induced scission of single-walled carbon

nanotubes. *J. Phys. Chem. B*, **111**, 1932–1937.

72 Tchoul, M.N., Ford, W.T., Lolli, G., Resasco, D.E., and Arepalli, S. (2007) Effect of mild nitric acid oxidation on dispersability, size, and structure of single-walled carbon nanotubes. *Chem. Mater.*, **19**, 5765–5772.

73 Marsh, D.H., Rance, G.A., Zaka, M.H., Whitby, R.J., and Khlobystov, A.N. (2007) Comparison of the stability of multiwalled carbon nanotube dispersions in water. *Phys. Chem. Chem. Phys.*, **9**, 5490–5496.

74 Chen, Z., Kobashi, K., Rauwald, U., Booker, R., Fan, H., Hwang, W.-F., and Tour, J.M. (2006) Soluble ultra-short single-walled carbon nanotubes. *J. Am. Chem. Soc.*, **128**, 10568–10571.

75 Li, J. and Zhang, Y. (2006) Cutting of multi walled carbon nanotubes. *Appl. Surf. Sci.*, **252**, 2944–2948.

76 Liu, Y., Gao, L., Sun, J., Zheng, S., Jiang, L., Wang, Y., Kajiura, H., Li, Y., and Noda, K. (2007) A multi-step strategy for cutting and purification of single-walled carbon nanotubes. *Carbon*, **45**, 1972–1978.

77 Aitchison, T.J., Ginic-Markovic, M., Matisons, J.G., Simon, G.P., and Fredericks, P.M. (2007) Purification, cutting, and sidewall functionalization of multiwalled carbon nanotubes using potassium permanganate solutions. *J. Phys. Chem. C*, **111**, 2440–2446.

78 Ziegler, K.J., Gu, Z., Shaver, J., Chen, Z., Flor, E.L., Schmidt, D.J., Chan, C., Hauge, R.H., and Smalley, R.E. (2005) Cutting single-walled carbon nanotubes. *Nanotechnology*, **16**, S539–S544.

79 Gu, Z., Peng, H., Hauge, R.H., Smalley, R.E., and Margrave, J.L. (2002) Cutting single-wall carbon nanotubes through fluorination. *Nano Lett.*, **2**, 1009–1013.

80 Hamon, M.A., Hu, H., Bhowmik, P., Niyogi, S., Zhao, B., Itkis, M.E., and Haddon, R.C. (2001) End-group and defect analysis of soluble single-walled carbon nanotubes. *Chem. Phys. Lett.*, **347**, 8–12.

81 Okpalugo, T.I.T., Papakonstantinou, P., Murphy, H., McLaughlin, J., and Brown, N.M.D. (2004) High resolution XPS characterization of chemical functionalized MWCNTs and SWCNTs. *Carbon*, **43**, 153–161.

82 Jung, A., Graupner, R., Ley, L., and Hirsch, A. (2006) Quantitative determination of oxidative defects on single walled carbon nanotubes. *Phys. Stat. Sol.(b)*, **243**, 3217–3220.

83 Coleman, J.N., Dalton, A.B., Curran, S., Rubio, A., Davey, A.P., Drury, A., McCarthy, B., Lahr, B., Ajayan, P.M., Roth, S., Barklie, R.C., and Blau, W.J. (2000) Phase separation of carbon nanotubes and turbostratic graphite using a functional organic polymer. *Adv. Mater.*, **12**, 213–216.

84 Hu, H., Bhowmik, P., Zhao, B., Hamon, M.A., Itkis, M.E., and Haddon, R.C. (2001) Determination of the acidic sites of purified single-walled carbon nanotubes by acid–base titration. *Chem. Phys. Lett.*, **345**, 25–28.

85 Marshall, M.W., Popa-Nita, S., and Shapter, J.G. (2006) Measurement of functionalized carbon nanotube carboxylic acid groups using a simple chemical process. *Carbon*, **44**, 1137–1141.

86 Mawhinney, D.B., Naumenko, V., Kuznetsova, A., Yates, J.T., Liu, J., and Smalley, R.E. (2000) Surface defect site density on single walled carbon nanotubes by titration. *Chem. Phys. Lett.*, **324**, 213–216.

87 Liu, J., Rinzler, A.G., Dai, H., Hafner, J.H., Bradley, R.K., Boul, P.J., Lu, A., Iverson, T., Shelimov, K., Huffman, C.B., Rodriguez-Macias, F., Shon, Y.-S., Lee, T.R., Colbert, D.T., and Smalley, R.E. (1998) Fullerene pipes. *Science*, **280**, 1253–1256.

88 Chen, J., Hamon, M.A., Hu, H., Chen, Y., Rao, A.M., Eklund, P.C., and Haddon, R.C. (1998) Solution properties of single-walled carbon nanotubes. *Science*, **282**, 95–98.

89 Wang, Y., Iqbal, Z., and Mitra, S. (2005) Microwave-induced rapid chemical functionalization of single-walled carbon nanotubes. *Carbon*, **43**, 1015–1020.

90 Li, S., Qin, Y., Shi, J., Guo, Z.-X., Li, Y., and Zhu, D. (2005) Electrical properties of soluble carbon nanotube/polymer

composite films. *Chem. Mater.*, **17**, 130–135.

91 Ford, W.E., Jung, A., Hirsch, A., Graupner, R., Scholz, F., Yasuda, A., and Wessels, J.M. (2006) Urea-melt solubilization of single-walled carbon nanotubes. *Adv. Mater.*, **18**, 1193–1197.

92 Hu, N., Dang, G., Zhou, H., Jing, J., and Chen, C. (2007) Efficient direct water dispersion of multi-walled carbon nanotubes by functionalization with lysine. *Mater. Lett.*, **61**, 5285–5287.

93 Gabriel, G., Sauthier, G., Fraxedas, J., Moreno-Manas, M., Martinez, M.T., Miravitlles, C., and Casabo, J. (2006) Preparation and characterization of single-walled carbon nanotubes functionalized with amines. *Carbon*, **44**, 1891–1897.

94 Shen, J., Huang, W., Wu, L., Hu, Y., and Ye, M. (2007) Study on amino-functionalized multiwalled carbon nanotubes. *Mater. Sci. Eng. A*, **A464**, 151–156.

95 Hong, S.Y., Tobias, G., Ballesteros, B., El Oualid, F., Errey, J.C., Doores, K.J., Kirkland, A.I., Nellist, P.D., Green, M.L.H., and Davis, B.G. (2007) Atomic-scale detection of organic molecules coupled to single-walled carbon nanotubes. *J. Am. Chem. Soc.*, **129**, 10966–10967.

96 Bergeret, C., Cousseau, J., Fernandez, V., Mevellec, J.-Y., and Lefrant, S. (2008) Spectroscopic evidence of carbon nanotubes' metallic character loss induced by covalent functionalization via nitric acid purification. *J. Phys. Chem. C*, **112**, 16411–16416.

97 Alvaro, M., Atienzar, P., De la Cruz, P., Delgado, J.L., Troiani, V., Garcia, H., Langa, F., Palkar, A., and Echegoyen, L. (2006) Synthesis, photochemistry, and electrochemistry of single-wall carbon nanotubes with pendent pyridyl groups and of their metal complexes with zinc porphyrin. Comparison with pyridyl-bearing fullerenes. *J. Am. Chem. Soc.*, **128**, 6626–6635.

98 Stephenson, J.J., Hudson, J.L., Leonard, A.D., Price, B.K., and Tour, J.M. (2007) Repetitive functionalization of water-soluble single-walled carbon nanotubes.

Addition of acid-sensitive addends. *Chem. Mater.*, **19**, 3491–3498.

99 D'Este, M., De Nardi, M., and Menna, E. (2006) A Co-functionalization approach to soluble and functional single-walled carbon nanotubes. *Eur. J. Org. Chem.*, 2517–2522.

100 Li, H., Kose, M.E., Qu, L., Lin, Y., Martin, R.B., Zhou, B., Harruff, B.A., Allard, L.F., and Sun, Y.-P. (2007) Excited-state energy transfers in single-walled carbon nanotubes functionalized with tethered pyrenes. *J. Photochem. Photobiol. A*, **185**, 94–100.

101 Burghard, M. (2005) Asymmetric end-functionalization of carbon nanotubes. *Small*, **1**, 1148–1150.

102 Chopra, N., Majumder, M., and Hinds, B.J. (2005) Bifunctional carbon nanotubes by sidewall protection. *Adv. Funct. Mater.*, **15**, 858–864.

103 Hu, J., Shi, J., Li, S., Qin, Y., Guo, Z.-X., Song, Y., and Zhu, D. (2005) Efficient method to functionalize carbon nanotubes with thiol groups and fabricate gold nanocomposites. *Chem. Phys. Lett.*, **401**, 352–356.

104 Wildgoose, G.G., Banks, C.E., and Compton, R.G. (2006) Metal nanoparticles and related materials supported on carbon nanotubes. Methods and applications. *Small*, **2**, 182–193.

105 Georgakilas, V., Gournis, D., Tzitzios, V., Pasquato, L., Guldi, D.M., and Prato, M. (2007) Decorating carbon nanotubes with metal or semiconductor nanoparticles. *J. Mater. Chem.*, **17**, 2679–2694.

106 Pan, B., Cui, D., He, R., Gao, F., and Zhang, Y. (2006) Covalent attachment of quantum dot on carbon nanotubes. *Chem. Phys. Lett.*, **417**, 419–424.

107 Li, W., Gao, C., Qian, H., Ren, J., and Yan, D. (2006) Multiamino-functionalized carbon nanotubes and their applications in loading quantum dots and magnetic nanoparticles. *J. Mater. Chem.*, **16**, 1852–1859.

108 Gao, C., Li, W., Morimoto, H., Nagaoka, Y., and Maekawa, T. (2006) Magnetic carbon nanotubes: synthesis by electrostatic self-assembly approach and application in biomanipulations. *J. Phys. Chem. B*, **110**, 7213–7220.

109 Hwang, S.-H., Moorefield, C.N., Wang, P., Jeong, K.-U., Cheng, S.Z.D., Kotta, K.K., and Newkome, G.R. (2006) Dendron-tethered and templated CdS quantum dots on single-walled carbon nanotubes. *J. Am. Chem. Soc.*, **128**, 7505–7509.

110 Sheeney-Haj-Ichia, L., Basnar, B., and Willner, I. (2005) Efficient generation of photocurrents by using CdS/carbon nanotube assemblies on electrodes. *Angew. Chem. Int. Ed.*, **44**, 78–83.

111 Jung, M.-S., Jung, S.-O., Jung, D.-H., Ko, Y.K., Jin, Y.W., Kim, J., and Jung, H.-T. (2005) Patterning of single-wall carbon nanotubes via a combined technique (chemical anchoring and photolithography) on patterned substrates. *J. Phys. Chem. B*, **109**, 10584–10589.

112 Jung, M.-S., Ko, Y.K., Jung, D.-H., Choi, D.H., and Jung, H.-T. (2005) Electrical and field-emission properties of chemically anchored single-walled carbon nanotube patterns. *Appl. Phys. Lett.*, **87**, 013114/013111–013114/013113.

113 Yu, J., Losic, D., Marshall, M., Bocking, T., Gooding, J.J., and Shapter, J.G. (2006) Preparation and characterisation of an aligned carbon nanotube array on the silicon (100) surface. *Soft Matter*, **2**, 1081–1088.

114 Yu, J., Shapter, J.G., Quinton, J.S., Johnston, M.R., and Beattie, D.A. (2007) Direct attachment of well-aligned single-walled carbon nanotube architectures to silicon (100) surfaces: a simple approach for device assembly. *Phys. Chem. Chem. Phys.*, **9**, 510–520.

115 Yu, J., Shapter, J.G., Johnston, M.R., Quinton, J.S., and Gooding, J.J. (2007) Electron-transfer characteristics of ferrocene attached to single-walled carbon nanotubes (SWCNT) arrays directly anchored to silicon(100). *Electrochim. Acta*, **52**, 6206–6211.

116 Flavel, B.S., Yu, J., Shapter, J.G., and Quinton, J.S. (2007) Patterned attachment of carbon nanotubes to silane modified silicon. *Carbon*, **45**, 2551–2558.

117 Flavel, B.S., Yu, J., Shapter, J.G., and Quinton, J.S. (2007) Patterned ferrocenemethanol modified carbon nanotube electrodes on silane modified silicon. *J. Mater. Chem.*, **17**, 4757–4761.

118 Yu, J., Mathew, S., Flavel, B.S., Johnston, M.R., and Shapter, J.G. (2008) Ruthenium porphyrin functionalized single-walled carbon nanotube arrays – a step toward light harvesting antenna and multibit information storage. *J. Am. Chem. Soc.*, **130**, 8788–8796.

119 Feldman, A.K., Steigerwald, M.L., Guo, X., and Nuckolls, C. (2008) Molecular electronic devices based on single-walled carbon nanotube electrodes. *Acc. Chem. Res.*, **41**, 1731–1741.

120 Guo, X., Small, J.P., Klare, J.E., Wang, Y., Purewal, M.S., Tam, I.W., Hong, B.H., Caldwell, R., Huang, L., O'Brien, S., Yan, J., Breslow, R., Wind, S.J., Hone, J., Kim, P., and Nuckolls, C. (2006) Covalently bridging gaps in single-walled carbon nanotubes with conducting molecules. *Science*, **311**, 356–359.

121 Whalley, A.C., Steigerwald, M.L., Guo, X., and Nuckolls, C. (2007) Reversible switching in molecular electronic devices. *J. Am. Chem. Soc.*, **129**, 12590–12591.

122 Guo, X., Whalley, A., Klare, J.E., Huang, L., O'Brien, S., Steigerwald, M., and Nuckolls, C. (2007) Single-molecule devices as scaffolding for multicomponent nanostructure assembly. *Nano Lett.*, **7**, 1119–1122.

123 Guldi, D.M., Rahman, G.M.A., Zerbetto, F., and Prato, M. (2005) Carbon nanotubes in electron donor–acceptor nanocomposites. *Acc. Chem. Res.*, **38**, 871–878.

124 Guldi, D.M., Rahman, G.M.A., Sgobba, V., and Ehli, C. (2006) Multifunctional molecular carbon materials – from fullerenes to carbon nanotubes. *Chem. Soc. Rev.*, **35**, 471–487.

125 Aprile, C., Martin, R., Alvaro, M., Scaiano, J.C., and Garcia, H. (2008) Functional macromolecules from single-walled carbon nanotubes: synthesis and photophysical properties of short single-walled carbon nanotubes functionalized

with 9,10-diphenylanthracene. *Chem. Eur. J.*, **14**, 5030–5038.

126 Ju, S.-Y. and Papadimitrakopoulos, F. (2008) Synthesis and redox behavior of flavin mononucleotide-functionalized single-walled carbon nanotubes. *J. Am. Chem. Soc.*, **130**, 655–664.

127 Alvaro, M., Aprile, C., Ferrer, B., and Garcia, H. (2007) Functional molecules from single wall carbon nanotubes. Photoinduced solubility of short single wall carbon nanotube residues by covalent anchoring of 2,4,6-triarylpyrylium units. *J. Am. Chem. Soc.*, **129**, 5647–5655.

128 Carmeli, I., Mangold, M., Frolov, L., Zebli, B., Carmeli, C., Richter, S., and Holleitner, A.W. (2007) A photosynthetic reaction center covalently bound to carbon nanotubes. *Adv. Mater.*, **19**, 3901–3905.

129 Herranz, M.A., Martin, N., Campidelli, S., Prato, M., Brehm, G., and Guldi, D.M. (2006) Control over electron transfer in tetrathiafulvalene-modified single-walled carbon nanotubes. *Angew. Chem. Int. Ed.*, **45**, 4478–4482.

130 Alvaro, M., Aprile, C., Atienzar, P., and Garcia, H. (2005) Preparation and photochemistry of single wall carbon nanotubes having covalently anchored viologen units. *J. Phys. Chem. B*, **109**, 7692–7697.

131 Baskaran, D., Mays, J.W., Zhang, X.P., and Bratcher, M.S. (2005) Carbon nanotubes with covalently linked porphyrin antennae: photoinduced electron transfer. *J. Am. Chem. Soc.*, **127**, 6916–6917.

132 Umeyama, T., Fujita, M., Tezuka, N., Kadota, N., Matano, Y., Yoshida, K., Isoda, S., and Imahori, H. (2007) Electrophoretic deposition of single-walled carbon nanotubes covalently modified with bulky porphyrins on nanostructured SnO₂ electrodes for photoelectrochemical devices. *J. Phys. Chem. C*, **111**, 11484–11493.

133 Liu, Z.-B., Tian, J.-G., Guo, Z., Ren, D.-M., Du, F., Zheng, J.-Y., and Chen, Y.-S. (2008) Enhanced optical limiting effects in porphyrin-covalently

functionalized single-walled carbon nanotubes. *Adv. Mater.*, **20**, 511–515.

134 Cho, E.S., Hong, S.W., and Jo, W.H. (2008) A new pH sensor using the fluorescence quenching of carbon nanotubes. *Macromol. Rapid Commun.*, **29**, 1798–1803.

135 Hong, C.-Y. and Pan, C.-Y. (2008) Functionalized carbon nanotubes responsive to environmental stimuli. *J. Mater. Chem.*, **18**, 1831–1836.

136 Yu, B., Zhou, F., Liu, G., Liang, Y., Huck, W.T.S., and Liu, W. (2006) The electrolyte switchable solubility of multi-walled carbon nanotube/ionic liquid (MWCNT/IL) hybrids. *Chem. Commun.*, 2356–2358.

137 Park, M.J., Lee, J.K., Lee, B.S., Lee, Y.-W., Choi, I.S., and Lee, S.-g. (2006) Covalent modification of multiwalled carbon nanotubes with imidazolium-based ionic liquids: effect of anions on solubility. *Chem. Mater.*, **18**, 1546–1551.

138 Rodriguez-Perez, L., Teuma, E., Falqui, A., Gomez, M., and Serp, P. (2008) Supported ionic liquid phase catalysis on functionalized carbon nanotubes. *Chem. Commun.*, 4201–4203.

139 Park, M.J. and Lee, S.-G. (2007) Palladium catalysts supported onto the ionic liquid-functionalized carbon nanotubes for carbon-carbon coupling. *Bull. Korean Chem. Soc.*, **28**, 1925–1926.

140 Klink, M. and Ritter, H. (2008) Supramolecular gels based on multi-walled carbon nanotubes bearing covalently attached cyclodextrin and water-soluble guest polymers. *Macromol. Rapid Commun.*, **29**, 1208–1211.

141 Liang, P., Zhang, H.-Y., Yu, Z.-L., and Liu, Y. (2008) Solvent-controlled photoinduced electron transfer between porphyrin and carbon nanotubes. *J. Org. Chem.*, **73**, 2163–2168.

142 Delgado, J.L., de la Cruz, P., Urbina, A., Lopez Navarrete, J.T., Casado, J., and Langa, F. (2007) The first synthesis of a conjugated hybrid of C60-fullerene and a single-wall carbon nanotube. *Carbon*, **45**, 2250–2252.

143 Urbina, A., Miguel, C., Delgado, J.L., Langa, F., Diaz-Paniagua, C., and Batallan, F. (2008) Isolated rigid rod behavior of functionalized single-wall

carbon nanotubes in solution determined via small-angle neutron scattering. *Phys. Rev. B*, **78**, 045420/045421–045420/045425.

144 Wu, W., Zhu, H., Fan, L., and Yang, S. (2008) Synthesis and characterization of a grapevine nanostructure consisting of single-walled carbon nanotubes with covalently attached [60]fullerene balls. *Chem. Eur. J.*, **14**, 5981–5987.

145 Giordani, S., Colomer, J.-F., Cattaruzza, F., Alfonsi, J., Meneghetti, M., Prato, M., and Bonifazi, D. (2009) Multifunctional hybrid materials composed of [60] fullerene-based functionalized-single-walled carbon nanotubes. *Carbon*, **47**, 578–588.

146 Coleman, J.N., Khan, U., Blau, W.J., and Gun'ko, Y.K. (2006) Small but strong: a review of the mechanical properties of carbon nanotube-polymer composites. *Carbon*, **44**, 1624–1652.

147 Lin, Y., Meziani, M.J., and Sun, Y.-P. (2007) Functionalized carbon nanotubes for polymeric nanocomposites. *J. Mater. Chem.*, **17**, 1143–1148.

148 Coleman, J.N., Khan, U., and Gun'ko, Y.K. (2006) Mechanical reinforcement of polymers using carbon nanotubes. *Adv. Mater.*, **18**, 689–706.

149 Daniel, S., Rao, T.P., Rao, K.S., Rani, S.U., Naidu, G.R.K., Lee, H.-Y., and Kawai, T. (2007) A review of DNA functionalized/ grafted carbon nanotubes and their characterization. *Sens. Actuators B*, **B122**, 672–682.

150 Lu, F., Gu, L., Meziani, M.J., Wang, X., Luo, P.G., Veca, L.M., Cao, L., and Sun, Y.-P. (2009) Advances in bioapplications of carbon nanotubes. *Adv. Mater.*, **21**, 139–152.

151 Reilly, R.M. (2007) Carbon nanotubes: potential benefits and risks of nanotechnology in nuclear medicine. *J. Nucl. Med.*, **48**, 1039–1042.

152 Lacerda, L., Bianco, A., Prato, M., and Kostarelos, K. (2006) Carbon nanotubes as nanomedicines: from toxicology to pharmacology. *Adv. Drug Deliv. Rev.*, **58**, 1460–1470.

153 Ramanathan, T., Fisher, F.T., Ruoff, R.S., and Brinson, L.C. (2005) Amino-functionalized carbon nanotubes for

binding to polymers and biological systems. *Chem. Mater.*, **17**, 1290–1295.

154 Gromov, A., Dittmer, S., Svensson, J., Nerushev, O.A., Perez-Garcia, S.A., Licea-Jimenez, L., Rychwalski, R., and Campbell, E.E.B. (2005) Covalent amino-functionalization of single-wall carbon nanotubes. *J. Mater. Chem.*, **15**, 3334–3339.

155 Coleman, K.S., Chakraborty, A.K., Bailey, S.R., Sloan, J., and Alexander, M. (2007) Iodination of single-walled carbon nanotubes. *Chem. Mater.*, **19**, 1076–1081.

156 Veloso, M.V., Souza Filho, A.G., Mendes Filho, J., Fagan, S.B., and Mota, R. (2006) *Ab initio* study of covalently functionalized carbon nanotubes. *Chem. Phys. Lett.*, **430**, 71–74.

157 Park, H., Zhao, J., and Lu, J.P. (2006) Effects of sidewall functionalization on conducting properties of single wall carbon nanotubes. *Nano Lett.*, **6**, 916–919.

158 Zhang, Z.Q., Liu, B., Chen, Y.L., Jiang, H., Hwang, K.C., and Huang, Y. (2008) Mechanical properties of functionalized carbon nanotubes. *Nanotechnology*, **19**, 395702/395701–395702/395706.

159 Valentini, L., Puglia, D., Armentano, I., and Kenny, J.M. (2005) Sidewall functionalization of single-walled carbon nanotubes through CF4 plasma treatment and subsequent reaction with aliphatic amines. *Chem. Phys. Lett.*, **403**, 385–389.

160 Van Lier, G., Ewels, C.P., Zuliani, F., De Vita, A., and Charlier, J.-C. (2005) Theoretical analysis of fluorine addition to single-walled carbon nanotubes: functionalization routes and addition patterns. *J. Phys. Chem. B*, **109**, 6153–6158.

161 Lee, Y.-S. (2007) Syntheses and properties of fluorinated carbon materials. *J. Fluorine Chem.*, **128**, 392–403.

162 Valentini, L., Armentano, I., Mengoni, F., Puglia, D., Pennelli, G., and Kenny, J.M. (2005) Chemical gating and photoconductivity of CF4 plasma-functionalized single-walled carbon nanotubes with adsorbed butylamine. *J. Appl. Phys.*, **97**, 114320/114321–114320/114325.

163 Valentini, L., Macan, J., Armentano, I., Mengoni, F., and Kenny, J.M. (2006) Modification of fluorinated single-walled carbon nanotubes with aminosilane molecules. *Carbon*, **44**, 2196–2201.

164 Oki, A., Adams, L., Khabashesku, V., Edigin, Y., Biney, P., and Luo, Z. (2008) Dispersion of aminoalkylsilyl ester or amine alkyl-phosphonic acid side wall functionalized carbon nanotubes in silica using sol–gel processing. *Mater. Lett.*, **62**, 918–922.

165 Oki, A., Adams, L., Luo, Z., Osayamen, E., Biney, P., and Khabashesku, V. (2007) Functionalization of single-walled carbon nanotubes with N-[3-(trimethoxysilyl) propyl]ethylenediamine and its cobalt complex. *J. Phys. Chem. Solids*, **69**, 1194–1198.

166 Dillon, E.P., Crouse, C.A., and Barron, A.R. (2008) Synthesis, characterization, and carbon dioxide adsorption of covalently attached polyethyleneimine-functionalized single-wall carbon nanotubes. *ACS Nano*, **2**, 156–164.

167 Zeng, L., Zhang, L., and Barron, A.R. (2005) Tailoring aqueous solubility of functionalized single-wall carbon nanotubes over a wide pH range through substituent chain length. *Nano Lett.*, **5**, 2001–2004.

168 Pulikkathara, M.X., Kuznetsov, O.V., and Khabashesku, V.N. (2008) Sidewall covalent functionalization of single wall carbon nanotubes through reactions of fluoronanotubes with urea, guanidine, and thiourea. *Chem. Mater.*, **20**, 2685–2695.

169 Zhang, L., Zhang, J., Schmandt, N., Cratty, J., Khabashesku, V.N., Kelly, K.F., and Barron, A.R. (2005) AFM and STM characterization of thiol and thiophene functionalized SWNTs: pitfalls in the use of chemical markers to determine the extent of sidewall functionalization in SWNTs. *Chem. Commun.*, 5429–5431.

170 Zhang, J., Zhang, L., Khabashesku, V.N., Barron, A.R., and Kelly, K.F. (2008) Self-assembly of sidewall functionalized single-walled carbon nanotubes by scanning tunneling microscopy. *J. Phys. Chem. C*, **112**, 12321–12325.

171 Pekker, S., Salvetat, J.P., Jakab, E., Bonard, J.M., and Forro, L. (2001) Hydrogenation of carbon nanotubes and graphite in liquid ammonia. *J. Phys. Chem. B*, **105**, 7938–7943.

172 Borondics, F., Bokor, M., Matus, P., Tompa, K., Pekker, S., and Jakab, E. (2005) Reductive functionalization of carbon nanotubes. *Fullerenes, Nanotubes, Carbon Nanostruct.*, **13**, 375–382.

173 Wunderlich, D., Hauke, F., and Hirsch, A. (2008) Preferred functionalization of metallic and small-diameter single walled carbon nanotubes via reductive alkylation. *J. Mater. Chem.*, **18**, 1493–1497.

174 Miller, G.P., Kintigh, J., Kim, E., Weck, P.F., Berber, S., and Tomanek, D. (2008) Hydrogenation of single-wall carbon nanotubes using polyamine reagents: combined experimental and theoretical study. *J. Am. Chem. Soc.*, **130**, 2296–2303.

175 Lu, X., Yuan, Q., and Zhang, Q. (2003) Sidewall epoxidation of single-walled carbon nanotubes: a theoretical prediction. *Org. Lett.*, **5**, 3527–3530.

176 Ogrin, D., Chattopadhyay, J., Sadana, A.K., Billups, W.E., and Barron, A.R. (2006) Epoxidation and deoxygenation of single-walled carbon nanotubes: quantification of epoxide defects. *J. Am. Chem. Soc.*, **128**, 11322–11323.

177 Ogrin, D. and Barron, A.R. (2006) Highly oxygenated fullerenes by catalytic epoxidation of C60 and single walled carbon nanotubes with methyltrioxorhenium-hydrogen peroxide. *J. Mol. Catal. A Chem.*, **244**, 267–270.

178 Liu, M., Yang, Y., Zhu, T., and Liu, Z. (2005) Chemical modification of single-walled carbon nanotubes with peroxytrifluoroacetic acid. *Carbon*, **43**, 1470–1478.

179 Liu, M., Yang, Y., Zhu, T., and Liu, Z. (2007) A general approach to chemical modification of single-walled carbon nanotubes with peroxy organic acids and its application in polymer grafting. *J. Phys. Chem. C*, **111**, 2379–2385.

180 Chen, Y., Haddon, R.C., Fang, S., Rao, A.M., Eklund, P.C., Lee, W.H., Dickey, E.C., Grulke, E.A., Pendergrass, J.C.,

Chavan, A., Haley, B.E., and Smalley, R.E. (1998) Chemical attachment of organic functional groups to single-walled carbon nanotube material. *J. Mater. Res.*, **13**, 2423–2431.

181 Hu, H., Zhao, B., Hamon, M.A., Kamaras, K., Itkis, M.E., and Haddon, R.C. (2003) Sidewall functionalization of single-walled carbon nanotubes by addition of dichlorocarbene. *J. Am. Chem. Soc.*, **125**, 14893–14900.

182 Kamaras, K., Itkis, M.E., Hu, H., Zhao, B., and Haddon, R.C. (2003) Covalent bond formation to a carbon nanotube metal. *Science*, **301**, 1501.

183 Lu, J., Wang, D., Nagase, S., Ni, M., Zhang, X., Maeda, Y., Wakahara, T., Nakahodo, T., Tsuchiya, T., Akasaka, T., Gao, Z., Yu, D., Ye, H., Zhou, Y., and Mei, W.N. (2006) Evolution of the electronic properties of metallic single-walled carbon nanotubes with the degree of CCl_2 covalent functionalization. *J. Phys. Chem. B*, **110**, 5655–5658.

184 Holzinger, M., Vostrowsky, O., Hirsch, A., Hennrich, F., Kappes, M., Weiss, R., and Jellen, F. (2001) Sidewall functionalization of carbon nanotubes. *Angew. Chem. Int. Ed.*, **40**, 4002–4005.

185 Holzinger, M., Abraham, J., Whelan, P., Graupner, R., Ley, L., Hennrich, F., Kappes, M., and Hirsch, A. (2003) Functionalization of single-walled carbon nanotubes with (*R*-)oxycarbonyl nitrenes. *J. Am. Chem. Soc.*, **125**, 8566–8580.

186 Moghaddam, M.J., Taylor, S., Gao, M., Huang, S., Dai, L., and McCall, M.J. (2004) Highly efficient binding of DNA on the sidewalls and tips of carbon nanotubes using photochemistry. *Nano Lett.*, **4**, 89–93.

187 Pastine, S.J., Okawa, D., Kessler, B., Rolandi, M., Llorente, M., Zettl, A., and Frechet, J.M.J. (2008) A facile and patternable method for the surface modification of carbon nanotube forests using perfluoroarylazides. *J. Am. Chem. Soc.*, **130**, 4238–4239.

188 Yinghuai, Z., Peng, A.T., Carpenter, K., Maguire, J.A., Hosmane, N.S., and Takagaki, M. (2005) Substituted carborane-appended water-soluble single-wall carbon nanotubes: new approach to boron neutron capture therapy drug delivery. *J. Am. Chem. Soc.*, **127**, 9875–9880.

189 Li, Z., Dong, Y., Haeussler, M., Lam, J.W.Y., Dong, Y., Wu, L., Wong, K.S., and Tang, B.Z. (2006) Synthesis of, light emission from, and optical power limiting in soluble single-walled carbon nanotubes functionalized by disubstituted Polyacetylenes. *J. Phys. Chem. B*, **110**, 2302–2309.

190 Lu, X., Tian, F., and Zhang, Q. (2003) The [2 + 1] cycloadditions of dichlorocarbene, silylene, germylene, and oxycarbonylnitrene onto the sidewall of armchair (5,5) single-wall carbon nanotube. *J. Phys. Chem. B*, **107**, 8388–8391.

191 Chen, Z., Nagase, S., Hirsch, A., Haddon, R.C., Thiel, W., and Schleyer, P.v.R. (2004) Side-wall opening of single-walled carbon nanotubes (SWCNTs) by chemical modification: a critical theoretical study. *Angew. Chem. Int. Ed.*, **43**, 1552–1554.

192 Bettinger, H.F. (2004) Effects of finite carbon nanotube length on sidewall addition of fluorine atom and methylene. *Org. Lett.*, **6**, 731–734.

193 Zhao, J., Chen, Z., Zhou, Z., Park, H., Schleyer, P.v.R., and Lu, J.P. (2005) Engineering the electronic structure of single-walled carbon nanotubes by chemical functionalization. *ChemPhysChem*, **6**, 598–601.

194 Chu, Y.-Y. and Su, M.-D. (2004) Theoretical study of addition reactions of carbene, silylene, and germylene to carbon nanotubes. *Chem. Phys. Lett.*, **394**, 231–237.

195 Zhang, C., Li, R.-F., Liang, Y.-X., Shang, Z.-F., Wang, G.-C., Xing, Y., Pan, Y.-M., Cai, Z.-S., Zhao, X.-Z., and Liu, C.-B. (2006) The nitrene cycloaddition on the sidewall of armchair single-walled carbon nanotubes. *THEOCHEM*, **764**, 33–40.

196 Bettinger, H.F. (2006) Addition of carbenes to the sidewalls of single-walled carbon nanotubes. *Chem. Eur. J.*, **12**, 4372–4379.

197 Lee, Y.-S. and Marzari, N. (2006) Cycloaddition functionalizations to preserve or control the conductance of

carbon nanotubes. *Phys. Rev. Lett.*, **97**, 116801/116801–116801/116804.

198 Coleman, K.S., Bailey, S.R., Fogden, S., and Green, M.L.H. (2003) Functionalization of single-walled carbon nanotubes via the Bingel reaction. *J. Am. Chem. Soc.*, **125**, 8722–8723.

199 Ashcroft, J.M., Hartman, K.B., Mackeyev, Y., Hofmann, C., Pheasant, S., Alemany, L.B., and Wilson, L.J. (2006) Functionalization of individual ultra-short single-walled carbon nanotubes. *Nanotechnology*, **17**, 5033–5037.

200 Umeyama, T., Tezuka, N., Fujita, M., Matano, Y., Takeda, N., Murakoshi, K., Yoshida, K., Isoda, S., and Imahori, H. (2007) Retention of intrinsic electronic properties of soluble single-walled carbon nanotubes after a significant degree of sidewall functionalization by the Bingel reaction. *J. Phys. Chem. C*, **111**, 9734–9741.

201 Huang, W., Lin, Y., Taylor, S., Gaillard, J., Rao, A.M., and Sun, Y.-P. (2002) Sonication-assisted functionalization and solubilization of carbon nanotubes. *Nano Lett.*, **2**, 231–234.

202 Lu, J., Nagase, S., Zhang, X., Maeda, Y., Wakahara, T., Nakahodo, T., Tsuchiya, T., Akasaka, T., Yu, D., Gao, Z., Han, R., and Ye, H. (2005) Structural evolution of [2 + 1] cycloaddition derivatives of single-wall carbon nanotubes: from open structure to closed three-membered ring structure with increasing tube diameter. *THEOCHEM*, **725**, 255–257.

203 Hemraj-Benny, T. and Wong, S.S. (2006) Silylation of single-walled carbon nanotubes. *Chem. Mater.*, **18**, 4827–4839.

204 Zhang, W. and Swager, T.M. (2007) Functionalization of single-walled carbon nanotubes and fullerenes via a dimethyl acetylenedicarboxylate-4-dimethylaminopyridine zwitterion approach. *J. Am. Chem. Soc.*, **129**, 7714–7715.

205 Georgakilas, V., Kordatos, K., Prato, M., Guldi, D.M., Holzinger, M., and Hirsch, A. (2002) Organic functionalization of carbon nanotubes. *J. Am. Chem. Soc.*, **124**, 760–761.

206 Li, J. and Grennberg, H. (2006) Microwave-assisted covalent sidewall functionalization of multiwalled carbon nanotubes. *Chem. Eur. J.*, **12**, 3869–3875.

207 Brunetti, F.G., Herrero, M.A., De Munoz, J., Giordani, S., Diaz-Ortiz, A., Filippone, S., Ruaro, G., Meneghetti, M., Prato, M., and Vazquez, E. (2007) Reversible microwave-assisted cycloaddition of aziridines to carbon nanotubes. *J. Am. Chem. Soc.*, **129**, 14580–14581.

208 Brunetti, F.G., Herrero, M.A., Munoz, J.D.M., Diaz-Ortiz, A., Alfonsi, J., Meneghetti, M., Prato, M., and Vazquez, E. (2008) Microwave-induced multiple functionalization of carbon nanotubes. *J. Am. Chem. Soc.*, **130**, 8094–8100.

209 Prato, M., Kostarelos, K., and Bianco, A. (2008) Functionalized carbon nanotubes in drug design and discovery. *Acc. Chem. Res.*, **41**, 60–68.

210 Bianco, A., Kostarelos, K., Partidos, C.D., and Prato, M. (2005) Biomedical applications of functionalized carbon nanotubes. *Chem. Commun.*, 571–577.

211 Campidelli, S., Klumpp, C., Bianco, A., Guldi, D.M., and Prato, M. (2006) Functionalization of CNT: synthesis and applications in photovoltaics and biology. *J. Phys. Org. Chem.*, **19**, 531–539.

212 Tagmatarchis, N., Prato, M., and Guldi, D.M. (2005) Soluble carbon nanotube ensembles for light-induced electron transfer interactions. *Physica E*, **29**, 546–550.

213 Campidelli, S., Sooambar, C., Lozano Diz, E., Ehli, C., Guldi, D.M., and Prato, M. (2006) Dendrimer-functionalized single-wall carbon nanotubes: synthesis, characterization, and photoinduced electron transfer. *J. Am. Chem. Soc.*, **128**, 12544–12552.

214 Garcia, A., Herrero, M.A., Frein, S., Deschenaux, R., Munoz, R., Bustero, I., Toma, F., and Prato, M. (2008) Synthesis of dendrimer-carbon nanotube conjugates. *Phys. Stat. Sol. (A)*, **205**, 1402–1407.

215 Ballesteros, B., Campidelli, S., de la Torre, G., Ehli, C., Guldi, D.M., Prato, M., and Torres, T. (2007) Synthesis, characterization and photophysical properties of a SWNT–phthalocyanine hybrid. *Chem. Commun.*, 2950–2952.

216 Ballesteros, B., De la Torre, G., Ehli, C., Rahman, G.M.A., Agullo-Rueda, F., Guldi, D.M., and Torres, T. (2007) Single-wall carbon nanotubes bearing covalently linked phthalocyanines – photoinduced electron transfer. *J. Am. Chem. Soc.*, **129**, 5061–5068.

217 Georgakilas, V., Bourlinos, A.B., Zboril, R., and Trapalis, C. (2008) Synthesis, characterization and aspects of superhydrophobic functionalized carbon nanotubes. *Chem. Mater.*, **20**, 2884–2886.

218 Georgakilas, V., Bourlinos, A., Gournis, D., Tsoufis, T., Trapalis, C., Mateo-Alonso, A., and Prato, M. (2008) Multipurpose organically modified carbon nanotubes: from functionalization to nanotube composites. *J. Am. Chem. Soc.*, **130**, 8733–8740.

219 Bae, J.H., Shanmugharaj, A.M., Noh, W.H., Choi, W.S., and Ryu, S.H. (2007) Surface chemical functionalized single-walled carbon nanotube with anchored phenol structures: physical and chemical characterization. *Appl. Surf. Sci.*, **253**, 4150–4155.

220 Tumpane, J., Karousis, N., Tagmatarchis, N., and Norden, B. (2008) Alignment of carbon nanotubes in weak magnetic fields. *Angew. Chem. Int. Ed.*, **47**, 5148–5152.

221 Pastorin, G., Wu, W., Wieckowski, S., Briand, J.-P., Kostarelos, K., Prato, M., and Bianco, A. (2006) Double functionalization of carbon nanotubes for multimodal drug delivery. *Chem. Commun.*, 1182–1184.

222 Lacerda, L., Pastorin, G., Wu, W., Prato, M., Bianco, A., and Kostarelos, K. (2006) Luminescence of functionalized carbon nanotubes as a tool to monitor bundle formation and dissociation in water: the effect of plasmid–DNA complexation. *Adv. Funct. Mater.*, **16**, 1839–1846.

223 Ménard-Moyon, C., Izard, N., Doris, E., and Mioskowski, C. (2006) Separation of semiconducting from metallic carbon nanotubes by selective functionalization with azomethine ylides. *J. Am. Chem. Soc.*, **128**, 6552–6553.

224 Lu, X., Tian, F., Wang, N., and Zhang, Q. (2002) Organic functionalization of the sidewalls of carbon nanotubes by Diels–Alder reactions: a theoretical prediction. *Org. Lett.*, **4**, 4313–4315.

225 Mercuri, F. and Sgamellotti, A. (2009) First-principles investigations on the functionalization of chiral and non-chiral carbon nanotubes by Diels–Alder cycloaddition reactions. *Phys. Chem. Chem. Phys.*, **11**, 563–567.

226 Delgado, J.L., de la Cruz, P., Langa, F., Urbina, A., Casado, J., and Lopez Navarrete, J.T. (2004) Microwave-assisted sidewall functionalization of single-wall carbon nanotubes by Diels–Alder cycloaddition. *Chem. Commun.*, 1734–1735.

227 Zhang, L., Yang, J., Edwards, C.L., Alemany, L.B., Khabashesku, V.N., and Barron, A.R. (2005) Diels–Alder addition to fluorinated single walled carbon nanotubes. *Chem. Commun.*, 3265–3267.

228 Menard-Moyon, C., Dumas, F., Doris, E., and Mioskowski, C. (2006) Functionalization of single-wall carbon nanotubes by tandem high-pressure/Cr (CO)$_6$ activation of Diels–Alder cycloaddition. *J. Am. Chem. Soc.*, **128**, 14764–14765.

229 Menard-Moyon, C., Gross, J.M., Bernard, M., Turek, P., Doris, E., and Mioskowski, C. (2007) Unexpected outcome in the reaction of triazolinedione with carbon nanotubes. *Eur. J. Org. Chem.*, 4817–4819.

230 Gergely, A., Telegdi, J., Meszaros, E., Paszti, Z., Tarkanyi, G., Karman, F.H., and Kalman, E. (2007) Modification of multi-walled carbon nanotubes by Diels–Alder and Sandmeyer reactions. *J. Nanosci. Nanotechnol.*, **7**, 2795–2807.

231 Sakellariou, G., Ji, H., Mays, J.W., Hadjichristidis, N., and Baskaran, D. (2007) Controlled covalent functionalization of multiwalled carbon nanotubes using [4 + 2] cycloaddition of benzocyclobutenes. *Chem. Mater.*, **19**, 6370–6372.

232 Hayden, H., Gun'ko, Y.K., and Perova, T.S. (2007) Chemical modification of multi-walled carbon nanotubes using a

tetrazine derivative. *Chem. Phys. Lett.*, **435**, 84–89.

233 Penicaud, A., Poulin, P., Derre, A., Anglaret, E., and Petit, P. (2005) Spontaneous dissolution of a single-wall carbon nanotube salt. *J. Am. Chem. Soc.*, **127**, 8–9.

234 Anglaret, E., Dragin, F., Penicaud, A., and Martel, R. (2006) Raman studies of solutions of single-wall carbon nanotube salts. *J. Phys. Chem. B*, **110**, 3949–3954.

235 Billups, W.E., Liang, F., Chattopadhyay, J., and Beach, J.M. (2007) Uses of single wall carbon nanotube salts in organic syntheses. *ECS Trans.*, **2**, 65–76.

236 Martinez-Rubi, Y., Guan, J., Lin, S., Scriver, C., Sturgeon, R.E., and Simard, B. (2007) Rapid and controllable covalent functionalization of single-walled carbon nanotubes at room temperature. *Chem. Commun.*, 5146–5148.

237 Borondics, F., Jakab, E., and Pekker, S. (2007) Functionalization of carbon nanotubes via dissolving metal reductions. *J. Nanosci. Nanotechnol.*, **7**, 1551–1559.

238 Liang, F., Sadana, A.K., Peera, A., Chattopadhyay, J., Gu, Z., Hauge, R.H., and Billups, W.E. (2004) A Convenient route to functionalized carbon nanotubes. *Nano Lett.*, **4**, 1257–1260.

239 Gu, Z., Liang, F., Chen, Z., Sadana, A., Kittrell, C., Billups, W.E., Hauge, R.H., and Smalley, R.E. (2005) *In situ* Raman studies on lithiated single-wall carbon nanotubes in liquid ammonia. *Chem. Phys. Lett.*, **410**, 467–470.

240 Chattopadhyay, J., Chakraborty, S., Mukherjee, A., Wang, R., Engel, P.S., and Billups, W.E. (2007) SET mechanism in the functionalization of single-walled carbon nanotubes. *J. Phys. Chem. C*, **111**, 17928–17932.

241 Chattopadhyay, J., Sadana, A.K., Liang, F., Beach, J.M., Xiao, Y., Hauge, R.H., and Billups, W.E. (2005) Carbon nanotube salts. Arylation of single-wall carbon nanotubes. *Org. Lett.*, **7**, 4067–4069.

242 Liang, F., Alemany, L.B., Beach, J.M., and Billups, W.E. (2005) Structure analyses of dodecylated single-walled carbon nanotubes. *J. Am. Chem. Soc.*, **127**, 13941–13948.

243 Stephenson, J.J., Sadana, A.K., Higginbotham, A.L., and Tour, J.M. (2006) Highly functionalized and soluble multiwalled carbon nanotubes by reductive alkylation and arylation: the billups reaction. *Chem. Mater.*, **18**, 4658–4661.

244 Chattopadhyay, J., de Cortez, F., Chakraborty, S., Slater, N.K.H., and Billups, W.E. (2006) Synthesis of water-soluble PEGylated single-walled carbon nanotubes. *Chem. Mater.*, **18**, 5864–5868.

245 Chattopadhyay, J., de Jesus Cortez, F., Chakraborty, S., Slater, N.K.H., and Billups, W.E. (2007) Application of lithium nanotube salts in organic syntheses: water-soluble pegylated single walled carbon nanotubes. *Pepr. Pap.-Am. Chem. Soc., Div. Fuel Chem.*, **52**, 124–125.

246 Moniruzzaman, M., Chattopadhay, J., Billups, W.E., and Winey, K.I. (2007) Tuning the mechanical properties of SWNT/Nylon 6,10 composites with flexible spacers at the interface. *Nano Lett.*, **7**, 1178–1185.

247 Liang, F., Beach, J.M., Kobashi, K., Sadana, A.K., Vega-Cantu, Y.I., Tour, J.M., and Billups, W.E. (2006) *In situ* polymerization initiated by single-walled carbon nanotube salts. *Chem. Mater.*, **18**, 4764–4767.

248 Mukherjee, A., Combs, R., Chattopadhyay, J., Abmayr, D.W., Engel, P.S., and Billups, W.E. (2008) Attachment of nitrogen and oxygen centered radicals to single-walled carbon nanotube salts. *Chem. Mater.*, **20**, 7339–7343.

249 Anderson, R.E. and Barron, A.R. (2007) Solubilization of single-wall carbon nanotubes in organic solvents without sidewall functionalization. *J. Nanosci. Nanotechnol.*, **7**, 3436–3440.

250 Garcia-Gallastegui, A., Obieta, I., Bustero, I., Imbuluzqueta, G., Arbiol, J., Miranda, J.I., and Aizpurua, J.M. (2008) Reductive functionalization of single-walled carbon nanotubes with lithium metal catalyzed by electron carrier additives. *Chem. Mater.*, **20**, 4433–4438.

251 Peng, H., Reverdy, P., Khabashesku, V.N., and Margrave, J.L. (2003) Sidewall functionalization of single-walled carbon

nanotubes with organic peroxides. *Chem. Commun.*, 362–363.

252 Umek, P., Seo, J.W., Hernadi, K., Mrzel, A., Pechy, P., Mihailovic, D.D., and Forro, L. (2003) Addition of carbon radicals generated from organic peroxides to single wall carbon nanotubes. *Chem. Mater.*, **15**, 4751–4755.

253 Liang, F., Beach, J.M., Rai, P.K., Guo, W., Hauge, R.H., Pasquali, M., Smalley, R.E., and Billups, W.E. (2006) Highly exfoliated water-soluble single-walled carbon nanotubes. *Chem. Mater.*, **18**, 1520–1524.

254 Ying, Y., Saini, R.K., Liang, F., Sadana, A.K., and Billups, W.E. (2003) Functionalization of carbon nanotubes by free radicals. *Org. Lett.*, **5**, 1471–1473.

255 Liao, S.-H., Yen, C.-Y., Hung, C.-H., Weng, C.-C., Tsai, M.-C., Lin, Y.-F., Ma, C.-C.M., Pan, C., and Su, A. (2008) One-step functionalization of carbon nanotubes by free-radical modification for the preparation of nanocomposite bipolar plates in polymer electrolyte membrane fuel cells. *J. Mater. Chem.*, **18**, 3993–4002.

256 Tasis, D., Papagelis, K., Prato, M., Kallitsis, I., and Galiotis, C. (2007) Water-soluble carbon nanotubes by redox radical polymerization. *Macromol. Rapid Commun.*, **28**, 1553–1558.

257 Papagelis, K., Kalyva, M., Tasis, D., Parthenios, J., Siokou, A., and Galiotis, C. (2007) Covalently functionalized carbon nanotubes as macroinitiators for radical polymerization. *Phys. Stat. Sol. (B)*, **244**, 4046–4050.

258 Peng, H., Alemany, L.B., Margrave, J.L., and Khabashesku, V.N. (2003) Sidewall carboxylic acid functionalization of single-walled carbon nanotubes. *J. Am. Chem. Soc.*, **125**, 15174–15182.

259 Long, B., Wu, T.M., and Stellacci, F. (2008) Ultra-fast and scalable sidewall functionalisation of single-walled carbon nanotubes with carboxylic acid. *Chem. Commun.*, 2788–2790.

260 Engel, P.S., Billups, W.E., Abmayr, D.W., Jr., Tsvaygboym, K., and Wang, R. (2008) Reaction of single-walled carbon nanotubes with organic peroxides. *J. Phys. Chem. C*, **112**, 695–700.

261 Wang, X., Li, S., Xu, Y., Wan, L., You, H., Li, Q., and Wang, S. (2007) Radical functionalization of single-walled carbon nanotubes with azo (bisisobutyronitrile). *Appl. Surf. Sci.*, **253**, 7435–7437.

262 Gunji, T., Akazawa, M., Arimitsu, K., and Abe, Y. (2006) Chemical modification of SWCNT and VGCF by radical addition reaction. *Chem. Lett.*, **35**, 630–631.

263 Chang, J.-Y., Wu, H.-Y., Hwang, G.L., and Su, T.-Y. (2008) Ultrasonication-assisted surface functionalization of double-walled carbon nanotubes with azobis-type radical initiators. *J. Mater. Chem.*, **18**, 3972–3976.

264 Liu, J., Zubiri, M.R.I., Dossot, M., Vigolo, B., Hauge, R.H., Fort, Y., Ehrhardt, J.-J., and McRae, E. (2006) Sidewall functionalization of single-wall carbon nanotubes (SWNTs) through aryl free radical addition. *Chem. Phys. Lett.*, **430**, 93–96.

265 Liu, J., Rodriguez i Zubiri, M., Vigolo, B., Dossot, M., Fort, Y., Ehrhardt, J.-J., and McRae, E. (2007) Efficient microwave-assisted radical functionalization of single-wall carbon nanotubes. *Carbon*, **45**, 885–891.

266 Wei, L. and Zhang, Y. (2007) Covalent sidewall functionalization of single-walled carbon nanotubes via one-electron reduction of benzophenone by potassium. *Chem. Phys. Lett.*, **446**, 142–144.

267 Wei, L. and Zhang, Y. (2007) Covalent sidewall functionalization of single-walled carbon nanotubes: a photoreduction approach. *Nanotechnology*, **18**, 495703/495701–495703/495705.

268 Barthos, R., Mehn, D., Demortier, A., Pierard, N., Morciaux, Y., Demortier, G., Fonseca, A., and Nagy, J.B. (2005) Functionalization of single-walled carbon nanotubes by using alkyl halides. *Carbon*, **43**, 321–325.

269 Liu, Y., Yao, Z., and Adronov, A. (2005) Functionalization of single-walled carbon nanotubes with well-defined polymers by radical coupling. *Macromolecules*, **38**, 1172–1179.

270 Oh, S.B., Kim, H.L., Chang, J.H., Lee, Y.-W., Han, J.H., An, S.S.A., Joo, S.-W., Kim, H.-K., Choi, I.S., and Paik, H.-J. (2008) Facile covalent attachment of well-defined poly(*t*-butyl acrylate) on carbon nanotubes via radical addition reaction. *J. Nanosci. Nanotechnol.*, **8**, 4598–4602.

271 Nakamura, T., Ohana, T., Ishihara, M., Hasegawa, M., and Koga, Y. (2007) Chemical modification of single-walled carbon nanotubes with sulfur-containing functionalities. *Diamond Relat. Mater.*, **16**, 1091–1094.

272 Nakamura, T., Ohana, T., Ishihara, M., Tanaka, A., and Koga, Y. (2006) Sidewall modification of single-walled carbon nanotubes with sulfur-containing functionalities and gold nanoparticle attachment. *Chem. Lett.*, **35**, 742–743.

273 Li, W., Bai, Y., Zhang, Y., Sun, M., Cheng, R., Xu, X., Chen, Y., and Mo, Y. (2005) Effect of hydroxyl radical on the structure of multi-walled carbon nanotubes. *Synth. Met.*, **155**, 509–515.

274 Basiuk, E.V., Monroy-Pelaez, M., Puente-Lee, I., and Basiuk, V.A. (2004) Direct solvent-free amination of closed-cap carbon nanotubes: a link to fullerene chemistry. *Nano Lett.*, **4**, 863–866.

275 Basiuk, E.V., Gromovoy, T.Y., Datsyuk, A.M., Palyanytsya, B.B., Pokrovskiy, V.A., and Basiuk, V.A. (2005) Solvent-free derivatization of pristine multi-walled carbon nanotubes with amines. *J. Nanosci. Nanotechnol.*, **5**, 984–990.

276 Yokoi, T., Iwamatsu, S.-I., Komai, S.-i., Hattori, T., and Murata, S. (2005) Chemical modification of carbon nanotubes with organic hydrazines. *Carbon*, **43**, 2869–2874.

277 Chochos, C.L., Stefopoulos, A.A., Campidelli, S., Prato, M., Gregoriou, V.G., and Kallitsis, J.K. (2008) Immobilization of oligoquinoline chains on single-wall carbon nanotubes and their optical behavior. *Macromolecules*, **41**, 1825–1830.

278 Stefopoulos, A.A., Chochos, C.L., Prato, M., Pistollis, G., Papagelis, K., Petraki, F., Kennou, S., and Kallitsis, J.K. (2008) Novel hybrid materials consisting of regioregular poly(3-octylthiophene)s

covalently attached to single-wall carbon nanotubes. *Chem. Eur. J.*, **14**, 8715–8724.

279 Bahr, J.L., Yang, J., Kosynkin, D.V., Bronikowski, M.J., Smalley, R.E., and Tour, J.M. (2001) Functionalization of carbon nanotubes by electrochemical reduction of aryl diazonium salts: a bucky paper electrode. *J. Am. Chem. Soc.*, **123**, 6536–6542.

280 Bahr, J.L. and Tour, J.M. (2001) Highly functionalized carbon nanotubes using *in situ* generated diazonium compounds. *Chem. Mater.*, **13**, 3823–3824.

281 Dyke, C.A. and Tour, J.M. (2003) Unbundled and highly functionalized carbon nanotubes from aqueous reactions. *Nano Lett.*, **3**, 1215–1218.

282 Dyke, C.A. and Tour, J.M. (2003) Solvent-free functionalization of carbon nanotubes. *J. Am. Chem. Soc.*, **125**, 1156–1157.

283 Price, B.K., Hudson, J.L., and Tour, J.M. (2005) Green chemical functionalization of single-walled carbon nanotubes in ionic liquids. *J. Am. Chem. Soc.*, **127**, 14867–14870.

284 Price, B.K. and Tour, J.M. (2006) Functionalization of single-walled carbon nanotubes\on water\. *J. Am. Chem. Soc.*, **128**, 12899–12904.

285 Stephenson, J.J., Hudson, J.L., Azad, S., and Tour, J.M. (2006) Individualized single walled carbon nanotubes from bulk material using 96% sulfuric acid as solvent. *Chem. Mater.*, **18**, 374–377.

286 Davis, V.A., Ericson, L.M., Parra-Vasquez, A.N.G., Fan, H., Wang, Y., Prieto, V., Longoria, J.A., Ramesh, S., Saini, R.K., Kittrell, C., Billups, W.E., Adams, W.W., Hauge, R.H., Smalley, R.E., and Pasquali, M. (2004) Phase Behavior and rheology of SWNTs in superacids. *Macromolecules*, **37**, 154–160.

287 Hudson, J.L., Casavant, M.J., and Tour, J.M. (2004) water-soluble, exfoliated, nonroping single-wall carbon nanotubes. *J. Am. Chem. Soc.*, **126**, 11158–11159.

288 Chen, X., Wang, J., Zhong, W., Feng, T., Yang, X., and Chen, J. (2008) A scalable route to highly functionalized multi-walled carbon nanotubes on a large scale. *Macromol. Chem. Phys.*, **209**, 846–853.

289 Hudson, J.L., Jian, H., Leonard, A.D., Stephenson, J.J., and Tour, J.M. (2006) Triazenes as a stable diazonium source for use in functionalizing carbon nanotubes in aqueous suspensions. *Chem. Mater.*, **18**, 2766–2770.

290 Liu, J., Zubiri, M.R., Vigolo, B., Dossot, M., Humbert, B., Fort, Y., and McRae, E. (2007) Microwave-assisted functionalization of single-wall carbon nanotubes through diazonium. *J. Nanosci. Nanotechnol.*, **7**, 3519–3523.

291 Nelson, D.J., Rhoads, H., and Brammer, C. (2007) Characterizing covalently sidewall-functionalized SWNTs. *J. Phys. Chem. C*, **111**, 17872–17878.

292 Peng, Z., Holm, A.H., Nielsen, L.T., Pedersen, S.U., and Daasbjerg, K. (2008) Covalent sidewall functionalization of carbon nanotubes by a \formation–degradation\approach. *Chem. Mater.*, **20**, 6068–6075.

293 Ellison, M.D. and Gasda, P.J. (2008) Functionalization of single-walled carbon nanotubes with 1,4-benzenediamine using a diazonium reaction. *J. Phys. Chem. C*, **112**, 738–740.

294 Burghard, M., Maroto, A., Balasubramanian, K., Assmus, T., Forment-Aliaga, A., Lee, E.J.H., Weitz, R.T., Scolari, M., Nan, F., Mews, A., and Kern, K. (2007) Electrochemically modified single-walled carbon nanotubes. *Phys. Stat. Sol.(b)*, **244**, 4021–4025.

295 Biju, V., Itoh, T., Makita, Y., and Ishikawa, M. (2006) Close-conjugation of quantum dots and gold nanoparticles to sidewall functionalized single-walled carbon nanotube templates. *J. Photochem. Photobiol. A*, **183**, 315–321.

296 Graff, R.A., Swanson, T.M., and Strano, M.S. (2008) Synthesis of nickel–nitrilotriacetic acid coupled single-walled carbon nanotubes for directed self-assembly with polyhistidine-tagged proteins. *Chem. Mater.*, **20**, 1824–1829.

297 Guo, Z., Du, F., Ren, D., Chen, Y., Zheng, J., Liu, Z., and Tian, J. (2006) Covalently porphyrin-functionalized single-walled carbon nanotubes: a novel photoactive and optical limiting donor–acceptor

nanohybrid. *J. Mater. Chem.*, **16**, 3021–3030.

298 Gomez-Escalonilla, M.J., Atienzar, P., Garcia Fierro, J.L., Garcia, H., and Langa, F. (2008) Heck reaction on single-walled carbon nanotubes. Synthesis and photochemical properties of a wall functionalized SWNT–anthracene derivative. *J. Mater. Chem.*, **18**, 1592–1600.

299 Li, H., Cheng, F., Duft, A.M., and Adronov, A. (2005) Functionalization of single-walled carbon nanotubes with well-defined polystyrene by click coupling. *J. Am. Chem. Soc.*, **127**, 14518–14524.

300 Campidelli, S., Ballesteros, B., Filoramo, A., Diaz, D.D., de la Torre, G., Torres, T., Rahman, G.M.A., Ehli, C., Kiessling, D., Werner, F., Sgobba, V., Guldi, D.M., Cioffi, C., Prato, M., and Bourgoin, J.-P. (2008) Facile decoration of functionalized single-wall carbon nanotubes with phthalocyanines via\click chemistry\. *J. Am. Chem. Soc.*, **130**, 11503–11509.

301 Kumar, I., Rana, S., Rode, C.V., and Cho, J.W. (2008) Functionalization of single-walled carbon nanotubes with azides derived from amino acids using click chemistry. *J. Nanosci. Nanotechnol.*, **8**, 3351–3356.

302 Cheng, F. and Adronov, A. (2006) Suzuki coupling reactions for the surface functionalization of single-walled carbon nanotubes. *Chem. Mater.*, **18**, 5389–5391.

303 Flatt, A.K., Chen, B., and Tour, J.M. (2005) Fabrication of carbon nanotube–molecule–silicon junctions. *J. Am. Chem. Soc.*, **127**, 8918–8919.

304 Chen, B., Flatt, A.K., Jian, H., Hudson, J.L., and Tour, J.M. (2005) Molecular grafting to silicon surfaces in air using organic triazenes as stable diazonium sources and HF as a constant hydride-passivation source. *Chem. Mater.*, **17**, 4832–4836.

305 Tulevski, G.S., Hannon, J., Afzali, A., Chen, Z., Avouris, P., and Kagan, C.R. (2007) Chemically assisted directed assembly of carbon nanotubes for the fabrication of large-scale device arrays. *J. Am. Chem. Soc.*, **129**, 11964–11968.

306 Buffa, F., Hu, H., and Resasco, D.E. (2005) Side-wall functionalization of single-walled carbon nanotubes with 4-hydroxymethylaniline followed by polymerization of e-caprolactone. *Macromolecules*, **38**, 8258–8263.

307 Yi, B., Rajagopalan, R., Foley, H.C., Kim, U.J., Liu, X., and Eklund, P.C. (2006) Catalytic polymerization and facile grafting of poly(furfuryl alcohol) to single-wall carbon nanotube: preparation of nanocomposite carbon. *J. Am. Chem. Soc.*, **128**, 11307–11313.

308 Wang, S., Liang, Z., Liu, T., Wang, B., and Zhang, C. (2006) Effective amino-functionalization of carbon nanotubes for reinforcing epoxy polymer composites. *Nanotechnology*, **17**, 1551–1557.

309 Vigolo, B., Mamane, V., Valsaque, F., Le, T.N.H., Thabit, J., Ghanbaja, J., Aranda, L., Fort, Y., and McRae, E. (2009) Evidence of sidewall covalent functionalization of single-walled carbon nanotubes and its advantages for composite processing. *Carbon*, **47**, 411–419.

310 Wang, G.-J., Huang, S.-Z., Wang, Y., Liu, L., Qiu, J., and Li, Y. (2007) Synthesis of water-soluble single-walled carbon nanotubes by RAFT polymerization. *Polymer*, **48**, 728–733.

311 Wu, W., Tsarevsky, N.V., Hudson, J.L., Tour, J.M., Matyjaszewski, K., and Kowalewski, T. (2007) Hairy single-walled carbon nanotubes prepared by atom transfer radical polymerization. *Small*, **3**, 1803–1810.

312 Matrab, T., Chancolon, J., L'Hermite, M.M., Rouzaud, J.-N., Deniau, G., Boudou, J.-P., Chehimi, M.M., and Delamar, M. (2006) Atom transfer radical polymerization (ATRP) initiated by aryl diazonium salts: a new route for surface modification of multiwalled carbon nanotubes by tethered polymer chains. *Colloids Surf. A*, **287**, 217–221.

313 Strano, M.S., Dyke, C.A., Usrey, M.L., Barone, P.W., Allen, M.J., Shan, H., Kittrell, C., Hauge, R.H., Tour, J.M., and Smalley, R.E. (2003) Electronic structure control of single-walled carbon nanotube functionalization. *Science*, **301**, 1519–1522.

314 Wang, C., Cao, Q., Ozel, T., Gaur, A., Rogers, J.A., and Shim, M. (2005) Electronically selective chemical functionalization of carbon nanotubes: correlation between Raman spectral and electrical responses. *J. Am. Chem. Soc.*, **127**, 11460–11468.

315 Nair, N., Usrey, M.L., Kim, W.-J., Braatz, R.D., and Strano, M.S. (2006) Estimation of the (n,m) concentration distribution of single-walled carbon nanotubes from photoabsorption spectra. *Anal. Chem.*, **78**, 7689–7696.

316 Nair, N., Kim, W.-J., Usrey, M.L., and Strano, M.S. (2007) A structure–reactivity relationship for single walled carbon nanotubes reacting with 4-hydroxybenzene diazonium salt. *J. Am. Chem. Soc.*, **129**, 3946–3954.

317 Fantini, C., Usrey, M.L., and Strano, M.S. (2007) Investigation of electronic and vibrational properties of single-walled carbon nanotubes functionalized with diazonium salts. *J. Phys. Chem. C*, **111**, 17941–17946.

318 Doyle, C.D., Rocha, J.-D.R., Weisman, R.B., and Tour, J.M. (2008) Structure-dependent reactivity of semiconducting single-walled carbon nanotubes with benzenediazonium salts. *J. Am. Chem. Soc.*, **130**, 6795–6800.

319 Schmidt, G., Gallon, S., Esnouf, S., Bourgoin, J.-P., and Chenevier, P. (2009) Mechanism of the coupling of diazonium to single-walled carbon nanotubes and its consequences. *Chem. Eur. J.*, **15**, 2101–2110.

320 Usrey, M.L., Lippmann, E.S., and Strano, M.S. (2005) Evidence for a two-step mechanism in electronically selective single-walled carbon nanotube reactions. *J. Am. Chem. Soc.*, **127**, 16129–16135.

321 Dyke, C.A., Stewart, M.P., and Tour, J.M. (2005) Separation of single-walled carbon nanotubes on silica gel. Materials morphology and Raman excitation wavelength affect data interpretation. *J. Am. Chem. Soc.*, **127**, 4497–4509.

322 Kim, W.-J., Usrey, M.L., and Strano, M.S. (2007) Selective functionalization and free solution electrophoresis of single-walled carbon nanotubes: separate enrichment of metallic and semiconducting SWNT. *Chem. Mater.*, **19**, 1571–1576.

323 Toyoda, S., Yamaguchi, Y., Hiwatashi, M., Tomonari, Y., Murakami, H., and Nakashima, N. (2007) Separation of semiconducting single-walled carbon nanotubes by using a long-alkyl-chain benzenediazonium compound. *Chem. Asian J.*, **2**, 145–149.

324 Tagmatarchis, N., Georgakilas, V., Prato, M., and Shinohara, H. (2002) Sidewall functionalization of single-walled carbon nanotubes through electrophilic addition. *Chem. Commun.*, 2010–2011.

325 Choi, J.H., Oh, S.B., Chang, J., Kim, I., Ha, C.-S., Kim, B.G., Han, J.H., Joo, S.-W., Kim, G.-H., and Paik, H.-J. (2005) Graft polymerization of styrene from single-walled carbon nanotube using atom transfer radical polymerization. *Polym. Bull.*, **55**, 173–179.

326 Xu, Y., Wang, X., Tian, R., Li, S., Wan, L., Li, M., You, H., Li, Q., and Wang, S. (2008) Microwave-induced electrophilic addition of single-walled carbon nanotubes with alkylhalides. *Appl. Surf. Sci.*, **254**, 2431–2435.

327 Balaban, T.S., Balaban, M.C., Malik, S., Hennrich, F., Fischer, R., Roesner, H., and Kappes, M.M. (2006) Polyacylation of single-walled carbon nanotubes under Friedel–Crafts conditions: an efficient method for functionalizing, purifying, decorating, and linking carbon allotropes. *Adv. Mater.*, **18**, 2763–2767.

328 Lee, H.-J., Han, S.-W., Kwon, Y.-D., Tan, L.-S., and Baek, J.-B. (2008) Functionalization of multi-walled carbon nanotubes with various 4-substituted benzoic acids in mild polyphosphoric acid/phosphorous pentoxide. *Carbon*, **46**, 1850–1859.

329 Viswanathan, G., Chakrapani, N., Yang, H., Wei, B., Chung, H., Cho, K., Ryu Chang, Y., and Ajayan Pulickel, M. (2003) Single-step *in situ* synthesis of polymer-grafted single-wall nanotube composites. *J. Am. Chem. Soc.*, **125**, 9258–9259.

330 Chen, S., Chen, D., and Wu, G. (2006) Grafting of poly(tBA) and PtBA-b-PMMA onto the surface of SWNTs using carbanions as the initiator. *Macromol. Rapid Commun.*, **27**, 882–887.

331 Blake, R., Gun'ko, Y.K., Coleman, J., Cadek, M., Fonseca, A., Nagy, J.B., and Blau, W.J. (2004) A generic organometallic approach toward ultra-strong carbon nanotube polymer composites. *J. Am. Chem. Soc.*, **126**, 10226–10227.

332 Blake, R., Coleman, J.N., Byrne, M.T., McCarthy, J.E., Perova, T.S., Blau, W.J., Fonseca, A., Nagy, J.B., and Gun'ko, Y.K. (2006) Reinforcement of poly(vinyl chloride) and polystyrene using chlorinated polypropylene grafted carbon nanotubes. *J. Mater. Chem.*, **16**, 4206–4213.

333 Graupner, R., Abraham, J., Wunderlich, D., Vencelova, A., Lauffer, P., Roehrl, J., Hundhausen, M., Ley, L., and Hirsch, A. (2006) Nucleophilic-alkylation-reoxidation: a functionalization sequence for single-wall carbon nanotubes. *J. Am. Chem. Soc.*, **128**, 6683–6689.

334 Mueller, M., Maultzsch, J., Wunderlich, D., Hirsch, A., and Thomsen, C. (2007) Raman spectroscopy of pentyl-functionalized carbon nanotubes. *Phys. Stat. Sol. RRL*, **1**, 144–146.

335 Mueller, M., Maultzsch, J., Wunderlich, D., Hirsch, A., and Thomsen, C. (2007) Raman spectroscopy on chemically functionalized carbon nanotubes. *Phys. Stat. Sol.(b)*, **244**, 4056–4059.

336 Wunderlich, D., Hauke, F., and Hirsch, A. (2008) Preferred functionalization of metallic and small-diameter single-walled carbon nanotubes by nucleophilic addition of organolithium and -magnesium compounds followed by reoxidation. *Chem. Eur. J.*, **14**, 1607–1614.

337 Chen, S., Shen, W., Wu, G., Chen, D., and Jiang, M. (2005) A new approach to the functionalization of single-walled carbon nanotubes with both alkyl and carboxyl groups. *Chem. Phys. Lett.*, **402**, 312–317.

338 Chen, S., Wu, G., and Chen, D. (2006) An easy approach to hydroxyethylated SWNTs and the high thermal stability of the inner grafted hydroxyethyl groups. *Nanotechnology*, **17**, 2368–2372.

339 Roubeau, O., Lucas, A., Pénicaud, A., and Derre, A. (2007) Covalent functionalization of carbon nanotubes through organometallic reduction and electrophilic attack. *J. Nanosci. Nanotechnol.*, **7**, 3509–3513.

340 Syrgiannis, Z., Hauke, F., Röhrl, J., Hundhausen, M., Graupner, R., Elemes, Y., and Hirsch, A. (2008) Covalent sidewall functionalization of SWNTs by nucleophilic addition of lithium amides. *Eur. J. Org. Chem.*, 2544–2550.

7
Carbon-Based Nanomaterial Applications in Biomedicine

Prabhpreet Singh, Tatiana Da Ros, Kostas Kostarelos, Maurizio Prato, and Alberto Bianco

7.1
Introduction

Nanotechnology refers to structures in the 1–100 nm size in at least one dimension. Using nanotechnology, it may be possible to achieve improved delivery of poorly water-soluble drugs, targeted delivery of drugs in a cell- or tissue-specific manner, codelivery of two or more drugs, visualization of sites of drug delivery by combining therapeutic agents with imaging modalities, and many more [1]. The development of nanocarriers for gene and drug delivery has become possible due to advances in nanotechnology. Currently, liposomes are used clinically as drug carriers while polymer micelles, and silica nanoparticles are potential carriers under development [2]. Indeed, there is continuous demand for novel delivery systems that are capable of protecting, transporting, and releasing active molecules (i.e., drugs, antigens, antibodies, nucleic acids) to specific sites of action [3–8]. This is of fundamental importance, particularly, in cancer therapy. Carbon nanomaterial can be considered as novel and innovative tools in the development of alternative methodologies for the delivery of therapeutic molecules [9]. This chapter will describe the structural characteristics and potential of carbon nanomaterials, including carbon nanotubes (CNTs), nanohorns (CNHs), and nanodiamonds (NDs) in biomedicine.

7.2
Carbon Nanotubes

7.2.1
Structures, Characteristics, and Derivatization of Carbon Nanotubes

Carbon nanotubes resemble graphene sheets rolled up to form cylinders, with different characteristics and peculiarities depending on the number of concentric cylinders and on the rolling up angle [10]. The main distinction among single-, double- and multiwalled carbon nanotubes, SWCNTs, DWCNTs, and MWCNTs, respectively, is the presence of different rolling curvatures, which confer the metallic

Carbon Nanotubes and Related Structures. Edited by Dirk M. Guldi and Nazario Martín
Copyright © 2010 WILEY-VCH Verlag GmbH & Co. KGaA, Weinheim
ISBN: 978-3-527-32406-4

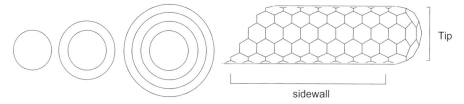

Figure 7.1 (Left) Frontal vision of SWCNT, DWCNT, and MWCNT; (right) CNT schematic representation, with tip and sidewall.

or semiconductor characteristics to the tubes (Figure 7.1, left). These objects are almost monodimensional, presenting diameters in the order of nanometers and lengths up to several microns. Improvements in CNT production are still under development, since till date every methodology has disadvantages, mainly represented by the presence of metallic nanoparticles and carbonaceous impurities in the production soot. The induction of selective chirality of CNTs is still far from being reached, the control of the rolling up angle has still not been obtained, and the separation of metallic or semiconducting CNTs is difficult and imprecise. The control of length and diameter, although better, is still not achieved as well.

The possibility of applications is wide, spanning in various fields of science from electronics to medicine. In some cases CNTs can be used as produced (pristine CNTs), but often a derivatization is necessary, especially to improve their manipulations and solubility characteristics.

From the reactivity point of view, it is possible to distinguish two different parts: the tip and the sidewall (Figure 7.1, right). The former resembles the fullerene hemisphere in curvature and reactivity, while the latter presents a monodirectional distortion from planarity and is much less reactive than the tip, due to lower strain energy [11].

The chemistry of CNTs is evolving quite rapidly and different approaches have been realized since the initial studies on their chemical modification. It is possible to distinguish three main methodologies: (a) the supramolecular derivatization, (b) the covalent functionalization, and (c) the filling of the internal cavities of the tubes themselves [12–14]. The first two methodologies are applied on the external CNT surface and can change some characteristics such as conductivity or solubility. The latter is quite interesting for producing CNT-based reservoirs, but is not effective in assisting the solubilization. Solubility is really poor in all the solvents for pristine CNTs that form tight bundles difficult to disperse. Considering that many properties are typical for individual CNTs and can be attenuated in the presence of ropes, the importance of good dispersions or solution appears to be fundamental not only for biological purposes but also in the materials chemistry.

The supramolecular approach (Scheme 7.1) exerts great importance when CNT use is related to the preservation of the intact aromatic structure, as in the case of metallic CNTs, where the introduction of defects on the carbon surface diminished their intrinsic capability of conducting electrons. Carbon nanotubes interact commonly with polymers and biopolymers. Poly(*m*-phenylenevinylene-*co*-2,5-dioctoxy-*p*-

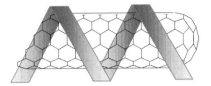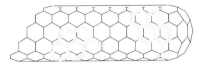

Scheme 7.1 Supramolecular complexation of CNTs by means of (bio)polymers and aromatic compounds.

phenylene vinylene) was the first to show stable π–π interactions with MWNTs [15]. Nakashima *et al.* reported the favorable interaction to CNTs of aromatic compounds such as pyrene, properly functionalized with a positive charge to induce water solubility [16, 17].

The covalent functionalization approach can be performed using two different methodologies: (i) amidation and esterification of oxidized SWCNTs; and (ii) sidewall functionalization [12–14, 18, 19];

i. **Amidation and esterification of oxidized SWCNTs**: The treatment with strong oxidizing acids and sonication modifies CNT surface with the introduction of carboxylic, carbonyl, and hydroxyl functions. The tips are removed and the opening of the tubes leads to the purification from metallic nanoparticles during this process. The carboxylic acids, which are the main type of the induced functionalizations, can be activated by thionyl or oxalyl chloride or in milder conditions, and then allowed to react with alcohols or amines (Scheme 7.2).

ii. **Sidewall functionalization of SWCNTs**: Many different methods have been proposed and developed for the sidewall derivatization of CNTs. Fluorination, radical, electrophilic, and nucleophilic additions, together with cycloaddition reactions have been widely used, obtaining a good variety of sidewall functionalities (Scheme 7.3).

7.2.2
Biological Applications of CNTs

Many applications of CNTs in medicine and related fields are envisaged and many research groups are involved in these efforts today. One of the major driving forces for biomedical exploration is the CNT capability to penetrate into cells. It is necessary to underline that apparently contradictory results can be reported, mainly because the structural differences among the studied materials are enormous. For example, some macroscopic differences, such as the use of SWCNTs or MWCNTs, type of surface modification, and degree of bundling, need to be reported. However often diameter

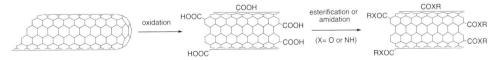

Scheme 7.2 Oxidation and covalent functionalization of CNTs.

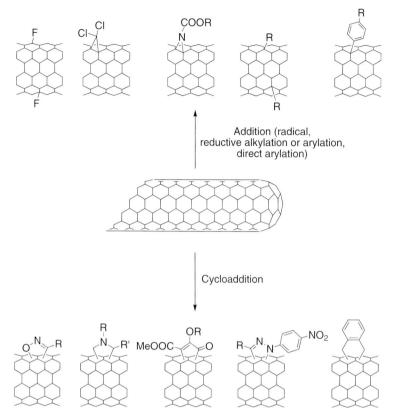

Scheme 7.3 Sidewall derivatization of CNTs.

and length values are neglected and, when reported, they represent an estimated average. The following sections report the representative examples of biological knowledge generated today according to the field of application and not taking into account the different CNT materials used.

7.2.2.1 Cell Penetration

It has already been accepted that CNTs can easily translocate into a wide range of cells [20]. The possible mechanisms of this process are phagocytosis, endocytosis, or membrane piercing, and the differences between the used CNT materials are probably due to the different surface coatings on the CNT backbone. Endocytosis has been reported for DNA–CNT complexes by Kam *et al.* [21, 22] and Heller *et al.* [23], while the membrane piercing [20] has been suggested for block copolymer-coated noncovalently functionalized MWCNTs [24] and oxidized, water-soluble CNTs [25].

Liu *et al.* investigated the possibility for carbon nanotubes to penetrate the cell wall and cell membrane of intact plant cells (BY-2 cells), using oxidized SWCNTs noncovalently labeled with fluorescein isothiocyanate (Ox-SWCNT/FITC) or pristine SWCNT complexed with FITC-labeled ssDNA (SWCNT/ssDNA). In both cases the results were really encouraging; in fact the CNTs enter cells without toxicity. Cells preserved their morphology, cytoplasmic fluidity, and proliferation rate, and, interestingly the ssDNA delivered alone was not founded in the BY-2 cells. The distribution in the cells is different: the Ox-SWCNT/FITC concentrated in the vacuoles while the SWCNT/ssDNA was seen in the cytoplasmic strands. The internalization was time- and temperature dependent, suggesting the involvement of endocytosis that in plant cells is blocked by decrease in the temperature [26].

7.2.2.2 Drug Delivery

As already mentioned, the possibility to fill the internal cavity of CNTs represents a new approach to create drug reservoirs. Hampel *et al.* explored this field using carboplatin as a filling drug [27]. They used oxidized MWCNTs with different diameters (ranging from 10 to 30 nm) and lengths (10–30 μm). The oxidation process was performed for opening the tubes. The incorporation seems to be dependent on the temperature. With different techniques (such as X-ray photoelectron spectroscopy) the authors confirmed the drug retention into CNTs, although also traces of external deposition of carboplatin were found. The biological tests performed on bladder cancer cells found that the incubation with carboplatin-MWCNTs inhibits cell growth, while MWCNTs alone do not exert any effects.

A novel strategy to functionalize carbon nanotubes with two different molecules using the 1,3-dipolar cycloaddition of azomethine ylides was presented by Pastorin *et al.*, who covalently linked methotrexate and FITC to MWCNTs (Figure 7.2, left). The penetration of this new derivative was studied by epifluorescence and confocal microscopy on human Jurkat T lymphocytes, demonstrating a fast accumulation in the cytoplasm [28].

McDevitt *et al.* prepared CNT derivatives bearing at the same time tumor-specific monoclonal antibodies, radiometal-ion chelates, or fluorescent probes. Water-soluble compounds, obtained by dipolar cycloaddition were treated with DOTA-NCS and LC-SMCC, and then ^{111}In was chelated (Figure 7.2, right). The conjugates were studied *in vitro* by flow cytometry and cell-based immunoreactivity assays were performed, while biodistribution was studied *in vivo* in mice with xenografted lymphoma [29].

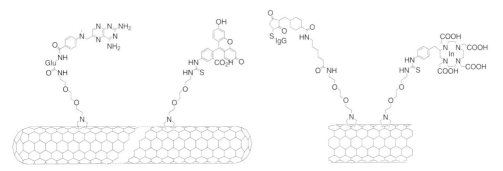

Figure 7.2 Molecular structures of bifunctional CNT derivatives.

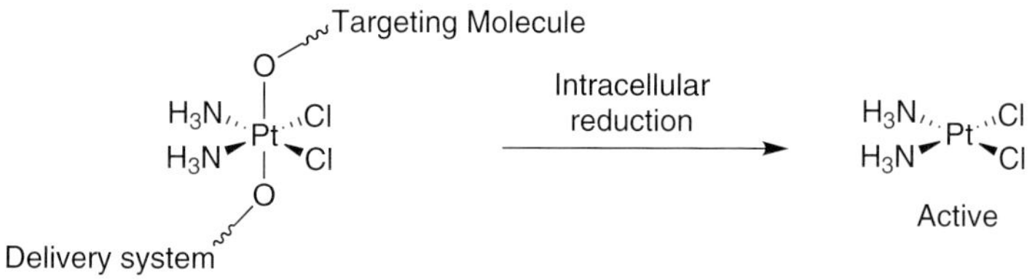

Targeting Molecule

Intracellular reduction

Active

Delivery system

Figure 7.3 Cisplatin release mechanism.

SWCNTs coated with modified phospholipids (PL-PEG-NH$_2$) allowed to link a platinum(IV) derivative through the amine residue to one of axial Pt ligands [30]. This molecule itself is almost nontoxic to testicular cancer cells, but becomes significantly cytotoxic when attached to the surface of amine-functionalized soluble SWCNTs. This is due to the fact that the presence of CNTs induces the prodrug cellular uptake, which takes place and confines the complex into endosomes, where the pH is lower than in the cell incubation medium. The acid environment facilitates the reduction from Pt(IV) to Pt(II) and the release of active cisplatin, with loss of axial ligands (Figure 7.3).

More recently the same authors described a new complex in which one platinum axial ligand is linked to folic acid, as targeting moiety (Figure 7.4, left) [31]. The preparation of high molecular weight complexes should also increase the blood circulation time. They investigated Folate Receptor-positive [FR(+)] human choriocarcinoma (JAR) and human nasopharyngeal carcinoma (KB) cell lines, using as negative control FR-negative [FR(−)] testicular carcinoma cells (NTera-2),

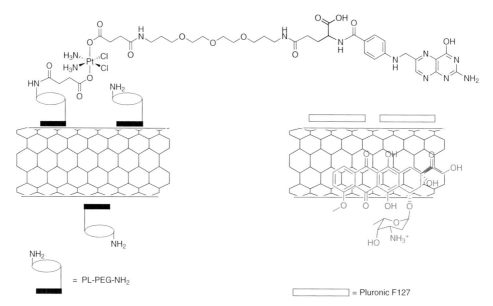

Figure 7.4 Noncovalent attachment of antitumoral drugs to CNTs.

demonstrating that FR(+) cells are more affected. Moreover, the so-prepared prodrug resulted to be much more active than cisplatin itself, forming, after intracellular reduction, cisplatin 1,2-d(GpG) intrastrand cross-links with nuclear DNA.

It has been recently demonstrated that doxorubicin adsorbed on MWCNT surface (Figure 7.4, right) by means of π–π interaction results to be more efficient than the free one, probably because of ameliorate delivery into cells [32].

7.2.2.3 Gene Delivery

One of the first approaches toward genetic material delivery was performed by Lu *et al.* who complexed RNA polymer poly(rU) to SWCNTs by nonspecific interactions [33]. The translocation across MCF7 breast cancer cells took place and the genetic material was found across the cellular and the nuclear membranes.

Covalent derivatized SWCNTs, obtained by amidation of the nanotube carboxylic groups with a chain presenting a terminal ammonium group, were studied as telomerase reverse transcriptase small interfering RNA (TERT-siRNA) delivery system into cells [34]. The effect was assessed taking into account the proliferation and the growth of tumor cells (Lewis lung cancer cells and HeLa cells), both *in vitro* and *in vivo*. So the silencing of the targeted gene is the evidence of a good intracellular delivery of siRNA, which is locally released from nanotubes and can induce its effect.

Dai and coworkers delivered siRNA to human T cells and peripheral blood mononuclear cells. In this case SWCNTs were again coated with PL-PEG-NH$_2$. Differences in efficacy were found to depend on functionalization and degree of hydrophilicity, related to the length of the introduced PEG chain in the PL-PEG construct [35]. Short SWCNTs, noncovalently functionalized with PL-PEG, were used. In this case a cleavable bond was introduced, linking by disulfide bond the PL-PEG unit and the delivering siRNA [36]. The siRNA functionality obtained in this model resulted to be more potent than lipofectamine used as control transfection agent.

Oxidized SWCNTs, bearing positive charges able to interact with fluorescein-labeled dsDNA (dsDNA-FAM), were also coated with folic acid modified phospholipids (PL-PEG-NH$_2$), by wrapping the alkyl chains (Figure 7.5). The multifunctional system was selectively uptaken by tumor cells, in which folic receptors are overexpressed, driving the oligonucleotide into cells, as demonstrated by fluorescence imaging [37].

SWCNTs were utilized as carriers for ssDNA probe into cells, demonstrating an increased resistance of the oligonucleotides toward nuclease digestion. In fact the ssDNA is protected from enzymatic cleavage and interference from nucleic acid binding proteins, still exerting its function of targeting mRNA [38]. This increases the potentiality of CNTs as oligonucleotide vectors, as already hypothesized by Kateb *et al.* [24]. They studied the possibility of using CNTs as DNA or RNA delivery systems in brain tumors, thus demonstrating the ability of microglia to efficiently internalize MWCNTs coated with Pluronic F108 as compared to glioma cells.

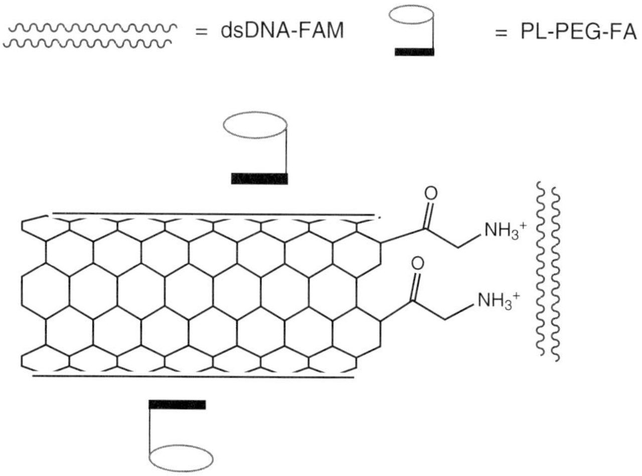

Figure 7.5 CNT-based multifunctional DNA delivery system.

7.2.2.4 Other Anticancer Approaches

The potential of CNTs as a genetic material delivery system is great, with application in gene therapy, considering the general biocompatibility of CNTs themselves. As mentioned already, nanotubes easily bind macromolecules, such as nucleic acids. Li *et al.* found that oxidized SWCNTs can inhibit DNA duplex association [39]. Moreover it has been demonstrated that they selectively induce human telomeric i-motif DNA formation, binding to the 5′-end major groove and directly stabilizing the charged CC$^+$ base pairs. These findings can be utilized in designing new compounds for cancer therapy, where the modulation of human telomeric DNA can play a fundamental role, although the biological effects of i-motif structure induction is not yet clear.

A polyadenylic acid [poly(rA)] tail is present in eukaryotic cells and human poly(rA) polymerase (PAP) and can be considered a tumor-specific target, and compounds interfering with this structure have interesting potential in the therapy. Zhao *et al.* studied oxidized SWCNTs and reported that CNTs induce single-stranded poly(rA) to self-structure into duplex but the mechanism is not understood and should involve the characteristics of the oxidized CNTs considering that amino-derivatized SWCNTs, surprisingly, do not induce self-structuation [40].

Antitumor immunotherapy is actually not very effective. In the past, Pantarotto *et al.* [41] have found that viral peptides conjugated to CNTs can improve the antipeptide immune response. Meng *et al.* follow this paved way to improve immunotherapy. Tumor lysate protein, readily obtainable from most solid tumors, has been linked to MWCNTs (Figure 7.6) and administered as a possible tumor-cell vaccine in a mouse model bearing the H22 liver cancer, with positive effects on the cure rate and cellular antitumor immune reaction [42].

Phospholipid–polyethylene glycol chains linked to a folic acid residue were linked by means of hydrophobic interactions to CNTs. The presence of folate should drive toward a selective uptake of the complex into cells overexpressing folate receptors, as it happens in some cancer cells (Scheme 7.4). After internalization, cell death was

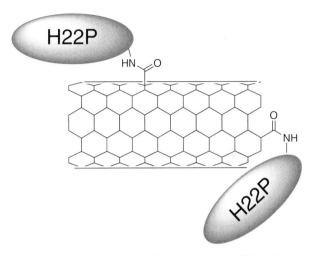

Figure 7.6 Example of tumor-cell vaccine based on CNT conjugates.

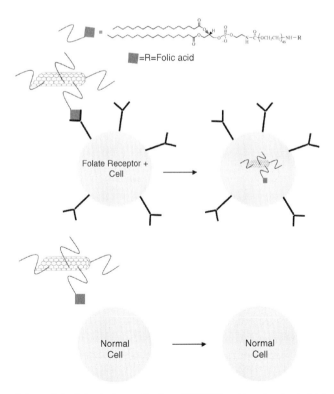

Scheme 7.4 Selective internalization of CNT-folic acid conjugates in folate receptor positive cells.

induced irradiating with near-infrared (NIR) light. This action proved to be selective because the presence of CNTs in cells increases the temperature, thereby inducing thermal ablation when irradiated [21].

Analogously, CNTs bearing two specific monoclonal antibodies (insulin-like growth factor 1 receptor, IGF1R and human endothelial receptor 2, HER2), prepared by supramolecular approach with properly functionalized pyrene units, have been used to kill breast cancer cells with NIR irradiation [43]. The main problem of near infrared irradiation application is due to its poor tissue penetration capacity, which allows the treatment of only superficial cancer lesions. Radiofrequency waves, on the contrary, penetrate more into tissues, and their interaction with internalized CNTs produces an increase in temperature of cells, thereby inducing cell death. Gannon *et al.* proposed the use of these waves, but in the reported case the limitation is due to the direct injection into tumor of SWCNT coated with polyphenylene-ethynylene polymer [44].

Another atypical approach to cancer treatment is related to angiogenesis. In fact, its inhibition by interfering in the growth factor balance can obstacle the insurgence of diseases, among which tumors can be mentioned. A good model to study angiogenesis is the chick chorioallantoic membrane (CAM). A study performed by Murugesan *et al.* employed carbon materials (such as CNT, fullerene, and graphite) to test their capability to inhibit FGF2- or VEGF-induced vessel formation in the CAM model. All the materials showed significant effects in the induced angiogenesis, acting more efficiently in comparison to VEGF, while they did not exert any effect on the basal process in absence of the already-mentioned growth factors. The mechanism involved in this process is not yet clarified but it is clear that this antiangiogenic action could be exploited in cancer treatment [45].

7.2.2.5 Neuron Interactions with CNTs

Few years ago, Lovat *et al.* reported the possibility of improving the neural signal transfer by growing neuronal circuits on CNT substrates, relating this activity to the high electrical conductivity of CNTs [46]. Later on, the same authors developed an integrated SWCNT–neuron system to test whether electrical stimulation delivered via SWCNTs can induce neuronal signaling. The patch clamped hippocampal cells, cultured on SWCNT substrates, have shown that they can be stimulated via the SWCNT layers. Any resistive coupling between biomembranes and SWCNTs was qualitatively distinguishable from a coupling between SWCNTs and the patch pipette through the patchseal path to ground [47]. Using single-cell electrophysiology techniques, electron microscopy analysis, and theoretical modeling, the same authors demonstrated that CNTs enhance the responsiveness of neurons, due to the formation of tight contacts between CNTs and the cell membranes. These could favor electrical shortcuts between the proximal and distal compartments of the neuron [48].

While in the already mentioned works, the CNTs were deposed on glass cover-slips for growing cells, a different approach has been used by Keefer *et al.* using

CNTs to electrochemically coat tungsten and stainless steel wire electrodes. This treatment led to an increase in recording and electrical stimulation of neuronal cultures [49]. Kotov and coworkers also prepared a layer-by-layer composite with SWCNTs and laminin, an essential glycoprotein of human extracellular matrix. The use of this surface for cellular culture induced neural stem cells (NSC) differentiation and the creation of functional neural network. Also in this case, the application of current through CNTs stimulated the action potentials [50]. Carbon nanotubes have also been arranged in vertical alignment to the gate insulator of an ion-sensitive field-effect transistor to act as electrical interfaces to neurons, to study the interactions with them, and an enhance efficiency has been recorded in neuronal electrical activity [51].

Water-soluble SWCNTs grafted to polyethylene glycol have been used in the culturing medium for neural growth. These derivatives demonstrated their ability to block the stimulated membrane endocytosis in neurons and this could justify the evidenced extended neurite length [52], while MWCNTs coated with Pluronic F127 (PF127) injected in mouse cerebral cortex do not induce degeneration of the neurons close to the injection site, while PF127 itself can induce apoptosis of mouse primary cortical neurons, implying that the presence of MWCNTs inhibits PF127-induced apoptosis [53].

All these successful works performed on neuronal growth and stimulation led to the hope of using CNTs as material for the reconstruction of neural injured tissues in the near future, paving the way for an application in spinal cord disease resolution and neurodegeneration restoration.

7.2.2.6 Antioxidant Properties of CNTs

CNTs, when instilled in lungs, induced inflammatory and fibrotic response, which was supposed to be due to oxidative stress derived from free radicals, although no real evidence of ROS generation from CNTs was observed. Moreover Fenoglio *et al.* reported that MWCNTs do not generate oxygen or carbon-based free radicals in the presence of H_2O_2 or formate, respectively, but, on the contrary, when radicals were induced, CNTs act as radical scavengers [54]. Very recently the antioxidant properties of CNTs have been reported by Tour and coworkers. SWCNTs and ultrashort SWCNTs (US-SWCNTs) were derivatized with butylated hydroxytoluene (BHT), using two different approaches as covalent attachment of triazene to the sidewalls of pluronic-wrapped SWCNT or amidation of carboxylic residues in the case of US-SWCNT derivatives. The oxygen radical scavenging ability of the different compounds has been evaluated leading to really interesting results, although the different compounds react differently. In the first case, the pluronic-coated SWCNTs were more efficient than the corresponding BHT-derivatives, while in the case of oxidized US-SWCNTs, higher the loading of BHT residues, better was the antioxidant activity. This new finding paves way for SWCNT application as novel medical therapeutics in the antioxidant field [55]. However, it is necessary to remember that a recent report presents contradictory results indicating MWCNT toxicity in A549 cells, which can be reduced by pretreatment with antioxidants to decrease ROS production and interleukin-8 gene expression [56].

7.2.2.7 Imaging using Carbon Nanotubes

SWCNTs are endowed of fluorescent properties that can be exploited in *in vitro* and *in vivo* imaging, and that have been used to determine the uptake of nanotubes in macrophages [57], the CNT elimination in rabbits [58], and their toxicity in fruitfly larvae [59]. This surprising use of CNTs for imaging purposes is only possible when some properties of the nanotubes are preserved, that is, at least 100 nm of the tubes must be unmodified and the CNTs should not be in bundles, conditions in which the fluorescence is quenched.

More recently Welsher *et al.* used semiconducting SWCNTs wrapped with poly-ethyleneglycol as near-infrared fluorescent tags. The polyethylene glycol was conjugated to antibodies for the selective recognition for CD20 cell surface receptor on B-cells or for HER2/neu positive breast cancer cells. The intrinsic NIR photoluminescence allowed the detection of the binding to the cells, in spite of the presence of a low autofluorescence in different cells [60].

MRI is another important imaging technique used in medicine. After the discovery of the great performances of $Gd@C_{60}$, $Gd@C_{82}$ and their derivatives as contrast agents [61–63], the evolution toward the analogous use of CNT has been done by Wilson and coworkers (Figure 7.7, left) [64]. The tubes, shortened by fluorination down to 20–80 nm, were loaded with $GdCl_3$. The obtained results demonstrated that these CNTs have a relaxivity (r1) 40 times greater than any current Gd^{3+} ion-based clinical agents. The aggregation, that in the case of fullerene endohedrals can vary the relaxivity, does not affect this parameter in the case of CNTs. However, more soluble compounds have been prepared by Ashcroft *et al.* [65]. The same authors demonstrated the stability of gadonanotubes in buffers, serum, at variation of pH and temperature. Moreover they discovered a great dependence of r1 on the pH, with ultrasensitivity in a range of pH 7.0–7.4. This suggests the possibility of using the gadonanotubes for the early detection of cancer cells in which the pH can dramatically decrease [66].

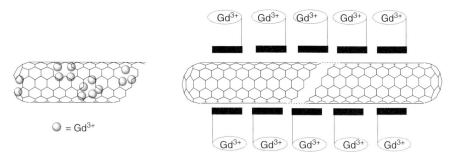

Figure 7.7 Different CNT derivatives bearing gadolinium.

A novel study on the biodistribution and the effect of SWCNTs (raw and superpurified, raw-SWCNT and SP-SWCNT, respectively) after *in vivo* exposure has been recently reported. The combination of ^{3}He and ^{1}H magnetic resonance imaging (MRI) was used in a rat model. Hyperpolarized gases such as ^{3}He acting as a contrast agent diffuses rapidly in lungs, permitting the determination of ventilated airways and alveolar spaces. The presence of metal impurity in the raw-SWCNTs was sufficient to induce a drop in magnetic field homogeneity detected in ^{3}He MRI, while no significant variation was observed after SP-SWCNT exposition [67].

Proton MRI was used to follow raw-SWCNTs after intravenous injection, finding them in spleen and kidneys. The absence of metal nanoparticles in the SP-SWCNT excluded signal modifications. When this technique was used to determine the fate of CNT after pulmonary instillation, no signal changes in liver, spleen, and kidney were detected, implying the absence of systemic circulation of CNTs after inhalation. This result was also confirmed by histological analysis, establishing the possibility of using noninvasive methodology to detect CNT presence, if associated with a proper iron impurity concentration.

A different approach has been reported by Richard *et al.* The authors, instead of filling the CNTs with the Gd derivative, used an amphiphilic gadolinium(III) chelate, absorbed on MWCNT external surface, in different concentrations (Figure 7.7, right). The obtained suspension resulted to be stable and r1 was measured in different conditions, finding a dependence on the Gd-chelate concentration. On the contrary, the transversal water proton relaxivity (T2) was independent from Gd concentration and frequency [68].

7.2.2.8 Various Applications of Carbon Nanotubes
CNTs are incredible sorbent materials for different compounds due to their large surface areas (BET value is about $\geq 600\,m^2/g$, depending on the CNT types taken into account). They have been already used as contaminant remover for water pollution in laboratory experiments [69, 70] and also for dioxin [71]. Chen *et al.* studied the sorption of americium and thorium on CNT surface, their kinetics and pH dependence [72, 73] but different efficiency probably due to different pretreatment of CNTs, found by other authors [74], and the possibility to use this new form of carbon in nuclear waste management cannot be excluded.

CNTs, both single- and multiwalled, were also used to capture bacteria, as *Streptococcus* mutans. The simple mixture of CNT bundles with bacteria leads to the formation of a precipitate, which shows bacterial adhesion to nanotubes, that depends on the CNT diameters [75].

7.2.3
Carbon Nanotube Toxicity

A big debate and many concerns about CNT toxicity arouse some years ago and up to now the information are fragmentary and apparently contradictory. Actually it is rather difficult to stigmatize definitive assumptions, considering that the performed studies cover a wide spectrum of derivatives presenting different characteristics. In general it is possible to affirm that toxicity decreases proportionally to the increase in dispersion and/or solubility degree, but it is necessary to go much further in this field [76]. A systematic study on the effect of structural defects of the MWCNT carbon cages on lung toxicity has been performed by Fenoglio *et al.*, who examined CNTs (i) with introduced structural defects, (ii) with reduced oxygenated carbon functionalities and metallic oxides, (iii) in a purified annealed sample, and (iv) in metal-deprived carbon framework with introduced structural defects [77]. The hydroxyl radical scavenging activities of modified CNTs were studied by EPR demonstrating that the original ground material exhibited an activity toward hydroxyl radicals, which disappeared in the purified annealed sample, but was detected again in metal-deprived carbon framework with introduced structural defects. *In vitro* experiments on rat lung epithelial cells were performed to assess the genotoxic potential of the different CNT preparations. Acute pulmonary toxicity and the genotoxicity of CNTs were reduced upon heating but restored upon grinding, indicating that the intrinsic toxicity of CNTs is mainly mediated by the presence of defective sites in their carbon framework [78]. The scavenging activity is related to the presence of defects and seems to go paired with the genotoxic and inflammatory potential of CNT.

Tabet *et al.* used dispersions of MWCNT in dipalmitoyl lecithin, ethanol, and phosphate-buffered saline (PBS), to study the CNT toxicity on human epithelial cell line A549. The presence of PBS induces agglomerate formation on top of the cells, but in all cases MWCNTs decrease the cellular metabolism without permeabilization of the cell membrane or apoptosis, while asbestos fibers penetrate into the cells and increase apoptosis [79]. Wick *et al.* investigated CNTs with different agglomeration degrees to determine their cytotoxicity on human MSTO-211H cells. Well-dispersed CNTs (obtained by using polyoxyethylene sorbitan monooleate) were less toxic than asbestos, and rope-like agglomerates induced higher cytotoxic effects than asbestos fibers [80].

Recently Guo *et al.* reported the modification of MWCNTs with glucosamine and decylamine by γ-ray irradiation (g-MWCNT and d-MWCNT), and they used these derivatives to perform cytotoxicity test on *Tetrahymena pyriformis*. The decylamine derivative exhibits a dose-dependent grown inhibition, attributed to the amine toxicity, driven into cells by the nanotubes. The comparison between purified MWCNTs and g-MWCNTs demonstrated growth stimulation by the latter. It seems that the hydrophilic compound is able to complex peptone, present in the medium, and to transfer it into cells. This means that any nonspecific interaction between CNTs and components of the culture medium must be considered in evaluating the cytotoxicity of CNTs [81]. Considering that almost nothing is known about CNT impact on natural ecosystems, the ciliated protozoan *Tetrahymena thermophila* has been also studied because of its role in the regulation of microbial populations through the ingestion and digestion of bacteria. It is, in fact, an important organism in wastewater treatment and, moreover an indicator of sewage effluent quality. SWCNTs have been found in these microorganisms, inducing protozoa aggregations with a consequent inability to interact with bacteria [82].

Salmonella typhimurium and *Escherichia coli* strains were used to perform mutagenicity test with MWCNTs. The mutagenic activity did not appear, even in presence of the metabolic activators [83].

Different tests devoted to establish the genotoxic potential of MWCNTs has been performed both *in vivo* in rats and *in vitro* on rat lung epithelial cells or human epithelial cells. Three days after carbon nanotube intratracheal administration, type II pneumocytes present micronuclei with a dose-dependent increase. The same behavior was reported for the *in vitro* experiments, proving the possibility of MWCNTs to induce clastogenic and aneugenic events (respectively increase the rate of genetic mutation due to DNA breaks and loss or gain of whole chromosomes) [84]. Often the clastogenicity is related to ROS generation and the possible presence of metallic nanoparticles (Fe or Co) would justify this action, but the MWCNT antioxidant behavior (see also Section 7.2.2.6) suggests different pathways and indicates the necessity to explore these systems more in details.

7.3
Carbon Nanohorns

7.3.1
Structure, Characteristics and Functionalization of SWCNHs

Single-walled carbon nanohorns (SWCNHs) were first observed by Iijima in 1999, during CO_2 laser ablation of graphite at room temperature without a metal catalyst [85]. SWCNHs are spherical aggregates (diameter 80–100 nm) of graphitic tubes (Figure 7.8). These tubular structures mainly consist of tubes that have closed

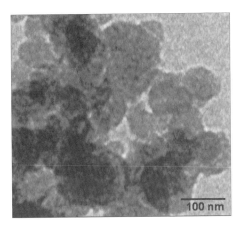

100 nm

Figure 7.8 Transmission electron microscopy photograph of pristine SWCNH.

ends with cone shaped caps (horns) with an average cone angle of 120°. Each individual tube has a diameter of 2–3 nm and an average length in the range of 30–50 nm [85].

Due to their peculiar geometry, reminiscent of a sponge, carbon nanohorns can be exploited for their capacity to adsorb most types of molecules [86]. Single-walled carbon nanohorns are very little soluble in organic solvents and basically insoluble in aqueous solution. A prerequisite for biomedical application is the increase in the solubility of such material. So, it has to be stressed that functionalization of carbon nanohorns is obligatory for rendering this material biocompatible. Indeed, carbon nanohorns behave like carbon nanotubes and can be dispersed or solubilized into physiological or water solutions, provided they are modified at their surface with suitable functional groups. Contrary to carbon nanotube, the reaction with SWCNHs proceeds well due to better dispersion of nanohorns in the examined solvents because the presence of rough surface structures of nanohorn aggregates prohibits the increase in the aggregate–aggregate contact area, leading to their weak interactions via van der Waals forces.

SWCNHs were fluorinated using elemental fluorine in the temperature range of 303–473 K without destroying the nanohorns [87]. The nature of C—F bonds in F-SWCNHs changed from semiionic to covalent with an increase in the fluorination temperature. After fluorination at 473 K, several nanoscale windows were produced on the walls of SWCNHs. Tagmatarchis *et al.* reported organic covalent functionalization on the sidewalls of SWCNHs, via the 1,3-dipolar cycloaddition of azomethine ylides generated *in situ* upon thermal condensation of different aldehydes and α-aminoacids (Figure 7.9). The functionalized nanohorns (**1–2**) were soluble in several organic solvents and were characterized by NMR, UV/Vis-NIR, IR, TEM, and EDX spectroscopy [88]. Prato and coworkers also reported similar 1,3-dipolar cycloaddition of azomethine ylides on SWCNH. Functionalized nanohorns (**1 and 3**) were soluble in DMF, chloroform, and dichloromethane but insoluble in THF, diethyl ether, and methanol [89]. Tagmatarchis and coworkers also reported the functionalization procedure of SWCNH that is based on the opening of their conical and highly strained end on treatment with molecular O_2 at 550 °C. The as-generated carboxylic groups terminated nanohorns, converted to the corresponding acyl chlorides, and reacted with a variety of amines, alcohols, and thiols, possesses short and long hydrophobic alkyl chains, polar oligoethylene units, aromatic chromophores such as pyrene or anthracene groups to give corresponding SWCNH-based amide (**4**), ester (**5**), and thioester (**6**) material [90]. Recently they also reported covalent functionalization of SWCNHs utilizing *in situ* generated aryl diazonium compounds. Due to grafting of aryl moieties on the skeleton of SWCNH (**7**), the solubility is significantly enhanced. Water-soluble cationic aryl ammonium functionalized SWCNHs are expected to be useful in biotechnological applications such as drugs and gene delivery systems [91]. Recently, we also reported the organic functionalization of CNHs by direct addition of diamine moiety to pristine CNHs that can be further modified with fluorescent probe (**8**). The resulting functionalized SWCNHs were biocompatible and were

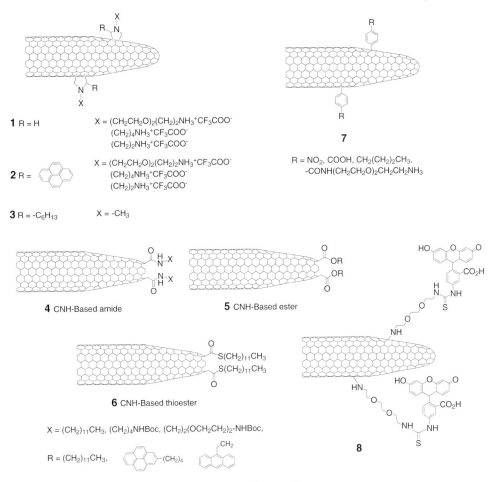

Figure 7.9 Structures of functionalized single-walled carbon nanohorns.

evaluated on the basis of their capacity to be integrated into a cellular system [92]. In a noncovalent modification approach, amphiphillic agents (i.e., molecules having both hydrophobic and hydrophilic moieties) have been used as surfactants to improve the solubility of the SWCNHs.

7.3.2
Biomedical Applications of Carbon Nanohorns

SWCNHs, with their unique flower-like structure, are promising substrates for supporting materials for drug delivery systems and radiographic contrast agents. Although SWCNHs is one of the most attractive nanomaterials, their biomedical applications have not been satisfactorily achieved [93].

7.3.2.1 Carbon Nanohorn as Potent Laser Therapeutic Agent

Dai *et al.* [36] have described a new cancer therapy, which utilizes carbon nanomaterial, Fe/Co graphitic cell nanocrystal, and near-infrared laser irradiation. Biological systems are transparent to the NIR light region (700–1100 nm) [21]. Carbon nanomaterial, when exposed to NIR radiation shows NIR-laser driven exothermy because optically stimulated electronic excitation of SWCNHs is transferred rapidly to molecular vibration energies causing heating in a similar way to photoinduced CNTs. For selective elimination of microorganisms, molecular recognition element (MRE) [wheat germ agglutinin (WGA) for *S. Cerevisiae* and antilipopolysaccharide core antibody (LPS core) for *E. coli*], noncovalently linked to SWCNHs, has recently been synthesized and studied by Miyako *et al.* [94]. They used SWCNH-COOH functionalized noncovalently with polyethyleneglycol carbamyl diasteroyl phosphatidiylethanol amine (PEG-PL) complex to attach noncovalently, respective MRE units. The MRE-PEG-PL-CNH complex binds specifically to microbial surface and after irradiation of CNH–microbe complex using NIR laser, microbial cells were destroyed due to NIR laser-triggered exothermy of the CNH complexes. Similarly, for removal of virus, they reported a complex containing T7 tag antibody noncovalently linked with SWCNH-PEG-PL. This complex binds with the model T7 bacteriophage virus and on interaction with NIR light, the T7 phage capsid protein becomes thermally denatured due to the photo-exthermocity of the NIR laser driven CNH complex (Figure 7.10) [95]. Recently, Zhang *et al.* reported Zinc phthalocyanine-oxidized SWCNH-bovine serum albumine (ZnPC-oxSWCNH-BSA) hybrid molecule-based cancer phototherapy system that uses a single laser. In this system, ZnPc acts as photodynamic therapy (PDT) agent and SWCNHox as the photohyperthermia (PHT) agent. When ZnPC-oxSWCNH-BSA complex was injected into tumors that were subcutaneously transplanted into mice, subsequent 670 nm laser irradiation caused the tumor to disappear, whereas on injecting either ZnPc or SWCNHox–BSA, the tumor continued to grow [96].

7.3.2.2 Carbon Nanohorn for Drug Delivery

SWCNHs with a diameter of 80–100 nm are particularly suitable as drug carriers to the tumor tissue because their size fulfils the condition for achieving the enhanced permeability and retention (EPR) effect. They can permeate through the damaged vessels in tumor tissue and remain there because of little lymphatic drainage. In addition to an extensive surface area, carbon nanohorns have a high number of interstices, which allow the adsorption of a large amount of guest molecules. As single-walled carbon nanohorns have tubes at the end of spherical aggregates with diameter 2–5 nm, holes can be generated at the tips of the tubes and can be exploited to insert different therapeutic agents into their empty space. Finally, SWCNHs do not exhibit cytotoxicity, making them potentially applicable in drug delivery systems.

Functionalized carbon nanohorns have been proposed for controlled drug release of anti-inflammatory and anticancer agents including dexamethasone (DEX), doxorubicin, and cisplatin. Shiba and coworkers reported the binding and release of the anti-inflammatory glucocorticoid dexamethasone by oxSWCNHs [97]. DEX--oxSWCNH complex exhibits sustained release of DEX into PBS (pH 7.4) at 37 °C and

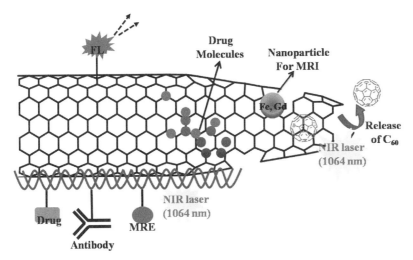

Figure 7.10 Schematic representation of noncovalent/covalent functionalization and adsorption of different therapeutic agents on/in carbon nanohorns for biomedical applications. The right end showing the release of C_{60} molecule (as a model drug) on exposure to NIR laser; adsorption of Fe and Gd nanoparticles for magnetic resonance imaging; adsorption of drugs (CDDP and DXR) molecules, whereas lower left end showing the solubilization of SWCNHs with PEG (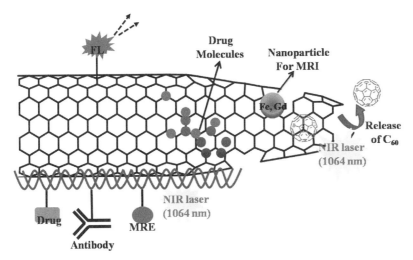) through noncovalent functionalization. Different therapeutic agents like drugs, antibodies, and MRE (molecular recognition element for microbes and yeast) can be covalently linked with PEG before being mixed with SWCNHs and the upper left end showing the covalent attachment of fluorescent moieties (**FI**) for fluorescence imaging in the cells.

more rapid release in the culture medium. DEX–oxSWCNHs activated GR-mediated transcription in mouse bone marrow stromal ST2 cells and inducted alkaline phosphatase in mouse osteoblastic MC3T3-E1 cells. The GR-mediated transcription activation obtained with 0.1 μg/ml DEX–oxSWCNHs was comparable with that obtained with 0.01–0.05 μM free DEX.

The same group also reported noncovalent modification of oxSWCNHs using an amide-linked polyethyleneglycol–doxorubicin (PEG–DXR) conjugates that results in a good dispersion [98]. Although the preparation of the conjugates required the use of organic solvents, which are not compatible with biological moieties (such as cell culture) but can eventually be eliminated using chromatographic separation, equilibrated with water. It has been demonstrated that carbon nanohorns adsorb PEG–doxorubicin via the doxorubicin moiety. The complexes, which have a diameter of 160 nm, contain more than 250 mg of PEG–doxorubicin per gram of nanohorns. Preliminary *in vitro* test have shown that complex exhibit DXR-dependent apoptotic activity against cancer cells. However the incidence of apoptosis obtained with 0.2 mg/ml PEG–DXR–oxSWCNHs was lower than that obtained with 10 ng/ml DXR. It is probable that PEG–doxorubicin is retained on the surface of the nanohorns, thus reducing its therapeutic effect. The authors could not exclude the possibility that some amount of free doxorubicin remained in their preparation and was responsible for apoptotic activity.

In another approach by the same group, cisplatin, *cis*-(NH$_3$)$_2$PtCl$_2$ (CDDP), an anticancer drug, was incorporated inside the SWCNHs. Carbon nanohorns do not alter the structure of the anticancer agent, which was slowly released in an aqueous solution [99]. Following the liberation of the drug, cell viability of human lung cancer cells was monitored for 48 h. The anticancer activity of the nanohorns containing cisplatin was almost comparable to the drug alone, while nanohorns used as controls presented no cytotoxic effects. Although carbon nanohorns can be easily dispersed in water, they have been shown to aggregate by their tendency to form clusters in the highly ionic and protein-rich cell culture media. The presence of aggregates in the micrometer scale formed by both oxidized and cisplatin-containing nanohorns is a major concern with this approach that will need to be overcome to achieve *in vivo* applications. Further, authors also investigated the effect of the hole-edge structure of the carbon nanohorns on cisplatin incorporation and release. They found that 70% of the cisplatin was released from SWCNHs having holes with hydrogen-terminated edges (NHh) and only 15% was released from SWCNHs having holes with oxygen containing functional groups at the hole edge (NHox), when they were immersed in phosphate-buffered saline [100]. Although SWCNHs having holes with hydrogen-terminated edges show high release, unfortunately, the hydrophobic property of an NH hinders the dispersion in PBS and thus cannot be used for the biological applications, whereas the release of CDDP from inside NHox in PBS was considerably suppressed because sodium ions in PBS replaced hydrogen of the oxygen containing functional group at the hole edge, which results in the plugging of the holes, hindering the CDDP release. For CDDP-incorporating NHox to be used as a CDDP-releasing carrier *in vivo*, the hole-size enlargement and control of the number of the functional groups at the hole edges of NHox might be effective in producing the necessary CDDP releasing quantities at a slower rate. So, optimization of hole opening of SWCNHs, by heat treatment from room temperature to 500–550 °C, in flowing dry air, with an increase rate of 1 °C/min, results in an 80% release of CDDP in PBS that greatly improved from the previous value of 15% [101].

Recently, the same authors have devised an alternative methodology to maintain well-dispersed nanohorns in physiological conditions or cell culture media [102]. They used peptide aptamer that specifically binds to the surface of the nanohorns as the SWCNH-binding block (NHBP-1) and synthesized a PEG-peptide aptamer conjugate for the dispersal of the SWCNHs. It was observed that modifications of the surfaces of the oxSWCNHs by PEG-NHBP do not interfere with the ability to load and release the cisplatin drug. In addition, this complex cisplatin@oxSWCNH-PEG-NHBP shows good dispersion in both PBS and culture medium, and exerted a potent cytotoxic effect against cancer cells. This is probably the approach to be followed to avoid the incapacitating aggregation phenomena described previously.

Carbon nanohorns modified with Gum Arabic can also be used for intracellular delivery because several biological cargoes could be linked with Gum Arabic [103]. Zhang *et al.* covalently attached rhodamine B to the polypeptide chain of Gum Arabic and through *in vitro* studies confirmed that modified SWCNHs can readily enter the cell, probably through the endocytotic pathway and were nontoxic.

Alternatively, carbon nanohorns have been modified with magnetite and administered *in vivo* for MRI applications. Miyawaki *et al.* reported a simple method for attaching Fe_3O_4 nanoparticles (6 nm in diameter) to hole opened SWCNHs [104], by depositing first $Fe(OAc)_2$ on oxSWCNHs that thermally decompose at 400 °C to yield super paramagnetic magnetite Fe_3O_4 strongly attached to oxSWCNHs. The attached magnetite nanoparticles induced remarkable darkening of the MR images, which allowed *in vivo* visualization of the accumulative behavior of SWCNHs in spleen and kidneys. Similarly, the same group reported that Gd-acetate clusters inside SWCNHs can be transformed into ultrafine Gd_2O_3 nanoparticles of 2–3 nm in average diameter and they retain their particle size even after treatment at 700 °C. Gd_2O_3 nanoparticle thus obtained can be useful to magnetic resonance imaging [105].

7.3.2.3 Toxicity of Carbon Nanohorns

Miyawaki *et al.* carried out an extensive *in vivo* and *in vitro* toxicological assessment of the as-grown SWCNHs for various exposure pathways, showing that as-grown SWCNHs have low toxicities. SWCNHs were found to be nonirritant and nondermal senstizer through skin primary and conjunctival irritation test and skin sensitization test. They are not carcinogenic. The acute peroral toxicity of SWCNHs was found to be quite low. Intratracheal instillation test revealed that SWCNHs rarely damaged rat lung tissue for a 90-day test period [106]. Isobe *et al.* reported the aminative deagglomerization of SWCNH agglomerates resulted in the formation of homogenous aqueous solution of primary particles of uniform size distribution. The amino group containing SWCNHs is taken up by mammalian cells but does not show significant cytotoxicity. In fact the cytotoxicity is much lower than that of the quartz microparticles [107]. Alternatively, the organic functionalization of SWCNHs permits to render such material soluble in physiological conditions. Pristine SWCNHs modified with a diamine moiety further functionalized with a fluorescent probe resulted biocompatible. Such functionalized SWCNHs were effectively phagocytozed by primary murine macrophages without affecting cell survival. However, some signs of activation, which could lead to an inflammatory status *in vivo*, were visible, such as a clear oxidative burst and IL6 production. The rapid uptake of this type of carbon nanoparticles can be particularly advantageous if a strategy targeting macrophages is envisaged. The delivery of biologically active agents using functionalized SWCNHs, which internalize into cells and contemporarily promote the activation of the immune system, is a very promising concept in vaccine development [92].

7.4
Carbon Nanodiamonds

7.4.1
Introduction

Nanodiamonds (NDs) have been already described in the 1960s [108], but in recent years, they have been widely investigated for their interesting properties. Diamonds are commonly known as stable and inert materials [109]. They are very difficult to manipulate being practically insoluble in any solvent. As a consequece, it was very difficult to imagine their use in nanomedicine. However recent findings have shown that if the dimensions of diamonds are reduced to the level of nanometers or microns, they can be treated as molecules that can be eventually functionalized at the surface [110–112]. This possibility increases their solubility and facilitates their manipulation remarkably. Shock wave transformation of graphite into sintered nanodiamond [113], or detonations of certain explosives in a closed container [114] are synthetic methods for production of nanodiamond besides commonly used CVD method (Figure 7.11) [115]. Most of the applications are based on polycrystalline diamond films or diamond particles with a relative large grain size produced by CVD, whereas, nanodiamond, powder-based particles with a much smaller size (<10 nm in diameter) can be readily generated by the detonation technique.

Nanodiamond can be functionalized with different chemical addends [116] like diazonium salts, azo-perfluoro alkyl group, fluorination, chlorination/amination, silylation and radical reaction with acyloxy, -aryl, -alkyl peroxides to name a few. Oxidized nanodiamond films (containing hydroxyl and carboxyl groups) can be also used for the covalent attachment of bioactive moioties, whereas, organic and biological molecules, such as apo-obelin and luciferase [117], cytochrome C [118], lysozyme [119] can also be noncovalently immobilized on the oxidized nanodiamond surface because oxygen-containing surface groups participate in hydrogen bonding and other interaction with the adsorbed species.

Nanodiamonds can be useful in a variety of biological applications such as carriers of drugs, genes or proteins, novel imaging probes, coatings for implantable

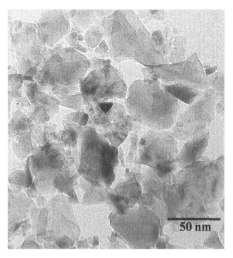

Figure 7.11 Transmission electron microscopy photograph of nonfunctionalized nanodiamonds obtained by high pressure/high temperature. (Courtesy of Christelle Mansuy.)

materials, biosensors, and biomedical nanorobots. Moreover, high adsorption capacity of nanodiamonds makes them good candidates for medical application. They can be used in protein chemistry as new adsorbents for effective extraction and purification of proteins. Due to the presence of hydrophobic and hydrophilic interaction, nanodiamonds possess high affinity for the proteins and, thus, can be used to capture the protein in dilute solution that can be analyzed by MALDI-time of flight (TOF) mass spectrometry. The promise of this method for clinical proteomics research is demonstrated with an application to human blood serum [120]. It was demonstrated that the immobilization of the protein on the nanodiamonds did not alter its stability and confirmation. Huang *et al.* [121] reported the immobilization of anti-*Salmonella* and anti-*Staphylococcus aureus* antibodies on hydrophobic and hydrophilic nanodiamond.

7.4.2
Carbon Nanodiamond as Delivery Vehicle

Huang *et al.* [122] described the adsorption of doxorubicin hydrochloride (DXR) on nanodiamond (2–8 nm diameter) that was highly functionalized with hydroxyl and carboxylic groups. Noncovalent complexes were formed with the addition of NaCl, whereas reversible release of the drug from the nanodiamonds was achieved by reducing the concentration of chloride ions (salt effect). The nanoparticles were able to enter into cells alone or complexed to DXR. To prove the capacity of these nanoobjects to pass the cell membrane, nanodiamonds were coated with a fluorescent polylysine derivative and then localized inside the cytoplasm. The same nanodiamonds were highly biocompatible, as demonstrated by the fact that cell viability was not reduced. The complexes with DXR were uptaken and apoptosis was assessed as a consequence of the liberation of the drug from the complex. The effects of doxorubicin-induced cell death were tested in comparison to the drug alone. Nanodiamonds sequestered DXR for a longer time, decreasing the efficacy compared to the drug alone, but they were proposed as an alternative technology for a delayed and time-controlled drug release, prolonging efficacy during the treatment (Figure 7.12).

Kossovsky *et al.* [123] reported surface modified diamond nanoparticles that can be used as effective antigen delivery vehicles. The particles consist of diamond substrate, coated with a carbohydrate film and the mussel adhesive protein (MAP). It is believed that diamond nanoparticles provide conformational stabilization as well as high degree of surface exposure to protein antigens resulting in a strong and specific antibody response in rabbits.

Liu *et al.* [124] proposed a model of visual system by carboxylated nanodiamond (cND)-conjugated alpha bungarotoxin (α-BTX), a neurotoxin that preserves the physiological activity of blocking the function of α7-nAChR (alpha7-nicotinic acetylcholine receptor) while targeting cells. It has been shown that α7-nAChRs located on the cellular membrane-mediated cation influx, particularly Ca^{2+} in flux. When cND–α-BTX complexes were used, they bound to α7-nAChR, located on the cell

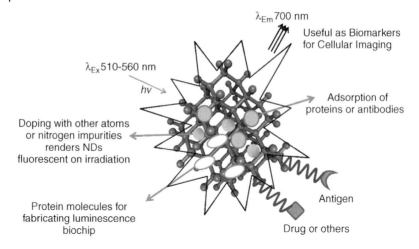

Nanodiamond particle

Figure 7.12 Schematic representation of noncovalent/covalent functionalization and adsorption of different therapeutic agents on/in carbon nanodiamonds for biomedical applications. The right part showing the adsorption of various proteins and antibodies, whereas, drugs or antigen can also be noncovalently attached to nanodiamonds. The left end showing that doping with other atoms or presence of nitrogen impurities renders the nanodiamond fluorescent and emits at 700 nm on irradiation with light at 510–560 nm, and can be useful for biomarker for cellular imaging. A luminescent biochip can also be made by attaching proteins molecule, noncovalently to nanodiamond.

membrane of oocytes and human lung A549 cancer cells, and blocked choline-evoked α7-nAChR-mediated downstream signaling pathway.

Puzyr *et al.* [117] studied the use of detonation nanodiamond particles along with light emitting-obelin and luciferase protein, for fabricating a luminescence biochip. The former bioluminescent system can be used for determining Ca^{2+} in the biological liquids, whereas the later system exhibits catalytic activity both in the bioenzyme reaction and with photoreduced flavin mononucleotide.

7.4.3
Carbon Nanodiamond as Biomarker for Cellular Imaging

NDs can also be doped with other atoms or can be modified by inducing defects and holes into their structure to render them fluorescent and are therefore extremely useful as cellular biomarkers for imaging purposes (Figure 7.12). Indeed, we can exploit the fluorescence properties of the nanodiamonds functionalized at their surface with specific ligands to target cancer cells with an exceptionally high sensitivity in the detection, which is fundamental for early tumor diagnosis. These NDs can exhibit two types of fluorescence: (i) bright red fluorescence and (ii) natural green fluorescence. Nanodiamonds can become fluorescent either by covalent attachment of fluorescent tags or by noncovalent adsorption of fluorescently labeled bioactive molecules. The most common defect is the presence of a negatively charged nitrogen vacancy center in the nanodiamond structure that can be created by irradiation of NDs having nitrogen impurities with beams of proton (3 MeV) or electron (2 MeV), followed by thermal annealing at ~800 °C. This defect center strongly adsorbs at 560 nm and emits fluorescence at 700 nm. Since the nitrogen atom is confined to an inert matrix, photobleaching is greatly reduced if not completely eliminated, and no fluorescence blinking takes place, thus rendering such nanodiamonds that are very useful as biomarkers for imaging [125, 126]. The fluorescence quantum efficiency is ~1, with a lifetime of 11.6 ns at room temperature. In a 10 nm size nanodiamond roughly 100 defects can be deposited without a very high energy transfer among them. Chang *et al.* [127] reported that fluorescent nanodiamond can also be produced by irradiation of type Ib diamond powders using 40 keV He$^+$ ions. The advantage of He$^+$ instead of proton includes first chemical inertness of He$^+$ and second, 40 keV of He$^+$ can produce 40 vacancies in comparison to only 13 vacancies using 3 MeV proton. The fluorescence is sufficiently bright and stable, suggesting fluoro-nanodiamonds as an ideal probe for long-term tracking and imaging. Zhao *et al.* [128] prepared fluorescent carbon nanocrystals by electrooxidation of graphite in aqueous solution, which possess no photobleaching and low cytotoxicity and thus can be used in biological labeling, imaging, and disease diagnosis. Borjanovic *et al.* [129] described that nanodiamonds when incorporates, into poly(dimethyl-siloxanes) matrix, show much stronger and more stable photoluminescence when irradiated with proton flux. The maximum emission occurred at 520 nm on excitation at 425 nm. Another defect is the H3 defect center (N-V-N) of a type Ia diamond that produces green emission (531 nm) [126]. Aharonovich *et al.* [130] studied the formation of Ni-related color center in CVD grown NDs by Ni implantation into the substrate prior to the growth. The photoluminescence peak appeared at >800 nm, proved to be a single photon emitter and did not interfere with other photoluminescence peaks arising from other defects.

Chang *et al.* [131] reported the preparation of fluorescent magnetic nanodiamond (FMND) for cellular imaging. These FMND possess water solubility of 2.1 mg/ml. The authors found that these water-soluble FMNDs easily entered into HeLa cells via nonreceptor-mediated endocytosis and remained in cytoplasm without entering into the nucleus and showing low cytotoxicity toward HeLa cells. Similarly, Faklaris *et al.* [132] studied the internalization of NDs in HeLa cancer cells and found that they mainly localize into the extra-endosomal region thus, providing a positive indication of their suitability for application in drug delivery. Vial *et al.* [133] analyzed the grafting of fluorescently labeled thiolated peptide onto NDs, which were previously coated by silanization or with polyelectrolyte layer. These modified NDs were nontoxic and readily enter the mammalian cells.

Besides fluorescence, nanodiamond unique spectroscopic Raman signal from sp^3 carbon structure could also be used as a biomarker for studying the molecular interaction at the single cell level via confocal Raman mapping. In recent reports, growth hormone receptor molecules in A549 human lung epithelial cells were observed by nanodiamond labeling. When cND alone interacts with cells, the cND Raman signals overlap with the cells suggesting that cND resides near the nucleus of the cell. However, when nanodiamonds were attached to growth hormone, the whole complex resides only on the surface of the cell, suggesting the presence of growth hormone receptors at the cell surface [134]. Chao *et al.* [135] described protein targeted cell interaction by first attaching the lysozyme to a carboxylated nanodiamond via physical adsorption using electrostatic interaction and then interaction of cND–lysozyme complex with *E. coli*. The intrinsic Raman signal from the cND–lysozyme complex was detected and used as a marker to locate the position of the interacting protein.

7.4.4
Biocompatibility and Toxicity

Schrand *et al.* [136] found that NDs were more biocompatible than carbon fiber, and pristine, carbon nanotubes. They studied neuroblastoma and alveolar macrophage cell lines, at concentrations ranging from 25 to 100 µg/ml for 24 h, and found that NDs do not disturb the mitochondrial membrane and do not generate ROS, while carbon nanotubes can cause membrane leakage and generate ROS. Further, the same authors reported that nanodiamond (diameter 2–10 nm), with or without surface modifications by acid or base, are biocompatible with a variety of cells including neuroblastoma, macrophages, keratinocytes, and PC-12 cells at concentrations ranging between 5 and 100 µg/ml at pH between 7.2 and 7.6. Other studies confirmed that cells can grow on nanodiamond-coated substrate without undergoing morphological changes [137]. Liu *et al.* [138] independently studied the biocompatibility of the nanodiamonds and carbon nanotubes on human lung A549 epithelial cells and HFL-1 normal fibroblasts. The ND particles, accumulated in A549 cells, were observed by AFM and confocal spectroscopy. Treatment of CND particles with 5 or 100 nm diameter at a concentration range of 0.1–100 µg/ml did not reduce cell viability or alter the protein expression profile in these cells whereas carboxylated carbon nanotubes induced cytotoxicity to these cells. Yu *et al.* [126] investigated the biocompatibility of synthetic abrasive diamond powders of large diameter (100 nm) in cell cultures and reported very low cytotoxicity in kidney cells. Puzyr *et al.* [139] also showed that the complete replacement of water in the experimental animal diet with modified nanodiamonds (MNDs) hydrosols for 6 months did not cause the death of the mice, nor affected the growth and weight dynamics of their organs. There were no indications of the destruction of their blood cells after the intravenous administration of MNDs.

7.5
Conclusions

This chapter describes the biomedical applications of three different forms of carbon-based nanomaterials, which are currently under investigation. Carbon nanotubes, nanohorns, and nanodiamonds can be functionalized with different types of therapeutic molecules following alternative chemical approaches. These novel nanomaterials can be modified to facilitate their manipulation and render them biocompatible. Such soluble/dispersible carbon-based nanoobjects in physiological conditions are then able to penetrate into the cells or can be administered *in vivo* to deliver their cargo molecules, which eventually display the desired activity.

Acknowledgments

This work was supported by the CNRS and the Agence Nationale de la Recherche (grant ANR-05-JCJC-0031-01), the University of Trieste, Italian Ministry of Education MUR (cofin Prot. 2006034372), Regione Friuli Venezia-Giulia, and The School of Pharmacy, University of London. Partial support is also acknowledged by the European Union FP6 NEURONANO (NMP4-CT-2006-031847), NINIVE (NMP4-CT-2006-033378), and FP7 ANTICARB (HEALTH-2007-201587) programs. A.B. and P.S. wish to thank the French-Indian CEFIPRA collaborative project (Project no. 3705-2).

References

1 Farokhzad, O.C. and Langer, R. (2009) Impact of nanotechnology on drug delivery. *ACS Nano*, **3**, 16–20.

2 Hollinger, R.V. (2004) *Drug delivery systems*, 2nd edn, CRC, Boca Raton, FL.

3 Allen, T.M. and Cullis, P.R. (2004) Drug delivery systems: entering the mainstream. *Science*, **303**, 1818–1822.

4 Lavan, D.A., McGuire, T., and Langer, R. (2003) Small-scale systems for *in vivo* drug delivery. *Nat. Biotechnol.*, **21**, 1184–1191.

5 Langer, R. (1998) Drug delivery and targeting. *Nature*, **392** (Suppl.), 5–10.

6 Duncan, R. (2003) The dawning era of polymer therapeutics. *Nat. Rev. Drug Discov.*, **2**, 347–360.

7 Varde, N.K. and Pack, D.W. (2004) Microspheres for controlled release drug delivery. *Exp. Opin. Biol. Ther.*, **4**, 35–51.

8 Boas, U. and Heegaard, P.M. (2004) Dendrimers in drug research. *Chem. Soc. Rev.*, **33**, 43–63.

9 Martin, C.R. and Kohli, P. (2003) The emerging field of nanotube biotechnology. *Nat. Rev. Drug Discov.*, **2**, 29–37.

10 Iijima, S. (1991) Helical microtubules of graphitic carbon. *Nature*, **354**, 56–58.

11 Niyogi, S., Hamon, M.A., Hu, H., Zhao, B., Bhowmik, P., Sen, R., Itkis, M.E., and Haddon, R.C. (2002) Chemistry of single-walled carbon nanotubes. *Acc. Chem. Res.*, **35**, 1105–1113.

12 Hirsch, A. (2002) Functionalization of single-walled carbon nanotubes. *Angew. Chem. Int. Ed.*, **41**, 1853–1859.

13 Pastorin, G., Kostarelos, K., Prato, M., and Bianco, A. (2005) Functionalized carbon nanotubes: Towards the delivery of therapeutic molecules. *J. Biomed. Nanotechnol.*, **1**, 133–142.

14 Tasis, D., Tagmatarchis, N., Bianco, A., and Prato, M. (2006) Chemistry of carbon nanotubes. *Chem. Rev.*, **106**, 1105–1136.

15 Curran, S.A., Ajayan, P.M., Blau, W.J., Carroll, D.L., Coleman, J.N., Dalton, A.B., Davey, A.P., Drury, A., McCarthy, B., Maier, S., and Strevens, A. (1998) A composite from poly(*m*-phenylenevinylene-*co*-2,5-dioctoxy-*p*-phenylenevinylene) and carbon nanotubes: a novel material for molecular

optoelectronics. *Adv. Mater.*, **10**, 1091–1093.

16 Nakashima, N., Tomonari, Y., and Murakami, H. (2002) Water-soluble single-walled carbon nanotubes via noncovalent sidewall-functionalization with a pyrene-carrying ammonium ion. *Chem. Lett.*, **31**, 638–639.

17 Wu, P., Chen, X., Hu, N., Tam, U.C., Blixt, O., Zettl, A., and Bertozzi, C.R. (2008) Biocompatible carbon nanotubes generated by functionalization with glycodendrimers. *Angew. Chem. Int. Ed.*, **47**, 1–5.

18 Hirsch, A. and Vostrowsky, O. (2005) Functionalization of carbon nanotubes, in "Functional molecular nanostructures". *Top. Curr. Chem.*, **245**, 193–237.

19 Peng, X. and Wong, S.S. (2009) Functional covalent chemistry of carbon nanotube surfaces. *Adv. Mater.*, **21**, 625–642.

20 Kostarelos, K., Lacerda, L., Pastorin, G., Wu, W., Wieckowski, S., Luangsivilay, J., Godefroy, S., Pantarotto, D., Briand, J.P., Muller, S., Prato, M., and Bianco, A. (2007) Cellular uptake of functionalized carbon nanotubes is independent of functional group and cell type. *Nat. Nanotechnol.*, **2**, 108–113.

21 Kam, N.W.S., O'Connell, M., Wisdom, J.A., and Dai, H.J. (2005) Carbon nanotubes as multifunctional biological transporters and near-infrared agents for selective cancer cell destruction. *Proc. Natl. Acad. Sci. USA*, **102**, 11600–11605.

22 Kam, N.W.S., Liu, Z.A., and Dai, H.J. (2006) Carbon nanotubes as intracellular transporters for proteins and DNA: an investigation of the uptake mechanism and pathway. *Angew. Chem. Int. Ed.*, **45**, 577–581.

23 Heller, D.A., Jeng, E.S., Yeung, T.K., Martinez, B.M., Moll, A.E., Gastala, J.B., and Strano, M.S. (2006) Optical detection of DNA conformational polymorphism on single-walled carbon nanotubes. *Science*, **311**, 508–511.

24 Kateb, B., Van Handel, M., Zhang, L.Y., Bronikowski, M.J., Manohara, H., and Badie, B. (2007) Internalization of MWCNTs by microglia: possible application in immunotherapy of brain tumors. *NeuroImage*, **37**, S9–S17.

25 Rojas-Chapana, J., Troszczynska, J., Firkowska, I., Morsczeck, C., and Giersig, M. (2005) Multi-walled carbon nanotubes for plasmid delivery into *Escherichia coli* cells. *Lab Chip*, **5**, 536–539.

26 Liu, Q., Chen, B., Wang, Q., Shi, X., Xiao, Z., Lin, J., and Fang, X. (2009) Carbon nanotubes as molecular transporters for walled plant cells. *Nano Lett.*, **9**, 1007–1010.

27 Hampel, S., Kunze, D., Haase, D., Krämer, K., Rauschenbach, M., Ritschel, M., Leonhardt, A., Thomas, J., Oswald, S., Hoffmann, V., and Büchner, B. (2008) Carbon nanotubes filled with a chemotherapeutic agent: a nanocarrier mediates inhibition of tumor cell growth. *Nanomedicine*, **3**, 175–182.

28 Pastorin, G., Wu, W., Wieckowski, S., Briand, J.P., Kostarelos, K., Prato, M., and Bianco, A. (2006) Double functionalisation of carbon nanotubes for multimodal drug delivery. *Chem. Commun.*, 1182–1184.

29 McDevitt, M.R., Chattopadhyay, D., Kappel, B.J., Jaggi, J.S., Schiffman, S.R., Antczak, C., Njardarson, J.T., Brentjens, R., and Scheinberg, D.A. (2007) Tumor targeting with antibody-functionalized, radiolabeled carbon nanotubes. *J. Nucl. Med.*, **48**, 1180–1189.

30 Feazell, R.P., Nakayama-Ratchford, N., Dai, H., and Lippard, S.J. (2007) Soluble single-walled carbon nanotubes as longboat delivery systems for platinum (IV) anticancer drug design. *J. Am. Chem. Soc.*, **129**, 8438–8439.

31 Dhar, S., Liu, Z., Thomale, J., Dai, H., and Lippard, S.J. (2008) Targeted single-wall carbon nanotube-mediated Pt(IV) prodrug delivery using folate as a homing device. *J. Am. Chem. Soc.*, **130**, 11467–11476.

32 Ali-Boucetta, H., Al-Jamal, K.H., McCarthy, D., Prato, M., Bianco, A., and Kostarelos, K. (2008) Multiwalled carbon nanotube–doxorubicin supramolecular complexes for cancer therapeutics. *Chem. Commun.*, 459–461.

33 Lu, Q., Moore, J.M., Huang, G., Mount, A.S., Rao, A.M., Larcom, L.L., and Chun

Ke, P. (2004) RNA polymer translocation with single-walled carbon nanotubes. *Nano Lett.*, **4**, 2473–2477.

34 Zhang, Z., Yang, X., Zhang, Y., Zeng, B., Wang, S., Zhu, T., Roden, R.B.S., Chen, Y., and Yang, R. (2006) Delivery of telomerase reverse transcriptase small interfering RNA in complex with positively charged single-walled carbon nanotubes suppresses tumor growth. *Clin. Cancer Res.*, **12**, 4933–4939.

35 Liu, Z., Winters, M., Holodniy, M., and Dai, H. (2007) siRNA Delivery into human T cells and primary cells with carbon-nanotube transporters. *Angew. Chem. Int. Ed.*, **46**, 2023–2027.

36 Shi Kam, N.W., Liu, Z., and Dai, H. (2005) Functionalization of carbon nanotubes via cleavable disulfide bonds for efficient intracellular delivery of siRNA and potent gene silencing. *J. Am. Chem. Soc.*, **127**, 12492–12493.

37 Yang, X., Zhang, Z., Liu, Z., Ma, Y., Yang, R., and Chen, Y. (2008) Multi-functionalized single-walled carbon nanotubes as tumor cell targeting biological transporters. *J. Nanopart. Res.*, **10**, 815–822.

38 Wu, Y., Phillips, J.A., Liu, H., Yang, R., and Tan, W. (2008) Carbon nanotubes protect DNA strands during cellular delivery. *ACS Nano*, **2**, 2023–2028.

39 Li, X., Peng, Y., Ren, J., and Qu, X. (2006) Carboxyl-modified single-walled carbon nanotubes selectively induce human telomeric i-motif formation. *Proc. Nat. Acad. Sci. USA*, **103**, 19658–19663.

40 Zhao, C., Peng, Y., Song, Y., Ren, J., and Qu, X. (2008) Self-assembly of single-stranded RNA on carbon nanotube: polyadenylic acid to form a duplex structure. *Small*, **4**, 656–661.

41 Pantarotto, D., Partidos, C.D., Hoebeke, J., Brown, F., Kramer, E., Briand, J.P., Muller, S., Prato, M., and Bianco, A. (2003) Immunization with peptide-functionalized carbon nanotubes enhances virus-specific neutralizing antibody responses. *Chem. Biol.*, **10**, 961.

42 Meng, J., Meng, J., Duan, J., Kong, H., Li, L., Wang, C., Xie, S., Chen, S., Gu, N., Xu, H., and Yang, X.-D. (2008) Carbon nanotubes conjugated to tumor lysate protein enhance the efficacy of an antitumor immunotherapy. *Small*, **4**, 1364–1370.

43 Shao, N., Lu, S., Wickstrom, E., and Panchapakesan, B. (2007) Integrated molecular targeting of IGF1R and HER2 surface receptors and destruction of breast cancer cells using single wall carbon nanotubes. *Nanotechnology*, **18**, 315101–315110.

44 Gannon, C.J., Cherukuri, P., Yakobson, B.I., Cognet, L., Kanzius, J.S., Kittrell, C., Weisman, R.B., Pasquali, M., Schmidt, H.K., Smalley, R.E., and Curley, S.A. (2007) Carbon nanotube-enhanced thermal destruction of cancer cells in a noninvasive radiofrequency field. *Cancer*, **111**, 2654–2665.

45 Murugesan, S., Shaker, A., Mousa, S.A., O'Connor, L.J., Lincoln, D.W., II, and Linhardt, R.J. (2007) Carbon inhibits vascular endothelial growth factor- and fibroblast growth factor-promoted angiogenesis. *FEBS Lett.*, **581**, 1157–1160.

46 Lovat, V., Pantarotto, D., Lagostena, L., Cacciari, B., Grandolfo, M., Righi, M., Spalluto, G., Prato, M., and Ballerini, L. (2005) Carbon nanotube substrates boost neuronal electrical signaling. *Nano Lett.*, **5**, 1107–1110.

47 Mazzatenta, A., Giugliano, M., Campidelli, S., Gambazzi, L., Businaro, L., Markram, H., Prato, M., and Ballerini, L. (2007) Interfacing neurons with carbon nanotubes: electrical signal transfer and synaptic stimulation in cultured brain circuits. *J. Neurosci.*, **27**, 6931–6936.

48 Cellot, G., Cilia, E., Cipollone, S., Rancic, V., Sucapane, A., Giordani, S., Gambazzi, L., Markram, H., Grandolfo, M., Scaini, D., Gelain, F., Casalis, L., Prato, M., Giugliano, M., and Ballerini, L. (2009) Carbon nanotubes might improve neuronal performance by favouring electrical shortcuts. *Nat. Nanotechnol.*, **4**, 126–133.

49 Keefer, E.W., Botterman, B.R., Romero, M.I., Rossi, A.F., and Gross, G.W. (2008) Carbon nanotube coating improves neuronal recordings. *Nat. Nanotechnol.*, **3**, 434–439.

50 Kam, N.W.S., Jan, E., and Kotov, N.A. (2009) Electrical stimulation of neural stem cells mediated by humanized carbon nanotube composite made with extracellular matrix protein. *Nano Lett.*, **9**, 273–278.

51 Massobrio, G., Massobrio, P., and Martinoia, S. (2008) Modeling the neuron-carbon nanotube-ISFET junction to investigate the electrophysiological neuronal activity. *Nano Lett.*, **8**, 4433–4440.

52 Malarkey, E.B., Reyes, R.C., Zhao, B., Haddon, R.C., and Parpura, V. (2008) Water soluble single-walled carbon nanotubes inhibit stimulated endocytosis in neurone. *Nano Lett.*, **8**, 3538–3542.

53 Bardi, G., Tognini, P., Ciofani, G., Raffa, V., Costa, M., and Pizzorusso, T. (2009) Pluronic-coated carbon nanotubes do not induce degeneration of cortical neurons *in vivo* and *in vitro*. *NNMB*, **5**, 96–104.

54 Fenoglio, I., Tomatis, M., Lison, D., Muller, J., Fonseca, A., Nagy, J.B., and Fubini, B. (2006) Reactivity of carbon nanotubes: free radical generation or scavenging activity? *Free Radic. Biol. Med.*, **40**, 1227–1233.

55 Lucente-Schultz, R.M., Moore, V.C., Leonard, A.D., Price, B.K., Kosynkin, D.V., Lu, M., Partha, R., Conyers, J.L., and Tour, J.M. (2009) Antioxidant single-walled carbon nanotubes. *J. Am. Chem. Soc*, **131**, 3934–3941.

56 Ye, S.F., Wu, Y.H., Hou, Z.Q., and Zhang, Q.Q. (2009) ROS and NF-kappaB are involved in upregulation of IL-8 in A549 cells exposed to multi-walled carbon nanotubes. *Biochem. Biophys. Res. Commun.*, **379**, 643–648.

57 Cherukuri, P., Bachino, S.M., Litovsky, S.H., and Weisman, R.B. (2004) Near-infrared fluorescence microscopy of single-walled carbon nanotubes in phagocytic cells. *J. Am. Chem. Soc.*, **126**, 15638–15639.

58 Cherukuri, P., Gannon, C.J., Leeuw, T.K., Schmidt, H.K., Smalley, R.E., Curley, S.A., and Weisman, R.B. (2006) Mammalian pharmacokinetics of carbon nanotubes using intrinsic near-infrared fluorescence. *Proc. Natl. Acad. Sci. USA*, **103**, 18882–18886.

59 Leeuw, T.K., Reith, R.M., Simonette, R.A., Harden, M.E., Cherukuri, P., Tsyboulski, D.A., Beckingham, K.M., and Weisman, R.B. (2007) Single-walled carbon nanotubes in the intact organism: near-IR imaging and biocompatibility studies in Drosophila. *Nano Lett.*, **7**, 2650–2654.

60 Welsher, K., Liu, Z., Daranciang, D., and Dai, H. (2008) Selective probing and imaging of cells with single-walled carbon nanotubes as near-infrared fluorescent molecules. *Nano Lett.*, **8**, 586–590.

61 Zhang, S., Sun, D., Li, X., Pei, F., and Liu, S. (1997) Synthesis and solvent enhanced relaxation property of water-soluble endohedral metallofullerenes. *Fullerene Sci. Tech.*, **5**, 1635–1643.

62 Bolskar, R.D. and Alford, J.M. (2003) Chemical oxidation of endohedral metallofullerenes: identification and separation of distinct classes. *Chem. Commun.*, **11**, 1292–1293.

63 Bolskar, R.D., Benedetto, A.F., Husebo, L.O., Price, R.E., Jackson, E.F., Wallace, S., Wilson, L.J., and Alford, J.M. (2003) First soluble M@C_{60} derivatives provide enhanced access to metallofullerenes and permit *in vivo* evaluation of Gd@C_{60}[C(COOH)$_2$]$_{10}$ as a MRI contrast agent. *J. Am. Chem. Soc.*, **125**, 5471–5478.

64 Sitharaman, B., Kissell, K.R., Hartman, K.B., Tran, L.A., Baikalov, A., Rusakova, I., Sun, Y., Khant, H.A., Ludtke, S.J., Chiu, W., Laus, S., Tóth, E., Helm, L., Merbach, A.E., and Wilson, L.J. (2005) Superparamagnetic gadonanotubes are high-performance MRI contrast agents. *Chem. Commun.*, **31**, 3915–3917.

65 Ashcroft, J.M., Hartman, K.B., Kissell, K.R., Mackeyev, Y., Pheasant, S., Young, S., Van der Heide, P.A.W., Mikos, A.G., and Wilson, L.J. (2007) Single-molecule I$_2$@US-tube nanocapsules: A new X-ray contrastagent design. *Adv. Mater.*, **19**, 573–576.

66 Hartman, K.B., Laus, S., Bolskar, R.D., Muthupillai, R., Helm, L., Tóth, E., Merbach, A.E., and Wilson, L.J. (2008) Gadonanotubes as ultrasensitive pH-smart probes for magnetic resonance imaging. *Nano Lett.*, **8**, 415–419.

67 Faraj, A.A., Cieslar, K., Lacroix, G., Gaillard, S., Canet-Soulas, E., and Cremillieux, Y. (2009) *In vivo* imaging of carbon nanotube biodistribution using magnetic resonance imaging. *Nano Lett.*, **9**, 1023–1027.

68 Richard, C., Doan, B.-T., Beloeil, J.-C., Bessodes, M., Tóth, E., and Scherman, D. (2008) Noncovalent functionalization of carbon nanotubes with amphiphilic Gd^{3+} chelates: Toward powerful T1 and T2 MRI contrast agents. *Nano Lett.*, **8**, 232–236.

69 Stafiej, A. and Pyrzynska, K. (2007) Adsorption of heavy metal ions with carbon nanotubes. *Sep. Purif. Technol.*, **58**, 49–52.

70 Li, Y.H., Wang, S.G., Wei, J.Q., Zhang, X.F., Xu, C.L., Luan, Z.K., Wu, D.H., and Wei, B.Q. (2002) Lead adsorption on carbon nanotubes. *Chem. Phys. Lett.*, **357**, 263–266.

71 Long, R.Q. and Yang, R.T. (2001) Carbon nanotubes as superior sorbent for dioxin removal. *J. Am. Chem. Soc.*, **123**, 2058–2059.

72 Wang, X., Chen, C., Hu, W., Ding, A., Xu, D., and Zhou, X. (2005) Sorption of Am(III) to multiwall carbon nanotubes. *Environ. Sci. Technol.*, **39**, 2856–2860.

73 Chen, C., Li, X., Zhao, D., Tan, X., and Wang, X. (2007) Adsorption kinetic, thermodynamic and desorption studies of Th(IV) on oxidized multi-wall carbon nanotubes. *Colloids Surf. A*, **302**, 449–454.

74 Belloni, F., Kütahyali, C., Rondinella, V.V., Carbol, P., Wiss, T., and Mangione, A. (2009) Can carbon nanotubes play a role in the field of nuclear waste management? *Environ. Sci. Technol.*, **43**, 1250–1255.

75 Akasaka, T. and Watari, F. (2009) Capture of bacteria by flexible carbon nanotubes. *Acta Biomater.*, **5**, 607–612.

76 Wick, P., Manser, P., Limbach, L.K., Dettlaff-Weglikowskab, U., Krumeich, F., Roth, S., Stark, W.J., and Bruinink, A. (2007) The degree and kind of agglomeration affect carbon nanotube cytotoxicity. *Toxicol. Lett.*, **168**, 121–131.

77 Fenoglio, I., Greco, G., Tomatis, M., Muller, J., Raymundo-Piñero, E., Béguin, F., Fonseca, A., Nagy, J.B., Lison, D., and Fubini, B. (2008) Structural defects play a major role in the acute lung toxicity of multiwall carbon nanotubes: Physicochemical aspects. *Chem. Res. Toxicol.*, **21**, 1690–1697.

78 Muller, J., Huaux, F., Fonseca, A., Nagy, J.B., Moreau, N., Delos, M., Raymundo-Piñero, E., Béguin, F., Kirsch-Volders, M., Fenoglio, I., Fubini, B., and Lison, D. (2008) Structural defects play a major role in the acute lung toxicity of multiwall carbon nanotubes: toxicological aspects. *Chem. Res. Toxicol.*, **21**, 1698–1705.

79 Tabet, L., Bussy, C., Amara, N., Setyan, A., Grodet, A., Rossi, M.J., Pairon, J.C., Boczkowski, J., and Lanone, S. (2009) Adverse effects of industrial multiwalled carbon nanotubes on human pulmonary cells. *J. Toxicol. Environ. Health A*, **72**, 60–73.

80 Wick, P., Manser, P., Limbach, L.K., Dettlaff-Weglikowskab, U., Krumeich, F., Roth, S., Stark, W.J., and Bruinink, A. (2007) The degree and kind of agglomeration affect carbon nanotube cytotoxicity. *Toxicol. Lett.*, **168**, 121–131.

81 Guo, J., Zhang, X., Zhang, S., Zhu, Y., and Li, W. (2008) The different bio-effects of functionalized multi-walled carbon nanotubes on *Tetrahymena pyriformis*. *Curr. Nanosci.*, **4**, 240–245.

82 Ghafari, P., St-Denis, C.H., Power, M.E., Jin, X., Tsou, V., Mandal, H.S., Bols, N.C., and Tang, X.S. (2008) Impact of carbon nanotubes on the ingestion and digestion of bacteria by ciliated protozoa. *Nat. Nanotechnol.*, **3**, 347–351.

83 Di Sotto, A., Chiaretti, M., Carru, G.A., Bellucci, S., and Mazzanti, G. (2009) Multi-walled carbon nanotubes: lack of mutagenic activity in the bacterial reverse mutation assay. *Toxicol. Lett.*, **184**, 192–197.

84 Muller, J., Decordier, I., Hoet, P.H., Lombaert, N., Thomassen, L., Huaux, F., Lison, D., and Kirsch-Volders, M. (2008) Clastogenic and aneugenic effects of multi-wall carbon nanotubes in epithelial cells. *Carcinogenesis*, **29**, 427–433.

85 Iijima, S., Yudasaka, M., Yamada, R., Bandow, S., Suenaga, K., Kokai, F., and Takahashi, K. (1999) Nano-aggregates of single-walled graphitic carbon nano-horns. *Chem. Phys. Lett.*, **309**, 165–170.

86 Yudasaka, M., Fan, J., Miyawaki, J., and Iijima, S. (2005) Studies on the adsorption of organic materials inside thick carbon nanotubes. *J. Phys. Chem. B*, **109**, 8909–8913.

87 Hattori, Y., Kanoh, H., Okino, F., Touhara, H., Kasuya, D., Yudasaka, M., Iijima, S., and Kaneko, K. (2004) Direct thermal fluorination of single wall carbon nanohorns. *J. Phys. Chem. B*, **108**, 9614–9618.

88 Tagmatarchis, N., Maigne, A., Yudasaka, M., and Iijima, S. (2006) Functionalization of carbon nanohorns with azomethine ylides: towards solubility enhancement and electron transfer processes. *Small*, **2**, 490–494.

89 Cioffi, C., Campidelli, S., Brunetti, F.G., Meneghetti, M., and Prato, M. (2006) Functionalisation of carbon nanohorns. *Chem. Commun.*, 2129–2131.

90 Pagona, G., Tagmatarchis, N., Fan, J., Yudasaka, M., and Iijima, S. (2006) Cone-End functionalization of carbon nanohorns. *Chem. Mater.*, **18**, 3918–3920.

91 Pagona, G., Karousis, N., and Tagmatarchis, N. (2008) Aryl diazonium functionalization of carbon nanohorns. *Carbon*, **46**, 604–610.

92 Lacotte, S., Garcia, A., Decossas, M., Al-Jamal, W.T., Li, S., Kostarelos, K., Muller, S., Prato, M., Dumortier, H., and Bianco, A. (2008) Interfacing functionalized carbon nanohorns with primary phagocytic cells. *Adv. Mater.*, **20**, 2421–2426.

93 Bianco, A., Kostarelos, K., and Prato, M. (2008) Opportunities and challenges of carbon-based nanomaterials for cancer therapy. *Exp. Opin. Drug Deliv.*, **5**, 331–342.

94 Miyako, E., Nagata, H., Hirano, K., Makita, Y., Nakayama, K.-I., and Hirotsu, T. (2007) Near-infrared laser-triggered carbon nanohorns for selective elimination of microbes. *Nanotechnology*, **18**, 475103–475109.

95 Miyako, E., Nagata, H., Hirano, K., Sakamoto, K., Makita, Y., Nakayama, K.-I., and Hirotsu, T. (2008) Photoinduced antiviral carbon nanohorns. *Nanotechnology*, **19**, 075106–075111.

96 Zhang, M., Murakami, T., Ajima, K., Tsuchida, K., Sandanayaka, A.S.D., Ito, O., Iijima, S., and Yudasaka, M. (2008) Fabrication of ZnPc/protein nanohorns for double photodynamic and hyperthermic cancer phototherapy. *Proc. Natl. Acad. Sci. USA*, **105**, 14773–14778.

97 Murakami, T., Ajima, K., Miyawaki, J., Yudasaka, M., Iijima, S., and Shiba, K. (2004) Drug-loaded carbon nanohorns: adsorption and release of dexamethasone *in vitro*. *Mol. Pharm.*, **1**, 399–405.

98 Murakami, T., Fan, J., Yudasaka, M., Iijima, S., and Shiba, K. (2006) Solubilization of single-wall carbon nanohorns using a PEG-Doxorubicin conjugate. *Mol. Pharm.*, **3**, 407–414.

99 Ajima, K., Yudasaka, M., Murakami, T., Maigne, A., Shiba, K., and Iijima, S. (2005) Carbon nanohorns as anticancer drug carriers. *Mol. Pharm.*, **2**, 475–480.

100 Ajima, K., Yudasaka, M., Maigne, A., Miyawaki, J., and Iijima, S. (2006) Effect of functional groups at hole edges on cisplatin release from inside single-wall carbon nanohorns. *J. Phys. Chem. B*, **110**, 5773–5778.

101 Ajima, K., Maigne, A., Yudasaka, M., and Iijima, S. (2006) Optimum hole-opening condition for cisplatin incorporation in single-wall carbon nanohorns and its release. *J. Phys. Chem. B*, **110**, 19097–19099.

102 Matsumara, S., Ajima, K., Yudasaka, M., Iijima, S., and Shiba, A. (2007) Dispersion of Cisplatin-loaded carbon nanohorns with a conjugate comprised of an artificial peptide aptamer and polyethylene glycol. *Mol. Pharm.*, **4**, 723–729.

103 Fan, X., Tan, J., Zhang, G., and Zhang, F. (2007) Isolation of carbon nanohorn assemblies and their potential for intracellular delivery. *Nanotechnology*, **18**, 195103–195108.

104 Miyawaki, J., Yudasaka, M., Imai, H., Yorimitsu, H., Isobe, H., Nakamura, E., and Iijima, S. (2006) *In vivo* magnetic resonance imaging of single-walled carbon nanohorns by labeling with magnetite nanoparticles. *Adv. Mater.*, **18**, 1010–1014.

105 Miyawaki, J., Yudasaka, M., Imai, H., Yorimitsu, H., Isobe, H., Nakamura, E., and Iijima, S. (2006) Synthesis of ultrafine Gd_2O_3 nanoparticles inside single-wall carbon nanohorns. *J. Phys. Chem. B*, **110**, 5179–5181.

106 Miyawaki, J., Yudasaka, M., Azami, T., Kubo, Y., and Iijima, S. (2008) Toxicity of single-walled carbon nanohorns. *ACS Nano*, **2**, 213–226.

107 Isobe, H., Tanaka, T., Maeda, R., Noiri, E., Solin, N., Yudasaka, M., Iijima, S., and Nakamura, E. (2006) Preparation, purification, characterization, and cytotoxicity assessment of water-soluble, transition-metal-free carbon nanotube aggregates. *Angew. Chem. Int. Ed.*, **45**, 6676–6680.

108 William-Andrew Publishing (2006) *Ultracrystalline Diamond* (eds O.A. Shenderova and D.M. Gruen), Norwich, UK.

109 Ferro, S. (2002) Synthesis of diamond. *J. Mater. Chem.*, **12**, 2843–2855.

110 Schreiner, P.R., Fokina, N.A., Tkachenko, B.A., Hausmann, H., Serafin, M., Dahl, J.E.P., Liu, S., Carlson, R.M.K., and Fokin, A.A. (2006) Functionalised nanodiamonds: triamantane and [121]tetramantane. *J. Org. Chem.*, **71**, 6709–6720.

111 Fokin, A.A., Tkachenko, B.A., Gunchenko, P.A., Gusev, D.V., and Schreiner, P.R. (2005) Functionalised nanodiamonds part 1. An experimental assessment of diamantane and computational predictions for higher diamondoids. *Chem. Eur. J.*, **11**, 7091–7101.

112 Liu, Y., Gu, Z., Margrave, J.L., and Khabashesku, V.N. (2004) Functionalization of nanoscale diamond powder: fluoro-, alkyl-, amino-, and amino acid–nanodiamond derivatives. *Mater. Chem.*, **16**, 3924–3930.

113 Novikov, N.V. (1999) New trends in high-pressure synthesis of diamond. *Diamond Relat. Mater.*, **8**, 1427–1432.

114 Shenderova, O.A., Zhirnov, V.V., and Brenner, D.W. (2002) Carbon nanostructures. *Crit. Rev. Solid State Mater. Sci.*, **27**, 227–356.

115 Frenklach, M., Kematick, R., Huang, D., Howard, W., Spear, K.E., Phelps, A.W., and Koba, R. (1989) Homogeneous nucleation of diamond powder in the gas phase. *J. Appl. Phys.*, **66**, 395–399.

116 Krueger, A. (2008) New carbon materials: biological applications of functionalized nanodiamond materials. *Chem. Eur. J.*, **14**, 1382–1390. and references therein.

117 Puzyr, A.P., Pozdnyakova, I.O., and Bondar, V.S. (2004) Design of luminescent biochip with nanodiamonds and bacterial luciferase. *Phys. Solid State*, **46**, 761–763.

118 Huang, L.-C.L. and Chang, H.-C. (2004) Adsorption and immobilization of cytochrome C on nanodiamonds. *Langmuir*, **20**, 5879–5884.

119 Chung, P.-H., Perevedentseva, E., Tu, J.-S., Chang, C.C., and Cheng, C.-L. (2006) Spectroscopic study of bio-functionalized nanodiamonds. *Diamond Relat. Mater.*, **15**, 622–625.

120 Kong, X.L., Huang, L.C.L., Hsu, C.-M., Chen, W.-H., Han, C.-C., and Chang, H.-C. (2005) High-affinity capture of proteins by diamond nanoparticles for mass spectrometric analysis. *Anal. Chem.*, **77**, 259–265.

121 Huang, T.S., Tzeng, Y., Liu, Y.K., Chen, Y.C., Walker, K.R., Guntupalli, R., and Liu, C. (2004) Immobilization of antibodies and bacterial binding on nanodiamond and carbon nanotubes for biosensor applications. *Diamond Relat. Mater.*, **13**, 1098–1102.

122 Huang, H., Pierstorff, E., Osawa, E., and Ho, D. (2007) Active nanodiamond hydrogels for chemotherapeutic delivery. *Nano Lett.*, **7**, 3305–3314.

123 Kossovsky, N., Gelman, A., Hnatyszyn, H.J., Rajguru, S., Garell, R.L., Torbati, S., Freitas, S.S.F., and Chow, G.-M. (1995) Surface-modified diamond nanoparticles as antigen delivery vehicles. *Bioconjugate Chem.*, **6**, 507–511.

124 Liu, K.-K., Chen, M.-F., Chen, P.-Y., Lee, T.J.F., Cheng, C.-L., Chang, C.-C., Ho, Y.-P., and Chao, J.-I. (2008) Alpha-bungarotoxin binding to target cell in a developing visual system by carboxylated nanodiamond. *Nanotechnology*, **19**, 205102.

125 Fu, C.-C., Lee, H.-Y., Chen, K., Lim, T.-S., Wu, H.-Y., Lin, P.-K., Wei, P.-K.,

Tsao, P.-H., Chang, H.-C., and Fann, W. (2007) Characterization and application of single fluorescent nanodiamonds as cellular biomarkers. *Proc. Natl. Acad. Sci. USA*, **104**, 727–732.

126 Yu, S.-J., Kang, M.-W., Chang, H.-C., Chen, K.-M., and Yu, Y.-C. (2005) Bright fluorescent nanodiamonds: No photobleaching and low cytotoxicity. *J. Am. Chem. Soc.*, **127**, 17604–17605.

127 Chang, Y.-R., Lee, H.-Y., Chen, K., Chang, C.-C., Tsai, D.-S., Fu, C.-C., Lim, T.-S., Tzeng, Y.-K., Fang, C.-Y., Han, C.-C., Chang, H.-C., and Fann, W. (2008) Mass production and dynamic imaging of fluorescent nanodiamonds. *Nat. Nanotechnology*, **3**, 284–288.

128 Zhao, Q.-L., Zhang, Z.-L., Huang, B.-H., Peng, J., Zhang, M., and Pang, D.-W. (2008) Facile preparation of low cytotoxicity fluorescent carbon nanocrystals by electrooxidation of graphite. *Chem. Commun.*, 5116–5118.

129 Borjanovic, V., Lawrence, W.G., Hens, S., Jaksic, M., Zamboni, I., Edson, C., Vlasov, I., Shenderova, O., and McGuire, G.E. (2008) Effect of proton irradiation on photoluminescent properties of PDMS-nanodiamond composites. *Nanotechnology*, **19**, 455701.

130 Aharonovich, I., Zhou, C., Stacey, A., Treussart, F., Roch, J.-F., and Prawer, S. (2008) Formation of color centers in nanodiamonds by plasma assisted diffusion of impurities from the growth substrate. *Appl. Phys. Lett.*, **93**, 243112.

131 Chang, I.P., Hwang, K.C., and Chiang, C.-S. (2008) Preparation of fluorescent magnetic nanodiamonds and cellular imaging. *J. Am. Chem. Soc.*, **130**, 15476–15481.

132 Faklaris, O., Garrot, D., Joshi, V., Druon, F., Boudou, J.-P., Sauvage, T., Georges, P., Curmi, P.A., and Treussart, F. (2008) Detection of single photoluminescent diamond nanoparticles in cells and study of the internalization pathway. *Small*, **4**, 2236–2239.

133 Vial, S., Mansuy, C., Sagan, S., Irinopoulou, T., Burlina, F., Boudou, J.-P., Chassaing, G., and Lavielle, S. (2008) Peptide-Grafted nanodiamonds: Preparation, cytotoxicity and uptake in cells. *ChemBioChem*, **9**, 2113–2119.

134 Cheng, C.-Y., Perevedentseva, E., Tu, J.-S., Chung, P.-H., Cheng, C.-L., Liu, K.-K., Chao, J.-I., Chen, P.-H., and Chang, C.-C. (2007) Direct and *in vitro* observation of growth hormone receptor molecules in A549 human lung epithelial cells by nanodiamond labeling. *Appl. Phys. Lett.*, **90**, 163903.

135 Chao, J.-I., Perevedentseva, E., Chung, P.-H., Liu, K.-K., Cheng, C.-Y., Chang, C.-C., and Cheng, C.-L. (2007) Nanometer-Sized diamond particle as a probe for biolabeling. *Biophys. J.*, **93**, 2199–2208.

136 Schrand, A.M., Dai, L., Schlager, J.J., Hussain, S.M., and Osawa, E. (2007) Differential biocompatibility of carbon nanotubes and nanodiamonds. *Diamond Relat. Mater.*, **16**, 2118–2123.

137 Schrand, A.M., Huang, H., Carlson, C., Schlager, J.J., Osawa, E., Hussain, S.M., and Dai, L. (2007) Are diamond nanoparticles cytotoxic? *J. Phys. Chem. B*, **111**, 2–7.

138 Liu, K.-K., Cheng, C.-L., Chang, C.-C., and Chao, J.-I. (2007) Biocompatible and detectable carboxylated nanodiamond on human cell. *Nanotechnology*, **18**, 325102.

139 Puzyr, A.P., Baron, A.V., Purtov, K.V., Bortnikov, E.V., Skobelev, N.N., Mogilnaya, O.A., and Bondar, V.S. (2007) Nanodiamonds with novel properties: A biological study. *Diamond Relat. Mater.*, **16**, 2124–2128.

8
Ground and Excited State Charge Transfer and its Implications

Vito Sgobba and Dirk M. Guldi

8.1
Introduction

Research on carbon nanotube (CNT) – a new form of carbon, configurationally equivalent to two-dimensional graphene sheets rolled into a tube – has been established as a highly interdisciplinary field. In recent years, many potential applications have been proposed for CNT, including conductive and high strength composites; energy storage and energy conversion devices; sensors; field emission displays and radiation sources; hydrogen storage media; and nanometer sized semiconductor devices, probes, and interconnects. Some of these applications are now realized in products [1–9].

Conceptionally, single-walled carbon nanotubes (SWNT) are considered as small strips of graphene sheets that are rolled up to form perfect seamless single-walled nanocylinders (see Figure 8.1). SWNT are usually described using the chiral vector, which connects two crystallographically equivalent sites on a graphene sheet. The way the graphene sheets are wrapped varies largely and is represented by a pair of indices (n, m). These integers relate the structure of each SWNT to both its diameter and chirality. The diameter of most SWNT is about 1 nm, while their length reaches to the order of centimeters [10]. A multiwalled carbon nanotube (MWNT) is similarly considered to be a coaxial assembly of cylinders of SWNT, like a Russian doll, one within another. The separation between tubes is almost equal to that between the layers in natural graphite. The simplest representative of a MWNT is a double-walled carbon nanotube (DWNT). Hence, CNT are one-dimensional objects with a well-defined direction along the nanotube axis that is analogous to the in-plane directions of graphite.

Importantly, CNT are either electrically conductive or semiconductive – depending on their helicity – leading to nanoscale wires and electrical components [11]. The electronic properties of a SWNT vary in periodic ways between metallic and semiconducting. If $(n - m)$ is a multiple of 3, then SWNT exhibit metallic behavior (i.e., finite value of carriers in the density of states at the Fermi energy). If $(n - m)$ is,

Carbon Nanotubes and Related Structures. Edited by Dirk M. Guldi and Nazario Martín
Copyright © 2010 WILEY-VCH Verlag GmbH & Co. KGaA, Weinheim
ISBN: 978-3-527-32406-4

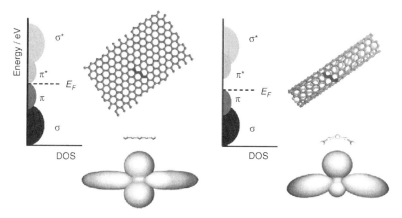

Figure 8.1 From a graphene sheet to CNT.

on the other hand, *not* a multiple of 3, then SWNT exhibit semiconducting behavior (i.e., no charge carriers in the density of states at the Fermi energy). The electrical transport in good quality metallic SWNT is ballistic, that is, electrons do not suffer from any scattering over a length scale of several micrometers and/or from any electromigration, even at room temperature. As a consequence, they may carry current densities 1000 times that of a typical copper wire [12]. Notably, even for semiconducting SWNT the charge transport is ballistic, but only in dimensions of a few hundred nanometers.

Early progress in understanding the optical characteristics of SWNT has been hampered by their aggregation – insoluble bundles and/or ropes [13]. To this end, mixing of energy states of different CNT results. In fact, the high polarizability and the smooth surface of SWNT are the inception to strong van der Waals interactions [14]. However, the preparation of functional CNT-based electron donor–acceptor nanocomposites necessitates the disentanglement of bundles and/or ropes. In this context, the chemical functionalization of CNT has attracted growing interest as a potent means to debundle and/or disperse them [15–19]. Nevertheless, debundling by ultrasonication is the most widely applied method [20]. It is implicit that the controlled conditions are the key requisites for avoiding/limiting damages to the sidewalls that are inevitable during such treatment [21]. Consequently, a number of techniques have been developed that allow the systematic debundling, separation, and characterization of different SWNT without the use of ultrasonication [22–24].

In the context of chemical reactivity, CNT are regarded as either sterically bulky π-conjugated ligands or as electron-deficient alkenes. Their shells of sp^2-hybridized carbons form highly aromatic hexagonal networks that are susceptible to a wide range of chemical reactions [25–31]. The smaller the SWNT diameter is, the more reactive they are [32–36]. A simple structural consideration suggests that CNT might be far more reactive at their ends than along their sidewalls. It is mainly the increased curvature at the ends leads to such a conclusion [27, 37]. Notwithstanding, upon closer inspection, the sidewalls show some fairly reactive sites – carboxylic groups and other oxygenated sites – since CNT are not as perfect as they were once thought to be. Defects such as pentagons, heptagons, vacancies, or dopants are found to drastically modify the electronic properties of these nanosystems [38–44]. Such functional defects are formed during the growth of CNT or introduced during the posttreatment applications. The posttreatment is primarily thought to remove graphite and other carbon particles from CNT. Nevertheless, depending on the actual conditions, the number of defects might be as high as 1–3% of all the carbon atoms present [45]. Topological defects and structural imperfections augment the curvature along the sidewalls [46–48]. The fascinating physical and chemical properties of CNT are tailored by the presence of defects, dopants, and so on [49–52]. A leading example is the rectifying behavior that originates exclusively from naturally occurring defects [53].

A comprehensive understanding of CNT chemistry, electrochemistry, and photophysics evolves not only as a necessity but also as a central fundament for applications in artificial photosynthesis and renewable energies [54–63].

8.2
Ground and Excited State Features

Employing absorption spectroscopy of SWNT suspensions shed light on their electronic states. In this context, the fact that a 1 nm diameter SWNT has a bandgap of approximately 1 eV is important. Such a value is close to that of silicon. But unlike silicon, semiconducting SWNT are direct bandgap materials and, as such, they are known to directly absorb and/or emit light.

As mentioned already, a common feature of CNT is that they tend to aggregate when synthesized and, thereby, limit optical studies to probe bundles of SWNT rather than individual SWNT. As a matter of fact, severe inhomogeneous broadening dominates the absorption spectra of bundles, since energy states of different CNT structures mix [64]. Such problems are circumvented by individualizing CNT through, for example, encasing them in micellar assemblies. Ultrasonic agitation, that is, dispersing SWNT in water/sodium dodecyl sulfate (SDS), followed by a centrifugation step – to remove bundles, ropes, and residual catalyst – has emerged as a probate means [19].

Three sets of bands, which all arise from electronic transitions between different van Hove singularities (see Figure 8.2) are normally discernable in SWNT suspensions. Bands centered in the 950–2000 nm range are ascribed to transitions that arise between the first (S_{11}) singularities in semiconducting SWNT. To the same extent, bands in the 700–950 nm range correspond to transitions between the second (S_{22}) singularities in semiconducting SWNT. Metallic SWNT, on the other hand, exhibit bands in the 400–700 nm. The origin of the latter are transitions between first (M_{11}) singularities [65, 66]. Despite these unambiguous characteristics, the absorption spectra are best described as the superposition of narrower absorption bands corresponding to the response of individual SWNT of different chiral indices and/or of different diameters [67, 68].

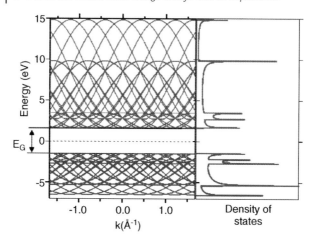

Figure 8.2 Density of states for semiconducting SWNT.

Electrostatic electron/hole interaction energies in the form of exciton binding energies are significant in SWNT – in the order of 0.3–0.5 eV at bandgap energies of approximately 1 eV [69–72]. An immediate consequence of this strong electron/hole attraction is that CNT photoexcited states are regarded as excitonic (i.e., electron/hole pairs) rather than as uncoupled electrons. In SWNT, excitons are characterized by electron/hole separations (i.e., Bohr radius) of approximately 2.5 nm [71, 73].

In SWNT bundles, fluorescence is hardly ever observed. Metallic SWNT that are statistically present in bundles are the inception for photoexcited carriers in semiconducting SWNT to relax along nonradiative pathways [74]. On the contrary, fluorescence measurements with individual SWNT – water/SDS – have revealed distinct fluorescing transitions for more than 30 different SWNT as shown in Figure 8.3 [75]. Nevertheless, inhomogeneous broadening and spectral overlap between fluorescence from different CNT structures hinder the recording of time-resolved fluorescence spectra. Interestingly, the fluorescence quantum yields are very low (i.e., 10^{-4}) – a finding that has been rationalized on the basis of multiple dark excitonic states that are situated below the lowest lying bright excitonic states [76]. However, a mixture of few specific semiconducting SWNT were recently suspended in toluene by using commercial poly[9,9-dioctylfluorenyl-2,7-diyl] as a dispersing/stabilizing agent [77]. Under these conditions, the fluorescence quantum yields are as high as 1% [78].

Resonance Raman spectroscopy emerged as another powerful technique for characterizing SWNT. Also notable is that it allowed optical transitions between different van Hove singularities and/or excitonic transitions leading to a marked resonance enhancement of the Raman scattering. By combining data that were acquired through fluorescence and resonance Raman investigations, each optical transition has been mapped out and assigned [75, 79].

In complementary work, which focused on the fluorescence of SWNT in water/polyacrylic acid, the excited states were shown to decay on the time scale of 10 ps.

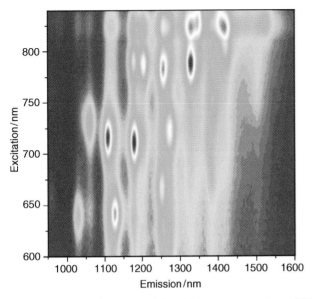

Figure 8.3 3D steady-state near-infrared fluorescence spectrum of SWNT/SDBS in D_2O – with increasing intensity from blue to green to yellow and to red.

There is little, if any, doubt that such an intrinsically fast relaxation must be associated with nonradiative processes. Taking this into consideration, the low fluorescence quantum yields are explainable. From the decay rates and the fluorescence quantum yields a radiative lifetime of 110 ns was deduced [80].

The optical features of SWNT are susceptible to environmental changes. To this end, the same SWNT exhibit different peak positions and line width – absorptive and emissive – when dissimilar surfactants are utilized. Further alterations may arise when trap states come into play as they may evolve from structural and/or chemical defects. Moreover, the fluorescence tends to shift to the red when the temperature is increased and is also affected by mechanical strains electric and magnetic fields [81–86]. Notably, the fluorescence of individual SWNT is typified by the absence of spectral and intensity fluctuations – unlike almost all other known emitters, which exhibit an emission intermittency or on–off blinking behavior for all excitation intensities on time scales that span many orders of magnitude [74].

The outstanding photostability of SWNT, namely resistance to photobleaching, together with their versatile wavelength tunability – absorbing light between 800 and 1100 nm and emitting light between 1300 and 1500 nm – render them as ideal single molecule fluorophores [87, 88]. Electrical measurements show strong electroluminescence in the near-infrared, which originates from radiative recombination of electrons and holes that are simultaneously injected into the undoped SWNT [89].

Ultrafast transient absorption spectroscopic measurements complement the fluorescence studies and have been used to clarify the time scales and nature of ground state recoveries in SWNT and to extract information about excitonic lifetimes. For relaxation from the excited state, an omnipresent fast decay component (i.e., 300–500 fs) is likely due to the presence of bundled SWNT and/or metallic SWNT. A much slower decay component (i.e., 100–130 ps) appears only when semiconducting SWNT are probed and corresponds likely to the intrinsic excited state lifetime of photoexcited excitons. Lifetimes in the 100–130 ps range were corroborated by photoluminescence lifetime and by correlated single photon counting spectroscopic measurements [79, 80].

Finally, it should be mentioned that if the excitation intensity is increased to photoexcited SWNT, multiple electron/hole pairs are generated. However, these electron/hole pairs were shown to deactivate in less than 3 ps. The energy released by the annihilated exciton is used to excite a second exciton to a higher energy level.

8.3
Ground State Charge Transfer – CNT as Electron Acceptors

The electron-acceptor properties of CNT and the factors that control these, need careful considerations as they impact the reactivity of the reduced state, oxidized state, and excited state. They also tie to electron storage capabilities and may answer questions on how the electron storage changes the overall energetics. Finally, aspects on how to integrate them into photonic, electrochemical, and electronic devices should not be overlooked.

8.3.1
Chemical Reduction

Reduction of SWNT bundles is achieved by exposure to reductants with different redox potentials (i.e., intercalation with alkali metals [90] and/or anion radicals) [91, 92]. The course of these reactions – filling the density of states and, in turn, modifying the conducting nature of individual SWNT – is typically monitored by visible/near-infrared absorption and *in situ* Raman measurements [90, 93, 94]. In the visible/near-infrared absorption spectrum, the most profound changes are seen as bleaching of the optical transitions that are associated with the filling of the corresponding electronic states. Actually, a gradual disappearance is noted for the S_{11}, S_{22}, and M_{11} transitions. Moreover, this process is accompanied by a blueshift [91–93, 95]. For instance, exposing SWNT to fluorenone/lithium and anthraquinone/lithium exclusively attenuates the S_{11} bands. On the contrary, reaction with naphthalene/lithium yields heavily reduced SWNT, in which the M_{11} have disappears completely. Concomitant with this disappearance is the gradual growth of new features that are positioned between those of S_{11} and S_{22}. However, the nature of these newly evolving bands is still unclear at this point [90, 95, 96].

Owing to their unique electronic structure, SWNT show characteristic Raman spectra, which are understood in terms of resonance enhancement in one-dimensional conductors with van Hove singularities in the electronic density of states. Figure 8.4 displays a characteristic spectrum. An important standard is the radial breathing mode (RBM), which is seen in the range between 160 and 365 cm^{-1}.

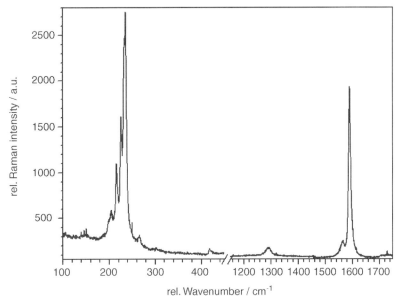

Figure 8.4 Raman spectrum of HiPCO SWNT/SDBS in H_2O ($\lambda_{excitation} = 800\,nm$).

The utilization of different laser energies assisted in resolving this part of the spectrum for several different SWNT deposited onto conducting electrodes – see Kataura plot [65, 97]. At first glance, when SWNT are reduced, the RBM intensities decreases, due to an inherent loss of the resonant conditions (i.e., bleaching of the S_{11}, S_{22}, and M_{11} transitions) [98, 99]. When strongly reductive conditions are applied, an additional shift of these frequencies to the blue is discerned. This is ascribed to charge induced C–C bond contractions and to charge induced SWNT debundling [100]. A similar intensity loss characterizes the tangential vibrational modes (HEM). Again, a likely rationale is the loss of resonance enhancement. Overall HEM are more susceptible – relative to RBM – in terms of altered frequencies and shapes upon reduction [101]. Electrons that are added to the conduction bands of SWNT are expected to weaken the sp^2 character and, in turn, inflict redshifted HEM [102].

8.3.2
Electrochemical Reduction

The electronic structure may also be tuned electrochemically by controlling the interfacial potential of a SWNT film as a working electrode. Here, SWNT films – pure buckypaper or cast on an inert back electrical contact (i.e., Pt, Au, or Hg) [94] – are simply immersed in an electrochemical cell in contact with an electrolyte solution, a Pt counterelectrode, and a Ag/AgCl reference electrode. It is implicit that control over the electrochemical reduction is much easier and more precise when compared

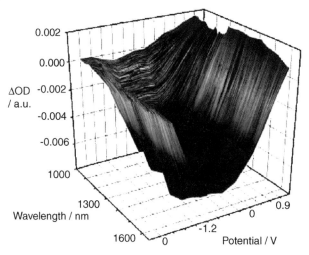

Figure 8.5 3D plot of the differential results from spectroelectrochemistry of SWNT/SDBS in DMSO (0.1 M TBA PF$_6$ supporting electrolyte).

with the chemical reduction. In order to increase the electrochemical window, especially in the cathodic range, the use of Hg electrodes assists in minimizing the evolution of H$_2$. Typical cyclic voltammograms display only a monotonous charge injection behavior over the entire scanning range – from -0.5 to $+0.7$ V. Double layer charging and overlapping peaks of faradaic charge transfers to individual SWNT of varying redox potential lead to this feature.

In situ spectroelectrochemistry, where bulk spectroscopy and phase boundary electrochemistry are complementarily combined, is, nevertheless, better suited to gather details about the reductive chemistry of SWNT (see Figure 8.5). For example, individual SWNT, which were deposited from water/SDS suspensions were tested by *in situ* micro-Raman spectroelectrochemistry [103, 104]. Aprotic electrolytes like acetonitrile [67, 105, 106], ethylene carbonate/dimethylcarbonate, tetrahydrofuran [106], and especially ionic liquids such as fluoroborate 1-butyl-3-methylimidazolium salts allow to explore a broad window of electrochemical potentials [107]. Overall, the spectroelectrochemical results confirm that charging induced bleaching of transitions between van Hove singularities. The bleaching of the optical transitions is mirrored by the quenching of resonance Raman scattering in the region of RBM and HEM. An additional benefit of this method is the establishment of the reversibility of the reduction and the rate of reduction [108, 109].

Recently, the redox properties of alkali metal reduced SWNT were determined in dry and deoxygenated DMSO solution containing tetrabutylammonium hexafluorophosphate as the supporting electrolyte [91, 96, 109]. The potentials were kept below -0.6 V to avoid additional reduction of sodium. Voltammetric investigations and visible/near-infrared absorption measurements showed a 35 meV redshift of the absorption peaks when compared with water/SDS suspensions and a redshift of

15 meV when compared with metal-free DMSO suspensions [110]. An accurate analysis of this data enabled to calculate the average standard reduction potentials of several individual SWNT. The latter were found to be in good agreement with the experimental values that were reported elsewhere [111, 112].

8.3.3
Reduction by Doping

SWNT modified with $AgNO_3$ – into the SWNT cores or into the interstitial channels of the SWNT lattice – were characterized by resonance Raman spectroscopy. A significant charge transfer induced perturbation of the SWNT electronic structure – downshift in the HEM [113, 114].

Figure 8.6 exemplifies substitutional doping, that is, the selective replacement of carbon atoms by boron or nitrogen atoms. Subsequently, strongly localized electronic features are introduced in the valence or conduction bands. Also the number of electronic states at the Fermi level is enhanced, depending on the location and concentration of the dopants [115]. One possible structure of N-SWNT is a three-coordinated nitrogen atom within the sp^2-hybridized network, which induces sharp localized states above the Fermi level due to the presence of additional electrons. These N-SWNT exhibit n-type conduction. The second type of substitutional nitrogen leaves two-coordinated (pyridine-like) nitrogens in the SWNT lattice – provided that an additional carbon atom is removed from the framework.

N-MWNT with low concentrations of nitrogen, that is, less than 1%, were synthesized by arc discharge in the copresence of melamine, nickel, and yttrium [116] or via pyrolysis of pyridine and methylpyrimidine [117]. The doped nanotubes exhibit strong features in the conduction band close to the Fermi level. Using tight binding and *ab initio* calculations, pyridine-like nitrogens are responsible for introducing donor states close to the Fermi level. The first observation of aligned arrays of N-MWNT involved the pyrolysis of aminodichlorotriazine and triaminotriazine over laser-etched cobalt thin films at 1050 °C [118, 119]. Importantly, controlling the concentration of substitutional nitrogen atoms in the carbon lattice should provide a way to tune the conducting properties of single-walled carbon nanotubes by *in situ* doping. Long strands of N-SWNT bundles were also produced by pyrolyzing ferrocene in an argon atmosphere in ethanol solutions with benzylamine at 950 °C.

    (a)   (b)   (c)

Figure 8.6 Structure of N-SWNT with (a) substitutional N, (b) pyridine-type N, and (c) B-SWNT.

They give rise to an electron conduction, quite different from that of pure SWNT, especially at temperatures lower than 20 K [120].

8.3.4
Miscellaneous

In light of the chemical reactivity of reduced CNT, it is interesting to note that the corresponding anions are versatile precursor states to perform sidewall functionalization. In fact, numerous contributions have demonstrated the great success on using them *en route* toward hydrogenated, alkylated, arylated, or carboxylated CNT. Powerful alternatives to produce reduced SWNT and MWNT precursors by chemical means are the uses of sodium, lithium, or potassium dissolved in liquid ammonia [121, 122]. The resulting CNT salts were treated with ethanol, methanol, or with oxygen to afford hydrogenated CNT [123, 124] or with alkyl or aryl halides to afford CNT grafted with alkyl or aryl groups, respectively [125–127]. Like fullerenes [128, 129], organolithium compounds react with CNT sidewalls. Reoxidation with air or oxygen yields alkylated and/or arylated SWNT [122, 130]. SWNT carbanions are also reactive toward carbon dioxide forming functionalized SWNT that bear carboxyl groups [131]. A recent report took the reductive alkylation reaction even further. When utilizing sodium/butyl iodide, a selectivity toward smaller diameter and metallic SWCNT was demonstrated. However, no pronounced difference in reactivity was seen during the hydrogenation with ethanol [132].

8.4
Ground State Charge Transfer – CNT as Electron Donors

8.4.1
Chemical Oxidation

Oxidation has been among the first reactions that were ever tested with CNT [133] and is still a key step in their purification (i.e., removal of Fe/Co/Ni catalysts and amorphous/graphitic carbon impurities) [134, 135]. Under oxidative conditions, opening the hemispherical caps often goes along with the shortening and chemical functionalization of SWNT.

Bundles – composed of up to hundreds of individual tubes – are the predominant form of SWNT. This directs their oxidation to the tips and to the defect sites. However, the reactivity is limited to the outer layer of the bundles [136]. Similar principles are applicable to the oxidation of MWNT, turning it into a powerful technique for MWNT thinning through the systematic removal of the outer layers [137].

O_2 – triplet ground state – does not oxidize CNT. Instead, O_2 physisorbs quite strongly to the sidewalls of CNT and transforms the semiconducting character into metallic one at room temperature. Upon photoexcitation, physisorbed O_2 desorbs from the CNT sidewalls [138]. $^1O_2$ – singlet excited state – on the other hand, reacts promptly to produce different cycloadducts (i.e., [2 + 2] or [4 + 4]) made of metallic SWNT. A great benefit of this reaction is that upon thermal activation, the cycloadducts readily undergo O−O bond cleavage [139]. This procedure was conveniently adapted to separate CNT of different chiralities [140].

Oxidative treatment of CNT was performed under a wide variety of experimental conditions. Hereby, sonications in oxygen containing acids (i.e., HNO_3, H_2SO_4, H_2SO_4 plus $K_2S_2O_8$, HNO_3 plus H_2SO_4, HNO_3 plus supercritical water, trifluoromethanesulfonic and chlorosulfonic acid, peroxytrifluoroacetic acid, H_2SO_4 plus $KMnO_4$, H_2SO_4 plus H_2O_2, $HClO_4$, H_2SO_4 plus $K_2Cr_2O_7$) play the most dominant roles. As a consequence of such treatments, carboxylate groups are generated and are distributed over the entire CNT. In the presence of H_2SO_4, sulfate, ketone, phenol, alcohol, and ether groups were also registered. Other often utilized oxidants are dilute ceric sulfate, H_2O_2, O_2 plasma, and RuO_4 [7].

Oxidation leads to the loss of structure in the visible/near-infrared absorption spectrum as a consequence of the conversion of a large number of sp^2-hybridized carbons, which leads to altered electronic structures of CNT [30]. The concentrations of carboxylic acid that are formed during the oxidative treatment were determined by evaluating the CO and CO_2 concentrations at high temperatures [141] or measuring the atomic oxygen percentage with calibrated energy-dispersive X-ray spectroscopy [142]. Chemical titration assays emerged as alternative tests to macroscopically estimate the defect density in CNT. Titrations of the purified SWNT with NaOH and $NaHCO_3$ solutions were used to determine the total percentage of acidic sites and carboxylic acid groups, respectively [45].

8.4.2
Electrochemical Oxidation

The effect of CNT chemical and electrochemical oxidation has been extensively studied with electrochemistry, photoelectrochemistry, and *in situ* visible/near-infrared and Raman spectroelectrochemistry. Oxidation is commonly accompanied by gradually disappearing S_{11}, S_{22}, and M_{11} transitions and concomitantly growing transitions in the range between S_{11} and S_{22} [98]. Raman spectroelectrochemistry, on the other hand, inflicts a reversible drop of the RBM and HEM intensities [98, 107]. In addition, strong blueshifts are seen for the RBM when potentials between 0.5 and 1 V are applied. The HEM blueshifts too – stiffening is connected to the introduction of holes in the π-band [143]. A common structural characteristic for both oxidation [105] and reduction is a steadily and irreversibly increasing D-band intensity [106]. Going beyond 1.2 V, photoanodic decomposition of SWNT triggers an irreversible breakdown [98].

8.4.3
Oxidation by Doping

There has been considerable interest in doping SWNT to manipulate their electronic properties. In the context of oxidation, doping reactions, namely, intercalation of iodine and bromine into the interstitial channels of SWNT was performed in the vapor phase and studied by Raman spectroscopy. The results – HEM shifts substantially to higher frequencies – are consistent with oxidizing CNT. Moreover, bromine leads, for example, to a conductivity decrease of up to 30 times [93].

Recently, air stable, iodine-doped SWNT/polymer composites were prepared. Such a composite material revealed electrical conductivities that are enhanced by a factor of 2–5 when compared to undoped SWNT/polymer composites [144]. Furthermore, Lewis acids (i.e., $FeCl_3$ [145], WF_6, CrO_3, $SOCl_2$) and also Brönsted acids (i.e., sulfuric, nitric, hydrochloric, N-hydroxyheptafluorobutyramide (HFBA), perfluorododecanoic acid (PFDA), 2-methyl acrylamidopropane sulfonic acid (AMPS), aminobutyl phosphonic acid (ABPA), and hexadecyl phosphonic acid (HDPA)) were investigated for CNT p-doping [62].

Boron, for example, once incorporated into the SWNT framework generates sharp localized states below the Fermi level. Consequently, the substitution of boron in the carbon lattice (see Figure 8.6) is expected to increase the hole-type charge carriers. Boron doped SWNT (B-SWNT) with boron contents up to 3% were synthesized using laser ablation of Co/Ni/B-doped carbon targets [146]. Higher atomic concentrations of boron, that is, between 10 and 20%, were also achieved, but required a post-treatment of laser ablation samples in corrosive ammonia environments [147]. Microwave conductivity studies on bulk B-MWNT suggest that these structures are intrinsically metallic [148].

Lastly, filling SWNT should be considered [31, 149–151]. The endohedral guest are viewed as dopants that, in turn, help to gain control over the SWNT properties such as conductance and/or electronic bandgap [152, 153]. It has been demonstrated that, for example, periodic arrays of C_{60} in SWNT give rise to a new hybrid electronic band [154]. The resulting increase in CNT conductivity is based on charge transfer that occurs from SWNT to C_{60} leaving holes as the primary charge carriers behind [155]. Important insights into the electronic structure of the encapsulated C_{60}, C_{82}, $La@C_{82}$, and $Gd@C_{82}$ [156] were lent from Raman spectroscopy. In particular, dramatic changes relative to empty SWNT are seen [157]. A sharp and intense line at $142\,cm^{-1}$ has been interpreted as a polymerization signature of the encapsulated metallofullerenes. Additional signals appeared at 400, 520, and $640\,cm^{-1}$. The upshift and the stiffening of HEM suggest also in this case, a charge transfer process from SWNT to the encapsulated fullerenes [158].

Additionally, a variety of organic molecules with different electron affinities (i.e., anthracene, tetracene, pentacene, tetracyanoquinodimethane (TCNQ), tetrafluoro-tetracyano-p-quinodimethane (F_4TCNQ), tetrakis(dimethylamino)ethylene (TDAE), 3,5-dinitrobenzonitrile) were encapsulated within SWNT. Remarkable is that the resistivity of TCNQ@SWNT and TDAE@SWCNT was approximately 50% of that found for pristine SWNT. However, implicit is that sizeable intratubular charge transfer interactions evolve, which contrasts, the lack of charge transfer activity in C_{60}@SWCNT. Note that TCNQ and C_{60} have approximately the same electron affinity [159].

8.5
Excited State Charge Transfer – CNT as Excited State Electron Acceptor

8.5.1
Covalent Electron Donor–Acceptor Conjugates

Via the cycloaddition of azomethine ylides several electron donors were attached to the sidewalls of CNT. In this context, ferrocene constitutes one of the first examples that was brought forward [130]. Spectroscopic and kinetic analyzes of the photophysical properties of SWNT-Fc (Figure 8.7) were interpreted in terms of intramolecular charge separation that evolves from photoexcited SWNT. The charge separation dynamics are very fast $(3.6 \times 10^9 \, s^{-1})$, whereas the charge recombination kinetics are very slow $(9.0 \times 10^5 \, s^{-1})$. As the correct identification of the product was deemed to be critical, additional time resolved pulse radiolytic and steady state electrolytic experiments were carried out to establish the characteristic fingerprints of the reduced SWNT. All techniques gave similar broad absorptions in the visible range and confirmed the formation of reduced SWNT.

In a modified strategy, namely purifying, shortening, and endowing SWNT with carboxylic groups electron donors like tetrathiafulvalene (TTF) and extended tetrathiafulvalene (exTTF) were tested [161]. Photophysical investigation supported the occurrence of photoinduced charge transfer processes in SWNT–TTF and SWNT–exTTF (Figure 8.8) and helped to identify the reduced SWNT and oxidized TTF and/or exTTF as metastable states. Overall, remarkable lifetimes – in the range of hundreds of nanoseconds – are noted. Most important is that we succeeded in controlling the rate of charge recombination by either systematically altering the

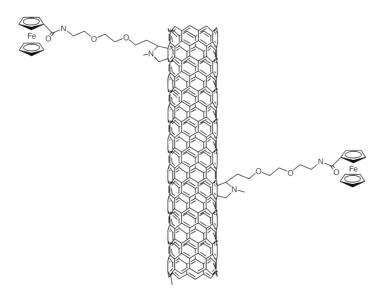

Figure 8.7 SWNT-Fc donor–acceptor conjugate.

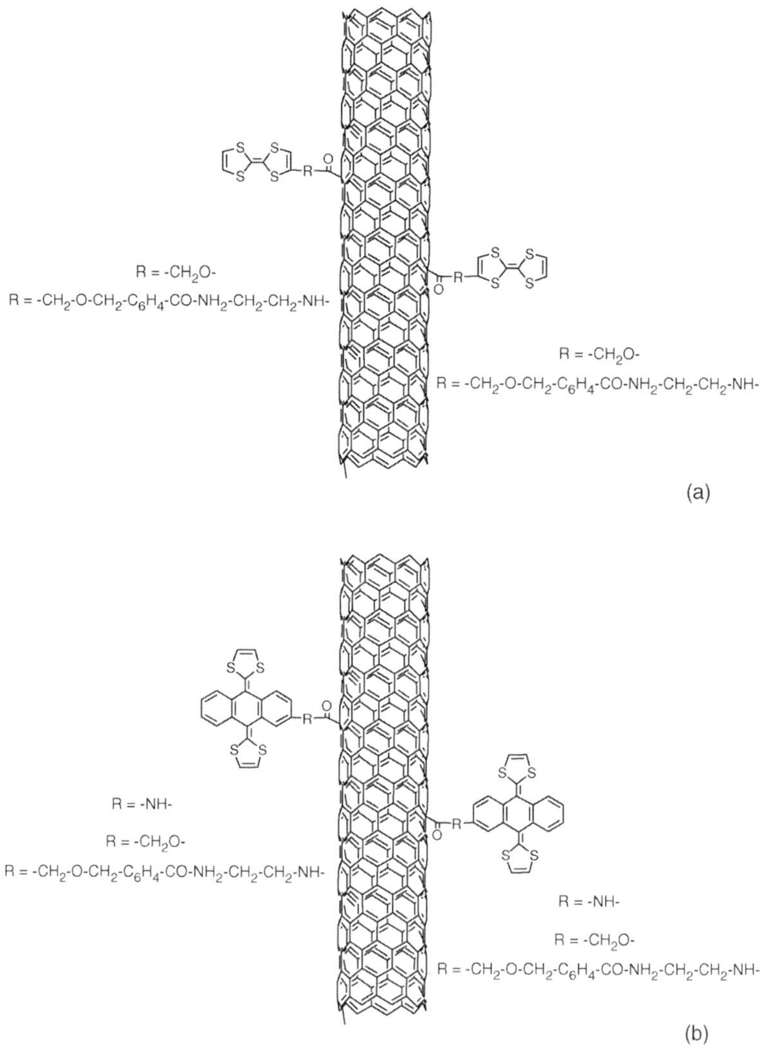

R = -CH₂O-
R = -CH₂-O-CH₂-C₆H₄-CO-NH₂-CH₂-CH₂-NH-

R = -CH₂O-
R = -CH₂-O-CH₂-C₆H₄-CO-NH₂-CH₂-CH₂-NH-

(a)

R = -NH-
R = -CH₂O-
R = -CH₂-O-CH₂-C₆H₄-CO-NH₂-CH₂-CH₂-NH-

R = -NH-
R = -CH₂O-
R = -CH₂-O-CH₂-C₆H₄-CO-NH₂-CH₂-CH₂-NH-

(b)

Figure 8.8 (a) SWNT–TTF and (b) SWNT–exTTF donor–acceptor conjugates.

relative donor–acceptor separations ($6.3 \times 10^6 - 2.9 \times 10^6 \, \text{s}^{-1}$) or integrating different electron donors ($5.2 \times 10^6 \, \text{s}^{-1}$ versus $3.6 \times 10^6 \, \text{s}^{-1}$) [162].

8.5.2
Noncovalent Electron Donor–Acceptor Hybrids

As a complement to the covalent approach, π–π interactions were pursued to anchor exTTF to the surface of SWNT by using a pyrene tether [163]. Nevertheless, π–π interactions between, for example, the concave hydrocarbon skeleton of exTTF and the convex surface of SWNT add further strength and stability to SWNT/pyrene–exTTF. In this context, for the first time a complete and concise characterization

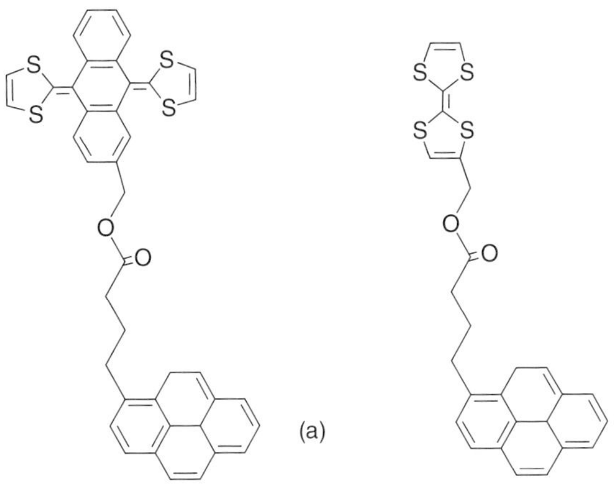

Figure 8.9   (a) Pyrene–exTTF and (b) pyrene–TTF.

of the radical ion pair state has been achieved, especially in light of injecting electrons into the conduction band of SWNT. The close proximity between exTTF and the electron-accepting SWNT leads to very rapid charge transfer $(1.1 \times 10^{12}\,\text{s}^{-1})$ that affords, in turn, a short lived radical ion pair state $(3.3 \times 10^{6}\,\text{s}^{-1})$.

In a follow-up work, we have taken this concept further by scrutinizing electron donor–acceptor interactions with different types of CNT and functionalized pyrene–TTF (Figure 8.9) [164]. The goal was to identify the best nanohybrids – in terms of processability and photoinduced charge transfer behavior – based on interacting SWNT, DWNT, and various MWNT as electron acceptors with an organic electron donor, namely TTF. Photophysical investigations evidence the formation of stable radical ion pair states only in the case of MWNT, while shorter lifetimes are obtained when considering SWNT. Most probably, the large number of concentric tubes – providing different acceptor levels – help in stabilizing the transient radical ion pairs.

8.6
Excited State Charge Transfer – CNT as Ground State Electron Acceptor

8.6.1
Covalent Electron Donor–Acceptor Conjugates

More powerful is the concept of integrating porphyrins [165] or phthalocyanines that serve as visible light-harvesting chromophores/electron donors [166, 167]. A first example involved the efficient covalent tethering of SWNT with porphyrins through the esterification of SWNT bound carboxylic acid groups. The work started with two porphyrin derivatives containing terminal hydroxyl groups, that is, [5-(p-hydroxyhexyloxyphenyl)-10,15,20-tris(p-hexadecyloxyphenyl)-21H,23H-porphine] and[5-(p-hydroxymethylphenyl)-10,15,20-tris(p-hexadecyloxyphenyl)-21H,23H-porphine]. In the corresponding SWNT–H$_2$P conjugates the photoexcited porphyrins deactivate through a transduction of excited state energy. Interestingly, the rates and

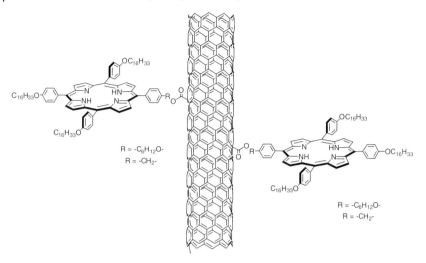

Figure 8.10 H$_2$P–SWNT donor–acceptor conjugate – linked via esterification of the carboxylic acid groups.

efficiencies of the excited state transfer depend on the length of the tether that links the porphyrins with the SWNT: the shorter tethers gave rise to weaker fluorescence quenching[168, 169] (Figure8.10).

Extending the scope of the aforementioned work, the carboxylic functionalities proved to be convenient for linking 5-*p*-hydroxyphenyl-10,15,20-tritolylporphyrin to either SWNT or MWNT (Figure 8.11). According to thermogravimetric analyzes the amount of grafted porphyrins was estimated to range from 8 to 22%, depending on the initially employed porphyrin concentrations.

Similarly, unsymmetrically substituted aminophthalocyanines ZnPc were linked to SWNT through a reaction with the terminal carboxylic acid groups of shortened SWNT (Figure 8.12). However, the resulting materials proved to be nearly insoluble in organic solvents [170, 171]. Reasonable alternatives involve a straightforward cycloaddition reaction with *N*-octylglycine and a formyl-containing ZnPc, a stepwise approach that involves cycloaddition of azomethine ylides to the double bonds of SWNT using *p*-formyl benzoic acid followed by esterification with an appropriate ZnPc [172, 173] (see Figure 8.13) or functionalization of SWNT with 4-(2-trimethyl-silyl)ethynylaniline and the subsequent ZnPc attachment using the *Huisgen* 1,3-dipolar cycloaddition [174]. The occurrence of charge transfer from photoexcited ZnPc to SWNT is observed in transient absorption experiments, which reflect the absorption of the one electron oxidized ZnPc cation and the concomitant bleaching of the van Hove singularities of SWNT. Charge separation (2.0×10^{10} s^{-1}) and charge recombination (7×10^{-5} s^{-1}) dynamics reveal a notable stabilization of the radical ion pair product (Figure 8.14).

The approach that involves placing of pyridyl isoxazolino functionalities along the sidewalls of short SWNT is quite different [175]. The synthesis is based on the cycloaddition of a nitrile oxide onto SWNT. The resulting SWNT–pyridine forms

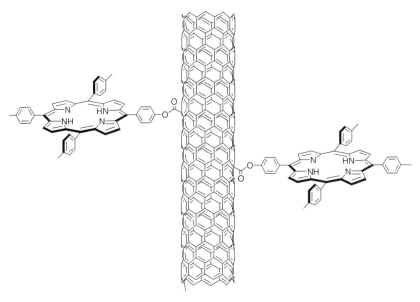

Figure 8.11 5-*p*-Hydroxyphenyl-10,15,20-tritolylporphyrin–SWNT donor–acceptor conjugate.

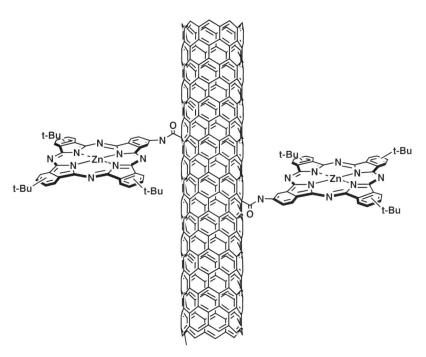

Figure 8.12 ZnPc–SWNT donor–acceptor conjugate – linked via esterification of the carboxylic acid groups.

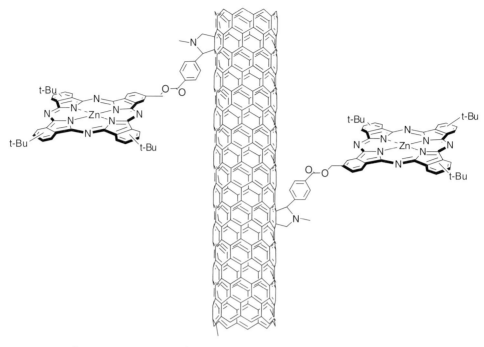

Figure 8.13   ZnPc–SWNT donor–acceptor conjugate – linked via the Prato reaction.

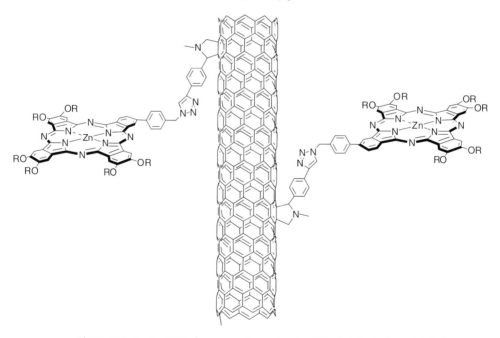

Figure 8.14   ZnPc–SWNT donor–acceptor conjugate – linked via the *Huisgen* 1,3-dipolar cycloaddition reaction.

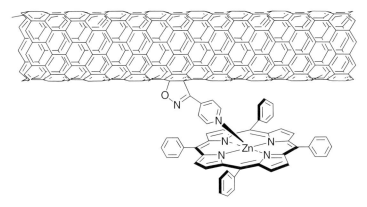

Figure 8.15 SWNT–pyridine/ZnP donor–acceptor conjugate.

complexes with ZnP and/or RuP [175, 176]. Formation of the SWNT–pyridine/ZnP complex was firmly established by a detailed electrochemical study. However, photochemical excitation of SWNT–pyridine/ZnP does not lead to charge transfer/generation of radical ion pair states. Instead, fluorescence and transient absorption studies indicate that the main process is energy transfer from the singlet excited state of ZnP to SWNT–pyridine (Figure 8.15).

SWNT functionalization is also the basis for applying Suzuki coupling reactions (Figure 8.16) [177]. Recent work documents that this type of coupling reactions

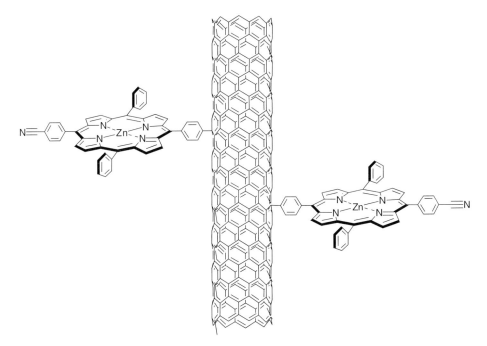

Figure 8.16 ZnP–SWNT donor–acceptor conjugate – linked via the Suzuki coupling reaction.

represents an efficient method for introducing ZnP onto the SWNT sidewalls. Despite all perceptions, covalently functionalized SWNT were found to serve as efficient quenchers even after their perfectly conjugated sidewall structure has been disrupted.

Functionalization with aryl diazonium salts, generated *in situ* is another effective method for CNT sidewall grafting [16, 178–180]. What is particularly interesting is that metallic SWNT seem to react more rapidly than semiconducting SWNT. The selectivity is dictated by the availability of electrons near the Fermi level to stabilize a charge transfer transition state preceding the bond formation. This opens up the way for a selective solubilization and separation. To this end, *in situ* generated tetraphenylporphyrin diazonium salts react directly with SWNT. These functionalized SWNT show superior optical limiting effects when compared with pristine SWNT [181].

An important consideration when associating SWNT with electron donors is to preserve the unique electronic structures of SWNT. A versatile approach involves grafting SWNT with polymers, such as poly(styrenesulfonate) (PSS^{n-}) [182], poly(4-vinylpyridine) (PVP) [183], and poly((vinylbenzyl)trimethylammonium chloride) ($PVBTA^{n+}$) [184] to form highly dispersable SWNT–PSS^{n-}, SWNT–PVP, and SWNT–$PVBTA^{n+}$, respectively. In the next step, coulomb complex formation was achieved with SWNT–PSS^{n-} and 5,15-bis-[2′,6′-bis-{2″,2″-bis-(carboxy)-ethyl}methyl-4′-*tert*-butyl-phenyl]-10,20-bis-(4′-*tert*-butyl-phenyl)porphyrin (H_2P^{8+}). Likewise a donor–acceptor nanohybrid has been prepared using electrostatic/van der Waals interactions between SWNT–$PVBTA^{n+}$ and 5,15-bis-[2′,6′-bis-{2″,2″-bis-(carboxy)-ethyl}-methyl-4′-*tert*-butyl-pheny]-10,20-bis-(4′-*tert*-butyl-phenyl)porphyrin (H_2P^{8-}). Several spectroscopic techniques like absorption, fluorescence, and TEM were used to monitor the complex formations. Importantly, photoexcitation of H_2P^{8+} or H_2P^{8-} in the newly formed nanohybrid structure results in efficient charge separation ($3.3 \times 10^9 \, s^{-1}$), which leads, subsequently, to the radical ion pair formation. In SWNT–PSS^{n-}/H_2P^{8+} (Figure 8.17) the newly formed radical ion pair exhibits a remarkably long lifetime ($7.1 \times 10^4 \, s^{-1}$), which constitutes one of longest reported for any CNT ensemble found so far. In SWNT–$PVBTA^{n+}$/H_2P^{8-} (Figure 8.18) the charge separation tends to be slightly faster ($4.5 \times 10^5 \, s^{-1}$). Differently, SWNT–PVP (Figure 8.19) were assayed in coordination tests with ZnP. Kinetic and spectroscopic evidence corroborates the successful formation of SWNT–PVP/ZnP nanohybrids in solutions. Within this SWNT–PVP/ZnP nanohybrid, static charge transfer quenching ($2.0 \times 10^9 \, s^{-1}$) converts the photoexcited ZnP chromophore into a microsecond-lived radical ion pair state, that is, one electron oxidized ZnP and reduced SWNT.

A series of SWNT that are functionalized with PAMAM dendrimers were described recently. As the dendrimers are linked directly to the SWNT surface via a divergent methodology, it allows to increase the number of functional groups without implementing significant damages to the conjugated π-system of SWNT and comes as a reasonable alternative to the SWNT-H_2P conjugate (Figure 8.21) [185]. Photophysical investigations reveal that in SWNT–$(H_2P)_n$ (Figure 8.20) some

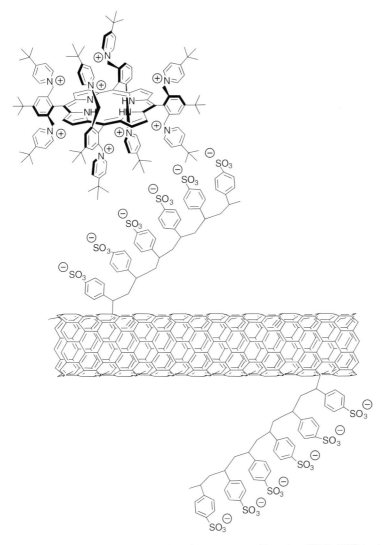

Figure 8.17 SWNT grafted with poly(sodium-4-styrenesulfonate) – SWNT–PSS^{n-} – that complexes H_2P^{8+}.

porphyrin units interact with SWNT while others do not; those that react (2.5×10^{10} s^{-1}) form a radical ion pair state that decays to ground state with kinetics of 2.9×10^5 s^{-1}.

CNT – bearing amides or esters – evolve as important precursors in the realization of bioinspired materials involving peptide nucleic acids [186], desoxyribonucleic acid [187], enzymes [188, 189], sugars [190], proteins [191], bacteria [192], viruses [193], biotin [194], and various nonporphyrin/nonphthalocyanine dyes [195–197]

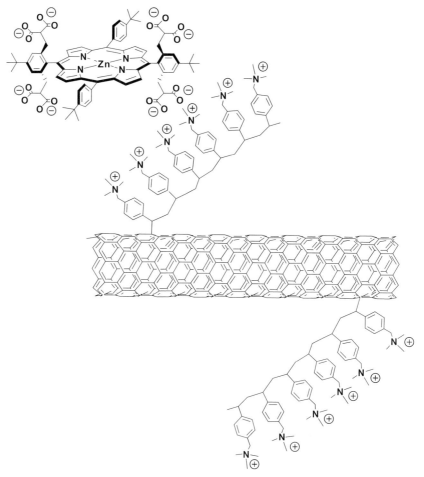

Figure 8.18 SWNT grafted with poly((vinylbenzyl)trimethylammonium chloride) – SWNT–PVBTA^{n+} – that complexes ZnP^{8-}.

8.6.2
Noncovalent Electron Donor–Acceptor Hybrids

Nevertheless, covalently modified CNT may not be suitable for applications, which are based on the high conductivity and/or mechanical strength of pristine SWNT. Noncovalent approaches might offer a solution to preserve the electronic and structural integrity of SWNT, permitting the use of their conductivity as well as their strength properties in future applications. Triggered by these incentives the noncovalent integration of a wide range of functional groups onto CNT emerged as a viable alternative [27, 198, 199].

As a leading example our recent success in the facile supramolecular association of pristine SWNT with linearly polymerized porphyrin polymers should be considered.

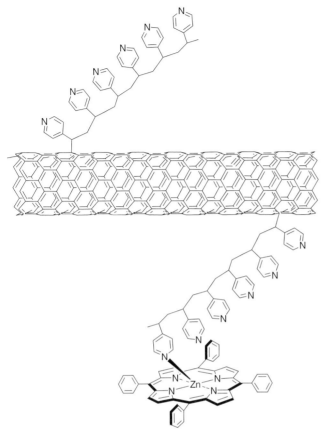

Figure 8.19 SWNT grafted with poly(4-vinylpyridine) – SWNT–PVP – that complexes ZnP.

The target SWNT nanohybrids, which are dispersable in organic media, were realized by the use of soluble and redox-inert poly(methylmethacrylate) (PMMA) bearing surface immobilized porphyrins (i.e., H_2P–polymer). Conclusive evidence for H_2P–polymer/SWNT interactions came from absorption spectroscopy: the finger-prints of SWNT and H_2P–polymer are discernable throughout the UV, visible, and near-infrared part of the spectrum. A similar conclusion was also derived from TEM and AFM. AFM images illustrate, for example, the debundling of individual SWNT. An additional feature of H_2P–polymer/SWNT (Figure 8.22) is an intrahybrid charge separation, which has been shown to be long lived ($4.7 \times 10^5 \, s^{-1}$) [200].

Stable H_2P/SWNT composites were also realized by condensating tetraformylpor-phyrins and diaminopyrenes on SWNT. The degree of interaction between SWNT and H_2P was evaluated by UV-visible absorption and fluorescence spectra, and chemical removal of H_2P from SWNT. In the H_2P/SWNT nanohybrids, the Soret and Q-bands of H_2P moieties were significantly broadened and their fluorescence was almost completely quenched. Apparent extinction coefficients at the Soret bands were decreased to about 20% [201].

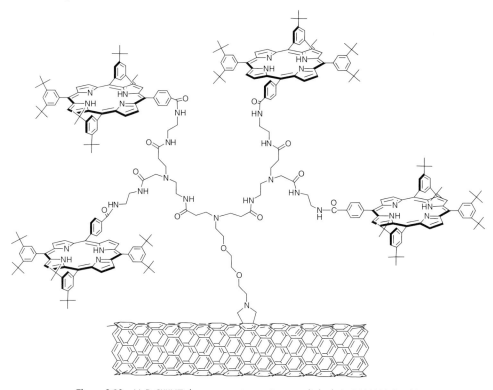

Figure 8.20 H$_2$P–SWNT donor–acceptor conjugate – linked via PAMAM dendrimers.

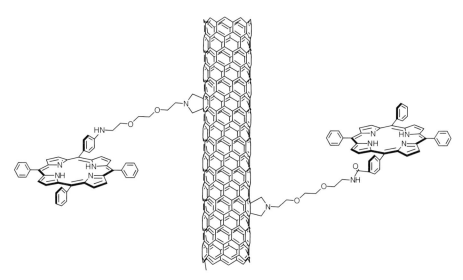

Figure 8.21 H$_2$P–SWNT donor–acceptor conjugate – linked via 1,3 dipolar cycloaddition.

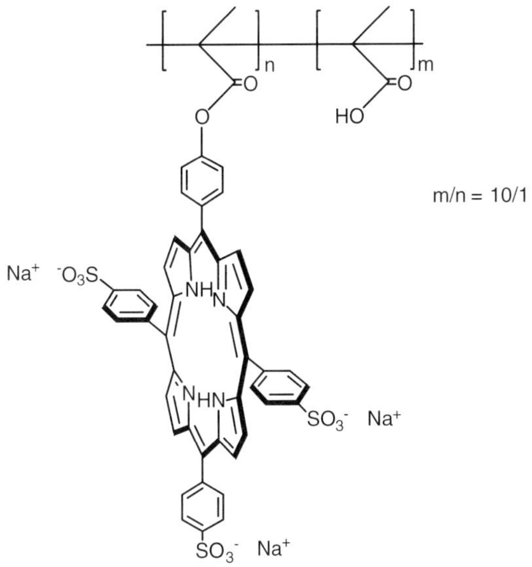

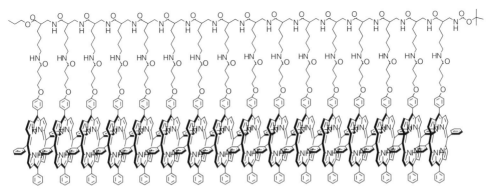

$m/n = 10/1$

Figure 8.22 Partial structure of H$_2$P–polymer.

SWNT/PVBTA^{n+} are water suspendable SWNT, in which PVBTA^{n+} are non-covalently wrapped around them – similar to Figure 8.18. Versatile nanohybrids have been prepared using electrostatic/van der Waals interactions between SWNT/PVBTA^{n+} nanohybrids and porphyrins (H$_2$P^{8-} and/or ZnP^{8-}). Several spectroscopic, microscopic, transient, and photoelectrochemical measurements were employed to characterize the resulting supramolecular complexes. The photoexcitation of the nanohybrids afforded long-lived radical ion pairs with lifetimes as long as 2.2 μs [184].

Strategies for SWNT separation are critical for developing nanotubes as useful nanoscale building blocks. Herein, diameter-selective dispersion of SWNT has been accomplished through noncovalent complexation with a flexible porphyrinic polypeptide P(H$_2$P)$_{16}$ bearing 16 porphyrin units in DMF (Figure 8.23). Supramolecular

Figure 8.23 Structure of H$_2$P–peptide hexadecamer.

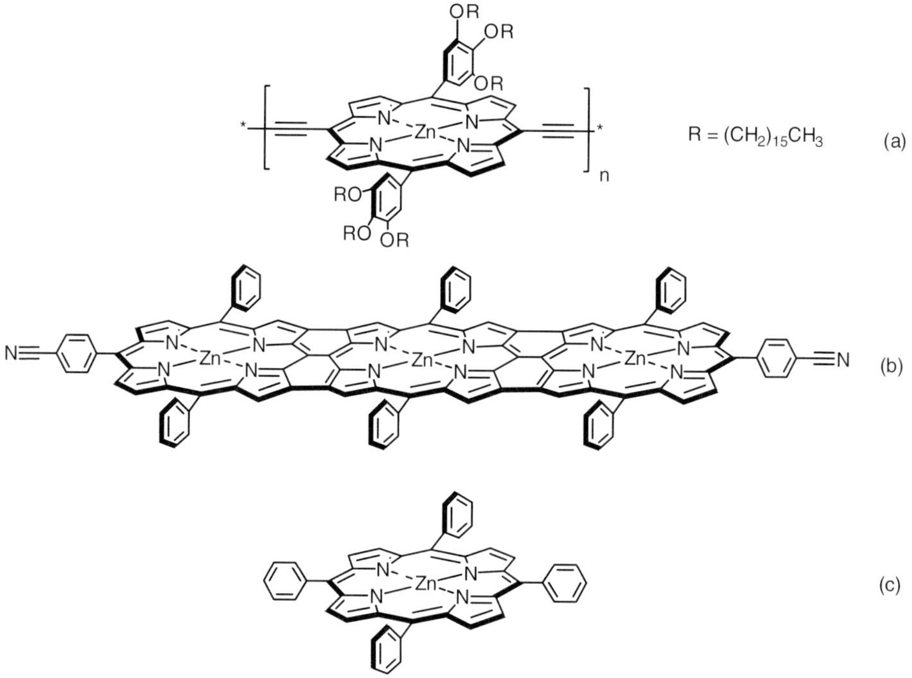

R = (CH$_2$)$_{15}$CH$_3$ (a)

(b)

(c)

Figure 8.24   (a) Partial structure of highly soluble ZnP–polymer, (b) a triply fused ZnP-trimer, and (c) ZnP.

formation occurs through wrapping of the peptide backbone and π–π interactions between the porphyrins and SWNT to extract the large-diameter nanotubes (about 1.3 nm) as revealed by visible/near-infrared absorption and Raman spectroscopy as well as high-resolution transmission electron microscopy. Like in the aforementioned case, photoexcitation of P(H$_2$P)$_{16}$ affords a slowly decaying radical ion pair state (2.7 × 10^6 s^{-1}) [202].

Following a similar strategy, SWNT were found to strongly interact with a highly soluble, conjugated ZnP–polymer (Figure 8.24 [203], a novel triply fused ZnP–trimer [204], and just ZnP. Likewise a rare earth phthalocyanine (i.e., HErPc$_2$) [205] and a tetraformylporphyrin/diaminopyrene polymer (Figure 8.25) were also tested. Successful complexation with, for example, the ZnP–polymer was manifested in a 127 nm bathochromic shift of the Q-band absorption. Additional evidence for

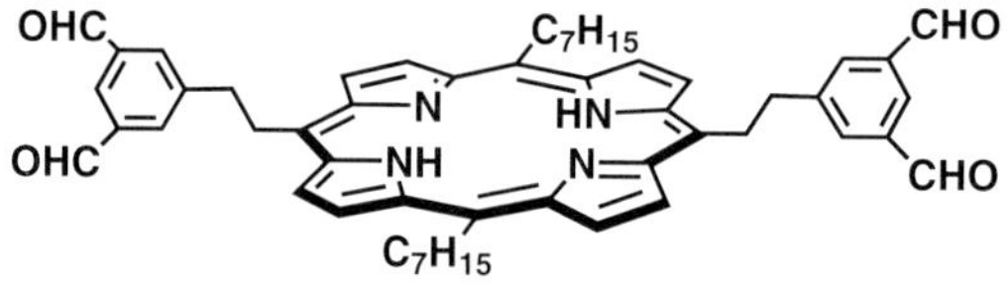

Figure 8.25   Tetraformylporphyrin/diaminopyrene.

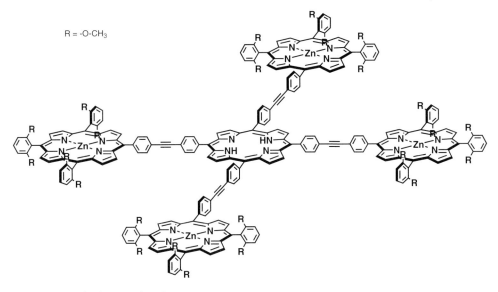

Figure 8.26 A dendritic porphyrinic pentamer.

interactions between ZnP–polymer, ZnP–trimer and ZnP and SWNT was obtained from fluorescence spectra, where the ZnP fluorescence was significantly quenched. The fluorescence quenching has been ascribed to energy transfer between the photoexcited porphyrin and SWNT.

Supramolecular structures consisting of dendritic porphyrins (Figure 8.26) and SWNT have been prepared and characterized as an efficient donor/acceptor system. Noncovalent interactions enable the pair system to produce suitable charge transfer through a process occurring from the dendritic porphyrin core to the graphenic wall of carbon nanotubes. Important is structure/architecture correlation based on photoelectrical investigations [185].

Generally, SWNT interactions with H_2P are appreciably stronger than what is typically seen when employing ZnP. AFM images of, for example, SWNT/H_2P revealed smaller bundle sizes than those recorded for SWNT/ZnP with even some individual specimen (i.e., 1.5 ± 0.2 nm).

The trend of stability was further corroborated with a water-soluble H_2P derivative, namely [meso-(tetrakis-4-sulfonatophenyl)porphine dihydrochloride] (Figure 8.27). In these SWNT suspensions, which were shown to be stable for several weeks, eminent interactions protect H_2P against protonations to the corresponding diacid. Moreover, these strongly interacting hybrids have been successfully aligned onto hydrophilic polydimethylsiloxane surfaces by flowing SWNT solutions along a desired direction and then transferred to silicon substrates by stamping [206].

Quite interesting is the following work: SWNT were effectively worked up with H_2P–[5,10,15,20-tetrakis(hexadecyloxyphenyl)-21H,23H-porphine] (Figure 8.27. The insoluble and recovered SWNT were separated from H_2P by treatment with

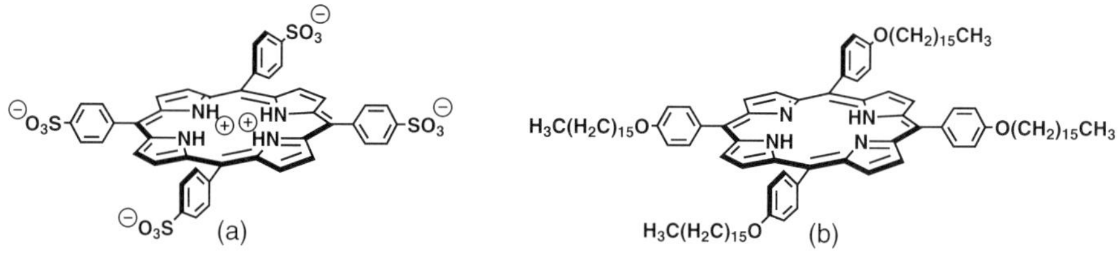

Figure 8.27 (a) [Meso-(tetrakis-4-sulfonatophenyl)porphine dihydrochloride] and (b) [5,10,15,20-tetrakis(hexadecyloxyphenyl)-21*H*,23*H*-porphine].

acetic acid and vigorous centrifugation. After heating the recovered SWNT and the free SWNT to 800 °C in a nitrogen atmosphere, the spectroscopic analysis showed that the semiconducting SWNT and the free SWNT are enriched in the recovered and metallic SWNT, respectively. We must assume selective interactions between H_2P and semiconducting SWNT as the inception to a successful separation of metallic and semiconducting nanotubes [207].

An impressive SWNT solubility has also been achieved upon treatment with zinc [5,10,15,20-tetraphenyl-21*H*,23*H*-porphine] in DMF. TEM and AFM images revealed the existence of high density SWNT and exfoliation of individual SWNT as a result of the tedious work-up procedure. The SWNT diameter ranged typically from 0.9 to 1.5 nm, underlining the successful debundling properties of ZnP. Evidence of interactions between porphyrins and SWNT was obtained from fluorescence spectra, where the porphyrin fluorescence was significantly quenched compared to the porphyrin alone. The fluorescence quenching has been ascribed to energy transfer between the photoexcited porphyrin and SWNT [208].

Solubilization/dispersion of SWNT was also accomplished with protoporphyrin IX and tetraphenylporphyrin iron(III) chloride (Figure 8.28) [209]. Transmission electron and atomic force microscopies, as well as UV-visible/near-infrared spectroscopy, revealed that the protoporphyrins IX dissolved SWNT in polar solvents under mild conditions of sonication using a bath-type sonicator, followed by centrifugation at $1000 \times g$. The absorption and fluorescence measurements of the octaalkylporphyrins in dimethylformamide containing 1 vol% THF with various amounts of SWNT

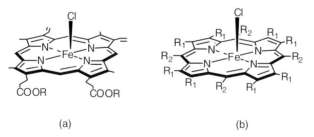

(a) (b)

Figure 8.28 (a) Protoporphyrin IX and (b) tetraphenylporphyrin iron(III) chloride.

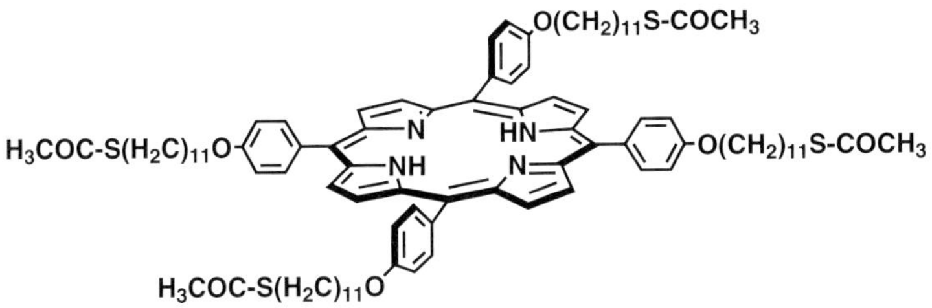

Figure 8.29 5,10,15,20-Tetraphenyl-21H,23H-porphine.

showed that the absorption maxima of the octaalkylporphyrins decreased with an increase in the concentration of SWNT without wavelength shift, and the fluorescence of the porphyrins was quenched. These spectral behaviors have been used to extract information about the SWNT/porphyrin interactions.

SWNT rings have been synthesized via noncovalent hybridization of SWNT and porphyrins (Figure 8.29). Optical absorption spectroscopy, scanning electron microscopy, and transmission electron microscopy results showed that the formation of the SWNT rings is strongly affected by the molecular architecture of the porphyrins, the hybrid concentration, the SWNT-to-porphyrin mass ratio, and the substrate on which the hybrid films were coated. The electron donation features of the substituents – para-positions of the phenyl rings in 5,10,15,20-tetraphenyl-21H,23H-porphine – and the structural regularity of the substituent groups facilitate the formation of J-aggregates [210].

Instead of immobilizing ZnP, H$_2$P, and so on directly onto the SWNT, ionic pyrene derivatives (i.e., 1-(trimethylammonium acetyl)pyrene (pyrene$^+$) (Figure 8.30)) were used to solubilize SWNT through π–π interactions [211, 212]. In fact, SWNT/pyrene$^+$ emerged as a versatile platform to perform van der Waals and electrostatic interactions. To this end, water-soluble porphyrins 5,15-bis-[2',6'-bis-{2'',2''-bis-(carboxy)-ethyl}-methyl-4'-tert-butyl-pheny]-10,20-bis-(4'-tert-butyl-phenyl) porphyrin octasodium (H$_2$P^{8-}) salt and the related zinc complex (ZnP^{8-}) were selected as ideal candidates. In SWNT/pyrene$^+$/ZnP^{8-}, fluorescence and transition absorption studies provided support for a rapid charge transfer (5.0×10^9 s^{-1}). MWNT interact similarly to SWNT with pyrene$^+$ and produce stable MWNT/pyrene$^+$. Interesting is the fact that a better delocalization of electrons in MWNT/pyrene$^+$/ZnP^{8-} helps to significantly enhance the stability of the radical ion pair state (1.7×10^5 s^{-1}) relative to SWNT/pyrene$^+$/ZnP^{8-} (2.5×10^6 s^{-1}). Percolation of the charge inside the concentric wires in MWNT decelerates the decay dynamics that are associated with the charge recombination [213].

Applying similar π–π interactions, SWNT were integrated with a series of negatively charged pyrene derivatives (pyrene$^-$) – 1-pyreneacetic acid, 1-pyrenecarboxylic acid, 1-pyrenebutyric acid, 8-hydroxy-1,3,6-pyrenetrisulfonic acid. But none of the resulting SWNT/pyrene$^-$ showed the tremendous stability that was seen in SWNT/pyrene$^+$. Still, a series of water soluble, positively charged 5,15-bis-[2',6'-bis-{2'',2''-bis-(carboxy)-ethyl}methyl-4'-tert-butyl-phenyl]-10,20-bis-(4'-tert-butyl-phenyl) porphyrins were shown to form photoactive SWNT/pyrene$^-$/ZnP^{8+}, and so on [214].

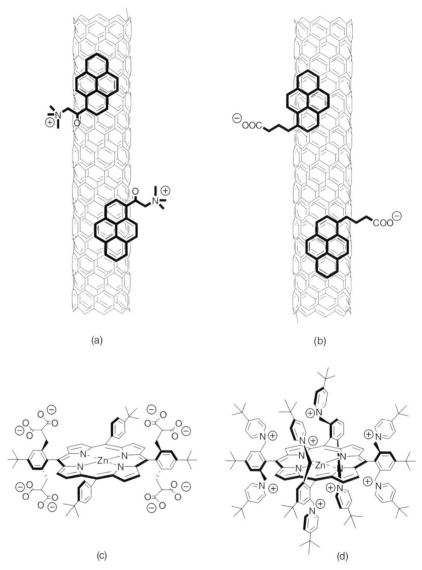

Figure 8.30 (a) SWNT/pyrene$^+$ hybrid that complexes (c) ZnP^{8-} and (b) SWNT/pyrene$^-$ hybrid that complexes (d) ZnP^{8+}.

Considering the broad adaptability and the stability of the SWNT/pyrene motif, π–π stacking with pyrene-imidazole (Figure 8.31), pyrene-*pyro*-pheophorbide *a* (Figure 8.32), pyrene-NH$_3^+$ (Figure 8.33), and pyrene-CdSe Figure 8.34) was demonstrated to solubilize SWNT [215–217, 220]. The case of pyrene-imidazole deserves special mention – through the use of the imidazole ligand naphthalocyanine (ZnNc) and ZnP were axially coordinated. Steady-state and time-resolved emission

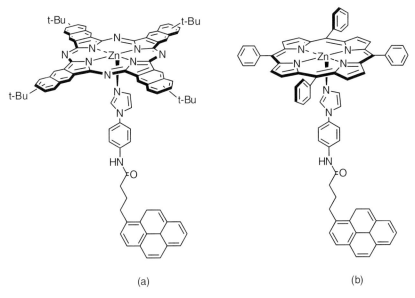

(a)   (b)

Figure 8.31 (a) ZnNc/pyrene and (b) ZnP/pyrene.

studies revealed efficient fluorescence quenching of the donor entities. Nanosecond transient absorption spectra revealed that the photoexcitation of ZnNc or ZnP resulted in the one-electron oxidation of the donor unit with a simultaneous one-electron reduction of SWNT. Both radical ion pair states decay on the nanosecond time scale (ZnPc: $1.7 \times 10^7 \, s^{-1}$; ZnP: $1.1 \times 10^7 \, s^{-1}$) to repopulate the ground state [215].

Viable alternatives to porphyrins or phthalocyanines emerged around excited state electron donors such as copolymers of unsubstituted thiophene and 7-(thien-3-ylsulfanyl)heptanoic acid or size-quantized thioglycolic acid stabilized CdTe@TGA^{n-} nanoparticles [218, 219]. They share in common their water solubility and their successful combination with SWNT/pyrene$^+$ via electrostatic forces. In the resulting nanohybrids strong electronic interactions were noted. In particular, poly-

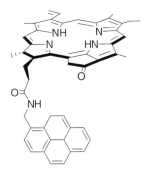

Figure 8.32 Pyrene-*pyro*-pheophorbide *a*.

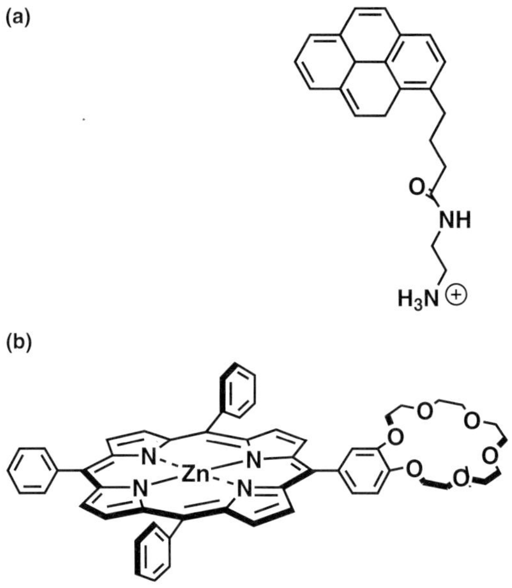

(a)

(b)

Figure 8.33 (a) Ammonium pyrene that complexes (b) ZnTCP.

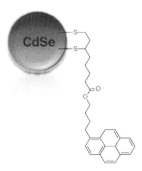

Figure 8.34 CdSe–pyrene.

thiophene and/or CdTe tend to donate excited state electrons to SWNT in the ground state, which slowly recombine – $3.5 \times 10^7 \, s^{-1}$ in SWNT/pyrene$^+$/CdTe@TGA^{n-}.

8.7
Excited State Charge Transfer – CNT as Ground State Electron Donor

8.7.1
Noncovalent Electron Donor–Acceptor Hybrids

In stark contrast to the wealth of information on charge transfer involving SWNT as electron acceptors, information on their use as electron donors is scarce. One of the

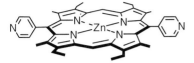

Figure 8.35   [5,15-Bis(4-pyridyl)-2,8,12,18-tetraethyl-3,7,13,17-tetramethyl zinc porphyrin].

first examples, in which SWNT act as electron donor is based on integrating [5,15-bis (4-pyridyl)-2,8,12,18-tetraethyl-3,7,13,17-tetramethyl zinc porphyrin] (Figure 8.35) – as electron acceptor – onto SWNT [221]. Field effect transistor characteristics suggest that the rate and magnitude of charge transfer depends on both the wavelength and the intensity of the applied light, with a maximum value of 0.37 electrons per porphyrin upon 420 nm illumination with $100 \, W/m^2$. The authors postulate that the direction and magnitude of the photoinduced charge transfer in these systems may be controllable.

Likewise, charge transfer proceeds in a self-assembled $SWNT/C_{60}$ that is photo-excited. It has been successfully demonstrated that SWNT act as an electron donor, while the function of C_{60} is that of an electron acceptor. Toward this, first, SWNT were noncovalently functionalized using pyrene-NH_3^+ (Figure 8.36). Pyrene-NH_3^+ not only interacts with SWNT, but also complexes benzo-18-crown-6. Such ammonium/ crown ether interactions were used to associate C_{60} (i.e., crown-C_{60}) to yield stable $SWNT/pyrene-NH_3^+/crown-C_{60}$. Steady state and time-resolved absorption spectroscopy prompted a photoinduced charge transfer, during which SWNT and C_{60} were oxidized and reduced, respectively. The rates of charge separation and charge recombination were found to be 3.5×10^9 and $1.0 \times 10^7 \, s^{-1}$, respectively [222].

Sapphyrins shown in Figure 8.37 are well-characterized pentapyrrolic macrocycles that are particularly attractive candidates for complexation to SWNT as they present

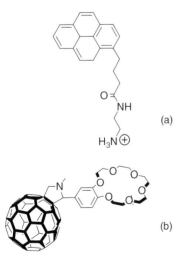

Figure 8.36   (a) Ammonium pyrene derivative that complexes (b) crown-C_{60}.

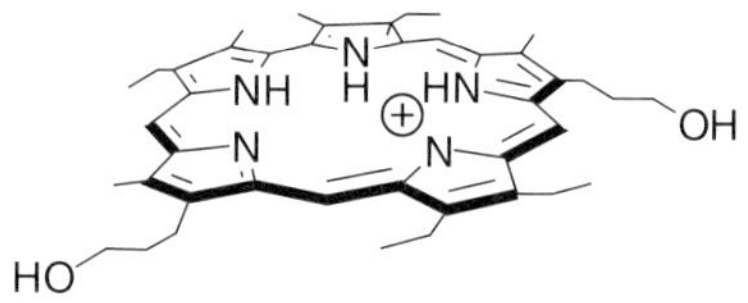

Figure 8.37 Sapphyrin.

larger π surfaces than typical tetrapyrrolic macrocycles. In particular, it was found that SWNT serve to template the supramolecular association of sapphyrins under two very different sets of solvent conditions. The resulting modified nanotubes undergo photoexcited intramolecular charge transfer from SWNT to the sapphyrin moiety upon photoexcitation, for which proof comes from a combination of steady-state investigations, femtosecond transient absorption spectroscopies, and pulse radiolysis experiments [223].

The objective of oxidizing SWNT has also been achieved using a perylene derivative (Figure 8.38), which combines excellent electron-accepting features and a reasonably sized π-system. The complementary use of microscopy and spectroscopy has shed light for the first time onto the mutual interactions between SWNT and a strong electron acceptor. In particular, microscopy demonstrated the benefits of tightly interacting π-systems toward the successful debundling and suspension of individual SWNT. Spectroscopy – including electrochemical, Raman, fluorescence, and transient absorption measurements – confirmed charge transfer to augment the communication in the ground state. More important is that distinct excited state interactions prevail and that a kinetically and spectroscopically well-characterized radical ion pair state is formed that recombines rapidly ($3.3 \times 10^9 \, \mathrm{s}^{-1}$).

8.7.2
Charge Transfer Interactions – CNT and Polymers

Lately, much attention has been directed to the optical and electronic features of CNT/conjugated polymers composites that stem from strongly interacting π-electrons of CNT and π-electrons that correlate with the lattice of the polymer skeleton. CNT wrapped with conducting polymers might – among many novel features – exhibit significantly improved solubilities. Conjugated poly(*m*-phenylenevinylene-*co*-2,5-dioctoxy-*p*-phenylenevinylene) (PmPV-*co*-DOctOPV) (Figure 8.39), whose structure is a simple variation of the more common PPV, is one of most widely tested polymer that forms novel nanohybrids with CNT. Here the coiled conformation

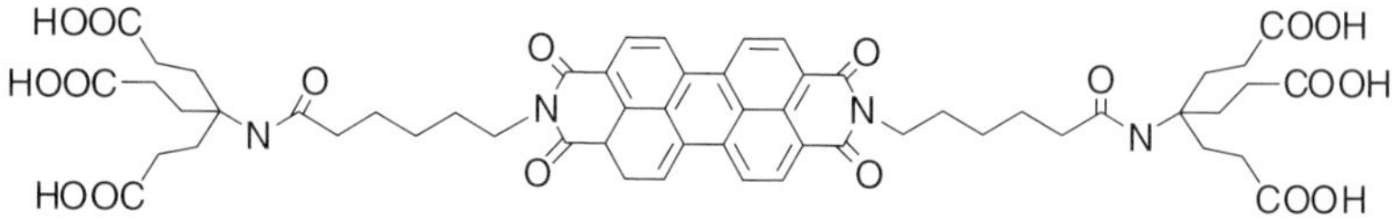

Figure 8.38 Perylene derivative.

Figure 8.39 (a) PmPV-*co*-DOctOPV, (b) PEI, (c) PVP, (d) PANI, (e) PAE, (f) PPY, and (g) P3MT.

allowed the polymer to wrap around the surface of CNT through simple π–π interactions [224, 225].

Complementary work demonstrated that PmPV helps in dispersing SWNT in organic media [226]. Interestingly, several spectroscopies performed with SWNT/PmPV attest to significantly broadened features, which correlate with π–π interacting forces as they are effective between SWNT and PmPV. Average diameters of aggregated SWNT are found around 7.1 nm – as discernable from AFM measurements. This indicates that noncovalent modification is a potent and powerful means to disrupt the inherently strong van der Waals interactions among individual CNT. Noteworthy is the observation that when higher polymer concentrations are used even individual SWNT are exfoliated from the initial bundles. The optoelectronic properties, namely a photoamplification of nearly thousand times, indicate that the polymer must be in intimate electrical contact with the SWNT.

Charge transfer interactions, often seen to dominate between amines and CNT, are the inception to CNT that interact with poly(ethylenimine) (PEI) and poly(4-vinylpyridine) [227]. Here, advantage has been taken of the high affinity of amines to interact with SWNT to assemble into SWNT/polymer nanohybrids. Microscopy of the films revealed that small bundles and even individual CNT were incorporated within multilayered films via the sequential adsorption of PEI and PVP followed by SWNT onto silicon substrates.

Along the same line, charge transfer reactivity enhances the electrical properties when polyaniline (PANI, Figure 8.39) interacts with MWNT [228]. This rationale explains why the electrical conductivity of MWNT/PANI increased by an order of magnitude relative to neat PANI. Alternatively, doping effects of MWNT should be considered. Following this fundamental characterization, electrochemical polymerization of PANI onto CNT electrodes was shown to be quite effective in dispersing conjugated polymeric films [229]. Absorption spectroscopy and AFM studies confirmed that, under ambient conditions, SWNT and MWNT adhere strongly to aniline by forming charge transfer complexes rather than undergoing weak van der Waals interactions [230]. In general, MWNT/PANI films tend to have improved electron and ion transfer properties – relative to pure PANI films. The charge transport upon oxidizing SWNT/PANI planar films is about four times faster than upon oxidation of planar films of PANI [231].

Impressive solubilizations of SWNT have been established in a novel wrapping approach using short and rigid functional conjugated polymers, poly(aryleneethynylene)s (PAE, Figure 8.39). This technique not only enables the dissolution of various types of CNT in organic solvents, but also assists in introducing numerous functional groups onto the SWNT surfaces [232, 233].

SWNT reinforced polyimide nanocomposites were synthesized by *in situ* polymerization of monomers [234, 235]. Quite considerably, the resulting blends showed optical transport and electrical conductivity enhancement at relatively low loadings (i.e., 0.1 vol%). Actually, the differences are as large as 10 orders of magnitude. Totally aromatic polyimides carrying sulfonates are capable of solubilizing a large amount of SWNT individually in solution [236]. Higher concentrations of SWNT in polyimide solutions form a gel composed of individually dissolved SWNT.

When applying electrochemical polymerization conditions, MWNT were codeposited with polypyrrole (PPY, Figure 8.39) and poly(3-methylthiophene) (P3MT, Figure 8.39) to form nanoporous composite films [237]. In the absence of supporting electrolytes, an increase in the MWNT concentration at the polymerizing electrode decreases the thickness of the polymer coating on each MWNT. Obviously, the ionic diffusion distances were greatly minimized, which improves the electrolyte access within the nanoporous MWNT/PPY films. This eventually amplifies the capacitance relative to similarly prepared PPY films.

8.8
Implications of Ground State Charge Transfer

8.8.1
Conducting Electrode Materials

The market for transparent conductors used in display, photovoltaic, and lighting markets will reach approximately $9.4 billion by 2015. Many materials that have been put forward as ITO replacements fall well below the standards set for conductivity and transparency achieved by ITO itself. In photovoltaics, the target is a surface resistivity of $\leq 50\,\Omega/sq$ with a trasmittance greater than 85% throughout the solar spectrum [238]. While extensively developed and optimized, ITO bears a number of deficiencies. One disadvantage results from its high deposition temperature of around 600 °C, which renders its compatibility with, for example, flexible substrates more than just problematic. In addition, cracks appear after repeated bending or strain, and are not resistant to acid [239]. What is even more puzzling is that traditional ITO has poor mechanical properties. More of an economical concern is the high costs of high quality ITO.

In this context, one of the first tests with CNT was the preparation of a composite of MWNT and poly(p-phenylenevinylene) (PPV). Thin and uniform MWNT films were obtained by spin coating the highly concentrated MWNT dispersion, which had been stabilized by chemical oxidation. Composites were realized by introducing the polymer precursor onto the surface of the MWNT film, followed by conversion at high temperature. A photovoltaic device, which was prepared from the composite by using MWNT as a hole-collecting electrode, gave rise to quantum efficiency that was nearly twice that of a standard device with a bear ITO electrode. It has been considered that the high efficiency arises from a complex interpenetrating network of PPV chains with MWNT and the relatively high work function of MWNT films (5.1 eV) [240].

Following this pioneering work, several methods have been probed to fabricate CNT thin films. The key is to disperse CNT in solutions with the aid of surfactants or surface derivatization. The resulting CNT suspensions were sprayed [241], spin coated [240, 242], drop casted [243] and deposited by LbL [244], Langmuir–Blodgett method [245], or by electrophoresis [246]. Still, the most suitable procedure for obtaining flexible and optically transparent films involves the transfer from a filter membrane to the transparent support by utilizing a solvent [247] or in the form of an adhesive film [248]. Such procedures are superior, since they guarantee better dispersions and the possibility of removing the solubilization agents by a simple washing/rinsing step. With the same technique – utilizing metallic SWNT sorted by density gradient ultracentrifugation – films with 75% transmittance in the visible and $231\,\Omega/sq$ conductivity were achieved. Similar results were obtained upon spin coating a mixture of metallic SWNT (i.e., DMSO) and PEDOT:PSS (i.e., aqueous solution 10% w/w) [249].

An alternative CNT film fabrication method involves dry state spinning of MWNT yarns from MWNT forests grown by chemical vapor deposition, followed by dipping in alcohol to induce the densification of the films [250]. Postdeposition treatments like doping with $SOCl_2$ [251], and with Nafion followed by nitric acid treatment [252], are worth mentioning.

Theoretical and experimental studies have established that the work function of SWNT networks is in the range between 4.8 and 4.9 eV. In other words, the SWNT work function should be sufficiently high to ensure efficient hole collection [253]. Hall effect measurements at room temperature corroborate the p-type nature of the conductivity [254]. Nevertheless, for optimal performance the CNT electrodes must be coated with a thin layer of the conducting polymer poly(3,4-ethylenedioxythiophene) (PEDOT) doped with poly(styrenesulfonate), which further reduces the resistance and helps to planarize the CNT films [250]. Early results confirm the proof of this principle by attesting that photovoltaic devices with SWNT thin films as transparent and conducting electrodes for hole collection are, in fact, competitive with comparable devices fabricated using ITO (i.e., device efficiencies between 1 and 2.5%) [253, 255, 256].

Finally, a recent example of flexible organic photovoltaics should be discussed, in which transparent and conducting SWNT thin films were implemented as current collecting electrodes on plastic substrates. As photoactive layer – forming a bulk heterojunction – zinc oxide nanowires and poly(3-hexylthiophene) (P3HT) were combined. The bulk heterojunctions for exciton dissociation were created by utilizing hydrothermally grown semialigned ZnO nanowires electron acceptors on top of solution deposited SWNT thin films and spin-coated P3HT polymers [257].

8.8.2
Counter Electrodes for DSSC

A high performance DSSC requires the counter electrode to be highly catalytic and highly conductive. Hence, platinum, which is a good catalyst for the reduction of redox species, such as triiodide/iodide, is usually used as the counter electrode of the DSSC. Notable, SWNT – as a substitute for platinum – reveal conversion efficiency of 3.5 and 4.5% when drop casted on FTO glass and on a Teflon membrane, respectively [258]. SWNT are good catalysts for the reduction of triiodide, and its catalytic activity is enhanced when exposing them to ozone as defects are introduced [259]. In addition, the sheet resistance of SWNT/Teflon is with 1.8 Ω/sq four times lower than that of a FTO glass [258]. Finally, an additional benefit of CNT is that they do not degrade under conditions, at which platinum was found to degrade, that is, in contact with an iodide/triiodide liquid electrolyte [260].

A slight modification implies MWNT/conducting poly(3,4-ethylenedioxythiophene):poly(styrenesulfonate) composites as DSSC counter electrodes. The devices exhibited high performance with energy conversion efficiencies of 6.5%, short circuit currents of 15.5 mA/cm^2, open circuit voltages of 0.66 V, and fill factors of 0.63. This performance is close to the devices using conventional platinum as the counter electrode and is significantly higher than the ones using a thin film of MWNT/poly(styrenesulfonic acid) composite as the counter electrode [261].

8.9
Implications of Excited State Charge Transfer

8.9.1
Active Component in Photoactive Layer

In recent years the continuing interest in alternative photovoltaic technologies, especially with respect to lowering production costs, triggered extensive research in the fields of organic photovoltaic devices. Compared with inorganic materials, organic materials offer great incentives including device production at exceptionally low costs, lightweight, and flexible devices that enable versatile applications [262, 263]. Organic materials absorb light in most of the visible and near-infrared regions of the solar spectrum to generate conduction band electrons and valence band holes. Coulombic interactions between the spatially confined charges result in strongly bound excitons. However, these excitons must be separated at the interphases between the electron donors and the electron acceptors or between the photoactive materials and semiconducting electrodes, before they recombine or relax in an alternative fashion to the ground state [264, 265]. Finally, the resulting free charge carriers should be transported to an external circuit, while minimizing the recombination pathways. To this end, the main obstacles to realize widespread applicability of organic semiconductor for photovoltaics are high exciton binding energy, low charge carrier mobility, and susceptibility to degradation.

Relative to known organic materials, CNT bear the great advantage to be structurally robust, chemically stable, and highly conductive. They can improve exciton dissociation in the presence of an external field at the heterointerfaces. In addition, their high surface areas – of about $1500 \, m^2/g$ – favor large donor–acceptor contact surfaces within the photoactive layer [266]. All these aspects are very important for separating the aforementioned excitons. Also the tubular shape of CNT appears interesting. This, in particular, should enable a lower percolation threshold for the acceptor phase and consequently a more efficient charge transport to the external circuit [267].

Much of the current research efforts on thin film photovoltaics have been directed toward composite materials made of SWNT [242, 268, 269] or MWNT [270, 271] and conjugated polymer – sometimes in the presence of C_{60} [272–274], CdSe [275, 276], or phthalocyanines [277].

Typical work-up procedures include the dispersion of CNT in solutions of poly(3-hexylthiophene) or poly(3-octylthiophene) (P3OT) and the spin coating onto transparent conductive electrodes. Postfabrication treatment such as heating to the point beyond the glass transition temperature of either P3HT or P3OT is another key point, especially for manipulating the phase separation of the blend. As the polymers are microcrystalline systems, they also affect the ordering of the polymeric chains. Overall, this annealing improves charge transfer, charge transport, and charge collection throughout the device [278]. As a consequence of such ordering, the hole mobility, and hence the power efficiency of the CNT/polymer device, increases significantly [279]. So far, following this strategy a maximum power conversion efficiency of 0.22% has been obtained [280]. Although the polymer CNT cells give rise to an exceptionally high open circuit voltage, namely of up to 1 V, as well as an advantageous near-infrared light harvesting [281], the corresponding photocurrents are quite low [282]. Weak interactions and incomplete exciton dissociations as they result especially at the utilized CNT concentration – 1 wt% for SWNT [283] and 5 wt% for MWNT [284] – are mainly responsible for this tendency. Higher CNT percentages, on the other hand, generate short circuits, since the CNT length is not compatible with the overall thickness of the photoactive layer.

Some promising reports on photoelectrochemical devices integrating SWNT as electron acceptor – on ITO – are based on layer-by-layer assembly strategies [285]. As a first step, the ITO electrodes are covered with a base layer of polyelectrolytes (i.e., poly(diallyl-dimethylammonium)chloride ($PDDA^{n+}$) or sodium polystyrene-4-sulfonate (PSS^{n-})) through hydrophobic-hydrophobic forces [218, 219, 286, 287]. In the next step, SWNT/pyrene$^+$ and finally porphyrin salts (i.e., H_2P^{8-} or ZnP^{8-}) are assembled electrostatically. This approach is applicable to other CNT materials such as DWNT, MWNT, and thin-MWNT. Importantly, the best performance was seen when thin-MWNT were used as electron acceptor layer material with mono-chromatic power conversion efficiencies of up to 10.7% for a device that contains a PDDA/SWNT/pyrene$^+$/ZnP^{8-} stack [287].

The best performance was seen when electrostatic/van der Waals interactions were used to assemble SWNT/PVBTA^{n+}/ZnP^{8-}. In fact, photocurrent measurements gave remarkable internal photon-to-current efficiencies of 9.90% when a potential of 0.5 V was applied [288].

Alternatively, PSCOOH (Figure 8.40) or CdTe quantum dots (QD) (Figure 8.41) were integrated together with SWNT/pyrene$^+$ onto ITO electrodes by a combination of van der Waals and electrostatic interactions. In the devices, PSCOOH and CdTe function as the light-harvesting chromophore, in which both donate electrons to SWNT. Quite remarkably, the monochromatic power conversion efficiencies for a SWNT/pyrene$^+$/PSCOOH stack and a SWNT/pyrene$^+$/CdTe stack were deter-mined as 1.2 and 2.3%, respectively [289, 290].

Electrophoresis is yet another interesting assembly strategy [291]. In particular, exposing SWNT that were suspended with the assistance of tetraoctylammonium bromide in tetrahydrofuran to an electrophoretic field emerged as a valuable approach to deposit them onto ITO covered with SnO$_2$ [292]. In fact, considerable photoconversion efficiencies were achieved when SWNT were deposited and as-sembled with light-harvesting CdS and porphyrins with values of 0.5 and 6.5%, respectively [293, 294]. Recently, a new methodology for the self-organization of C$_{60}$ on the sidewall of SWNT for the use in photoelectrochemical devices used SnO$_2$ has been developed. The photocurrent generation efficiency – 18% at 400 nm under

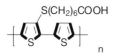

Figure 8.40 PSCOOH.

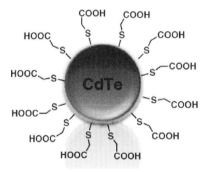

Figure 8.41 CdTe QD stabilized with thioglycolic acid.

an applied potential of 0.05 V versus SCE – is the highest value among CNT-based photoelectrochemical devices, in which CNT are deposited electrophoretically, electrostatically, or covalently onto semiconducting electrodes [295].

SWNT support networks can be incorporated into mesoscopic TiO_2/Ru(II) trisbipyridyl complex films to improve the charge transport in dye-sensitized solar cells. While no net increase in power conversion efficiency is seen, an increase in photon to current efficiency represents the benefits of the SWNT as a conducting scaffold to facilitate charge separation and charge transport in nanostructured semiconductor films [296]. Utilizing CdS QD, the presence of a SWNT sublayer increased the short circuit current under irradiation condition and also reduced the charge recombination process under dark condition. The power conversion efficiency of CdS/TiO_2 on ITO increased 50% in the presence of SWNT due to the improved charge collecting efficiency and reduced recombination [297].

8.9.2
Gas Sensors

Recently, CNT-based gas sensors have received considerable attention because of their outstanding properties such as faster response, higher sensitivity, lower operating temperature, and wider variety of gases that may be detected compared with the other types of gas sensors. CNT-based gas sensing utilizes a change in an electrical property due to adsorption of gas molecules as the output signal [298].

The CNT electrical resistance, thermoelectric power, and local density of states, as determined by transport measurements and scanning tunneling spectroscopy are extremely sensitive to air or oxygen exposure. Such parameters are reversibly tuned by surprisingly small concentrations of adsorbed gases. An impressing demonstration of such alterations is that an intrinsically semiconducting nanotube converts into an apparent metallic one once exposed to air/oxygen [299]. Annealing the contacts in vacuum [300] or in an inert gas [301], removes the adsorbed oxygen and reconverts the device from p- to n-type. n-type doping of SWNT by ammonia adsorption was also observed [302]. The effects are smaller when SWNT are in contact with H_2, He, and N_2. Simply, weaker interactions are responsible for such trends [303].

References

1 Reich, S., Thomsen, C., and Maultzsch, J. (2004) *Carbon Nanotubes: Basic Concepts and Physical Properties*, Wiley-VCH Verlag GmbH, Weinheim.

2 Dresselhaus, M.S., Dresselhaus, G., and Avouris, P. (2001) *Carbon Nanotubes: Synthesis, Structure, Properties and Application*, Springer, Berlin.

3 Roth, S. and Carroll, D. (2004) *One-Dimensional Metals: Conjugated Polymers, Organic Crystals, Carbon Nanotube*, Wiley-VCH Verlag GmbH, Weinheim.

4 Meyyappan, M. (2006) *Carbon Nanotubes, Science and Application*, Wiley-VCH Verlag GmbH, Weinheim.

5 Haddon, R.C. (2002) Carbon Nanotubes. *Acc. Chem. Res.*, **35**, 997–1113.

6 Sgobba, V., Rahman, G.M.A., Ehli, C., and Guldi, D. (2007) Chapter 11, in *Fullerenes – Principles and Applications* (eds F. Langa and J.F. Nierengarten), RSC, Cambridge, UK.

7 Sgobba, V., Rahman, G.M.A., and Guldi, D.M. (2006) Chapter 6, in *Chemistry of Carbon Nanotubes* (eds V.A. Basiuk and E.V. Basiuk), American Scientific Publishers, North Lewis Way, CA, USA.

8 Carlson, L.J. and Krauss, T.D. (2008) Photophysics of individual single-walled carbon nanotubes. *Acc. Chem. Res.*, **41**, 235–243.

9 Rotkin, S.V. and Subramoney, S. (2005) *Applied Physics of Carbon Nanotubes*, Springer-Verlag, Berlin.

10 Tans, S.J., Devoret, M.H., Dal, H., Thess, A., Smalley, R.E., Geerligs, L.J., and Dekker, C. (1997) Individual single-wall carbon nanotubes as quantum wires. *Nature*, **386**, 474–477.

11 Ajayan, P.M. and Zhou, O.Z. (2001) Applications of carbon nanotubes. *Carbon Nanotubes*, **80**, 391–425.

12 Robertson, J. (2004) Realistic applications of CNTs. *Mater. Today*, **7**, 46–52.

13 Thess, A., Lee, R., Nikolaev, P., Dai, H., Petit, P., Robert, J., Xu, C., Lee, Y.H., Kim, S.G., Rinzler, A.G., Colbert, D.T., Scuseria, G.E., Tomanek, D., Fischer, J.E., and Smalley, R.E. (1996) Unraveling nanotubes: field emission from an atomic wire. *Science*, **273**, 1550–1553.

14 Girifalco, L.A., Hodak, M., and Lee, R.S. (2000) Carbon nanotubes, buckyballs, ropes, and a universal graphitic potential. *Phys. Rev. B: Condens. Matter*, **62**, 13104–13110.

15 Tasis, D., Tagmatarchis, N., Georgakilas, V., and Prato, M. (2003) Soluble carbon nanotubes. *Chem. Eur. J.*, **9**, 4000–4008.

16 Dyke, C.A. and Tour, J.M. (2004) Overcoming the insolubility of carbon nanotubes through high degrees of sidewall functionalization. *Chem. Eur. J.*, **10**, 812–817.

17 Melle-Franco, M., Marcaccio, M., Paolucci, D., Paolucci, F., Georgakilas, V., Guldi, D.M., Prato, M., and Zerbetto, F. (2004) Cyclic voltammetry and bulk electronic properties of soluble carbon nanotubes. *J. Am. Chem. Soc.*, **126**, 1646–1647.

18 Guldi, D.M., Holzinger, M., Hirsch, A., Georgakilas, V., and Prato, M. (2003) First comparative emission assay of single-wall carbon nanotubes-solutions and dispersions. *Chem. Commun.*, 1130–1131.

19 O'Connell, M.J., Bachilo, S.M., Huffman, C.B., Moore, V.C., Strano, M.S., Haroz, E.H., Rialon, K.L., Boul, P.J., Noon, W.H., Kittrell, C., Ma, J.P., Hauge, R.H., Weisman, R.B., and Smalley, R.E. (2002) Band gap fluorescence from individual single-walled carbon nanotubes. *Science*, **297**, 593–596.

20 Shen, K., Curran, S., Xu, H., Rogelj, S., Jiang, Y., Dewald, J., and Pietrass, T. (2005) Single-walled carbon nanotube purification, pelletization, and surfactant-assisted dispersion: a combined TEM and resonant micro-Raman spectroscopy study. *J. Phys. Chem. B*, **109**, 4455–4463.

21 Monthioux, M., Smith, B.W., Burteaux, B., Claye, A., Fischer, J.E., and Luzzi, D.E. (2001) Sensitivity of single-wall carbon nanotubes to chemical processing: an electron microscopy investigation. *Carbon*, **39**, 1251–1272.

22 Strano, M.S., Dyke, C.A., Usrey, M.L., Barone, P.W., Allen, M.J., Shan, H., Kittrell, C., Hauge, R.H., Tour, J.M., and

Smalley, R.E. (2003) Electronic structure control of single-walled carbon nanotube functionalization. *Science*, **301**, 1519–1522.

23 Zhang, G., Qi, P., Wang, X., Lu, Y., Li, X., Tu, R., Bangsaruntip, S., Mann, D., Zhang, L., and Dai, H. (2006) Selective etching of metallic carbon nanotubes by gas-phase reaction. *Science*, **314**, 974–977.

24 Krupke, R., Hennrich, F., Löhneysen, H.V., and Kappes, M.M. (2003) Separation of metallic from semiconducting single-walled carbon nanotubes. *Science*, **301**, 344–347.

25 Balasubramanian, K. and Burghard, M. (2005) Chemically functionalized carbon nanotubes. *Small*, **1**, 180–182.

26 Banerjee, S., Hemraj-Benny, T., and Wong, S.S. (2005) Covalent surface chemistry of single-walled carbon nanotubes. *Adv. Mater.*, **17**, 17–29.

27 Hirsch, A. (2002) Functionalization of single-walled carbon nanotubes. *Angew. Chem. Int. Ed.*, **41**, 1853–1859.

28 Banerjee, S., Kahn, M.G.C., and Wong, S.S. (2003) Rational chemical strategies for carbon nanotube functionalization. *Chem. Eur. J.*, **9**, 1898–1908.

29 Lu, X. and Chen, Z. (2005) Curved pi-conjugation, aromaticity, and the related chemistry of small fullerenes ($<C_{60}$) and single-walled carbon nanotubes. *Chem. Rev.*, **105**, 3643–3696.

30 Bahr, J.L. and Tour, J.M. (2002) Covalent chemistry of single-wall carbon nanotubes. *J. Mater. Chem.*, **12**, 1952–1958.

31 Tasis, D., Tagmatarchis, N., Bianco, A., and Prato, M. (2006) Chemistry of carbon nanotubes. *Chem. Rev.*, **106**, 1105–1136.

32 Nikolaev, P., Bronikowski, M.J., Bradley, R.K., Rohmund, F., Colbert, D.T., Smith, K.A., and Smalley, R.E. (1999) Gas-phase catalytic growth of single-walled carbon nanotubes from carbon monoxide. *Chem. Phys. Lett.*, **313**, 91–97.

33 Mylvaganam, K. and Zhang, L.C. (2004) Chemical bonding in polyethylene-nanotube composites: A quantum mechanics prediction. *J. Phys. Chem. B*, **108**, 5217–5220.

34 Lu, X., Tian, F., Xu, X., Wang, N., and Zhang, Q. (2003) A theoretical exploration of the 1,3-dipolar cycloadditions onto the sidewalls of (n, n) armchair single-wall carbon nanotubes. *J. Am. Chem. Soc.*, **125**, 10459–10464.

35 Bettinger, H.F. (2003) Experimental and computational investigations of the properties of fluorinated single-walled carbon nanotubes. *ChemPhysChem*, **4**, 1283–1289.

36 Zhou, W., Ooi, Y.H., Russo, R., Papanek, P., Luzzi, D.E., Fischer, J.E., Bronikowski, M.J., Willis, P.A., and Smalley, R.E. (2001) Structural characterization and diameter-dependent oxidative stability of single wall carbon nanotubes synthesized by the catalytic decomposition of CO. *Chem. Phys. Lett.*, **350**, 6–14.

37 Zhao, J. and Balbuena, P.B. (2006) Structural and reactivity properties of finite length cap-ended single-wall carbon nanotubes. *J. Phys. Chem. A*, **110**, 2771–2775.

38 Hashimoto, A., Suenaga, K., Gloter, A., Urita, K., and Iijima, S. (2004) Direct evidence for atomic defects in graphene layers. *Nature*, **430**, 870–873.

39 Watts, P.C.P., Hsu, W.-K., Kroto, H.W., and Walton, D.R.M. (2003) Are bulk defective carbon nanotubes less electrically conducting? *Nano Lett.*, **3**, 549–553.

40 Charlier, J.-C. (2002) Defects in carbon nanotubes. *Acc. Chem. Res.*, **35**, 1063–1069.

41 Ouyang, M., Huang, J.L., Cheung, C.L., and Lieber, C.M. (2001) Atomically resolved single-walled carbon nanotube intramolecular junctions. *Science*, **291**, 97–100.

42 Louie, S.G. (2001) Electronic properties, junctions, and defects of carbon nanotubes. *Top. Appl. Phys.*, **80**, 113–145.

43 Ebbesen, T.W. and Takada, T. (1995) Topological and sp^3 defect structures in nanotubes. *Carbon*, **33**, 973–978.

44 Charlier, J.C., Ebbesen, T.W., and Lambin, P. (1996) Structural and electronic properties of pentagon–heptagon pair defects in carbon nanotubes. *Phys. Rev. B: Condens. Matter*, **53**, 11108–11113.

45 Hu, H., Bhowmik, P., Zhao, B., Hamon, M.H., Itkis, M.E., and Haddon, R.C.

(2001) Determination of the acidic sites of purified single-walled carbon nanotubes by acid–base titration. *Chem. Phys. Lett.*, **345**, 25–28.

46 Stone, A.J. and Wales, D.J. (1986) Theoretical studies of icosahedral footballene sixty-carbon-atom molecules and some related species. *Chem. Phys. Lett.*, **128**, 501–503.

47 Bettinger, H.F. (2005) The reactivity of defects at the sidewalls of single-walled carbon nanotubes: the stone-wales defect. *J. Phys. Chem. B.*, **109**, 6922–6924.

48 Lu, X., Chen, Z., and Schleyer, P.V.R. (2005) Are stone-wales defect sites always more reactive than perfect sites in the sidewalls of single-wall carbon nanotubes? *J. Am. Chem. Soc.*, **127**, 20–21.

49 Dai, H. (2000) Controlling nanotube growth. *Phys. World*, **13**, 43–47.

50 Ouyang, M., Huang, J.-L., and Lieber, C.M. (2002) Fundamental electronic properties and applications of single-walled carbon nanotubes. *Acc. Chem. Res.*, **35**, 1018–1025.

51 Grujicic, M., Cao, C., and Singh, R. (2003) The effect of topological defects and oxygen adsorption on the electronic transport properties of single-walled carbon-nanotubes. *Appl. Surf. Sci.*, **211**, 166–183.

52 Freitag, M. (2008) Doped defects tracked down. *Nat. Mater.*, **7**, 840–841.

53 Yao, Z., Postma, H.W.C., Balents, L., and Dekker, C. (1999) Carbon nanotube intramolecular junctions. *Nature*, **402**, 273–274.

54 Collings, A.F. and Critchley, C. (2005) *Artificial Photosynthesis: From Basic Biology to Industrial Application*, Wiley-VCH Verlag GmbH, Weinheim.

55 Balzani, V. (2001) *Electron Transfer in Chemistry*, Wiley-VCH Verlag GmbH, Weinheim.

56 Guldi, D.M. (2002) Fullerene-porphyrin architectures; photosynthetic antenna and reaction center models. *Chem. Soc. Rev.*, **31**, 22–36.

57 Deisenhofer, J. and Norris, J.R. (1993) *The Photosynthetic Reaction Center*, Academic Press, San Diego, USA.

58 Gust, D., Moore, T.A., and Moore, A.L. (1993) Molecular mimicry of photosynthetic energy and electron transfer. *Acc. Chem. Res.*, **26**, 198.

59 Pelizzetti, E. and Schiavello, M. (1997) *Photochemical Conversion and Storage of Solar Energy*, Kluwer, Dordrecht, The Netherlands.

60 Nelson, J. (2003) *The Physics of Solar Cells (Properties of Semiconductor Materials)*, Imperial College Press, London.

61 Brabec, C., Dyakonov, V., Parisi, J., and Sariciftci, S. (2003) *Organic Photovoltaics: Concepts and Realization, Springer Series in Materials Science*, Springer, Berlin.

62 Sgobba, V. and Guldi, D.M. (2008) Carbon nanotubes-electronic/electrochemical properties and application for nanoelectronics and photonics. *Chem. Soc. Rev.*, **38**, 165–184.

63 Brabec, C., Dyakonov, V., and Scherf, U. (2008) *Organic Photovoltaic*, Wiley-VCH Verlag GmbH, Weinheim.

64 Hartschuh, A., Pedrosa, H.N., Peterson, J., Huang, L., Anger, P., Qian, H., Meixner, A.J., Steiner, M., Novotny, L., and Krauss, T.D. (2005) Single carbon nanotube optical spectroscopy. *ChemPhysChem*, **6**, 577–582.

65 Kataura, H., Kumazawa, Y., Maniwa, Y., Umezu, I., Suzuki, S., Ohtzuka, Y., and Achiba, Y. (1999) Optical properties of single-wall carbon nanotubes. *Synth. Met.*, **103**, 2555–2558.

66 Hamon, M.A., Itkis, M.E., Niyogi, S., Alvarez, T., Kuper, C., Menon, M., and Haddon, R.C. (2001) Effect of rehybridization on the electronic structure of single-walled carbon nanotubes. *J. Am. Chem. Soc.*, **123**, 11292–11293.

67 Kazaoui, S., Minami, N., Matsuda, H.K.N., and Achiba, Y. (2001) Electrochemical tuning of electronic states in single-wall carbon nanotubes studied by *in situ* absorption spectroscopy and ac resistance. *Appl. Phys. Lett.*, **78**, 3433–3435.

68 Nair, N., Usrey, M.L., Kim, W.-J., Braatz, R.D., and Strano, M.S. (2006) Estimation of the (n, m) concentration distribution of single-walled carbon nanotubes from

photoabsorption spectra. *Anal. Chem.*, **78**, 7689–7696.

69 Perebeinos, V., Tersoff, J., and Avouris, P. (2004) Scaling of excitons in carbon nanotubes. *Phys. Rev. Lett.*, **92**, 257402/1–257402/4.

70 Zhao, H. and Mazumdar, S. (2004) Electron-electron interaction effects on the optical excitations of semiconducting single-walled carbon nanotubes. *Phys. Rev. Lett.*, **93**, 157402/1–157402/4.

71 Spataru, C.D., Ismail-Beigi, S., Benedict, L.X., and Louie, S.G. (2004) Excitonic effects and optical spectra of single-walled carbon nanotubes. *Phys. Rev. Lett.*, **92**, 077402/1–077402/4.

72 Wang, Z., Zhao, H., and Mazumdar, S. (2006) Quantitative calculations of the excitonic energy spectra of semiconducting single-walled carbon nanotubes within a π-electron model. *Phys. Rev. B: Condens. Matter*, **74**, 195406/1–195406/6.

73 Wang, F., Dukovic, G., Brus, L.E., and Heinz, T.F. (2005) The optical resonances in carbon nanotubes arise from excitons. *Science*, **308**, 838–841.

74 Hartschuh, A., Pedrosa, H.N., Novotny, L., and Krauss, T.D. (2003) Simultaneous fluorescence and Raman scattering from single carbon nanotubes. *Science*, **301**, 1354–1357.

75 Bachilo, S.M., Strano, M.S., Kittrell, C., Hauge, R.H., Smalley, R.E., and Weisman, R.B. (2002) Structure-assigned optical spectra of single-walled carbon nanotubes. *Science*, **298**, 2361–2366.

76 Dresselhaus, S.M., Dresselhaus, G., Saito, R., and Jorio, A. (2007) Exciton photophysics of carbon nanotubes. *Ann. Rev. Phys. Chem.*, **58**, 719–747.

77 Chen, F., Wang, B., Chen, Y., and Li, L.J. (2007) Toward the extraction of single species of single-walled carbon nanotubes using fluorene-based polymers. *Nano Lett.*, **7**, 3013–3017.

78 Lebedkin, S., Hennrich, F., Kiowski, O., and Kappes, M.M. (2008) Photophysics of carbon nanotubes in organic polymer-toluene dispersions: Emission and excitation satellites and relaxation pathways. *Phys. Rev. B: Condens. Matter*, **77**, 165429/1–165429/8.

79 Jones, M., Engtrakul, C., Metzger, W.K., Ellingson, R.J., Nozik, A.J., Heben, M.J., and Rumbles, G. (2005) Analysis of photoluminescence from solubilized single-walled carbon nanotubes. *Phys. Rev. B: Condens. Matter*, **71**, 115426/1–115426/9.

80 Wang, F., Dukovic, G., Brus, L.E., and Heinz, T.F. (2004) Time-resolved fluorescence of carbon nanotubes and its implication for radiative lifetimes. *Phys. Rev. Lett.*, **92**, 177401/1–177401/4.

81 Cronin, S.B., Yin, Y., Walsh, A., Capaz, R.B., Stolyarov, A., Tangney, P., Cohen, M.L., Louie, S.G., Swan, A.K., Ünlü, M.S., Goldberg, B.B., and Tinkham, M. (2006) Temperature dependence of the optical transition energies of carbon nanotubes: The role of electron–phonon coupling and thermal expansion. *Phys. Rev. Lett.*, **96**, 127403/1–127403/4.

82 Shan, G. and Bao, S. (2006) The effect of deformations on electronic structures and optical properties of carbon nanotubes. *Physica E*, **35**, 161–167.

83 Walsh, A.G., Vamivakas, A.N., Yin, Y., Cronin, S.B., Ünlü, M.S., Goldberg, B.B., and Swan, A.K. (2007) Screening of excitons in single, suspended carbon nanotubes. *Nano Lett.*, **7**, 1485–1488.

84 Woodside, M.T. and McEuen, P.L. (2002) Scanned probe imaging of single-electron charge states in nanotube quantum dots. *Science*, **296**, 1098–1101.

85 McEuen, P.L., Bockrath, M., Cobden, D.H., Yoon, Y.G., and Louie, S.G. (1999) Disorder, pseudospins, and backscattering in carbon nanotubes. *Phys. Rev. Lett.*, **83**, 5098–5101.

86 Zaric, S., Ostojic, G.N., Kono, J., Shaver, J., Moore, V.C., Strano, M.S., Hauge, R.H., Smalley, R.E., and Wei, X. (2004) Optical signatures of the Aharonov–Bohm phase in single-walled carbon nanotubes. *Science*, **304**, 1129–1132.

87 Barone, P.W., Parker, R.S., and Strano, M.S. (2005) *In vivo* fluorescence detection of glucose using a single-walled carbon nanotube optical sensor: design, fluorophore properties, advantages, and disadvantages. *Anal. Chem.*, **77**, 7556–7562.

88 Barone, P.W., Baik, S., Heller, D.A., and Strano, M.S. (2005) Near-infrared optical sensors based on single-walled carbon nanotubes. *Nat. Mater.*, **4**, 86–92.

89 Misevich, J.A., Martel, R., Avouris, P., Tsang, J.C., Heinze, S., and Tersoff, J. (2003) Electrically induced optical emission from a carbon nanotube FET. *Science*, **300**, 783–786.

90 Kazaoui, S., Minami, N., Jacquemin, R., Kataura, H., and Achiba, Y. (1999) Amphoteric doping of single-wall carbon-nanotube thin films as probed by optical absorption spectroscopy. *Phys. Rev. B: Condens. Matter*, **60**, 13339–13342.

91 Petit, P., Mathis, C., Journet, C., and Bernier, P. (1999) Tuning and monitoring the electronic structure of carbon nanotubes. *Chem. Phys. Lett.*, **305**, 370–374.

92 Jouguelet, E., Mathis, C., and Petit, P. (2000) Controlling the electronic properties of single-wall carbon nanotubes by chemical doping. *Chem. Phys. Lett.*, **318**, 561–564.

93 Rao, A.M., Eklund, P.C., Bandow, S., Thess, A., and Smalley, R.E. (1997) Evidence for charge transfer in doped carbon nanotube bundles from Raman scattering. *Nature*, **388**, 257–259.

94 Kavan, L. and Dunsch, L. (2007) Spectroelectrochemistry of carbon nanostructures. *ChemPhysChem*, **8**, 974–998.

95 Jacquemin, R., Kazaoui, S., Yu, D., Hassanien, A., Minami, N., Kataura, H., and Achiba, Y. (2000) Doping mechanism in single-wall carbon nanotubes studied by optical absorption. *Synth. Met.*, **115**, 283–287.

96 Bendiab, N., Anglaret, E., Bantignies, J.L., Zahab, A., Sauvajol, J.L., Petit, P., and Mathis, C. (2001) Stoichiometry dependence of the Raman spectrum of alkali-doped single-wall carbon nanotubes. *Phys. Rev. B*, **64**, 245424/1–245424/6.

97 Son, H., Reina, A., Samsonidze, G.G., Saito, R., Jorio, A., Dresselhaus, M.S., and Kong, J. (2006) Selection rules for one- and two-photon absorption by excitons in carbon nanotubes. *Phys. Rev. B: Condens. Matter*, **74**, 073406/1–241406/4.

98 Kavan, L., Rapta, P., Dunsch, L., Bronikowski, M.J., Willis, P., and Smalley, R.E. (2001) Electrochemical tuning of electronic structure of single-walled carbon nanotubes: *in situ* Raman and Vis–NIR study. *J. Phys. Chem. B*, **105**, 10764–10771.

99 Gupta, S., Hughes, M., Windle, A.H., and Robertson, J. (2004) Charge transfer in carbon nanotube actuators investigated using *in situ* Raman spectroscopy. *J. Appl. Phys.*, **95**, 2038–2048.

100 Kavan, L., Kalbáč, M., Zukalová, M., and Dunsch, L. (2005) Electrochemical doping of chirality-resolved carbon nanotubes. *J. Phys. Chem. B*, **109**, 19613–19619.

101 Rafailov, P.M., Maultzsch, J., Thomsen, C., and Kataura, H. (2005) Electrochemical switching of the Peierls-like transition in metallic single-walled carbon nanotubes. *Phys. Rev. B*, **72**, 045411/1–045411/7.

102 Dresselhaus, M.S. and Dresselhaus, G. (1981) Intercalation compounds of graphite. *Adv. Phys.*, **30**, 139–326.

103 Murakoshi, K. and Okazaki, K. (2005) Electrochemical potential control of isolated single-walled carbon nanotubes on gold electrode. *Electrochim. Acta*, **50**, 3069–3075.

104 Okazaki, K., Nakato, Y., and Murakoshi, K. (2003) Absolute potential of the Fermi level of isolated single-walled carbon nanotubes. *Phys. Rev. B*, **68**, 035434/1–035434/5.

105 Corio, P., Santos, P.S., Brar, V.W., Samsonidze, G.G., Chou, S.G., and Dresselhaus, M.S. (2003) Potential dependent surface Raman spectroscopy of single wall carbon nanotube films on platinum electrodes. *Chem. Phys. Lett.*, **370**, 675–682.

106 Claye, A., Rahman, S., Fischer, J.E., Sirenko, A., Sumanasekera, G.U., and Eklund, P.C. (2001) *In situ* Raman scattering studies of alkali-doped single wall carbon nanotubes. *Chem. Phys. Lett.*, **333**, 16–22.

107 Kavan, L. and Dunsch, L. (2003) Ionic liquid for *in situ* Vis/NIR and Raman spectroelectrochemistry: doping of carbon nanostructures. *ChemPhysChem*, **4**, 944–950.

108 Cambedouzou, J., Sauvajol, J.L., Rahmani, A., Flahaut, E., Peigney, A., and Laurent, C. (2005) Raman spectroscopy of iodine-doped double-walled carbon nanotubes. *Phys. Rev. B*, **69**, 235422/1–235422/6.

109 Pénicaud, A., Poulin, P., Derré, A., Anglaret, E., and Petit, P. (2005) Spontaneous dissolution of a single-wall carbon nanotube salt. *J. Am. Chem. Soc.*, **127**, 8–9.

110 Paolucci, D., Franco, M.M., Iurlo, M., Marcaccio, M., Prato, M., Zerbetto, F., Pénicaud, A., and Paolucci, F. (2008) Singling out the electrochemistry of individual single-walled carbon nanotubes in solution. *J. Am. Chem. Soc.*, **130**, 7393–7399.

111 Wang, Z., Pedrosa, H., Krauss, T., and Rothberg, L. (2007) Determination of the exciton binding energy in single-walled carbon. Reply. *Phys. Rev. Lett.*, **98**, 019702/1.

112 Dukovic, G., Wang, F., Song, D., Sfeir, M.Y., Heinz, T.F., and Brus, L.E. (2005) Structural dependence of excitonic optical transitions and band-gap energies in carbon nanotubes. *Nano Lett.*, **5**, 2314–2318.

113 Fagan, S.B., Souza Filho, A.G., Mendes Filho, J., Corio, P., and Dresselhaus, M.S. (2005) Electronic properties of Ag- and CrO$_3$-filled single-wall carbon nanotubes. *Chem. Phys. Lett.*, **456**, 54–59.

114 Corio, P., Santos, A.P., Temperini, M.L.A., Brar, V.W., Pimenta, M.A., and Dresselhaus, M.S. (2004) Characterization of single wall carbon nanotubes filled with silver and with chromium compounds. *Chem. Phys. Lett.*, **383**, 475–480.

115 Terrones, M., Jorio, A., Endo, M., Rao, A.M., Kim, Y.A., Hayashi, T., Terrones, H., Charlier, J.C., Dresselhaus, G., and Dresselhaus, M.S. (2004) New direction in nanotube science. *Mater. Today*, **7**, 30–45.

116 Glerup, M., Steinmetz, J., Samaille, D., Stephan, O., Enouz, S., Loiseau, A., Roth, S., and Bernier, P. (2004) Synthesis of N-doped SWNT using the arc-discharge procedure. *Chem. Phys. Lett.*, **387**, 193–197.

117 Sen, R., Satishkumar, B.C., Govindaraj, S., Harikumar, K.R., Renganathan, M.K., and Rao, C.N.R. (1997) Nitrogen-containing carbon nanotubes. *J. Mater. Chem.*, **7**, 2335–2337.

118 Terrones, M., Grobert, N., Olivares, J., Zhang, J.P., Terrones, H., Kordatos, K., Hsu, W.K., Hare, J.P., Townsend, P.D., Prassides, K., Cheetham, A.K., Kroto, H.W., and Walton, D.R.M. (1997) Controlled production of aligned-nanotube bundles. *Nature*, **388**, 52–55.

119 Terrones, M., Redlich, P., Grobert, N., Trasobares, S., Hsu, W.K., Terrones, H., Zhu, Y.Q., Hare, J.P., Reeves, C.L., Cheetham, A.K., Ruhle, M., Kroto, H.W., and Walton, D.R.M. (1999) Carbon nitride nanocomposites. Formation of aligned CxNy nanofibers. *Adv. Mater.*, **11**, 655–658.

120 Villalpando-Paez, F., Zamudio, A., Elias, A.L., Son, H., Barros, E.B., Chou, S., Kim, Y.A., Muramatsu, H., Hayashi, T., Kong, J., Terrones, H., Dresselhaus, G., Endo, M., Terrones, M., and Dresselhaus, M.S. (2006) Synthesis and characterization of long strands of nitrogen-doped single-walled carbon nanotubes. *Chem. Phys. Lett.*, **424**, 345–352.

121 Birch, A.J. (1950) The reduction of organic compounds by metal–ammonia solutions. *Quart. Revs.*, **4**, 69–93.

122 Stephenson, J.J., Sadana, A.K., Higginbotham, A.L., and Tour, J.M. (2006) Highly functionalized and soluble multiwalled carbon nanotubes by reductive alkylation and arylation: the billups reaction. *Chem. Mater.*, **18**, 4658–4661.

123 Chen, Y., Haddon, R.C., Fang, S., Rao, A.M., Lee, W.H., Dickey, E.C., Grulke, E.A., Pendergrass, J.C., Chavan, A., Haley, B.E., and Smalley, R.E. (1998) Chemical attachment of organic functional groups to single-walled carbon nanotube material. *J. Mater. Res.*, **13**, 2423–2431.

124 Pekker, S., Salvetat, J.-P., Jakab, E., Bonard, J.-M., and Forro, L. (2001)

Hydrogenation of carbon nanotubes and graphite in liquid ammonia. *J. Phys. Chem. B.*, **105**, 7938–7943.

125 Liang, F., Sadana, A.K., Peera, A., Gu, J., Chattopadhyay, Z., Hauge, R.H., and Billups, W.E. (2004) A convenient route to functionalized carbon nanotubes. *Nano Lett.*, **4**, 1257–1260.

126 Borondics, F., Jakab, E., and Pekker, S. (2007) Functionalization of carbon nanotubes via dissolving metal reductions. *J. Nanosci. Nanotechnol.*, **7**, 1551–1559.

127 Chattopadhyay, J., Sadana, A.K., Liang, F., Beach, J.M., Xiao, Y., Hauge, R.H., and Billups, W.E. (2005) Carbon nanotube salts. Arylation of single-wall carbon nanotubes. *Org. Lett.*, **7**, 4067–4069.

128 Liang, F., Alemany, L.B., Beach, J.M., and Billups, W.E. (2005) Structure analyses of dodecylated single-walled carbon nanotubes. *J. Am. Chem. Soc.*, **127**, 13941–13948.

129 Hirsch, A., Soi, A., and Karfunkel, H.R. (1992) Titration of C_{60}: a synthesis of organofullerenes. *Angew. Chem. Int. Ed. Engl.*, **31**, 766–768.

130 Graupner, R., Abraham, J., Wunderlich, D., Vencelova, A., Lauffer, P., Roehrl, J., Hundhausen, M., Ley, L., and Hirsch, A. (2006) Nucleophilic-alkylation–reoxidation: a functionalization sequence for single-wall carbon nanotubes. *J. Am. Chem. Soc.*, **128**, 6683–6689.

131 Chen, S., Shen, W., Wu, G., Chen, D., and Jiang, M. (2005) A new approach to the functionalization of single-walled carbon nanotubes with both alkyl and carboxyl groups. *Chem. Phys. Lett.*, **402**, 312–317.

132 Wunderlich, D., Hauke, F., and Hirsch, A. (2008) Preferred functionalization of metallic and small-diameter single walled carbon nanotubes via reductive alkylation. *J. Mater. Chem.*, **18**, 1493–1497.

133 Ajayan, P.M., Ebbesen, T.W., Ichihashi, T., Iijima, S., Tanigaki, K., and Hiura, H. (1993) Opening carbon nanotubes with oxygen and implications for filling. *Nature*, **362**, 522–525.

134 Sen, R., Rickard, S.M., Itkis, M.E., and Haddon, R.C. (2003) Controlled

purification of single-walled carbon nanotube films by use of selective oxidation and near-IR spectroscopy. *Chem. Mater.*, **15**, 4273–4279.

135 Huang, S. and Dai, L. (2002) Plasma etching for purification and controlled opening of aligned carbon nanotubes. *J. Phys. Chem. B.*, **106**, 3543–3545.

136 Dujardin, E., Ebbesen, T.W., Treacy, A., and Krishnan, M.M.J. (1998) Purification of single-shell carbon nanotubes. *Adv. Mater.*, **10**, 611–613.

137 Ajayan, P.M. and Iijima, S. (1993) Capillarity-induced filling of carbon nanotubes. *Nature*, **361**, 333–334.

138 Collins, P.G., Bradley, K., Ishigami, M., and Zettl, A. (2000) Extreme oxygen sensitivity of electronic properties of carbon nanotubes. *Science*, **287**, 1801–1804.

139 Saini, R.K., Chiang, I.W., Peng, H., Smalley, R.E., Billups, W.E., Hauge, R.H., and Margrave, J.L. (2003) Covalent sidewall functionalization of single wall carbon nanotubes. *J. Am. Chem. Soc.*, **125**, 3617–3621.

140 Kawai, T. and Miyamoto, Y. (2008) Chirality-dependent C-C bond breaking of carbon nanotubes by cyclo-addition of oxygen molecule. *Chem. Phys. Lett.*, **453**, 256.

141 Mawhinney, D.B., Naumenko, V., Kuznetsova, A., Yates, J.T., Jr., Liu, J., and Smalley, R.E. (2000) Surface defect site density on single walled carbon nanotubes by titration. *Chem. Phys. Lett.*, **324**, 213–216.

142 Chen, J., Rao, A.M., Lyuksyutov, S., Itkis, M.E., Hamon, M.A., Hu, H., Cohn, R.W., Eklund, P.C., Colbert, D.T., Smalley, R.E., and Haddon, R.C. (2001) Dissolution of full-length single-walled carbon nanotubes. *J. Phys. Chem. B*, **105**, 2525–2528.

143 Pietronero, L. and Strassler, S. (1981) Bond-length change as tool to determine charge transfer and electron-phonon coupling in graphite intercalation compounds. *Phys. Rev. Lett.*, **47**, 593–596.

144 Sankapal, B.R., Setyowati, K., Chen, J., and Liu, H. (2007) Electrical properties of air-stable, iodine-doped carbon-nanotube–polymer composites. *Appl. Phys. Lett.*, **91**, 173103/1–173103/3.

145 Baxendale, M., Mordkovich, V.Z., Yoshimura, S., Chang, R.P.H., and Jansen, A.G.M. (1998) Metallic conductivity in bundles of FeCl₃-intercalated multiwall carbon nanotubes. *Phys. Rev. B: Condens. Matter*, **57**, 15629–15632.

146 Gai, P., Stephan, O., McGuire, K., Rao, A.M., Dresselhaus, M.S., Dresselhaus, G., and Colliex, C. (2004) Structural systematics in boron-doped single wall carbon nanotubes. *J. Mater. Chem.*, **14**, 669–675.

147 Borowiak-Palen, E., Pichler, T., Fuentes, G.G., Graff, A., Kalenczuk, R.J., Knupfer, M., and Fink, J. (2003) Efficient production of B-substituted single-wall carbon nanotubes. *Chem. Phys. Lett.*, **378**, 516–520.

148 Terrones, M., Hsu, W.K., Schilder, A., Terrones, H., Grobert, N., Hare, J.P., Zhu, Y.Q., Schwoerer, M., Prassides, K., Kroto, H.W., and Walton, D.R.M. (1998) Novel nanotubes and encapsulated nanowires. *Appl. Phys. A: Mater.*, **66**, 307–317.

149 Monthioux, M. (2002) Filling single-wall carbon nanotubes. *Carbon*, **40**, 1809–1823.

150 Khlobystov, A.N., Britz, D.A., and Briggs, G.A.D. (2005) Molecules in carbon nanotubes. *Acc. Chem. Res.*, **38**, 901–909.

151 Vostrowsky, O. and Hirsch, A. (2004) Molecular peapods as supramolecular carbon allotropes. *Angew. Chem. Int. Ed.*, **43**, 2326–2329.

152 Chiu, P.W., Yang, S.F., Yang, S.H., Gu, G., and Roth, S. (2003) Temperature dependence of conductance character in nanotube peapods. *Appl. Phys. A*, **76**, 463–467.

153 Lee, J., Kim, H., Kahng, S.-J., Kim, G., Son, Y.-W., Ihm, J., Kato, H., Wang, Z.W., Okazaki, T., Shinohara, H., and Kuk, Y. (2002) Bandgap modulation of carbon nanotubes by encapsulated metallofullerenes. *Nature*, **415**, 1005–1008.

154 Okada, S., Saito, S., and Oshiyama, A. (2001) Energetics and electronic structures of encapsulated C₆₀ in a carbon nanotube. *Phys. Rev. Lett.*, **86**, 3835–3838.

155 Vavro, J., Llaguno, M.C., Satishkumar, B.C., Luzzi, D.E., and Fischer, J.E. (2002) Electrical and thermal properties of C₆₀-filled single-wall carbon nanotubes. *Appl. Phys. Lett.*, **80**, 1450–1452.

156 Hornbaker, D.J., Kahng, S.-J., Misra, S., Smith, B.W., Johnson, A.T., Mele, E.J., Luzzi, D.E., and Yazdani, A. (2002) Mapping the one-dimensional electronic states of nanotube peapod structures. *Science*, **295**, 828–831.

157 Cho, Y., Han, S., Kim, G., Lee, H., and Ihm, J. (2003) Orbital hybridization and charge transfer in carbon nanopeapods. *Phys. Rev. Lett.*, **90**, 106402/1–106402/4.

158 Debarre, A., Julien, R., Nutarelli, D., Richard, A., and Tchenio, P. (2003) Specific Raman signatures of a dimetallofullerene peapod. *Phys. Rev. Lett.*, **91**, 085501/1–085501/4.

159 Takenobu, T., Takano, T., Shiraishi, M., Murakami, Y., Ata, M., Kataura, H., Achiba, Y., and Iwasa, Y. (2003) Stable and controlled amphoteric doping by encapsulation of organic molecules inside carbon nanotubes. *Nat. Mater.*, **2**, 683–688.

160 Guldi, D.M., Marcaccio, M., Paolucci, D., Paolucci, F., Tagmatarchis, N., Tasis, D., Vázquez, E., and Prato, M. (2003) Single-wall carbon nanotube-ferrocene nanohybrids: observing intramolecular electron transfer in functionalized SWNTs. *Angew. Chem. Int. Ed.*, **42**, 4206–4209.

161 Segura, J.L. and Martin, N. (2001) New concepts in tetrathiafulvalene chemistry. *Angew. Chem. Int. Ed.*, **40**, 1372–1409.

162 Herranz, M.Á., Martin, N., Campidelli, S., Prato, M., Brehm, G., and Guldi, D.M. (2006) Control over electron transfer in tetrathiafulvalene-modified single-walled carbon nanotubes. *Angew. Chem. Int. Ed.*, **45**, 4478–4482.

163 Herranz, M.A., Ehli, C., Campidelli, S., Gutierrez, M., Hug, G.L., Ohkubo, K., Fukuzumi, S., Prato, M., Martin, N., and Guldi, D.M. (2008) Spectroscopic characterization of photolytically generated radical ion pairs in single-wall carbon nanotubes bearing surface-immobilized tetrathiafulvalenes. *J. Am. Chem. Soc.*, **130**, 66–73.

164 Ehli, C., Guldi, D.M., Angeles Herranz, M., Martin, N., Campidelli, S., and Prato, M. (2008) Pyrene-tetrathiafulvalene supramolecular assembly with different types of carbon nanotubes. *J. Mater. Chem.*, **18**, 1498–1503.

165 (a) Kalyanasundaram, K. (1997) *Photochemistry of Polypyridine and Porphyrin Complexes*, Academic Press, London;(b) Kadish, K.M., Smith, K.M., and Guilard, R. (2003) *The Porphyrin Handbook*, Academic Press, New York.

166 Leznoff, C.C. and Lever, A.B.P. (1996) *Phthalocyanines: Properties and Applications*, vol. **1–4**, VCH, Weinheim.

167 de la Torre, G., Vázquez, P., Agulló-López, F., and Torres, T. (2004) Role of structural factors in the nonlinear optical properties of phthalocyanines and related compounds. *Chem. Rev.*, **104**, 3723–3750.

168 Baskaran, D., Mays, J.W., Zhang, X.P., and Bratcher, M.S. (2005) Carbon nanotubes with covalently linked porphyrin antennae: photoinduced electron transfer. *J. Am. Chem. Soc.*, **127**, 6916–6917.

169 Li, H., Martin, R.B., Harruff, B.A., Carino, R.A., Allard, L.F., and Sun, Y.-P. (2004) Single-walled carbon nanotubes tethered with porphyrins: synthesis and photophysical properties. *Adv. Mater.*, **16**, 896–900.

170 de la Torre, G., Blau, W., and Torres, T. (2003) A survey on the functionalization of single-walled nanotubes. The chemical attachment of phthalocyanine moieties. *Nanotechnology*, **14**, 765–771.

171 Xu, H.-B., Chen, H.-Z., Shi, M.-M., Bai, R., and Wang, M. (2005) A novel donor–acceptor heterojunction from single-walled carbon nanotubes functionalized by erbium bisphthalocyanine. *Mater. Chem. Phys.*, **94**, 342–346.

172 Ballesteros, B., Campidelli, S., de la Torre, G., Ehli, C., Guldi, D.M., Prato, M., and Torres, T. (2007) Synthesis, characterization and photophysical properties of a SWNT–phthalocyanine hybrid. *Chem. Commun.*, **28**, 2950–2952.

173 Ballesteros, B., de la Torre, G., Ehli, C., Rahman, G.M.A., Agullo-Rueda, F., Guldi, D.M., and Torres, T. (2007) Single-wall carbon nanotubes bearing covalently linked phthalocyanines-photoinduced electron transfer. *J. Am. Chem. Soc.*, **129**, 5061–5068.

174 Campidelli, S., Ballesteros, B., Filoramo, A., Díaz, D.D., de la Torre, G., Torres, T., Rahman, G.M.A., Ehli, C., Kiessling, D., Werner, F., Sgobba, V., Guldi, D.M., Cioffi, C., Prato, M., and Bourgoin, J.-P. (2008) Facile decoration of functionalized single-wall carbon nanotubes with phthalocyanines via click chemistry. *J. Am. Chem. Soc.*, **130**, 11503–11509.

175 Alvaro, M., Atienzar, P., de la Cruz, P., Delgardo, J.L., Troiani, V., Garcia, H., Langa, F., Palkar, A., and Echegoyen, L. (2006) Synthesis, photochemistry, and electrochemistry of single-wall carbon nanotubes with pendent pyridyl groups and of their metal complexes with zinc porphyrin. Comparison with pyridyl-bearing fullerenes. *J. Am. Chem. Soc.*, **128**, 6626–6635.

176 Yu, J., Mathew, S., Flavel, B.S., Johnston, M.R., and Shapter, J.G. (2008) Ruthenium porphyrin functionalized single-walled carbon nanotube arrays – a step toward light harvesting antenna and multibit information storage. *J. Am. Chem. Soc.*, **130**, 8788–8796.

177 Cheng, F. and Adronov, A. (2006) Suzuki coupling reactions for the surface functionalization of single-walled carbon nanotubes. *Chem. Mater.*, **18**, 5389–5391.

178 Stephenson, J.J., Hudson, J.L., Azad, S., and Tour, J.M. (2006) Individualized single walled carbon nanotubes from bulk material using 96% sulfuric acid as solvent. *Chem. Mater.*, **18**, 374–377.

179 Price, B.K., Hudson, J.L., and Tour, J.M. (2005) Green chemical functionalization of single-walled carbon nanotubes in ionic liquids. *J. Am. Chem. Soc.*, **127**, 14867–14870.

180 Chen, B., Flatt, A.K., Jian, H., Hudson, J.L., and Tour, J.M. (2005) Molecular grafting to silicon surfaces in air using organic triazenes as stable diazonium sources and HF as a constant hydride-passivation source. *Chem. Mater.*, **17**, 4832–4836.

181 Guo, Z., Du, F., Ren, D., Chen, Y., Zheng, J., Liu, Z., and Tian, J. (2006) Covalently porphyrin-functionalized single-walled carbon nanotubes: a novel photoactive and optical limiting donor–acceptor nanohybrid. *J. Mater. Chem.*, **16**, 3021–3030.

182 Guldi, D.M., Rahman, G.M.A., Ramey, J., Marcaccio, M., Paolucci, D., Paolucci, F., Qin, S., Ford, W.T., Balbinot, D., Jux, N., Tagmatarchis, N., and Prato, M. (2004) Donor–acceptor nanoensembles of soluble carbon nanotubes. *Chem. Commun.*, 2034–2035.

183 Guldi, D.M., Rahman, G.M.A., Qin, S., Tchoul, M., Ford, W.T., Marcaccio, M., Paolucci, D., Paolucci, F., Campidelli, S., and Prato, M. (2006) Versatile coordination chemistry towards multifunctional carbon nanotube nanohybrids. *Chem. Eur. J.*, **12**, 2152–2161.

184 Rahman, G.M.A., Troeger, A., Sgobba, V., Guldi, D.M., Jux, N., Tchoul, M.N., Ford, W.T., Mateo-Alonso, A., and Prato, M. (2008) Improving photocurrent generation: supramolecularly and covalently functionalized single-wall carbon nanotubes-polymer/porphyrin donor–acceptor nanohybrids. *Chem. Eur. J.*, **14**, 8837–8846.

185 Campidelli, S., Sooambar, C., Diz, E.L., Ehli, C., Guldi, D.M., and Prato, M. (2006) Dendrimer-functionalized single-wall carbon nanotubes: Synthesis, characterization, and photoinduced electron transfer. *J. Am. Chem. Soc.*, **128**, 12544–12552.

186 Williams, K.A., Veenhuizen, P.T.M., de la Torre, B.G., Eritja, R., and Dekker, C. (2002) Nanotechnology: carbon nanotubes with DNA recognition. *Nature*, **420**, 761.

187 Wong, N., Kam, S., Liu, Z., and Dai, H. (2005) Functionalization of carbon nanotubes via cleavable disulfide bonds for efficient intracellular delivery of siRNA and potent gene silencing. *J. Am. Chem. Soc.*, **127**, 12492–12493.

188 Wang, Y., Iqbal, Z., and Malhotra, S.V. (2005) Functionalization of carbon nanotubes with amines and enzymes. *Chem. Phys. Lett.*, **402**, 96–101.

189 Zhang, Y., Shen, Y., Li, J., Niu, L., Dong, S., and Ivaska, A. (2005) Electrochemical functionalization of single-walled carbon nanotubes in large quantities at a room-temperature ionic liquid supported three-dimensional network electrode. *Langmuir*, **21**, 4797–4800.

190 Kandimalla, V.B. and Ju, H. (2006) Binding of acetylcholinesterase to multiwall carbon nanotube-cross-linked chitosan composite for flow-injection amperometric detection of an organophosphorous insecticide. *Chem. Eur. J.*, **12**, 1074–1080.

191 Sun, Y.P., Fu, K., Lin, Y., and Huang, W. (2002) Functionalized carbon nanotubes: properties and applications. *Acc. Chem. Res.*, **35**, 1096–1104.

192 Elkin, T., Jiang, X., Taylor, S., Lin, Y., Gu, L., Yang, H., Brown, J., Collins, S., and Sun, Y.P. (2005) Immuno-carbon nanotubes and recognition of pathogens. *Chem. Biol. Chem.*, **6**, 640–643.

193 Portney, N.G., Singh, K., Chaudhary, S., Destito, G., Schneemann, A., Manchester, M., and Ozkan, M. (2005) Organic and inorganic nanoparticle hybrids. *Langmuir*, **21**, 2098–2103.

194 Li, M., Dujardin, E., and Mann, S. (2005) Programmed assembly of multi-layered protein/nanoparticle-carbon nanotube conjugates. *Chem. Commun.*, 4952–4954.

195 Kam, N.W.S., Jessop, T.C., Wender, P.A., and Dai, H. (2004) Nanotube molecular transporters: internalization of carbon nanotube-protein conjugates into mammalian cells. *J. Am. Chem. Soc.*, **126**, 6850–6851.

196 Alvaro, M., Aprile, C., Atienzar, P., and Garcia, H. (2005) Preparation and photochemistry of single wall carbon nanotubes having covalently anchored viologen units. *J. Phys. Chem. B*, **109**, 7692–7697.

197 Wang, P., Moorefield, C.N., Li, S., Hwang, S.H., Shreiner, C.D., and Newkome, G.R. (2006) Terpyridine CuII-mediated reversible nanocomposites of single-wall carbon nanotubes: towards metallo-nanoscale architectures. *Chem. Commun.*, 1091–1093.

198 Star, A., Stoddart, J.F., Steuerman, D., Diehl, M., Boukai, A., Wong, E.W., Yang, X., Chung, S.W., Choi, H., and Heath, J.R. (2001) Preparation and properties of polymer-wrapped single-walled carbon nanotubes. *Angew. Chem. Int. Ed.*, **40**, 1721–1725.

199 Guldi, D.M., Rahman, G.M.A., Jux, N., Tagmatarchis, N., and Prato, M. (2004) Integrating single-wall carbon nanotubes into donor–acceptor nanohybrids. *Angew. Chem. Int. Ed.*, **43**, 5526–5530.

200 Guldi, D.M., Taieb, H., Rahman, G.M.A., Tagmatarchis, N., and Prato, M. (2005) Novel photoactive single-walled carbon nanotube–porphyrin polymer wraps: efficient and long-lived intracomplex charge separation. *Adv. Mater.*, **17**, 871–875.

201 Satake, A., Miyajima, Y., and Kobuke, A. (2005) Porphyrin-carbon nanotube composites formed by noncovalent polymer wrapping. *Chem. Mater.*, **17**, 716–724.

202 Saito, K., Troiani, V., Qiu, H., Solladie, N., Sakata, T., Mori, H., Ohama, M., and Fukuzumi, S. (2007) Nondestructive formation of supramolecular nanohybrids of single-walled carbon nanotubes with flexible porphyrinic polypeptides. *J. Phys. Chem. C*, **111**, 1194–1199.

203 Cheng, F. and Adronov, A. (2006) Noncovalent functionalization and solubilization of carbon nanotubes by using a conjugated Zn–porphyrin polymer. *Chem. Eur. J.*, **12**, 5053–5059.

204 Cheng, F., Zhang, S., Adronov, A., Echegoyen, L., and Diederich, F. (2006) Triply fused Zn(II)–porphyrin oligomers: synthesis, properties, and supramolecular interactions with single-walled carbon nanotubes (SWNTs). *Chem. Eur. J.*, **12**, 6062–6070.

205 Cao, L., Chen, H.Z., Zhou, H.B., Zhu, L., Sun, J.Z., Zhang, X.B., Xu, J.M., and Wang, M. (2003) Carbon-nanotube-templated assembly of rare-earth phthalocyanine nanowires. *Adv. Mater.*, **15**, 909–913.

206 Chen, J. and Collier, C.P. (2005) Noncovalent functionalization of single-walled carbon nanotubes with water-soluble porphyrins. *J. Phys. Chem. B*, **109**, 7605–7609.

207 Li, H., Zhou, B., Lin, Y., Gu, L., Wang, W., Fernando, K.A.S., Kumar, S., Allard, L.F., and Sun, Y.-P. (2004) Selective interactions of porphyrins with semiconducting single-walled carbon nanotubes. *J. Am. Chem. Soc.*, **126**, 1014–1015.

208 Murakami, H., Nomura, T., and Nakashima, N. (2003) Noncovalent porphyrin-functionalized single-walled carbon nanotubes in solution and the formation of porphyrin–nanotube nanocomposites. *Chem. Phys. Lett*, **378**, 481–485.

209 Murakami, H., Nakamura, G., Nomura, T., Miyamoto, T., and Nakashima, N. (2007) Noncovalent porphyrin-functionalized single-walled carbon nanotubes: solubilization and spectral behaviors. *J. Porph. Phthal.*, **11**, 418–427.

210 Geng, J., Ko, Y.K., Youn, S.C., Kim, Y.H., Kim, S.A., Jung, D.H., and Jung, H.T. (2008) Synthesis of SWNT rings by noncovalent hybridization of porphyrins and single-walled carbon nanotubes. *J. Phys. Chem. C*, **112**, 12264–12271.

211 Nakashima, N., Tomonari, Y., and Murakami, H. (2002) Water-soluble single-walled carbon nanotubes via noncovalent sidewall-functionalization with a pyrene-carrying ammonium ion. *Chem. Lett.*, **6**, 638–639.

212 Ehli, C., Rahman, G.M.A., Jux, N., Balbinot, D., Guldi, D.M., Paolucci, F., Marcaccio, M., Paolucci, D., Melle-Franco, M., Zerbetto, F., Campidelli, S., and Prato, M. (2006) Interactions in single wall carbon nanotubes/pyrene/porphyrin nanohybrids. *J. Am. Chem. Soc.*, **128**, 11222–11231.

213 Guldi, D.M., Rahman, G.M.A., Jux, N., Balbinot, D., Tagmatarchis, N., and Prato, M. (2005) Multiwalled carbon nanotubes in donor–acceptor nanohybrids-towards long-lived electron transfer products. *Chem. Commun.*, 2038–2040.

214 Guldi, D.M., Rahman, G.M.A., Jux, N., Balbinot, D., Hartnagel, U., Tagmatarchis, N., and Prato, M. (2005) Functional single-wall carbon nanotube nanohybrids-associating SWNTs with

water-soluble enzyme model systems. *J. Am. Chem. Soc.*, **127**, 9830–9838.

215 Chitta, R., Sandanayaka, A.S.D., Schumacher, A.L., D'Souza, L., Araki, Y., Ito, O., and D'Souza, F. (2007) Donor–acceptor nanohybrids of zinc naphthalocyanine or zinc porphyrin noncovalently linked to single-wall carbon nanotubes for photoinduced electron transfer. *J. Phys. Chem. C*, **111**, 6947–6955.

216 Kavakka, J.S., Heikkinen, S., Kilpelaeinen, I., Mattila, M., Lipsanen, H., and Helaja, J. (2007) Noncovalent attachment of pyro-pheophorbide a to a carbon nanotube. *Chem. Commun.*, 519–521.

217 D'Souza, F., Chitta, R., Sandanayaka, A.S.D., Subbaiyan, N.K., D'Souza, L., Araki, Y., and Ito, O. (2007) Self-assembled single-walled carbon nanotube:zinc-porphyrin hybrids through ammonium ion–crown ether interaction: construction and electron transfer. *Chem. Eur. J.*, **13**, 8277–8284.

218 Rahman, G.M.A., Guldi, D.M., Cagnoli, R., Mucci, A., Schenetti, L., Vaccari, L., and Prato, M. (2005) Combining single wall carbon nanotubes and photoactive polymers for photoconversion. *J. Am. Chem. Soc.*, **127**, 10051–10057.

219 Guldi, D.M., Rahman, G.M.A., Sgobba, V., Kotov, N.A., Bonifazi, D., and Prato, M. (2006) CNT–CdTe versatile donor–acceptor nanohybrids. *J. Am. Chem. Soc.*, **128**, 2315–2323.

220 Hu, L., Zhao, Y.L., Ryu, K., Zhou, C., Stoddart, J.F., and Gruner, G. (2008) Light-induced charge transfer in pyrene/CdSe–SWNT hybrids. *Adv. Mater.*, **20**, 939–946.

221 Hecht, D.S., Ramirez, R.J.A., Briman, M., Artukovic, E., Chichak, K.S., Stoddart, J.F., and Gruener, G. (2006) Bioinspired detection of light using a porphyrin-sensitized single-wall nanotube field effect transistor. *Nano Lett.*, **6**, 2031–2036.

222 D'Souza, F., Chitta, R., Sandanayaka, A.S.D., Subbaiyan, N.K., D'Souza, L., Araki, Y., and Ito, O. (2007) Supramolecular carbon nanotube-fullerene donor–acceptor hybrids for photoinduced electron transfer. *J. Am. Chem. Soc.*, **129**, 15865–15871.

223 Boul, P.J., Cho, D.G., Rahman, G.M.A., Marquez, M., Ou, Z., Kadish, K.M., Guldi, D.M., and Sessler, J.L. (2007) Sapphyrin-nanotube assemblies. *J. Am. Chem. Soc.*, **129**, 5683–5687.

224 Ryan, K.P., Lipson, S.M., Drury, A., Cadek, M., Ruether, M., O'Flaherty, S.M., Barron, V., McCarthy, B., Byrne, H.J., Balu, W.J., and Coleman, J.N. (2004) Carbon-nanotube nucleated crystallinity in a conjugated polymer based composite. *Chem. Phys. Lett.*, **391**, 329–333.

225 Coleman, J.N., Dalton, A.B., Curran, S., Rubio, A., Davey, A.P., Drury, A., McCarthy, B., Lahr, B., Ajayan, P.M., Roth, S., Barllie, R.C., and Blau, W.J. (2000) Phase separation of carbon nanotubes and turbostratic graphite using a functional organic polymer. *Adv. Mater.*, **12**, 213–216.

226 Star, A., Stoddart, J.F., Steuerman, D., Diehl, M., Boukai, A., Wong, E.W., Yang, X., Chung, S.W., Choi, H., and Heath, J.R. (2001) Preparation and properties of polymer-wrapped single-walled carbon nanotubes. *Angew. Chem. Int. Ed.*, **40**, 1721–1725.

227 Rouse, H., Lillehei, P.T., Sanderson, J., and Iochi, E.J. (2004) Polymer/single-walled carbon nanotube films assembled via donor–acceptor interactions and their use as scaffolds for silica deposition. *Chem. Mater.*, **16**, 3904–3910.

228 Zengin, H., Zhou, W., Jin, J., Czerw, R., Smith, D.W., Echegoyen, L., Carroll, D.L., Foulger, S.H., and Ballato, J. (2002) Carbon nanotube doped polyaniline. *Adv. Mater.*, **14**, 1480–1483.

229 Bavastrello, V., Carrara, S., Ram, M.K., and Nicolini, C. (2004) Optical and electrochemical properties of poly(o-toluidine) multiwalled carbon nanotubes composite Langmuir–Schaefer films. *Langmuir*, **20**, 969–973.

230 Li, X.H., Wu, B., Huang, J.E., Zhang, J., Liu, Z.F., and Li, H.I. (2003) Fabrication and characterization of well-dispersed single-walled carbon nanotube/polyaniline composites. *Carbon*, **41**, 1670–1673.

231 Granot, E., Basnar, B., Cheglakov, Z., Katz, E., and Willner, I. (2006) Enhanced bioelectrocatalysis using single-walled carbon nanotubes (SWCNTs)/polyaniline hybrid systems in thin-film and microrod structures associated with electrodes. *Electroanalysis*, **18**, 26–34.

232 Rice, N.A., Soper, K., Zhou, N., Merschrod, E., and Zhao, Y. (2006) Dispersing as-prepared single-walled carbon nanotube powders with linear conjugated polymers. *Chem. Commun.*, 4937–4939.

233 Chen, J., Liu, H., Weimer, W.A., Halls, M.D., Waldeck, D.H., and Walker, G.C. (2002) Noncovalent engineering of carbon nanotube surfaces by rigid, functional conjugated polymers. *J. Am. Chem. Soc.*, **124**, 9034–9035.

234 Lillehei, P.T., Park, C., Rouse, J.H., and Siochi, E.J. (2002) Imaging carbon nanotubes in high performance polymer composites via magnetic force microscopy. *Nano Lett.*, **2**, 827–829.

235 Park, C., Crooks, R.E., Siochi, E.J., Harrison, J.S., Evans, N., and Kenik, E. (2003) Adhesion study of polyimide to single-wall carbon nanotube bundles by energy-filtered transmission electron microscopy. *Nanotechnology*, **14**, L11–L14.

236 Wise, K.E., Park, C., Siochi, E.J., and Harrison, J.S. (2004) Stable dispersion of single wall carbon nanotubes in polyimide: the role of noncovalent interactions. *Chem. Phys. Lett.*, **391**, 207–211.

237 Hughes, M., Chen, G.Z., Shaffer, M.S.P., Fray, D.J., and Windle, A.H. (2004) Controlling the nanostructure of electrochemically grown nanoporous composites of carbon nanotubes and conducting polymers. *Comp. Sci. Technol.*, **64**, 2325–2331.

238 Geng, H.Z., Kim, K.K., So, K.P., Lee, Y.S., Chang, Y., and Lee, Y.H. (2007) Effect of acid treatment on carbon nanotube-based flexible transparent conducting films. *J. Am. Chem. Soc.*, **129**, 7758–7759.

239 Hu, L., Gruner, G., Li, D., Kaner, R.B., and Cech, J. (2007) Patternable transparent carbon nanotube films for electrochromic devices. *J. Appl. Phys.*, **101**, 016102/1–016102/3.

240 Ago, H., Petrisch, K., Shaffer, M.S.P., Windle, A.H., and Friend, R.H. (1999) Composites of carbon nanotubes and conjugated polymers for photovoltaic devices. *Adv. Mater.*, **11**, 1281–1285.

241 Barnes, T.M., van de Lagemaat, J., Levi, D., Rumbles, G., Coutts, T.J., Weeks, C.L., Britz, D.A., Levitsky, I., Peltola, J., and Glatkowski, P. (2007) Optical characterization of highly conductive single-wall carbon-nanotube transparent electrodes. *Phys. Rev. B: Condens. Matter*, **75**, 235410/1–235410/10.

242 Kymakis, E., Alexandrou, I., and Amaratunga, G.A. (2003) High open-circuit voltage photovoltaic devices from carbon-nanotube-polymer composites. *J. Appl. Phys.*, **93**, 1764–1768.

243 Zhao, J., Park, H., Han, J., and Lu, J.P. (2004) Electronic properties of carbon nanotubes with covalent sidewall functionalization. *J. Phys. Chem. B*, **108**, 4227–4420.

244 Jung, M.S., Choi, T.L., Joo, W.J., Kim, J.Y., Han, I.T., and Kim, J.M. (2007) Transparent conductive thin films based on chemically assembled single-walled carbon nanotubes. *Synth. Met.*, **157**, 997–1003.

245 Li, X., Zhang, L., Wang, X., Shimoyama, I., Sun, X., Seo, W.S., and Dai, H. (2007) Langmuir–Blodgett assembly of densely aligned single-walled carbon nanotubes from bulk materials. *J. Am. Chem. Soc.*, **129**, 4890–4891.

246 Lima, M.D., de Andrade, M.J., Bergmann, C.P., and Roth, S. (2008) Thin, conductive, carbon nanotube networks over transparent substrates by electrophoretic deposition. *J. Mater. Chem.*, **18**, 776–779.

247 Wu, Z., Chen, Z., Du, X., Logan, J.M., Sippel, J., Nikolou, M., Kamaras, K., Reynolds, J.R., Tanner, D.B., Hebard, A.F., and Rinzler, A.G. (2004) Transparent, conductive carbon nanotube films. *Science*, **305**, 1273–1277.

248 Cao, Q., Hur, S.H., Zhu, Z.T., Sun, Y., Wang, C., Meitl, M.A., Shim, M., and Rogers, J.A. (2006) Highly bendable, transparent thin-film transistors that use carbon-nanotube-based conductors and semiconductors with elastomeric dielectrics. *Adv. Mater.*, **18**, 304–309.

249 Wang, W., Fernando, K.A.S., Lin, Y., Meziani, M.J., Veca, L.M., Cao, L., Zhang, P., Kimani, M.M., and Sun, Y.P. (2008) Metallic single-walled carbon nanotubes for conductive nanocomposites. *J. Am. Chem. Soc.*, **130**, 1415–1419.

250 Zhang, M., Fang, S., Zakhidov, A.A., Lee, S.B., Aliev, A.E., Williams, C.D., Atkinson, K.R., and Baughman, R.H. (2005) Strong, transparent, multifunctional, carbon nanotube sheets. *Science*, **309**, 1215–1219.

251 Zhang, D., Ryu, K., Liu, X., Polikarpov, E., Ly, J., Tompson, M.E., and Zhou, C. (2006) Transparent, conductive, and flexible carbon nanotube films and their application in organic light-emitting diodes. *Nano Lett.*, **6**, 1880–1886.

252 Geng, H.Z., Kim, K.K., Song, C., Xuyen, N.T., Kim, S.M., Park, K.A., Lee, D.S., An, K.H., Lee, Y.S., Chang, Y., Lee, Y.J., Choi, J.Y., Benayad, A., and Lee, Y.H. (2008) Doping and de-doping of carbon nanotube transparent conducting films by dispersant and chemical treatment. *J. Mater. Chem.*, **18**, 1261–1266.

253 van de Lagemaat, J., Barnes, T.M., Rumbles, G., Shaheen, S.E., Coutts, T.J., Weeks, C., Levitsky, I., Peltola, J., and Glatkowski, P. (2006) Organic solar cells with carbon nanotubes replacing In$_2$O$_3$: Sn as the transparent electrode. *Appl. Phys. Lett.*, **88**, 233503/1–233503/3.

254 Li, Z., Kandel, H.R., Dervishi, E., Saini, V., Xu, Y., Biris, A.R., Lupu, D., Salamo, G.J., and Biris, A.S. (2008) Comparative study on different carbon nanotube materials in terms of transparent conductive coatings. *Langmuir*, **24**, 2655–2662.

255 Rowell, M.W., Topinka, M.A., McGehe, M.D., Prall, H.J., Dennler, G., Sariciftci, N.S., Hu, L., and Gruner, G. (2006) Organic solar cells with carbon nanotube network electrodes. *Appl. Phys. Lett.*, **88**, 233506/1–233506/3.

256 Du Pasquier, A., Unalan, H.E., Kanwal, A., Miller, S., and Chhowalla, M. (2005) Conducting and transparent single-wall carbon nanotube electrodes for

polymer-fullerene solar cells. *Appl. Phys. Lett.*, **87**, 203511/1–203511/3.

257 Unalan, H.E., Hiralal, P., Kuo, D., Parekh, B., Amaratunga, G., and Chhowalla, M. (2008) Flexible organic photovoltaics from zinc oxide nanowires grown on transparent and conducting single walled carbon nanotube thin films. *J. Mater. Chem.*, **18**, 5909–5912.

258 Suzuki, K., Yamaguchi, M., Kumagai, M., and Yanagida, S. (2003) Application of carbon nanotubes to counter electrodes of dye-sensitized solar cells. *Chem. Lett.*, **32**, 28–29.

259 Trancik, J.E., Calabrese, B.S., and Hone, J. (2008) Transparent and catalytic carbon nanotube films. *Nano Lett.*, **8**, 982–987.

260 Koo, B.K., Lee, D.L., Kim, H.J., Lee, W.J., Song, J.S., and Kim, H.J. (2006) Seasoning effect of dye-sensitized solar cells with different counter electrodes. *J. Electroceram.*, **V17**, 79–82.

261 Fan, B., Mei, X., Sun, K., and Ouyang, J. (2008) Conducting polymer/carbon nanotube composite as counter electrode of dye-sensitized solar cells. *Appl. Phys. Lett.*, **93**, 143103/1–143103/3.

262 Yu, G., Gao, J., Hummelen, J.C., Wudl, F., and Heeger, A.J. (1995) Polymer photovoltaic cells: enhanced efficiencies via a network of internal donor–acceptor heterojunctions. *Science*, **270**, 1789–1791.

263 Granstrom, M., Petritsch, K., Arias, A.C., Lux, A., Andersson, M.R., and Friend, R.H. (1998) Laminated fabrication of polymeric photovoltaic diodes. *Nature*, **395**, 257–260.

264 Lien, D.H., Zan, H.W., Tai, N.H., and Tsai, C.H. (2006) Photocurrent amplification at carbon nanotube-metal contacts. *Adv. Mater.*, **18**, 98–103.

265 Sun, J.L., Wei, J., Zhu, J.L., Xu, D., Liu, X., Sun, H., Wu, D.H., and Wu, N.L. (2006) Photoinduced currents in carbon nanotube/metal heterojunctions. *Appl. Phys. Lett.*, **88**, 131107/1–131107/3.

266 Cinke, M., Li, J., Chen, B., Cassell, A., Delzeit, L., Han, J., and Meyyappan, M. (2002) Pore structure of raw and purified HiPco single-walled carbon nanotubes. *Chem. Phys. Lett.*, **365**, 69–74.

267 Sgobba, V. and Guldi, D.M. (2008) Carbon nanotubes as integrative materials for

organic photovoltaic devices. *J. Mater. Chem.*, **18**, 153–157.

268 Landi, J., Raffaelle, R.P., Castro, S.L., and Bailey, S.G. (2005) Single-wall carbon nanotube-polymer solar cells. *Prog. Photovoltaics*, **13**, 165–172.

269 Xu, Z., Wu, Y., Hu, B., Ivanov, I.N., and Geohegan, D.B. (2005) Carbon nanotube effects on electroluminescence and photovoltaic response in conjugated polymers. *Appl. Phys. Lett.*, **87**, 263118/1–263118/3.

270 Miller, A.J., Hatton, R.A., Ravi, P., and Silva, S. (2006) Water-soluble multiwall-carbon-nanotube-polythiophene composite for bilayer photovoltaics. *Appl. Phys. Lett.*, **89**, 123115/1–123115/3.

271 Miller, A.J., Hatton, R.A., Silva, S., and Ravi, P. (2006) Interpenetrating multiwall carbon nanotube electrodes for organic solar cells. *Appl. Phys. Lett.*, **89**, 133117/1–133117/3.

272 Pradhan, B., Batabyal, S.K., and Pal, A.J. (2006) Functionalized carbon nanotubes in donor/acceptor-type photovoltaic devices. *Appl. Phys. Lett.*, **88**, 093106/1–093106/3.

273 Li, C., Chen, Y., Wang, Y., Iqbal, Z., Chhowalla, M., and Mitra, S. (2007) A fullerene-single wall carbon nanotube complex for polymer bulk heterojunction photovoltaic cells. *J. Mater. Chem.*, **17**, 2406–2411.

274 Kalita, G., Adhikari, S., Aryal, H.R., Umeno, M., Afre, R., Soga, T., and Sharon, M. (2008) Fullerene (C_{60}) decoration in oxygen plasma treated multiwalled carbon nanotubes for photovoltaic application. *Appl. Phys. Lett.*, **92**, 063508/1–063508/3.

275 Landi, B.J., Castro, S.L., Ruf, H.J., Evans, C.M., Bailey, S.G., and Raffaelle, R.P. (2005) CdSe quantum dot-single wall carbon nanotube complexes for polymeric solar cells. *Sol. Energy Mater. Sol. Cells*, **87**, 733–746.

276 Rud, J.A., Lovell, L.S., Senn, J.W., Qiao, Q., and McLeskey, J.T., Jr. (2005) Water soluble polymer/carbon nanotube bulk heterojunction solar cells. *J. Mater. Sci.*, **40**, 1455–1458.

277 Kymakis, E. and Amaratunga, G.A.J. (2003) Photovoltaic cells based on dye-sensitisation of single-wall carbon nanotubes in a polymer matrix. *Sol. Energy Mater. Sol. Cell*, **80**, 465–472.

278 Kim, Y., Choulis, S.A., Nelson, J., Bradley, D.D.C., Cook, S., and Durrant, J.R. (2005) Device annealing effect in organic solar cells with blends of regioregular poly(3-hexylthiophene) and soluble fullerene. *Appl. Phys. Lett.*, **86**, 063502/1–063502/3.

279 Dittmer, J.J., Marseglia, E.A., and Friend, R.H. (2000) Electron trapping in dye/polymer blend photovoltaic cells. *Adv. Mater.*, **12**, 1270–1274.

280 Kymakis, E., Koudoumas, E., Franghiadakis, I., and Amaratunga, G.A.J. (2006) Post-fabrication annealing effects in polymer-nanotube photovoltaic cells. *J. Phys. D: Appl. Phys.*, **39**, 1058–1062.

281 Kazaoui, S., Minami, N., Nalini, B., Kim, Y., and Hara, K. (2005) Near-infrared photoconductive and photovoltaic devices using single-wall carbon nanotubes in conductive polymer films. *J. Appl. Phys.*, **98**, 084314/1–084314/6.

282 Geng, J. and Zeng, T. (2006) Influence of single-walled carbon nanotubes induced crystallinity enhancement and morphology change on polymer photovoltaic devices. *J. Am. Chem. Soc.*, **128**, 16827–16833.

283 Kymakis, E. and Amaratunga, G.A.J. (2005) Carbon nanotubes as electron acceptors in polymeric photovoltaics. *Rev. Adv. Mater. Sci.*, **8**, 300–305.

284 Canestraro, C.D., Schnitzler, M.C., Zarbin, A.J.G., da Luz, M.G.E., and Roman, L.S. (2006) Carbon nanotubes based nanocomposites for photocurrent improvement. *Appl. Surf. Sci.*, **252**, 5575–5578.

285 Srivastava, S. and Kotov, N.A. (2008) Composite layer-by-layer (LBL) assembly with inorganic nanoparticles and nanowires. *Acc. Chem. Res.*, **41**, 1831–1841.

286 Guldi, D.M., Rahman, G.M.A., Prato, M., Jux, N., Qin, S., and Ford, W.T. (2005) Single-wall carbon nanotubes as integrative building blocks for solar-energy conversion. *Angew. Chem. Int. Ed.*, **44**, 2015–2018.

287 Guldi, D.M., Rahman, G.M.A., Sgobba, V., Campidelli, S., and Prato, M. (2006) Supramolecular assemblies of different carbon nanotubes for photoconversion processes. *Adv. Mater.*, **18**, 2264–2269.

288 Rahman, G.M.A., Troeger, A., Sgobba, V., Guldi, D.M., Jux, N., Tchoul, M.N., Ford, W.T., Mateo-Alonso, A., and Prato, M. (2008) Improving photocurrent generation: supramolecularly and covalently functionalized single-wall carbon nanotubes-polymer/porphyrin donor–acceptor nanohybrids. *Chem. Eur. J.*, **14**, 8837–8846.

289 Rahman, G.M.A., Guldi, D.M., Cagnoli, R., Mucci, A., Schenetti, L., Vaccari, L., and Prato, M. (2005) Combining single wall carbon nanotubes and photoactive polymers for photoconversion. *J. Am. Chem. Soc.*, **127**, 10051–10057.

290 Guldi, D.M., Rahman, G.M.A., Sgobba, V., Kotov, N.A., Bonifazi, D., and Prato, M. (2006) CNT–CdTe versatile donor–acceptor nanohybrids. *J. Am. Chem. Soc.*, **128**, 2315–2323.

291 Itoh, F., Suzuki, I., and Miyairi, K. (2005) Field emission from carbon-nanotube-dispersed conducting polymer thin film and its application to photovoltaic devices. *Jpn. J. Appl. Chem.*, **44**, 636–640.

292 Barazzouk, S., Hotchandani, S., Vinodgopal, K., and Kamat, P.V. (2004) Single-wall carbon nanotube films for photocurrent generation. a prompt response to visible-light irradiation. *J. Phys. Chem. B*, **108**, 17015–17018.

293 Robel, I., Bunker, B.A., and Kamat, P.V. (2005) Single-walled carbon nanotube-CdS nanocomposites as light-harvesting assemblies: Photoinduced charge-transfer interactions. *Adv. Mater.*, **17**, 2458–3263.

294 Hasobe, T., Fukuzumi, S., and Kamat, P.V. (2006) Organized assemblies of single wall carbon nanotubes and porphyrin for photochemical solar cells: charge injection from excited porphyrin into single-walled carbon nanotubes. *J. Phys. Chem. B*, **110**, 25477–25484.

295 Umeyama, T., Tezuka, N., Fujita, M., Hayashi, S., Kadota, N., Matano, Y., and Imahori, H. (2008) Clusterization, electrophoretic deposition, and photoelectrochemical properties of fullerene-functionalized carbon nanotube composites. *Chem. Eur. J.*, **14**, 4875–4885.

296 Brown, P., Takechi, K., and Kamat, P.V. (2008) Single-walled carbon nanotube scaffolds for dye-sensitized solar cells. *J. Phys. Chem. C*, **112**, 4776–4782.

297 Lee, W., Lee, J., Lee, S., Yi, W., Han, S.H., and Cho, B.W. (2008) Enhanced charge collection and reduced recombination of CdS/TiO$_2$ quantum-dots sensitized solar cells in the presence of single-walled carbon nanotubes. *Appl. Phys. Lett.*, **92**, 153510/1–153510/3.

298 Zhang, T., Mubeen, S., Myung, N.V., and Deshusses, M.A. (2008) Recent progress in carbon nanotube-based gas sensors. *Nanotechnology*, **19**, 332001/1–332001/14.

299 Collins, P.G., Bradley, K., Ishigami, M., and Zettl, A. (2000) Extreme oxygen sensitivity of electronic properties of carbon nanotubes. *Science*, **287**, 1801–1804.

300 Derycke, V., Martel, R., Appenzeller, J., and Avouris, P. (2001) Carbon nanotube inter-and intramolecular logic gates. *Nano Lett.*, **1**, 453–456.

301 Martel, R., Derycke, V., Lavoie, C., Appenzeller, J., Chan, K., Tersoff, J., and Avouris, P. (2001) Ambipolar electrical transport in semiconducting single-wall carbon nanotubes. *Phys. Rev. Lett.*, **87**, 256805/1–256805/4.

302 Kang, D., Park, N., Ko, J., Bae, E., and Park, W. (2005) Oxygen-induced p-type doping of a long individual single-walled carbon nanotube. *Nanotechnology*, **16**, 1048–1052.

303 Jhi, S.H., Louie, S.G., and Cohen, M.L. (2000) Electronic properties of oxidized carbon nanotubes. *Phys. Rev. Lett.*, **85**, 1710–1713.

9
Photovoltaic Devices Based on Carbon Nanotubes and Related Structures

Emmanuel Kymakis

9.1
Introduction

Carbon nanotubes (CNTs) have been integrated into organic photovoltaic cell devices both as the electron acceptor material [1] and as the transparent electrode [2]. The CNTs are attractive since they are not only efficient e-acceptors but also provide a high field at the polymer/nanotube interfaces favoring exciton dissociation. In this chapter, a review of the research conducted in this exciting field is presented.

Carbon nanotubes are allotropes of carbon with a nanostructure that can have a length-to-diameter ratio of 28 000 000 : 1 [3], which is significantly larger than any other material. They are unique one-dimensional tubular structures, basically concentric cylinders of graphene with one or more layers capped by roughly hemispherical graphite structures [4] (Figure 9.1).

These cylindrical carbon molecules have novel properties that make them potentially useful in many applications in nanotechnology, electronics, optics, and other fields of materials science, as well as potential uses in architectural fields. These exceptional properties include ballistic transport, high conductivity, which is, in fact, comparable to those of copper and/or silicon and exceeds those of any conducting polymer by several orders of magnitude [5], strong mechanical [6], thermal [7] and environmental resistance [8], and finally, optical polarizability [9]. They exhibit extraordinary strength and unique electrical properties, and are efficient conductors of heat. Their final usage, however, may be limited by their potential toxicity.

Nanotubes are members of the fullerene structural family, which also includes the spherical buckyballs. The ends of a nanotube might be capped with a hemisphere of the buckyball structure. Their name is derived from their size since the diameter of a nanotube is on the order of a few nanometers (approximately 1/50 000th of the width of a human hair), while they can be up to several millimeters in length (as of 2008). Nanotubes are categorized as single-walled nanotubes (SWNTs) and multiwalled nanotubes (MWNTs).

Carbon Nanotubes and Related Structures. Edited by Dirk M. Guldi and Nazario Martín
Copyright © 2010 Wiley-VCH Verlag GmbH & Co. KGaA, Weinheim
ISBN: 978-3-527-32406-4

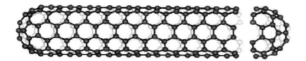

Figure 9.1 Structural formula of a carbon nanotube.

Carbon nanotubes are usually made by carbon arc discharge, laser ablation of carbon, or chemical vapor deposition [10]. Depending on the method of their synthesis, impurities in the form of catalyst particles, amorphous carbon, and nontubular fullerenes are also produced. Thus, subsequent purification steps are required to separate the tubes from other forms of nontubular carbon. This involves chemical processes such as acid reflux, filtration, centrifugation, and repeated washes with solvents and water. Typical nanotube diameters range from 0.4 to 3 nm for SWNTs and from 1.4 to at least 100 nm for MWNTs. Therefore, it is evident that the nanotube properties can be tuned by changing the diameter.

The SWNTs are presently produced only on a small scale and are extremely expensive. High-purity samples cost about \$500/g, and samples containing substantial amounts of impurities cost about \$60/g. On the other hand, MWNTs are quite inexpensive, about \$300/kg.

Many researchers have founded spin-off companies, such as Carbon Nanotechnology, Inc. (CNI) that was started by the deceased Professor Smalley of Rice University. Designed for large-scale production of high-pressure carbon monoxide (HiPCO) nanotubes, CNI produces around 10 kg a day. Other spin-offs include Professor Eklund's Carbolex, which can produce up to 35 g/day of SWNTs by arc discharge, and Professor Resasco's SouthWest NanoTechnologies. The past few years have seen a substantial increase in the number of companies producing nanotubes on commercial scale and speculative forecasting of their possible uses. For example, Nanocyl in Belgium is capable of producing 10 kg/day, NanoLedge in France produces 120 g/day, Nanothinx in Greece produces 100 g/day, and Nanocarblab in Russia is capable of 3 g/day. A moderate level of optimism, coupled with informed speculation and current market trend, suggests that nanotubes prices will significantly reduce in coming years [11].

9.2
Photovoltaic Cells Based on Carbon Nanotubes

9.2.1
Carbon Nanotubes as Electron Acceptors in Organic PVs

Most of the conjugated polymers used in organic PV devices transport holes preferentially and can be considered as p-type materials. Combining the physical and chemical characteristics of conjugated polymers with high conductivity along the tube axis of carbon nanotubes provides a great deal of incentives to disperse CNTs into the photoactive layer in order to obtain more efficient OPV devices. The interpenetrating bulk donor–acceptor heterojunction in these devices can achieve charge separation and collection because of the existence of a bicontinuous network. Along this network, electrons and holes can travel toward their respective contacts through the electron acceptor and the polymer hole donor. Photovoltaic efficiency enhancement is proposed to be due to the introduction of internal polymer/nanotube junctions within the polymer matrix. The high electric field at these junctions can split up the excitons, while the SWNT can act as a pathway for the electrons.

Since the discovery of photoinduced charge transfer between conjugated polymers (as donors) and buckminsterfullerene C_{60} and its derivatives (as acceptors), several efficient photovoltaic systems based on the donor–acceptor principle using a combination of polymer and fullerenes have been fabricated [12, 13]. Practically, the solubility and stability of both the donor and the acceptor are critically important. In general, the most successful OPV cells are those with BHJ architecture based on soluble poly(3-hexylthiophene) (P3HT) and poly(3-octylthiophene) (P3OT) as the donor and PCBM as the acceptor [14, 15]. But PCBM is not necessarily the optimum structure for solution-processed OPV materials, and the power efficiency of these OPV devices is still low compared to that of conventional inorganic devices [16].

These blends are then spin coated onto a transparent conductive electrode with thicknesses that vary from 60 to 120 nm. These conductive electrodes are usually glass covered with indium tin oxide (ITO) and a 40 nm sublayer of poly(3,4-ethylenedioxythiophene) (PEDOT) and poly(styrenesulfonate) (PSS). PEDOT and PSS help smooth the ITO surface, decreasing the density of pinholes and stifling current leakage that occurs along shunting paths. Through thermal evaporation or sputter coating, a 20–70 nm thick layer of aluminum, and sometimes an intermediate layer of lithium fluoride, is then applied onto the photoactive material.

This charge transfer mechanism in a polymer matrix containing fullerene provides the motivation for investigating the use of the longest fullerene molecule, the carbon nanotubes, as the electron transport material. The nanotubes consist of one or more sheets of graphite wrapped around each other in concentric cylinders. Individually, they could be metallic or semiconducting depending on their chirality (spiral conformation) and diameter, making them ideal reinforcing fillers in composite materials.

Multiple investigations with both multiwalled carbon nanotubes and single-walled carbon nanotubes integrated into the photoactive material have been completed [17–20]. First, Ago *et al.* [21] investigated the use of MWNTs as the cathode electrode in poly(*p*-phenylene vinylene) (PPV) photodiodes. The polymer–nanotube composite was prepared by spin coating a highly concentrated MWNT dispersion. An atomic force image showed that the PPV covers the surface of MWNTs and forms a well-mixed composite. The quantum efficiency of the device was significantly higher than the device having ITO as the electrode, suggesting that MWNT materials can act as a good cathode electrode due to the formation of a complex interpenetrating network with polymer chains. The polymer–nanotube interface quenches the radiative recombination and hence enhances the photocurrent.

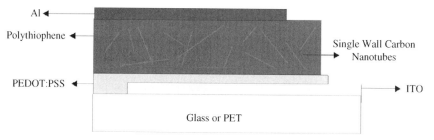

Figure 9.2 Polymer–nanotube photovoltaic cell structure.

The first report of using carbon nanotubes as e-acceptors in bulk heterojunction solar cells was from Kymakis and Amaratunga in 2002, who blended SWNTs with polythiophenes and observed an increase in the photocurrent of two orders of magnitude [17]. This effect was attributed to the formation of internal polymer/ nanotube junctions resulting in a better exciton dissociation and balance bipolar transport through the entire volume of the polymer composite.

Typical devices can be fabricated by spin coating the photoactive polymer–nanotube film from a chloroform solution onto a transparent conductive oxide (ITO) substrate, followed by evaporation of a metallic back contact, such as Al; the device structure is shown in Figure 9.2. Although it is still difficult to take their full advantage since the pristine CNTs lack solubility in any solvent, covalent and noncovalent modifications of the outer surface of CNTs have been proved to be successful methods to achieve soluble functional CNTs. The noncovalent modification methods include the wrapping of CNTs with conjugated polymers and small organic dyes [22, 23]. The covalent modification of the convex surfaces of CNTs by aliphatic molecules, organic functional molecules, atom transfer radical polymerization (ATRP) of vinyl monomers, and polyetherimide has been well documented in the literature [24–26].

The contribution of SWNTs in the photovoltaic characteristics has been thought to enhance charge separation and transport. At low doping levels, percolation pathways are established due to the high aspect ratio and surface area of SWNTs, which can provide the means for high carrier mobility and efficient charge transfer. Moreover, not only the SWNTs are good electron acceptors but they can also enhance the electron mobility in the bulk of the polymer by transporting electrons along their axis. Similar results were obtained from Landi *et al.* [27], Rodríguez *et al.* [28], and Arranz-Andrés and Blau [29], using the same device configuration.

So far, the maximum power conversion efficiency reached with these systems is about 0.22% with postfabrication thermal annealing at 120 °C for 5 min [30]. The postfabrication annealing temperature dependence on the principal cell parameters of the device such as power conversion efficiency (η), the open-circuit voltage (V_{oc}), short-circuit current density (J_{sc}), and fill factor (FF) are plotted in Figure 9.3. It is clear that the postfabrication annealing temperature of the device has a strong influence on both the photocurrent and the fill factor.

The J_{sc} of the polymer–nanotube devices is 0.25 mA/cm^2 for postfabrication temperatures between 20 and 40 °C, while postfabrication annealing at temperatures

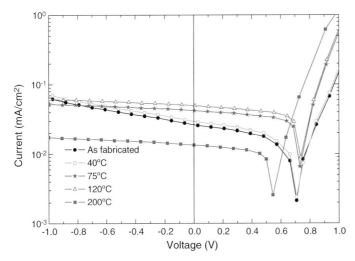

Figure 9.3 The *J–V* curves under AM 1.5 illumination for a ITO/PEDOT:PSS/P3OT-SWNT cell with different postfabrication annealing temperatures.

above 100 °C gives higher current densities of about 0.5 mA/cm². It is worth noticing that the device, which is annealed at very high temperatures, tends to deprecate and its performance, governed by J_{sc} and V_{oc}, is inferior to the nonannealed device. A similar trend is obtained for the FF dependence on temperature. The FF increases with increasing temperature in the whole temperature range, in discrepancy with J_{sc} and V_{oc}, which significantly decrease at high temperatures. The best photovoltaic characteristics are obtained with postfabrication annealing at 120 °C. The enhancement in the photovoltaic characteristics of the device can be attributed to an enhanced movement of the polymer chains (polymer mobility) and polymer structure and to a better electrical contact at the PEDOT:PSS/P3OT–SWNT interface, thereby reducing recombination losses.

Nevertheless, the major limitation so far, with respect to the fullerene-based counterparts, is the existence of an upper limit in the SWNT concentration (around 1 wt.%) for an optimal photovoltaic cell device performance [1]. Figure 9.4 depicts the short-circuit current (I_{sc}) and the open-circuit voltage of the devices for various SWNT concentrations. Average experimental values are plotted since there is a scattering in the values, possibly due to the random distribution of the nanotubes in the polymer matrix, resulting in a partial discontinuity of the donor or acceptor phases. As can be seen, the photocurrent increases very efficiently with SWNT concentration, attains a maximum at about 1% and then decreases very rapidly for even higher SWNTs concentrations. On the other hand, the behavior of V_{oc} is quite different. V_{oc} slightly increases for concentrations up to 1%, tending to saturate at even higher concentrations. Moreover, the variation in the photocurrent with concentration is quite large, a factor of 500, while the variation in V_{oc} is only at a factor of 2. In any case, the maximum efficiency was obtained from the composite containing 1% of carbon nanotubes. For higher concentrations, the photocurrent is

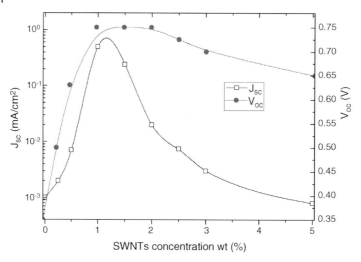

Figure 9.4 Dependence of the short-circuit current and the open-circuit voltage on the SWNT concentration for the polymer–nanotube composite.

believed to be mainly limited by a lower photogeneration rate since the exciton generation takes place only in the polymer. The maximum intensity of the solar spectrum occurs at a wavelength of about 555 nm, which falls within the band of green light. Upon increase in nanotube concentration, the distance between individual nanotubes becomes smaller than 555 nm, which results in a significant decrease in the absorption, and thus in the photogeneration rate. It can be concluded that there are sufficient interfaces to ensure efficient exciton dissociation and that there are bicontinuous conducting paths to provide percolation of electrons and holes to the appropriate electrode. However, as the nanotube content further increases, the photocurrent decreases, confirming the proportion of collected photons decreases, indicating that the nanotubes do not contribute to the photocurrent.

In our recent work, the photocurrent dependence on the SWNT concentration was directly correlated with the hole blocking process observed in the dark I–V characteristics of the photovoltaic cells [31]. As the SWNT concentration increased from 0 to 1%, the photocurrent was improved by more than two orders of magnitude; this enhancement was attributed to the introduction of internal polymer/nanotube junctions within the polymer matrix. This results in an increase in the electron mobility, and hence a balance of the hole and electron mobilities at 1%, where the maximum photocurrent was observed. The same trend was observed in the effective hole mobility where the maximum value is also obtained at 1%, indicating a direct link between the hole mobility and the photocurrent. Therefore, it can be concluded that above 1%, the efficiency of the cells is limited due to reduced photogeneration rate and hole blocking, mainly caused by the metallic nanotubes. This hole blocking arises from the presence of metallic nanotubes in the SWNTs, which enhance recombination due to their lack of direct bandgap.

Furthermore, upon increasing the SWNT concentration, it is plausible that SWNTs will be in the form of crystalline ropes that consists of bundles of nanotubes clumped in a snug parallel association due to van der Waals attractive forces. In this way, the negative impact of the metallic nanotubes is boosted since in a given bundle only one metallic nanotube is adequate for transforming the entire bundle to a quasimetallic state.

Advantages of both nanotubes and fullerenes can be harnessed with the introduction of nanotubes at low concentrations in a polymer/fullerene matrix. In such a device, an improvement of the photovoltaic performance can be obtained by the enhancement of the dissociation of excitons at the polymer/fullerene and polymer/nanotube junctions, as well as the electron transport through the nanotubes. Several groups have realized the concept by efficiently incorporating carbon nanotubes in conventional polymer–fullerene donor–acceptor photovoltaic cells [32–34].

9.2.2
Hole Collecting Electrodes

Indium tin oxide is a transparent metal, commonly used as the hole collecting electrode in optoelectronic devices such as flat-panel displays, touch screens, organic light-emitting diodes, and photovoltaic cells. Although extensively developed and optimized, ITO bears, nevertheless, a number of deficiencies. One disadvantage results from its high deposition temperature of around 600 °C, which is not compatible with roll-to-roll manufacturing methods on flexible substrates. On top of that, ITO is comparatively expensive ($400–500/kg) since it suffers from the low availability of indium, and indium is a toxic material. Also, the nonoptimal conductivity of the ITO strongly affects the series resistance of the device, limiting the efficiency. To overcome these restrictions, ITO can be replaced with a hole transport polymer, such as the poly(3,4-ethylene-dioxythiophene)–poly(styrene sulfonate) (PEDOT–PSS) [35, 36]. However, the PEDOT:PSS carrier mobility and density are quite low, resulting in conductivities not adequate for charge extraction. This problem can be effectively tackled by doping the polymers with carbon nanotubes resulting in an enhancement of conductivity by several orders of magnitude for low doping ratios [37]. Recently, several groups have come up with the idea to use a carbon nanotube mat as the transparent conductive electrode [2, 38, 39] replacing the ITO as the hole collecting electrode. The pure nanotube mats demonstrated high conductivity and transparency, and the photovoltaic characteristics were superior to the ITO.

Recent research and technological improvements made by several academic and research groups and companies have resulted in the formation of very thin layers of highly transparent conductive films based on carbon nanotube dispersions that can achieve performance comparable to most of today's commercial sources of sputtered ITO on plastic film [40, 41]. In addition, the CNT films exhibit the same degree of mechanical reliability as that of ITO and can be formed using low-temperature printing techniques, making them an attractive alternative for applications requiring much lower cost of fabrication and affording flexibility.

A wide range of methods have been probed to fabricate CNT thin films, such as spraying [42], spin coating [21], and Langmuir–Blodgett deposition [43]. The most appropriate procedure to obtain flexible and optically transparent films involves, however, the transfer from a filter membrane to the transparent support by using a solvent [44], or in the form of an adhesive film [45].

In particular, the use of CNTs as hole collecting transparent electrodes in polymer–fullerene photovoltaic cells was first demonstrated in 2006 by Du Pasquier *et al.* [2]. The photovoltaic characteristics of the devices with the CNTs were superior to the ITO ones due to the 3D connection between the nanotube percolation and the photoactive composite.

Also, most recently the groups of McGehee, Sariciftci, and Grüner [46] in a joint effort replaced the ITO with a printed CNT network. The success of this technology lies in the fact that the fabrication method combines the high quality attainable from vacuum filtration processes with the potentially low-cost patterning technique of PDMS stamp transfer resulting in transparent films with high mechanical flexibility and conductivity [47].

Unidym was founded to capitalize this promising technology targeting in a wide variety of optoelectronic applications, such as touch screens, flat-panel displays, and solar cells. Furthermore, Eikos, Inc. has developed high-transparent CNT inks for use as efficient electrodes, to impart electrical conductivity while maintaining high optical transparency in a variety of polymeric films and coatings [48]. The company claimed that they can produce CNT films comprising a polymer binder with 90% visible light transmittance and about $200\,\Omega$/sq sheet resistance. Also, cycle testing, CNT films showed no signs of cracking until about 27 000 cycles, with failure observed at about 32 000 cycles. To put the CNT performance into perspective, 25 000 cycles are roughly equivalent to 7 years of use for a rollout flexible display device flexed 10 times per day. In comparison, ITO films developed cracks that led to catastrophic failure (open circuit) at about 6000 cycles.

Last but not the least, Baughman's team at the University of Texas at Dallas engineered a way to spin carbon nanotubes into thin, ultralight sheets [49]. After chemically growing trillions of carbon nanotubes into "forests," the team in Dallas then brings them together to create larger threads, which are then woven into strips of fabric. Although very light, the scientists claimed that an acre of the material would only weigh about 4 oz, the material is very strong because of its carbon-only structure.

9.3
Related Structures

Novel materials for both donor and acceptor phases with better HOMO/LUMO matching, stronger light absorption, and higher charge mobility with good stability are much needed. However, some drawbacks, such as their insolubility, impurities, and bundling structure, have greatly hindered the device performance of carbon nanotubes.

Graphene is one of the most common allotropes of carbon. Unlike diamond, graphite is an electrical conductor, and can be used, for instance, as the material in the electrodes of an electrical arc lamp. Graphite holds the distinction of being the most stable form of carbon under standard conditions. Therefore, it is used in thermochemistry as the standard state for defining the heat of formation of carbon compounds.

Graphene is able to conduct electricity due to delocalization of the pi bond electrons above and below the planes of the carbon atoms. These electrons are free to move, so are able to conduct electricity. However, the electricity is conducted only along the plane of the layers. In diamond, all four outer electrons of each carbon atom are "localized" between the atoms in covalent bonding. The movement of electrons is restricted and diamond does not conduct an electric current. In graphene, each carbon atom uses only three of its four outer energy level electrons in covalently bonding to three other carbon atoms in a plane. Each carbon atom contributes one electron to a delocalized system of electrons that is also a part of the chemical bonding. The delocalized electrons are free to move throughout the plane. For this reason, graphene conducts electricity along the planes of carbon atoms but does not conduct in a direction at right angles to the plane.

Graphene as a very recent rising star in materials science with two-dimensional (2D) structure consisting of sp^2-hybridized carbon exhibits remarkable electronic and mechanical properties that qualify it for application in future optoelectronic devices [50]. It is a gapless semiconductor with unique electronic properties and its electron mobility reaches $10.000 \, cm^2/(V \, s)$ at room temperature [51]. Its one-atom thickness and large 2D plane lead to a large specific area, and therefore very large interfaces can be formed when it is added to a polymer matrix. Recently, such graphene thin films were used as the transparent electrode in dye-sensitized solar cells [52] and organic photovoltaic cells [53, 54].

Graphene films were used as window anode electrodes in solid-state dye-sensitized solar cells. The graphene films were prepared via either top-down exfoliation of graphite oxide flakes followed by thermal reduction or bottom-up construction of extremely large PAHs, nanographene molecules followed by thermal fusion. Furthermore, the graphene electrode compared to conventional ITO film shows high conductivity, good transparency in both the visible and the near-infrared regions, ultrasmooth surface with tunable wettability, and high chemical and thermal stabilities.

Also, the use of graphene as an electron acceptor material in organic bulk heterojunction photovoltaic cells was demonstrated [55]. Functionalized graphene was dispersed in a polythiophene matrix, and an optimum power conversion efficiency of 1.4% was observed at a graphene content of 10% upon annealing.

In a semiconductor-based device structure, electromagnetic radiation incident on the semiconductor gives rise to an optical transition rate that is proportional to the duration of the interaction between the photon and the semiconductor, that is, to the transit time of the photon per unit thickness of the semiconductor. In comparison, for an electromagnetic field in a semiconductor arising from a surface-plasmon excitation in a proximate metal nanoparticle, the duration of the interaction between the field and the semiconductor will be determined by the lifetime of the surface plasmon excitation. Recently, plasmonic enhancement of conventional P3HT:PCBM system was demonstrated in a spin-cast device with an incorporated thin Ag nanoparticle film [56, 57]. The incorporation of a plasmon-active material is shown to enhance device efficiency by enhancing the short-circuit current. Improvement in the external quantum efficiency at red wavelengths was observed to have mainly resulted from the enhanced absorption of the photoactive conjugated polymer due to the high electromagnetic field strength in the vicinity of the excited surface plasmons.

9.4
Future Directions

Although the efficiencies of the polymer solar cells significantly increase upon the introduction of nanotubes, they are, as yet, far from rivaling those of fullerene-based devices. The low photocurrent is mainly caused by the mismatch of the optical absorption of the polymer and the solar spectrum, and the fact that the nanotubes do not contribute to the photogeneration process, as the fullerenes do. This problem can be overcome by incorporating an organic dye with high absorption coefficient at the polymer–nanotube junctions, a process known as sensitization. By choosing dyes that absorb radiation over a large range, the area of photoexcitation of the diode can be broadened beyond the usual range of the polymer and can cover a high proportion of the available solar energy and at the same time introduce photoconductivity properties to the nanotubes.

Depending on SWNT length and degree of bundling, their percolation threshold can be as high as 11 wt.% [58]. In contrast, optimal SWNT solar cell performance was achieved with only 1 wt.% SWNTs [1]. This might arise from the presence of 33% metallic SWNTs within the sample, forcing us to adequately dilute them in order to minimize their undesired contribution. In photovoltaic devices, one would expect only semiconducting SWNTs at a concentration as high as the percolation threshold (for a given length and bundling level) to achieve the optimum interpenetrating network resulting in a high photocurrent. The presence of metallic nanotubes is believed to negatively impact the photovoltaic properties by providing unwanted pathways for electron/hole recombination due to its lack of direct bandgap. Therefore, employing SWNT samples enriched with semiconducting nanotubes is crucial. This will minimize the probability of finding a metallic nanotube within a given bundle of semiconducting SWNTs, which transform the entire bundle to a quasi-metallic state. Furthermore, this will enable us to increase the amount of nanotubes within the polymer matrix and although it partially reduces the interface area within the polymer and SWNTs, we improve bundling among nanotubes and enhance carrier conduction.

In this context, an effective separation route involves the selective precipitation of met- from sem-SWNTs from THF dispersions of SWNT functionalized with octadecylamine [59]. Successive precipitation of the resolubilization cycles is employed to enrich the sem-SWNTs content to a level of greater than 97%. More recently,

Papadimitrakopoulos team has used molecules from vitamin B_2 and attached them to nanotubes in such a way that could distinguish different types of nanotubes [60].

Another significant limiting factor is the random distribution of the nanotubes in the polymer matrix. As it was seen in previous investigations, for an exciton to be dissociated, it should be within a distance of 27 nm from the junction [1]. In order to tackle the random distribution network problem, an ideal device architecture for polymer/SWNT dispersed heterojunction solar cell can be made by building a vertically aligned (VA) network (instead of mixing a polymer with SWNTs in a solution to cast a film), in which the junctions are uniformly dispersed in a way that a junction exists every 27 nm within the polymer matrix. In this way, all the photogenerated excitons can be theoretically dissociated at the polymer–nanotube junctions. The percolation threshold of aligned nanotubes could be even lower than that of randomly mixed ones without having any isolated electron trapping sites. In addition, each component of the dispersed heterojunction makes contact with only one electrode, which prevents shunting between positive and negative electrodes.

References

1 Kymakis, E. and Amaratunga, G.A.J. (2005) *Rev. Adv. Mater. Sci.*, **10**, 300–305.

2 Pasquier, A.D., Unalan, H.E., Kanwal, A., Miller, S., and Chhowalla, M. (2005) *Appl. Phys. Lett.*, **87**, 203511.

3 Dresselhaus, M.S., Dresselhaus, G., and Avouris, P.M. (2001) *Carbon Nanotubes: Synthesis, Structure, Properties and Applications*, Springer-Verlag, Berlin.

4 Baughman, R.H., Zakhidov, A.A., and de Heer, W.A. (2002) *Science*, **297**, 787.

5 Mamedov, A.A., Kotov, N.A., Prato, M., Guldi, D.M., Wicksted, J.P., and Hirsch, A. (2002) *Nat. Mater.*, **1**, 190.

6 Cataldo, F. (2002) *Fullerenes, Nanotubes, Carbon Nanostruct.*, **10**, 293.

7 Sun, C.Q., Bai, H.L., Tay, B.K., Li, S., and Jiang, E.Y. (2003) *J. Phys. Chem.*, **107**, 7544.

8 Dresselhaus, M.S., Dai, H. (2004) Advances in carbon nanotubes. *MRS Bull.*, **29** (4).

9 Ajayan, P.M. (1999) *Chem. Rev.*, **99** (7), 1787.

10 Harris, P.J.F. (1999) *Carbon Nanotubes and Related Structures*, Cambridge University Press.

11 Robichaud, C.O., Tanzie, D., Weilenmann, U., and Wiesner, M.R. (2005) *Environ. Sci. Technol.*, **39** (22), 8985–8994.

12 Sariciftci, N.S., Smilowitz, L., Heeger, A.J., and Wudl, F. (1992) *Science*, **258**, 1474–1478.

13 Halls, J.J.M., Pichler, K., Friend, R.H., Moratti, S.C., and Holmes, A.B. (1996) *Appl. Phys. Lett.*, **68**, 3120–3122.

14 Kim, J.Y., Lee, K., Coates, N.E., Moses, D., Nguyen, T.Q., Dante, M., and Heeger, A.J. (2007) *Science*, **317**, 222.

15 Li, G., Shrotriya, V., Huang, J.S., Yao, Y., Moriarty, T., Emery, K., and Yang, Y. (2005) *Nat. Mater.*, **4**, 864.

16 Backer, S.A., Sivula, K., Kavulak, D.F., and Frechet, J.M.J. (2007) *Chem. Mater.*, **19**, 2927.

17 Kymakis, E. and Amaratunga, G.A.J. (2002) *Appl. Phys. Lett.*, **80** (1), 112–114.

18 Raffaelle, R.P., Landi, B.J., Castro, S.L., Ruf, H.J., Evans, C.M., and Bailey, S.G. (2005) *Sol. Energy Mater. Sol. Cells*, **87** (1–4), 733–746.

19 Kazaoui, S., Minami, N., Nalini, B., Kim, Y., and Hara, K. (2005) *J. Appl. Phys.*, **98** (8), 84314-1–84314-6.

20 Guldi, D.M., Rahman, A., Sgobba, V., and Ehli, C. (2006) *Chem. Soc. Rev.*, **35**, 165–184.

21 Ago, H., Petritsch, K., Shaffer, M.S.P., Windle, A.H., and Friend, R.H. (1999) *Adv. Mater.*, **11** (15), 1281–1285.

22 Coleman, J.N., Dalton, A.B., Curran, S., Rubio, A., Davey, A.P., Drury, A. *et al.* (2000) *Adv. Mater.*, **12**, 213–216.

23 Steuerman, D.W., Star, A., Narizzano, R., Choi, H., Ries, R.S., Nicolini, C. *et al.* (2002) *J. Phys. Chem. B*, **106**, 3124–3130.

24 Chen, J., Liu, H.Y., Weimer, W.A., Halls, M.D., Waldeck, D.H., and Walker, G.C. (2002) *J. Am. Chem. Soc.*, **124**, 9034–9035.

25 Gao, C., Vo, C.D., Jin, Y.Z., Li, W.W., and Armes, S.P. (2005) *Macromolecules*, **38**, 8634–8648.

26 Tasis, D., Tagmatarchis, N., Bianco, A., and Prato, M. (2006) *Chem. Rev.*, **106**, 1105–1136.

27 Landi, B.J., Raffaelle, R.P., Castro, S.L., and Bailey, S.G. (2005) *Prog. Photovoltaics Res. Appl.*, **13** (2), 165–172.

28 Rodríguez, L.M., Ruz, F., de la Cruz, P., Delgado, J.L., Langa, F., and Urbina, A. (2005) Proceedings of the 20th Photo Solar Energy Conference, p. 338.

29 Arranz-Andrés, J. and Blau, W.J. (2008) *Carbon*, **46** (15), 2067–2075.

30 Kymakis, E., Koudoumas, E., Franghiadakis, I., and Amaratunga, G.A.J. (2006) *J. Phys. D: Appl. Phys.*, **39**, 1058–1062.

31 Kymakis, E., Servati, P., Tzanetakis, P., Koudoumas, E., Kornilios, N., Rompogiannakis, I., Franghiadakis, Y., and Amaratunga, G.A.J. (2007) *Nanotechnology*, **18**, 435702.

32 Berson, S., De Bettignies, R., Bailly, S., Guillerez, S., and Jousselme, B. (2007) *Adv. Funct. Mater.*, **17** (16), 3363–3370.

33 Li, C., Chen, Y., Wang, Y., Iqbal, Z., Chhowalla, M., and Mitra, S. (2007) *J. Mater. Chem.*, **17** (23), 2406–2411.

34 Kymakis, E., Kornilios, N., and Koudoumas, E. (2008) *J. Phys. D: Appl. Phys.*, **41**, 165110.

35 Arias, A.C., Granstrom, M., Petritsch, K., and Friend, R.H. (1999) *Synth. Met.*, **102**, 953.

36 Zhang, F., Johansson, M., Andersson, M.R., Hummelen, J.C., and Inganäs, O. (2002) *Adv. Mater.*, **14**, 662.

37 Coleman, J.N., Curran, S., Dalton, A.B. *et al.* (1998) *Phys. Rev. B*, **58**, R7492.

38 Wu, Z., Chen, Z., Du, X., Logan, J.M., Sippel, J., Nikolou, M., Kamaras, K., Reynolds, J.R., Tanner, D.B., Hebard, A.F., and Rinzler, A.G. (2004) *Science*, **305**, 1273.

39 Kymakis, E., Stratakis, E., and Koudoumas, E. (2007) *Thin Solid Films*, **515**, 8598–8600.

40 Wu, Z.C., Chen, Z.H., Du, X., Logan, J.M., Sippel, J., Nikolou, M., Kamaras, K., Reynolds, J.R., Tanner, D.B., Hebard, A.F., and Rinzler, A.G. (2004) *Science*, **305**, 1273.

41 Hu, L., Hecht, D.S., and Grüner, G. (2004) *Nano Lett.*, **4** (12), 2513.

42 Barnes, T.M., van de Lagemaat, J., Levi, D., Rumbles, G., Coutts, T.J., Weeks, C.L., Britz, D.A., Levitsky, I., Peltola, J., and Glatkowski, P. (2007) *Phys. Rev. B*, **75**, 235410.

43 Li, X., Zhang, L., Wang, X., Shimoyama, I., Sun, X., Seo, W.-S., and Dai, H. (2007) *J. Am. Chem. Soc.*, **129**, 4890.

44 Hu, L., Hecht, D.S., and Gruner, G. (2004) *Nano Lett.*, **4**, 2513.

45 Cao, Q., Hur, S.-H., Zhu, Z.-T., Sun, Y., Wang, C., Meitl, M.A., Shim, M., and Rogers, J.A. (2006) *Adv. Mater.*, **18**, 304.

46 Rowell, M.W., Topinka, M.A., McGehee, M.D., Prall, H.-J., Dennler, G., Sariciftci, N.S., Hu, L., and Gruner, G. (2006) *Appl. Phys. Lett.*, **88**, 233506.

47 Zhou, Y., Hu, L.B., and Gruner, G. (2006) *Appl. Phys. Lett.*, **88**, 123109.

48 Van De Lagemaat, J., Barnes, T.M., Rumbles, G., Shaheen, S.E., Coutts, T.J., Weeks, C., Levitsky, I., and Glatkowski, P. (2006) *Appl. Phys. Lett.*, **88** (23), JA-590-39472.

49 Zhang, M., Fang, S., Zakhidov, A.A., Lee, S.B., Aliev, A.E., Williams, C.D., Atkinson, K.R., and Baughman, R.H. (2005) *Science*, **309** (5738), 1215.

50 Geim, A.K. and Novoselov, K.S. (2007) *Nat. Mater.*, **6**, 183.

51 Novoselov, K.S., Geim, A.K., Morozov, S.V., Jiang, D., Zhang, Y., Dubonos, S.V., Grigorieva, I.V., and Firsov, A.A. (2004) *Science*, **306**, 666.

52 Wang, X., Zhi, L., and Müllen, K. (2008) *Nano Lett.*, **8** (1), 323–327.

53 Wu, J., Becerril, H.A., Bao, Z., Liu, Z., Chen, Y., and Peumans, P. (2008) *Appl. Phys. Lett.*, **92**, 263302.

54 Eda, G., Lin, Y.Y., Miller, S., Chen, C.W., Su, W.F., and Chhowalla, M. (2008) *Appl. Phys. Lett.*, **92**, 233305.

55 Liu, Z., Liu, Q., Huang, Y., Ma, Y., Yin, S., Zhang, X., Sun, W., and Chen, Y. (2008) *Adv. Mater.*, **20** (20), 3924–3930.

56 Chen, X., Zhao, C., Rothberg, L., and Ng, M.-K. (2008) *Appl. Phys. Lett.*, **93** (12), 123302.

57 Morfa, A.J., Rowlen, K.L., Reilly, T.H., II, Romero, M.J., and Van De Lagemaat, J. (2008) *Appl. Phys. Lett.*, **92** (1), 013504.

58 Kymakis, E. and Amaratunga, G.A.J. (2006) *J. Appl. Phys.*, **99** (8), 084302.

59 Chattopadhyay, D., Galeska, I., and Papadimitrakopoulos, F. (2003) *J. Am. Chem. Soc.*, **125**, 3370–3375.

60 Ju, S.-Y., Doll, J., Sharma, I., and Papadimitrakopoulos, F. (2008) *Nat. Nanotechnol.*, **3** (6), 356–362.

10
Layer-by-Layer Assembly of Multifunctional Carbon Nanotube Thin Films

Bong Sup Shim and Nicholas A. Kotov

10.1
Introduction

Carbon nanotubes (CNTs), as part of large discipline of nanotechnology [1], receive a lot of attention from diverse spectrum of scientists due to their unique properties and potentially revolutionary applications. However, one of the key challenges for practical utilization of carbon nanotubes is their processing into macroscale composite materials retaining these unique properties specific to individual rolled graphene sheets in the resulting material to be used in the final product. This necessitates development of new methods of composite production.

Macroscale assembly of the nanomaterials requires special techniques with molecular-scale component manipulation, which is distinguished from conventional composite processing techniques such as mix-and-molding and prepreg layering. This nanoprocessing will intimately explore the chemical functionalities of the building blocks and enable nanometer-level control of organization in the super-structures. Among the different approaches that are being explored, the layer-by-layer (LBL) assembly technique stands out as one of the simplest and most versatile methods. In simplest of the cases, the technique is a method of alternating deposition of oppositely charged components from dilute solutions or dispersions on a suitable substrate. Since the first demonstration of LBL assembly for oppositely charged microparticles by Iler [2] and later on by Decher *et al.* in the 1990s for oppositely charged polyelectrolytes [3, 4], the LBL field has experienced rapid growth. The technique has soon become one of the most popular and well-established methods for the preparation of multifunctional thin films thanks not only to its simplicity but also to its robustness and versatility [5]. Introduction of hybrid organic/inorganic films has further enriched the functionality and applicability of LBL. Nearly any type of macromolecular species, including inorganic molecular clusters [6], nanoparticles [7], nanotubes and nanowires [8, 9], nanoplates [10], organic dyes [11], organic nanocrystals [12, 13], dendrimers [14], porphyrin [15], polysaccharides [16, 17], polypeptides [18], nucleic acids and DNA [19], proteins [20, 21], and viruses [22]

Carbon Nanotubes and Related Structures. Edited by Dirk M. Guldi and Nazario Martín
Copyright © 2010 Wiley-VCH Verlag GmbH & Co. KGaA, Weinheim
ISBN: 978-3-527-32406-4

can be successfully used as assembly components [23]. Remarkable versatility has further led to a number of novel designs and applications, such as superhydrophobic surfaces [24, 25], chemical sensors and semipermeable membranes [23, 26–28], drug and biomolecules delivery systems [23, 29, 30], memory devices [31], optically active and responsive films [13, 32–34], cell and protein adhesion-resistant coatings [23, 35], fuel cells and photovoltaic materials [36], biomimetic and bioresponsive coatings [25, 37], semiconductors [38, 39], catalysts [40, 41], and magnetic devices [42, 43] and many more [5, 23]. The technique has opened the door to an unlimited number of structural and functional combinations of colloids and macromolecules.

While the LBL field clearly covers a vast number of molecular species and architectures, in this review we concentrate on the state of the art in synthesis and properties of multilayer hybrid films based on CNTs. These building blocks possess unique structural and physicochemical properties, thus enabling preparation of a variety of functional composites. Moreover, carbon nanotubes are some of the only few nanomaterials that can provide general comparative standards about the efficiencies of transferring the properties of nanoblocks into organized LBL composites. The review is accordingly divided into two sections covering the two building blocks, with each section further divided into subsections covering different research areas and applications of the resulting multilayers: (i) structure and properties of CNTs; (ii) techniques for preparation of nanofilms; (iii) organization of the LBL films; (iv) organization of the adsorbed molecules; (v) functionalities of the films; (vi) and conclusions.

10.2
Structure and Properties of CNTs

CNTs are rolled-up structures of a perfect hexagonal carbon crystal molecular sheet as a tubular cylinder. As the carbon crystal sheet, a graphene, is the strongest known material and a zero-bandgap semiconductor, CNTs have shown many unique mechanical and electrical properties by the rolling direction types and the number of walls: single-walled and multiwalled carbon nanotubes (SWNTs, MWNTs). Stiffness and toughness of CNTs are arguably known as about up to 1 TPa and 300 GPa, respectively [46]. SWNTs are tough due to inward collapse and plastic deformation [47], whereas MWNTs have shown unique "sword-in-sheath" breakage pattern [46].

Electrically, a SWNT can be either semiconducting or metallic, which can be calculated by the chiral vector indices (n, m) of graphene layer [48]. Although separation of each type and production techniques of one structural kind are actively under development, overall one-third of bulk SWNTs are metallic and the rest are semiconducting. Besides the properties of each CNT, the type distribution of CNTs collectively affects the macroscale performance of composites. Thus, both the identification of CNT type and the control of the distribution ratio are critical research factors for the assembly of CNT multilayers. Raman spectroscopy is widely used to extract the information of individual CNT structures and types [49, 50]

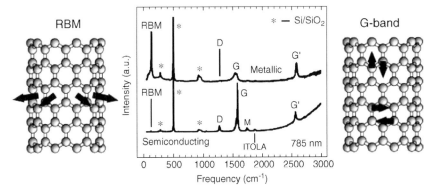

Figure 10.1 Representative image of Raman spectroscopy measurements. Left shows the radial breathing mode (RBM) and right depicts the graphene corresponding band (G-band). The middle-top and bottom spectrograms indicate the metallic and the semiconducting SWNTs, respectively [48].

(Figure 10.1). Furthermore, theoretical predictions of each type of CNTs are well documented for electronic, photonic, and phonon dispersion states in a single CNT [48, 51, 52].

Usually, the length of CNTs varies in range from micrometer to centimeter so that the extremely high aspect ratios of CNTs in nanoscale dimensions further enrich the application potentials. However, given the hydrophobic graphene and smooth crystalline walls, dispersion of CNTs still remains a daunting challenge. Collective van der Waals forces, which exert on a CNT bundle and prevent exfoliation, dramatically increase as the length increases. Thus, at present only less than 100 µm long CNTs are employed for solution processing, which is a prerequisite of LBL assemblies. Furthermore, two typical dispersion techniques of CNTs, oxidation of CNTs and interfacing with stabilizers, severely degrade or at least modify the properties of CNTs. Overall, the tipping point of LBL assembly of CNTs originates from the widespread difficulties of uniform dispersion and precise controlled structural casting into a solid composite by conventional composite processing techniques (Figure 10.2).

10.3
Structural Organization in Multilayers of Carbon Nanotubes

The primary advantages of LBL assembly of CNTs include (1) uniform dispersion of CNTs into a composite that is enabled by direct adsorption of CNTs from a solution to a solid state without phase segregation; (2) tunable multifunctional properties of a composite that is enabled by accurately controlled multicomponent nanolayers; and (3) simple, robust, and versatile processibility of CNT nanothin composite coating. The early pioneering work on CNT LBL assembly is the successful conquest over

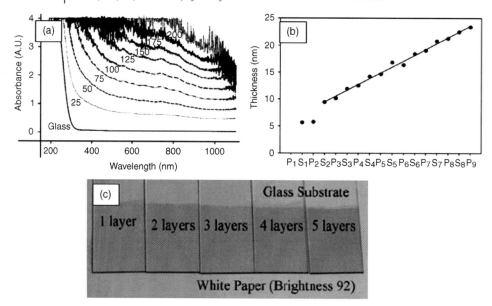

Figure 10.2 Typical LBL layering process of SWNTs monitored by (a) UV-vis light absorbance spectroscopy and (b) thickness estimation by ellipsometry. (c) Optical transparency changes by adding SWNT layers. The bumpy peaks in a spectrogram of (a) indicate van Hove singularities of SWNTs, which are an indirect evidence of both good exfoliation and dispersion [53].

dispersion challenges of CNTs in polymer composites. Mamedov *et al.* [8] reported that exceptionally uniform dispersion of SWNTs in nanothin-layered structures showed great potential in their mechanical properties even with weak polymers, which has drawn broad attention from various disciplines. Notable mechanical functionalities of the SWNT LBL composite are originated by not only the uniform dispersion but also by the high loading of CNTs and functional activated interfacial bondings between CNTs and polyelectrolyte matrix, poly(ethylene imine) (PEI). Successive experimental reports confirmed that LBL assembly of SWNTs indeed allowed exceptional exfoliation and homogeneous dispersion in a polymeric composite [54], and nanoscale stepwise deposition of CNTs showed tremendous potential for development of a wide range of functional materials [55–58]. The loading of CNTs in LBL-assembled composites can be controlled in the range from 10 [53]–75% [59], which depends on many variables of LBL assembly such as LBL polymer matrix, stabilizer of CNTs, process conditions, and so on. The layered growth by dipping is usually linear with constant slope [60, 61], and the surface roughness of each layer relies on the layer growth rate and the thickness of each layer [61] although it is limited within nanoscale range, which allows precise organization controls during CNT LBL assembly. The adsorption interaction between CNTs and polymers can be conventional electrostatic force [62], hydrogen bonds [60], or van der Waals interactions [63] and later modified to strong covalent bonds [60]. The features of LBL assembly were also effective in carbon nanomaterials of various sizes including MWNTs [55], vapor-

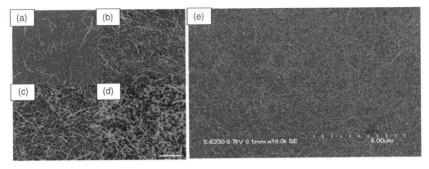

Figure 10.3 Typical microscopic images of CNT LBL assembly. (a–d) Tapping-mode AFM images of (b) 1, (b) 3, (c) 6, and (d) 9 layers of [PDDA/SWNT] LBL. The scale bar and z-heights represent 1.25 μ and 50 nm, respectively. (e) SEM images of nine layers of [PDDA/SWNT] LBL film [54].

grown hollow carbon fibers with a diameter of 50–150 nm [64], and exfoliated graphene nanoplatelet [65] (Figure 10.3).

The unusual properties of CNTs are modulated in LBL-assembled nanocomposites by mix-matching with complementary polymeric LBL partners. For their use as sensors, cell scaffolds, drug delivery carrier, fuel cell membranes, and sensors, a judicious selection of polymeric matrix in LBL assembly is required for enhancing their performance in application. The scope of macromolecules as a CNT LBL counterpart extends from conventional LBL polymers such as poly(vinyl alcohol) (PVA) [53], poly(allylamine hydrocholoride) (PAH) [66], poly(ethylene imine) (PEI) [8, 67], poly(diallyldimethyl ammonium choloride) (PDDA), and poly(acrylic acid) (PAA) [68], to functional materials such as poly(3,4-ethylenedioxythiphene)/poly (styrene sulfonate) (PEDOT/PSS) [69], poly(anilne) (PANI) [70], polypyrrole (PPy) [71, 72], light sensitive diazoresin (DR) [60], poly(viologen) derivatives [73, 74], porphyrin [74], and prussian blue (PB) [75, 76], and to biorelated materials such as blood compatible poly(lactic acid-*co*-glycolic acid) (PLGA) [77], biopolymer chitosan [76], antimicrobial lysozyme (LSZ) [78], and antisense oligodeoxynucleotide (ASODN) enzymes [79]. In the following sections, detailed examples of CNT LBL assembly classified by applications are introduced.

10.4
Electrical Conductor Applications

The combination of exceptional nanoorganization of LBL assembly and unique electrical properties of CNTs has opened up innumerable research opportunities. In 2005, Kovtyukhova *et al.* reported that LBL composites of densely distributed SWNTs have highly anisotropic electrical properties because of their layered structures [56]. The conductivity differences between in-plane and out-of-plane directions are more than a factor of 3 so that great potential for property modulation was suggested with the control of ultrathin film thickness [56, 80]. The same research

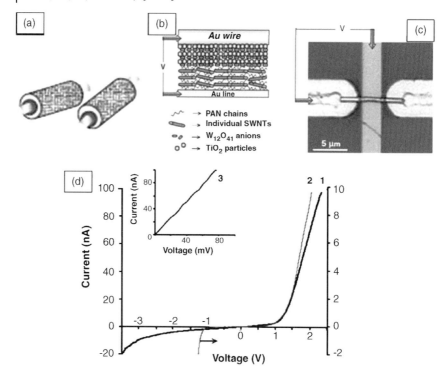

Figure 10.4 Tube-shell type p–n junction LBL on a Au nanowire. (a) schematic models of tube-shell p–n junction, (b) structural p–n junction layers, and (c) microscope image of test setup. SWNT LBL thin film provides p-type conductor. (d) *I–V* characteristics of (1) p- and n-type layer coating, (2) only n-type coating, and (3) only p-type coating on a Au nanowire [57].

team further demonstrated p–n heterojunction diodes on Au nanowires with high rectifying efficiency, which was fabricated by mixed LBL assembly of SWNTs, conductive polymer (PANI), and semiconducting nanoparticles on Au nanowires [57]. In this research, the features of SWNT LBL films were highly p-type conductive, ultrathin to coat over Au nanowires, and stable to provide junction performances even after alumina membrane dissolution. The test schematics and performances are shown in Figure 10.4. As a way to improve the electrical properties of LBL-assembled CNT nanocomposites, Shim *et al.* reported that electrical conductivities depend on micro-/nanoscale continuous conductive path specific to molecular structures of LBL composites [53]. Following this analysis, heat treatment techniques at 200 and 300 °C were suggested to form tighter SWNT connections whose *in situ* measurements of nanostructural changes were demonstrated. These treatments improved the electrical conductivities of SWNT LBL film by two orders of magnitude. Furthermore, they revealed that LBL-assembled SWNT composites were up to 10 times stronger than SWNT-only mats. The films were suggested to be used as high-performance transparent conductors (TCs), which justified the advantages of SWNT LBL composites as new class electronic materials [53]. These SWNT LBL films

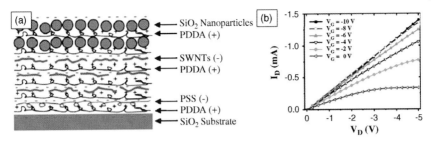

Figure 10.5 (a) Schematic presentation of molecular assembly structure of TFT with SWNT semiconducting layers and SiO$_2$ nanoparticle dielectric layers. (b) Examples of TFT performances that showed drain current–voltage characteristics by varying the gate voltage from −10 to 0 V [81].

were further demonstrated as low-cost, high-efficiency polymeric thin-film transistors (TFTs) by Xue *et al.* [81]. The same team also suggested that these micropatterned electronic devices could be fabricated on a highly flexible substrate by LBL assembly [82] (Figure 10.5).

Among the applications of these properties, TCs provide direct commercialization opportunities using precise nanoscale organization of LBL assembly of SWNTs. Yu *et al.* reported a SWNT/PDDA LBL composite with 2.5 kΩ/sq at 86.5% light transmittance [83], and they formed the similar LBL coating as a transparent surface electrode on a PVDF actuator with an excellent audio speaker performance [84]. Ham *et al.* used conductive PEDOT/PSS with SWNTs to form TCs by LBL assembly [69] although the TC performances are lower than SWNT-only mats, bucky papers. The competitive TC performances of SWNT LBL composites compared to conventional ITOs were reported by Shim *et al.*, whose TC coating with vigorous acid doping demonstrated lower than 100 Ω/sq at 80% light transmittance. Unlike other processes for SWNT nanomats, bucky papers, simple and easy-to-scale-up processibility of LBL assembly and the high mechanical integrity of SWNT LBL composites make them to be a viable alternative to TC materials of choice. As an extension of TC properties, Jain *et al.* reported that durable electrochromic devices, which usually require high transparency, high contrast, low surface roughness, accurate thickness control, and fast ion conductivity, were fabricated by LBL assembly of poly[2-(3-thienyl)ethoxy-4-butylsulfonate]/poly(allylamine hydrochloride) on SWNT electrodes [85].

10.5
Sensor Applications

A great deal of SWNT nanocomposites fabricated by LBL assembly technique were applied to physical/chemical sensors by using (1) electrical property change of CNTs by chemical species, (2) electrochemical responses of counter LBL partner of CNTs, (3) structural deformation-induced electrical resistance changes of a composite, or (4) especially designed micro-/nanoelectrical mechanical system (MEMS/NEMS)

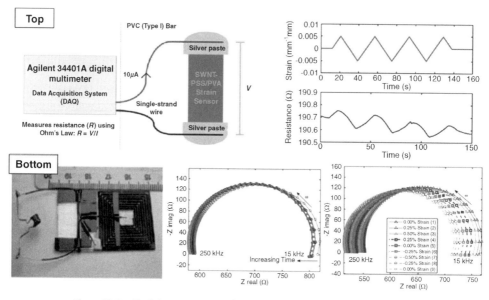

Figure 10.6 (Top) Strain-sensor application of SWNT LBL films. (Bottom) Patterned SWNT LBL films as an inductively coupled antenna used in an RFID system and their sensor performances [86].

devices. Typical examples of monitoring physical and chemical states by CNT LBL composites are demonstrated by Loh *et al.* (Figure 10.6). They reported that a strain sensor by tailored piezoresistive response [86] and a pH sensing strips sensitive enough to monitor environmental changes caused by corrosion of metal structures [70] were constructed by SWNT/PVA and SWNT/PANI LBL composites, respectively. Detection targets of SWNT LBL composites are limited only by our imagination. Biological materials such as DNA [87], glucose [71, 72, 76, 88–93], dopamine [94, 95], uric acid [94], and toxic materials such as arsenic [96] and phenols [72] were often detected by amperometric measurements of CNT LBL nanocomposites using the combinations of electrocatalytic activities, electrochemical sensitivities of CNTs, and various materials' immobilization by versatile selection of LBL assembly. Environmental changes such as humidity are easily monitored by CNT LBL composites [97].

The integration of LBL-assembled composites into MEMS/NEMS devices is a new fusion technology of top-down and bottom-up nanoprocess [44]. CNTs are also leading these frontiers owing to their unique electrical and mechanical properties. The LBL-assembled SWNT nanocomposites were patterned as flexible microcantilever arrays whose movements were demonstrated by Xue and Cui [98]. This CNT LBL microcantilever can find future applications in biosensors and microvalve for microfluidic structure channels. As another recent important progress, highly efficient nanomembrane microsensor platforms by MEMS devices were developed by SWNT LBL composites [45] following ultrathin membrane design concepts [99]. The thickness of freely suspended sensors could vary from 7 to 26 nm due to strong mechanical properties of SWNT LBL composites. Kang *et al.* claim that this freely

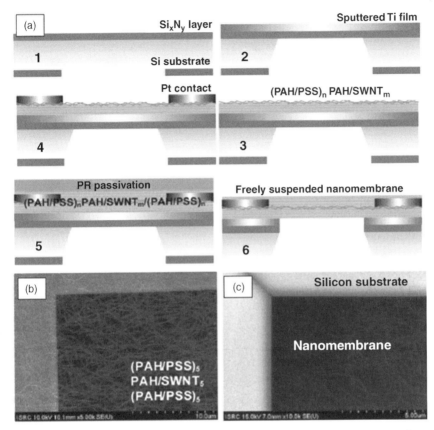

Figure 10.7 (a) Fabrication process of SWNT LBL nanomembrane arrays. (b) Top view and (c) bottom view of SEM images of MEMS structure integrated nanomembranes [116].

suspended nanomembrane array has paved the way for new nanomembrane microsensors with its exceptional sensitivity, versatile multicomponent functionality, and extreme stability (Figure 10.7).

10.6
Fuel Cell Applications

Combined with unique merit of LBL assemblies such as nanothin organized structures, controlled porosity, and free selection of ionic matrix, fuel cell proton exchange membranes (PEMs) were constructed by LBL-assembled CNT nanocomposites as highly conductive and chemically durable electrodes [100–103]. Michel *et al.* reported that CNTs wrapped with Nafion and Pt catalysts showed unusual fuel cell performances when they are LBL assembled to a fuel cell PEM, which is featured with nanoscale optimized electron/proton movements, simple, low-cost production, and efficient utilization of Pt catalysts along with mechanical strength, high electrical and high thermal conductivities, and stable chemical durability [100]. This LBL scheme is further adapted as a blending technique between proton exchange Nafion and CNTs to produce nanolayer controlled structures, to improve proton conduction efficiency at high operating temperature [104]. Interestingly, biofuel cell applications were also developed by CNT/poly-l-(lysine)/fungal laccase LBL assembly, which features electrical potential generation by stable enzyme immobilization [105].

10.7
Nano-/Microshell LBL Coatings and Biomedical Applications

The LBL assembly principles are effective not only on bulk planar surfaces but also on micro-/nanocomplex objects that may have potential applications such as in bio-medicine and biocompatible microprosthetic devices. For these complex coatings, two design schemes associated with CNTs are introduced. The first is assembling CNT LBL films over microspheres. Starting from Sano *et al.*'s initial success reports of hollow carbon microvessels in 2002 [106], efforts have been directed to improve the properties of microspherical shells such as strength, permeability, and porosity [107, 108], and to modify production process by changing types of microsphere templates [107, 109] or by adding calcination process to remove polymers [108]. Various types of microobjects can be used as templates for LBL assemblies. Coatings over soft microcapsules were developed to protect liposome by CNT-reinforced LBL composites [66]. Pan *et al.* reported that electrospun polystyrene (PS) microfibers were employed as a template of MWNT LBL composites [110]. By dissolving of PS, hollow microtubular structures of MWNT LBL nanocomposites were easily obtained.

The other types of nanoshell designs are direct LBL assemblies over CNT strands such as templates. Artyukhin *et al.* reported that SWNTs attached with ionic pyrene derivatives were successfully wrapped by polymeric LBL assemblies [111]. This technique has also been applied to produce porous indium oxide nanotubes that may be useful as toxic gas detectors [112]. Du *et al.* reported that nanotubular structures of indium oxide produced by LBL assembly over CNT strands and purified by calcinations greatly expanded their active surface areas and simultaneously improved sensitivity to NH_3 gas [112]. As other examples, stable biomolecule coatings over MWNTs were demonstrated by Liu *et al.* [113]. They formed a thick and molecular brick-like rigid β-cyclodextrin layer over an MWNT by LBL assembly techniques. Certainly, this biomolecule wrapping of CNTs was proved to be an effective tool for intracellular delivery of biomedicine [79]. Jia *et al.* reported that anticancer enzymes, ASODNs, and fluorescent labeling quantum dots coated on MWNTs by LBL assembly were successfully injected into tumor cells [79]. Thus, further development of LBL assembly over CNT nanoobjects may be needed for the future biomedicine or tissue engineering application.

Along with above-mentioned structural controls, biomedical functionality and cell interface compatibility tests were performed with CNT LBL nanocomposites. Following cellular adhesion experiments [114], Gheith *et al.* reported that neural cells and their neurite growth were enhanced by electrical stimulus through CNT LBL

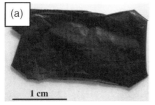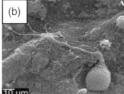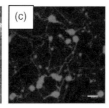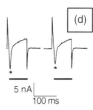

Figure 10.8 (a) Example of freestanding LBL nanofilms used for evaluating neural excitation. (b) SEM and (c) confocal microscopy images of an NG1-8-15 neural cell adhered to SWNT LBL film. (d) Two independent representative current traces (over time) from NG108-15 cells stimulated by extrinsic current passed through the SWNT film (100 ms, 1 Hz) and recorded in whole-cell voltage clamp mode. The current through SWNT film results in rapid inward currents (asterisks) through neural membrane indicative of Na^+ ion currents specific to neural excitation and action potential.

composites [68], which was the first observation of cellular interactions with CNT materials by electrical potential (Figure 10.8). Furthermore, these neural cell tests were extended to the differentiation of embryonic neural stem cells on CNT LBL nanocomposites [67]. These live cell interface experiments corroborate that CNT LBL composites are indeed one of the viable functional material options for biomedical implant, prosthetic devices, and stem cell growth platforms.

Another type of biocompatible and blood compatible surface tests were performed on LBL-assembled CNT/PLGA nanocomposites for thromboresistance suppression in foreign artificial blood prostheses [77]. In contrast to this biocompatible, antifouling or antimicrobial coating provides good application examples of LBL assembly. Nepal *et al.* designed LBL assembly of antimicrobial lysozyme with CNTs for improving mechanical properties of surfaces [78]. The nanoindentation measurement, however, needs to be made with care while determining the Young's modulus and hardness of LBL film because the substrate effect is quite significant in a thin film. The suggestion of valid measurement condition is no dipper than 10% of the total thickness as indicated by Pavoor *et al.* [115].

10.8
Conclusions

LBL technology offers an interesting possibility to control the structure of composite materials on nanometer scale with a high degree of accuracy. This is the pathway to attaining the record properties of materials and some of them have already been demonstrated. This fact has fundamental significance because it allows one to establish the design rules for composite structures initially on a small scale and then to replicate them on a larger scale probably with other methods of composite manufacturing. Some of the future challenges that we see in this field include the following: (1) scale-up of the LBL materials and development of new methods of deposition that combine a high degree of structural control and speed. Substantial effort in this direction is being undertaken. (2) Realization of greater mechanical properties in CNT composites. At present, the ideal stress transfer was realized only for clay platelets [116]. CNTs will require much greater research effort to achieve such conditions. This challenge will probably imply a substantial change in understanding how the deformations should be transferred through different interfaces. (3) Wider utilization of the properties of the described composites in the biological arena. At present, we are just in the beginning of the development of a new generation of composite biomaterials made by LBL technique that are likely to provide enabling technologies for applications in medicine, from implants to imaging.

References

1 Talapin, D.V. (2008) *ACS Nano*, **2**, 1097.
2 Iler, R.K. and Colloid, J. (1966) *J. Colloid Interface Sci.*, **21**, 569.
3 Decher, G. and Schmitt, J. (1992) *Prog. Colloid Polym. Sci.*, **89**, 160.
4 Decher, G. (1997) *Science*, **277**, 1232.
5 Hammond, P.T. (2004) *Adv. Mater.*, **16**, 1271.
6 Ingersoll, D., Kulesza, P.J., and Faulkner, L.R. (1994) *J. Electrochem. Soc.*, **141**, 140.
7 Kotov, N.A., Dekany, I., and Fendler, J.H. (1995) *J. Phys. Chem.*, **99**, 13065.
8 Mamedov, A.A., Kotov, N.A., Prato, M., Guldi, D.M., Wicksted, J.P., and Hirsch, A. (2002) *Nat. Mater.*, **1**, 190.
9 Jiang, C., Ko, H., and Tsukruk, V.V. (2005) *Adv. Mater.*, **17**, 2127.
10 Keller, S.W., Kim, H.N., and Mallouk, T.E. (1994) *J. Am. Chem. Soc.*, **116**, 8817.
11 Cooper, T.M., Campbell, A.L., and Crane, R.L. (1995) *Langmuir*, **11**, 2713.
12 Podsiadlo, P., Choi, S.Y., Shim, B., Lee, J., Cuddihy, M., and Kotov, N.A. (2005) *Biomacromolecules*, **6**, 2914.
13 Podsiadlo, P., Sui, L., Elkasabi, Y., Burgardt, P., Lee, J., Miryala, A., Kusumaatmaja, W., Carman, M.R., Shtein, M., Kieffer, J., Lahann, J., and Kotov, N.A. (2007) *Langmuir*, **23**, 7901.
14 He, J.A., Valluzzi, R., Yang, K., Dolukhanyan, T., Sung, C., Kumar, J., Tripathy, S.K., Samuelson, L., Balogh, L., and Tomalia, D.A. (1999) *Chem. Mater.*, **11**, 3268.
15 Araki, K., Wagner, M.J., and Wrighton, M.S. (1996) *Langmuir*, **12**, 5393.
16 Lvov, Y., Onda, M., Ariga, K., and Kunitake, T. (1998) *J. Biomater. Sci., Polym. Ed.*, **9**, 345.

17 Richert, L., Lavalle, P., Vautier, D., Senger, B., Stoltz, J.F., Schaaf, P., Voegel, J.C., and Picart, C. (2002) *Biomacromolecules*, **3**, 1170.
18 Boulmedais, F., Ball, V., Schwinte, P., Frisch, B., Schaaf, P., and Voegel, J.C. (2003) *Langmuir*, **19**, 440.
19 Lvov, Y., Decher, G., and Sukhorukov, G. (1993) *Macromolecules*, **26**, 5396.
20 Hong, J.D., Lowack, K., Schmitt, J., and Decher, G. (1993) *Prog. Colloid Polym. Sci.*, **93**, 98.
21 Lvov, Y., Ariga, K., and Kunitake, T. (1994) *Chem. Lett.*, 2323.
22 Yoo, P.J., Nam, K.T., Qi, J., Lee, S.K., Park, J., Belcher, A.M., and Hammond, P.T. (2006) *Nat. Mater.*, **5**, 234.
23 Tang, Z., Wang, Y., Podsiadlo, P., and Kotov, N.A. (2006) *Adv. Mater.*, **18**, 3203.
24 Zhai, L., Cebeci, F.C., Cohen, R.E., and Rubner, M.F. (2004) *Nano Lett.*, **4**, 1349.
25 Zhai, L., Berg, M.C., Cebeci, F.C., Kim, Y., Milwid, J.M., Rubner, M.F., and Cohen, R.E. (2006) *Nano Lett.*, **6**, 1213.
26 Ellis, D.L., Zakin, M.R., Bernstein, L.S., and Rubner, M.F. (1996) *Anal. Chem.*, **68**, 817.
27 Constantine, C.A., Mello, S., Dupont, V.A., Cao, X., Santos, D., Jr., Oliveira, O.N., Jr., Strixino, F.T., Pereira, E.C., Cheng, T., Defrank, J.J., and Leblanc, R.M. (2003) *J. Am. Chem. Soc.*, **125**, 1805.
28 Koktysh, D.S., Liang, X., Yun, B.G., Pastoriza-Santos, I., Matts, R.L., Giersig, M., Serra-Rodriguez, C., Liz-Marzan, L.M., and Kotov, N.A. (2002) *Adv. Funct. Mater.*, **12**, 255.
29 Wood, K.C., Chuang, H.F., Batten, R.D., Lynn, D.M., and Hammond, P.T. (2006) *Proc. Natl. Acad. Sci. USA*, **103**, 10207.

30 Jewell, C.M., Zhang, J., Fredin, N.J., and Lynn, D.M. (2005) *J. Control Release*, **106**, 214.

31 Lee, J.S., Cho, J., Lee, C., Kim, I., Park, J., Kim, Y.M., Shin, H., Lee, J., and Caruso, F. (2007) *Nat. Nanotechnol.*, **2**, 790.

32 Hiller, J., Mendelsohn, J.D., and Rubner, M.F. (2002) *Nat. Mater.*, **1**, 59.

33 DeLongchamp, D.M. and Hammond, P.T. (2004) *Adv. Funct. Mater.*, **14**, 224.

34 Moriguchi, I. and Fendler, J.H. (1998) *Chem. Mater.*, **10**, 2205.

35 Heuberger, R., Sukhorukov, G., Voeroes, J., Textor, M., and Moehwald, H. (2005) *Adv. Funct. Mater.*, **15**, 357.

36 Tokuhisa, H. and Hammond, P.T. (2003) *Adv. Funct. Mater.*, **13**, 831.

37 Zhang, J., Senger, B., Vautier, D., Picart, C., Schaaf, P., Voegel, J.C., and Lavalle, P. (2005) *Biomaterials*, **26**, 3353.

38 Mamedov, A.A., Belov, A., Giersig, M., Mamedova, N.N., and Kotov, N.A. (2001) *J. Am. Chem. Soc.*, **123**, 7738.

39 Wang, D., Rogach, A.L., and Caruso, F. (2002) *Nano Lett.*, **2**, 857.

40 Liu, J., Cheng, L., Song, Y., Liu, B., and Dong, S. (2001) *Langmuir*, **17**, 6747.

41 Shen, Y., Liu, J., Jiang, J., Liu, B., and Dong, S. (2003) *J. Phys. Chem. B*, **107**, 9744.

42 Nolte, A.J., Rubner, M.F., and Cohen, R.E. (2004) *Langmuir*, **20**, 3304.

43 Mamedov, A.A. and Kotov, N.A. (2000) *Langmuir*, **16**, 5530.

44 Hua, F., Cui, T.H., and Lvov, Y.M. (2004) *Nano Lett.*, **4**, 823.

45 Kang, T.J., Cha, M., Jang, E.Y., Shin, J., Im, H.U., Kim, Y., Lee, J., and Kim, Y.H. (2008) *Adv. Mater.*, **20** (16), 3131–3137.

46 Yu, M.F., Lourie, O., Dyer, M.J., Moloni, K., Kelly, T.F., and Ruoff, R.S. (2000) *Science*, **287**, 637.

47 Calvert, P. (1999) *Nature*, **399**, 210.

48 Dresselhaus, M.S., Dresselhaus, G., and Jorio, A. (2004) *Annu. Rev. Mater. Res.*, **34**, 247.

49 Jorio, A., Saito, R., Hafner, J.H., Lieber, C.M., Hunter, M., McClure, T., Dresselhaus, G., and Dresselhaus, M.S. (2001) *Phys. Rev. Lett.*, **86**, 1118.

50 Rao, A.M., Richter, E., Bandow, S., Chase, B., Eklund, P.C., Williams, K.A., Fang, S., Subbaswamy, K.R., Menon, M., Thess, A., Smalley, R.E., Dresselhaus, G., and Dresselhaus, M.S. (1997) *Science*, **275**, 187.

51 Fantini, C., Jorio, A., Souza, M., Strano, M.S., Dresselhaus, M.S., and Pimenta, M.A. (2004) *Phys. Rev. Lett.*, **93**.

52 Dresselhaus, M.S. and Eklund, P.C. (2000) *Adv. Phys.*, **49**, 705.

53 Shim, B.S., Tang, Z., Morabito, M.P., Agarwal, A., Hong, H., and Kotov, N.A. (2007) *Chem. Mater.*, **19**, 5467.

54 Rouse, J.H. and Lillehei, P.T. (2003) *Nano Lett.*, **3**, 59.

55 Olek, M., Ostrander, J., Jurga, S., Moehwald, H., Kotov, N., Kempa, K., and Giersig, M. (2004) *Nano Lett.*, **4**, 1889.

56 Kovtyukhova, N.I. and Mallouk, T.E. (2005) *J. Phys. Chem. B*, **109**, 2540.

57 Kovtyukhova, N.L. and Mallouk, T.E. (2005) *Adv. Mater.*, **17**, 187.

58 Rouse, J.H., Lillehei, P.T., Sanderson, J., and Siochi, E.J. (2004) *Chem. Mater.*, **16**, 3904.

59 Xue, W. and Cui, T.H. (2007) *Nanotechnology*, **18**.

60 Shi, J.H., Qin, Y.J., Luo, H.X., Guo, Z.X., Woo, H.S., and Park, D.K. (2007) *Nanotechnology*, **18**.

61 Paloniemi, H., Lukkarinen, M., Aaritalo, T., Areva, S., Leiro, J., Heinonen, M., Haapakka, K., and Lukkari, J. (2006) *Langmuir*, **22**, 74.

62 Shen, J.F., Hu, Y.Z., Qin, C., and Ye, M.X. (2008) *Langmuir*, **24**, 3993.

63 Moya, S.E., Ilie, A., Bendall, J.S., Hernandez-Lopez, J.L., Ruiz-Garcia, J., and Huck, W.T.S. (2007) *Macromol. Chem. Phys.*, **208**, 603.

64 Shim, B.S., Starkovich, J., and Kotov, N. (2006) *Compos. Sci. Technol.*, **66**, 1174.

65 Hendricks, T.R., Lu, J., Drzal, L.T., and Lee, I. (2008) *Adv. Mater.*, **20**, 2008.

66 Angelini, G., Boncompagni, S., De Maria, P., De Nardi, M., Fontana, A., Gasbarri, C., and Menna, E. (2007) *Carbon*, **45**, 2479.

67 Jan, E. and Kotov, N.A. (2007) *Nano Lett.*, **7**, 1123.

68 Gheith, M.K., Pappas, T.C., Liopo, A.V., Sinani, V.A., Shim, B.S., Motamedi, M., Wicksted, J.P., and Kotov, N.A. (2006) *Adv. Mater.*, **18**, 2975.

69 Ham, H.T., Choi, Y.S., Chee, M.G., Cha, M.H., and Chung, I.J. (2008) *Polym. Eng. Sci.*, **48**, 1.

70 Loh, K.J., Kim, J., Lynch, J.P., Kam, N.W.S., and Kotov, N.A. (2007) *Smart Mater. Struct.*, **16**, 429.

71 Shirsat, M.D., Too, C.O., and Wallace, G.G. (2008) *Electroanalysis*, **20**, 150.

72 Korkut, S., Keskinler, B., and Erhan, E. (2008) *Talanta*, **76**, 1147.

73 Wang, X., Huang, H.X., Liu, A.R., Liu, B., Wakayama, T., Nakamura, C., Miyake, J., and Qian, D.J. (2006) *Carbon*, **44**, 2115.

74 Wang, X., Huang, H.X., Liu, A.R., Liu, B., Chen, M., and Qian, D.J. (2008) *Thin Solid Films*, **516**, 3244.

75 Wang, L., Guo, S.J., Hu, X.O., and Dong, S.J. (2008) *Colloid Surf. A: Physicochem. Eng. Aspects*, **317**, 394.

76 Zou, Y.J., Xian, C.L., Sun, L.X., and Xu, F. (2008) *Electrochim. Acta*, **53**, 4089.

77 Koh, L.B., Rodriguez, I., and Zhou, J.J. (2008) *J. Biomed. Mater. Res. A*, **86A**, 394.

78 Nepal, D., Balasubramanian, S., Simonian, A.L., and Davis, V.A. (2008) *Nano Lett.*, **8**, 1896.

79 Jia, N.Q., Lian, Q., Shen, H.B., Wang, C., Li, X.Y., and Yang, Z.N. (2007) *Nano Lett.*, **7**, 2976.

80 Palumbo, M., Lee, K.U., Ahn, B.T., Suri, A., Coleman, K.S., Zeze, D., Wood, D., Pearson, C., and Petty, M.C. (2006) *J. Phys. D: Appl. Phys.*, **39**, 3077.

81 Xue, W., Liu, Y., and Cui, T.H. (2006) *Appl. Phys. Lett.*, **89**.

82 Xue, W. and Cui, T.H. (2008) *Sens. Actuators A Phys.*, **145**, 330.

83 Yu, X., Rajamani, R., Stelson, K.A., and Cui, T. (2008) *Surf. Coat. Technol.*, **202**, 2002.

84 Yu, X., Rajamani, R., Stelson, K.A., and Cui, T. (2006) *Sens. Actuators A Phys.*, **132**, 626.

85 Jain, V., Yochum, H.M., Montazami, R., Heflin, J.R., Hu, L.B., and Gruner, G. (2008) *J. Appl. Phys.*, **103**.

86 Loh, K.J., Lynch, J.P., Shim, B.S., and Kotov, N.A. (2008) *J. Intell. Mater. Syst. Struct.*, **19**, 747.

87 Ma, H.Y., Zhang, L.P., Pan, Y., Zhang, K.Y., and Zhang, Y.Z. (2008) *Electroanalysis*, **20**, 1220.

88 Yan, X.B., Chen, X.J., Tay, B.K., and Khor, K.A. (2007) *Electrochem. Commun.*, **9**, 1269.

89 Wu, B.Y., Hou, S.H., Yin, F., Zhao, Z.X., Wang, Y.Y., Wang, X.S., and Chen, Q. (2007) *Biosens. Bioelectron.*, **22**, 2854.

90 Sun, Y.Y., Wang, H.Y., and Sun, C.Q. (2008) *Biosens. Bioelectron.*, **24**, 22.

91 Liu, Y., Wu, S., Ju, H.X., and Xu, L. (2007) *Electroanalysis*, **19**, 986.

92 Qu, F.L., Yang, M.H., Chen, J.W., Shen, G.L., and Yu, R.Q. (2006) *Anal. Lett.*, **39**, 1785.

93 Zhang, J., Feng, M., and Tachikawa, H. (2007) *Biosens. Bioelectron.*, **22**, 3036.

94 Zhang, Y.Z., Pan, Y., Sit, S., Zhang, L.P., Li, S.P., and Shao, M.W. (2007) *Electroanalysis*, **19**, 1695.

95 Zhang, M.N., Gong, K.P., Zhang, H.W., and Mao, L.Q. (2005) *Biosens. Bioelectron.*, **20**, 1270.

96 Liu, Y.X. and Wei, W.Z. (2008) *Electrochem. Commun.*, **10**, 872.

97 Yu, H.H., Cao, T., Zhou, L.D., Gu, E.D., Yu, D.S., and Jiang, D.S. (2006) *Sens. Actuators B Chem.*, **119**, 512.

98 Xue, W. and Cui, T.H. (2007) *Sens. Actuators A Phys.*, **136**, 510.

99 Jiang, C., Markutsya, S., Pikus, Y., and Tsukruk, V.V. (2004) *Nat. Mater.*, **3**, 721.

100 Michel, M., Taylor, A., Sekol, R., Podsiadlo, P., Ho, P., Kotov, N., and Thompson, L. (2007) *Adv. Mater.*, **19**, 3859.

101 Yuan, J.H., Wang, Z.J., Zhang, Y.J., Shen, Y.F., Han, D.X., Zhang, Q., Xu, X.Y., and Niu, L. (2008) *Thin Solid Films*, **516**, 6531.

102 Shi, J., Hu, Y.Q., and Hua, Y.X. (2008) *Electroanalysis*, **20**, 1483.

103 Wang, L., Guo, S.J., Huang, L.J., and Dong, S.J. (2007) *Electrochem. Commun.*, **9**, 827.

104 Chen, W.F., Wu, J.S., and Kuo, P.L. (2008) *Chem. Mater.*, **20**, 5756.

105 Deng, L., Shang, L., Wang, Y.Z., Wang, T., Chen, H.J., and Dong, S.J. (2008) *Electrochem. Commun.*, **10**, 1012.

106 Sano, M., Kamino, A., Okamura, J., and Shinkai, S. (2002) *Nano Lett.*, **2**, 531.

107 Ji, L.J., Ma, J., Zhao, C.G., Wei, W., Ji, L.J., Wang, X.C., Yang, M.S., Lu, Y.F., and Yang, Z.Z. (2006) *Chem. Commun.*, 1206.

108 Shi, J.H., Chen, Z.Y., Qin, Y.J., and Guo, Z.X. (2008) *J. Phys. Chem. C*, **112**, 11617.

109 Kim, B.S., Kim, B., and Suh, K.D. (2008) *J. Polym. Sci. A: Polym. Chem.*, **46**, 1058.

110 Pan, C., Ge, L.Q., and Gu, Z.Z. (2007) *Compos. Sci. Technol.*, **67**, 3271.

111 Artyukhin, A.B., Bakajin, O., Stroeve, P., and Noy, A. (2004) *Langmuir*, **20**, 1442.

112 Du, N., Zhang, H., Chen, B.D., Ma, X.Y., Liu, Z.H., Wu, J.B., and Yang, D.R. (2007) *Adv. Mater.*, **19**, 1641.

113 Liu, K.S., Fu, H.G., Xie, Y., Zhang, L.L., Pan, K., and Zhou, W. (2008) *J. Phys. Chem. C*, **112**, 951.

114 Gheith, M.K., Sinani, V.A., Wicksted, J.P., Matts, R.L., and Kotov, N.A. (2005) *Adv. Mater.*, **17**, 2663.

115 Pavoor, P.V., Bellare, A., Strom, A., Yang, D.H., and Cohen, R.E. (2004) *Macromolecules*, **37**, 4865.

116 Podsiadlo, P., Kaushik, A.K., Arruda, E.M., Waas, A.M., Shim, B.S., Xu, J., Nandivada, H., Pumplin, B.G., Lahann, J., Ramamoorthy, A., and Kotov, N.A. (2007) *Science*, **318**, 80.

11
Carbon Nanotubes for Catalytic Applications

Eva Castillejos and Philippe Serp

11.1
Introduction

Filamentous carbons have been produced for long, the first reports dating back to more than a century ago, from hydrocarbons in metallic crucibles that were unintentionally acting as catalysts [1]. Thus, catalysis constitutes an important aspect of carbon nanotube synthesis (see Chapter 1), and scientists from the catalysis community have been involved in the mechanisms of filamentous carbon formation for decades. Indeed, fibrous carbon was considered an unwanted by-product of reactions involving conversion of carbon-containing gases and all efforts were focused on how to prevent their formation, which was identified as one of the reasons of catalyst deactivation and reactor degradation in the process involving carbon monoxide or hydrocarbon decomposition [2]. In addition, studies on vapor-grown carbon fibers have fixed the basis of carbon nanotube (CNT) catalytic growth mechanisms [3]. Basically, catalytic filamentous carbons can be classified into two categories, the tubes, both single-walled (SWCNTs) and multi-walled (MWCNTs), which present a hollow cavity, and the carbon nanofibers (CNFs), which do not. In this chapter, we will discuss CNTs possessing a tubular morphology with one or several concentric graphene layers and also CNFs for the other structures.

The use of CNTs for catalysis, as catalyst support or directly as catalyst, is a relatively new topic of investigation since the first report in 1993 [4]. Since then, attention has been paid to the preparation of metal particles on CNTs or CNFs [5, 6], and specific activities/selectivity of CNT or carbon nanofiber-supported catalysts were reported [7–11]. At present, the main area of interest in this domain are (i) the involvement of CNT/CNF-supported catalysts in liquid-phase reactions for fine chemicals; (ii) electrocatalysis for fuel cells; and (iii) the production of specific materials for gas sensors [12].

In this chapter, we will highlight the main features and specific characteristics that make carbon nanotubes attractive for heterogeneous catalysis and will exemplify this with the most relevant studies in the literature.

Carbon Nanotubes and Related Structures. Edited by Dirk M. Guldi and Nazario Martín
Copyright © 2010 WILEY-VCH Verlag GmbH & Co. KGaA, Weinheim
ISBN: 978-3-527-32406-4

11.2
Macroscopic shaping of CNTs

Pure CNTs are generally obtained as a "black fluffy powder," composed of microscopic aggregates of entangled carbon nanotubes, presenting a high mesoporosity, a low bulk density, and low mechanical strength [13]. Such a mesoporous support offers interesting opportunities for the dispersion of a catalytically active phase, and for both gas-phase and slurry-phase catalysis. SEM and TEM micrographs of a Pt/CNT powder are shown in Figure 11.1. However, CNTs in a powder form suffer from some drawbacks; for example, their handling is hazardous in large-scale use. For slurry-phase operations, the dispersion, agglomeration, and filtration of CNT agglomerates can drastically impact the catalytic performances. For gas-phase applications, their low-density results in a low catalyst mass/reactor volume ratio and can cause uncontrolled pressure drop along the catalyst bed. It is also important that supported catalysts do not fracture or attrite during use because such fragments may become entrained in the reaction stream and must then be separated from the reaction mixture. Different solutions have been proposed in order to overcome these problems.

The first approach is based on the shaping of the CNT aggregates. CNT mats possessing interesting mechanical property, thermal conductivity, and gas permeability consists of self-supporting entangled assemblies of CNTs that are prepared by careful filtration and drying of CNT suspensions [14]. Rigid porous structures, such as pellets, can be produced by extruding a paste-like suspension of functionalized nanotubes or a mixture of as-made aggregates and gluing agent followed by a calcination step to drive off conveying liquids and either cross-linking the functionalized nanotubes or pyrolizing the gluing agent [15]. Interestingly, the metal dispersion and specific surface area in the CNT extrudates is equal to or greater

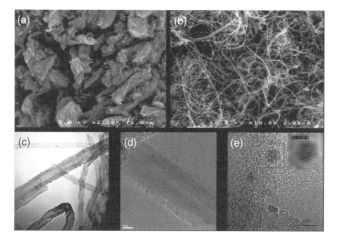

Figure 11.1 SEM and TEM micrographs of Pt/CNT powders.

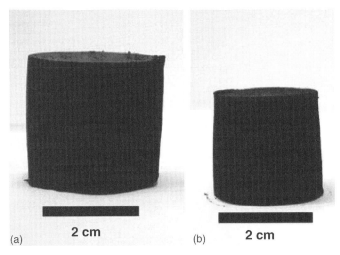

Figure 11.2 Self-supported MWCNT monolith (a) after wetting and (b) after drying. Reprinted with permission from Ref. [17]. Copyright 2006 Elsevier.

than that in the original metal-loaded carbon nanotubes. Alternatively, strong and electrically conducting CNT aerogels can be produced from wet gel precursor [16].

The second approach does not involve postproduction treatment and is rather based on specific CNT production processes. CNT monoliths with macroscopic shape (Figure 11.2) have been produced by constraint synthesis; allowing a significant increase in the apparent density from 20 kg/m^3 for CNT powder to 200 kg/m^3 for the self-supported CNTs [17]. Aggregates with controlled particle sizes (0.6–0.8 mm), high bulk density (500–900 kg/m^3), and high mechanical strength (crushing strength 1.25 MPa) have been prepared from high-loading Ni/SiO$_2$ catalysts. The so obtained macroscopic particles consist of submillimetric granules characterized by dense CNF networks, which can be purified by removing the Ni/SiO$_2$ catalyst [13, 18]. Dense arrays of vertically aligned CNTs have been prepared by CVD on spherical supports (Figure 11.3) [19, 20]. The use of macroscopic supports, such as graphite felt (Figure 11.4) [21], carbon cloth [22], carbon fiber fabric [23], carbon foam [24], woven glass [25] or silica microfibers [26], Ni foams [27], and cordierite monoliths wash-coated with alumina [28], for the growth of CNTs has also been explored to prepare structured and resistant catalyst supports. Finally, the design of structured (micro) reactors incorporating aligned CNTs offers some interesting perspectives on chemical synthesis [29, 30]. In particular, improvements in catalytic activity and stability have been reported for Pt/CNT-incorporated microreactor [30].

11.3
Specific Metal–Support Interaction

In catalysis, the electrical conductivity of CNTs, the presence of structural defects, the CNT helicity, the curvature of the surface of these materials, and the presence of

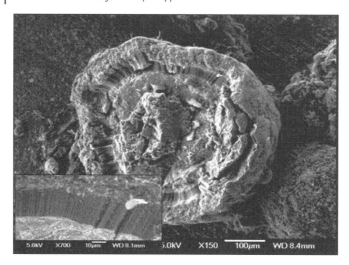

Figure 11.3 SEM images of aligned CNTs prepared by CVD on spherical supports.

an inner cavity (see Section 11.7) are expected to affect the metal–support interaction, possibly in a different manner if compared to classical carbon support such as activated carbon (AC) or graphite.

Thus, theoretical studies have shown that the curvature of the surface has an impact on both the nature of the binding sites for the metal [31, 32] and the surface diffusion of metallic adatoms [33, 34]. Studies conducted on nickel have shown that the nature of the most stable anchoring sites (Figure 11.5) varies sensibly passing from graphite (atop site) to SWCNTs (bridge site) due to the different curvature of the surfaces where the active species are deposited [31]. For a single platinum atom, DFT analysis shows that Pt binds stronger to the outer wall of small-diameter CNTs, compared to larger nanotubes [32]. Theoretical calculations have also demonstrated

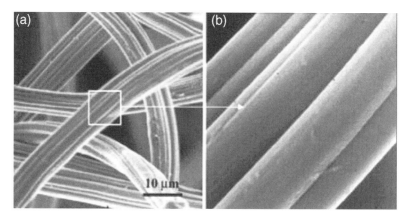

Figure 11.4 SEM images of the microfilaments that constitute the starting macroscopic graphite felt. Reprinted with permission from Ref. [21]. Copyright 2005 Elsevier.

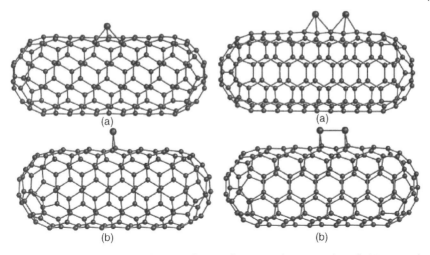

Figure 11.5 The two stable binding sites for a single Ni on carbon nanotube wall: (a) atop and atop–atop site, (b) bridge and bridge–bridge site. Reprinted with permission from Ref. 31. Copyright 2000 Elsevier.

that adatoms surface diffusion is affected both by the curvature and the helicity of SWCNTs; adatoms diffuse more easily along the axial direction for armchair SWCNTs and/or on CNT concave surface than for zigzag SWCNTs and/or on CNT convex surface [34]. The formation of silver clusters from silver adatoms on CNTs has been experimentally studied to infer the curvature effect on surface diffusion [33]. The stronger dependence of curvature effect has been noticed for small-diameter CNTs.

The shape of metallic nanoparticles (NPs) also depends on the adsorption, and it has been proposed that metal nanoparticles grown on CNT surfaces may undergo a bending deformation along the transverse direction to the CNT because of the nanoparticle–CNT interaction and the large curvature of small-diameter CNTs [35]. Finally, it has been demonstrated that mechanically bent SWCNTs possess kink sites that are chemically more reactive than the others [36].

If these latter studies consider adsorption or diffusion of metals on defect-free CNTs, other studies have emphasized the crucial role of surface defects, such as Stone Wales defects, vacancies or derivative point defects, in metal adsorption [37–39], or of support morphology in surface diffusion [40]. Thus, alkali metal atoms preferably interact with the pentagonal and heptagonal carbon rings of Stone Wales defects than with the perfect graphene [39]. Other theoretical calculations have shown that the binding energy between 3d transition metals and CNTs is significantly enhanced when vacancy sites are present, compared to defect-free CNTs [37, 40] or Stone Wales defects [40]. The mobility, morphology, and melting of Pt nanoparticles have been compared theoretically for a single graphite sheet and a bundle of SWCNTs [40]. The structure of Pt nanoparticles was more disordered on SWCNTs, and the predicted diffusion coefficient was the lowest on this very support, indicating that nanotube-supported nanoparticles may undergo less sintering than those on graphite. The wetting of the carbon surface by a transition metal is also determined by the concentration and nature of the defects in the carbon material [41]. Thus, SWCNTs, which contain a low concentration of defects, are difficult to be wetted, whereas MWCNTs or AC are much easily wetted by transition metals. The chemical reactivity of open-tip CNTs has also to be considered, as stable tip structures can be obtained by putting in contact transition metals with the open ends of SWCNTs [42]. Similarly, on CNFs for which the graphene layers stacked obliquely with respect to the fiber axis [43], strong metal–support interaction and thus high dispersion are usually obtained. Concerning the chemical nature of graphene edges, present in high density in CNFs, computational analyses have shown that under ambient conditions, a significant fraction of the oxygen-free edge sites are neither H-terminated nor free radicals but carbene- or carbine-like sites [44].

Of course, the nature of the transition metal has to be taken into account when considering metal–support interaction. Thus, although strong interactions have been measured for palladium and platinum, such interactions are weaker for gold [45]. Similarly, Ti and Ni strongly interact with the sidewall of CNTs through a partial covalent bonding [46, 47], whereas Au and Al interact weakly through van der Waals forces [46]. These results should be correlated with a fast surface diffusion process in the case of gold and a low diffusion rate for titanium. In addition, if the binding energy of a single metal atom is much higher for Pt than for Pd, the binding energy of a cluster can be lowered due to strong metal–metal binding within Pt [45].

As for graphite [48, 49], the nature of the metal–support interaction may also be tuned by the introduction of chemical species on the surface of CNTs [50]. The general approach consists in oxidizing the carbon surface, which strongly impacts the metal–support interaction through the formation of direct chemical bonds between the surface oxygen groups and the metal.

For catalytic applications, specific metal–support interactions often lead to specific reactivity. Thus, studies on ethylene, but-1-ene and buta-1,3-diene, hydrogenation have been conducted on nickel catalysts supported on different types of CNFs, γ-alumina and AC [51–53]. The authors state that the nickel crystallite activity and selectivity can undergo important modifications by interacting with the support. Indeed, it was found that the catalyst supported on CNFs gives higher conversion than those observed employing γ-Al$_2$O$_3$ or AC, even though the metallic particle size was larger (6.4–8.1 nm on CNFs, 5.5 nm on AC, and 1.4 nm on γ-Al$_2$O$_3$). These results point to the fact that catalytic hydrogenation might be extremely sensitive to the nature of the metal–support interaction. HREM studies have been performed to get a deeper insight into metallic particle morphology: upon CNF supports, the deposited crystallites were found to adopt very thin, hexagonal structures, and the relevant growth pathways are generally believed to be followed in situations in which strong metal–support interactions are present to cause the metal to spread on the support surface. By contrast, a globular particle geometry was observed when Ni was supported on γ-Al$_2$O$_3$, providing a somewhat weaker metal–support interaction.

Finally, the high electrical conductivity of CNTs may also favor charge transfer phenomena between the support and the active phase [54]. Thus, charge transfer, resulting in SWCNT hole doping, was evident during the room-temperature spontaneous reduction, or galvanic displacement, of Au(III) and Pt(II) ions on the surface of SWCNTs [55]. Similar results were already known in the case of highly orientated pyrolytic graphite (HOPG) [56]. However, for Ni adatom interacting with SWCNT or graphite, it was shown that the intensity and the direction of the charge transfer were different for these two supports [31]. The electron transfer between CNTs and TiO_2 nanoparticles, which consists in the transfer of photoinduced electrons (e^-) into the conduction band of TiO_2, might be an explanation of the excellent performances of TiO_2/CNT composites for photocatalytic reactions [57].

11.4
Dispersion of the Active Phase

The preparation of a solid catalyst constitutes an entire field of research for which the ultimate goal is to produce a material possessing the desired activity, selectivity, and lifetime [58]. As the synthesis is a crucial step that affects a catalyst's performance, a fundamental understanding of catalyst synthesis is important. If we focus our attention on carbon (AC, graphite, and carbon black) supported catalysts, the most applied synthesis techniques are liquid-phase impregnation (excess solution or incipient wetness), ion exchange, and deposition precipitation [59, 60]. For CNTs or CNFs, since supported nanoparticles could find applications other than in catalysis, a plethora of materials have been deposited on their surface and many methods have been developed to attain such deposition [5, 6], from classical ones such as liquid-phase impregnation to more sophisticated ones using supercritical CO_2 [61] or biphasic mixtures [62]. Generally speaking, in order to control the dispersion of the metallic phase on carbon materials, several aspects have to be taken into account [63, 64]: (i) the porous structure of activated carbons; (ii) the oxygen groups present on activated carbons or introduced on the surface of high surface area graphite and carbon blacks [65]; and (iii) the specific interactions of metal precursors with surface defects or graphite edges for graphitized carbon blacks. Thus, a common aspect to most of the carbon materials is that a surface activation or functionalization, meaning the creation of active groups on the carbon surface, is essential to attain a small metal particle size, which means high dispersion of the metallic phase.

11.4.1
Surface Area and Porosity of CNT

Nitrogen adsorption studies performed on CNTs have highlighted the porous nature of these materials. For SWCNTs, nitrogen isotherms clearly showed the microporous nature of this type of CNTs. Most experiments show that the specific surface area of SWCNTs is often larger than that of MWCNTs or CNFs (Table 11.1). For an MWCNT

Table 11.1 Specific surface area of SWCNTs, MWCNTs, and CNFs.

	SWCNTs	MWCNTs	CNFs
Diameter (nm)	0.5–2 (generally 1–1.5)	5–200 (generally 10–60)	10–500 (generally 50–100)
Length	Few μm to 20 cm	Few to hundreds of μm	Few to hundreds of μm
d_{002} (Å)	—	3.39–3.48	3.36–3.44
S_{BET} (m^2/g)	400–900	150–450	10–250
Porosity (cm^3/g)	Microporous V_{micro}: 0.15–0.3	Mesoporous V_{meso}: 0.5–2	Mesoporous V_{meso}: 0.2–2

aggregate, pores can be mainly divided into inner hollow cavities of small diameter (narrowly distributed, mainly 3–10 nm), external walls, and aggregated pores (widely distributed, 20–100 nm) formed by interaction of isolated MWCNTs. Thus, MWCNTs are essentially a mesoporous material that also presents some macroporosity (pores >50 nm). For CNFs, which is also a mesoporous material, the specific surface area is generally lower than for MWCNTs. Although it is well known, in the case of activated carbon support, that the porosity and specific surface area of the support have a direct impact on metal dispersion [66], no specific study has yet been conducted on that matter for CNTs.

11.4.2
CNT Surface Activation to Improve Particle Dispersion

From a general point of view, both experimental and theoretical results conclude that surface activation of CNT is necessary when a high dispersion is aimed. Indeed, as-produced CNTs do not possess a high amount of functional groups on their surface, and only surface defects such as vacancies, dangling bonds at open ends, or Stone Wales defects can be considered as anchoring sites for metals (see Section 11.3). On such CNTs, the result is often a poor dispersion of the metallic phase and the impossibility to reach high metal loadings. Thus, on nonfunctionalized MWCNTs, the preparation of a 5% w/w Pt/MWCNT sample from [Pt(CH$_3$)$_2$(cod)] (cod = 1,5-cyclooctadiene) leads to the effective metal loading of 2.0% w/w and the deposition of large (14 nm mean diameter) Pt particles (Figure 11.6).

The broad spectrum of CNT and CNF applications has allowed the development of reliable methods for their functionalization, and today chemical reactions on these materials receive an increasing attention [67, 68]. The most common methods used for metal particle anchoring involve treatment with strong acids or oxidizing agents such as HNO$_3$, H$_2$SO$_4$/HNO$_3$, H$_2$SO$_4$/H$_2$O$_2$, KMnO$_4$, and O$_3$ [69]. For example, nitric acid functionalization provides carboxylic functionalities on CNT surface [50], which can act as anchoring sites for the metal. These drastic treatments should be carefully monitored in order to control surface oxygen group's concentration and to avoid damage to CNT walls and increase in the specific surface area after prolonged reflux. On HNO$_3$-treated MWCNTs, the preparation of a 5% w/w Pt/MWCNT sample

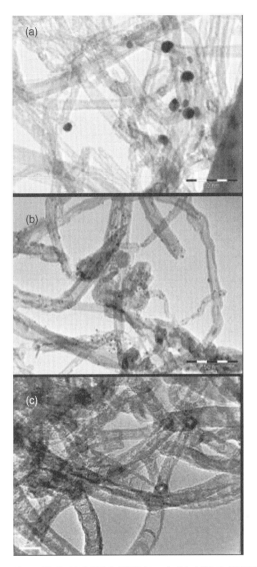

Figure 11.6 (a) 2.1% Pt/CNT (4 nm), (b) 4.8% Pt/CNT-COOH (5 nm), (c) 10% Pt/CN$_x$ (2 nm).

from [Pt(CH$_3$)$_2$(cod)] leads to the effective metal loading of 4.8% w/w of Pt and the deposition of small (<5 nm mean diameter) Pt particles (Figure 11.6). XPS studies have shown that, for oxygen plasma-treated MWCNTs, no mixed Pt-C phase is formed and that the formation of C−O−Pt bonds at the nanoparticle–CNT interface may contribute to reduce the electronic interaction between Pt nanoparticles and the CNT surface [70]. Even though oxidative surface treatments on the supports seemingly improve the dispersion of the metallic phase, it is important to stress that a better knowledge of the concentration and distribution of the attached functional moieties will help to understand the dependence of CNT and CNF reactivity on their structure. Of course, the difference in dispersion obtained on functionalized and not functionalized CNTs will have a direct impact on the catalytic performances of the catalysts. Thus, in comparison with cobalt particles on unoxidized nanotubes (Co mean particle size 100 nm), the cobalt nanoparticles deposited on the oxidized nanotubes have much smaller size and higher dispersion (Co mean particle size <5 nm), resulting in a significantly better catalytic performance in dehydrogenation of cyclohexanol to cyclohexanone [71]. One should also keep in mind the importance of the surface chemistry of the support with regard to the final performance of the catalysts. Indeed, if anchoring sites are needed to reach high dispersion, an excess of this group may impact the catalytic performance. Thus, for Ru/CNF, the crucial role of cinnamaldehyde adsorption on the support has been reported, with CNFs with low amounts of oxygen-containing groups allowing increased activity and higher selectivity [72]. Similarly, the use of PtRu/MWCNT for the same hydrogenation reaction leads to high activities in hydrogenation and a selectivity to cinnamyl alcohol >90%, after heating the catalyst to remove the oxygen-containing groups [73].

Interestingly, high metal dispersion can be obtained without surface functionalization on nitrogen-doped carbon nanotubes [74]. In that case, it is likely that the nitrogen functionalities introduced during CNT synthesis act as anchoring groups for the metal. Figure 11.6 shows 1–2 nm Pt nanoparticles deposited from [Pt(CH$_3$)$_2$(COD)] on pristine CN$_x$ nanotubes.

11.4.3
Specific Interactions of Metal Precursors with Surface Defects of CNTs and CNFs

Owing to their specific structure, CNTs and CNFs can be considered as tunable and polyfunctional ligands for metal coordination (Figure 11.7). The different orientation of the graphene layer in CNTs and CNFs may induce differences in dispersion of the active phase. For PtRu nanoparticles (2% w/w) deposited on oxidized MWCNTs and CNFs, a smaller mean particle size has been measured for the PtRu/CNF sample [73]. If the different metal dispersions might be explained by different amounts of oxygen-containing surface groups, the higher this amount, the better the dispersion, it is not possible to rule out the fact that the higher dispersion obtained on CNF-supported catalysts might also be due to the specific orientation of the graphene layers in this support, which are likely more favorable to nanoparticles' anchorage and stabilization [75].

11.4.4
Influence of Catalyst Preparation Procedure on Metal Loading and Dispersion

Various experimental procedures have been developed for coating nanotubes [5, 6], the choice of the coating method depending largely on the application of the coated nanotubes, the scale of operation, and cost considerations. Basically, the synthesis procedures can be classified into four groups: (i) deposition methods from solution

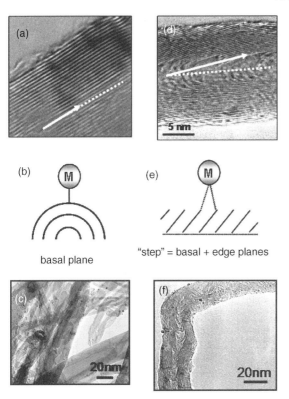

Figure 11.7 Different orientation of the graphene layer in (a, b) CNTs, (d, e) CNFs. Different dispersion of the active phase in (c) CNTs and (f) CNFs.

(wetness impregnation, homogeneous deposition precipitation method, polyol process, electroless deposition, and sol–gel); (ii) self-assembly methods; (iii) electro- and electrophoretic deposition; and (iv) deposition from gas phase (electron beam evaporation, atomic layer deposition, and CVD).

We can see from Table 11.2 that several successful experimental methods have been developed to decorate the CNT surface with different metals or oxides. As already stated, a prior functionalization or activation of the CNT surface is necessary to achieve uniform coatings. Wetness impregnation methods, though multistep processes, are most commonly used due to their simplicity and suitability for large-scale operation. In general, relatively high dispersions are obtained with this process (Table 11.2). Gas-phase methods have been adopted mainly for the *in situ* fabrication of electronic devices. For catalyst preparation, these latter methods have two limitations: (i) if it is conducted by successive steps (atomic layer deposition), high metal loadings are difficult to reach while keeping high dispersion; and (ii) it is difficult to scale up since deposition must be carried out in fluidized bed, and CNTs and CNFs are difficult to fluidize without external activation such as vibration. Electrochemical methods are becoming increasingly attractive for application in electrocatalytic

Table 11.2 Elements that have been deposited on CNTs or CNFs.

Coating method	Element	Particle size (nm)	Support	References
Deposition from solution				
Wetness impregnation	Pt	1–1.5	MWCNT	[76]
Deposition precipitation	Pt	5	CNF	[77]
Polyol process	Pt	2–3	SWCNT	[78]
Sol–gel	SiO$_2$	Film	MWCNT	[79]
Electroless deposition	Pt, Au	2 (wires)	SWCNT	[80]
Self-assembly methods				
Prefabricated metal nanoparticles	Au	10	MWCNT	[81]
Layer-by-layer electrostatic adsorption	Pt	6–10	MWCNT	[82]
Microemulsion-templated deposition	Pd, Rh	2–10	MWCNT	[83]
Electro- and electrophoretic deposition				
Electrochemical coating of nanotube surfaces	Pt	30–70	MWCNT	[84]
Growth mechanisms in electrochemical deposition	Pt	—	SWCNT	[85]
Tip-selective electrochemical deposition	Pd	1–3	MWCNT	[86]
Deposition from gas phase				
Atomic layer deposition	Al$_2$O$_3$	Film	MWCNT	[87]
CVD coating	Pt	2–3	MWCNT	[88]

reactions and fuel cells since high metal loadings can be reached. However, the electrodeposition method presents limitations due to the difficulty in controlling precisely the metal loading because of the concurrent reduction in protons and to the difficulty in reaching small particle sizes (Table 11.2). Until now, coated CNTs have not been prepared on an industrial scale and progresses in the upscaling of preparation methods and in the definition of reliable methods are needed.

11.5
Electrically and Thermally Conductive Supports

11.5.1
Electrical Conductive Supports

Graphite is a semimetal with a layered crystal structure. The model graphite crystal, that is, highly ordered pyrolytic graphite, is known for its high anisotropy among its electrical properties, the electrical conductivity being much higher within the graphene layer (*a*-axis) than in the direction normal to the graphene layers (*c*-axis). Grain boundaries, one of the most commonly occurring extended defects in HOPG because of its polycrystalline character, may impact the electronic properties of this material from a metallic character for the best ordered materials to semiconductor [89, 90]. The electronic structure of SWCNTs depends on their diameter and how the graphene is rolled. In particular, the electronic structure strongly depends on the *chirality* vector [91] defining the type of nanotube: (*n, n*) tubes ("armchair" type) are predicted to be metallic, while (*n, m*) tubes with $n \neq m$ are wide-gap or narrow-gap semiconductors, depending on the particular *m* and *n*. If $2n + m$ or $n + 2m$ is an integer multiple of 3, the SWCNT is predicted to be a narrow-gap semiconductor with good room-temperature conductivity. However, these theoretically predicted electronic properties are often modified by the presence of defects such as pentagons, heptagons, vacancies, or impurities [92], and production techniques do not allow the selective production of only one type of SWCNTs, so the final purity of the obtained material, that is, after purification steps, is often far from being satisfactory. In MWCNTs, built from parallel concentric SWCNTs, each graphene may present electrical properties similar to an isolated SWCNT. The helicity of each shell is random during growth, and there is little coupling to neighboring shells. Since the bandgap of a SWCNT is proportional to ~$1/D$, it decreases as the tube diameter increases. As most MWCNTs are larger than 5 nm in diameter, they typically have a vanishing bandgap at 300 K, and are metallic in nature, as could be a nanoscale HOPG cylinder. Herringbone CNFs are nearly 100% metallic wires with linear *I–V* properties [93]. A temperature-dependent study over a range of 4–300 K revealed that the electron transport along the CNF can be modeled by combining graphite *a*-axis and *c*-axis resistivity [93], the intrinsic resistance of the CNF being dominated by electron hopping between graphitic layers similar to graphite *c*-axis transport. Even though there is scope to improve the intrinsic conductivity of these materials, it appears to be sufficient for catalytic applications as conductive support. Finally, it is worth noting that the modulation of electronic properties of CNTs can be envisaged by intercalation processes (alkali metals), surface modification (adsorption of amines, polymers), or doping (nitrogen or boron).

The unique electrical and structural properties of CNTs make them ideal materials for use in gas and vapor sensors [94], batteries [95], electrochemical systems [96–98], and in fuel cell construction, where they are used in fabricating the bipolar plate, the gas diffusion layer, and can also act as a support for the active metal in the catalyst layer. Thus, one of the fields to which the specific features of CNTs, and particularly their electric conduction, could bring the most significant advancements is probably fuel cell electrocatalysis [99], where CNTs are being investigated for use as catalyst support. This could lead to the production of more stable, high-activity catalysts, with low platinum loadings (<0.1 mg/cm^2) and therefore low cost. In order to reach the DOE cost reduction targets, among other component is cost reductions, the total Pt catalyst loading must be reduced to 0.03 mg/cm^2 [99].

In fuel cell electrocatalysis, the structure and properties of the carbon support, which constitutes the catalyst layer, have a direct impact on the performance of fuel cells. This material must have (i) a high electronic conductivity; (ii) a high meso-porosity to attain high metal dispersion at reasonable loading; and (iii) a good hydrophobicity for water management. In comparison to the more widely used carbon black support, CNTs have significantly higher electronic conductivities and present higher mesoporous volumes for comparable or higher surface areas.

Among the numerous studies published on the use of CNTs as catalyst supports for direct methanol and proton exchange membrane fuel cells, the most studied reactions are methanol oxidation, followed by oxygen reduction, and, to a lesser extent, hydrogen oxidation. Platinum [89] is the most used metal, followed by Pt-Ru systems [100], and only few studies have been performed with cheaper metals such as Pd [101]. All kinds of CNTs and CNFs have been used for these reactions such as herringbone CNFs [102], MWCNTs [99], SWCNTs [103], DWCNTs [103], and nitrogen-doped CNTs [104]. In most of these works, high dispersions of the metallic phase, typically 2–5 nm mean particle size, have been obtained. Although comparison between these studies constitute a difficult issue because of the different origin of CNTs, the general tendency observed is that the catalysts prepared on CNTs are more active and sometimes present a higher resistance to poisoning [105] than those prepared on conventional carbon black supports. In some cases, it is also possible to obtain similar or better performances with significant reduction of metal loadings [106]. When compared to the commonly used carbon black support, the increase in the power density of a single stack is between 20–40%, even if higher values [107] have been reached. It has also been reported that the degradation rate of Pt/CNTs was nearly two times lower than that of Pt/C under the same testing conditions, which was attributed to the specific interaction between Pt and CNTs and to the higher resistance of the CNTs to electrochemical oxidation [108]. Only few studies concern comparison between the different types of CNTs [103], and if we consider the electronic conductivity of the supports, we would expect the following order: SWCNTs > DWCNTs > MWCNTs > CNFs. In general, and this result could be rationalized on the basis of the electronic conductivity of the support, the activity of metal-supported MWCNT systems is superior to CNF-based catalysts, even if high metal dispersions are reached on CNFs. For DWCNTs, which present a higher specific surface area than MWCNTs, better performances have been obtained [103]. Contrasting results have been reported for SWCNTs, and more work is needed to shed light on the potential of these materials.

The advances made with the use of CNTs and CNFs as support for fuel cell electrocatalysis are generally attributed to (i) the possibility to reach high metal dispersions and high electroactive surface area; for carbon black, the catalyst particles can sink into the microporosity; (ii) the peculiar 3D mesoporous network formed by these materials, which improves the mass transport (see Section 11.6); and (iii) their excellent conducting properties that improve electron transfer. Many studies have also pointed out that the morphology and the activity of metal supported on CNT catalysts are strongly affected by their synthesis methods [109], and innovative catalyst syntheses are still needed for metal loading reduction and performance optimization.

11.5.2
Thermally Conductive Supports

Since the thermal control on nanocatalysts becomes increasingly important as the size of the system diminishes, another important feature that ought to be taken into account is the thermal conductivity of the support. Therefore, for exothermic reactions the thermal conductivity of CNTs or CNFs should play a crucial role in controlling the catalytic performance. Despite the importance of thermal management, there has been little progress, mainly due to technical difficulties, in measuring the thermal conductivity of individual CNTs. The measured thermal conductivities of graphite along the basal plane range from 940 to 2000 W/(K m), and those perpendicular to the basal plane are two or three orders of magnitude lower. For CNTs, a similar behavior is very likely to be observed for their axial and transverse conductivities. Although experimental data and theoretical predictions are scattered by one order of magnitude, most of the studies concerning CNT thermal conductivity agree about very high values comparable to diamond and to in-plane graphite sheet. Consequently, they can withstand a high heat transfer for use in highly exothermic reactions, that is, Fischer–Tropsch [110] or selective oxidation of H_2S into elemental sulfur [111].

11.6
Mass Transfer Limitations

Among the different stages involved in a catalytic reaction, we should distinguish the diffusion (physical) steps and the adsorption/desorption (chemical) steps. When using porous-supported catalysts, molecular diffusion, often called simply (internal) diffusion, and external diffusion (Figure 11.8) may govern the rate of the catalytic reaction if microporous materials are used to support the active phase or if working at elevated temperatures. In particular, when active sites are accessible within the particle, internal mass transfer (molecular diffusion) has a tremendous influence on the rate of reaction within the catalyst.

For carbon materials, and particularly for activated carbon, the pore size distribution consists mainly in micropores, with low meso- or macropore volume. Even if the metal is well dispersed in the microporosity, mass transfer limitations decrease the apparent activity and modify the selectivity of the catalysts [59]. For filamentous carbon the situation is different. First, the structure of a filamentous carbon pellet is different from that of a conventional support (Figure 11.9). Indeed, for MWCNT aggregates, aggregated pores formed by interaction of isolated MWCNTs, and external walls are often the most important for adsorption issues and metal dispersion since most of the time MWCNTs and CNFs are closed. Figure 11.10 shows SEM and TEM micrographs of a PtRu/MWCNT catalyst at different magnifications: from the MWCNT aggregate to the PtRu nanoparticle anchored on CNT wall. Thus, the open structure of filamentous carbon (CNTs and CNFs) and their high mesoporosity (Table 11.1) should allow to significantly restrict mass transfer limitations and the pore blocking by metal particle deposition. Furthermore, the flexible nature of CNTs makes the catalyst pellet a structurally dynamic support presenting a porosity that can evolve according to external stimulation.

However, if aligned and both ends opened CNTs are used to build a membrane (Figure 11.9), gas and liquid transport measurements show substantial flow

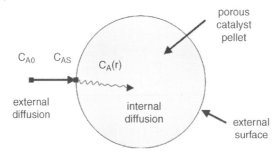

Figure 11.8   Mass transfer and reaction steps for a catalyst pellet.

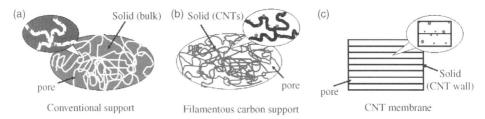

Figure 11.9   Macrostructure and dispersion of the active phase in (a) a conventional support, (b) a filamentous carbon pellet, and (c) an aligned and both ends opened CNT membrane.

enhancements with respect to real Knudsen gas diffusion [112]. Alternative transport mechanisms have been proposed that emphasize transfer limitations at the tube entrance. In any case, the very efficient transport in the tubes makes the kinetics of surface transfer more important. It was also shown that in that case the classic model

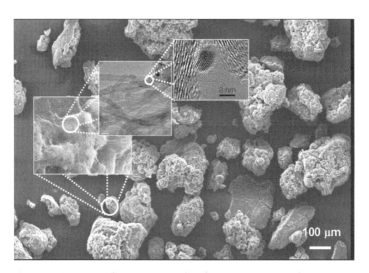

Figure 11.10   SEM and TEM micrographs of a PtRu/MWCNT catalyst.

for gas (Knudsen) transport does not properly describe diffusion through CNT membranes [113]. This is due to the ultralow friction of smooth CNT walls with molecules. Thus, CNTs are somewhat unique in this aspect because of the atomic planarity and high rigidity of the graphene and the nonpolar nature of the sp^2 carbon network.

Owing to the advantageous properties of CNTs as support, several studies have been carried out for different catalytic reactions [114]. Several studies report that for liquid-phase catalytic and also electrocatalytic reactions, the use of CNT or CNF supports possessing large mesoporosity gave better catalytic activities than microporous activated carbon for which mass transfer limitations were operating.

The peculiarity of CNFs has been advantageously used for hydrazine decomposition over Ir catalysts, where Ir/CNF presents far better performances than a commercial Ir/Al$_2$O$_3$ catalyst [115]. It has been proposed that part of the injected hydrazine was trapped inside the micropores of alumina containing no Ir particles and thus could not be decomposed. The absence of any detrimental microporosity in the CNF-supported iridium catalyst is therefore of great interest. The high performance of the catalytic decomposition of hydrazine compared to the iridium/CNF catalyst was also attributed to the high external surface area of the carbon support, which allows a high dispersion of the active phase. This high external surface area was observed because of the entangled carbon nanofiber structure.

The selective hydrogenation of cinnamaldehyde has been studied on bimetallic PtRu catalysts supported on MWCNTs and activated carbon. The use of PtRu/MWCNT (175 m^2/g, PtRu nanoparticles 1.8 nm) leads to activities higher than with PtRu/AC (700 m^2/g, PtRu nanoparticles 2.4 nm), and a selectivity to cinnamyl alcohol >90%, was reached after heating the catalysts to remove the oxygen-containing groups [73]. Palladium systems have also been investigated since they are very active in the selective hydrogenation of the C=C bonds [116]. The catalytic results obtained over the Pd supported on CNFs and a commercial Pd/C catalyst show the superiority in terms of reaction rate of the CNF-based catalyst. This high catalytic activity was attributed to the high external surface area (high surface-to-volume ratio) of the carbon nanofibers compared to the charcoal grains that improves the mass transfer during the reaction. In addition, the presence of a large amount of micropores (>60% of the total surface area) in the commercial catalyst can be at the origin of diffusion phenomena, slowing down the apparent rate of the reaction. The total absence of microporosity in the nanofibers avoided such a drawback.

Composite electrodes consisting of Pt nanoparticles supported on MWCNTs grown directly on carbon paper have been compared with a commercial Pt on carbon black catalyst for the electrooxidation of methanol [117]. Compared to standard Pt/C electrode (0.10 mg$_{Pt}$/cm^2, Pt nanoparticles 2–3 nm), the Pt/CNT/carbon paper composite electrodes (0.11 mg$_{Pt}$/cm^2, Pt nanoparticles 2 nm) exhibit higher electrocatalytic activity. This has been essentially attributed to the 3D structure of CNT-based electrodes. Indeed, in spite of the high surface area of the carbon black particles, due to their dense structure, the carbon black-based support presents significant mass transfer limitations, leading to a very low Pt utilization.

11.7
Confinement Effect

The study at the mesoscopic or microscopic scale of the impact of confinement of the active phase on the catalytic performances is a subject of current interest [118, 119]. Indeed, confinement at the mesoscopic scale may (i) induce variation in the physical properties of the confined fluids (viscosity, density, solubility, and adsorption), (ii) induce wall or curvature effects, which may modify adsorption or electronic properties or induce concentration gradients, and (iii) influence the energetics and dynamics of catalytic processes [118]. At the microscopic scale, the confinement may influence the chemical reaction through (i) shape–catalytic effect, (ii) physical "soft" effects (electrostatic interactions), or (iii) chemical "hard" interactions (covalent bonding) [119].

The opportunity of using the inner cavity of CNTs as a nanoreactor of few nanometers in diameter and only some micrometers in length has lead, only a few years after their discovery, to the first studies dealing with CNT opening, filling, or wetting [120–122]. If the complete filling of CNTs, which allows to produce nanowires [123, 124], is now relatively controlled [125], the selective confinement of isolated nanoparticles in the CNT cavity is still a synthetic challenge, which should offer interesting perspectives for performing chemistry in a confined space while exploiting CNT properties.

At present, three main routes exist to introduce NPs inside CNTs. The first one is based on impregnation techniques and consists in the filling of oxidized CNTs by capillarity with a solution containing a metal salt [126–133] or preformed NPs [134, 135]. Although the precise quantification of the percentage of NPs located inside CNTs is difficult to establish, TEM, XPS [136], and particularly 3D TEM [128] studies have allowed to determine the fraction of metal present in the inner cavity of CNTs between 15 and 50%. For very low metal loadings (1.2% w/w Rh/CNT), it was claimed that 80% of the rhodium NPs were selectively confined inside CNTs (Figure 11.11) [127]. A more selective filling of very large diameter (≥ 300 nm) carbon filaments by preformed polystyrene NPs has been achieved (Figure 11.12) [135]. As the wettability of CNT surface is a critical issue for selective confinement, the use of supercritical fluids (zero surface tension) has been investigated for polymer coating [137] or nanoparticle transport [138] in CNTs. Another strategy consists in the opening of MWCNT by catalytic oxidation (Ag or Fe catalysts) and in the subsequent confinement using simple impregnation methods and ultrasonication [139]. A simple and selective procedure to confine, even at high metal loadings, high proportion (>80%) of preformed nanoparticles involves the use of surface chemistry to favor nanoparticles/inner surface of the tubes interaction and nanoparticles/outer surface of the tubes repulsion (Figure 11.13). 3D TEM analyses have shown that this procedure allows to confine 80% of PtRu NPs (2 nm) in the inner cavity of 40 nm internal diameter CNTs (Figure 11.14), even at metal loading as high as 40% w/w [140].

The second one is based on the sublimation of a metal precursor, which can be used to introduce it in the CNT cavity [141, 142]. For these two routes, in some

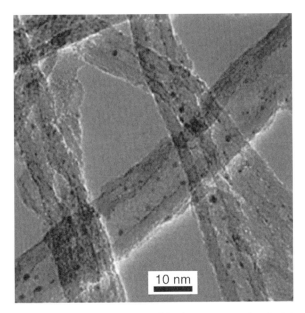

Figure 11.11 TEM micrograph of rhodium nanoparticle selectivity confined inside CNTs. Reprinted by permission from Ref. [127]. Copyright 2007 Macmillan Publishers Ltd.

cases, an enhancement in the confinement selectivity can be obtained if an extra selective washing step is performed to remove NPs deposited on the outer surface [143–145].

The third multistep approach consists in producing CNTs inside anodic aluminum oxide membranes, filling the resulting template with a solution of NPs and dissolving

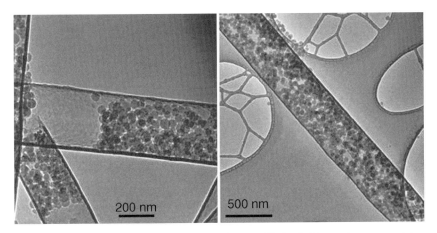

Figure 11.12 TEM of CVD-CNTs of 200–300 nm diameter filled with 51 nm polystyrene particles. Reprinted with permission from Ref. [17]. Copyright 2008 Royal Society of Chemistry.

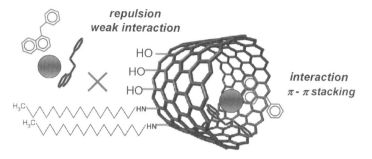

Figure 11.13 Strategy to confine NPs in the inner cavity of CNTs.

the alumina template [146–148]. By this way, 10 nm iron nanoparticles have been confined in 250 nm diameter CNTs (Figure 11.15).

Owing to synthesis difficulties, relatively few experimental studies provide reliable results on confinement effects on CNTs. For example, the possibility of hydrogen generation via splitting of H_2O confined in SWCNTs by the use of a camera flash has been reported [149, 150]. For this reaction, the camera flash induces an ultraphotothermal effect that allows the temperature to rise high and causes water molecules to dissociate into H_2 and O_2. In addition, the ultraefficient transport of water and gas through the hydrophobic cavity of CNTs makes CNT membranes promising for many applications [151]. Theoretical calculations have shown that water [152] and also gold NP [153] diffusion is slower in zigzag than in armchair SCWNTs. These results can be rationalized in terms of topology of potential energy surface (water diffusion) or of surface corrugation (NP diffusion). Temperature-programmed reduction/oxidation experiments have shown that the reactivity of iron oxide NPs (5–8 nm) is increased (easy reduction) while that of the metallic iron NPs is decreased (difficult reoxidation) when confined inside CNTs [154]. Dimensionally confined phase transitions have

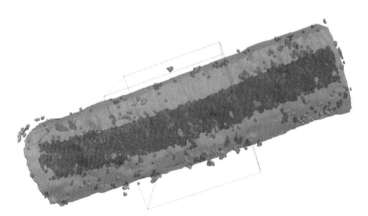

Figure 11.14 Three-dimensional TEM confinement of PtRu NPs in the inner cavity of CNTs. Courtesy of Dr. Ovidiu Ersen, IPCMS, Strasbourg, France.

(a)

(b)

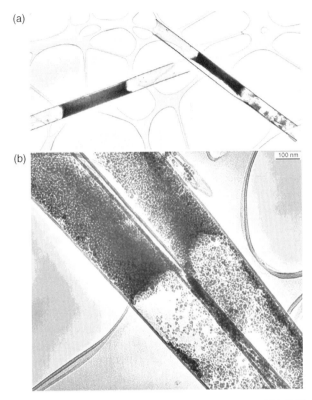

100 nm

Figure 11.15 Filling of released nanotubes. TEM image of (a) CNTs with open ends, filled with magnetic particles from water-based ferrofluid EMG 508, and (b) part of a branched CNT. Reprinted with permission from Ref. [146]. Copyright 2005 American Chemical Society.

been reported for water [155] and ionic liquids [156] entrapped within CNTs. It has also been shown that heptene molecules confined in SWCNTs have a reduced reactivity toward atomic hydrogen compared to the same molecules adsorbed on the external surface [157].

As far as catalytic studies are concerned, some reports give evidence of an effect of confinement on the performances of the catalysts. Rh NPs (4–5 nm) confined in CNTs (4–8 nm inner diameter) are one order of magnitude more active for ethanol production from syngas than Rh NPs (1–2 nm) deposited on the convex CNT surface [127]. Temperature-programmed desorption experiments have demonstrated that a higher concentration of active hydrogen species is obtained when Rh NPs are located inside CNTs. For Fischer–Tropsch synthesis on Fe/C systems, the confinement of 4–8 nm iron NPs in the inner cavity of MWCNTs (4–8 nm inner diameter) permits reaching higher activity and selectivity in C_{5+} hydrocarbons compared to 6–10 nm Fe NPs deposited on the convex surface of CNTs (two times less active) or to 8–12 nm Fe NPs deposited on activated carbon (six times less active). These results have been rationalized by considering the modified redox properties of the confined catalyst, for which higher amounts of iron carbide are present [158]. For selective hydrogenation of cinnamaldehyde, an improvement in C=O selective hydrogenation compared to C=C and C=O hydrogenation has been observed when Pt NPs (5 nm) are located inside CNTs (60–100 nm inner diameter) rather than NPs (2–5 nm) outside MWCNTs (1–10 nm inner diameter) [159]. In the case of Pd NPs (5 nm) located mostly inside CNTs (20–40 nm inner diameter), very high selectivity toward C=C bond hydrogenation has been obtained [126]. In the NH_3 decomposition reaction to produce hydrogen, FeCo alloyed NPs (14 nm) located inside CNTs (20–50 nm inner diameter) give higher conversion (48% versus 24%) than FeCo NPs (15 nm) deposited on the external surface of the same CNTs [133]. Noticeably, for this latter reaction the mean NP size after catalysis does not change for the FeCo NPs inside (15 nm), whereas the NPs located outside CNTs have a tendency to sinter (30 nm). Finally, iron [160] and cobalt NPs [161] have been deposited inside SWCNTs and MWCNTs, respectively. Ferrocene was encapsulated inside 1.4 nm SWCNTs and the $FeCp_2$@SWCNT catalyst was used to prepare double-walled CNTs, for which the inner diameter is around 0.7 nm [160]. On 5–15 nm Co NPs, carbon nanofibers have been grown in the inner cavity of MWCNTs (20–80 nm inner diameters) to produce CNT@CNFs composites [161].

11.8
Conclusion

Carbon nanotubes offer a range of attractive properties and can constitute a new family of highly efficient catalyst support. They are produced on semi-industrial scale and the current barriers to mass-market use of these materials include the availability of high-quality materials at low prices and in commercial quantities. In addition, the key issues concerning standardization and toxicity should be carefully addressed.

Catalytic studies conducted on CNT- or CNF-based catalysts have shown promising results in terms of performances that can be correlated to (i) the use of a high-purity support; (ii) the possibility to reach high dispersion of the metallic phase; (iii) the flexibility of these supports, which allows through the orientation of the graphene layers to tune the metal–support interactions; (iv) the open structure of these supports that helps avoid mass transfer limitation; (v) the exceptional mechanical, thermal, and electrical properties of CNTs; (vi) the possibility to operate in a confined space when the active phase is selectively deposited in CNT inner cavity; and (vii) the possibility of macroscopic shaping.

Acknowledgment

E.C. acknowledges the support of the French Agence Nationale de la Recherche (ANR), under grant Nanoconficat (PNANO-05-NANO-030) "Confinement de nano-particules métalliques dans des nanotubes de carbone: vers des catalyseurs haute-ment sélectifs."

References

1 Hughes, T.V. and Chambers, C.R. (1889) U.S. Patent 405, 480

2 Bartholomew, C.H. (1982) *Catal. Rev. - Sci. Eng.*, **16**, 67.

3 Biró, L.P., Bernardo, C.A., Tibbetts, G.G., and Lambin, P. (eds) (2009) *Carbon Filaments and Nanotubes: Common Origins, Differing Applications?* Kluwer Academic Publishers, Dordrecht, 2001. NATO Science Series E. Edition Collection Sous-collection 372, pp. 315–630.

4 Moy, D. and Hoch, R., US Patent 5,569,635 (1996), Hyperion Catalysts International, Inc.

5 Wildgoose, G.G., Banks, C.E., and Compton, R.G. (2006) *Small*, **2**, 182.

6 Georgakilas, V., Gournis, D., Tzitzios, V., Pasquato, L., Guldi, D.M., and Prato, M. (2007) *J. Mater. Chem.*, **17**, 2679.

7 De Jong, K.P. and Geus, J.W. (2000) *Cat. Rev.-Sci. Eng.*, **42**, 481.

8 Goldshleger, N.F. (2001) *Fullerene Sci. Technol.*, **9**, 255.

9 Serp, P., Corrias, M., and Kalck, P. (2003) *Appl. Catal. A*, **173**, 337.

10 Zhang, H.B., Lin, G.D., and Yuan, Y.Z. (2005) *Curr. Top. Catal.*, **4**, 1.

11 Vairavapandian, D., Vichchulada, P., and Lay, M.D. (2008) *Anal. Chim. Acta*, **626**, 119.

12 Kauffman, D.R. and Star, A. (2008) *Angew. Chem., Int. Ed.*, **47**, 6550.

13 Bitter, J.H. in *Nanocatalysis* (ed. D. Murzin), Research Signpost, Kerala, India, 2006, pp. 99–125.

14 Smajda, R., Kukovecz, Á., Kónya, Z., and Kiricsi, I. (2007) *Carbon*, **45**, 1176.

15 Jun, M., David, M., Asif, C., and Jun, Y. U.S. Patent Application 20060142149, 2006.

16 Mateusz, B., Bryning, B., Milkie, D.E., Islam, M.F., Hough, L.A., Kikkawa, J.M., and Yodh, A.G. (2007) *Adv. Mater.*, **19**, 661.

17 Amadou, J., Begin, D., Nguyen, P., Tessonnier, J.P., Dintzer, T., Vanhaecke, E., Ledoux, M.J., and Pham-Huu, C. (2006) *Carbon*, **44**, 2587.

18 van der Lee, M.K., van Dillen, A.J., Geus, J.W., de Jong, K.P., and Bitter, J.H. (2006) *Carbon*, **44**, 629.

19 Philippe, R., Caussat, B., Plee, D., Kalck, P., and Serp, P. French Patent, Arkem-Institut National Polytechnique de Toulouse, 52711, 2008.

20 Xiang, R., Luo, G.H., Qian, W.Z., Wang, Y., Wei, F., and Li, Q. (2007) *Chem. Vap. Deposition*, **13**, 533.

21 Ledoux, M.J. and Pham-Huu, C. (2005) *Catal. Today*, **102–103**, 2.

22 Chen, C.C., Chen, C.F., Hsu, C.H., and Li, I.H. (2005) *Diamond Relat. Mater.*, **14**, 770.

23 Tzeng, S.-S., Hung, K.-H., and Ko, T.-H. (2006) *Carbon*, **44**, 859.

24 Wenmakers, P.W.A.M., van der Schaaf, J., Kuster, B.F.M., Schouten, J.C., and Foam, H. (2008) *J. Mater. Chem.*, **18**, 2393.

25 Keller, N., Rebmann, G., Barraud, E., Zahraa, O., and Keller, V. (2005) *Catal. Today*, **101**, 323.

26 Ochoa-Fernández, E., Zhixin Yu, D.C., Tøtdal, B., Rønning, M., and Holmen, A. (2005) *Catal. Today*, **102–103**, 45.

27 Chinthaginjala, J.K., Thakur, D.B., Seshan, K., and Lefferts, L. (2008) *Carbon*, **46**, 1638.

28 García-Bordejé, E., Kvande, I., Chen, D., and Rønning, M. (2006) *Adv. Mater.*, **18**, 1589.

29 Janowska, I., Winé, G., Ledoux, M.J., and Pham-Huu, C. (2007) *J. Mol. Catal. A: Chem.*, **267**, 92.

30 Ishigami, N., Ago, H., Motoyama, Y., Takasaki, M., Shinagawa, M., Takahashi, K., Ikuta, T., and Tsuji, M. (2007) *Chem. Commun.*, 1626.

31 Menon, M., Andriotis, A.N., and Froudakis, G.E. (2000) *Chem. Phys. Lett.*, **320**, 425.

32 Chen, G. and Kawazoe, Y. (2006) *Phys. Rev. B*, **73**, 125410.

33 Wu, M.-C., Li, C.-L., Hu, C.-K., Chang, Y.-C., Liaw, Y.-H., Huang, L.-W., Chang, C.-S., Tsong, T.-T., and Hsu, T. (2006) *Phys. Rev. B*, **74**, 125424.

34 Shu, D.J. and Gong, X.G. (2001) *J. Chem. Phys.*, **114**, 10922.

35 Zhong, J. and Stocks, G.M. (2005) *Appl. Phys. Lett.*, **87**, 133105.

36 Park, S., Srivastava, D., and Cho, K. (2003) *Nano Lett.*, **3**, 1273.

37 Park, Y., Lahaye, R.J.W.E., and Lee, Y.H. (2007) *Comput. Phys. Commun.*, **177**, 46.

38 Mpourmpakis, G. and Froudakis, G. (2006) *J. Chem. Phys.*, **125**, 204707.

39 Morrow, B.H. and Striolo, A. (2008) *Nanotechnology*, **19**, 195711.

40 Mpourmpakis, G., Froudakis, G.E., Andriotis, A.N., and Menon, M. (2005) *Appl. Phys. Lett.*, **87**, 193105.

41 Zhong, Z., Liu, B., Sun, L., Ding, J., Lin, J., and Tan, K.L. (2002) *Chem. Phys. Lett.*, **362**, 135.

42 Mpourmpakis, G., Froudakis, G.E., Andriotis, A.N., and Menon, M.E. (2005) *Appl. Phys. Lett.*, **87**, 193105.

43 Zhu, Y.A., Sui, Zh.J., Zhao, T.J., Dai, Y.Ch., Cheng, Zh.M., and Yuan, W.K. (2005) *Carbon*, **43**, 1593.

44 Radovic, L.R. and Bockrath, B. (2005) *J. Am. Chem. Soc.*, **127**, 5917.

45 Maiti, A. and Ricca, A. (2004) *Chem. Phys. Lett.*, **395**, 7.

46 Zhang, Y., Franklin, N.W., Chen, R.J., and Dai, H. (2000) *Chem. Phys. Lett.*, **331**, 35.

47 Zhang, Y. and Dai, H. (2000) *Appl. Phys. Lett.*, **77**, 3015.

48 Yang, D.-Q. and Sache, E. (2008) *J. Phys. Chem. C*, **112**, 4075.

49 Guerrero-Ruiz, A., Badenes, P., and Rodríguez-Ramos, I. (1998) *Appl. Catal. A*, **173**, 313.

50 Solhy, A., Machado, B.F., Beausoleil, J., Kihn, Y., Gonçalves, F., Pereira, M.F.R., Órfão, J.J.M., Figueiredo, J.L., Faria, J.L., and Serp, P. (2008) *Carbon*, **46**, 1194.

51 Park, C. and Baker, R.T.K. (1999) *J. Phys. Chem. B*, **103**, 2453.

52 Chambers, A., Nemes, T., Rodriguez, N.M., and Baker, R.T.K. (1998) *J. Phys. Chem. B*, **102**, 2251.

53 Park, C. and Baker, R.T.K. (1998) *J. Phys. Chem. B*, **102**, 5168.

54 Iurlo, M., Paolucci, D., Marcaccio, M., and Paolucci, F. (2008) *Chem. Commun.*, 4867.

55 Choi, H.C., Shim, M., Bangsaruntip, S., and Dai, H. (2002) *J. Am. Chem. Soc.*, **124**, 9059.

56 Shen, P., Chi, N., Chan, K.-Y., and Phillips, D.L. (2001) *Appl. Surf. Sci.*, **172**, 159.

57 Wang, W.D., Serp, P., Kalck, P., and Faria, J.L. (2005) *Appl. Catal. B*, **56**, 305.

58 Ertl, G., Knözinger, H., Schüth, F., and Weitkamp, J. (eds) (2007) *The Prediction of*

Solid Catalyst Performance, Wiley-VCH Verlag GmbH, Weinheim.

59 Radovic, L.R. and Rodriguez-Reinoso, F. (1997) in *Chemistry, Physics of Carbon*, vol. 25 (ed. P.A. Thrower), Marcel Dekker, New York, p. 2438.

60 Bitter, J.H. and de Jong, K.P. (2009) in *Carbon Materials for Catalysis* (eds P. Serp and J.L. Figueiredo), John Wiley & Sons, Inc., Hoboken, NJ.

61 Bayrakceken, A., Kitkamthorn, U., Aindow, M., and Erkey, C. (2007) *Scr. Mater.*, **56**, 101.

62 Lee, K.Y., Kim, M., Lee, Y.W., Lee, J.-J., and Han, S.W. (2007) *Chem. Phys. Lett.*, **440**, 249.

63 Auer, E., Freund, A., Pietsch, J., and Tacke, T. (1998) *Appl. Catal. A*, **173**, 259.

64 Cerro-Alarcón, M., Maroto-Valiente, A., Rodríguez-Ramos, I., and Guerrero-Ruiz, A. (2005) *Carbon*, **43**, 2711.

65 Yu, X. and Ye, S. (2007) *J. Power Sources*, **172**, 133.

66 Cabiac, A., Cacciaguerra, T., Trens, P., Durand, R., Delahay, G., Medevielle, A., Plée, D., and Coq, B. (2008) *Appl. Catal. A*, **340**, 229.

67 Tasis, D., Tagmatarchis, N., Bianco, A., and Prato, M. (2006) *Chem. Rev.*, **106**, 1105.

68 Khabashesku, V.N. and Pulikkathara, M.X. (2006) *Mendeleev Commun.*, **16**, 61–66.

69 Vairavapandian, D., Vichchulada, P., and Lay, M.D. (2008) *Anal. Chim. Acta*, **626**, 119.

70 Bittencourt, C., Hecq, M., Felten, A., Pireaux, J.J., Ghijsen, J., Felicissimo, M.P., Rudolf, P., Drube, W., Ke, X., and Van Tendeloo, G. (2008) *Chem. Phys. Lett.*, **462**, 260.

71 Liu, Z.J., Yuan, Z.Y., Zhou, W., Peng, L.M., and Xu, Z. (2001) *Phys. Chem. Chem. Phys.*, **3**, 2518.

72 Toebes, M.L., Prinsloo, F.F., Bitter, J.H., van Dillen, A.J., and de Jong, K.P. (2003) *J. Catal.*, **214**, 78.

73 Vu, H., Gonçalves, F., Philippe, R., Lamouroux, E., Corrias, M., Kihn, Y., Plée, D., Kalck, P., and Serp, P. (2006) *J. Catal.*, **240**, 18.

74 Lepró, X., Terrés, E., Vega-Cantú, Y., Rodríguez-Macías, F.J., Muramatsu, H.,

Kim, Y.A., Hayahsi, T., Endo, M., Torres, M., and Terrones, R.M. (2008) *Chem. Phys. Lett.*, **463**, 124.

75 Zhang, Y., Toebes, M.L., van der Eerden, A., O'Grady, W.E., de Jong, K.P., and Koningsberger, D.C. (2004) *J. Phys. Chem. B*, **108**, 1850919.

76 Satishkumar, B.C., Vogl, E.M., Govindaraj, A., and Rao, C.N.R. (1996) *J. Phys. D: Appl. Phys.*, **29**, 3173–3176.

77 Li, X. and Hsing, I.-M. (2006) *Electrochim. Acta*, **51**, 5250–5258.

78 Kongkanand, A., Vinodgopal, K., Kuwabata, S., and Kamat, P.V. (2006) *J. Phys. Chem. B*, **110**, 16185–16191.

79 Bottini, M., Tautz, L., Huynh, H., Monosov, E., Bottini, N., Dawson, M.I., Bellucci, S., and Mustelin, T. (2005) *Chem. Commun.*, 758–760.

80 Choi, H.C., Shim, M., Bangsaruntip, S., and Dai, H. (2002) *J. Am. Chem. Soc.*, **124**, 9058–9059.

81 Jiang, K., Eitan, A., Schadler, L.S., Ajayan, P.M., Siegel, R.W., Grobert, N., Mayne, M., Reyes-Reyes, M., Terrones, H., and Terrones, M. (2003) *Nano Lett.*, **3**, 275–277.

82 Yuan, U., Wang, Z., Zhang, Y., Shen, Y., Han, D., Zhang, Q., Xu, X., and Niu, L. (2008) *Thin Solid Films*, **516**, 6531–6535.

83 Yoon, B. and Ma, C. (2005) *J. Am. Chem. Soc.*, **127**, 17174–17175.

84 Tang, H., Chen, J.H., Huang, Z.P., Wang, D.Z., Ren, Z.F., Nie, L.H., Kuang, Y.F., and Yao, S.Z. (2004) *Carbon*, **42**, 191–197.

85 Day, T.M., Unwin, P.R., Wilson, N.R., and Macpherson, J.V. (2005) *J. Am. Chem. Soc.*, **127**, 10639–10647.

86 Bradley, J.-C., Babu, S., and Ndungu, P. (2005) *Fullerenes, Nanotubes, Carbon Nanostruct.*, **13**, 227–237.

87 Lee, J.S., Min, B., Cho, K., Kim, S., Park, J., Lee, Y.T., Kim, N.S., Lee, M.S., Park, S.O., and Moon, J.T. (2003) *J. Cryst. Growth*, **254**, 443–448.

88 Serp, P., Feurer, R., Kihn, Y., Kalck, P., Faria, J.L., and Figueiredo, J.L. (2002) *J. Phys. IV France*, **12**, 29–36.

89 Pantin, V., Avila, J., Valbuena, M.A., Esquinazi, P., Dávila, M.E., and Asensio, M.C. (2006) *J. Phys. Chem. Solids*, **67**, 546.

90 Cervenka, J. and Flipse, C.F.J. http://arxiv.org/abs/0810.5653v1.

91 Monthioux, M., Serp, P., Flahaut, E., Laurent, C., Peigney, A., Razafinimanana, M., Bacsa, W., and Broto, J.-M. (2007) in *Springer Handbook of Nanotechnology*, second revised and extended Edition (ed. B. Bhushan), Springer-Verlag, Heidelberg, Germany, pp. 43–112.

92 Charlier, J.C. (2002) *Acc. Chem. Res.*, **35**, 1063.

93 Ngo, Q., Cassell, A.M., Austin, A.J., Li, J., Krishnan, S., Meyyappan, M., and Yang, C.Y. (2006) *IEEE Electron Device Lett.*, **27**, 221.

94 Kauffman, D.R. and Star, A. (2008) *Angew. Chem., Int. Ed.*, **47**, 6550.

95 Zou, L., Lv, R., Kang, F., Gan, L., and Shen, W. (2008) *J. Power Sources*, **184**, 566.

96 Gooding, J.J. (2005) *Electrochim. Acta*, **50**, 3049.

97 Gong, K., Chakrabarti, S., and Dai, L. (2008) *Angew. Chem., Int. Ed.*, **47**, 5446.

98 Banks, C.E., Davies, T.J., Wildgoose, G.G., and Compton, R.G. (2005) *Chem. Commun.*, 829.

99 Lee, K., Zhang, J., Wang, H., and Wilkinson, D.P. (2006) *J. Appl. Electrochem.*, **36**, 507.

100 Ostojic, G.N., Ireland, J.R., and Hersam, Mark C. (2008) *Langmuir*, **24**, 9784.

101 Grigoriev, S., Lyutikova, E.K., Martemianov, S., and Fateev, V.N. (2007) *Int. J. Hydrogen Energy*, **32**, 4438.

102 Kvande, I., Briskeby, S.T., Tsypkin, M., Rønning, M., Sunde, S., Tunold, R., and Chen, D. (2007) *Top. Catal.*, **45**, 81.

103 Li, W., Wang, X., Chen, Z., Waje, M., and Yan, Y. (2006) *J. Phys. Chem. B*, **110**, 15353.

104 Maiyalagan, T. (2008) *Appl. Catal. B*, **80**, 286.

105 Li, L., Wu, G., and Xu, B.-Q. (2006) *Carbon*, **44**, 2973.

106 Wee, J.-H., Lee, K.-Y., and Kim, S.H. (2007) *J. Power Sources*, **165**, 667.

107 Shao, Y.Y., Yin, G.P., Gao, Y.Z., and Shi, P.F. (2006) *J. Electrochem. Soc.*, **153**, 1093.

108 Tavasoli, A., Abbaslou, R.M.M., Trepanier, M., and Dalai, A.K. (2008) *Appl. Catal. A*, **345**, 134.

109 Gong, K., Chakrabarti, S., and Dai, L. (2008) *Angew. Chem., Int. Ed.*, **47**, 5446.

110 Ledoux, M.-J. and Pham-Huu, C. (2005) *Catal. Today*, **102**, 2.

111 Shinkarev, V.V., Fenelonov, V.B., and Kuvshinov, G.G. (2003) *Carbon*, **41**, 295.

112 Verweij, H., Schillo, M.C., and Li, J. (2007) *Small*, **3**, 1996.

113 Jakobtorweihen, S., Lowe, C.P., and Keil, F.J. (2007) *J. Chem. Phys.*, **127**, 024904.

114 Serp, P. (2009) in *Carbon Materials for Catalysis* (eds P. Serp and J.L. Figueiredo), John Wiley & Sons, Inc., Hoboken, NJ.

115 Viera, R., Pham-Huu, C., Keller, N., and Ledoux, M.J. (2002) *Chem. Commun.*, 954.

116 Pham-Huu, C., Keller, N., Charbonniere, L.J., Ziessel, R., and Ledoux, M.J. (2000) *Chem. Commun.*, 1871.

117 Saha, M.S., Li, R., and Sun, X. (2008) *J. Power Sources*, **177**, 314.

118 Goettmann, F. and Sanchez, C. (2007) *J. Mater. Chem.*, **17**, 24.

119 Kondratyuk, P. and Yates, J.T. (2007) *J. Am. Chem. Soc.*, **129**, 8736.

120 Ajayan, P.M. and Iijima, S. (1993) *Nature*, **361**, 333.

121 Tsang, C., Chen, Y.K., Harris, P.J.F., and Green, M.L.H. (1994) *Nature*, **372**, 159.

122 Dujardin, E., Ebbesen, T.W., Hiura, H., and Tanigaki, K. (1994) *Science*, **265**, 1850.

123 Chen, S., Wu, G., Sha, M., and Huang, S. (2007) *J. Am. Chem. Soc.*, **129**, 2416.

124 Kondratyuk, P. and Yates, J.T., Jr. (2007) *J. Am. Chem. Soc.*, **129**, 8736.

125 Monthioux, M. (2002) *Carbon*, **40**, 1809.

126 Tessonier, J.-P., Pesant, L., Ehret, G., Ledoux, M.J., and Pham-Huu, C. (2005) *Appl. Catal. A*, **288**, 203.

127 Pan, X., Fan, Z., Chen, W., Ding, Y., Luo, H., and Bao, X. (2007) *Nat. Mater.*, **6**, 507.

128 Ersen, O., Werckmann, J., Houllé, M., Ledoux, M.J., and Pham-Huu, C. (2007) *Nano Lett.*, **7**, 1898.

129 Winter, F., Leendert Bezemer, G., vand der Spek, C., Meeldijk, J.D., van Dillen, A.J., Geus, J.W., and de Jong, K.P. (2005) *Carbon*, **43**, 327–332.

130 Wu, H.-Q., Wei, X.-W., Shao, M.-W., Gu, J.-S., and Qu, M.-Z. (2002) *J. Mater. Chem.*, **12**, 1919–1921.

131 Zhang, A.M., Dong, J.L., Xu, Q.H., Rhee, H.K., and Li, X.L. (2004) *Catal. Today*, **93–95**, 347–352.

132 Zhang, J., Hu, Y.-S., Tessonnier, J.-P., Weinberg, G., Maier, J., Schlögl, R., and Su, D.S. (2008) *Adv. Mater.*, **20**, 1450.

133 Zhang, J., Muller, J.-O., Zheng, W., Wang, D., Su, D., and Schlogl, R. (2008) *Nano Lett.*, **8**, 2738.

134 Kim, B.M., Qian, S., and Bau, H.H. (2005) *Nano Lett.*, **5**, 873.

135 Bazilevsky, A.V., Sun, K., Yarin, A.L., and Megaridis, C.M. (2008) *J. Mater. Chem.*, **18**, 696.

136 Winter, F., Leendert Bezemer, G., van der Spek, C., Meeldijk, J.D., van Dillen, A.J., Geus, J.W., and de Jong, K.P. (2005) *Carbon*, **43**, 327.

137 Lock, E.H., Merchan-Merchan, W., D'Arcy, J., Saveliev, A.V., and Kennedy, L.A. (2007) *J. Phys. Chem. C*, **111**, 13655.

138 Chang, J.-Y., Mai, F.-D., Lo, B., Chang, J.-J., Tzing, S.-H., Ghule, A., and Ling, Y.-C. (2003) *Chem. Commun.*, 2362.

139 Wang, C., Guo, S., Pan, X., Chen, W., and Bao, X. (2008) *J. Mater. Chem.*, **18**, 5782.

140 Castillejos, E., Deboutière, P.J., Solhy, A., Martinez, V., Kihn, Y., Ersen, O., Philippot, K., Chaudret, B., and Serp, P. *Angew. Chem., Int. Ed.*, 2009, **48**, 2529–2533.

141 Costa, P.M.F.J., Sloan, J., Rutherford, T., and Green, M.L.H. (2005) *Chem. Mater.*, **17**, 6579.

142 Schulte, K., Swarbrick, J.C., Smith, N.A., Bondino, F., Magnano, E., and Khlobystov, A.N. (2007) *Adv. Mater.*, **19**, 3312.

143 Jain, D. and Wilhelm, R. (2007) *Carbon*, **45**, 602.

144 Capobianchi, A., Foglia, S., Imperatori, P., Notargiacomo, A., Giamatteo, M., Del Buono, T., and Palange, E. (2007) *Carbon*, **45**, 2205.

145 Fu, Q., Weinberg, G., and Su, D.-S. (2008) *New Carbon Mater.*, **23**, 17.

146 Korneva, G., Ye, H., Gogotsi, Y., Halverson, D., Friedman, G., Bradley, J.-C., and Kornev, K.G. (2005) *Nano Lett.*, **5**, 879.

147 Wang, X.-H., Orikasa, H., Inokuma, N., Yang, Q.-H., Hou, P.-X., Oshima, H., Iyoh, K., and Kyotani, T. (2007) *J. Mater. Chem.*, **17**, 986.

148 Orikasa, H., Inokuma, N., Ittisanronnachai, S., Wang, X.H.,

Kitakami, O., and Kyotani, T. (2008) *Chem. Commun.*, 2215.

149 Guo, D.Z., Zhang, G.M., Zhang, Z.X., Xue, Z.Q., and Gu, Z.N. (2006) *J. Phys. Chem. B*, **110**, 1571.

150 Srinivasan, C. (2006) *Curr. Sci.*, **90**, 756.

151 Noy, A., Park, H.G., Fornasiero, F., Holt, J.K., Grigoropoulos, C.P., and Bakajin, O. (2007) *Nano Today*, **2**, 22.

152 Liu, Y.-C., Shen, J.-W., Gubbins, K.E., Moore, J.D., Wu, T., and Wang, Q. (2008) *Phys. Rev. B*, **77**, 125438.

153 Schoen, P.A.E., Walther, J.H., Poulikakos, D., and Koumoutsakos, P. (2007) *Appl. Phys. Lett.*, **90**, 253116.

154 Chen, W., Pan, X., and Bao, X. (2007) *J. Am. Chem. Soc.*, **129**, 7421.

155 Koga, K., Gao, G.T., Tanaka, H., and Zeng, X.C. (2001) *Nature*, **412**, 802.

156 Chen, S., Wu, G., Sha, M., and Huang, S. (2007) *J. Am. Chem. Soc.*, **129**, 2416.

157 Kondratyuk, P. and Yates, J.T., Jr. (2007) *J. Am. Chem. Soc.*, **129**, 8736.

158 Chen, W., Fan, Z., Pan, X., and Bao, X. (2008) *J. Am. Chem. Soc.*, **130** (29), 9414.

159 Ma, H., Wang, L., Chen, L., Dong, C., Yu, W., Huang, T., and Qian, Y. (2007) *Catal. Commun.*, **8**, 452.

160 Shiozawa, H., Pichler, T., Grüneis, A., Pfeiffer, R., Kuzmany, H., Liu, Z., Suenaga, K., and Kataura, H. (2008) *Adv. Mater.*, **20**, 1443.

161 Zhang, J., Hu, Y.-S., Tessonnier, J.-P., Weinberg, G., Maier, J., Schlögl, R., and Su, D.S. (2008) *Adv. Mater.*, **20**, 1450.

12
Carbon Nanotubes as Containers

Thomas W. Chamberlain, Maria del Carmen Gimenez-Lopez, and Andrei N. Khlobystov

12.1
Introduction

The capacity of carbon nanotubes to serve as containers for a wide range of molecular and atomic compounds continues to stimulate the scientific imagination ever since it was discovered that the molecules can be inserted into the quasi-1D channels of nanotubes just over a decade ago [1]. The main attraction of nanotubes as nanosized containers is their extremely high mechanical stability, their relative chemical inertness, and the availability of different diameters from subnanometer to hundreds of nanometers. Often molecules and atoms encapsulated in nanotubes become arranged in geometrically regular 1D arrays, some of which do not exist outside carbon nanotubes and thus can be regarded as products of confinement at the nanoscale. The structural and dynamic properties of some materials encapsulated in nanotubes drastically change as a result of this confinement, and there is a growing body of evidence that the chemical reactivity of encapsulated compounds can also be altered inside carbon nanotubes. The effects of the encapsulated species on the physicochemical properties of nanotubes are not fully understood, this may be related to the intrinsic complexity of nanotube properties or the polydispersity of nanotube samples. However, careful experimentation involving a combination of local probe (microscopy) and spectroscopy techniques is expected to shed new light on this problem.

In this chapter, we review the current state of the area of filled carbon nanotubes (from 2005 onward). Most of the examples described in the chapter involve single-walled (SWNT) and multiwalled (MWNT) carbon nanotubes with internal diameters below 100 nm. Tubular structures with diameters above 100 nm are technically considered to be in the category of *micro*tubes and should be distinguished from the genuine examples of nanoscale confinement. Unfortunately, in the current literature, the distinction between nanotubes and microtubes is not sufficiently clear, which may lead to some confusion in this area in the long term. Also, we have limited the scope of this chapter to examples only where the nanotubes were filled after their formation (i.e., cases where carbon nanotubes coformed or formed around the noncarbon material are not included) and where the presence of guest species

Carbon Nanotubes and Related Structures. Edited by Dirk M. Guldi and Nazario Martín
Copyright © 2010 WILEY-VCH Verlag GmbH & Co. KGaA, Weinheim
ISBN: 978-3-527-32406-4

inside nanotubes is unambiguously confirmed by transmission electron microscopy (TEM), which still remains the only direct method of verification of nanotube filling.

This area of nanotube research has recently been reviewed in research journals [2–4] and a book [5], where the readers can find examples and references dating before 2005.

12.2
Mechanisms of Nanotube Filling

Depending on the nature of the compound to be inserted in carbon nanotubes (guest compound), one of the three general methods of nanotube filling can be used. For compounds that can be vaporized, a mixture of guest compound and carbon nanotubes with open termini is typically enclosed in a sealed container and heated at temperatures near or above the sublimation point of the compound. The mechanism of this process has been explored in detail for C_{60} fullerene as the guest molecule [6, 7]. It is generally accepted that prior to their insertion into the nanotubes, the guest molecules become adsorbed onto the nanotube surface (Figure 12.1b) but remain mobile and migrate freely along the nanotube eventually reaching the nanotube's terminus (Figure 12.1c). A moderate activation barrier (about 0.3 eV for C_{60}) may need to be overcome for the molecule to enter the nanotube, but this energetic penalty is richly compensated upon encapsulation of the molecule (Figure 12.1d) due to the significantly enhanced van der Waals interactions between the nanotube and the encapsulated molecule (i.e., nanotube surface wraps around the molecule maximizing the surface of the contact).

Clearly, the efficiency of encapsulation in this case depends critically on the ratio of the diameter of nanotube and the diameter of guest molecule [4]. For the optimum value of nanotube–guest molecule interactions, the distance between the surface of

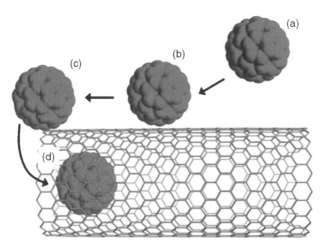

Figure 12.1 Stages of carbon nanotube filling with fullerene C_{60} via the gas-phase mechanism.

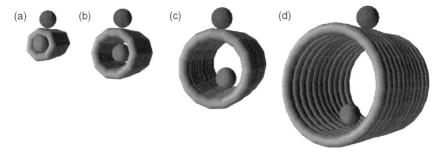

Figure 12.2 Schematic illustration of different ratios of guest molecule diameter to nanotube diameter.

the molecule and the inner surface of nanotube needs to be close to the van der Waals gap of $\sim0.3\,nm$, so that if the diameter of the molecule is n nm, the optimum diameter of carbon nanotube for hosting such molecule is $d_{NT}=n + 2 \times 0.3\,nm$ (Figure 12.2b). If the nanotube diameter is below this value, the guest molecule will not be able to enter the nanotube (Figure 12.2a); if the nanotube diameter is significantly larger than the optimum value (Figure 12.2d), the nanotube–molecule interaction will be less effective which could result in a reversible encapsulation.

Carbon nanotubes immersed in a liquid with surface tension below $200\,mN/m$ become spontaneously filled according to the Young–Laplace law ([8] and references therein). This filling mechanism is often referred to as capillary filling. However, it must be noted that classical continuum approximation of liquids inside nanotubes with diameters below 8–10 nm breaks down, so that the well-defined gas–liquid interface (meniscus) disappears, and the liquids confined in nanotubes exhibit anomalous behavior [8]. The ionic compounds have intrinsic low volatility and therefore most of them have been inserted in nanotubes via the capillary filling mechanism by immersing nanotubes into a pure molten salt or an eutectic mixture of salts. Ionic inorganic salts such as KI were among the first compounds inserted into nanotubes, and their presence and structural behavior in nanotubes have been unambiguously demonstrated by high-resolution TEM (HRTEM). Detailed theoretical investigations of the capillary filling process [9, 10] revealed that it has a stepwise mechanism. Initially, disordered cations and anions in the molten phase become ordered upon insertion into the nanotube as a result of confinement (Figure 12.3). Coordinatively unsaturated cations generated at the opening of carbon nanotube become a driving force for the filling as they maximize their coordination number by pulling in anions and by rotating the 1D crystallite around its axis (Figure 12.3) [9].

Encapsulation of compounds that neither melt nor sublime has been shown to be possible by immersion of nanotubes with open termini in solutions of these compounds. This method tends to be less efficient than others as the guest compound in this case is surrounded by a large number of solvent molecules that interfere with the filling process and have to be taken into account when considering the mechanism. No viable mechanism explicitly including and explaining the role of the solvent in this filling method has been formulated so far. The key role of the

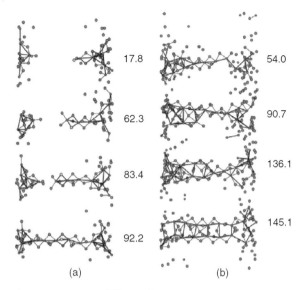

Figure 12.3 Stages of filling carbon nanotube with ionic compound LaCl₃. (Reproduced with permission from Ref. [9].)

solvent is to enable the transport of guest molecules from the bulk solution phase into carbon nanotubes. However, the fate of the thousands of solvent molecules that enter the nanotubes with each guest molecule (Figure 12.4) remains largely unclear. At present, only tentative attempts to provide a model for transport of fullerenes from conventional and supercritical fluid solutions into carbon nanotubes have been made [11], and there is a clear need for a solid theoretical basis describing the filling of nanotubes from solutions.

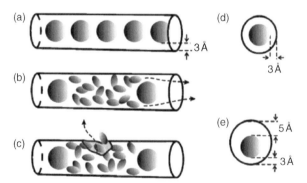

Figure 12.4 Schematic diagram illustrating the importance of solvent molecules (shown by oblong shapes) for nanotube filling in solution. Solvent molecules are often omitted from consideration in nanotube filling mechanisms. However, in most cases the physical size of solvent molecules is comparable with the size of encapsulated compound (round shape), and so escape routes of solvent from nanotubes are important to consider (b, c).

12.3
Fullerenes as Guest Molecules

The interactions of fullerenes with carbon nanotubes have been studied more than the interactions of any other molecule. Being structurally and chemically related to carbon nanotubes, fullerenes form extremely effective van der Waals interactions with the interior of nanotubes that can reach up to 3 eV per C_{60}, depending on the diameter of the nanotube. In addition, the distinct round shape of fullerene cages and their relatively high stability in the electron beam make them ideal molecules for observation in TEM, which is widely used as a direct characterization tool for 1D assemblies of fullerenes inside nanotubes.

12.3.1
Fullerene C_{60}

Buckminster fullerene, C_{60}, was the first molecule inserted and unambiguously identified in SWNT [1]. The facile nature of the processes with which C_{60} can be inserted into SWNT coupled with its relatively high abundance, compared to the higher fullerenes, makes C_{60} still the most comprehensively studied molecule within SWNT (Table 12.1). The increase in the number of encapsulation methods reported, ranging from sublimation at high temperatures to ultrasonication at room temperature in solution, and the improvement and increased availability of HRTEM facilities have led to a vast amount of work being published in recent times. The increased understanding of fullerene encapsulation has even led to its use as a diagnostic test for SWNT purity [12].

The unique mechanical properties of carbon nanotubes have led to their application in a number of strong, light-weight materials [13]. Conventional wisdom would predict that filling would increase the mechanical strength of a hollow structure; hence, work has been done to test and fully quantify the mechanical properties of C_{60}@SWNT peapod structures. Nanochemical resonance studies of SWNT (PLV method) peapod bundles reveal an increase in average bending modulus from 240 ± 105 GPa for empty bundles to 650 ± 105 GPa for C_{60}@SWNT bundles [14]. The lower bending modulus for empty SWNT bundles compared to theoretical values reported for individual SWNT is explained by the weak interaction and sliding effects that are known to exist between nanotube surfaces. The significant increase in stiffness of the nanotube after filling is due to an increase in the strain energy of the system upon the deformation of C_{60} chains [14]. Theoretical calculations predict a size-dependent effect upon the buckling strength of individual SWNT when C_{60} is encapsulated [15]; the linear chains of fullerenes in C_{60}@[10,10]SWNT are cylindrically symmetric that helps increase the buckling strength. However, for the wider [14,14]SNWT and [18,18]SWNT where the nonlinear fullerene packing breaks the nanotube symmetry, the buckling strength actually decreases due to an uneven distribution of applied force on the circumference of the SWNT [15].

Optical bandgap modification of SWNT is also reported as a result of fullerene encapsulation [16–19], proposing the development of SWNT peapod structures as molecular wires and sensors. Photoluminescence (PL) and the corresponding

Table 12.1 Fullerene guest compounds inserted in carbon nanotubes.

Compound	Type of nanotube[a]	Nanotube internal diameter (nm)	Method of filling	Reference
C_{60}	SWNT (AD)	1.4	Gas phase, 10^{-6} Torr, 650 °C, 6 h	[12]
C_{60}	SWNT (PLV)	1.4	Gas phase, 10^{-6} Torr, 650 °C, 3 h	[14]
C_{60}	SWNT (PLV)	1.3	Gas phase, 10^{-6} Torr, 500 °C, 48 h	[16]
C_{60}	DWNT (AD)	4–5	Gas phase	[18]
C_{60}	SWNT (AD)	1.3–1.4	Gas phase, 10^{-6} Torr, 500 °C, 48 h	[21]
C_{60}	SWNT (PLV)	?	Gas phase, 10^{-6} Torr, 500 °C, 48 h	[22]
C_{60}	SWNT (AD)	1.44–1.56	Gas phase, 10^{-6} Torr, 500 °C, 2 h	[23]
C_{60}	SWNT (CVD)	1.28	Gas phase, 500 °C, 2 days	[24]
C_{60}	SWNT/DWNT (CVD)	1–5	Gas phase, 10^{-6} Torr, 500–550 °C, 18–24 h	[25]
C_{60}	SWNT (LA)	1.39	Gas phase, 10^{-6} Torr, 650 °C, 3 h	[27]
C_{60}	SWNT (AD)	?	Gas phase, low pressure, 450 °C, 4 days	[28]
C_{60}	SWNT (AD)	1.3–1.4	Gas phase, low pressure, 440 °C, 2 days	[30]
C_{60}	SWNT/DWNT		Gas phase, 10^{-6} Torr, 650 °C, 9 h	[31]
C_{60}	SWNT (PLV)	1.2–1.3	Gas phase, 10^{-6} Torr, 500 °C, 48 h	[33]
C_{60}	SWNH	2–5	Solution (toluene), ultrasonication, 1 min, slow evaporation	[34]
$C_{60} + C_{70}$	DWNT (CVD)	2.5–3	Gas phase, 10^{-6} Torr, 500 °C, 2 days	[32]
$C_{60} + C_{70} + C_{60}H_{28}$	SWNT (AD)	1.22–1.6	Gas phase, 10^{-4} Torr, 420 °C, 10 days	[36]

C_{60} + C_{70}	SWNT (AD)	1.4–1.58	Gas phase, 10^{-6} Torr, 500 °C, 24 h	[37]
C_{80} and $Er_3N@C_{80}$	SWNT	?	Gas phase, 10^{-6} Torr, 500 °C, 2 days	[35]
$Dy_3N@C_{80}$	SWNT (LA)	?	Gas phase, 440 °C, 24 h	[38]
$N@C_{60}$	SWNT (AD)	1.4	Solution (toluene), ultrasonication, 2 h	[39]
$N@C_{60}$	SWNT (AD)	1.4, 2.5–3	Solution (toluene), ultrasonication, 2 h	[40]
$Sc@C_{82}$ + $La@C_{82}$	SWNT (AD)	?	Gas phase, 10^{-6} Torr, 440 °C, 4 days	[41]
$Tb_2@C_{92}$ + $Gd@C_{82}$ + $Gd_2@C_{92}$	SWNT (AD)	1.4–1.6		[42]
$Gd@C_{82}$ + $Ca@C_{82}$	SWNT (AD)	1.3–1.4	Gas phase, 10^{-6} Torr, 500 °C, 2 days	[43]
$La_2@C_{80}$	SWNT (AD)	1.4	Gas phase, 10^{-6} Torr, 450 °C, 24 h	[44]
$Gd@C_{82}$	SWNT (AD)	0.65–1.9, 5	Gas phase, 10^{-6} Torr, 500 °C, 2 days	[45]
$Gd@C_{82}$ + $Ti_2@C_{80}$ + $Ce_2@C_{80}$ + $Gd_2@C_{92}$	SWNT (LA)	1.3–1.6	Gas phase, 10^{-4} Torr, 450–500 °C, 48 h	[46]
$Gd@C_{82}$	SWNT (AD)	?	Gas phase, 10^{-6} Torr, 500 °C, 2 days	[47]
C_{60}-alkyl	SWNT (AD)	1.39–1.41	Supercritical fluid (scCO2), 50 °C, 100–150 bar, 5 days	[50]
C_{60}-aryl	SWNT (AD)	1.39–1.41	Supercritical fluid (scCO2), 50 °C, 100–150 bar, 5 days	[51]
C_{60}-retinal chromophore	SWNT (PLV)	1.38	Solution (chloroform), 65 °C, 3 h	[52]
C_{60}-$C_6H_{12}NO$	SWNT (AD)	1.39–1.41	Solution (carbon disulphide), ultrasonication, 30 min, supercritical fluid (scCO2), 50 °C, 100–150 bar, 5 days	[53]
$C_{59}N$	SWNT	1.4	Solution (toluene), ultrasonication, 10 min	[54]
C_{60}-C_8H_9O	SWNT (AD)	1.4	Solution (toluene), ultrasonication, 1 h	[55]
$C_{59}N$ + $(C_{59}N)_2$	SWNT (AD)		Gas phase, 10^{-6} Torr, 520 °C, 72 h; Supercritical fluid (scCO2), 50 °C, 100–150 bar, 5 days	[56]

a) When known, the method of nanotube growth is given in brackets: AD: arc discharge; CVD: chemical vapor deposition; PLV: pulsed laser vaporization; LA: laser ablation.

excitation spectroscopy reveal that the optical transition energies shift upon encapsulation of C_{60} molecules strongly depends on the chirality and diameter of the SWNT [16]. This is understood to be a result of the local strain effects of the nanotube and hybridization between the encapsulated fullerenes and the outer SWNT. The decrease of 36 meV for the bandgap of $C_{60}@[14,14]SWNT$ is in agreement with previous studies, which observed a downshift in the first van Hove gap of 60 meV for a similar system [20].

$C_{60}@SWNT$ peapod structures have been used as the channel in field-effect transistors (FETs) fabricated to explore their exciting conductance properties [18].

Incorporation of metal atoms within the $C_{60}@SWNT$ structure was attempted in an effort to tune the SWNT conductance properties, metal doping of $C_{60}@SWNT$ with potassium and cesium as well as a general method for inclusion of any alkali metal into peapod structures has been reported [21–23]. By exposing the formed $C_{60}@SWNT$ to potassium vapor at 200 °C for 50 h, potassium atoms were introduced into the gaps between pairs of fullerene cages and the nanotube sidewall [21]. Cesium was introduced by first forming the fulleride anion with cesium as the counterion using chemical reduction. The resultant salt was then encapsulated inside the C_{60} by sublimation methods leading to an ordered Cs_2C_{60} structure observed by HRTEM (Figure 12.5) [23]. Replacement of the Cs with any alkali metal can be achieved by forming the appropriate fulleride salt precursor. The metal atoms within the structure were confirmed to be cationic using electron energy loss spectroscopy (EELS) with a suitable alkali metal salt (such as CsCl) for comparison. Cs^+ ions inside the nanotube can migrate only in the limited space between C_{60} cages and their arrangement is sensitive to elastic deformation of the outer nanotube [23]. The behavior of the doped chemical species is therefore significantly affected by the distortion and defects of the nanotube.

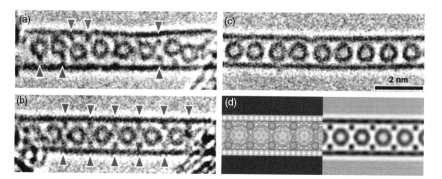

Figure 12.5 (a–c) TEM images of $C_{60}@SWNT$ peapods after doping with cesium. Cs^+ ions inserted inside the peapods are indicated by triangles. (d) Schematic illustration of the ordered Cs_2C_{60} structure inside a SWNT (left) and its simulated TEM image (right). The middle of the nanotube wall thickness of the model in (d) corresponds to the limit of the dark contrast given by the nanotube wall in the simulated TEM image. (Reproduced with permission from Ref. [23].)

Electrical transport measurements of cesium-doped C_{60}@DWNT show the structures to exhibit both n- and p-type semiconducting behavior; it is demonstrated that by controlling the metal concentration it is possible to form p–n junctions [18]. Both these metal-doped peapod structures show great potential as building blocks for electronic nanodevices [18].

The fabrication of nanodevices from C_{60}@SWNT has been aided by the development of a method to form peapod structures directly onto silicon substrates [24, 25]. Chemical vapor deposition of vaporized methanol over metal catalyst particles leads to high-purity/high-density SWNT on a SiO_2 wafer, the ends of the SWNT were opened by oxidation in air and C_{60} vapor was passed over the wafer. The resultant peapods confirmed by HRTEM and the samples were analyzed using Raman spectroscopy where a downshift in the radial breathing mode (RBM) of the SWNT was attributed to the encapsulation of C_{60} [24]. Peapod structures have also been synthesized directly onto TEM grids that allow instant analysis of the structure formed [25].

Formation of DWNTs by thermal treatment of peapod structure of SWNT containing fullerenes was first observed in 2001 [26]. Recently, work has been carried out to understand the growth mechanism of this process; *in situ* Raman spectroscopy and HRTEM studies and the use of X-ray diffraction (XRD) have been reported (Figure 12.6) [27, 28]. The Raman spectra of a C_{60}@SWNT undergoing thermal treatment show that the C_{60} molecules break up significantly faster than the inner tube is formed, the discrepancy in time constant would imply the existence of an intermediate phase. Comparison of the Raman data with XRD gives evidence that the intermediate phase that is formed is solid and has the linear periodicity of the starting material [27]. It is also observed that the formation of inner tubes in SWNT peapod structures with a larger diameter occurs up to twice as slowly as the intermediate phase decays; this is attributed to the larger number of Stone–Wales transformations that are required to form these larger nanotubes [27].

An *in situ* HRTEM study shows a time series of images depicting five neighboring C_{60} fullerenes coalescing into a large tubular C_{300} molecule over the course of 25 min, the coalescence is attributed to the electron beam irradiation [28]. The C_{300} molecule

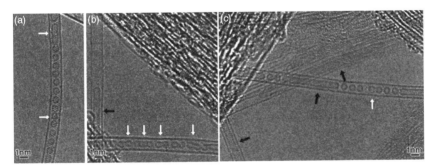

Figure 12.6 HRTEM images of peapods transformed at 1250 °C after (a) 5 min, (b) 15 min, and (c) 25 min. The white and black arrows indicate modulated peapods (intermediate phase) and DWNTs, respectively. (Reproduced with permission from Ref. [27].)

translates along the nanotube displaying a unique corkscrew motion. A separate XRD study details the progressive polymerization of C_{60} within SWNT at a pressure of 4 GPa and temperature of 1023 °C where the final system is stable at ambient conditions and still contains a 1D chain structure with a periodicity of 0.87 nm and no DWNT formation [29]. By immersing C_{60}@SWNT peapods into molten iodine, it is possible to introduce iodine into the empty area between fullerene cages in larger SWNT [30]. HRTEM was used to observe the iodine molecules as dark spots with up to five atoms in a row depicted. The system was then heated at 550 °C (250 °C below the previously reported minimum temperature required for fullerene coalescence), DWNTs were formed and characterized using HRTEM and Raman spectroscopy. The iodine is thought to assist the coalescence of the fullerene cages although no mechanism is postulated. HRTEM also provides evidence that iodine atoms are trapped between the original nanotube and the newly formed inner tube, potentially providing a novel method of synthesizing DWNT containing intercalated guest species [30].

The efficient conversion of C_{60}@SWNT into DWNT was also used in an electron paramagnetic resonance (EPR) study [31]. Two small lines are observed in the EPR spectra of both C_{60}@SWNT and DWNT with temperature dependence that implies "super-Curie" paramagnetic behavior. The linewidth of the two lines is attributed to the inner species, and the outer walls are metallic in nature and the outer nanotube's resonance broadens upon growth of the inner nanotubes. A sudden decrease in relaxation rate below 20 K is attributed to the opening of a spin gap [31]. The thermal stability of C_{60}@DWNT peapods, formed by vapor deposition, was studied using thermo gravimetric analysis (TGA) [32]. TGA of the as-prepared DWNT peapods showed that they had a higher thermal stability (burning temperature of 636.2 °C) in comparison to pristine DWNTs (581.8 °C). It is predicted that the increase in burning temperature is caused by the rearrangement of the electronic structure between C_{60} and nanotube resulting in a slightly negative charged area between fullerenes and the inner tube and a positive charged area on the surface of the outer tube. This adversely affects the oxidation reaction for the peapods compared to the pristine DWNT and allows the filling yield of the DWNT to be calculated from the TGA data [32].

PL is a useful tool for characterizing the unique electronic properties of SWNTs; little work has been possible on DWNT. It was expected that the inner tube is protected from the environment and its intrinsic optical properties will be preserved; however, it is difficult to assign PL to the inner tubes of DWNT as samples of DWNT made by CVD also contain SWNT impurities with diameters identical to the inner tube of the DWNT. By studying DWNT synthesized by thermal treatment of C_{60}@SWNT structures, any SWNT impurities will have significantly larger diameters than the inner tube of the resultant DWNT. This was exploited in Ref. [33] where the first unambiguous PL study of the inner tubes of DWNT is presented. The PL is severely suppressed as a consequence of an interlayer interaction between the inner and outer tubes that efficiently quenches the PL signals [33].

Fullerene release from carbon nanostructures by near-infrared (NIR) laser irradiation has also been achieved [34]. Single-wall carbon nanohorns (SWNHs) can encapsulate C_{60} in an analogous fashion to SWNT to form peapod structures, when irradiated with an NIR laser fullerene is released into the surrounding solution. This procedure is proposed to have potential applications in controlled drug delivery [34].

12.3.2
Higher Fullerenes

Significantly less work has been dedicated to the study of higher fullerenes C_n $(n > 60)$ in recent years, and those reported are concentrated mainly on the second most abundant fullerene, C_{70}. This is perhaps due to the fact that fullerenes other than C_{60} and C_{70} are less readily available commercially in a pure form in sufficient quantities. Another problem is that many of the higher fullerenes, such as C_{78}, C_{80}, C_{82}, or C_{92} can exist in many different isomers that are difficult to separate and distinguish unambiguously. Different isomers of fullerenes have different shapes, some of which substantially deviate from a spheroidal shape of C_{60}. The different shapes of these fullerenes lead to disorder when encapsulated in SWNT that makes controlling the precise orientations of the internal molecules in the nanotube impossible.

The problem of the presence of a number of isomers may be resolved by improving the aberration-corrected TEM. A detailed study of a single isomer of C_{80} inside a SWNT is presented in Ref. [35], showing the ellipsoidal D_{5d}-C_{80} in atomic resolution with the pyrene-like tetracyclic component of the cylindrical body unambiguously identified. Sequential images show the rotation of the fullerene cage within the SWNT (Figure 12.7) with a fullerene tilt angle of $30°$ observed experimentally as the

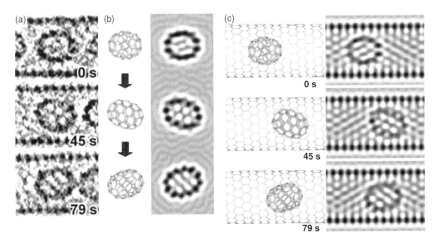

Figure 12.7 (a) FFT-processed TEM images of the D_{5d}-C_{80} fullerene. (b) Suggested orientations of the D_{5d}-C_{80} cage (left panel) and simulated TEM images (right panel). (c) Possible arrangement of the D_{5d}-C_{80} cage inside the (18,18) SWNT (left panels) and simulated TEM images (right panels). (Reproduced with permission from Ref. [35].)

orientation that gives the best contrast profile [35]. Time-resolved TEM studies show examples of nonspherical fullerene cage tumbling and rotating as they migrate along the length of the nanotube. The use of electron microscopy to analyse higher fullerenes encapsulated in SWNT to identify the different fullerene isomers enables the identification of a specific fullerene isomer. This could lead to the availability of high-purity samples of single isomers of the higher fullerenes, which is a crucial step if their chemistry is to be investigated fully.

A study of C_{70} is reported where C_{70} was encapsulated in SWNT of a variety of diameters, ranging from 1.22 to 1.6 nm, in order to probe the effect the encapsulated fullerenes have on the nanotube [36]. The systems were probed using HRTEM and UV–Vis–NIR and Raman spectroscopy, and it is observed that filling the SWNT with C_{70} markedly alters the electronic structure of the nanotube. Raman spectroscopy shows a redshift in the radial breathing mode of the SWNT for peapod structures formed from narrow nanotubes (1.22–1.3 nm in diameter), whereas for wider nanotubes (1.5–1.6 nm in diameter) a blueshift is observed. These variations in vibrational energy are attributed to changes in the electron density surrounding the carbon framework of the SWNT due to interaction with electron density upon the encapsulated fullerene cage. The uneven distribution of electron density upon the ellipsoidal fullerene cage means that, unlike spherical C_{60} peapod structures, changes in orientation of the fullerene cage within the nanotube greatly affect the size of the fullerene–nanotube interaction. As the diameter of the SWNT increases, the ellipsoidal cage rotates perpendicular to the nanotube axis to minimize the total energy of the peapod structure, and this change in orientation leads to a larger fullerene–nanotube interaction causing a blueshift in the energy of the RBM mode. Raman spectra at 633 and 1064 nm excitation wavelengths reveal that the RBM frequencies of C_{70} peapods are downshifted compared to empty nanotubes [36]. These results suggest that the average diameter of the electron cloud in SWNT is affected by the filling of C_{70}, and the magnitude and nature of the effect depend more upon the SWNT diameter than upon the filling of the spheroidal C_{60} [36].

The geometrical confinement of C_{70} inside SWNT was explored by subjecting C_{70} and C_{60} peapod structures in SWNT (diameter 1.4 nm) to high pressure and temperature conditions similar to those reported for polymerizing pure C_{60} in its crystalline form [37]. *Ex situ* X-ray diffraction (XRD) experiments highlight that although the encapsulated C_{60} molecules polymerize in 1D chains inside the SWNT, no polymerization of the C_{70} molecules is observed. This is attributed to the fact that only the double bonds in the poles of the C_{70} molecule can participate in the formation of covalent bonds between molecules and thus, unlike C_{60}, C_{70} molecules do not form linear polymer chains but are arranged in zigzag chains with a periodicity of 1.79 nm. The approximate value of the smallest diameter of nanotube that the zigzag chains can be accommodated in is 1.8 nm; it is predicted that the zigzag chains are not energetically stable inside the much narrower SWNT, so no polymerization occurs [37]. The unpredictable behavior of C_{70} peapods due to the much stricter constraints on the reactivity of the nonspheroidal C_{70} than the highly symmetrical C_{60} makes C_{70} peapods interesting systems as they exhibit a fullerene behavior different from the bulk phase.

12.3.3
Endohedral Fullerenes

Endohedral fullerenes are fullerenes that have additional atoms, ions, or clusters enclosed within their inner spheres. Two types of endohedral complexes exist: endohedral metallofullerenes and nonmetal endohedral fullerenes. Endohedral metallofullerenes are characterized by the fact that valence electrons of metal atom normally transfer to the fullerene cage and that the metal atom takes a position off-center in the cage. The amount of the electron transfer is not always easy to determine. Contrary to the endohedral metallofullerenes, in nonmetal endohedral fullerenes no charge transfer of the encapsulated atom to the fullerene cage takes place, and the endohedral atom is usually located in the center of the cage.

Endohedral metallofullerenes are generally inserted into SWNT using sublimation methods identical to those used for the gas-phase encapsulation of C_{60} (Table 12.1). The fullerene and the chemically or thermally opened nanotubes are sealed in an evacuated container and heated above the sublimation point of the fullerene [38]. No mechanistic studies of the route for this encapsulation have been reported, although it is thought to proceed in a similar fashion to C_{60}. Until recently, the high temperature of sublimation methods prevented the encapsulation of the thermally sensitive nonmetal endohedral fullerenes; however, the development of room-temperature methods such as solution and supercritical fluid filling has enabled peapod systems of the fullerenes to be formed in modest yields of 5–10% [39, 40].

Filling rates are usually calculated by analyzing the filled nanotube sample using HRTEM and from the images obtained estimating the ratio of filled and empty nanotubes [35, 41]. It is also possible to use photoemission spectroscopy to obtain estimates for the filling factor with an accuracy of $\pm 10\%$ (compared to HRTEM) [38]. Comparison of the emission spectra of the $(Dy_3N@C_{80})@SWNT$ peapod structure at 400 eV and that of pristine SWNT gives a quantifiable signal corresponding to the encapsulated fullerene [38]. Evidence of charge transfer between the $Dy_3N@C_{80}$ and the SWNT is also quantified by looking at the structured photoemission close to the Fermi level. Subtraction of the pristine SWNT spectra gives well-reproduced atomic multiplet profiles almost identical to those observed for $Dy_3N@C_{80}$. From the relative intensities of the two sets of peaks, it is possible to estimate the effective oxidation state of the Dy ions to be about 3.0, which is larger than 2.8 reported previously for the pristine $Dy_3N@C_{80}$ [38].

The improvement in resolution of transmission electron microscopes has led to increasingly detailed studies of the structure of endohedral fullerene peapods. Time-dependent orientational changes in $Er_3N@C_{80}$ fullerenes inside SWNT were demonstrated using an aberration-corrected transmission electron microscope at the real atomic level (Figure 12.8) [35]. Detailed structures of the individual fullerene molecules and the encapsulated Er atoms were unambiguously visualized for the first time. This allows precise determination of both the orientation and the position of each fullerene cage with respect to the SWNT and also the intercage orientation of the Er_3N cluster to be deduced [35].

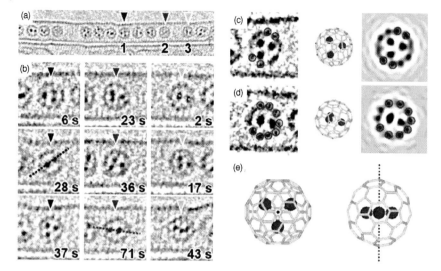

Figure 12.8 (a) A TEM image of the I_h-Er$_3$N@C$_{80}$ molecules aligned inside the (18,18) SWNT. (b) Sequential TEM images of the I_h-Er$_3$N@C$_{80}$ molecules **1**, **2**, and **3** (left, center, and right panels, respectively) marked with triangles in (a). (c) and (d) FFT-processed TEM images of the I_h-Er$_3$N@C$_{80}$ molecules **1** at 37 s (top panel) and **3** at 2 s (bottom panel). (e) Suggested orientations of the I_h-Er$_3$N@C$_{80}$ molecules (left) and simulated TEM images (right panels). (Reproduced with the permission from Ref. [35].)

In situ HRTEM allows to study the dynamic behavior of the fullerene cages; it is also possible to study the behavior of the metal ions within the fullerene cages that can give more insight into the cases of the energetic barriers to rotational and translational motion of the fullerene cages [41]. SWNTs were filled with the paramagnetic fullerene Sc@C$_{82}$ to form peapods, however in the HRTEM images taken using a 2 s exposure time, no Sc ions were observed. This is attributed to the constant rotation of the fullerene cage within the SWNTs so the Sc^{3+} ion is not stationary for a sufficiently long period of time to be visualized. The presence of Sc^{3+} was confirmed by encapsulating Sc@C$_{82}$ in MWNT, where it becomes stationary for a sufficiently long period of time to enable Sc^{3+} ions to be seen. In contrast, La@C$_{82}$ metallofullerenes in SWNT peapods have fixed orientations for extended periods of up to 50 s in SWNTs and La ions are observed in the HRTEM images, the La@C$_{82}$ is demonstrated to rapidly rotate between these fixed orientations [41]. Comparison of the behavior of the two endohedral fullerenes within the SWNT demonstrates that the intermetallofullerene interactions, which depend on charge transfer, play an important role in the dynamics of the systems along with the metallofullerene–nanotube interactions.

The effect the SWNT has on the rotational and translational motion of the encapsulated fullerene was also investigated [28]. A time series of HRTEM images presents the movement of a zigzag chain of encapsulated Sc$_3$C$_2$@C$_{80}$ endohedral fullerenes encapsulated in SWNT, the entire chain is observed to rotate under the influence of the electron beam inducing structural modification of the diameter and cross-sectional shape of the SWNT. The expansion and contraction of the SWNT are measured as 17% demonstrating nanoactuation in carbon peapods [28].

By increasing the irradiation time of the electron beam, it is possible to create defects in the carbon framework of the fullerene cages of metallofullerenes in peapod structures [42]. This enables the metal ion or ions to escape and migrate through the peapod structure; *in situ* microscopy enables the dynamic behavior of these metal ions to be precisely followed. $Tb_2@C_{92}$, $Gd@C_{82}$, and $Gd_2@C_{92}$ peapods were studied using HRTEM [42]. When exposed to the e-beam, a Tb^{3+} ion breaks out of the cage through an induced atomic defect that is a product of vacancy formation, the large size of a Tb^{3+} ion (0.185 nm) indicates that the fullerene must go through considerable deformation for the ion to escape. The fullerene reforms and Tb^{3+} ion is then observed to move through the peapod structure passing between fullerenes and the nanotube sidewall.

In attempts to observe metal ions being transferred between fullerene cages, a heteropeapod was formed containing both $Gd@C_{82}$ and $Gd_2@C_{92}$ metallofullerenes [42]. Despite the different affinity of two kinds of fullerenes, both were encapsulated within an SWNT forming a mixed peapod. During the HRTEM observation, three molecules are seen to coalesce and one Gd^{3+} ion moves through the atomic pathway interconnecting two neighboring cages. The fullerenes separate and an elongated fullerene containing three Gd^{3+} ions is formed. This is direct evidence of weak interactions between the metal ions and the encapsulating cage as intramolecular motion of the metal ions inside the fullerene readily occurs [42].

A complementary study compares peapod structures of $Gd@C_{82}$ and $Ca@C_{82}$ under identical electron beam irradiation using HRTEM *in situ* studies [43]. The Gd^{3+} is observed to break out of the cage through an irradiation-induced defect site and migrate through the peapod system (Figure 12.9a and b). The behavior of the Ca^{2+} ions inside the defective C_{82} is completely different, as the ion is seen to migrate to the defect site in the fullerene cage but is unable to escape even though radii of Gd^{3+} and Ca^{2+} are not significantly different (Figure 12.9d and e). Density functional theory predicts that the migratory behavior of the two ions in the defective cage is completely different due to the different interactions between vacant orbitals of the metal and the antibonding orbitals of the cage. The Gd^{3+} breakout is directly induced by the first carbon atom displacement, but the Ca^{2+} requires an additional bond cleavage with a larger activation energy (Figure 12.9c and f).

Owing to the absence of any metallic element, it is difficult to characterize the nonmetal endohedral fullerenes using HRTEM. Fortunately, the majority of work with nonmetal endohedral fullerenes is carried out using the spin active $N@C_{60}$ molecule. EPR has been used for $N@C_{60}$ peapods structures to probe the presence of the endohedral fullerene within the SWNT and the interactions within the nanotube, both interfullerene and between the fullerene and the nanotube sidewall [39].

A study used the EPR signal of the singlet nitrogen to show that the thermal stability of $N@C_{60}$ (the temperature at which the nitrogen escapes the fullerene cage) increased when encapsulated in SWNT to above 620 K compared to 500 K in its crystalline form. However, the temperature-dependent spin lattice relaxation time of the encapsulated molecule was shown to be significantly shorter (10 times) than the

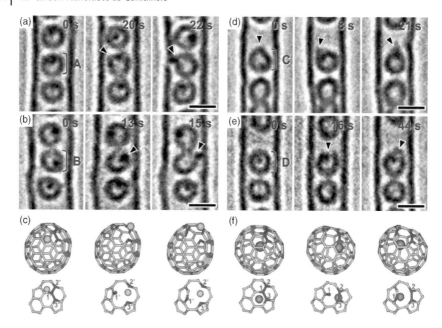

Figure 12.9 Sequential TEM images of Gd@C$_{82}$ (a and b) and Ca@C$_{82}$ (d and e) metallofullerenes aligned inside SWNTs. The Gd atoms in molecules (a) and (b) and the Ca atoms in (d) and (e) are detected as dark dots, and the defective parts of these fullerene cages are indicated by black arrows. Structures of Ca@C82 and Gd@C82 before and after defect formation (c and f, respectively) derived from defect evolution process. (Reproduced with permission from Ref. [43].)

crystalline material. This is explained by the presence of additional relaxation mechanisms within the SWNT. These include coupling of conduction electrons of the nanotube with the electron on nitrogen and paramagnetic relaxation from both the defects in the nanotubes and the transition metal catalyst particles that are inevitably present in nanotube samples [39]. The presence of dipolar coupling between the N@C$_{60}$ fullerene and electron spins in the nanotube is also observed in a separated study [40].

The coalescence of C$_{60}$ fullerenes inside SWNT to give DWNT has been studied in detail (Section 12.3.1), with various techniques used to study the formation and nature of these new systems; however, little work has been done on metallofullerenes. Recently, La$_2$@C$_{80}$ was encapsulated in SWNT and the resulting peapod structure was thermally treated at 1273 K for 48 h to coalesce the fullerenes within the nanotube [44]. This resulted in a DWNT with a narrow inner tube ($d_{NT} = 0.7$ nm) and the entrapped La ions form a linear arrangement of metal ions consisting of metal atom dimers orientated perpendicular to the direction of the nanotube. Detailed analysis revealed that the La–La distance in a metal atom dimer and the dimer–dimer separation measured from the center of the intensity maxima are 0.44 ± 0.02 and 0.48 ± 0.02 nm, respectively. Both these distances are larger than the spacing between La atoms in the crystal. It is anticipated that the newly formed atomic wires will have physical properties different from that of the endohedral fullerene or crystalline La.

This technique was taken further using $Gd@C_{82}$ encapsulated in SWNT of various diameters and thermally treating at high pressures to give a host of DWNTs with inner tube diameters from 0.65 to 5 nm filled with Gd nanowires [45]. These orientations of the metal ions inside the SWNTs that have been synthesized consist of a single Gd atomic chain, a one-dimensional alignment of Gd squares and nanowires that correspond to a one-dimensional segment of the bulk close-packed structure. It is proposed that by using this nanotemplate reaction, it is possible to fabricate various types of metal nanowires that are specific to the 1D space inside the SWNTs. The use of di- and trimetallofullerenes and mixtures of metallofullerenes containing different metals should allow the control of the structure of the resultant nanowires and also enable alloys to be formed [45].

Manipulating the bandgap of SWNT has been achieved by filling the nanotube with a wide variety of endohedral fullerenes [46]. By using the endohedral fullerene peapods as the channels of field-effect transistors, it is possible to fine-tune the size of the bandgap simply by changing the type of endohedral fullerene [46]. By varying the gate bias, it is shown that all metallofullerenes produce FETs exhibiting p- and n-type conductance, the so-called ambipolar carrier transportation. A comparison of the effects on nanotube bandgap induced by C_{60}, $Gd@C_{82}$, and $Gd_2@C_{92}$ showed direct reduction of the bandgap with an increase in charge on the fullerene cage. The size of the off-state region in FET decreases from 25 V for C_{60} to 15 and 0 V for $Gd@C_{82}$ and $Gd_2@C_{92}$. This is attributed to the degree of charge transfer between the endohedral fullerene and the SWNT directly governing the width of the off-state regions. This bandgap modulation is derived from the formation of new fullerene-derived states near the bandgap that are considered as the direct result of electron transfer between the metallofullerene and the SWNT [46].

A similar study with $Dy@C_{82}$ metallofullerene peapods used novel etching techniques to remove the peapod sample from the FET after transport measurements had been carried out [47]. HRTEM images of the very same metallofullerene peapod sample were then taken on which transport measurements had been carried out. It was found that some of the bundles contained few fullerenes while others had filling rates of over 95%. Both types of bundles behaved as semiconductors; however, the poorly filled bundle exhibited Coulomb blockade oscillations for low bias voltages, which is expected for empty nanotubes [47]. This technique allows the confirmation of the composition of peapod bundles after transport measurements have been performed.

12.3.4
Functionalized Fullerenes

Very few examples of the encapsulation of exohedrally functionalized fullerenes, fullerenes with functional groups attached to the outside of the cage, have been reported (Table 12.1). The decreased thermal stability of functionalized fullerenes

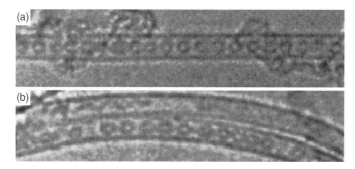

Figure 12.10 HRTEM of C_{60}@SWNT (a) and functionalized fullerene, $C_{61}(COOEt)_2$@SWNT (b). The icosahedral C_{60} cage appears as a perfect circle in micrographs (a), whereas functionalized fullerenes appear elongated (b) possibly due to a break in the symmetry and distortion of the fullerene cage upon addition of a functional group.

prohibits the use of the standard sublimation techniques for C_{60}; however, since the first encapsulation of functionalized fullerenes in SWNT was reported using low-temperature supercritical carbon dioxide [11], low temperature, solution, and supercritical fluid filling methods have led to the encapsulation of a number of different functionalized fullerenes [48–51]. The addition of functionality to the carbon framework of fullerene also decreases the stability of the molecule to the electron beam in HRTEM so the visualization of encapsulated molecules is difficult, the cages of functionalized fullerenes appear elongated in shape compared to the dark circles seen for the spherical C_{60} (Figure 12.10) [50].

High filling rates of SWNT with the functionalized fullerene, *N*-methyl-3,4-fulleropyrrolidine are reported using ultrasonication followed by refluxing in *n*-hexane for several hours [48, 49]. *n*-Hexane is considered to be small enough to escape from the peapod structure by passing between the fullerene and the nanotube sidewall leaving a tightly packed peapod structure [49]. Fullerene encapsulation was confirmed by Raman spectroscopy, EELS, and HRTEM. A series of HRTEM images taken at 2 s intervals show individual functional groups attached to the fullerene cage, these groups are seen to rotate with the functional groups always pointing toward the sidewall of the nanotube [48].

Functional group visualization is also reported in an encapsulation study of a fullerene functionalized with a retinal chromophore, the fullerene was inserted into the SWNT by refluxing the thermally opened SWNT and functionalized fullerene in chloroform for 3 h [52]. The functional group of the fullerene is observed in sequential HRTEM images [52].

Addition of functional groups to the fullerenes in peapod structures has been identified as a way of manipulating their properties; by attaching sterically bulky functional groups, attempts were made to control the spacing of fullerenes in SWNT [50, 51]. The ability to accurately control the intermolecular separation is instrumental in tuning interactions between the confined molecules within a 1D array and in modulating the intrinsic properties of the SWNT such as their electronic bandgap. Fullerenes with a variety of alkyl chains attached were encapsulated in

SWNT using supercritical carbon dioxide at high pressures but low temperature. The resulting peapods were analyzed by HRTEM and although no functional groups were observed, an increase in interfullerene separation of up to 0.8 nm was measured: the increase in separation and the elongated shape of the fullerene were taken as evidence of functionalized fullerene encapsulation [50]. The irregular nature of the peapods formed was attributed to the ability of the alkyl chains to pack inside the nanotube in a variety of ways, and it is predicted that, in contrast to previous studies [48], the alkyl functional group would prefer to form folded conformations rather than attaching itself to the aromatic nanotube sidewall [50]. The fact that alkyl chains appeared to be not visible in HRTEM may be related to high mobility of these groups that are known to be flexible.

A more recent study details the use of functionalized fullerenes with aromatic chains encapsulated in SWNT [51]. The peapods formed were analyzed by Raman spectroscopy that confirms fullerene encapsulation and the periodicity was characterized by XRD and HRTEM (Figure 12.11). A smaller but more regular separation of 0.2 nm was observed that cannot be fully explained even considering the shorter but more conformationally rigid spacer units used. Semiempirical molecular modeling of the peapods formed indicates the possibility of the aromatic spacers exhibiting a variety of π–π interactions within the SWNT that lead the aromatic chain to slip between the sidewall and carbon cage of a neighboring functionalized fullerene resulting in a shortening of observed interfullerene separation (Figure 12.11d) [51]. Recently, functionalized fullerenes consisting of aromatic chains terminated with bulky nonaromatic groups have been encapsulated in SWNT and led to an increased interfullerene separation of up to 0.5 nm (T.W. Chamberlain, J. Howells, A.N. Khlobystov, and N.R. Champness, unpublished results).

In an attempt to investigate the inner environment of SWNTs, a fulleropyrrolidine bearing a nitroxide free radical was inserted into SWNT using supercritical carbon dioxide techniques [53]. Electron paramagnetic resonance revealed limited aniso-

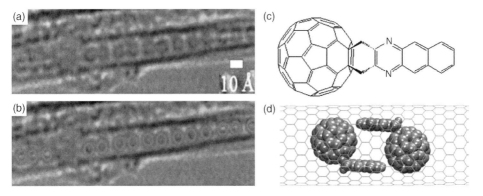

Figure 12.11 HRTEM of an aromatic chain-functionalized fullerene@SWNT peapod structure. (a) The same HRTEM with the circles overlaid on the fullerene positions (b). Structure of aromatic chain-functionalized fullerene (c) and molecular model (d) depicting possible orientation of two encapsulated functionalized fullerenes that explain the pairs of molecules observed in the HRTEM image (a).

tropic rotational freedom at room temperatures (300 K) indicating a strong interaction between the fullerene and the surrounding nanotube. The unaltered nature of fullerenes' EPR spectral characteristics including g-tensor and isotropic hyperfine constant suggest that the interaction involves the fullerene cage rather than the functional group [53].

Considerable work has been done to encapsulate the heterofullerene radical $C_{59}N^{\bullet}$ in SWNT, and EPR spectroscopy of the fullerene molecule probe inside the SWNT provides information on the internal environment [54]. The magnetic $C_{59}N^{\bullet}$ is an ideal candidate as it occupies the interior of the nanotube and unlike $N@C_{60}$, the unpaired electron is located on the fullerene cage. The first peapod systems formed were created by encapsulating a chemically inert functionalized $C_{59}N$ derivative, using toluene solution, then thermally treating the peapods in vacuum to remove the side group [40, 54, 55]. EPR studies are used to monitor the production of the spin active $C_{59}N^{\bullet}$ radicals with heating, the characteristic triplet signal from N nucleus can be used to quantify the amount of the radical present and therefore be used to obtain approximate filling yields [54]. A hindered, anisotropic rotation of $C_{59}N^{\bullet}$ was deduced from the temperature dependence of the electron spin resonance spectra near room temperature.

A comparative study explores the difference in structure obtained when using different filling techniques [56]. Using sublimation of the $(C_{59}N)_2$ dimer, the $C-C$ interfullerene bond dissociates generating two $C_{59}N^{\bullet}$ radicals; these enter the SWNT and three different scenarios are possible: stabilization of the radical in the interior of the SWNT, recombination of two radicals to form the bisazafullerene dimer $(C_{59}N)_2$, or the unpaired electron can be transferred to another fullerene cage initiating an oligomerization reaction. In the case of the oligomerization, the resultant chain of molecules would be forced by the confines of the SWNT to form linear chains compared to the branched chains observed in the bulk.

By a careful analysis of HRTEM data, measuring the interfullerene cage distances, it is possible to confirm that all three of these scenarios occur creating an uncontrollable mixture of dimers, oligomers, and $C_{59}N$ monomers. By using the milder supercritical carbon dioxide method, it is possible to insert the $(C_{59}N)_2$ dimer into the SWNT without dissociation and the resultant peapods consisting of purely the dimer in its pristine form. Potentially, this system can be used to generate the $C_{59}N^{\bullet}$ radical species [56].

The affinity of the fullerene cage for the interior of SWNT provides a strong energetic driving force for the insertion of fullerene molecules, and the ability to covalently link any functional group to the carbon framework of the cage means an almost endless list of chemically active groups could be introduced inside SWNTs. The 1D confinement imposed by the SWNT could then be used to control the direction of any subsequent chemical reactions between neighboring functional groups and could be a route to chemical transformations not possible under standard, unconfined conditions (i.e., solution, solid crystal, gas phase). The use of time-resolved *in situ* HRTEM to monitor these transformations could allow chemical reactions to be studied. The scope for encapsulating functionalized fullerenes in SWNT is large and may find applications in nanoelectronics, sensing, solar cells, or quantum computing in the future; however, more investigations into the mechanisms of solution and supercritical fluid filling need to be undertaken if this field is to advance.

12.4
Other Types of Molecules

Although other molecules may have less distinct shapes and only limited electron beam and thermal stabilities compared to fullerenes, the number of well-documented examples of nonfullerene molecules inserted in nanotubes is gradually increasing. In order to understand the true potential of carbon nanotubes as nanoscale containers, it is important to broaden the range of guest molecules beyond fullerenes. Traditional methods of vibrational and electronic spectroscopies can be readily used for studying structural and functional properties of 1D molecular chains inside nanotubes. A virtually infinite range of coordination and organometallic compounds allows selection of guest molecules with desired magnetic or optical properties that can be incorporated within the nanotube structure, which in turn can be integrated as part of a functional electronic device.

12.4.1
Molecules Without Metal Atoms

Octasiloxane ($Si_8H_8O_{12}$) is a cube-shaped molecule with 8 H atoms pointing outward from each corner of the cube. Since the hydrogen atoms of the octasiloxane come in direct contact with the nanotube surface when the molecule is nestled inside, the vibrational frequency of Si—H bond can be used to probe the interactions between the octasiloxane molecules and the nanotube. The vibrational frequency of this bond normally appears as a sharp band at $2277\,cm^{-1}$ (IR spectra) in solution. Upon insertion, the Si—H absorption band of the resultant $Si_8H_8O_{12}$@SWNT system is substantially broadened and its center is redshifted by approximately $15\,cm^{-1}$. The most likely explanation for the shift of ν(Si—H) in SWNTs is the elongation of Si—H bonds as a result of dispersion forces acting between $Si_8H_8O_{12}$ and the nanotube. The broadening may be attributed to different local environments for the individual Si—H bonds depending on the orientation of octasiloxane with regard to the nanotube and with each other [57].

A diatomic molecule, iodine, has been inserted in SWNT from its gas phase. Polymorphic structures of atomic chains and crystals of iodine have been studied at atomic level by HRTEM in SWNT with different diameters. It has been found that the number of atomic chains in the host strongly depends on the diameter of the nanotube. Single chains of iodine have been observed in thinner SWNTs with diameters of approximately 1.05 ± 0.05 nm. In wider SWNTs (diameter 1.30 ± 0.05 or 1.40 ± 0.05 nm), double or triple helical chains of iodine atoms have been found. No correlation between the helical structure of iodine chains and nanotube chirality has been noticed. Most interestingly, a phase transition between

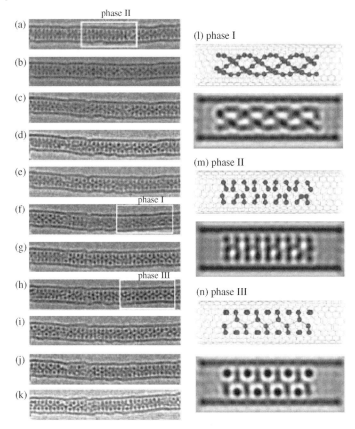

Figure 12.12 (a–k) A variety of packing patterns formed by I_2 in nanotubes. (Reproduced with permission from Ref. [58].)

the atomic chains and new crystalline phase of I_2 is observed by HRTEM inside nanotubes with diameters around the critical value of 1.45 nm (Figure 12.12). In nanotubes with diameters of 1.55 nm or above, a crystalline iodine phase identical to the bulk I_2 is observed [58].

Relatively, low molecular weight polymers, such as poly(ethylene oxide) (PEO, 600 kDa) and poly(caprolactone) (PCL, 80 kDa), were encapsulated in MWNTs [59] (Table 12.2). TEM images reveal foam and peapod-like patterns of the dried polymer in the nanotube channels. The encapsulation method relies on a diffusion mechanism driving polymers from dilute volatile solutions into nanotubes. The process takes place selectively depending on the size of the polymer macromolecules, as relatively small, flexible polymer molecules can conform to enter these nanotubes, but larger macromolecules (~1000 kDa) remain outside nanotubes. A transparent thermoplastic polymer, poly(methylmethacrylate) (PMMA), was also encapsulated by solvent evaporation in nanotubes [60]. Rayleigh instability, which is a result of confinement in nanotubes, is observed by TEM showing formation of compartmentalized PMMA nanorods surrounded by amorphous carbon nanotube after thermal treatment.

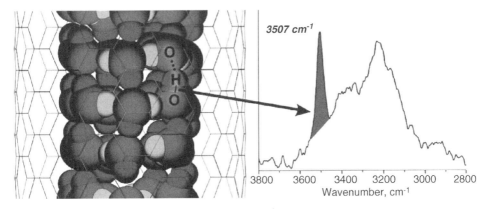

Figure 12.13 Packing of water molecules inside a (10,10)SWNT at 123 K, and an IR spectrum showing the OH stretching mode at 3507 cm^{-1}. (Reproduced with the permission from Ref. [65].)

Because of the importance of water in biology and medicine, a large number of studies have been dedicated to the interactions between H_2O and nanotubes under different conditions. Molecular dynamics simulations predict the spontaneous water filling of SWNTs and MWNTs and show that the nanotube sidewalls have a strongly hydrophobic nature (Figure 12.13). It was found that water molecules exhibit three different modes of nanotube filling depending on the nanotube diameter, which are called a "wire mode," "layered mode," and "bulk mode." The density of the confined water decreases with the nanotube diameter, while nanotube chirality seems to have no effect [61–64]. Although water in nanotubes has been extensively studied by a range of spectroscopy [65–67], diffraction [68, 69], and other techniques [70–73], the direct microscopy verification of the structural and dynamic behavior of water molecules in nanotubes still remains beyond reach.

Table 12.2 Molecules without metal atoms inserted in carbon nanotubes.

Compound	Type of nanotube	Nanotube internal diameter (nm)	Method of filling	Reference
$Si_8H_8O_{12}$	SW/MWNT (AD/HiPCO)	1.3–1.5/1–3	Gas phase (175 °C, vacuum, 2 days) or scCO$_2$ solution (50 °C, ~24 h)	[57]
Iodine	SWNT (AD/HiPCO)	1.2–1.6/0.6–1.3	Gas phase, 150 °C, vacuum, 24 h	[58]
PEO, PCL	MWNT	50–100	Solution (water, dichloromethane), RT, solvent evaporation	[59]
PMMA	NT/AAO composite	?	Solution (chloroform), RT, solvent evaporation followed by heating at 150–200 °C	[60]

Density functional theory calculations of the nuclear magnetic resonance spectroscopy of gas molecules (acetylene, ethylene, methane, CO, N_2, HF, and He) encapsulated within SWNTs have been reported. They showed that the encapsulation affects chemical shifts of the nuclei of guest molecules due to the ring currents generated in the nanotube. The chemical shift strongly depends on the chirality of the nanotube and less on the nanotube diameter. For example, in the zigzag nanotubes the chemical shift is typically 10–20 ppm greater than in armchair nanotubes [74].

12.4.2
Organometallic and Coordination Compounds

Insertion of molecules containing metal atoms into carbon nanotubes is of particular interest as it presents an opportunity to tune the physicochemical properties of nanotubes through controlling the electronic states of the metal atoms.

The filling of SWNTs with metallocenes has been carried out for cobaltocene $(Co(C_5H_5)_2)$ and its derivatives [75] and ferrocene $(Fe(C_5H_5)_2)$ [76]. Chains of metallocene molecules in nanotubes were observed by HRTEM as a series of dark contrast features lying between the nanotube sidewalls. Shapes of individual molecules could not be distinguished for metallocenes, but EDX spectroscopy confirmed the presence of metal atoms inside nanotubes. The C–H stretching vibration of ferrocene in $Fe(C_5H_5)_2@SWNT$ appears to be downshifted by $200\,cm^{-1}$ and its redox peaks are broadened as a result of interactions with carbon nanotubes [76].

Cobaltocene has been predicted to be an efficient n-dopant for SWNTs, transferring 0.61 electrons per molecule to the nanotube. Because of the relatively weak energy of interaction between cobaltocene and nanotubes, cobaltocene can be selectively inserted in SWNTs of specific diameter (0.935 nm) (Table 12.3). UV–Vis absorption spectroscopy indicated that the oxidation state of Co atom was $+3$ in Co $(C_5H_5)_2@SWNT$ and $Co(C_5H_4C_2H_5)_2@SWNT$ structures, thus confirming the electron transfer from the molecules to the nanotube. The positive charged cobaltocenium ions $[Co(C_5H_5)_2]^+$ nested in the nanotube induced a $\sim20\,meV$ redshift in the photoluminescence spectra of the nanotubes, whereas the spectra of empty (too narrow or too wide) nanotubes remained unchanged [75, 77].

Ferrocene molecules were encapsulated inside SWNTs and explored as molecular precursors for reactions inside nanotubes. In this process, $Fe(C_5H_5)_2$ acts as a catalyst and a carbon source for the growth of a secondary, inner nanotube, yielding DWNTs as the final product. By Raman spectroscopy it has been shown that DWNT formation is possible at temperatures as low as $600\,°C$. Moreover, the Fe 2p photoemission spectroscopy revealed that annealing at different temperatures leads to the selective growth of either iron-doped ($600\,°C$ for 2 h) or undoped ($1150\,°C$ for 1 h) DWNTs. More evidence for DWNT growth was obtained from HRTEM where clean DWNTs were observed after heating $Fe(C_5H_5)_2@SWNT$ at $1150\,°C$. However, for Fe $(C_5H_5)_2@SWNT$ heated at $600\,°C$, iron carbide (Fe_3C) nanocrystals were found inside the SWNTs. This iron carbide phase is thermodynamically metastable at

Table 12.3 Organometallic and coordination compounds inserted in carbon nanotubes.

Compound	Type of nanotube	Nanotube internal diameter (nm)	Method of filling	Reference
Co(C₅H₅)₂/Co (C₅H₄C₂H₅)₂	SWNT(HiPCO)	0.9–1.40	Gas phase, vacuum, 100 °C, 3 days	[75]
FeCp₂	SWNT(LA)	1.4	Gas phase, vacuum, 70–150 °C, several days	[78]
FeCp₂	SWNT	?	Gas phase (in vacuum), 180 °C	[79]
Fe(CO)₅	MWNT	60–200	Neat liquid, RT, 1 day	[80]
CoPc	MWNT	2.4	Gas phase (in vacuum), 375 °C, 3 days	[81]
[W₆O₁₉][nBu₄N]₂	DWNT/MWNT (AD)	1.21	Solution (ethanol), 78 °C, overnight	[82]

600 °C and gradually transforms into metallic iron that diffuses out of the nanotubes and aggregates into nanoparticles. The intensity distribution of the radial breathing mode of the DWNTs derived from $Fe(C_5H_5)_2$@SWNT was significantly different from the DWNTs grown from fullerene peapods, suggesting a larger interlayer separation in DWNTs grown from $Fe(C_5H_5)_2$@SWNT [78].

Ferrocene molecules were also encapsulated in SWNTs at 180 °C as precursors for magnetic Fe nanoparticles formed using a flash annealing process. Fe nanoparticles with a diameter distribution of 0.5–0.7 nm formed in nanotubes were imaged by TEM suggesting that each nanoparticle consists of only several Fe atoms. Moreover, the presence of Fe in nanotubes was confirmed by EDX. A distinct Raman peak at 197 cm^{-1} observed for $Fe(C_5H_5)_2$@SWNT was attributed to a diameter-selective filling process of nanotubes with diameters of 1.26 nm. However, a marked reduction in the intensity of the same peak was observed in the Raman spectra of the Fe@SWNTs. Magnetic measurements of Fe@SWNTs show ferromagnetic behavior at room temperature and low temperatures due to the small size of Fe nanoparticles. Also, Fe@SWNTs exhibit high-performance n-type FET behavior and excellent single-electron transistor characteristics at low temperatures [79].

Iron pentacarbonyl ($Fe(CO)_5$), a volatile organometallic compound, was encapsulated in MWNTs from solution and used to synthesize magnetic $\gamma\text{-}Fe_2O_3$@MWNT by vacuum thermolysis and subsequent oxidation. Raman spectra of the nanocomposite show three broadbands around 400, 490, and 600 cm^{-1}, which are in agreement with the $\gamma\text{-}Fe_2O_3$ spectrum reported earlier. HRTEM imaging reveals that ~11 nm $\gamma\text{-}Fe_2O_3$ nanoparticles are formed on the surface and inside MWNTs, which exhibit ferri-magnetic behavior at room temperature [80].

Cobalt phthalocyanine (CoPc), which consists of a planar macrocycle with a single cobalt ion in the middle, has been encapsulated in DWNTs and MWNTs. No internal order was observed by TEM measurements. Nitrogen K-edge in NEXAFS spectra revealed nonrandom stacking of the molecules inside of the nanotube, characterized by a stacking angle $\gamma = 36 \pm 5°$, which results from a mixture of phases in the nanotube with a majority component of α-CoPc ($\gamma = 26°$) [81].

The tungsten Lindqvist ion $[W_6O_{19}]^{2-}$ belongs to a family of cluster anions called polyoxometalates (POMs). $[W_6O_{19}][nBu_4N]_2$ salt was inserted into DWNTs and SWNTs from a saturated solution in ethanol and the encapsulated anions were imaged by HRTEM. Prolonged electron beam irradiation of these species gave rise to translational molecular motion in narrower DWNTs that was sterically modulated by the internal surface of the host nanotube as a result of the nonspherical shape of the anion. In wider nanotubes, this molecular motion is not observed and an irreversible agglomeration of the clusters takes place instead [82] (Table 12.3).

12.5
Ionic Compounds

Methods of insertion of ionic inorganic compounds in nanotubes are distinct from insertion of molecular species. Because of their ionic nature, capillary filling mechanisms from neat molten ionic compounds at high temperatures or their aqueous solutions at moderate temperatures are typically used. Both methods introduce ionic compounds on the surface as well as inside the nanotubes, which requires careful washing of the filled nanotubes so that the surface-adsorbed material is removed, but at the same time the encapsulated material remains intact. The key problem here is that the ionic compounds seem to have significantly lower affinity for the carbon nanotube interior compared to molecular species (such as fullerenes), which makes it likely that the ionic guest compounds become removed from the nanotube during washing or other postfilling processing.

12.5.1
Salts

Filling carbon nanotubes with halogenide salts MX_n, where $X = Cl$, Br, or I, is a well-established procedure. Some halogenides (or their eutectic mixtures) melt at reasonably low temperatures so that they can be inserted in nanotubes in a molten state (Table 12.4). The exact structure of the 1D ionic crystallite formed inside nanotube is determined by the internal diameter of carbon nanotube [83], but in most cases it is derived directly from the structure of the bulk (unconfined) metal salt. The crystallite structures can be observed in nanotubes at the atomic scale using conventional HRTEM, as individual or small numbers of halogen and metal ions provide sufficiently high TEM contrast. For example, KI and CsI crystallized in nanotubes from the molten phase show no significant structural deviation from their bulk structure, but a detailed TEM analysis shows a wide range of defects in the 1D crystallites including plane faults, small twists, and vacancies or interstitial defects [84]. CuI in nanotubes forms a hexagonal structure similar to that of the

Table 12.4 Inorganic guest compounds inserted into carbon nanotubes.

Compound	Type of nanotube[a]	Nanotube internal diameter (nm)	Method of filling	Reference
KI	DWNT (AD)	2–3	Molten salt, 620–650 °C, several hours	[83]
CsI	DWNT (AD)	2–3	Molten salt, 670–730 °C, several hours	[83]
CuI	SWNT (AD)	1.0–1.4	Molten salt, 705 °C, 6 h	[85]
LaI$_2$	SWNT (AD)	1.5	Molten salt, 802 °C, 1 h	[86]
HgTe	SWNT (AD)	1.36/1.49	Molten salt	[87]
PbI$_2$	SWNT (AD) DWNT (AD)	?	Molten salt, 450 °C, 5 h	[89]
KI	SWNT (HiPCO)	0.9–1.0	Molten salt	[91]
MnTe$_2$	SWNT (AD)	1.0–1.3	Molten salt, 835 °C, 6 h	[93]
AgNO$_3$	MWNT (CVD)	10–20	Aqueous solution, RT, 1 h	[95]
AgNO$_3$	MWNT (CVD)	?	Solution (water/isopropanol), 30 °C, ultrasound, 1 h	[96]
PdNO$_3$	MWNT	?	Aqueous solution, RT, several hours	[97]
Fe(NO$_3$)$_3$	MWNT (CVD)	4–8	Aqueous solution, ultrasound	[98]
CdCl$_2$	MWNT (CVD)	4–10	Aqueous solution, RT, reduced pressure (removal of solvent by sublimation)	[99]
Fe(NO$_3$)$_3$ + Sm(NO$_3$)$_3$	MWNT	?	Solution (water/NHO$_3$), 100 °C, 4.5 h	[100]
Fe(NO$_3$)$_3$ + Co(NO$_3$)$_2$	MWNT	~60	Solution (water/ethanol), <100 °C	[101]
CsOH	SWNT (CVD)	?	Molten phase, 300 °C, 4 h, vacuum	[102]
NaOH	SWNT (CVD)	?	Molten phase, 418 °C, 4 h, vacuum	[102]
KOH	SWNT (CVD)	?	Molten phase, 450 °C, 4 h, vacuum	[102]
Ba(OH)$_2$	SWNT (CVD)	?	Molten phase, 508 °C, 4 h, vacuum	[102]
Re$_2$O$_7$	SWNT (AD)	?	Molten phase, 250 °C, 3 h, vacuum	[103]
CrO$_3$	SWNT (AD)	?	Aqueous solution, HCl, RT, 1 day	[104]
S	SWNT (AD)	?	120 °C, 6 h, vacuum	[111]
Se	SWNT (AD)	?	320 °C, 6 h, vacuum	[111]
Te	SWNT (AD)	?	520 °C, 6 h, vacuum	[111]

a) When known, the method of nanotube growth is given in brackets.

bulk material but slightly distorted [85]. The [00l]-direction of CuI lattice coincides with the axis of the nanotube.

Sometimes, encapsulation of ionic salts produces unexpected results. For example, the stoichiometry of lanthanum iodide inserted in carbon nanotubes was found to be LaI_2 by HRTEM, which is different from the LaI_3 usually observed in the bulk [86]. This change in the salt composition upon encapsulation within a nanotube was attributed to the confinement effects permitting only two iodide anions per each metal cation and possible electronic interactions between the guest compound and the nanotube potentially stabilizing the unusual stoichiometry. Semiconducting HgTe exhibits an unusual coordination geometry of mercury cations and telluride anions as they adopt planar trigonal and trigonal pyramidal geometries, respectively, upon encapsulation in nanotubes [87, 88]. These can also be considered to be a result of the confinement imposed by the 1D nanotube channel. More subtle effects of nanotubes on the structure of guest compounds have been demonstrated for PbI_2 [89]. This salt is known to exist in different polymorphic forms in the bulk crystalline state; however, only one of these forms (2H polymorph) was observed inside carbon nanotubes (Figure 12.14). The distortion of the nanotube from the perfect cylindrical shape (Figure 12.14f) appears to be very important for the formation of ordered crystallites, as no crystalline phase of PbI_2 was observed in rigid narrow double-walled nanotubes ($d_{NT\ internal} < 2\,nm$) [89]. It is also interesting that no phonons corresponding to the encapsulated PbI_2 were observed in the Raman spectra of this composite [90].

Comparative Raman and optical spectroscopy measurements of nanotubes filled with KI and empty nanotubes indicated some effects of the ionic guest compound on the nanotube container, including polarization van der Waals interactions and possible charge transfer from KI to the nanotube [91]. These effects are more pronounced for narrower nanotubes. Electron transport measurements in nanotubes filled with KI showed negative differential resistance that could be related to the periodic electrostatic modulation potential induced by the permanent dipoles of

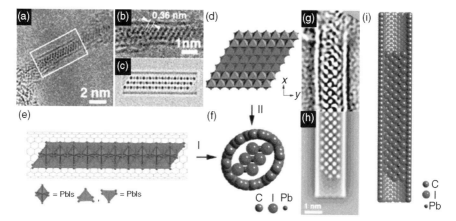

Figure 12.14 HRTEM micrographs (a)–(c) and structural diagrams (d)–(i) showing the atomic arrangement of PbI_2 when encapsulated in a nanotube. (Reproduced with permission from Ref. [89].)

KI-pairs inside the nanotube [92]. Photoluminescence excitation mapping of nano-tubes filled with $MnTe_2$ showed nanotube electronic bandgap reduction by up to 3.8% as a result of interaction with the encapsulated compound [93]. In this case, the greater effects were observed for wider nanotubes, which are exacted to contain a larger amount of $MnTe_2$ than narrower nanotubes. Nanotubes filled with AgBr or AgI appeared to be more susceptible to oxidation than corresponding empty nanotubes, which was explained by possible strain and/or defects induced by the guest compounds on the nanotube structure [94]. However, the fact that Ag^0/Ag^+ compounds can act as efficient oxidation catalysts may also contribute to the reduced oxidation stability of filled nanotubes.

Many ionic salts, such as nitrates or chlorides, have high solubility in water. Numerous recent examples demonstrate that MWNTs, whose internal diameters are significantly wider than that of SWNTs, can be filled when immersed in aqueous solutions of ionic salts at ambient temperature, sometimes with the help of ultrasound. It is possible that salts crystallize from supersaturated solutions onto the nanotube inner wall; however, this encapsulation is highly reversible, and the salts deposited inside wide MWNTs can be easily removed from nanotubes if the solution becomes too dilute or during the washing procedure. The filling yields tend to be quite low, and in most cases the soluble salts are converted into insoluble metals or metal oxides, which are more firmly attached to the MWNT interior. For example, $AgNO_3$ in MWNTs can be transformed into metallic silver either by reduction using hydrogen at elevated temperatures [95] or by reduction with γ-rays [96]. $PdNO_3$ or Fe $(NO_3)_3$ can be thermally decomposed to the corresponding oxides PdO [97] or Fe_2O_3 [98], which can then be reduced to metallic particles using hydrogen. Water-soluble $CdCl_2$ salt was successfully inserted into nanotubes using an unusual low-temperature sublimation technique for the solvent removal and care-fully designed washing procedure to avoid the loss of $CdCl_2$ from the nanotubes [99]. $CdCl_2$ was converted into insoluble CdS in MWNTs by treating it with H_2S [99]. Syntheses of intermetallic materials in nanotubes by immersing MWNTs in a solution containing a mixture of different metal nitrates and subsequent calcination at high temperature have been recently attempted [100, 101]. Although metal deposition inside the nanotube has been demonstrated in both cases, the interme-tallic nature of the compound formed inside MWNTs has to be yet proven by direct local probe methods, such as selected area e-beam diffraction, EDX analysis, and/or HRTEM.

12.5.2
Oxides and Hydroxides

There are a small number of studies on the encapsulation of metal oxides and hydroxides in carbon nanotubes. Some alkali metal hydroxides melt at relatively low temperatures without decomposition, which makes them suitable for insertion in nanotubes from a molten phase. CsOH, NaOH, KOH, and $Ba(OH)_2$ have been successfully inserted in SWNT at elevated temperatures [102]. As these compounds have extremely high solubility in water, they can be removed from nanotubes by washing. Metal oxides can also be filled in nanotubes using either a capillary method for low-melting point oxides, such as Re_2O_7 [103], or from aqueous solutions for water-soluble oxides, such as CrO_3 [104]. It is interesting that Re_2O_7 changes its stoichiometry upon encapsulation in nanotubes and forms discrete clusters in the nanotube cavity. These clusters can be reduced to metallic Re by hydrogen at 200 °C. There has been an attempt to insert the polyoxometallate compound $Cs_xPMo_{12}O_{40}$ in MWNTs, but it was found to be mostly deposited on the nanotube surface [105]. Although the effects of encapsulation on the physicochemical properties of metal oxides are not entirely understood, some studies suggest that the redox behavior of metal oxides may be significantly affected by the nanotube interior. Thus, Fe_2O_3 was found to be reduced to metallic iron inside MWNTs at lower temperatures than that required for the reduction of Fe_2O_3 adsorbed on the nanotube surface [106].

12.5.3
Other Inorganic Materials

Being effective electron acceptors, carbon nanotubes can be readily doped with alkali metals. However, most of the doping methods suffer from the lack of control over the amount of alkali metal inserted into the carbon nanotube. An unusual method for controlled filling of carbon nanotubes with Cs atoms using a plasma technique has been proposed [107]. Although, the presence of alkali metal ions inside nanotubes is not unambiguously proven, the doped nanotubes show some interesting electronic properties consistent with n-doping by Cs.

Electrolysis of water with Ag electrodes in the presence of MWNTs enables coating of nanotubes with silver and some deposition of silver in the nanotube cavities [108]. Intermetallic permalloy can be efficiently deposited into MWNTs in an electrochemical cell from a mixture of water and methanol [109]. Permalloy-filled nanotubes exhibit clear magnetic behavior. AFM measurements on Ag-filled SWNTs generated by *in situ* photodecomposition of AgCl showed that the effective dielectric function of SWNT is altered and the C—C bonds of the nanotube sidewalls are weakened as a result of the metallic silver imbedded in SWNTs [110]. However, photodecomposition of AgCl is also likely to produce significant defects in the nanotube sidewall (e.g., by released Cl atoms attacking the C—C bonds) that are difficult to take into account while elucidating the effects of the metallic filling.

Main group elements S, Se, and Te have been recently inserted in SWNTs using the molten-phase method [111]. The presence of the elements inside nanotubes resulted in shifts of the Raman lines of host nanotubes that are consistent with the electron-accepting character of the guest elements.

12.6
Nanoparticles in Nanotubes

Nanoparticles of different materials have been shown to have a high affinity for the nanotube sidewalls and adsorb readily on the nanotube exterior. However, examples where preformed nanoparticles enter the nanotube cavity (i.e., *not* formed inside nanotubes from molecular or ionic precursors) are quite rare. Most of these examples are reported for wide MWNTs or for carbon microtubes whose internal diameters are greater than 100 nm. In most cases, it is impossible to avoid the deposition of nanoparticles on the nanotube surface. Nanoparticles adsorbed on nanotubes sometimes can be mistaken for nanoparticles imbedded inside nanotubes, as the conventional TEM provides 2D projections of 3D objects. This ambiguity can be usually resolved by tilting the specimen and by taking a series of TEM images of the same nanotube as a function of the tilt angle. This series of images can be then reconstructed in a 3D model of nanotube, where the positions of nanoparticles can be reliably verified.

An efficient method of nanotube filling with Fe_3O_4 nanoparticles has been demonstrated for open MWNTs [112]. Nanotubes immersed in a suspension of Fe_3O_4 nanoparticles in water or organic solvent at room temperature over several hours become well filled with the magnetic nanoparticles (up to 80%). Fe_3O_4 was subsequently reduced to metallic iron by hydrogen (Figure 12.15). Carbon microtubes (average diameter $\sim$300 nm) can also be filled with a liquid dispersion of Fe_3O_4 nanoparticles [113]. Nanoparticles of Pd (diameter 3–4 nm) can be efficiently inserted into nanotubes with internal diameters above 30 nm from aqueous dispersions by a capillary mechanism [114]. However, nanotubes with internal diameters below 15 nm remained unfilled under the same conditions, which may indicate the importance of solvation shells around the nanoparticles during the encapsulation

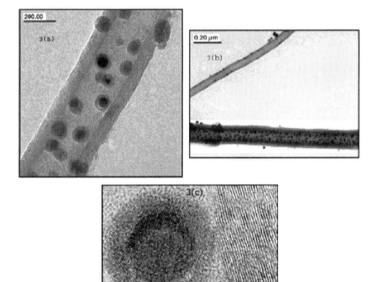

Figure 12.15 Nanoparticles of Fe_3O_4 were inserted in MWNTs and reduced by hydrogen into nanoparticles of α-Fe. (Reproduced with permission from Ref. [112].)

process. Organic colloidal nanoparticles of polystyrene can be inserted into carbon microtubes with average inner diameters ~400 nm [115] and ~500 nm [116] from water or organic solvents.

12.7
Concluding Remarks

Being nanosized containers, carbon nanotubes provide a unique tool for studying the chemistry and physics of molecules and ions confined in a quasi-1D channel. Initially, nanotubes filled with guest compounds were considered as a curiosity of the nanoworld. Over the past decade, the understanding of the capacity of nanotubes to confine materials has significantly advanced, and some important rules of the host nanotube/guest compound interactions have been discovered. On a practical side, the methodology of filling carbon nanotubes with a wide range of different materials has been developed and refined, so that filled nanotubes have become more accessible. We anticipate that in the nearest future these advances will give a momentum to the development of real applications for these nanostructures, ranging from nanoelectronics to drug delivery systems to nanosized reaction vessels.

References

1 Smith, B.W., Monthioux, M., and Luzzi, D.E. (1998) *Nature*, **396**, 323–324.

2 Monthioux, M. and Flahaut, E. (2007) *Mater. Sci. Eng.*, **C27**, 1096–1101.

3 Khlobystov, A.N., Britz, D.A., and Briggs, G.A.D. (2005) *Acc. Chem. Res.*, **38**, 901–911.

4 Britz, D.A. and Khlobystov, A.N. (2006) *Chem. Soc. Rev.*, **35**, 637–659.

5 Khlobystov, A.N. (2008) *Chemistry of Carbon Nanotubes* (eds V.A. Basiuk and E.V. Basiuk), American Scientific Publishers, USA.

6 Ulbricht, H., Moos, G., and Hertel, T. (2003) *Phys. Rev. Lett.*, **90**, 095501.

7 Girifalco, L.A. and Hodak, M. (2002) *Phys. Rev. B*, **65**, 125404.

8 Mattia, D. (2008) *Microfluid Nanofluid*, **5**, 289–305.

9 Wilson, M. and Friedrichs, S. (2006) *Acta Cryst.*, **A62**, 287–295.

10 Wilson, M. (2007) *Faraday Discuss.*, **134**, 283–295.

11 Khlobystov, A.N., Britz, D.A., O'Neil, S.A., Wang, J., Poliakoff, M., and Briggs,

G.A.D. (2004) *J. Mater. Chem.*, **14**, 2852–2857.

12 Kim, Y., Torrens, O.N., Kikkawa, J.M., Abou-Hamad, E., Goze-Bac, C., and Luzzi, D.E. (2007) *Chem. Mater.*, **19**, 2982.

13 Iijima, S., Brabec, C., Maiti, A., and Bernholc, J. (1996) *J. Chem. Phys.*, **104**, 2089.

14 Jaroenapibal, P., Chikkannanavar, S.B., and Luzzi, D.E. (2005) *J. Appl. Phys.*, **98**, 044301.

15 Zhou, L., Zhu, B.E., Wang, Y.X., and Zhu, J. (2007) *Nanotechnology*, **18**, 275709.

16 Okazaki, T., Okubo, S., Nakanishi, T., Joung, S.-K., Saito, T., Otani, M., Okada, S., Bandow, S., and Iijima, S. (2008) *J. Am. Chem. Soc.*, **130**, 4122.

17 Mizubayashi, J., Haruyama, J., Takesue, I., Okazaki, T., Shinohara, H., Harada, Y., and Awano, Y. (2008) *Microelectr. J.*, **39**, 222.

18 Li, Y.F., Hatakeyama, R., and Kaneko, T. (2007) *Appl. Phys. A*, **88**, 745.

19 Mizubayashi, J., Haruyama, J., Takesue, I., Okazaki, T., Shinohara, H., Harada, Y.,

and Awano, Y. (2007) *Phys. Rev. B*, **75**, 205431.

20 Hornbaker, D.J., Kahng, S.-J., Misra, S., Smith, B.W., Johnson, A.T., Mele, E.J., Luzzi, D.E., and Yardini, A. (2002) *Science*, **295**, 828.

21 Guan, L., Suenaga, K., Shi, Z., Gu, Z., and Ijima, S. (2005) *PRL*, **94**, 045502.

22 Sun, B.-Y., Sato, Y., Suenaga, K., Okazaki, T., Kishi, N., Sugai, T., Bandow, S., Ijima, S., and Shinohara, H. (2005) *J. Am. Chem. Soc.*, **127**, 17972.

23 Sato, Y., Suenaga, K., Bandow, S., and Ijima, S. (2008) *Small*, **4** (8), 1080.

24 Ohno, Y., Kurokawa, Y., Kishimoto, S., Mizutani, T., Shimada, T., Ishida, M., Okazaki, T., Shinohara, H., Murakami, Y., Maruyama, S., Sakai, A., and Hiraga, K. (2005) *Appl. Phys. Lett.*, **86**, 023109.

25 Chikkannanavar, S.B., Luzzi, D.E., Paulson, S., and Johnson, A.T., Jr. (2005) *Nano Lett.*, **5** (1), 151.

26 Bandow, S., Takizawa, M., Hirahara, K., Yudasaka, M., and Iijima, S. (2001) *Chem. Phys. Lett.*, **337**, 48.

27 Pfeiffer, R., Holzweber, M., Peterlik, H., Kuzmany, H., Liu, Z., Suenaga, K., and Kataura, H. (2007) *Nano Lett.*, **7** (8), 2428.

28 Warner, J.H., Ito, Y., Zaka, M., Ge, L., Akachi, T., Okimoto, H., Porfyrakis, K., Watt, A.A.R., Shinohara, H., and Briggs, G.A.D. (2008) *Nano. Lett.*, **8** (8), 2328.

29 Chorro, M., Rols, S., Cambedouzou, J., Alvarez, L., Almairac, R., and Sauvajol, J.-L. (2006) *Phys. Rev. B*, **74**, 205425.

30 Guan, L., Suenaga, K., Okazaki, T., Shi, Z., Gu, Z., Gu, Z., and Iijima, S. (2007) *J. Am. Chem. Soc.*, **129**, 8954.

31 Náfrádi, B., Nemes, N.M., Fehér, T., Forró, L., Kim, Y., Fischer, J.E., Luzzi, D.E., Simon, F., and Kuzmany, H. (2006) *Phys. Stat. Sol. (b)*, **243** (13), 3106.

32 Ning, G., Kishi, N., Okimoto, H., Shiraishi, M., Kato, Y., Kitaura, R., Sugai, T., Aoyagi, S., Nishibori, E., Sakata, M., and Shinohara, H. (2007) *Chem. Phys. Lett.*, **441**, 94.

33 Okazaki, T., Bandow, S., Tamura, G., Fujita, Y., Iakoubovski, K., Kazaoui, S., Minami, N., Saito, T., Suenaga, K., and Iijima, S. (2006) *Phys. Rev. B*, **74**, 153404.

34 Miyako, E., Nagata, H., Hirano, K., Makita, Y., and Hirotsu, T. (2008) *Chem. Phys. Lett.*, **456**, 220.

35 Sato, Y., Suenaga, K., Okubo, S., Okazaki, T., and Iijima, S. (2007) *Nano Lett.*, **7** (12), 3704.

36 Ryabenko, A.G., Kselev, N.A., Hutchison, J.L., Moroz, T.N., Bukalov, S.S., Mikhalityn, L.A., Loutfy, R.O., and Moravsky, A.P. (2007) *Carbon*, **45**, 1492.

37 Chorro, M., Cambedouzou, J., Iwasiewicz-Wabnig, A., Noé, L., Rols, S., Monthioux, M., Sundqvist, B., and Launois, P. (2007) *Lett. J. Explor. Front. Phys.*, **79**, 56003.

38 Shiozawa, H., Rauf, H., Pichler, T., Knupfer, M., Kalbac, M., Yang, S., Dunsh, L., Büchner, B., Batchelor, D., and Kataura, H. (2006) *Phys. Rev. B*, **73**, 205411.

39 Tóth, S., Quintavalle, D., Náfrádi, B., Korecz, L., Forró, L., and Simon, F. (2008) *Phys. Rev. B*, **77**, 214409.

40 Corzilius, B., Gembus, A., Weiden, N., Dinse, K.-P., and Hata, K. (2006) *Phys. Stat. Sol. (b)*, **243** (13), 3273.

41 Warner, J.H., Watt, A.A.R., Ge, L., Porftrakis, K., Akachi, T., Okimoto, H., Ito, Y., Ardavan, A., Montanari, B., Jefferson, J.H., Harrison, N.M., Shinohara, H., and Briggs, G.A.D. (2008) *Nano Lett.*, **8** (4), 1005.

42 Urita, K., Sato, Y., Suenaga, K., Gloter, A., Hashimoto, A., Ishida, M., Shimada, T., Shinohara, H., and Iijima, S. (2004) *Nano Lett.*, **4** (12), 2451.

43 Sato, Y., Yumura, T., Suenaga, K., Urita, K., Kataura, H., Kodama, T., Shinohara, H., and Iijima, S. (2006) *Phys. Rev. B*, **73**, 233409.

44 Guan, L., Suenaga, K., Okubo, S., Okazaki, T., and Iijima, S. (2008) *J. Am. Chem. Soc.*, **130**, 2162.

45 Kitaura, R., Imazu, N., Kobayashi, K., and Shinohara, H. (2008) *Nano Lett.*, **8** (2), 693.

46 Shimada, T., Ohno, Y., Suenaga, K., Okazaki, T., Kishimoto, S., Mizutani, T., Taniguchi, R., Kato, H., Cao, B., Sugai, T., and Shinohara, H. (2005) *Jpn. J. Appl. Phys.*, **44** (1A), 469.

47 Obergfell, D., Meyer, J.C., Haluska, M., Khlobystov, A.N., Yang, S., Fan, L., Liu,

D., and Roth, S. (2006) *Phys. Stat. sol. (b)*, **13**, 3430.

48 Mrzal, A., Hassanien, A., Liu, Z., Suenaga, K., Miyata, Y., Yanagi, K., and Kataura, H. (2006) *Surf. Sci.*, **601**, 5116.

49 Liu, Z., Koshino, M., Suenaga, K., Mrzel, A., Kataura, H., and Iijima, S. (2006) *PRL*, **96**, 088304.

50 Chamberlain, T.W., Camenisch, A., Champness, N.R., Briggs, G.A.D., Benjamin, S.C., Ardavan, A., and Khlobystov, A.N. (2007) *J. Am. Chem. Soc.*, **129**, 8609.

51 Chamberlain, T.W., Pfeiffer, R., Peterlik, H., Kuzmany, H., Zerbetto, F., Melle-Franco, M., Staddon, L., Champness, N.R., Briggs, G.A.D., and Khlobystov, A.N. (2008) *Small*, **4** (12), 2262.

52 Liu, Z., Yanagi, K., Suenaga, K., Kataura, H., and Iijima, S. (2007) *Nat. Nanotechnol.*, **2**, 422.

53 Campestrini, S., Corvaja, C., De Nardi, M., Ducati, C., Franco, L., Maggini, M., Meneghetti, M., and Menna, E. and Ruaro, G. (2008) *Small*, **4** (3), 350.

54 Simon, F., Kuzmany, H., Náfrádi, B., Fehér, T., Forró, L., Fülöp, F., Janossy, A., Korecz, L., Rockenbauer, A., Hauke, F., and Hirsch, A. (2006) *PRL*, **97**, 136801.

55 Simon, F., Kuzmany, H., Bernardi, J., Hauke, F., and Hirsch, A. (2006) *Carbon*, **44**, 1958.

56 Pagona, G., Rotas, G., Khlobystov, A.N., Chamberlain, T.W., Porfyrakis, K., and Tagmatarchis, N. (2008) *J. Am. Chem. Soc.*, **130**, 6062.

57 Wang, J., Kuimova, M.K., Poliakoff, M., Briggs, G.A.D., and Khlobystov, A.N. (2006) *Angew. Chem. Int. Ed.*, **45**, 5188–5191.

58 Guan, L., Suenaga, K., Shi, Z., Gu, Z., and Lijima, S. (2007) *Nano Lett.*, **7**, 1532–1535.

59 Bazilevsky, A.V., Sun, K., Yarin, A.L., and Megaridis, C.M. (2007) *Langmuir*, **23**, 7451–7455.

60 Chen, J.-T., Zhang, M., and Russell, T.P. (2007) *Nano Lett.*, **7** (1), 183–187.

61 Rana, M. and Chandra, A. (2008) *J. Chem. Sci.*, **207**, 367–376.

62 Meng, L., Li, Q., and Shuai, Z. (2008) *J. Chem. Phys.*, **128**, 134703–134709.

63 Alexiadis, A. and Kassinos, S. (2008) *Chem. Eng. Sci.*, **63**, 2047–2056.

64 Hennrich, F., Arnold, K., Lebedkin, S., Quintillá, A., Wenzel, W., and Kappes, M. (2007) *Phys. Stat. Sol. (b)*, **244**, 3896–3900.

65 Byl, O., Liu, J.C., Wang, Y., Yim, W.L., Johson, J.K., and Yates, J.T. (2006) *J. Am. Chem. Soc.*, **128**, 12090–12097.

66 Matsuda, K., Hibi, T., Kadowaki, H., Kataura, H., and Maniwa, Y. (2006) *Phys. Rev. B*, **74**, 073415.

67 Wenseleers, W., Cambré, S., Culin, J., Bouwen, A., and Goovarts, E. (2007) *Adv. Mater.*, **19**, 2274–2278.

68 de Souza, N.R., Kolesnikov, A.I., Burnham, D.J., and Loong, C.K. (2006) *J. Phys.: Condens. Matter*, **18**, S232.

69 Maniwa, Y., Katura, H., Abe, M., Udaka, E., Suzuki, S., Achiba, Y., Kira, H., Matsuda, K., Kadozaki, H., and Okabe, Y. (2005) *Chem. Phys. Lett.*, **401**, 534–538.

70 Zimmerli, U., Gonnet, P.G., Walther, J.H., and Koumoutsakos, P. (2005) *Nano Lett.*, **5**, 1017–1022.

71 Hummer, G. (2007) *Mol. Phys.*, **105**, 201–207.

72 Huang, B., Xia, Y., Zhao, M., Li, F., Liu, X., Ji, Y., and Song, C. (2005) *J. Chem. Phys.*, **122**, 084708.

73 Bekyarova, E., Hanzawa, Y., Kaneko, K., Silvestre-Albero, J., Sepulveda-Escribano, A., Rodriguez-Reinoso, F., Kasuya, D., Yudasaka, M., and Iijima, S. (2002) *Chem. Phys. Lett.*, **366**, 463–468.

74 Besley, N.A. and Noble, A. (2008) *J. Chem. Phys.*, **128**, 101102–101105.

75 Li, L.-J., Khlobystov, A.N., Wiltshire, J.G., Briggs, G.A.D., and Nicholas, R.J. (2007) *Nat. Mater.*, **4**, 481–485.

76 Guan, L., Shi, Z., Li, M., and Gu, Z. (2005) *Carbon*, **43**, 2780–2785.

77 Khlobystov, A.N., Britz, D.A., and Briggs, G.A.D. (2005) *Acc. Chem. Res.*, **38**, 901.

78 Shiozawa, H., Pichler, T., Grüneis, A., Pfeiffer, R., Kuzmany, H., Liu, Z., Suenaga, K., and Kataura, H. (2008) *Adv. Mater.*, **20**, 1443–1449.

79 Li, Y., Kaneko, T., Ogawa, T., Takahashi, M., and Hatakeyama, R. (2008) *Jpn. J. Appl. Phys.*, **47**, 2048–2055.

80 Tan, F., Fan, X., Zhang, G., and Zhang, F. (2007) *Mater. Lett.*, **61**, 1805–1808.

81 Schulte, K., Swarbrick, J.C., Smith, N.A., Bondino, F., Magnano, E., and Khlobystov, A.N. (2007) *Adv. Mater.*, **19**, 3312–3316.

82 Sloan, J., Matthewman, G., Dyer-Smith, C., Sung, A.-Y., Liu, Z., Suenaga, K., Kirkland, A.I., and Flahaut, E. (2008) *ACS Nano*, **2**, 966–976.

83 Wilson, M. (2006) *J. Chem. Phys.*, **124**, 124706.

84 Costa, P.M.F.J., Friedrichs, S., Sloan, J., and Green, M.L.H. (2005) *Chem. Mater.*, **17**, 3122–3129.

85 Chernysheva, M.V., Eliseev, A.A., Lukashin, A.V., Tretyakov, Y.D., Savilov, S.V., Kiselev, N.A., Zhigalina, O.M., Kumskov, A.S., Krestinin, A.V., and Hutchison, J.L. (2007) *Physica E*, **E 37**, 62–65.

86 Friedrichs, S., Falke, U., and Green, M.L.H. (2005) *ChemPhysChem*, **6**, 300–305.

87 Carter, R., Sloan, J., Kirkland, A.I., Meyer, R.R., Lindan, P.J.D., Lin, G., Green, M.L.H., Vlandas, A., Hutchison, J.L., and Harding, J. (2006) *Phys. Rev. Lett.*, **96**, 215501.

88 Sloan, J., Carter, R., Meyer, R.R., Vlandas, A., Kirkland, A.I., Lindan, P.J.D., Lin, G., Harding, J., and Hutchison, J.L. (2006) *Phys. Stat. Sol.*, **243**, 3257–3262.

89 Flahaut, E., Sloan, J., Friedrichs, S., Kirkland, A.I., Coleman, K.S., Williams, V.C., Hanson, N., Hutchison, J.L., and Green, M.L.H. (2006) *Chem. Mater.*, **18** (8), 2059–2069.

90 Gonzalez, J., Power, Ch., Belandria, E., Broto, J.M., Puech, P., Sloan, J., and Flahaut, E. (2007) *Phys. Stat. Sol.*, **244**, 136–141.

91 Ilie, A., Bendall, J.S., Roy, D., Philip, E., and Green, M.L.H. (2006) *J. Phys. Chem. B*, **110**, 13848–13857.

92 Ilie, A., Egger, S., Friedrichts, S., Kang, D.J., and Green, M.L.H. (2007) *Appl. Phys. Lett.*, **91**, 253124.

93 Li, L.J., Lin, T.W., Doig, J., Mortimer, I.B., Wiltshire, J.G., Taylor, R.A., Sloan, J., Green, M.L.H., and Nicholas, R.J. (2006) *Phys. Rev. B*, **74**, 245418.

94 Bendall, J.S., Ilie, A., Welland, M.E., Sloan, J., and Green, M.L.H. (2006) *J. Phys. Chem. B*, **110** (13), 6569–6573.

95 Zhao, D.-L., Li, X., and Shen, Z.-M. (2008) *Mater. Sci. Eng. B*, **150**, 105–110.

96 Yu, H., Peng, J., Zhai, M., Li, J., and Wei, G. (2008) *Physica E*, **40**, 2694–2697.

97 Tessonnier, J.-P., Pesant, L., Ehret, G., Ledoux, M.J., and Pham-Huu, C. (2005) *Appl. Catal. A: Gen.*, **288**, 203–210.

98 Chen, W., Pan, F.X., and Bao, X. (2008) *J. Am. Chem. Soc.*, **130**, 9414–9419.

99 Capobianchi, A., Foglia, S., Imperatori, P., Notargiacomo, A., Giammatteo, M., Del Buono, T., and Palange, E. (2007) *Carbon*, **45**, 2205–2208.

100 Seifu, D., Hijji, Y., Hirsch, G., and Karna, S.P. (2008) *J. Magn. Magn. Mater.*, **320**, 312–315.

101 Tessonnier, J.P., Winé, G., Estournés, C., Leuvrey, C., Ledoux, M.J., and Pham-Huu, C. (2005) *Catal. Today*, **102–103**, 29–33.

102 Shao, L., Tobias, G., Huh, Y., and Green, M.L.H. (2006) *Carbon*, **44**, 2849–2867.

103 Costa, P.M.F.J., Sloan, J., Rutherford, T., and Green, M.L.H. (2005) *Chem. Mater.*, **17**, 6579–6582.

104 Lota, G., Frackowiak, E., Mittal, J., and Monthioux, M. (2007) *Chem. Phys. Lett.*, **434**, 73–77.

105 Karina Cuentas-Gallegos, A., Martínez-Rosales, R., Rincón, M.E., Hirata, G.A., and Orozco, G. (2006) *Opt. Mater.*, **29**, 126–133.

106 Chen, W., Pan, X., Willinger, M.G., Su, D.S., and Bao, X. (2006) *J. Am. Chem. Soc.*, **128**, 3136–3137.

107 Hatakeyama, R. and Li, Y.F. (2007) *J. Appl. Phys.*, **102**, 034309.

108 Samant, K.M., Chaudhari, V.R., Kapoor, S., and Haram, S.K. (2007) *Carbon*, **45**, 2126–2139.

109 Wang, X.-H., Orikasa, H., Inokuma, N., Yang, Q.-H., Hou, P.-X., Oshima, H., Itoh, K., and Kyotani, T. (2007) *J. Mater. Chem.*, **17**, 986–991.

110 Ilie, A., Bendall, J.S., Kubo, O., Sloan, J., and Green, M.L.H. (2006) *Phys. Rev. B*, **74**, 235418.

111 Chernysheva, M.V., Kiseleva, E.A., Verbitskii, N.I., Eliseev, A.A., Lukashin, A.V., Tretyakov, Y.D.,

Savilov, S.V., Kiselev, N.A., Zhigalina, O.M., Kumskov, A.S., Krestinin, A.V., and Hutchison, J.L. (2008) *Physica E*, **40**, 2283–2288.

112 Jain, D. and Wilhelm, R. (2007) *Carbon*, **45**, 602–606.

113 Korneva, G., Ye, H., Gogotsi, Y., Halverson, D., Friedman, G., Bradley, J.-C., and Kornev, K.G. (2005) *Nano Lett.*, **5** (5), 879–884.

114 Ersen, O., Werckmann, J., Houllé, M., Ledoux, M.-J., and Pham-Huu, C. (2007) *Nano Lett.*, **7** (7), 1898–1907.

115 Bazilevsky, A.V., Sun, K., Yarin, A.L., and Megaridis, C.M. (2008) *J. Mater. Chem.*, **18**, 696–702.

116 Kim, B.M., Qian, S., and Bau, H.H. (2005) *Nano Lett.*, **5** (5), 873–878.

13
Carbon Nanohorn

Masako Yudasaka and Sumio Iijima

13.1
Introduction

Graphene-based tubular objects with nanometer diameters have some attractive properties; they are mechanically strong, chemically and thermally stable, and have unique optical and electrical properties. Multiwalled carbon nanotubes (MWNTs) [1] and single-walled carbon nanotubes (SWNTs) [2] are popular representative nanographene materials being widely studied by many researchers. There are several unpopular graphene nanotubules, one of them is single-walled carbon nanohorns (SWNHs), which were identified and named in 1998 [3]. The SWNHs have been obtained in large quantities at high purity without using catalysts, which has been advantageous for advancing studies on their properties and potential applications. In this chapter, we introduce the production, structure, and chemical functionalization of SWNHs and discuss recent research on their potential application in drug delivery systems and assessment of their toxicity.

13.2
Production

Without using any catalyst, SWNHs can be produced at room temperature by CO_2 laser ablation of graphite in Ar gas at 760 Torr [3]. The production process has been improved, achieving a production rate of 0.5–1 kg/day and a purity of 90–95% [4]. These are in contrast with most of the nanometer-sized graphene materials produced by using either catalysts or templates. For example, the SWNT formation requires metal catalysts with a size of 1–2 nm at a high temperature, typically 800–1000 °C, which makes large-scale production of high-purity material difficult. Recently, the graphene sheet has taken its place among potential nanomaterials. Graphene sheets are obtained using an old technology, peeling graphite crystals with adhesive tape [5]. Because this method does not fit large-scale production, the catalytic growth methods are actively addressed.

Carbon Nanotubes and Related Structures. Edited by Dirk M. Guldi and Nazario Martín
Copyright © 2010 WILEY-VCH Verlag GmbH & Co. KGaA, Weinheim
ISBN: 978-3-527-32406-4

13.3
Structure and Growth Mechanism

The SWNH is a single-graphene tubule of an irregular shape (nonuniform diameter of 2–5 nm and length of 40–50 nm) with a long horn-shaped tip (Figure 13.1) [3]. The tips often have cone angles of approximately 19°, indicating the existence of five pentagonal rings (Figure 13.1d) [3]. Approximately, 2000 SWNHs assemble to form a spherical aggregate with a diameter of 80–100 nm (Figure 13.1a and b) [3]. The separation of the aggregate into individual SWNHs has not yet been successful, suggesting that SWNHs are partially covalently bonded. It has also been reported that there are 10–20 nm caves near the center of the aggregates, which was inferred from the transmission electron microscopy (TEM) observation of Gd_2O_3 particles of that size located there [6].

Single-graphene objects with shapes similar to the horn-shaped tips of SWNHs were found in soot by several researchers, including the author (S.I.), in the mid-1990s [7–9], before the discovery of SWNHs and their spherical aggregates in 1998. Because the soot was made by the evaporation of carbon, it appears that the horn-shaped graphene is easily formed from carbon under high temperatures. Computer simulation supports this idea and indicates that the two-graphene sheets spontaneously roll up, accompanying the edge–edge bond creation, resulting in the horn-shape formation [10]. We consider this spontaneous change from graphene to horn-shape to be key in the mechanism of SWNH formation and it could occur in the course of cooling the hot liquid carbon droplets to solid graphite [11]. Similar processes are likely to occur in the arc discharges of graphite rod electrodes. Actually, SWNHs can also be formed by arc discharges [12–14]. More information about the growth methods and mechanisms were introduced in a previous paper [15].

13.4
Properties

Isotherm measurements of nitrogen adsorption at 77 K and high-pressure helium buoyancy at 303 K reveal that grown SWNHs are closed tubules with a specific surface area of approximately 300 m^2/g [16], total pore volume of 0.40 ml/g [17], and a particle density of 1.25 g/ml [16], which increases by opening the holes on the walls up to approximately 1450–1460 m^2/g [17, 18], 1.05 ml/g [17], and 2.05 g/ml [19], respectively. It has also been reported that there are three adsorption sites, the inter-SWNH micropore, the inside-wall surface, and the inside space, and their volume ratio is approximately 1 : 2 : 2 [19].

The holes are opened by oxidation using O_2 [18–21] (Figure 13.2), CO_2 [22], and oxidative acids [17, 23, 24]. The hole-opening method with the least carbonaceous dust generation is slow combustion, in which the temperature increases at the rate of 1 °C/min up to 400–550 °C in air [20]. An increase in the number and size of the holes under oxidative conditions are reflected in the increase of material adsorption

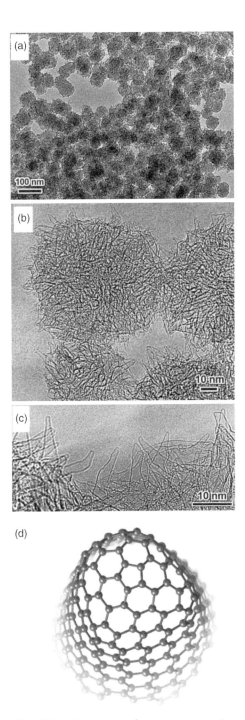

Figure 13.1 Transmission electron microscope images of SWNHs (a–c) [3], and a computer graphic image of the horn tip with a cone angle of 19° (d). The computer graphics are courtesy of NEC Fundamental Research Laboratory.

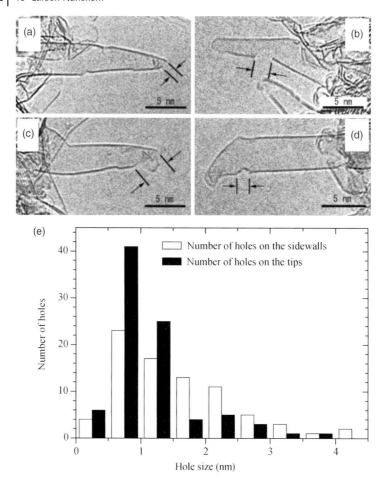

Figure 13.2 Transmission electron microscope images of SWNHs with holes opened by heating in oxygen at 575 °C (a–d). The holes are found at the tips and structural defects on the sidewalls of SWNHs. Hole sizes were measured as designated by the bars and arrows in (a)–(d). Histograms of the hole sizes are shown in (e). There is a tendency for large holes to open on the sidewalls [21].

quantities [17–20, 22–25], such as shown in Figure 13.3. The SWNHs with opened holes are referred to, hereafter, as SWNHox.

Applications of SWNHox that utilize the holes have been attempted. Molecular size-dependent entrance into hole-opened SWNHox has been reported [25, 26]; related to this, it has been clarified that the hole size could be optimized to make electrical condensers with high capacitance from SWNHox and organic electrolytes [27].

Hydrogen and methane adsorption capacities have been measured for possible applications. SWNHox does not adsorb much hydrogen [28], but abundant methane [29] reaches the DOE target for car fuel cells. Despite the low hydrogen adsorption capacity, it appears that the separation of H_2 and D_2 when using SWNHox is

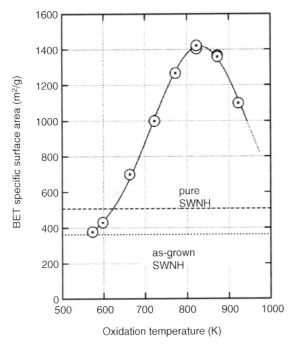

Figure 13.3 BET surface area of SWNHox at various treating temperatures in oxygen. "Pure SWNH" implies a SWNH without impurities such as amorphous carbon and graphitic particles [18].

potentially applicable because H_2/D_2 separation is possible at 77 K, utilizing the narrow spaces at the tips inside SWNHox [30].

For the chemical and biological applications, the adsorption of organic molecules or water by SWNH and SWNHox was examined. The adsorption behavior of benzene and *m*-xylene on SWNHox reflected the strong π–π electron interaction with graphene [31, 32]. It is interesting that water, with a low affinity for graphene, is also adsorbed by SWNHox. The water adsorption by SWNHox could be mediated by water pentamer formation, which is inferred from the hysteresis exhibited in adsorption–desorption isotherms [33].

Not much is known about the optical, electrical, and magnetic properties of SWNH and SWNHox as there has been a limited number of reports [3, 34–43].

13.5
Functionalization

13.5.1
Material Incorporation and Release

The incorporation of various molecules inside SWNHox is easy [44–56] and their release controllable [47–50, 55–57], suggesting that SWNHox is useful as a material carrier. For the large-scale production of material-incorporated SWNHox, facile incorporation methods carried out in solution at room temperature have been developed [44]. There are several variations in the methods depending on the affinity between the guest material, the graphene walls of SWNHox, and the solvent [44]. When the affinity between the guest molecules and graphene is high, the guest molecules enter SWNHox simply by immersion in the solution of guest molecules [47, 54]. A method that is often used is the solvent evaporation method ("nanoprecipitation"), in which the solvent is completely evaporated from the dispersion of SWNHox in solution [48, 50, 55, 56]. This method is applicable even when the guest molecules have a low affinity for graphenes. Notably, hydrophilic molecules with a low affinity for graphene, such as metal acetate, are incorporated inside SWNHox [45, 46, 48, 51–53] simply by immersion, where incorporation is considered to be assisted by the hydrophilic groups at the edges of the holes [45] (see Section 13.5.3) or multiple cluster formation, as inferred in the water adsorption inside SWNHox [33].

The release rate of the incorporated material is fast when immersed in good solvents of guest molecules, whereas it is extremely slow in a poor solvent. In toluene, C_{60}-incorporated SWNHox (C_{60}@SWNHox) corresponds to the former case, but in ethanol the latter [50]. However, the rate is not always fast when the solubility of the incorporated material is high, such as cisplatin@SWNHox in aqueous solutions [48, 49]. It is likely that the guest molecules gain certain stabilization energies by being confined inside the narrow spaces or that the guest molecules create stable clusters.

The incorporation and release in the gas phase is also possible. When C_{60}@SWNHox is exposed to liquid or vapor ethanol, a poor solvent of C_{60}, the C_{60} molecules are released and crystallize outside of SWNHox. The crystallized C_{60} again enters SWNHox upon exposure to toluene vapor, a good solvent of C_{60} [58].

If plugs could be put at the holes, the release of materials from SWNHox would be more easily controlled, which has been demonstrated using gadolinium oxide as the plug that is chemically attached to the oxygenated groups at the hole edges [59] (see Section 13.5.3).

The materials incorporated inside SWNHox can be perfectly confined inside SWNHox by thermally (1000–1200 °C) closing the holes. For this, the holes should be opened only at the tips of SWNHox but not on the sidewalls because the sidewalls cannot thermally close (Figure 13.4) [60, 61]. Such confinements were successful for nanoparticles of Gd_2O_3 [62].

13.5.2
Chemical Modification of Structure Defects

The SWNH tips are reactive, which is apparent from the preferential opening of tip-holes by oxidation [60]. The theoretical calculation also indicated a high reactivity [63]. Bumps and dips on the sidewalls are the next reactive sites in SWNH where the holes are opened by oxidation but at higher temperatures [60]. Various molecules are attached to the SWNHs, bonded at these structural defects [61–72]. Nakamura and

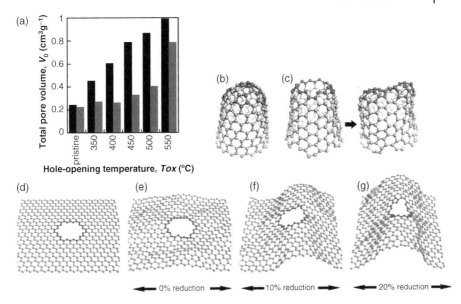

(a)

Total pore volume, V_0 (cm^{3}g^{-1})

Hole-opening temperature, *Tox* (°C)

(b) (c)

(d) (e) (f) (g)

◄─ 0% reduction ─► ◄─10% reduction ─► ◄─20% reduction ─►

Figure 13.4 The total pore volume of SWNHox increases with the hole-opening temperature, the heat treatment temperature in O$_2$ gas (Tox), which decreases after heat treatment in Ar at 1200 °C for 3 h. The decrease in pore volume was remarkable for SWNHox with a Tox of 350–500 °C because the holes at the tips close at 1200 °C (b, c). On the other hand, the pore volumes decreased only a little for SWNHox treated at a Tox of 550 °C because the holes on the sidewalls of SWNHox are not closed (d–g) [60].

coworkers attached amino groups directly to the tips of SWNHs using their original method [66], and the amino groups further reacted with triamide molecules, visible with TEM, supporting the premise that SWNHs were chemically modified as designed [72] (Figure 13.5).

13.5.3
Chemical Functionalization at Hole Edges

Functional molecules are attached to carboxylic groups at the edges of holes in SWNTs, which was first shown by Haddon and coworkers [73]. The carboxylic groups are created by oxidation, and their existence is demonstrated by infrared absorption spectroscopy, thermogravimetric analysis, and temperature-programmed mass spectroscopy.

Details of the functional groups at the edges of SWNHox holes were similarly studied, and it has been found that the number of carboxylic and other oxygenated groups upon hydrogen peroxide oxidation was larger than when oxidized with oxygen gas [24]. The existence of carboxylic groups at the graphene edges was confirmed by TEM by staining them with Pt compounds [74].

The oxygenated groups at the hole edges are chemically reactive [75]. The carboxylic groups react with the amino groups of various molecules by forming

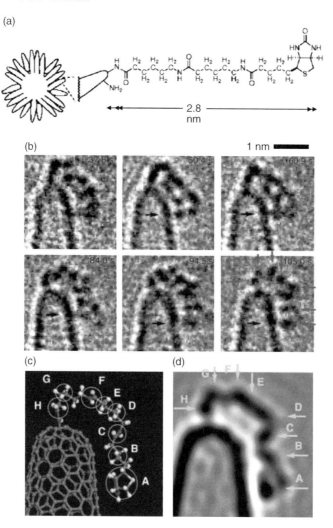

Figure 13.5 Molecular structure of triamide–SWNH conjugates (a) and TEM images (b). Conformation model of triamide (c) and the simulated TEM image (d) [72]. Gray arrows in (b) correspond to gray arrows in (d).

amide bonds. There have been increasing reports of SWNHox being modified in this way [24, 54, 76–79].

Coordination is another way of using the carboxylic groups to functionalize SWNHox. This process was demonstrated using Gd acetate (as shown in Figure 13.6a [45]). The Gd acetates attached to the hole edges function as plugs, which hinders the incorporation of the molecules inside SWNHox and their release [45, 59]. Here, the explanation of the material incorporation mechanisms is supplemented. The Gd acetate molecules attached to the edges of the holes

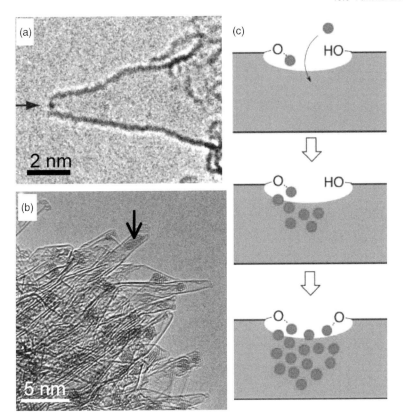

Figure 13.6 Transmission electron microscope images of SWNHox with an open hole tip, where a Gd acetate molecule (indicated by an arrow) is trapped (a), and SWNHox with Gd acetate clusters (indicated by an arrow) inside SWNHox (b). A proposed model for the Gd acetate (blue circle) trap at the hole edge and the successive accumulation inside the SWNHox, resulting in a cluster formation (c) [45].

trigger an accumulation of Gd acetate at the holes, which successively enlarges the Gd acetate cluster toward the inside of the SWNHox; thus, large Gd acetate clusters are formed near the holes inside SWNHox (Figure 13.6b and c) [45]. A similar incorporation mechanism was often found to that described in Section 13.5.1.

Another example of coordination is the attachment of (terpyridine)copper(II) to the carboxylic groups of SWNHox [80]. Microscopic structure analysis supported that the reaction proceeded as designed (Figure 13.7a–e) [81]. Due to their ionic properties, the obtained complexes dispersed well in water (Figure 13.7f). Using this system, interesting optical data were obtained revealing that the optically excited singlet electrons of (terpyridine)copper(II) groups were transferred to SWNHox [80].

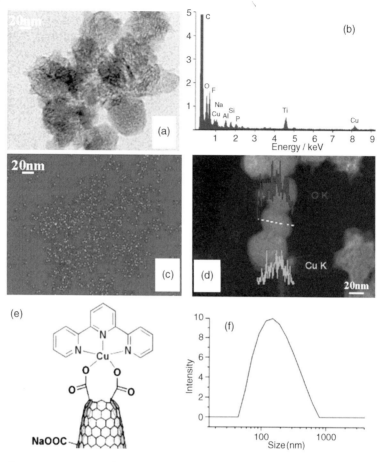

Figure 13.7 Structure of the (terpyridine) copper(II)–SWNHox complex. STEM image (a), EDX spectrum (b), Cu EDX-mapping for the same area with (a) (c), and line profiles of Cu and oxygen on a SWNHox aggregate (d). Molecular structure of (terpyridine)copper(II) (e) and the particle size distribution measured with the dynamic light scattering method (f) [80].

13.5.4
Physical Modification

The dispersion of SWNTs in solvents, especially water, is essential for their application, and the good dispersants for water dispersion are found in surfactants, polymers, and living materials, among others, most of which are hydrophilic–hydrophobic bifunctional molecules. Phospholipid-polyethylene glycol is one such compound [81] that disperses SWNTs in a physiological solution of phosphate-buffered saline, which is advantageous for the biological application of SWNTs [81].

For drug delivery applications of SWNHs, new dispersants have been developed. One interesting example is the use of a special peptide aptamer that specifically adsorbs on SWNHs [82] and is conjugated with polyethylene glycols [83, 84]. Polyethylene glycol conjugated with the anticancer drug doxorubicin greatly improves the dispersion of SWNH, which has enabled the *in vivo* testing of SWNHs [85, 86].

The physical adsorption of catalysts on SWNHs has also been investigated [46, 51, 87–89], and certain effects have been confirmed for organic material decomposition to generate hydrogen or protons.

13.6
Toxicity

The cytotoxicity of SWNHs and SWNHox and its functionalized forms has been tested using various cell lines, and no serious cell death was found [24, 47–49, 66, 83, 85, 90]. Tailored toxicity tests have shown no abnormal signs in animals (Table 13.1) [91]. In the histological studies of functionalized SWNH and SWNHox, intravenous injection did not result in appreciable abnormal changes in the tissues of most of the organs of mice after a half year [92].

Table 13.1 Toxicology testing of SWNHs [91].

Test	Test organism/animal	Dosage	Findings
Reverse mutation (Ames) test	*S. typhimurium* and *E. coli* strains	78–1250 µg/plate	No positive increase in re-vertants; no growth inhibitory effect
Chromosomal aberration test	Chinese-hamster lung fibroblast cell line	0.010–0.078 or 0.313–2.5 mg/ml	Negligible positive incidence of structural chromosomal aberrations or polyploidy
Skin primary irritation test	Rabbits	0.015 g/site	Primary irritation index (PII) = 0; no clinical signs of abnormalities; normal body weight gain
Eye irritation test	Rabbits	0.02 g/eye	Draize irritation score = 0; no clinical signs of abnormalities; normal body weight gain
Skin sensitization (adjuvant and patch) test	Guinea pigs	Induction: 0.02 g/site; challenge: 0.01 g/site	Mean response score = 0; no clinical signs of abnormalities; normal body weight gain
Peroral administration test	Rats	2000 mg/kg	No mortality; no clinical signs of abnormalities; normal body weight gain
Intratracheal instillation test	Rats	17.3 mg/kg	No mortality; rales for all animals, including control group; normal body weight gain; black lung spots and anthracosis; foamy macrophage in intra-alveolar spaces

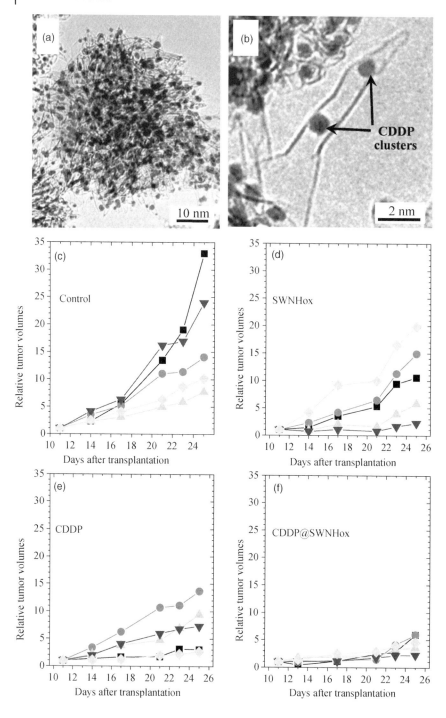

The SWNHs would be useful as a standard material for the assessment of nanomaterial toxicity. There have not been any nanomaterials with defined sizes, high purity, high dispersion in aqueous solutions, and low toxicity, except for functionalized SWNHs. In this context, Isobe *et al.* reported that amino-SWNH may serve as a standard material for the assessment of the toxicity of nanomaterials [66].

13.7
Drug Delivery Applications

Drug incorporation inside SWNHox first succeeded in dexamethasone [47], an anti-inflammatory drug. There are several medical materials that can be incorporated [47–49, 54, 55], among which cisplatin (CDDP), an anticancer drug, is interesting and discussed here to some extent [48, 49]. The incorporation of CDDP inside SWNHox was relatively easy: SWNHox was mixed with an aqueous solution of CDDP and left until the water evaporated. The CDDP-incorporated SWNHox (CDDP@SWNHox) obtained contained approximately 50% CDDP [49]. The anticancer efficiency of CDDP@SWNHox has been tested *in vitro* and *in vivo* [49] and shown, in both, to be higher than that of CDDP itself. This high anticancer effect was due to slow CDDP release from SWNHox and a tendency of SWNHox to attach to the cell surface *in vitro*, keeping the concentration of CDDP high near the cancer cells. Similar effects are expected *in vivo*, and, additionally, SWNHox tends to remain in the tumors when directly injected into subcutaneously transplanted tumors [86] (Figure 13.8).

The SWNTs are expected to be useful as photohyperthermia cancer therapy because they absorb light in the phototherapy window, a wavelength of approximately 700 nm, and transform the light energy into thermal energy, warming up the surroundings and inducing cell death *in vitro* [82]. This characteristic is the same for SWNHs [54]. To achieve a higher phototherapy efficiency, zinc phthalocyanine (ZnPc) was loaded inside and on the outside surfaces of SWNHox. The ZnPc is a potential drug for photodynamic therapy [94], generating reactive oxygen species by light absorption. The double phototherapy of photohyperthermia and photodynamic using ZnPc/SWNHox demonstrates the high phototherapy effects [54]. In *in vivo* tests, ZnPc/SWNHox–BSA (bovine serum albumin) was locally injected into tumors subcutaneously transplanted in mice. Here, BSA attachment to SWNHox was to increase the hydrophilicity of ZnPc/SWNHox [24, 54]. The tumors were irradiated with laser for 15 min/day for 10 days, which made the tumors disappear (Figure 13.9). This did not occur when ZnPc or SWNHox–BSA was used alone, suggesting that the enhanced efficiency is due to the double phototherapy (Figure 13.9) [54].

Figure 13.8 Transmission electron microscope images of cisplatin-incorporated SWNHox (CDDP@SWNHox) (a, b). Relative tumor sizes for each mouse, normalized on day 11, measured after transplantation (c–f). On days 11 and 15, the samples (c, saline; d, SWNHox; e, cisplatin (CDDP); f, CDDP@SWNHox) were injected into tumors that were subcutaneously transplanted into the mice on day 0. (dosages: CDDP, 0.5 mg/kg; SWNHox, 0.5 mg/kg) [49].

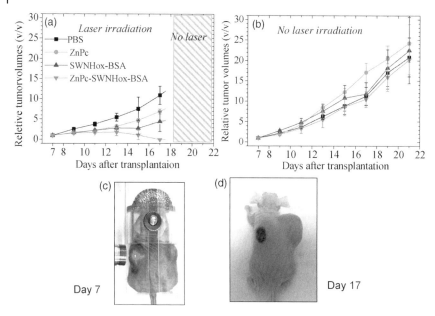

Day 7   Day 17

Figure 13.9 Average relative tumor size according to the number of days after transplantation. On day 7, samples were injected (dosages: ZnPc, roughly 0.3 mg/kg; SWNHox, roughly 1.5 mg/kg) into tumors that were subcutaneously transplanted into the mice on day 0. The tumors were laser-irradiated (wavelength, 670 nm; power, 160 mW; spot diameter, 5 mm) for 15 min every 24 h from days 7–17 and stopped on days 18–21 (a). The results for no laser irradiation are shown in (b). A picture of a mouse on day 7 with two tumors on the right and left flanks, into which ZnPc–SWNHox–BSA was intratumorally injected, and the left flank tumor was laser-irradiated (c). The mouse in (c) was position-fixed by a cage in a plastic tube. On day 17, the tumor on the left flank disappeared after the double phototherapy, whereas the tumor on the right flank increased in size (d) [54].

For the application of SWNHs in the drug delivery system for cancer therapy, SWNHs are better for use in intravenous injections. Intravenously injected SWNHs are known to be trapped by the reticuloendothelial system, a serious problem in cancer therapy in terms of targeting tumors. To solve these problems, key issues are controlling the size of SWNH aggregates and the chemical functionalization. If these problems are solved, SWNHs will be widely useful in living bodies.

13.8
Summary

In this section, we have introduced broad aspects of SWNH research. The SWNHs are easily produced in large quantity with high purity without metal catalysts and have various potential applications. Because we consider the functionalization of SWNH or SWNHox as crucial for any application being really useful, we have referred to as many references as possible to show the current research status of the functionalization of SWNHs. This content might be helpful for anybody who is interested in developing SWNH applications.

References

1 Iijima, S. (1991) Helical micro tubules of graphitic carbon. *Nature*, **354**, 56–58.

2 Iijima, S. and Ichihashi, T. (1993) Ingle-shell carbon nanotubes of 1-nm diameter. *Nature*, **363**, 603–605.

3 Iijima, S., Yudasaka, M., Yamada, R., Bandow, S., Suenaga, K., Kokai, F., and Takahashi, K. (1999) Nanoaggregates of single-walled graphitic carbon nano-horns. *Chem. Phys. Lett.*, **309**, 165.

4 Azami, T., Kasuya, D., Yuge, R., Yudasaka, M., Iijima, S., Yoshitake, T., and Kubo, Y. AT Large-scale production of single-wall carbon nanohorns with high purity. *J. Phys. Chem. C*, **112**, 1330–1334.

5 Yudasaka, M., Kikuchi, R., Matsui, T., Kamo, H., Ohki, Y., Yoshimura, S., and Ota, E. (1994) Graphite thin film formation by chemical vapor deposition, **64**, 842–844.

6 Yuge, R., Ichihashi, T., Miyaswaki, J., Yoshitake, T., Iijima, S., and Yudasaka, M. (2009) Hidden caves in an aggregate of single-wall carbon nanohorns found by using Gd_2O_3 probes. *J. Phys. Chem. C.*, **113**, 2741–2744.

7 Harris, P.J.F., Tsang, S.C., Claridge, J.B., and Green, M.L.H. (1994) High-resolution electron microscopy studies of a microporous carbon produced by arc-evaporation. *J. Chem. Soc. Faraday Trans.*, **90**, 2799–2802.

8 Saito, Y., Nishikubo, K., Kawabata, K., and Matsumoto, T. (1996) Carbon nanocapsules and single-layered nanotubes produced with platinum-group metals (Ru, Rh, Pd, Os, Ir, Pt) by arc discharge. *J. Appl. Phys.*, **80**, 3062–3067.

9 Iijima, S., Wkabayashi, T., and Achiba, Y. (1996) Structure of carbon soot prepared by laser ablation. *J. Phys. Chem.*, **100**, 5839–5843.

10 Kawai, T., Miyamoto, Y., Sugino, O., and Koga, Y. (2002) Nanotube and nanohorn nucleation from graphitic patches: tight-binding molecular-dynamics simulations. *Phys. Rev. B*, **66**, 033404.

11 Kokai, F., Takahashi, K., Yudasaka, M., and Iijima, S. (1999) Emission imaging spectroscopic and shadow graphic studies on the growth dynamics of graphitic carbon particles synthesized by CO_2 laser vaporization. *J. Phys. Chem. B*, **103**, 8686–8693.

12 Takikawa, H., Ikeda, M., Hirahara, K., Hibi, Y., Tao, Y., Ruiz, P.A., Jr., Sakakibara, T., Itoh, S., and Iijima, S. (2002) Fabrication of single-walled carbon nanotubes and nanohorns by means of a torch arc in open air. *Physica B*, **323**, 277–279.

13 Yamaguchi, T., Bandow, S., and Iijima, S. (2004) Synthesis of carbon nanohorn particles by simple pulsed arc discharge ignited between pre-heated carbon rods. *Chem. Phys. Lett.* **389**, 181–185.

14 Wang, H., Chhowalla, M., Sano, N., Jia, S., and Amaratunga, G.A.J. (2004) Large-scale synthesis of single-walled carbon nanohorns by submerged arc. *Nanotechnology*, **15**, 546–550.

15 Yudasaka, M., Iijima, S., and Crespi, V.H. (2008) Single-wall carbon nanohorns and nanocones, in *Topics in Applied Physics Volume 111: Carbon Nanotubes* (eds A. Jorio, M.S. Dresselhaus, and G. Dresselhaus), Springer-Verlag, Berlin, Heidelberg.

16 Murata, K., Kaneko, K., Kokai, F., Takahashi, K., Yudasaka, M., and Iijima, S. (2000) Pore structure of single-wall carbon nanohorn aggregates. *Chem. Phys. Lett.*, **331**, 14–20.

17 Yang, C.-M., Noguchi, H., Murata, K., Yudasaka, M., Hashimoto, A., Iijima, S., and Kaneko, K. (2005) Highly ultramicroporous single-walled carbon nanohorn assemblies. *Adv. Mater.*, **17**, 866–870.

18 Utsumi, S., Miyawaki, J., Tanaka, H., Hattori, Y., Ito, T., Ichikuni, N., Kanoh, H.,

Yudasaka, M., Iijima, S., and Kaneko, K. (2005) Opening mechanism of internal nanoporosity of single-wall carbon nanohorn. *J. Phys. Chem. B*, **109**, 14319.

19 Murata, K., Kaneko, K., Steele, W., Kokai, F., Takahashi, K., Kasuya, D., Hirahara, K., Yudasaka, M., and Iijima, S. (2001) Molecular potential structures of heat-treated single-wall carbon nanohorn assemblies. *J. Phys. Chem. B*, **105**, 10210–10216.

20 Fan, J., Yudasaka, M., Miyawaki, J., Ajima, K., Murata, K., and Iijima, S. (2006) Control of hole opening in single-wall carbon nanotubes and single-wall carbon nanohorns using oxygen. *J. Phys. Chem. B*, **110**, 1587–1591.

21 Ajima, K., Yudasaka, M., Suenaga, K., Kasuya, D., Azami, T., and Iijima, S. (2004) Materials storage mechanism in porous nanocarbons. *Adv. Mater.*, **16**, 397–401.

22 Bekyarova, E., Kaneko, K., Yudasaka, M., Kasuya, D., Iijima, S., Huirobro, A., and Rodriguez-Reinoso, F. (2003) Controlled opening of single-wall carbon nanohorns by heat treatment in carbon dioxide. *J. Phys. Chem. B*, **107**, 4479–4484.

23 Yang, C.-M., Kasuya, D., Yudasaka, M., Iijima, S., and Kaneko, K. (2004) Microporosity development of single-wall carbon nanohorn with chemically induced coalescence of the assembly structure. *J. Phys. Chem. B*, **108**, 17775–17782.

24 Zhang, M., Yudasaka, M., Ajima, K., Miyawaki, J., and Iijima, S. (2007) Light-assisted oxidation of single-wall carbon nanohorns for abundant creation of oxygenated groups that enables chemical modifications with proteins to enhance biocompatibility. *ACS Nano*, **1**, 265–272.

25 Murata, K., Hirahara, K., Yudasaka, M., Iijima, S., Kasuya, D., and Kaneko, K. (2002) Nanowindow-induced molecular sieving effect in a single-wall carbon nanohorn. *J. Phys. Chem. B*, **109**, 12668–12669.

26 Krungleviciute, V., Calbi, M.M., Wagner, J. A., Migoe, A.D., Yudasaka, M., and Iijima, S. (2008) Probing the structure of carbon nanohorn aggregates by adsorbing gases of different sizes. *J. Phys. Chem. C*, **112**, 5742–5746.

27 Yang, C.-M., Kim, Y.-J., Endo, M., Kanoh, H., Yudasaka, M., Iijima, S., and Kaneko, K. (2007) Nanowindow-regulated specific capacitance of supercapacitor electrodes of single-wall carbon nanohorns. *J. Am. Chem. Soc.*, **129**, 20–21.

28 Murata, K., Kaneko, K., Kanoh, H., Kasuya, D., Takahashi, K., Kokai, F., Yudasaka, M., and Iijima, S. (2002) Adsorption mechanism of supercritical hydrogen in internal and interstitial nanospaces of single-wall carbon nanohorns assembly. *J. Phys. Chem. B*, **106**, 11132–11138.

29 Bekyarova, E., Murata, K., Yudasaka, M., Kasuya, D., Iijima, S., Tanaka, H., Kanoh, H., and Knaneko, K. (2003) Single-wall nanostructured carbon for methane storage. *J. Phys. Chem. B*, **107**, 4682–4684.

30 Tanaka, H., Kanoh, H., Yudasaka, M., Iijima, S., and Knaeko, K. (2005) Quantum effects on hydrogen isotope adsorption on single-wall carbon nanohorns. *J. Am. Chem. Soc.*, **127**, 7511–7516.

31 Fan, J., Yudasaka, M., Kasuya, Y., Kasuya, D., and Iijima, S. (2004) Influence of water on desorption of benzene adsorbed within single-wall carbon nanohorns. *Chem. Phys. Lett.*, **397**, 5–10.

32 Yudasak, M., Fan, J., Miyawaki, J., and Iijima, S. (2005) Studies on the adsorption of organic materials inside thick carbon nanotubes. *J. Phys. Chem. B*, **109**, 8909–8913.

33 Bekyarova, E., Hanzawa, Y., Kaneko, K., Silvestre-Albero, J., Sepulveda-Escribano, A., Rodriguez-Reinoso, F., Kauysa, D., Yudasaka, M., and Iijima, S. (2002) Cluster-mediated filling of water vapor in intra-tube and interstitial nanospaces of single-wall carbon nanohorns. *Chem. Phys. Lett.*, **366**, 463–468.

34 Berber, S., Kwon, Y.-K., and Tomanek, D. (2000) Electronic and structural properties of carbon nanohorns. *Phys. Rev. B*, **62**, R2291.

35 Garaj, S., Thien-Nga, L., Gaal, G., Forro, L., Takahashi, K., Kokai, F., Yudasaka, M., and Iijima, S. (2000) Electronic properties of carbon nanohorns studied by ESR. *Phys. Rev. B*, **62**, 17115–17119.

36 Bandow, S., Kokai, F., Takahashi, K., and Yudasaka, M. (2001) Unique magnetism

observed in single-wall carbon nanohorns. *Cryst. Appl. Phys. A*, **73**, 281–285.

37 Imai, H., Babu, P.K., Oldfield, E., Wieckosdki, A., Kasuya, D., Azami, T., Shimakawa, Y., Yudasaka, M., Kubo, Y., and Iijima, S. (2006) 13C NNMR spectroscopy of carbon nanohorns. *Phys. Rev. B*, **73**, 125405.

38 Bandow, S., Rao, A., Sumanasekera, G., Eklund, P.C., Kokai, F., Takahashi, K., Yudasaka, M., and Iijima, S. (2000) Evidence for anomalously small charge transfer in doped single-wall carbon nanohorn aggregates with Li, K, and Br. *Appl. Phys. A*, **71**, 561–564.

39 Bonard, J.-M., Gaal, R., Garaj, S., Thien-Nga, L., Forro, L., Takahashi, K., Kokai, F., Yudasaka, M., and Iijima, S. (2002) Field emission properties of carbon nanohorn films. *J. Appl. Phys.*, **91**, 10107–10109.

40 Urita, K., Seki, S., Utsumi, S., Noguchi, D., Kanoh, H., Tanaka, H., Hattori, Y., Ochiai, Y., Aoki, N., Yudasaka, M., Iijima, S., and Kaneko, K. (2006) Effect of gas adsorption on the electrical conductivity of single-wall carbon nanohorns. *Nano Lett.*, **6**, 1325–1328.

41 Ustumi, S., Honda, H., Hattori, Y., Kanoh, H., Takahashi, K., Sakai, H., Abe, M., Yudasaka, M., Iijima, S., and Kaneko, K. (2007) Direct evidence on C−C single-bonding in single-wall carbon nanohorn aggregates. *J. Phys. Chem. C*, **111**, 5572–5575.

42 Urita, K., Seki, S., Tsuchiya, H., Honda, H., Ustumi, S., Hayakasa, C., Kanoh, H., Ohba, T., Tanaka, H., Yudasaka, M., Iijima, S., and Kaneko, K. (2008) Mechanochemically induced sp-bond-associated reconstruction of single-wall carbon nanohorns. *J. Phys. Chem. C*, **112**, 8759–8762.

43 Fujimori, T., Urita, K., Aoki, Y., Kanoh, H., Ohba, T., Yudasaka, M., Iijima, S., and Kaneko, K. (2008) Fine nanostructure analysis of single-wall carbon nanohorns by surface-enhanced Raman scattering. *J. Phys. Chem. C.*, **112**, 7552–7556.

44 Yudasaka, M., Ajima, K., Suenaga, K., Ichihashi, T., Hashimoto, A., and Iijima, S. (2003) Nano-extraction and nano-condensation for C_{60} incorporation into

single-wall carbon nanotubes in liquid phases. *Chem. Phys. Lett.*, **380**, 42–46.

45 Hashimoto, A., Yorimitsu, H., Ajima, K., Suenaga, K., Isobe, H., Miyawaki, J., Yudasaka, M., Iijima, S., and Nakamura, E. (2004) Selective deposition of a gadolinium(III) cluster in a hole opening of single-wall carbon nanohorn. *Proc. Natl. Acad. Sci.*, **101**, 8527–8530.

46 Yuge, R., Ichihashi, T., Shimakawa, Y., Kubo, U., Yudasaka, M., and Iijima, S. (2004) Preferential deposition of Pt nanoparticles inside single-wall carbon nanohorns. *Adv. Mater.*, **16**, 1420–1423.

47 Murakami, T., Ajima, K., Miyawaki, J., Yudasaka, M., Iijima, S., and Shiba, K. (2004) Drug-loaded carbon nanohorns: adsorption and release of dexamethasone *in vitro. Mol. Pharm.*, **1**, 399–405.

48 Ajima, K., Yudasaka, M., Murakami, T., Maigne, A., Shiba, K., and Iijima, S. (2005) Carbon nanohorns as anticancer drug carriers. *Mol. Pharm.*, **2**, 475–480.

49 Ajima, K., Murakami, T., Mizoguchi, Y., Tsuchida, K., Ichihashi, T., Iijima, S., and Yudasaka, M. (2008) Enhancement of *in vivo* anticancer effects of cisplatin by incorporation inside single-wall carbon nnanohorns. *ACSNano*, **2**, 2057–2064.

50 Yuge, R., Yudasaka, M., Miyawaki, J., Kubo, Y., Ichihashi, T., Imai, H., Nakamura, E., Isobe, H., Yorimitsu, H., and Iijima, S. (2005) Controlling the incorporation and release of C_{60} in nanometer-scale hollow spaces inside single-wall carbon nanohorns. *J. Phys. Chem. B*, **109**, 17861–17867.

51 Bekyarova, E., Hashimoto, A., Yudasaka, M., Hattori, Y., Murata, K., Kanoh, H., Kasuya, D., Iijima, S., and Kaneko, K. (2005) Palladium nanoclusters deposited on single-walled carbon nanohorns. *J. Phys. Chem. B*, **109**, 3711–3714.

52 Miyawaki, J., Yudasaka, M., Imai, H., Yorimitsu, H., Isobe, H., Nakamura, E., and Iijima, S. (2006) *In vivo* magnetic resonance imaging of single-wall carbon nanohorns by labeling with magnetite nanoparticles. *Adv. Mater.*, **18**, 1010–1014.

53 Miyawaki, J., Yudasaka, M., Imai, H., Yorimitsu, H., Isobe, H., Nakamura, E., and Iijima, S. (2006) Synthesis of ultrafine Gd$_2$O$_3$ nanoparticles inside single-wall carbon nanohorns. *J. Phys. Chem. B*, **110**, 5179–5181.

54 Zhang, M., Murakami, T., Ajima, K., Tsuchida, K., Sandanayaka, A.S.D., Ito, O., Iijima, S., and Yudasaka, M. (2008) Fabrication of ZnPc/protein nanohorns for double photodynamic and hyperthermic cancer phototherapy. *Proc. Natl. Acad. Sci.*, **105**, 14773–14778.

55 Xu, J., Yudasaka, M., Kouraba, S., Sekido, M., Yamamoto, Y., and Iijima, S. (2008) Single wall carbon nanohorn as a drug carrier for controlled release. *Chem. Phys. Lett.*, **461**, 189–192.

56 Yuge, R., Yudasaka, M., Maigne, A., Tomonari, M., Miyawaki, J., Kubo, Y., Imai, H., Ichihashi, T., and Iijima, S. (2008) Adsorption phenomena of tetracyano-*p*-quinodimethane adsorbed on single-wall carbon nanohorns. *J. Phys. Chem. C*, **112**, 5416–5422.

57 Miyako, E., Nagata, H., Hirano, K., Makita, Y., and Hirotsu, T. (2008) Photodynamic release of fullerenes from within carbon nanohorn. *Chem. Phys. Lett.*, **456**, 220–222.

58 Miyawaki, J., Yudasaka, M., Yuge, R., and Iijima, S. (2007) Organic-vapor-induced repeatable entrance and exit of C$_{60}$ into/from single-wall carbon nanohorns at room temperature. *J. Phys. Chem. C*, **111**, 9719–9722.

59 Yuge, R., Yudasaka, M., Miyawaki, J., Kubo, Y., Isobe, H., Yorimitsu, H., Nakamura, E., and Iijima, S. (2007) Plugging and unplugging of holes of single-wall carbon nanohorns. *J. Phys. Chem. C*, **111**, 7348–7351.

60 Miyawaki, J., Yuge, R., Kawai, T., Yudasaka, M., and Iijima, S. (2007) Evidence of thermal closing of atomic-vacancy holes in single-wall carbon nanohorns. *J. Phys. Chem. C*, **111**, 1553–1555.

61 Fan, J., Yuge, R., Miyawaki, J., Kawai, T., Iijima, S., and Yudasaka, M. (2008) Close-open-close evolution of holes at the tips of conical graphenes of single-wall carbon nanohorns. *J. Phys. Chem. C*, **112**, 8600–8603.

62 Yuge, R., Ichihashi, T., Miyawaki, J., Yoshitake, T., Iijima, S., and Yudasaka, M.

(2009) Hidden caves in an aggregate of single-wall carbon nanohorns found by using Gd$_2$O$_3$ probes. *J. Phys. Chem. C.*, **113**, 2741–2744.

63 Petsalakis, I.D., Pagona, G., Theodorakopoulos, G., Tagmatarchis, N., Yudasaka, M., and Iijima, S. (2006) Unbalanced strain-directed functionalization of carbon nanohorns: a theoretical investigation based on complementary methods. *Chem. Phys. Lett.*, **429**, 194–198.

64 Tagmatarchis, N., Maigne, A., Yudasaka, M., and Iijima, S. (2006) Functionalization of carbon nanohorns with azomethine ylides: towards solubility enhancement and electron-transfer processes. *Small*, **4**, 490–494.

65 Cioffi, C., Campidelli, S., Brunetti, F.G., Meneghetti, M., and Prato, M. (2006) Functionalization of carbon nanohorns. *Chem. Commun.*, 2129–2131.

66 Isobe, H., Tanaka, T., Maeda, R., Noiri, E., Solin, N., Yudasaka, M., Iijima, S., and Nakamura, E. (2006) Preparation, purification, characterization, and cytotoxicity assessment of water-soluble, transition-metal-free carbon nanotube aggregates. *Angew. Chem. Int. Ed.*, **45**, 6676–6680.

67 Pagona, G., Rotas, G., Patsalakis, I.D., Theodorakopolous, G., Fan, J., Maigne, A., Yudasaka, M., Iijima, S., and Tagmatarchis, N. (2007) Soluble functionalized carbon nanohorns. *J. Nanosci. Nanotechnol.*, **7**, 3468–3472.

68 Pagona, G., Fan, J., Maignè, A., Yudasaka, M., Iijima, S., and Tagmatarchis, N. (2007) Aqueous carbon nanohorn–pyrene–porphyrin nanoensembles: controlling charge-transfer interactions. *Diam Relat. Mater.*, **16**, 1150–1153.

69 Petsalakis, I.D., Pagona, G., Tagmatarchis, N., and Theodorakopoulos, G. (2007) Theoretical study in donor–acceptor carbon-based hybrids. *Chem. Phys. Lett.*, **448**, 115–120.

70 Cioffi, C., Campidelli, S., Sooambar, C., Marcaccio, M., Marcolongo, G., Meneghetti, M., Paolucci, D., Paolucci, F., Ehli, C., Rahman, G.M.A., Sgobba, V., Guldi, D.M., and Prato, M. (2007) Synthesis, characterization, and

photoinduced electron transfer in functionalized single wall carbon nanohorns. *J. Am. Chem. Soc.*, **129**, 3938–3945.

71 Pagona, G., Karousis, N., and Tagmatarchis, N. (2008) Aryl diazonium functionalization of carbon nanohorns. *Carbon*, **46**, 604–610.

72 Nakamura, E., Koshino, M., Tanaka, T., Niimi, Y., Harano, K., Nakamura, Y., and Isobe, H. (2008) Imaging of conformational changes of biotinylated triamide molecules covalently bonded to a carbon nanotube surface. *J. Am. Chem. Soc.*, **130**, 7808–7809.

73 Chen, G., Hamon, M.A., Hu, H., Chen, Y., Rao, A.M., Eklund, P.C., and Haddon, R.C. (1998) Solution properties of single-wall carbon nanotubes. *Science*, **282**, 95–98.

74 Yuge, R., Zhang, M., Tomonari, M., Yoshitake, T., Iijima, S., and Yudasaka, M. (2008) Site identification of carboxyl groups on graphene edges with Pt derivatives. *ACSNano*, **2**, 1865–1870.

75 Miyawaki, J., Yudasaka, M., and Iijima, S. (2004) Solvent effects on hole-edge structure for single-wall carbon nanotubes and single-wall carbon nanohorns. *J. Phys. Chem. B*, **108**, 10732–10735.

76 Pagona, G., Tagmatarchis, N., Fan, J., Yudasaka, M., and Iijima, S. (2006) Cone-end functionalization of carbon nanohorns. *Chem. Mater.*, **18**, 3918–3920.

77 Pagona, G., Sandanayaka, A.S.D., Araki, Y., Fan, J., Tagmatarchis, N., Charalambidis, G., Coustolelos, A.G., Boitrel, B., Yudasaka, M., Iijima, S., and Ito, O. (2007) Covalent functionalization of carbon nanohorns with porphyrins: nanohybrid formation and photoinduced electron and energy transfer. *Adv. Funct. Mater.*, **17**, 1705–1711.

78 Sandanayaka, A.S.D., Pagona, G., Fan, J., Tagmatarchis, N., Yudasaka, M., Iijima, S., Araki, Y., and Ito, O. (2007) Photoinduced electron-transfer processes of carbon nanohorns with covalently linked pyrene chromophores: charge-separation and electron-migration systems. *J. Mater. Chem.*, **17**, 2540–2546.

79 Pagona, G., Sandanayaka, A.S.D., Hosobe, T., Charalambidis, G., Coutsolelos, A.G.,

Yudasaka, M., Iijima, S., and Tagmatarchis, N. (2008) Characterization and photoelectrochemical properties of nanostructured thin film composed of carbon nanohorns covalently functionalized with porphyrins. *J. Phys. Chem. C*, **112**, 15735–15741.

80 Rotas, G., Sandanayaka, A.S.D., Tagmatarchis, N., Ichihashi, T., Yudasaka, M., Iijima, S., and Ito, O. (2008) (Terpyridine)copper(III)-carbon nanohorns: metallo-nanocomplexes for photoinduced charge separation. *J. Am. Chem. Soc.*, **130**, 4725–4731.

81 Kam, N.W.S., O'connel, M., Wisdom, J.A., and Dai, H. (2005) Carbon nanotubes as multifunctional biological transporters and near-infrared agents for selective cancer cell destruction. *Proc. Natl. Acad. Sci.*, **102**, 1160–11605.

82 Kase, D., Kulp, J.L., III, yudasaka, M., Evans, J.S., Iijima, S., and Shiba, K. (2004) Affinity selection of peptides phage libraries against single-wall carbon nanohorns indentifies a peptide aptamer with conformational variability. *Langmuir*, **20**, 8939–8941.

83 Matsumura, S., Ajima, K., Yudasaka, M., Iijima, S., and Shiba, K. (2007) Dispersion of cisplatin-loaded carbon nanohorns with a conjugate comprised of an artificial peptide aptamer and polyethylene glycol. *Mol. Pharm.*, **4**, 723–729.

84 Matsumura, S., Sato, S., Yudasaka, M., Tomida, A., Tsuruo, T., Iijima, S., and Shiba, K. (2009) Prevention of carbon nanohorn agglomeration using conjugate comprised of comb-shaped polyethylene glycol and a peptide aptamer. *Mol. Pharm.*, **6**, 441–447.

85 Murakami, T., Fan, J., Yudasaka, M., Iijima, S., and Shiba, K. (2006) Solubilization of single-wall carbon nanohorns using a PEG-doxorubicin conjugate. *Mol. Pharm.*, **3**, 407–414.

86 Murakami, T., Sawada, H., Tamura, G., Yudasaka, M., Iijima, S., and Tsuchida, K. (2008) Water-dispersed single-wall carbon nanohorns as drug carriers for local cancer chemotherapy. *Nanomedicine*, **3**, 453–463.

87 Yoshitake, T., Shimakawa, Y., Kuroshima, S., Kimura, H., Ichihashi, T., Kubo, Y., Kasuya, D., Takahashi, K., Kokai, F.,

Yudasaka, M., and Iijima, S. (2002) Preparation of fine platinum catalyst supported on single-wall carbon nanohorns for fuel cell application. *Physica B*, **323**, 124–126.

88 Murata, K., Hashimoto, A., Yudasaka, M., Kasuya, D., Kaneko, K., and Iijima, S. (2004) The use of charge transfer to enhance the methane-storage capacity of single-walled, nanostructured carbon. *Adv. Mater.*, **16**, 1520–1522.

89 Murata, K., Yudasaka, M., and Iijima, F S. (2006) Hydrogen production from methane and water at low temperature using EuPt supported on single-wall carbon nanohorns. *Carbon*, **44**, 799–823.

90 Lacotte, S., Garcia, A., Decossas, M., Al-Jamal, W.T., Li, S., Kostarelos, K., Muller, S., Prato, M., Dumortier, H., and Bianco, A. (2008) Interfacing functionalized carbon nanohorns with primary phagocytic cells. *Adv. Mater.*, **20**, 2421–2426.

91 Miyawaki, J., Yudasaka, M., Azami, T., Kubo, Y., and Iijima, S. (2008) Toxicity of single-walled carbon nanohorns. *ACSNano*, **2**, 213–226.

92 Miyawaki, J., Yudasaka, M., Zhang, M., and Iijima, S. (2007) Intravenous toxicity of single-walled carbon nanohorns. 34th Fullerene-Nanotube Symposium.

93 Shopova, M., Stoichkov, N., Milev, A., Georgiev, K., Gizbreht, A., Jori, G., and Ricchelli, F. (1995) Photodynamic therapy of experimental tumors with Zn(II)-phthalocyanine and pulsed laser irradiation. *Lasers in Medical Science*, **10**, 43–46.

14
Self-Organization of Nanographenes

Wojciech Pisula, Xinliang Feng, and Klaus Müllen

14.1
Introduction

14.1.1
Graphene, Graphene Nanoribbon, and Nanographene

Since centuries, graphite is mostly known as a thin stick of pigment in pencils allowing to write down a text on a piece of paper. Apart from diamond and fullerene, this mineral is one of the allotropes of carbon consisting of graphene layers in which the sp^2-bonded carbon atoms are arranged in a hexagonal lattice with separation of 0.142 nm, and the distance between sheets is 0.335 nm. Such perfect isolated two-dimensional (2D) crystals were believed to be unstable in free state. Few years ago, Novoselov and Geim proved the contrary. For the first time, large, planar and freestanding graphene sheets were prepared by mechanical exfoliation of graphite crystals [1, 2]. Today, it is supposed that this novel process occurs even during writing with the pencil when the graphite ink is transferred to paper. Such atomically flat sheets as infinitely large aromatic macromolecules reveal quite unique properties from most conventional three-dimensional materials and allow new and surprising insights into fundamental physics [3]. For instance, the fact that intrinsic graphene is a zero-gap semiconductor attracts remarkable attention from both experimental and theoretical communities and makes it highly promising for future applications in various electronic devices such as transparent electrodes, ultrasensitive sensors, and high-performance ballistic transistors [4–6]. Graphene owes its extraordinary behavior to its relatively large size exceeding 100 μm of the strict hexagonal lattice of the carbon atoms in relation to the unbonded edge. The cutting of "infinite" graphene sheets into smaller pieces, the so-called graphene nanoribbons (GNRs) with width less than 50 nm and lengths below 500 nm, changes this situation dramatically. It permits controlling the electronic structure of GNRs by varying the width, the bandgap by the crystallographic orientation of the GNR axis, the electronic structure by the nature of the edge passivation, among other important issues such as the effect of structural defects on the electronic and thermodynamic properties of GNRs. It is

Carbon Nanotubes and Related Structures. Edited by Dirk M. Guldi and Nazario Martín
Copyright © 2010 Wiley-VCH Verlag GmbH & Co. KGaA, Weinheim
ISBN: 978-3-527-32406-4

predicted that GNRs with armchair-shaped edges can be either metallic or semi-conducting depending on their widths, whereby GNRs with zigzag-shaped edges are metallic with peculiar edge states on both sides of the ribbon regardless of their width [10]. For the armchair GNRs, calculations indicate bandgap oscillations as a function of the GNRs width less than 3 nm [7]. The relation between bandgap, edge nature, and crystallographic orientation of GNR axis also suggests that, similar to carbon nanotubes (CNTs), optical spectra can be used as a powerful characterization tool for GNRs. Recent experiments on thin stripes of graphene fabricated by lithography, which are smaller sized GNRs of hundreds of nanometers in length and tens of nanometers width, reveal a decline of the resistivity with the decrease of the ribbon width and confirm the impact of edge states [8]. Indeed, the general trend of the semiconducting character for GNRs results from the increase of the energy gap with decreasing width and is consistent with theoretical predictions [9, 10]. However, this fabrication process does not provide sufficient control over the edge structure and overall crystallographic direction that is essential for the properties of GNRs. Therefore, it is believed that the interplay between precise width, edge orientation, edge structure, and chemical termination of the edges in GNRs remains a challenging area for future research. Apart from the exfoliation of graphite, mesoscopic graphene sheets can be produced by chemical reduction of graphite oxide [11] or epitaxial growth on silicon carbide or metal surfaces [12, 13]. The first approach using hydrazine as a reducing agent provides a stable dispersion of reduced graphite oxide in aqueous solution, leading to processable graphene sheets in large quantities. However, the harsh condition of this method does not allow producing graphene and GNR layers with well-defined edge shape. Very recently, Dai and colleagues developed a chemical exfoliation of expandable graphite, resulting in high-quality GNRs with width below 10 nm [14]. Measurements of the electronic transport exhibited, unlike the graphene sheets, a semiconducting character for all the sub-10 nm GNRs and afforded field-effect transistors with on/off ratios of 10^5 at room temperature, qualifying them for high switching speed electronics as silicon replacement. Nevertheless, the yield of this method is rather low and the edge of these GNRs cannot be clearly defined. Synthetic organic chemistry presents a different concept on GNRs. In this bottom-up approach, long and soluble graphene nanoribbons terminated by hydrogen or alkyl substituents and a discrete width are obtained in large quantities (Figure 14.1), which enables a facile solution processing and device fabrication [15, 16]. The synthesis is based on the versatile organic chemistry of polycyclic aromatic hydrocarbons (PAHs) [17, 18] and opens the door to a broad variety of even smaller aromatic structures, to the so-called nanographenes. Nanographene can be defined as a nanometer-sized (average diameter <10 nm) 2D flat hexagonal carbon ring network with open edges [19–22, 24, 32], which stands in contrast to the other members of the nanocarbon family with closed π-electron systems, such as fullerenes and carbon nanotubes. So far, the largest synthesized monodispersed PAH molecule consists of 222 carbon atoms and has 10 benzene rings with a disk diameter of 3 nm, while one of the smallest but most studied nanographene, hexa-*peri*-hexabenzocoronene (HBC), has a size of about 1.4 nm (Figure 14.1) [23].

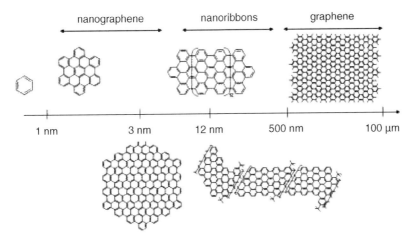

Figure 14.1 Nanographene, nanoribbons, and graphene defined by their size.

Identical to graphene, parent nanographenes are not soluble in common organic solvents and are not meltable as required for such facile processibility and device fabrication. As a great advantage, organic synthesis does not set limits to size and shape of the nanographenes, instead enables the integration of solubilizing substituents and further functionality [24]. The early synthesis of these molecules is based on the pioneering contributions by E. Clar and R. Scholl at the beginning of the last century [19]. Later on, more efficient and mild synthetic methods were developed leading to the expansion of this class of systems [25]. By employing the previous concept, large dendritic oligophenylene precursors with different size and shape were first designed by using the Diels–Alder reaction or cyclotrimerization of suitable building blocks. After planarization of the precursors, all kinds of large benzenoid PAHs with different molecular sizes, symmetries, and peripheries have become available. For example, large nanographenes with triangle shape (C60), linear ribbon shape (C78), cordite shape (C96), square shape (C132), and so on are also attainable (Figure 14.2) [26, 27]. The most recent progress in the synthesis of large PAHs is the inclusion of additional double bonds on the periphery of HBCs, which act as the "zigzag" armchair [28]. Up to now, HBCs with monozigzag, double zigzag, and trizigzag are well designed and synthesized (Figure 14.3). In addition to all-hydrocarbon-based nanographene building blocks, electron-rich and electron-poor heteroatoms such as sulfur and nitrogen have also been successfully incorporated into the nanographene units [29, 30]. Therefore, compared to graphene prepared from physical exfoliation and chemical reduction methods, the bottom-up synthetic chemical approach offers the great opportunity to tailor the molecular size, shape, edge, and composition of the nanographenes, and more important, together with their structure perfection. Obviously, the effect of size, shape, and edge structure on the nanographenes is more profound than on the GNRs. This is clearly reflected by their pristine color and optoelectronic properties. With a larger number of carbons including C42 (HBC), C60, and C222, the color of nanographene powders changes

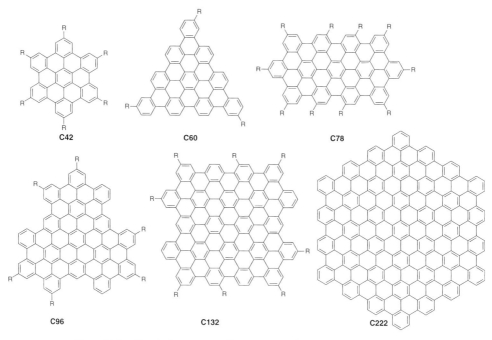

Figure 14.2 Chemical structures of large nanographenes.

from yellow to red to dark purple. This trend is also obvious from the optical absorption properties of nanographenes, for which the maximum wavelength of the lowest energy absorption band (alpha band) increases linearly and broadens with the number of "full" six-membered rings, suggesting the extension of π-conjugation and the reduction of the bandgap. This behavior is very similar to that of linear conjugated oligomers for which the absorption maximum redshifts with increased chain length below the effective conjugation length. The nature of the edge structure also plays an important role in their electronic properties and chemical reactivity [31]. According to Clar's aromatic sextet rule, all-benzenoid graphitic molecules with either "armchair" and/or "cove" peripheries normally show extremely high chemical stability. However, the nanographenes with zigzag peripheries have higher electronic energies and thus show higher chemical reactivity. The zigzag periphery with double-bond-like structure dramatically changes the electronic and optoelectronic properties of

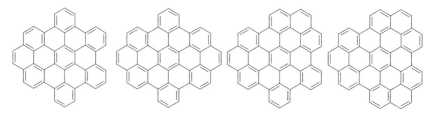

Figure 14.3 Nanographenes with different zigzag peripheries at the edge.

nanographenes. It is known that the UV–vis and fluorescence peaks of the zigzag nanographene series undergo a dramatic redshift with respect to the benzenoid analogues, indicating the reduction of electronic bandgap.

14.1.2
Organization of Nanographenes

Nanographenes can be used not only as model π-systems for modern synthetic and theoretical chemistry but also as building blocks to fabricate various electronic and optoelectronic devices, in which the nanographene units serve as transporting media for electrons [32]. Nanographenes have gained attraction in recent times as promising candidates owing to their semiconducting properties for the application in electronic and optoelectronic devices such as flexible E-paper that might replace in future partly the "traditional" writing with a pencil on cellulose-based paper. Since the late 1970s, π-conjugated systems have been in focus as active components for field-effect transistors (FETs), light-emitting diodes, and solar cells including small molecules and polymers [33]. Today, organic semiconductors exhibit charge carrier mobilities equal to the amorphous silicon, while their facile solution processability makes mass applications on flexible substrates possible by roll-to-roll printing [34]. Typically, the fabrication of organic electronic devices is mainly based on the deposition and solidification of the semiconductor on a substrate. The substance is first fluidized by melting at higher temperature or dissolved in a solvent and deposited on a surface. In the next step, the material is solidified by cooling or drying during which a self-assembly of the molecules takes place. In order to control and optimize the molecular organization, a fundamental understanding of the processes involved in solidification has to be established. Before this key point is discussed in more detail, the term "organization" needs a definition since it includes many aspects and will appear in the following sections in different contexts. Organization of nanographenes can occur over different length scales and under various conditions. For instance, ordered monolayers at solid–liquid interfaces obtained from diluted solutions allow the inspection of single sheets. If the solubility is increased, an initial structure formation occurs in the solvent since the molecules pack in larger aggregates. After deposition on a surface, interfacing effects become more dominant on the arrangement of the nanographenes toward the substrate. If the deposition is controlled through a processing technique, a homogeneous organization over macroscopic areas is achieved. In contrast, this kind of surface influence can be ignored for the organization in the bulk in which molecular interactions are most essential for the formation of columnar superstructures. Interestingly, these columnar stacks of nanographenes can be regarded as nanotube-like architectures with the aromatic layer direction perpendicular to the columnar axis. In contrary, the graphene planes in carbon nanotubes, which have a diameter of one to few nanometers and a length of even hundreds of micrometers, are oriented along the tube axis [35]. Introduction of aliphatic chains at the disk-shaped aromatic core provides solubility and leads to a distinct thermal behavior. Flexible substituents increase the overall disorder and result in the self-assembly into complex columnar superstructures with liquid crystalline (LC) properties. In such mesophases, self-healing of structural defect and grain boundaries, which decrease the migration of charges, can improve the device. Hexa-*peri*-hexabenzocoronene, the so-called "superbenzene", is a well-studied example for discotic LC nanographenes. Due to the orbital overlap of the large π-system, the charge carrier mobility along the stacking direction satisfactorily reaches high values for applications in field-effect transistors. The insulating aliphatic chains around the semiconductive aromatic stack provide a one-dimensional charge carrier transport that, however, is rather sensitive to structural bifurcation defects. In addition, the charge movement between electrodes strongly depends on the macroscopic order through the active layer. The self-assembly during solution processing is controlled by the introduction of adequate alkyl substituents. Besides the π-interactions between the aromatic parts, additional functional groups such as hydrogen bonds, and local dipole–dipole interactions enhance the organization. Investigations of the aggregation ability in solution and structure formation of nanographenes with different side chains on surfaces offer valuable information for the device fabrication and operation. This chapter provides an overview on the general molecular design toward discotic nanographenes of different geometry carrying various functional substituents that have a great impact on the interaction between the single blocks, the supramolecular organization, thermotropic behavior, and their processibility from solution and isotropic melt. This provides the opportunity to control the complex self-assembly on the surface during processing leading finally to the implementation of such molecules in field-effect transistors and photovoltaic cells.

14.2
Single Sheets of Nanographenes

Techniques for the isolation of single sheets of graphene-type systems are based on the size of the desired layers (Figure 14.4). As mentioned in the introduction, individual, freestanding graphene layers are obtained by mechanical exfoliation. The isolation of single nanographenes requires a different approach. Vacuum sublimation and physisorption at solid–liquid interfaces were successfully applied for realizing single sheets of HBC (**1**) on surfaces [36, 37]. The resulting organizations formed by both methods were inspected by scanning tunneling microscopy (STM) revealing an epitaxial growth of the HBC nanographenes into highly ordered, close-packed monolayers with a 2D crystal pattern corresponding to the underlying surface such as basal plane of highly oriented pyrolytic graphite (HOPG) (Figure 14.5a). The molecules were arranged with their aromatic planes parallel to the surface due to interactions between the conducting substrate and the molecule. The tunneling current of single nanographenes is measured by the STM tip that can be positioned precisely with respect to the molecules [38]. As alkyl-substituted HBCs and larger nanographenes cannot be sublimed nondestructively, monolayers of this kind of systems are obtained only at solid–liquid interface since even little amounts of material allow the layer formation. Monolayers of superbiphenyl (**3**) [39], C60

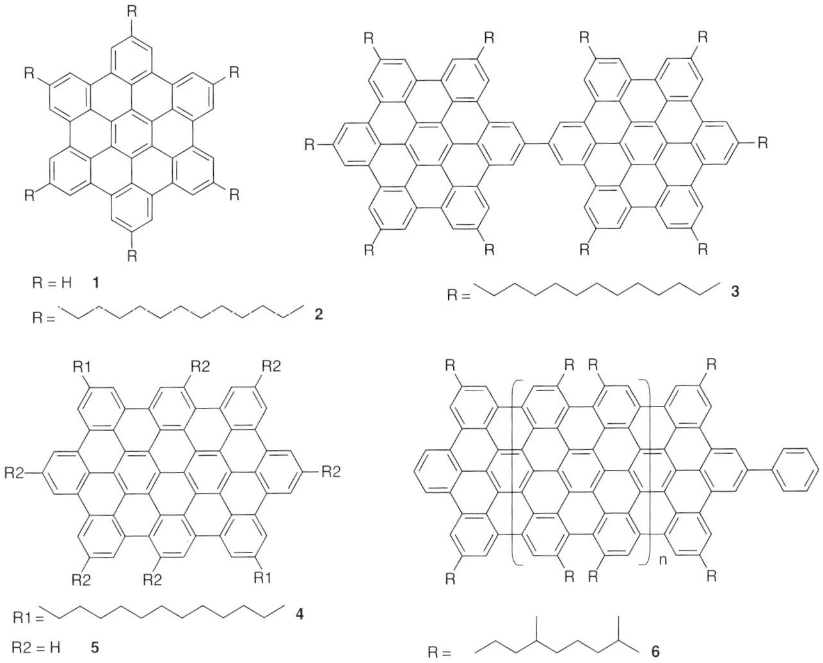

R = H **1**

R = ~~~~~~~~~ **2**

R = ~~~~~~~~~~~~~ **3**

R1 = ~~~~~~~~~ **4**

R2 = H **5**

R = ~~~~ **6**

Figure 14.4 Nanographenes forming monolayers of single sheets.

(**4** and **5**) [40], C132 (Figure 14.2) [41], and nanoribbons (**6**) [15] were successfully prepared (Figure 14.5). Therefore, the side chains play an important role for the solubility, the absorption on the conducting substrate, and the organization of the nanographene within the crystal lattice. For hexadodecyl-substituted HBC (**2**), the chains were found to be arranged along the main axis of the graphite substrate, while alkylphenyl-substituted HBC (**7**) revealed variable vertical displacements in the lattice from the substrate [42]. The type and size of the 2D crystal is mainly determined by the symmetry and substitution pattern of the nanographene. For

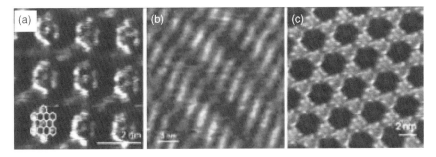

Figure 14.5 STM images of (a) HBC (**1**). Reprinted with permission from Ref. [32]. Copyright 2007 American Chemical Society. (b) Alkyl-substituted nanoribbon **6**. Reprinted with permission from Ref. [15]. Copyright 2008 American Chemical Society. (c) C_3 symmetric **10** with zigzag peripheries at the edge. Adapted from Ref. [43].

instance, C_3 symmetric nanographenes (**10**) with resolved triangle molecular shape packed in a highly regular and unique honeycomb lattice (Figure 14.5c), which was formed by weak intermolecular and interfacial forces, whereby the unit cell parameters indicated that it dissolved partially in the supernatant organic solution [43]. Functional groups at the nanographene periphery can be additionally employed to structure formation at interfaces. Lateral dimeric structures of nanographenes grow at solid–liquid interfaces if hydrogen bonds are introduced through one attached carboxylic acid [44]. As the next step, the functionalization of nanographenes by covalently linked acceptors such as anthraquinone leads to a different tunneling behavior of the subunits that can be exploited for molecular chemical field-effect transistors [45].

A diluted concentration and lack of aggregation are essential prerequisites for the formation of the above-described monolayers at solid–liquid interfaces. Even at these conditions, nanographenes assemble into bilayers due to aggregation. With increasing concentration, reaggregation of the molecules intensifies and the effect of the surface on the organization becomes negligible. After the molecules precipitate and a bulk state is established, the single building blocks arrange within the supramolecular organization due to specific molecular interactions. The assembly of nanographenes in the bulk, which can be regarded as multilayer structure, is related to another aspect of "organization" that is described in detail in the next section.

14.3
Organization in the Bulk State

14.3.1
Liquid Crystalline Columnar Phases

The previous section illustrated that nanographenes of different shape and size can carry various substitution patterns governing the monolayer formation. The substitution of nanographenes by aliphatic side chains has a great influence on the organization. In strong contrast to plain PAHs, the alkyl chain-decorated derivatives self-assemble into well-defined supramolecular structures. Typically, two types of organizations, columnar and nematic, can arise. While nematic phases reveal an orientational arrangement of molecules, but no long-range translational order, columnar structures are formed due to π-stacking interactions between individual rigid cores, which stack on top of each other, separated by a local nanophase between the aromatic and aliphatic fractions. The columns order in arrays that can be described by crystallographic 2D unit cells, whereby the alkyl side chains fill the intercolumnar space. The unit cell parameters are determined by the core size and the side chain length. For instance, the lateral packing characterized by the lattice constant of an intercolumnar hexagonal arrangement increases with the molecular masses of the core, forming the columnar stacks, and with the length of the side chains that fill the core periphery. For a model describing the relationship between the morphological and the molecular parameters for such columnar hexagonal arrangement of the discotic nanographenes with different core sizes, it is reasonable to regard the density inside the columns similar to graphite, which is 2.25 g/cm³, and the density of the alkyl side chains close to low molecular polyethylene at 0.9 g/cm³ [46]. Indeed, for nanographenes consisting of 132 carbon atoms, the unit cell parameters determined by X-ray scattering are in close agreement with this theoretical model taking into consideration a nanophase separation between nanographenes and the polyethylene-type matrix (Figure 14.6). This model confirms once again the graphitic character of the discotic nanographenes.

The attachment of flexible aliphatic side chains at the aromatic disk has also a pronounced impact on the thermal behavior of the materials. Since the side chains

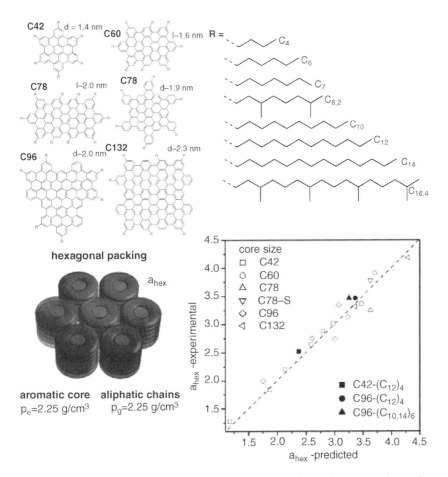

Figure 14.6 The effects of the core size, the side chain length, and the number of substituents on the supramolecular organization of nanographenes were investigated by X-ray scattering. In this example, the aromatic core size of the compounds varied between the hexa-*peri*-hexabenzocoronene core consisting of 42 carbon atoms and an enlarged aromatic core of 132 carbon atoms, whereas the length of the side chains extended up to 20 carbon atoms per chain. The unit cell parameters determined experimentally and theoretically are in close agreement taking into account the model based on the densities of graphite and low molecular polyethylene.

increase both the overall disorder in the system and the heterogeneity between the two nanophases, the enthalpic energy of the aromatic part is overcome by the entropic contribution of side-chain mobility. The entropic contribution depends on the steric demand and thus on the geometry of the substituents, while the enthalpic energy goes together with the π-orbital area of the aromatic core. The thermal behavior of discotic nanographenes can be separated into three main phases with characteristic features in organization and molecular dynamics. For the crystalline state, a reduced molecular dynamics of the alkyl-substituted nanographenes is typical with a high overall order. Side chains are fully or partially crystallized, while the aromatic cores are tilted with their planes with respect to the columnar axis in order to optimize orbital interactions between molecules [47]. Due to this molecular tilt inside the stacks, noncylindrical columns are formed resulting in rectangular or monoclinic arrays (Figure 14.7b). The increase in molecular dynamics can lead to the formation of (multiple) intermediate liquid crystalline states. These phases are assigned by lateral and longitudinal fluctuations of the nanographenes accompanied by molecular rotation around the columnar axis. In this state, the molecules pack in hexagonally arranged columns being most favorable for cylindrical structures [48]. Due to the molecular rotation in most of the cases, the aromatic plane is perpendicular to the columnar axis, whereby the aliphatic side chains adapt a fluid-like nature. This liquid crystalline state represents a unique and important property for the material since it allows a spontaneous self-healing of (structural) defects resulting

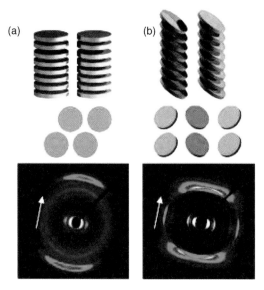

Figure 14.7 Schematic representation of the intra- and intercolumnar organization of alkyl-substituted discotic nanographenes in the (a) liquid crystalline phase and (b) crystalline. In the LC phase with an orthogonal packing in the columns, while in the crystalline state the molecules are tilted with respect to the columnar axis. The information regarding the structure at different temperatures is derived form the 2D patterns of the extruded fibers (the white arrow indicates the fiber axis). The 2D patterns presented are for HBC-C$_{12}$ (**2**).

in improvement for charge carrier transport [49]. Further heating leads to an amorphous melt in which the columnar structures break into monomeric species or small aggregates.

To understand the relation between the molecular structure and thermotropic properties of discotics nanographenes, detailed investigations are necessary. As presented in Figure 14.6, a close relationship between aromatic size, side chain length, and lattice parameters was found on the basis of X-ray scattering results. However, more information on the thermal properties, self-assembly in solution or during solidification from the isotropic melt, is needed to gain a full picture of the material. As mentioned above, X-ray scattering can provide detailed information of the organization in the bulk state. The application of a two-dimensional detector and sample preparation via fiber extrusion offer significantly more details on the organization of the molecules within the complex superstructure. For most of the discotic nanographenes, the extrusion is performed in the soft and deformable, low-viscosity liquid crystalline phase. A number of parameters control this mechanical processing such as die shape, cylinder and orifice diameters, and the piston velocity (Figure 14.8a). The X-ray diffraction experiments revealed that under these sample preparation conditions the columns are in most of the cases well oriented in the direction of the extrusion. For the measurement, the fiber is positioned vertically to the detector as schematically illustrated in Figure 14.8b. The resulting 2D pattern can be separated for the analysis into the equatorial and meridional planes, on which the reflections are positioned characteristically for a discotic nanographene (Figure 14.8c). Usually, the equatorial small-angle reflections are attributed to intercolumnar correlations confirming that the columnar stacks are oriented along the extrusion direction. Additional higher order scattering intensities in the same plane are related to the first main reflection and their positions with respect to each other assign the lattice type. Meridional reflections are attributed to the d-spacings along the columnar stacks such as the π-stacking distance between individual molecules. Figure 14.7a shows a typical 2D pattern for a hexadodecyl-substituted HBC (2) in its liquid crystalline phase that is formed at temperatures above 105 °C.

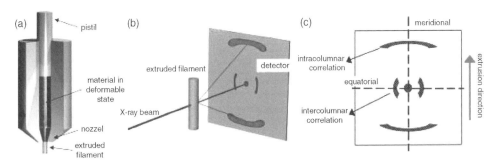

Figure 14.8 Schematic illustration of (a) the filament alignment by extrusion using a mini-extruder, (b) the position of the fiber sample toward the 2D detector, and (c) characteristic 2D wide-angle X-ray scattering pattern of an extruded filament based on discotic nanographenes. Adapted from Ref. [46].

The meridional reflections indicate that the molecules are arranged perpendicularly with their planes to the columnar axis with a π-stacking distance of 0.35 nm. The relation of the positions of the equatorial scattering intensities by 1: $\sqrt{2}$: 2 points toward a hexagonal intercolumnar arrangement (Figure 14.7a) that is most favorable for cylindrical structures and characteristic of the liquid crystalline state. In contrast to this, in the crystalline phase the nanographene molecules revealed a different organization in which the disks are tilted to the columnar axis as indicated by the so-called off-meridional reflections (Figure 14.7b). This intracolumnar packing leads to asymmetric columns that adapt a rectangular or monoclinic unit cell presented as an example in Figure 14.7b. The degree of order and molecular dynamics, both determine the type of the thermal phase, are closely related to each other. Therefore, besides X-ray scattering it is essential to perform very fast magic angle spinning (MAS) solid-state NMR spectroscopic measurements to gain a full picture of the thermotropic behavior of a discotic nanographene. This technique provides detailed information on the molecular packing and dynamics of these disk-like molecules in different phases [50]. In comparison to the above-discussed X-ray scattering results for HBC-C$_{12}$ (2), in the MAS solid-state NMR distinct aromatic resonances are observed in the crystalline phase due to different magnetic environments of the symmetry equivalent aromatic protons caused by the ring current effect of adjacent layers. On the other hand, the packing changes in the liquid crystalline phase, and an averaging of the multiple signals into a single resonance is observed due to an axial motion of the HBC disks in the column (Figure 14.9).

14.3.2
Helical Packing of Discotic Nanographenes

The organization of single building blocks into complex helical structure is displayed impressively in nature such as the double helix of DNA [51]. Inspired by this type of

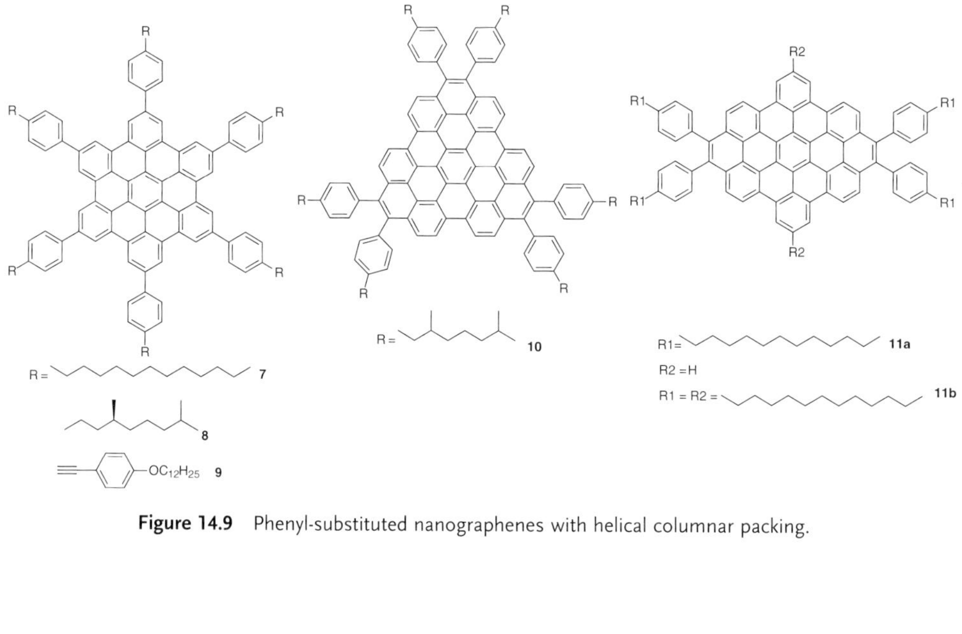

Figure 14.9 Phenyl-substituted nanographenes with helical columnar packing.

molecular arrangement and the resulting distinct properties of the material, supramolecular chemists attempt to understand the mechanism behind this self-organization process in order to adapt the principles for functional materials to gain desired properties. The helical packing of discotic molecules has been already reported for small PAHs such as hexahexylthiotriphenylene [52]. Indeed, time-of-flight (TOF) experiments for this compound in its helical phase showed the highest mobilities for discotics so far, which was related to both the lack of grain boundaries due to the liquid crystalline phase and the high crystalline-like helical order. Theoretical considerations indicated that these values are even significantly larger [53]. The formation of helical stacks not only does enhance the charge carrier transport, but it also leads to high stability of the supramolecular structures at elevated temperatures due to the improved interaction via molecular interlocking. As described above, in the crystalline phase alkyl-substituted nanographenes do not perform a rotation with respect to each other, but the aromatic cores are rather shifted to each other [54]. In contrast, bulky substituents attached directly at the core are rotated on the molecular plane and hinder in this way a direct face-to face packing of the molecules [55]. In order to overcome the steric hindrance at the core proximity, the building blocks rotate with respect to each other by a corresponding angle that depends on the number and position of substituents and the molecular symmetry. Hexaphenylalkyl-substituted HBCs (**7–8**) are one of the most prominent examples for helically packed nanographenes [56]. The structural analysis revealed that these disks are twisted by 20° in the columnar stacks to each other and form a helical pitch of 1 nm consisting of three molecules, whereby every fourth disk is in the same positional order (Figure 14.10a). These experimental observations were in good agreement with a structural model that was developed on the basis of accompanying quantum

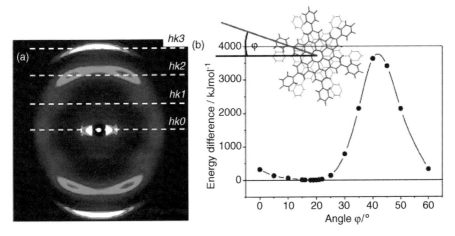

Figure 14.10 (a) 2D WAXS pattern of the extruded filament of HBC-PhC$_{12}$ (**7**) (dashed lines indicate Miller's indexes assigning the intracolumnar packing), (b) definition of the twisting angle in the columnar stack of HBC-Ph, and dependence of the energy of an HBC dimer on the twisting angle as obtained in quantum chemical calculations at the RIBP86/SVP level. Adapted from Ref. [56].

chemical calculations. This model indicated the out-of-plane rotation of the attached phenyls toward the aromatic core provoking this helical organization (Figure 14.10b). In the case of **7**, solid-state NMR experiments indicated that the phenyl substituents additionally perform a flipping from 20° to −20° toward the HBC plane [48]. On the other hand, the helical packing is independent of chiral alkyl substituents, while alkyl chirality generates only a uniform handedness of the twist. The superstructure of both compounds **7** and **8** fulfills the classification of a plastic columnar discotic that possesses three-dimensional order with minor fluctuations in positions. The appearance of plastic phases in columnar structures is rare and was also reported for triphenylenes [57, 58]. Typically, at higher temperatures the degree of helical organization decreases due to lateral and longitudinal dynamics of the discotic molecules. Furthermore, as for other plastic discotics annealing at ambient conditions improves the long-range arrangement of the disks along the columnar structures. This reorganization is attributed to a self-healing process of the plastic material. As mentioned above, this feature is desirable for the application of this kind of material as active layers in electronic devices. The helical packing of the molecules results finally in a considerable stability of the mesophase up to 500 °C, which is rather unique for liquid crystals. The introduction of more bulky substituents can change helical packing. For instance, the increase in the space filling between the columnar stacks by attaching diphenylacetylene arms tightens the intracolumnar packing of the molecules, reduces the rotation angle, and increases the helical order [59].

The helical packing was successfully controlled also for other core extended nanographenes (**10** and **11**), both possessing another disk symmetry, by attaching phenyl attachments of adequate steric demand. In both cases, the presence of bulky phenyls at the core proximity induced a helical arrangement in the columnar stacks following the same principles that were applied for the hexaphenyl-substituted HBCs. X-ray scattering results indicated for the C_3 symmetric **10** a rotation of 40° to each other and a helical pitch of 1 nm consisting of three molecules [43]. This specific intracolumnar arrangement was attributed to the C_3 molecular symmetry of **10**. Further development of the molecular design and the synthetic route allowed to introduce phenyl substituents at distinct positions at the aromatic core. Upon modification of the substitution pattern, the intracolumnar packing is controlled from helical stacking to a staggered arrangement [60]. For **11a** a helical period of 1 nm at every fourth unit suggesting a packing with a molecular rotation of 60° was found. In sharp contrast, compound **11b** revealed a staggered packing characterized by a corresponding period of 0.72 nm, twice the stacking distance, and a correlation of every second molecule along the columns. The lateral rotation of the ovalene-shaped nanographenes by 90° to each other is initiated not only by the hindrance of the phenyl rings at the aromatic core as for **11b** but also by the introduction of alkyl side chains at the *para* position. The different lateral rotation angles of the molecules **11a** and **11b** are attributed to a different peripheral space filling of the side chains. Since compound **11a** carries only four alkyl substituents in comparison to **11b** with six chains, the resulting rotation of **11a** is smaller to fill the peripheral space by the attached phenyls.

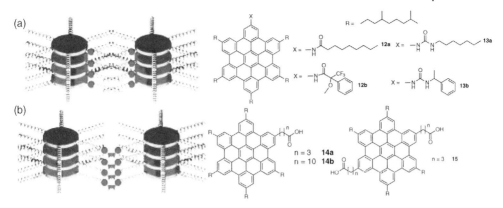

Figure 14.11 Schematic illustration of (a) intracolumnar and (b) intercolumnar hydrogen bonds.

14.3.3
Complementary Interactions

The introduction of hydrogen bonds at the aromatic core of a nanographene, with a strength ranging from 4 to 120 kJ/mol, besides already existing π-interactions (0–50 kJ/mol), is an additional instrument for the development of supramolecular architectures that is well known for a diversity of molecular systems [61, 62].

As one possibility, functional groups capable of forming hydrogen bonds are added either directly or through rigid linkers at the nanographene disk to enhance the degree of order within a single column (Figure 14.11a). On the other hand, hydrogen bonds between columns are formed if the functional group is separated from the aromatic core by a flexible spacer (Figure 14.11b). The attachment of amido (**12**) or ureido (**13**) groups to the HBC moieties is thereby especially effective due to their relatively strong and directional hydrogen bonds acting within the columns [63]. The structural investigations indeed indicated a strong influence of the hydrogen bonding interactions on stacking and especially on the thermotropic properties. Interestingly, after annealing of compounds **12a**, **13a**, and **13b** the phase transition of the liquid crystalline phase became irreversible. After annealing at high temperatures in the liquid crystalline state and slowly cooling down the sample to the initial temperature, only an irreversible endothermic transition during the heating cycle was observed and the crystalline phase with its characteristic organization could not be recovered. Instead, the high temperature LC supramolecular arrangement was maintained due to the intermolecular hydrogen bonding between side chains cementing the intracolumnar structure, while the order imposed by π-interactions remained. It has to be pointed out that this effect could be observed only for nanographenes with functional groups attached directly to the aromatic cores leading to intracolumnar hydrogen bonds that are established between disks within the same stack. In contrast to this, the position of carboxylic acids (**14** and **15**) or alcohol groups as hydrogen bonding functionalities terminated at long alkyl spacers far from the aromatic core resulting in the formation of intercolumnar hydrogen bonds [44]. The length of the tethers between the aromatic core and the functional groups appeared as a key factor for the organization. This strong influence was reflected by an unusually high mesophase transition temperature and nontilted stacking in a pseudocrystalline phase. It was observed that a short tether (**14a**) can stay longer in the hydrogen-bonded state due to the reduced freedom of movement. This leads to stronger intermolecular interactions and thus to an improved order in the bulk and higher transition temperatures. These additional noncovalent forces distort the preferred low temperature tilted arrangement toward a nontilted one. By adding a second short tethered carboxylic group to the *para* position of the nanographene (**15**), a complete distortion was achieved and a change in the intercolumnar lattice from hexagonal to orthorhombic was observed. Besides hydrogen bonds, dipole interactions serve as a tool to effectively control the self-assembly [64]. C_3-symmetric HBCs with alternating polar ester and apolar alkyl chains that are attached to the hexaphenyl HBCs either directly or through a spacer group indeed revealed a profound influence of even subtle structural changes. Compound **16**, which directly carries the ester groups at the hexaphenyl-substituted aromatic core [65], revealed a unique columnar alignment in the extruded filaments since its columnar superstructures were arranged perpendicular to the alignment direction what is only observed for high molecular weight main-chain discotic polymers consisting of covalently linked triphenylenes [66]. The packing mode of **16** was therefore attributed to the pronounced noncovalent dipole intercolumnar forces induced by ester groups coexisting with the π-stacking interactions as schematically illustrated in Figure 14.12. The alternating attachment of the polar ester groups and apolar alkyl chains resulted in a local phase separation of the substituents in the molecular periphery. The comparison to HBCs possessing a different molecular symmetry highlighted the crucial role of the disk design in the organization. The further structure formation during alignment is assumed to follow the scheme presented in Figure 14.12. First, a 2D in-plane hexagonal network is established due to the strong intermolecular dipole interactions between the ester groups of individual building blocks. This network possesses the necessary aspect ratio to be oriented with the molecular planes along the alignment direction. In the second step, the π-stacking interactions during alignment lead to a 3D hexagonal columnar array.

Dipole interaction can occur not only between columnar structures, as described above, but also within the stacks influencing the packing of individual nanographenes. In the case of C_3, symmetric *meta*-trimethoxy-substituted HBC **17** with three alternating methoxy and alkyl substituents, X-ray scattering experiments exposed a remarkable helical organization with a large intracolumnar correlations (Figure 14.13) [67]. This enhanced order was obtained after cooling the compound from the liquid crystalline state. The pronounced dipole forces were also indicated from the changed intracolumnar packing mode from a tilted to a nontilted arrangement after annealing as reported for nanographenes **12** and **13** with hydrogen bonds. The analysis of the X-ray pattern implied that in contrast to the nanographenes substituted by bulky phenyls, the steric hindrance is not the origin for the helical organization but local dipole moments between the C−O bonds of individual neighboring molecules in the columnar stacks. Therefore, the rotation angle was relatively small for **17**. The introduction of additional noncovalent forces does not only influence the interactions between molecules and their packing but also the self-assembly from solution leading in many cases to pronounced fibrous nanostructures several hundred micrometers in length. This essential aspect of supramolecular chemistry will be discussed later.

The stabilization of self-assembled columnar structures can also be achieved through chemical covalent cross-linking. This approach consists of cross-linking discotic nanographenes bearing a functional group to fix the molecules irreversibly into a superstructure network. HBC derivatives with reactive methacrylate or acrylate moieties were successfully cross-linked by both light and heat as external initiation to freeze *in situ* their organization in either in the crystalline phase or in the liquid crystalline state [68, 69]. Figure 14.14 presents the DSC traces together with X-ray scattering patterns observed during cross-linking of **18** in the LC phase that was

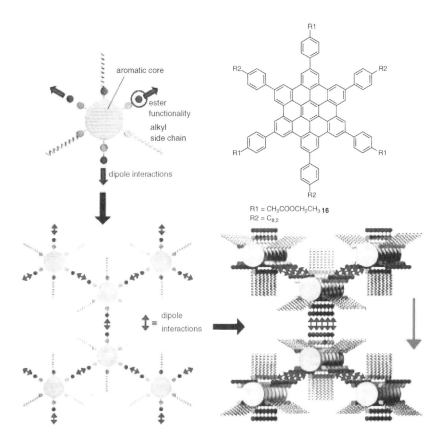

R1 = CH$_2$COOCH$_2$CH$_3$ **16**
R2 = C$_{8.2}$

Figure 14.12 Schematic illustration of self-assembly and orientation during mechanical processing of **16** that assembles first through dipole interactions into a 2D network and in a further step through π-stacking into a 3D hexagonal columnar organization. The red arrow indicates the alignment direction to which the columns are arranged perpendicularly. Adapted from Ref. [65].

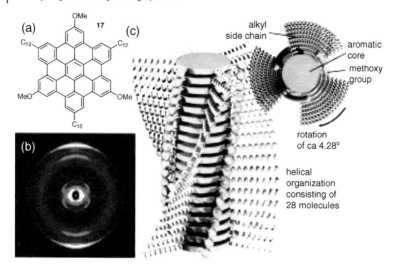

Figure 14.13 (a) C₃ symmetric HBC **17** with three alternating methoxy and alkyl substituents, (b) 2D-WAXS at 30 °C after annealing, and (c) schematic illustration of the intracolumnar arrangement of **17** at 30 °C after annealing into a helical organization with a disk rotation of about 4.28°. Adapted from Ref. [67].

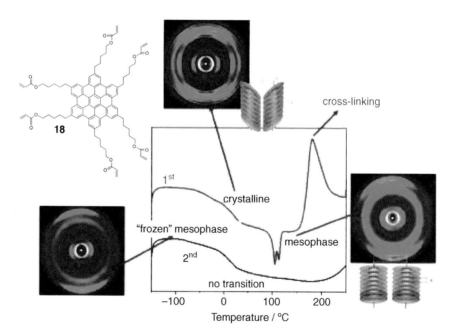

Figure 14.14 DSC heating traces and X-ray scattering patterns of the cross-linking process of **18**. The first heating shows a phase transition from crystalline to liquid crystalline phase and the characteristic organizations in the corresponding phases. At temperatures above 150 °C, a large exothermic peak in the DSC indicates the cross-linking. Back at low temperatures, the LC structure is maintained, while the phase transition cannot be recovered during the second heating.

maintained after cooling back the material. As an advantage in comparison to other discotics, the structural order was not influenced by the cross-linking process but rendered the material insoluble. Since grain boundaries and structural defects wane in the LC phase, the fixation of such state down to ambient temperatures is considered to be beneficial for charge transport in devices. Furthermore, photo-cross-linking of the acrylate-functionalized nanographenes can be applied for lithographic processes to produce nanostructured surfaces.

14.4
Charge Carrier Transport Along Nanographene Stacks

Discotic nanographenes possess extensive molecular π-orbitals in which the electrons are delocalized. Due to the interaction between the π-orbitals of adjacent aromatic cores, a one-dimensional pathway for charge migration along the columnar structures is created. The peripheral substitution of the aromatic cores with long aliphatic hydrocarbon chains provides an insulation of the conducting core as schematically illustrated in Figure 14.15a. The charge transport takes place via hopping between localized states of neighboring aromatic molecules with an assumed jump time between 75 and 750 fs for HBCs in the mesophase [70]. A detailed study of the charge transport mechanism in discotic liquid crystalline materials was performed by Warman and van de Craats by using the pulse-radiolysis time-resolved microwave conductivity (PR-TRMC) technique [71]. With using this method, it is possible to estimate the charge carrier mobility on a nanometer length scale (interface effects caused by electrodes are avoided and electric field effects are minimized), so traps caused by structural defects and impurities are far

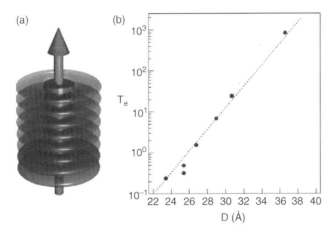

Figure 14.15 (a) Schematic illustration of the columnar 1D pathway for charge migration along a nanographene stack, whereby the conductive aromatic cores are insulated by the attached alkyl side chains; (b) semilogarithmic plot of the exponential conductivity decay time, versus the calculated disk diameter, D, for hexa-alkyl-substituted HBCs (≜).

less disturbing. Therefore, it is important to emphasize that the charge carrier mobilities determined by this technique provide information different from techniques such as TOF [72, 73] or FET. In the TOF measurement, the transition time of laser-generated charge carriers is recorded that migrate from one electrode to an opposite one under an externally applied electric field. The motion of the charges through the bulk material is highly defect controlled and is decreased by local and macroscopic traps. In strong contrast, PR-TRMC reveals only intrinsic or local charge carrier mobilities between several molecules, whereas the mobilities in the bulk obtained by the TOF and FET experiments are usually several orders of magnitude lower due to a higher degree of disorder and possible trapping sites at grain boundaries. The charge carrier mobility derived from the PR-TRMC technique can be considered as the maximum value. Indeed, for well-aligned triphenylenes it was possible to confirm the local PR-TRMC mobility by TOF measurements [74, 75]. It has to be stressed out that the higher the dimensionality of the charge transport in a semiconductor, the lower the influence of single defects. Since such a defect acts as significant scattering side in one-dimensional columnar pathways, structural attempts toward higher dimensionality are mandatory.

In general, the charge transfer along an ideal conjugated system is more efficient than the charge hopping between neighboring π-stacked aromatic molecules. Nevertheless, high mobilities at room temperature for discotic HBCs were obtained due to their pronounced self-organization propensity into highly ordered supramolecular structures [76], which might be considered to be an upper limit attainable for noncovalently bonded materials. It was found that the charge carrier mobility in the hexagonal liquid crystalline phase is more dependent on the size of the aromatic core than on the cofacial distance between neighboring cores within the columnar stack [77]. This relationship was explained by the increased π-orbital overlapping in the case of the larger aromatic cores leading to a more stable columnar structure thus increasing the intracolumnar charge transport. The mobility increased from $0.1 \, \text{cm}^2/(\text{V s})$ for triphenylenes [78] to $1.1 \, \text{cm}^2/(\text{V s})$ for HBCs [76], whereby the influence of the chain length on the charge carrier mobility became negligible for discotic nanographenes with larger aromatic [79] cores.

The transport of mobile charge carriers can be limited by their recombination and trapping as both decrease the device performance. Recombination can be controlled by the insulating effect of the alkyl side chains around the conductive columnar core. The lifetime determined for a series of linear and branched alkyl chain HBC derivatives by PR-TRMC provided information on this insulating nature of the alkyl mantel and was derived from the timescale for intercolumnar tunneling of electrons through the intervening aliphatic hydrocarbon chains [80]. This close relation is evident from the plot in Figure 14.15b presenting a significant increase in the timescale of the decay of the radiation-induced conductivity from a few hundred nanoseconds to close to a millisecond as the peripheral alkyl substituents increase in size from 8 to 24 carbon atoms with corresponding disk diameters, D, from 2.4 to 3.7 nm. The decay kinetics is a function only of the total number of peripheral carbon atoms with no evidence for specific effects of chain branching. On the other hand, charge carriers can be trapped by inorganic impurities or by structural defects. As for

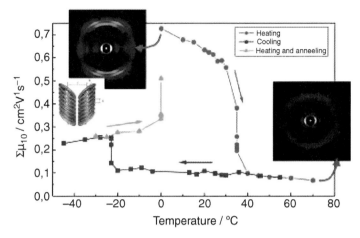

Figure 14.16 Temperature dependence of the PR-TRMC mobility of **20** for the first heating (full circles), cooling (open circles), and final heating back to room temperature (full triangles). Adapted from Ref. [81].

inorganic semiconductors, the mobility of charge carriers for discotic nanographenes is a function of intracolumnar order. Typically, in highly ordered crystalline phases the mobility reaches its maximum. Heating the discotic to above the liquid crystalline phase transition, the decrease of order is accompanied by a sharp decline in mobility [81]. For columnar disordered systems, as shown in Figure 14.16, the mobility might drop even close to zero. The reversible phase transition at lower temperatures allows to recover the high order together with pronounced mobility value.

14.5
Solution Aggregation and Fiber Formation

The self-assembly of nanographenes in solution or upon deposition from solution is an important aspect of processing these materials as organic semiconductors in thin-film electronic devices, and apart from monolayer and bulk, it is the third definition of organization. Organization in solution can be considered an initial structure formation upon surface deposition. The solution processing determines the macroscopic order and film morphology and effects therefore the charge carrier transport through the active layer. The fabrication parameters such as the nature of solvent, temperature, concentration, and surface treatment allow to control the structure formation. On the other hand, the molecular design of nanographenes, for example, shape and size of the aromatic core and type of substituents, also plays a crucial role for self-organization during device preparation. Therefore, the understanding of the relation between the molecular architecture and the self-assembly both in solution and on the surface during solution deposition is necessary for a successful device fabrication. The size and geometry of the alkyl substituents is the most effective

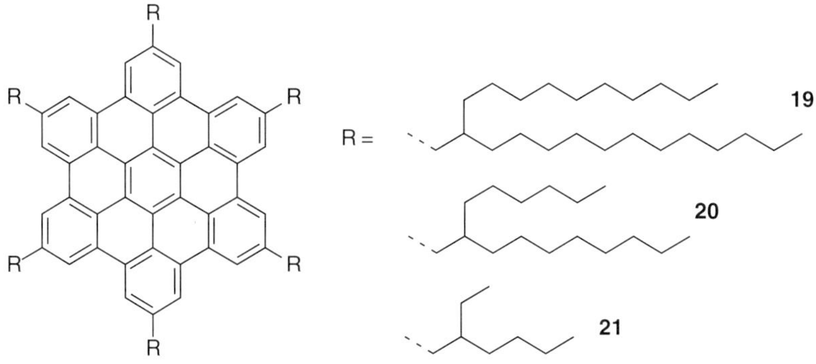

R =

19

20

21

Figure 14.17 HBCs with different side chains to control their solubility and self-assembly from solution.

instrument to control the interaction of molecules in solution (Figure 14.17). The noncovalent π-stacking interactions are disturbed by the steric hindrance of side chains, whereby the degree of steric demand is assigned by their length and bulkiness. Concentration-dependent ^{1}H NMR chemical shifts (Figure 14.18) were recorded for two HBC derivatives (**2** and **19**) with side chains of significantly different steric hindrance [82]. The derivative with linear dodecyl side chains HBC-C$_{12}$ (**2**) revealed strong aggregation at considerably lower concentrations in comparison to **19** as evidenced by the change in the chemical shift. As expected, compounds with less bulky substituents are well soluble and, as presented in Table 14.1, possess a higher self-association constant.

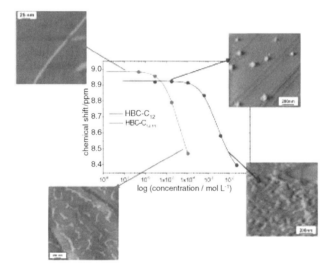

Figure 14.18 Concentration-dependent ^{1}H NMR chemical shifts of **2** and **19** recorded in 1,1,2,2-tetrachloroethane-d_2 (30 °C, 500 MHz) and fitted with the indefinite self-association model including next nearest-neighbor relations. The insets display AFM topologies of spin-coated films. Adapted from Ref. [82].

Table 14.1 Association constants for self-aggregation of the hexa-alkyl HBC derivatives (**2** and **19–21**) in 1,1,2,2-tetrachloroethane-d_2 at 30 °C.

Compound	K_a (l/mol)
2	898
19	2.9
20	1.8
21	13.4

Apart from the influence of the solution aggregation, the side chains show a great effect on the self-assembly during solvent evaporation. The insets in Figure 14.18 display different morphologies after spin coating the solution with the corresponding concentrations. For **19** with a high self-association constant, distinct fibers were formed. At low concentrations, it was possible to observe even single nanofibers. An additional structure analysis revealed that the columnar stacks are organized along these fibers. In strong contrast, at high concentrations only an amorphous film of **19** is observed in the AFM image, while at low concentrations single 60–80 nm wide blobs with a height of about 3 nm appeared, which were randomly distributed over the surface indicating poor aggregation.

Although few nanographenes exhibit a low association constant such as **21**, fiber formation from solution can be adjusted by changing the evaporation rate during processing. Compound **21** revealed significantly shorter fibers when drop cast from solvents such as THF (Figure 14.19a) in comparison to **2** that formed pronounced fibers from various solvents of different boiling points. Interestingly, changing to high-boiling point solvents with low evaporation rates centimeter-long, crystalline fibers of **21** were obtained. This concentration-dependent assembly of **21** indicates the important role of the kinetics on the structure formation that was attributed to the steric requirement of the dovetailed alkyl substituents. Nanographenes with this kind of kinetics can be macroscopically aligned by methods such as dip coating [83, 84].

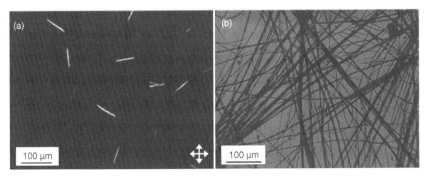

Figure 14.19 Polarized optical microscopy (POM) images of **21** drop cast from a solvent with (a) low and (b) high boiling point. Adapted from Ref. [82].

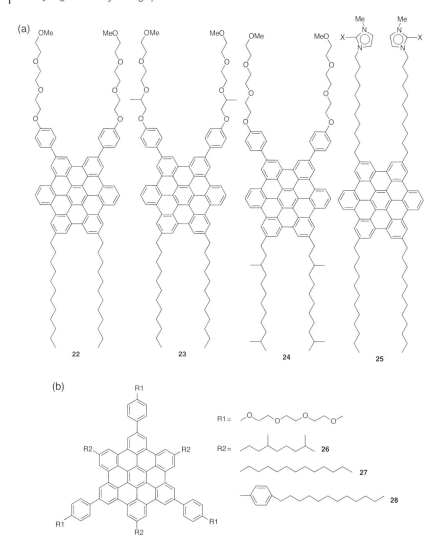

Figure 14.20 Amphiphilic HBC nanographenes with (a) Gemini shape and (b) alternating chain arrangement that organize into nanotubes and nanofibers.

During this dip coating process at higher concentrations and low solvent evaporation rates, a uniaxial orientation of **21** into surface layers was achieved [82] (Figure 14.20).

As described above, the π-stacking interactions from alkylated HBCs compete with intermolecular van der Waals interactions during the formation of long fibrous nanostructures. The introduction of additional noncovalent forces can lead to the growth of nano-objects with a more complex organization. Aida and coworkers developed Gemini-shaped amphiphilic HBCs with two hydrophobic dodecyl chains on one side of the core and two hydrophilic triethylene glycol (TEG) chains on the

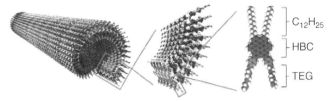

Figure 14.21 The hierarchical structure formation of a nanotube consisting of **22**. Reprinted with permission from Ref. [86]. Copyright 2008 American Chemical Society.

other (**22–24**) [85, 86]. Due to the incompatibility between the hydrophobic and hydrophilic parts of the molecules, a self-assembly in polar solvents takes place to form several tens of micrometer-long tubular nanotubes consisting of helically rolled-up bilayer tapes. The structure is presented in Figure 14.21, where a local phase separation occurred with the inner and outer layers being connected by interdigitation of the dodecyl side chains and the TEG chains being located on both sides of the tubular wall. The substitution of the ethylene glycol chains by chiral oxyalkylene side chains (**23**) results in one-handed helical chirality of the graphitic nanotubes, where the single helical sense was determined by the major enantiomer [87]. Even the incorporation of large hydrophobic groups, such as norbornene [88] and trinitrofluorenone [89] pendants, into the termini of the ethylene glycol chains hardly hinders the nanotube formation. On the other hand, no nanotube could be observed for HBC nanographenes with branched alkyl side chains (**24**) probably due to their pronounced steric demand similar to the fully alkylated HBCs [90]. In contrast to this tubular assembly, amphiphilic HBC carrying imidazolium ion pendants showed filled fibers [91]. The combined secondary forces such as π-stacking and ion–ion interactions between the imidazolium salts established a nanofiber organization with its size and shape sensitive to the evaporation rate of the corresponding solvents.

The amphiphilic character is also established by alternating the hydrophilic–hydrophobic substituents (**26–28**) around the aromatic core providing an additional control tool over the self-assembly of nanostructures [92]. The steric demand of the hydrophobic alkyl chains is essential for the tailoring of the liquid crystalline phase in bulk and self-assembly in solution and on the surface. Derivatives with linear alkyl side chains (**27**) formed after drop casting pronounced helical bundles with a diameter between 100 and 400 nm and the length of several tens of micrometers in which the columns were aligned along these structures (Figure 14.22a and b). It has to be noted that this organization was in contrast to the Gemini-shaped nanographenes. Compound **26** with bulky branched chains showed instead fibrous structures only small globular droplets composed of randomly deposited material. This corresponded to the low aggregation behavior of **26** in solution as demonstrated also by the UV and PL spectra and was in line with Aida's observation for the Gemini-shaped HBC (**24**). Local dipole interactions not only influence the organization in the bulk solid state as described above but also assist the nanofiber formation during solution deposition. In the case of **16**, fibrous structures with a diameter of around 100–200 nm and several micrometers in length were grown from a THF/methanol

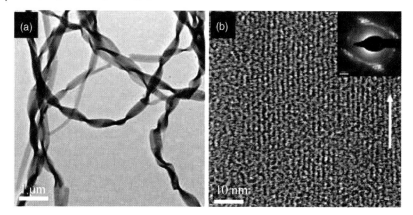

Figure 14.22 (a) Electron microscopy (EM) of **27** grown from MeOH:CHCl3 = 2 : 1 solution; (b) HRTEM image displaying columnar structures of **27** within such fiber (inset: electron diffraction pattern with reflections assigned to the π-stacking distance of 0.35 nm). Adapted from Ref. [92].

solvent mixture. The pronounced aggregation in this solvent mixture was indicated by significant broadening and tailing of the bands of the UV/vis and fluorescence. The structural analysis implied that the submicrometer-sized fibers consist of 50–100 bundles of stacked molecular wires that are oriented along fiber. On the other hand, the drop casting of **17** from different solvents including toluene, THF, and CHCl₃ displayed in all cases a similar morphology consisting of a fibrous network with fibers of an average diameter of 500 nm and a length up to a few hundred micrometers. High-resolution TEM displayed individual columns well oriented along the fiber direction suggesting a high supramolecular order. The distinct organization was related to the methoxy units at the *meta* positions and the high intermolecular interactions that were already observed for the bulk (Figure 14.13).

Local phase separation in the disk periphery can be induced not only between side chains of different nature as described for the amphiphilic nanographenes but also between different sections within the same substituent assisting self-organization into nanostructures. For instance, HBC derivatives with side chains consisting of an alkylated and a perfluoroalkylated section assemble into long monostranded molecular stacks from benzotrifluoride (**29**) [93, 94]. The alkylated section acted as a four-carbon spacer between the core structure and the six-carbon perfluorinated tail (Figure 14.23). The pronounced tendency toward long fiber formation was related to maximum flexibility between the rigid core structure and the semirigid perfluoroalkyl tail and to the prevention of crystalline alkyl chain–alkyl chain interactions.

The remarkable increase in performance of silicon electronics over the last few decades is based mainly on the downscaling of the circuits that is expected to run into fundamental limits in near future. Alternative technologies for further device miniaturization are molecular electronics that consists of single molecules, in mostly carbon nanotubes, bridging the electrodes [95, 96]. The growth of nanographenes into nano-objects enables their application in molecular electronics within the nanoscale. However, the investigation of individual nanoscale fibers in electronic devices is not straightforward since first of all, a sufficient contact of the nanostructures with metal

(a)

(b)

R1 = OC$_{12}$H$_{25}$; R2 = H **30**
R1 = R2 = OC$_{12}$H$_{25}$ **31**

R1 = H; R2 = COCl **32**
R1 = OC$_{12}$H$_{25}$; R2 = COCl **33**

Figure 14.23 (a) Perfluoroalkylated HBC and (b) hexa-*cata*-hexabenzocoronenes as nanographenes.

electrodes has to be ensured. The pick-and-place method of single fibers of **31** has been applied for the fabrication of FET devices (Figure 14.24a) [97]. After a mat of nano-stacks from solution is grown by slow evaporation, an elastomer stamp is pressed into this mat so that few wires are transferred to the stamp. This stamp is again pressed onto the device substrate to deposit these structures. Top contact transistor character-istics indicated a hole-transporting behavior of the nanofibers of **31**, whereby their length spanned the electrodes (Figure 14.24). The structure study revealed that the

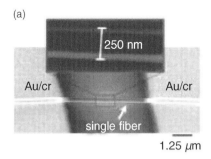

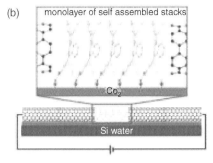

Figure 14.24 Nanodevices based on nanographenes: (a) image of an isolated fiber of **30** that spans electrodes in a FET device. Reprinted with permission from Ref. [97]. Copyright 2006 American Chemical Society. (b) Setup of a molecular sensor based on monolayers of **32** or **33** with groups capable to form covalent bondings to the silicon oxide surface. Reprinted with permission from Ref. [101]. Copyright 2006 National Academy of Sciences USA.

columnar stacks are oriented along the fiber direction. The carrier mobility in these objects was determined to be 0.02 cm^2/(V s). Identical values have been reported for spin-coated films of the fourfold alkoxy-substituted liquid crystalline **30** [98]. The aromatic core of **30** has two π-systems attributed to the inner radialene core, which allows a 1D charge transport, and the outer out of-plane alkoxyphenyl rings insulating the inner pathway [99]. The device performance of **30** was enhanced once again by using the so-called pick-up-stick transistor concept that is based on a thin-film device consisting of a bilayer structure with an array of carbon nanotubes as an underlayer and **30** overcoat. The improved device charge carrier mobility of 0.1 cm^2/(V s) in comparison to that of HBC without a CNT underlayer was attributed to the modulation by the CNT density [100]. Modified field-effect transistors also bear great potential as molecular sensors. Monolayers of functionalized nanographenes **32** and **33**, which covalently bond to silicon oxide surface of a transistor silicon wafer, were contacted by an individual metallic single-walled carbon nanotube to obtain ultrasmall point contacts that were separated by only a few nanometers as the source and drain electrodes (Figure 14.24b) [101]. The transistor channel in which the molecules assembled laterally into columns was reached through an oxidative cutting of nanotube. Transistors with large current modulation and high gate efficiency were fabricated that varied during exposure to electron-deficient molecules confirming the successful operation as sensors.

An alternative but exceptional approach toward nanoscopic one-dimensional fibers or tubes is based on templating the organic material in microporous inorganic membranes with a well-defined pore diameter [102]. The organic molecules are introduced into the pores from the isotropic melt or solution and self-assemble while wetting the pore walls (Figure 14.25a). The tubular or filled nanofibers with regular and well-defined shape are obtained after removing (dissolving) the template [103]. The shape of these objects is adjusted simply by the geometric confinement of the pores. Discotic nanographene **19** was deposited from the melt into an aluminum template, whereby in the pores during cooling the molecules assembled into highly ordered and macroscopically self-aligned organization [104]. X-ray scattering techniques revealed that the molecules arranged in the edge-on fashion at the pore walls, while the columnar stacks self-aligned along the pore. Identical orientation was reported for melt templated discotic triphenylenes [105, 106]. After removing the template, filled fibers were obtained with the structure that was found already in the pores (Figure 14.25b). Contrary to this melt processing, which resulted in filled wires, solution deposition allows the fabrication of hollow tubes. After filling the membranes from solution, compound **18** was thermally cross-linked through functional acrylate groups providing mechanical stability to the nanostructures [69]. After dissolution of the inorganic template, well-defined nanoscaled tubes were observed in which the columns were oriented along the axis (Figure 14.25c). As an additional possibility, the prealigned organic material can be directly pyrolyzed in the template yielding graphite nanotubes, where the graphitic sheets are perpendicular to the tube wall [107]. In contrast to common carbon nanotubes, in which the graphene layer is directed along the tube axis, the above-mentioned templated nanotubes based on nanographene reveal layers arranged perpendicular to the axis.

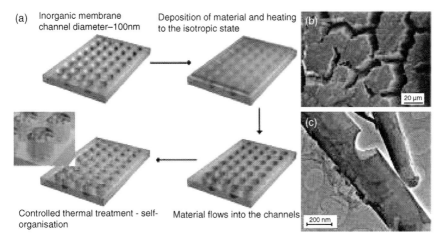

(a) Inorganic membrane channel diameter–100nm   Deposition of material and heating to the isotropic state   (b)

20 µm

(c)

Controlled thermal treatment - self-organisation   Material flows into the channels

200 nm

Figure 14.25 (a) Schematic illustration of the templating process using an inorganic membrane toward nanotubes and nanofibers, (b) nanofibers obtained from **19** by melting processing, and (c) nanotubes of cross-linked **18** as obtained from solution deposition into the membrane pores. Adapted from Refs [69, 104].

14.6
Solution Alignment on Surfaces

Disorder in the active semiconducting layer deposited between the electrodes, grain boundaries (defects), and metal interface resistance hinder the macroscopic charge carrier transport and thus decrease the device performance. For all organic semiconductors, macroscopic order and molecular alignment are essential for an improved electronic performance [108, 109]. To attain the high local charge carrier PR-TRMC mobilities of alkylated nanographenes in devices, it is required to span the columnar structures in a defect-free way between the electrodes. Nanographenes as disk-shaped molecules can be ordered in two different ways on the surface that finally determine the direction of the charge carrier transport with respect to the substrate. In the edge-on organization, the molecules are arranged with their molecular plane perpendicular to the surface (Figure 14.26a). Further stacking of the molecules results in columns that are lying on surface, where the charge carrier transport takes place parallel to the surface. This is the case in a field-effect transport in which the transport of charges between the source and the drain electrode occurs close to the interface under an applied gate voltage. The other possible molecular arrangement is the so-called "face-on" in which the disks are lying flat on the surface (Figure 14.26b). "Face-on" monolayers might induce further columnar growth and finally a homeotropic phase with columns staying on the substrate. This enables a charge transport vertical to the surface between the top and bottom electrodes that is considered to enhance the performance in photovoltaic cells.

Solution processable organic semiconductors are especially interesting for industrial applications due to their cheap and large area production by roll-to-roll printing

(a)   (b)

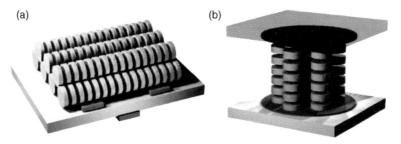

Figure 14.26 Schematic illustration of (a) edge-on orientation of the molecules desired in FETs and (b) face-on arrangement allowing a homeotropic alignment considered to favor the performance of photovoltaic cells.

on flexible substrates [110]. Techniques such as offset or screen printing do not allow to control the morphology and the orientation of the conjugated molecules on the surface. Apart from the development of new organic semiconductors, it is thus important to focus on alternative technologies for an efficient solution processing. The *zone-casting* technique was successfully applied for the macroscopic orientation of different discotic nanographenes from solution and their exploitation in electronic devices. The zone-casting apparatus is shown schematically in Figure 14.27a and consists of two main heating blocks. The device operates in the following way: the solution is deposited by the nozzle on the support moving at a defined speed. A meniscus is formed between the nozzle and the moving support in which a concentration gradient is established. During solvent evaporation a critical concentration is reached at which the material starts to nucleate on the moving support as an aligned thin layer. This technique has been used for the orientation of polymers and small semiconducting molecules proving its practicability for a broad range of organic systems [111–113]. Furthermore, this processing method is upscalable with potential for industrial applications and mass production. However, before the implementation of a new technology on a larger scale, first, a fundamental under-

(a)   (b)

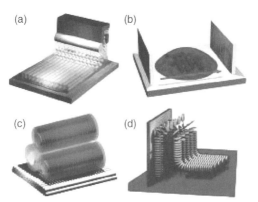

(c)   (d)

Figure 14.27 Schematic illustrations of different solution alignment techniques: (a) zone casting, (b) in magnetic or electric field, (c) on prealigned PTFE layer, and (d) Langmuir–Blodgett.

standing is required about the aggregation of discotic nanographenes in solution and self-assembly during solution deposition on a surface. This information is crucial for an optimized solution processing and improved alignment.

Especially nanographenes with pronounced solution aggregation and preaggregation such as **2** (HBC-C_{12}) were found to be appropriate for the zone-casting alignment. This derivative self-assembled during the processing into highly ordered surface layers [114]. A detailed structural study of the layers revealed closely packed columnar stacks uniaxially oriented in the deposition direction with exceptionally high length (Figure 14.28a and b). Diffraction methods confirmed "single-crystalline" like macroscopic domains with a high orientation of edge-on arranged nanographenes [115, 116]. The molecules stacked in the characteristic crystalline herringbone order having a molecular tilt of 45° with respect to the stacking direction. This crystalline order in the thin layer could be changed by heating the sample to its liquid crystalline state, in which the disks adapted the typical cofacial, nontilted arrangement. This transition in intracolumnar packing induced a change in the optical behavior from near net zero optical anisotropy between cross-polarizers for the crystalline phase to maximum anisotropy above the LC phase transition [117, 118]. Cooling the thin film back to its crystalline phase restored the optical isotropy verifying the reversibility of this effect.

Field-effect transistors of **2** zone-cast films were tested in the top contact geometry [119]. The device characteristics revealed mobilities up to 10^{-2} cm^2/(V s) and good on–off ratio of 10^4 along the alignment direction and thus in the direction of the columnar structures (Figure 14.28c). Measurements performed perpendicular to the columns showed mobilities two orders of magnitude lower implying the 1D character of the discotic columnar semiconductor. In contrast to this, drop-cast, unoriented fibers as shown in Figure 14.19b possess isotropic mobilities up to 10^{-4} cm^2/(V s) [120]. Despite the extraordinary high order and orientation, the local PR-TRMC charge carrier mobilities for **2** were not reached. This large discrepancy was attributed to local packing defects in single columns acting as charge carrier barriers. For instance, Y-shaped bifurcations are well known to appear in 1D

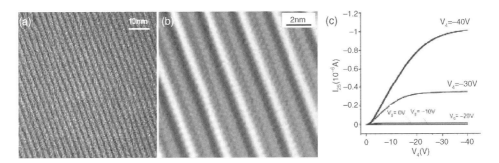

Figure 14.28 (a) High-resolution transmission electron microscopy of zone-cast HBC-C_{12} (**2**) with individual columns oriented in the zone-casting direction, (b) filtered inverse FFT (IFFT) image indicating intermolecular periodicity along the columns, and (c) *I–V* FET output characteristics for the zone-cast layer of **2** measured along the columnar alignment.

semiconducting systems limiting their performance, reflecting the importance for an increase in the dimensionality of the charge carrier pathways. Identical to the crystalline **2**, liquid crystalline nanographenes of different size and shape, HBC-PhC$_{12}$ (**7**) and C96 (Figure 14.2), were also successfully processed by zone casting into macroscopic orientation films [121]. This is especially interesting since liquid crystalline molecules are typically substituted by bulky side chains that decrease the molecular interaction in solution and lower the tendency for self-assembly during solution deposition. On the other hand, the liquid crystallinity state of thin films for HBC-PhC$_{12}$ (**7**) and C96 is attractive for device application due to self-healing of macroscopic defects such as grain boundaries. The structural study indicated an appropriate columnar organization and edge-on arrangement of the molecules for FET operation. The driving force for the orientation of molecules during zone casting is attributed to the concentration gradient and the moving direction of the substrate. Another possibility is the alignment along a magnetic or electric field. For a variety of systems, it was reported that aromatic molecules tend to arrange with their ring plane along the magnetic field [122, 123]. Such external driving force of a 20 T gradient was applied during solution deposition of **7** in a transistor channel [124]. In the large-area monodomains, the nanographenes were edge-on arranged with their planes along the applied magnetic field, while the columnar stacks assembled by 40° to the applied gradient. The film morphology consisted of individual fibers aligned perpendicular to the magnetic field direction. Charge carrier mobilities up to 10^{-4} cm^2/(V s) perpendicular to the aromatic planes were determined.

However, the orientation of HBC-PhC$_{12}$ (**7**) by using electric fields is also effective as reported for films drop cast on glass substrates or HOPG resulting in highly ordered structures after solvent evaporation (Figure 14.27b) [125]. Thereby, the electric field was applied parallel to the substrate surface during adsorption from solution and subsequent drying. The long-range uniaxially aligned columnar structures in extended domains were arranged perpendicular to the direction of the imposed electric field. As expected, in the absence of the field force, the molecules adopted a face-on orientation on the HOPG that is related to the strong intermolecular interactions between HOPG and HBC [126]. Such epitaxial influence of the substrate on the self-assembly of the molecules is well known for preoriented and friction-deposited poly(tetrafluoroethylene) (PTFE) first reported by Wittmann and coworkers [127, 128]. They showed for a large variety of materials that such PTFE surface layers are capable to induce macroscopic orientation when the molecules are deposited from solution or by sublimation. Indeed, solution processable discotic nanographenes were successfully aligned over large areas by this method. In the oriented films, the columnar stacks were found to be arranged parallel to the underlying PTFE chains, while the epitaxial influence of the PTFE layer was reflected on the organization of the 2D unit cell [129–131]. Finally, tests of these oriented films of **7** in FETs revealed mobilities of up to 5×10^{-4} cm^2/(V s) with maximum values along the columnar alignment [132]. As discussed above, amphiphilic interactions can lead to complex self-assembly as nanofibers during solution deposition. But amphiphilicity of nanographenes can be explored also for their alignment on water–air interfaces through Langmuir–Blodgett (LB) [133, 134]. For this method,

Gemini-shaped nanographenes are required to carry on the one side a hydrophobic and on the other one a hydrophilic substituent. After deposition onto a water surface, the HBC nanographene molecules arrange with their hydrophilic side toward the liquid. Such functionality can be introduced through an asymmetrically substituted and terminated by a carboxylic acid group (**14**) or a polar anthraquinone group that allows to establish a high columnar orientation after the transfer of the film from the water surface onto a substrate [135, 136]. During this processing, the lattice depends on the surface pressure applied on the film. Furthermore, the type of substrate plays an important role. The application of hydrophobic quartz substrates gives typically well-defined multilayer films after vertical dipping. Further improvement of the LB alignment was found when poly(ethylene imine)-functionalized substrates were used as anchor points for the first layer [137, 138]. In all reported cases, the assembled nanographene columns are oriented along the dipping direction with disk planes perpendicular to the columnar axes and stacked in a cofacial manner.

14.7
Thermal Processing

As discussed earlier, the introduction of the space-filling branched side chains significantly decreases the aggregation of nanographenes, whereby the molecules substituted by linear side chains possess a solubility more than three orders of magnitude lower. The variation in the length and geometry of aliphatic substituents not only controls the solubility but also affects the thermal properties of discotic nanographenes. The attachment of alkyl substituents to the nanographene allows to tune the phase transitions of the liquid crystalline state in the desired temperature range. Plain nanographenes do not melt, while solubilizing linear alkyl side chains leads to liquid crystalline phases, however, in most of the cases at higher temperature. For practical utilization, this gives the opportunity to maintain the highly crystalline order at low operating temperatures, while only in the liquid crystalline phase at high temperatures the material can undergo a healing process of structural defects. Simple extension of the linear chains does not significantly influence the thermal properties. HBCs with linear alkyl side chains possess an isotropic phase transition temperature (T_i) above 400 °C, which is too far from applicable temperatures for melt processing. In addition, at such temperatures thermal decomposition can occur. The introduction of long-branched, space-filling side chains with a higher rotational freedom is an alternative concept to dramatically lower the phase transitions by decreasing the π-stacking interactions of the aromatic cores to a large extent.

In order to analyze the influence of branched side chains in comparison to linear alkyl chains and chains containing methylene groups, the isotropization temperature was plotted as a function of the side-chain volume for hexasubstituted HBCs as in Figure 14.29b [139]. The linear and methylene-containing chains are considered as cylinders and the branched ones as cones. By plotting the side-chain volume on a logarithmic scale, a linear relationship of the isotropization temperature is evident for each type of the side chain. Particularly, the branched HBC derivatives follow strictly the linearity indicated by the dashed line. It is apparent that the number of atoms in the side chains is negligible, but the steric requirements, resulting from the high degree of rotational freedom at the branching site as verified by solid-state NMR, play a much more important role in the thermal behavior of the nanographenes [81]. The extended branching of the side chains leads to a dramatic lowering of the isotropization temperature in comparison to the substitution of linear alkyl chains. For instance, in order to obtain an identical isotropization temperature for an HBC derivative substituted by linear side chains as for a nanographene bearing the branched chains, the number of carbon atoms has to be significantly increased in order to reach the same side-chain volume. From Figure 14.29 it is also possible to derive that the position of branching is an additional factor for the determination of the thermal behavior. HBC derivatives substituted by side chains branched at the β-position (**19–21**) show a lower isotropization temperature than HBCs with ether linkages.

The substitution symmetry is a strong parameter that is not included in Figure 14.29. Highly asymmetric nanographenes such as **34** revealed a remarkable lowering of the isotropization temperature down to 190 °C that was attributed to reduced symmetry and the slight nonplanarity of the aromatic core (Figure 14.30) [140]. Interestingly, despite this "unwrapping" of the alky mantel, this compound still self-assembled into liquid crystalline columnar structures with well-arranged molecules. Furthermore, the reduction in the side chains around the nanographene core is expected to favor an intercolumnar charge carrier hopping and to lessen the influence of local intraco-

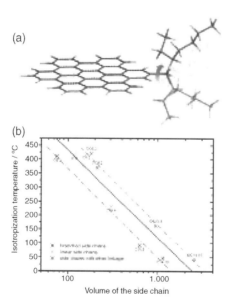

Figure 14.29 (a) Schematic illustration of the rotational freedom and the space filling of a branched C6,2 alkyl chain at different configurations in relation to the aromatic core (due to simplicity only one side chain is attached to the core), (b) relation between the isotropization temperature of hexasubstituted HBC nanographenes and the volume of side chains with different architectures.

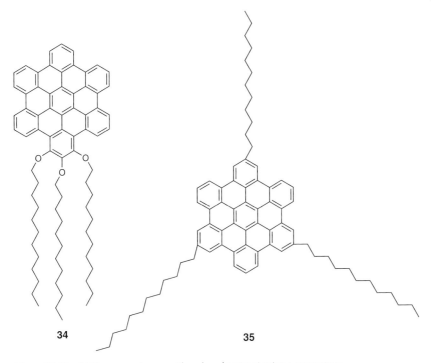

34   35

Figure 14.30 HBC nanographenes with reduced isotropization temperature.

lumnar packing defects. As a disadvantage, molecular asymmetry decreases the crystallinity that is beneficial for the charge transport. Surprisingly, D_3 symmetric HBC (Figure 14.30, **35**) with three dodecyl chains showed a transition temperature to the isotropic melt of only 171 °C maintaining the pronounced crystallinity as indicated by X-ray scattering. The high-substitution symmetry of **35** preserves a distinct supramolecular order crucial for the device applications, while the accessible phase transition makes thermal processing more facile. In general, the reduction of alkyl substituents results in a higher concentration of the active chromophore and increases the conductivity of the active material.

The role of the side chains is not limited to the control of the thermal properties. The introduction of such substituents changes the self-assembly and provokes specific morphologies with well-defined molecular orientations during thermal processing from the isotropic state. Such distinct effect is especially obvious by comparing **19** and **20** both carrying long branched side chains [141]. Cooling both compounds as thin surface layers from the melt results in extended domains but with contrary orientations of the columnar structures (Figure 14.31). The spherulitic domains of **19** revealed radially oriented columnar structures (Figure 14.31b), while for **20** the columnar growth took place around the nucleation center as indicated by synchrotron radiation microfocused X-ray scattering (Figure 14.31a). In both cases, the molecules were arranged in the edge-on fashion. This variation in the molecular organization resulted in different optical properties of the corresponding domains.

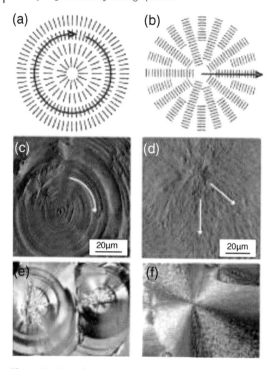

Figure 14.31 Schematic columnar orientation for (a) **20** and (b) **19** in the domains; AFM topographic images of (c) **20** and (d) **19** after cooling on a single surface; and corresponding POM images of (e) **20** and (f) **19**. Arrows indicate the columnar growth direction. Adapted from Ref. [141].

The application of a λ-plate led to a red-blue distribution in the spherulite image of **19** that is attributed to optically negative properties of these domains (Figure 14.31f) [142]. The textures of **20** exhibit an opposite distribution of the red-blue color due to optically positive behavior (Figure 14.31e). In consequence, the pronounced directed growth of **19** over large areas during simple cooling from the isotropic melt allowed to uniaxially align this compound along a temperature gradient through zone crystallization [142]. Thereafter, a thin film is moved at a defined speed from a hot plate, with a temperature above T_i, to a cold plate. At the gap between both plates, a temperature gradient is established along which the columnar structures are oriented with edge-on arranged molecules as demonstrated by X-ray scattering.

The homeotropic orientation of discotic nanographenes spontaneously occurs during processing from the isotropic melt. The type of surface or cooling rate can influence the degree of orientation; however, mainly the molecular design determines the ability toward this specific organization on surfaces. The mechanism taking place during alignment for discotics is not fully understood. Most discotics can be aligned only in the sandwiched geometry between two glass slides [143]. Only a few examples have been reported for the successful alignment on one surface [144]. The introduction of heteroatoms through ether linkages in the vicinity of the HBC core together with bulky branched side chains (**37–40**) resulted in the formation of

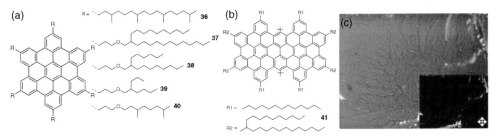

Figure 14.32 Nanographenes capable of aligning homeotropically (a) HBCs (**36–40**), (b) large C72 (**41**), and (c) POM image of homeotropic C72 (**41**) (inset with cross-polarizers).

large homeotropically aligned monodomains [145]. However, pure hydrocarbon nanographenes such as HBCs with branched side chains (**36**) or "supernaphthalene" consisting of 72 aromatic carbon atoms (**41**) showed for the first time that heteroatoms are not required for the formation of a homeotropic phase (Figure 14.32b and c) [146, 147]. It is assumed that the homeotropic alignment is thermodynamically favored for many discotic nanographenes and that in the first stage of the assembly noncovalent van der Waals interactions between the molecules and the polar surface play an important role. Typically, the homeotropic monodomains possess a dendritic morphology that appears black between cross-polarizers since the optical axis is aligned along the incident light beam. These dendritic textures are due to a density change and dewetting of some parts of the film while solidification during cooling from the isotropic melt [148].

The direction of the charge transport toward the surface depends on the arrangement of the nanographenes on the surface. Time-of-flight experiments highlight this close relation between charge transport and columnar orientation (Figure 14.33). In

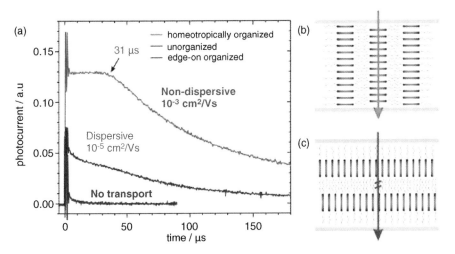

Figure 14.33 (a) Example of time-of-flight hole current transients of nanographenes organized differently on the surface: face-on, weakly organized, and edge-on arranged toward the surface as illustrated schematically (right). Adapted from Ref. [141].

these studies, the charge carrier drift perpendicular to the surface is monitored through the active material that is sandwiched between two electrodes. The edge-on oriented derivative did not allow to record any signal since the bulky alkyl corona that surrounds the aromatic core prohibited a charge carrier migration perpendicular to the stacking axis. When molecules capable to form homeotropic phases were processed too rapidly and small, disordered domains were obtained, an almost dispersive transport of charge carriers was displayed due to scattering sites for charges at grain boundaries and columnar defects. Well homeotropically ordered macroscopic monodomains provide an undisturbed pathway for charge carriers as evident from hole transients with a clear plateau appeared typical for a nondispersive transport (Figure 14.33).

14.8
Nanographenes in Heterojunctions for Solar Cells

The homeotropic alignment of discotic nanographenes is considered to favor the performance of photovoltaic cell. However, the application of organic semiconductors in this type of devices requires the mixture of two components with different electronic affinity to form the so-called heterojunctions that ensure efficient charge separation in the thin layer [149, 150]. Until now, it was not possible to achieve homeotropically aligned discotics in desired heterojunction structures. Furthermore, the solution processing, which is usually used for the fabrication of photovoltaics, does not lead to this characteristic alignment of molecules on the surface. Nevertheless, superior efficiencies were obtained for HBC-PhC$_{12}$ (**2**) and perylene diimide (PDI) mixtures that were spin coated from solution to form a controlled phase separation between vertical layers of both compounds in the thin film (Figure 14.34a). In this way, a large surface contact area between both materials led to high external quantum efficiency of 34% at 490 nm [151]. The operation of a photovoltaic cell strongly depends on the length of alkyl side chains of HBCs. The HBC chromophore is diluted and the ability to absorb light decreases with longer alkyl chains [152]. On the other hand, derivatives with shorter substituents are more crystalline and possess a significantly higher order. This pronounced crystallinity results in larger interfacial donor/acceptor separation and in an improved device performance by evaluating **19–21** (Figure 14.34b). In contrast, the derivatives with longer chains do not segregate well from the acceptor and assemble in columnar structures containing both compounds, a disadvantage for the charge transport along the stacks [153]. Longer alkyl side chains around the conducting aromatic core not only act as an insulating mantel for the 1D charge transport, but also these hinder the exciton polarization and separation in the photovoltaic cell decreasing the device performance. Interestingly, for triangle-shaped nanographenes, this tendency was contrary since the best performance was observed for molecules with the longest branched side chains (**42–43**, Figure 14.34c) [154]. These compounds own a broad liquid crystalline range due to their D$_{3h}$ symmetric extended graphene corona. The derivative with the longest side chains (**43**) revealed a significant structure improve-

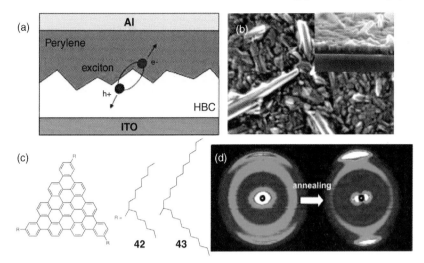

Figure 14.34 (a) Schematic illustration of a heterojunction structure with phase separation between vertical layers of the acceptor perylene and donor HBC, (b) scanning electron microscopy of a thin layer consisting of PDI and HBC, and (c) triangle-shaped nanographenes for photovoltaics, 2D WAXS patterns of **43** before and after annealing. Adapted from Refs [152, 154].

ment after annealing (Figure 14.34d). Especially, the intracolumnar packing was enhanced that is crucial for the charge transport. This annealing procedure of thin film in the photovoltaic device resulted in considerably better performance indicating self-healing of structural defects during thermal processing. Taking into account the insulating effect of the chains that dissolve the aromatic chromophore, the normalized EQE for **43** was one of the highest in comparison to other discotic nanographenes.

14.9
Processing of Nondiscotic Nanographenes

The PR-TRMC studies have proven that extended nanographenes with large delocalized π-orbitals might possess a faster charge carrier transport in comparison to small disks. Large PAHs have a strong π-stacking and a high order. To ensure processibility, long aliphatic side chains are required, which on the other hand have an insulating effect on the charge migration and leading to a defect-sensitive 1D transport. However, pure nanographenes, which might allow a charge transport over 2D or 3D, are in principle inprocessible since these compounds are not soluble in common organic solvent and are not meltable. Solvent-free matrix-assisted laser desorption/ionization (MALDI) soft landing is an alternative technique to overcome this problem allowing the deposition of large nanographenes onto surfaces in an ultrapure form [155]. In the case of deposited HBC (**1**) on HOPG and

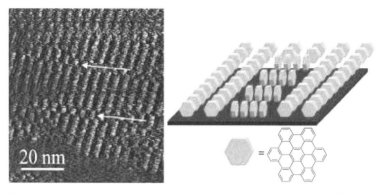

Figure 14.35 AFM image of ultrapure crystalline HBC (**1**) nanographene soft landed on HOPG. Reprinted with permission from Ref. [155]. Copyright 2007 Nature Publishing Group.

dielectrics, the generated layers consist of crystalline "edge-on" ordered nanographenes (Figure 14.35). This arrangement of molecular planes normal to the surface is attributed to the high electric field at the substrate surface during the soft-landing process and stands in contrast to the monolayer organization of individual sheets obtained by vacuum sublimation.

14.10
Conclusions

The bottom-up approach to the implementation of graphene-like structures in novel electronics is realized by the application of nanosized single building blocks based on large polyaromatic hydrocarbons. The well-designed synthesis enables the variation of their shape, size, and edge structure and allows to introduce different solubilizing substituents and functional groups that affect to a great extent the electronic properties, self-organization (in monolayers, in bulk, in solution, and in thin surface layers), processability, and device performance. This stands in a strong contrast to large graphene sheets whose properties and processing cannot be controlled in a straightforward way. Therefore, nanographenes offer a great potential beyond their single-sheet nature, namely, if the aggregation into the above-described multilayers occurs. Improved order and macroscopic alignment of nanographenes in the active layer can be obtained over their strong intermolecular interactions and over a variety of processing technologies controlling the arrangement of the molecules on surfaces. High molecular packing ensures a pronounced local charge carrier transport, while defect-free organization over long range is important for the migration of charge carriers between the electrodes in devices. The formation of distinct thermotropic properties of discotic nanographenes is an essential aspect of self-healing of local defects taking place even at ambient temperatures and improving the device performance. However, the maximum performance of nanographene-based devices has not yet been achieved. The one-dimensional charge transport character is a limiting factor despite the high order achieved in the devices. Therefore, for future work, an adjusted molecular design is in the focus for the increase in the dimensionality of the charge carrier transport that can decrease the influence of local structural defects. Two options are in the focus: (1) conjugated bridges between the columnar stacks also favoring an intercolumnar motion of charges and (2) nanoribbons of different size and shape that allow the transport along such ribbons and between π-stacked molecules. In the latter case, one can expect brick stone-type packing that might increase the dimensionality more. As a different parameter on the device operation, the self-assembly on surface during processing can be further improved by the introduction of functional groups capable of interfacial interactions leading to high macroscopic order.

For miniaturization from microscopic to nanoscopic device architectures, the processing of functional nanographenes into nanotubes and nanofibers does not only provide valuable information about their ability to self-assemble, but also make the fabrication molecular electronics, for example, molecular sensors, realistic. Furthermore, this kind of nanotubes containing self-assembled molecular building blocks represents an alterative to common carbon nanotubes. Molecularly assembled nanotubes offer a great range of variation over the chemical and physical properties. Moreover, self-assembly into individual nanostructures of functionalized nanoscaled graphenes paves the way for novel device applications in molecular scale.

References

1 Novoselov, K.S., Geim, A.K., Morozov, S.V., Jiang, D., Zhang, Y., Dubonos, S.V., Grigorieva, I.V., and Firsov, A.A. (2004) Electric field effect in atomically thin carbon films. *Science*, **306**, 666–669.

2 Novoselov, K.S., Jiang, D., Schedin, F., Booth, T.J., Khotkevich, V.V., Morozov, S.V., and Geim, A.K. (2005) Two-dimensional atomic crystals. *Proc. Natl Acad. Sci. USA*, **102**, 10451–10453.

3 Geim, A.K. and Novoselov, K.S. (2007) The rise of graphene. *Nat. Mater.*, **6**, 183–191.

4 Wang, X., Zhi, L.J., and Müllen, K. (2008) Transparent, conductive graphene electrodes for dye-sensitized solar cells. *Nano Lett.*, **8**, 323–327.

5 Chen, J.H., Jang, C., Xiao, S., Ishigami, M., and Fuhrer, M.S. (2008) Intrinsic and extrinsic performance limits of graphene devices on SiO2. *Nat. Nanotechnol.*, **3**, 206–209.

6 Lemme, M.C., Echtermeyer, T.J., Baus, M., and Kurz, H. (2007) A graphene field-effect device. *IEEE Electron Device Lett.*, **28**, 282–284.

7 Barone, V., Hod, O., and Scuseria, G.E. (2006) Electronic structure and stability of semiconducting graphene nanoribbons. *Nano Lett.*, **6**, 2748–2755.

8 Chen, Z., Lin, Y.-M., Rooks, M.J., and Avouris, P. (2007) Graphene nano-ribbon electronics. *Physica E*, **40**, 228–232.

9 Han, M.Y., Özyilmaz, B., Zhang, Y., and Kim, P. (2007) Energy band-gap engineering of graphene nanoribbons. *Phys. Rev. Lett.*, **98**, 206805.

10 Son, Y.W., Cohen, M.L., and Louie, S.G. (2006) Energy gaps in graphene nanoribbons. *Phys. Rev. Lett.*, **97**, 216803.

11 Stankovich, S., Dikin, D.A., Dommett, G.H.B., Kohlhaas, K.M., Zimney, E.J., Stach, E.A., Piner, R.D., Nguyen, S.T., and Ruoff, R.S. (2006) Graphene-based composite materials. *Nature*, **442**, 282–286.

12 Krishnan, A., Dujardin, E., Treacy, M.M.J., Hugdahl, J., Lynum, S., and Ebbesen, T.W. (1997) Graphitic cones and

the nucleation of curved carbon surfaces. *Nature*, **388**, 451–454.

13 Land, T.A., Michely, T., Behm, R.J., Hemminger, J.C., and Comsa, G. (1992) STM investigation of single layer graphite structures produced on Pt(111) by hydrocarbon decomposition. *Surf. Sci.*, **264**, 261–270.

14 Li, X., Wang, X., Zhang, L., Lee, S., and Dai, H. (2008) Chemically derived, ultrasmooth graphene nanoribbon semiconductors. *Science*, **319**, 1229–1232.

15 Yang, X., Dou, X., Rouhanipour, A., Zhi, L., Räder, H.J., and Müllen, K. (2008) Two-dimensional graphene nanoribbons. *J. Am. Chem. Soc.*, **130**, 4216–4217.

16 Wu, J., Gherghel, L., Watson, M.D., Li, J., Wang, Z., Simpson, C.D., Kolb, U., and Müllen, K. (2003) From branched polyphenylenes to graphite ribbons. *Macromolecules*, **36**, 7082–7089.

17 Clar, E. (1964) *Polycyclic Hydrocarbons*, Vol. I/II, Academic Press, New York.

18 Clar, E. (1972) *The Aromatic Sextet*, Wiley-VCH, London.

19 Enoki, T. and Kazuyuki, T. (2008) Unconventional electronic and magnetic functions of nanographene-based host–guest systems. *Dalton Trans.*, 3733–3781.

20 Enoki, T. and Kobayashi, Y. (2005) Magnetic nanographite: an approach to molecular magnetism. *J. Mater. Chem.*, **15**, 3999–4002.

21 Kusakabe, K. and Maruyama, M. (2003) Magnetic nanographite. *Phys. Rev. B*, **67**, 092406.

22 Maruyama, M. and Kusakabe, K. (2004) Theoretical prediction of synthesis methods to create magnetic nanographite. *J. Phys. Soc. Jpn.*, **73**, 656–663.

23 Simpson, C.D., Brand, J.D., Berresheim, A.J., Przybilla, L., Rader, H.J., and Müllen, K. (2002) Synthesis of a giant 222 carbon graphite sheet. *Chem. Eur. J.*, **6**, 1424–1429.

24 Watson, M.D., Fechtenkötter, A., and Müllen, K. (2001) Big is beautiful–"aromaticity" revisited from the viewpoint of macromolecular and supramolecular benzene chemistry. *Chem. Rev.*, **101**, 1267–1300.

25 Harvey, R.G. (1997) *Polycyclic Aromatic Hydrocarbons*, Wiley-VCH, New York.

26 Dötz, F., Brand, J.D., Ito, S., Gherghel, L., and Müllen, K. (2000) Synthesis of large polycyclic aromatic hydrocarbons: variation of size and periphery. *J. Am. Chem. Soc.*, **122**, 7707–7717.

27 Yang, X.Y., Dou, X., Rouhanipour, A., Zhi, L.J., Räder, H.J., and Müllen, K. (2008) Two-dimensional graphene nanoribbons. *J. Am. Chem. Soc.*, **130**, 4216–4217.

28 Kastler, M., Schmidt, J., Pisula, W., Sebastiani, D., and Müllen, K. (2006) From armchair zigzag peripheries in nanographenes. *J. Am. Chem. Soc.*, **128**, 9526–9534.

29 Draper, S.M., Gregg, D.J., and Madathil, R. (2002) Heterosuperbenzenes: a new family of nitrogen-functionalized, graphitic molecules. *J. Am. Chem. Soc.*, **124**, 3486–3487.

30 Takase, M., Enkelmann, V., Sebastiani, D., Baumgarten, M., and Müllen, K. (2007) Annularly fused hexapyrrolohexaazacoronenes: an extended pi system with multiple interior nitrogen atoms displays stable oxidation states. *Angew. Chem., Int. Ed.*, **46**, 5524–5527.

31 Stein, S.E. and Brown, R.L. (1987) π-Electron properties of large condensed polyaromatic hydrocarbons. *J. Am. Chem. Soc.*, **109**, 3721–3729.

32 Wu, J., Pisula, W., and Müllen, K. (2007) Graphenes as potential material for electronics. *Chem. Rev.*, **107**, 718–747.

33 Heeger, A.J. (2001) Semiconducting and metallic polymers: the fourth generation of polymeric materials. *Angew. Chem., Int. Ed.*, **40**, 2591–2611.

34 McCulloch, I., Heeney, M., Bailey, C., Genevicius, K., MacDonald, I., Shkunov, M., Sparrowe, D., Tierney, S., Wagner, R., Zhang, W., Chabinyc, M.L., Kline, R.J., Mcgehee, M.D., and Toney, M.F. (2006) Liquid-crystalline semiconducting polymers with high charge-carrier mobility. *Nat. Mater.*, **5**, 328–333.

35 Iijima, S. (1991) Helical microtubules of graphitic carbon. *Nature*, **354**, 56–58.

36 Samori, P., Severin, S., Simpson, C.D., Müllen, K., and Rabe, J.P. (2002) Epitaxial composite layers of electron donors and acceptors from very large polycyclic aromatic hydrocarbons. *J. Am. Chem. Soc.*, **124**, 9454–9457.

37 Samorí, P., Keil, M., Friedlein, R., Birgerson, J., Watson, M., Müllen, K., Salaneck, W.R., and Rabe, J.P. (2001) Growth of ordered hexakis-dodecyl-hexabenzocoronene layers from solution: a SFM and ARUPS study. *J. Phys. Chem. B*, **105**, 11114–11119.

38 Müllen, K. and Rabe, J. (2008) Nanographenes as active components of single-molecule electronics and how a scanning tunneling microscope puts them to work. *Acc. Chem. Res.*, **41**, 511–520.

39 Ito, S., Herwig, P.T., Böhme, T., Rabe, J.P., Rettig, W., and Müllen, K. (2000) Bis-hexa-perihexabenzocoronenyl: a superbiphenyl. *J. Am. Chem. Soc.*, **122**, 7698–7706.

40 Iyer, V.S., Yoshimura, K., Enkelmann, V., Epsch, R., Rabe, J.P., and Müllen, K. (1998) A soluble C60 graphite segment. *Angew. Chem., Int. Ed*, **37**, 2696–2699.

41 Samori, P., Severin, N., Simpson, C.D., Müllen, K., and Rabe, J.P. (2002) Epitaxial composite layers of electron donors and acceptors from very large polycyclic aromatic hydrocarbons. *J. Am. Chem. Soc.*, **124**, 9454–9457.

42 Samorí, P., Fechtenkötter, A., Jäckel, F., Böhme, T., Müllen, K., and Rabe, J.P. (2001) Supramolecular staircase via self-assembly of disklike molecules at the solid–liquid interface. *J. Am. Chem. Soc.*, **123**, 11462–11467.

43 Feng, X., Wu, J., Ai, M., Pisula, W., Zhi, L., Rabe, J.P., and Müllen, K. (2007) Triangle-shaped polycyclic aromatic hydrocarbons. *Angew. Chem., Int. Ed.*, **46**, 3033–3036.

44 Wasserfallen, D., Fischbach, I., Tchebotareva, N., Kastler, M., Pisula, W., Jäckel, F., Watson, M.D., Schnell, I., Rabe, J.P., Spiess, H.W., and Müllen, K. (2005) Influence of hydrogen bonds on the supramolecular order of hexa-*peri*-hexabenzocoronenes. *Adv. Funct. Mater.*, **15**, 1585–1594.

45 Jäckel, F., Wang, Z., Watson, M.D., Müllen, K., and Rabe, J.P. (2004) Nanoscale array of inversely biased molecular rectifiers. *Chem. Phys. Lett.*, **387**, 372–376.

46 Pisula, W., Tomovic, Z., Simpson, C., Kastler, M., Pakula, T., and Müllen, K. (2005) Relationship between core size, side chain length, and the supramolecular organization of polycyclic aromatic hydrocarbons. *Chem. Mater.*, **17**, 4296–4303.

47 Fischbach, I., Pakula, T., Minkin, P., Fechtenkötter, A., Müllen, K., Spiess, H.W., and Saalwachter, K. (2002) Structure and dynamics in columnar discotic materials: a combined X-ray and solid-state NMR study of hexabenzocoronene derivatives. *J. Phys. Chem. B*, **106**, 6408–6418.

48 Fischbach, I., Ebert, F., Spiess, H.W., and Schnell, I. (2004) Rotor modulations and recoupling strategies in ^{13}C solid-state magic-angle-spinning NMR spectroscopy: probing molecular orientation and dynamics. *ChemPhysChem*, **5**, 895–908.

49 Boden, N., Bushby, R.J., Clements, J., Movaghar, B., Donovan, K.J., and Kreouzis, T. (1995) Mechanism of charge transport in discotic liquid crystals. *Phys. Rev. B*, **52**, 13274–13280.

50 Brown, S.P., Schnell, I., Brand, J.D., Müllen, K., and Spiess, H.W. (1999) An investigation of π–π packing in a columnar hexabenzocoronene by fast magic-angle spinning and double-quantum ^1H solid-state NMR spectroscopy. *J. Am. Chem. Soc.*, **121**, 6712–6718.

51 Watson, J.D. and Crick, F.H. (1953) Molecular structure of nucleic acids: a structure for deoxyribose nucleic acid. *Nature*, **171**, 737–738.

52 Fontes, E., Heiney, P.A., and Dejeu, W.H. (1988) Liquid-crystalline and helical order in a discotic mesophase. *Phys. Rev. Lett.*, **61**, 1202–1205.

53 Cornil, J., Lemaur, V., Calbert, J.-P., and Brédas, J.-L. (2002) Charge transport in discotic liquid crystals: a molecular scale description. *Adv. Mater.*, **14**, 726–729.

54 Ochsenfeld, C., Brown, S.P., Schnell, I., Gauss, J., and Spiess, H.W. (2001) Structure assignment in the solid state by the coupling of quantum chemical calculations with NMR experiments: a columnar hexabenzocoronene derivative. *J. Am. Chem. Soc.*, **123**, 2597–2606.

55 Wu, J., Fechtenkötter, A., Gauss, J., Watson, M.D., Kastler, M., Fechtenkötter, C., Wagner, M., and Müllen, K. (2004) Controlled self-assembly of hexa-*peri*-hexabenzocoronenes in solution. *J. Am. Chem. Soc.*, **126**, 11311–11321.

56 Pisula, W., Tomovic, Z., Watson, M.D., Müllen, K., Kussmann, J., Ochsenfeld, C., Metzroth, T., and Gauss, J. (2007) Helical packing of discotic hexaphenyl hexa-*peri*-hexabenzocoronenes: theory and experiment. *J. Phys. Chem. B*, **111**, 7481–7487.

57 Glüsen, B., Kettner, A., and Wendorff, J.H. (1997) A plastic columnar discotic phase. *Mol. Cryst. Liq. Cryst.*, **303**, 115–120.

58 Glüsen, B., Heitz, W., Kettner, A., and Wendorff, J.H. (1996) A plastic columnar discotic phase Dp. *Liq. Cryst.*, **20**, 627–633.

59 Wu, J., Watson, M.D., Zhang, L., Wang, Z., and Müllen, K. (2004) Hexakis (4-iodophenyl)-*peri*-hexabenzocoronene – a versatile building block for highly ordered discotic liquid crystalline materials. *J. Am. Chem. Soc.*, **126**, 177–186.

60 Feng, X., Pisula, W., and Müllen, K. (2007) From helical to staggered stacking of zigzag nanographenes. *J. Am. Chem. Soc.*, **129**, 14116–14117.

61 Brunsveld, L., Folmer, B.J.B., Meijer, E.W., and Sijbesma, R.P. (2001) Supramolecular polymers. *Chem. Rev.*, **101**, 4071–4098.

62 Ciferri, A. (2002) Supramolecular polymerizations. *Macromol. Rapid Commun.*, **23**, 511–529.

63 Dou, X., Pisula, W., Wu, J., Bodwell, G.J., and Müllen, K. (2008) Reinforced self-assembly of hexa-*peri*-hexabenzocoronenes by hydrogen bonds: from microscopic aggregates to macroscopic fluorescent organogels. *Chem. Eur. J.*, **14**, 240–249.

64 Datta, A. and Pati, S.K. (2006) Dipolar interactions and hydrogen bonding in supramolecular aggregates: understanding cooperative phenomena for 1st hyperpolarizability. *Chem. Soc. Rev.*, **35**, 1305–1323.

65 Feng, X., Pisula, W., Zhi, L., Takase, M., and Müllen, K. (2008) Controlling the columnar orientation of C3-symmetric superbenzenes through alternating polar/apolar substitutents. *Angew. Chem., Int. Ed.*, **47**, 1703–1706.

66 Hsu, T.C., Hüser, B., Pakula, T., Spiess, H.W., and Stamm, M. (1990) Morphology and phase transition study of polymers with discotic mesogens. *Makromol. Chem.*, **191**, 1597–1609.

67 Feng, X., Pisula, W., Takase, M., Dou, X., Enkelmann, V., Wagner, M., Ding, N., and Müllen, K. (2008) Synthesis, helical organization, and fibrous formation of C3 symmetric methoxy-substituted discotic hexa-*peri*-hexabenzocoronene. *Chem. Mater.*, **20**, 2872–2874.

68 Brand, J.D., Kubel, C., Ito, S., and Müllen, K. (2000) Functionalized hexa-*peri*-hexabenzocoronenes: stable supramolecular order by polymerization in the discotic mesophase. *Chem. Mater.*, **12**, 1638–1647.

69 Kastler, M., Pisula, W., Davies, R.J., Gorelik, T., Kolb, U., and Müllen, K. (2007) Nanostructuring with a crosslinkable discotic material. *Small*, **3**, 1438–1444.

70 Piris, J. (2004) Optoelectronic properties of discotic materials for device, Dissertation, Applications Delft University of Technology, The Netherlands.

71 Warman, J.M. and van de Craats, A.M. (2003) Charge mobility in discotic materialsstudied by PR-TRMC. *Mol. Cryst. Liq. Cryst.*, **396**, 41–72.

72 Boden, N., Bushby, R.J., and Clements, J. (1993) Mechanism of quasi-one-dimensional electronic conductivity in discotic liquid crystals. *J. Chem. Phys.*, **98**, 5920–5931.

73 Simmerer, J., Glüsen, B., Paulus, W., Kettner, A., Schuhmacher, P., Adam, D.,

Etzbach, K.H., Siemensmeyer, K., Wendorff, J.H., Ringsdorf, H., and Haarer, D. (1996) Transient photoconductivity in a discotic hexagonal plastic crystal. *Adv. Mater.*, **8**, 815–819.

74 van de Craats, A.M., Siebbeles, L.D.A., Bleyl, I., Haarer, D., Berlin, Y.A., Zharikov, A.A., and Warman, J.M. (1998) Mechanism of charge transport along columnar stacks of a triphenylene dimer. *J. Phys. Chem. B*, **102**, 9625–9634.

75 van de Craats, A.M., Warman, J.M., de Haas, M.P., Adam, D., Simmerer, J., Haarer, D., and Schuhmacher, P. (1996) The mobility of charge carriers in all four phases of the columnar discotic material hexakis(hexylthio)triphenylene: combined TOF and PR-TRMC results. *Adv. Mater.*, **8**, 823–826.

76 van de Craats, A.M., Warman, J.M., Fechtenkötter, A., Brand, J.D., Harbison, M.A., and Müllen, K. (1999) Record charge carrier mobility in a room-temperature discotic liquid-crystalline derivative of hexabenzocoronene. *Adv. Mater.*, **11**, 1469–1472.

77 van de Craats, A.M. and Warman, J.M. (2001) The core-size effect on the mobility of charge in discotic liquid crystalline materials. *Adv. Mater.*, **13**, 130–133.

78 van de Craats, A.M., de Haas, M.P., and Warman, J.M. (1997) Charge carrier mobilities in the crystalline solid and discotic mesophases of hexakis-hexylthio and hexakis-hexyloxy triphenylene. *Synth. Met.*, **86**, 2125–2126.

79 Debije, M.G., Piris, J., de Haas, M.P., Warman, J.M., Tomović, Ž., Simpson, C.D., Watson, M.D., and Müllen, K. (2004) The optical and charge transport properties of discotic materials with large aromatic hydrocarbon cores. *J. Am. Chem. Soc.*, **126**, 4641–4645.

80 Warman, J.M., Piris, J., Pisula, W., Kastler, M., Wasserfallen, D., and Müllen, K. (2005) Charge recombination via intercolumnar electron tunneling through the lipid-like mantle of discotic hexabenzocoronenes. *J. Am. Chem. Soc.*, **127**, 14257–14262.

81 Pisula, W., Kastler, M., Wasserfallen, D., Mondeshki, M., Piris, J., Schnell, I., and Müllen, K. (2006) Relation between supramolecular order and charge carrier mobility of branched alkyl hexa-*peri*-hexabenzocoronenes. *Chem. Mater.*, **18**, 3634–3640.

82 Kastler, M., Pisula, W., Wasserfallen, D., Pakula, T., and Müllen, K. (2005) Influence of alkyl substituents on the solution aggregation of hexa-*peri*-hexabenzocoronene derivatives. *J. Am. Chem. Soc.*, **127**, 4286–4296.

83 Gao, P., Beckmann, D., Tsao, H.N., Feng, X., Enkelmann, V., Baumgarten, M., Pisula, W., and Müllen, K. (2009) Dithieno[2,3-d;2′,3′-d′]benzo[1,2-b;4,5-b′]dithiophene (DTBDT) as semiconductor for high-performance, solution-processed organic field-effect transistors. *Adv. Mater.*, **21**, 213–216.

84 Tsao, H.N., Cho, D., Andreasen, J.W., Rouhanipour, A., Breiby, D.W., Pisula, W., and Müllen, K. (2009) The influence of morphology on high performance polymer field-effect transistors. *Adv. Mater.*, **21**, 209–212.

85 Hill, J.P., Jin, W., Kosaka, A., Fukushima, T., Ichihara, H., Shimomura, T., Ito, K., Hashizume, T., Ishii, N., and Aida, T. (2004) Self-assembled hexa-*peri*-hexabenzocoronene graphitic nanotube. *Science*, **304**, 1481–1483.

86 Jin, W., Yamamoto, Y., Fukushima, T., Ishii, N., Kim, J., Kato, K., Takata, M., and Aida, T. (2008) Systematic studies on structural parameters for nanotubular assembly of hexa-*peri*-hexabenzocoronenes. *J. Am. Chem. Soc.*, **130**, 9434–9440.

87 Yamamoto, Y., Fukushima, T., Jin, W., Kosaka, A., Hara, T., Nakamura, T., Saeki, A., Seki, S., Tagawa, S., and Aida, T. (2006) *Adv. Mater.*, **18**, 1297–1300.

88 Yamamoto, T., Fukushima, T., Yamamoto, Y., Kosaka, A., Jin, W., Ishii, N., and Aida, T. (2006) Stabilization of a kinetically favored nanostructure: surface ROMP of self-assembled conductive nanocoils from a norbornene-appended hexa-*peri*-hexabenzocoronene. *J. Am. Chem. Soc.*, **128**, 14337–14340.

89 Yamamoto, Y., Fukushima, T., Suna, Y., Ishii, N., Saeki, A., Seki, S., Tagawa, S., Taniguchi, M., Kawai, T., and Aida, T.

(2006) Photoconductive coaxial nanotubes of molecularly connected electron donor and acceptor layers. *Science*, **314**, 1761–1764.

90 Jin, W., Fukushima, T., Niki, M., Kosaka, A., Ishii, N., and Aida, T. (2005) Self-assembled graphitic nanotubes with one-handed helical arrays of a chiral amphiphilic molecular grapheme. *Proc. Natl. Acad. Sci. U.S.A.*, **102**, 10801–10806.

91 El Hamaoui, B., Zhi, L., Pisula, W., Kolb, U., Wu, J., and Müllen, K. (2007) Self-assembly of amphiphilic imidazolium-based hexa-*peri*-hexabenzocoronenes into fibreous aggregates. *Chem. Commun.*, **23**, 2384–2386.

92 Feng, X., Pisula, W., Kudernac, T., Wu, D., Zhi, L., De Feyter, S., and Müllen, K. (2009) Controlled self-assembly of C$_3$-symmetric hexa-perihexabenzocoronenes with alternating hydrophilic and hydrophobic substituents in solution, in the bulk, and on a surface, *J. Am. Chem. Soc.*, **131**, 4439–4448.

93 Aebischer, O.F., Aebischer, A., Tondo, P., Alameddine, B., Dadras, M., Güdelb, H.-U., and Jenny, T.A. (2006) Self-aggregated perfluoroalkylated hexa-*peri*-hexabenzocoronene fibers observed by cryo-SEM and fluorescence spectroscopy. *Chem. Commun.*, 4221–4223.

94 Aebischer, O.F., Aebischer, A., Donnio, B., Alameddine, B., Dadras, M., Güdel, H.-U., Guillon, D., and Jenny, T.A. (2007) Controlling the lateral aggregation of perfluoroalkylated hexa-*peri*-hexabenzocoronenes. *J. Mater. Chem.*, **17**, 1262–1267.

95 Joachim, C., Gimzewski, J.K., and Aviram, A. (2000) Electronics using hybrid-molecular and mono-molecular devices. *Science*, **408**, 541–548.

96 Avouris, P. (2002) Molecular electronics with carbon nanotubes. *Acc. Chem. Res.*, **35**, 1026–1034.

97 Xiao, S., Tang, J., Beetz, T., Guo, X., Tremblay, N., Siegrist, T., Zhu, Y., Steigerwald, M., and Nuckolls, C. (2006) Transferring self-assembled, nanoscale cables into electrical devices. *J. Am. Chem. Soc.*, **128**, 10700–10701.

98 Xiao, S., Myers, M., Miao, Q., Sanaur, S., Pang, K., Steigerwald, M.L., and Nuckolls, C. (2005) Molecular wires from contorted aromatic compounds. *Angew. Chem., Int. Ed.*, **44**, 7390–7394.

99 Cohen, Y.S., Xiao, S., Steigerwald, M.L., Nuckolls, C., and Kagan, C.R. (2006) Enforced one-dimensional photoconductivity in core-cladding hexabenzocoronenes. *Nano Lett.*, **6**, 2838–2841.

100 Harris, K.D., Xiao, S., Lee, C.Y., Strano, M.S., Nuckolls, C., and Blanchet, G.B. (2007) High mobility, air-stable organic transistors from hexabenzocoronene/carbon nanotube bilayers. *J. Phys. Chem. C*, **111**, 17947–17951.

101 Guo, X., Myers, M., Xiao, S., Lefenfeld, M., Steiner, R., Tulevski, G.S., Tang, J., Baumert, J., Leibfarth, F., Yardley, J.T., Steigerwald, M.L., Kim, P., and Nuckolls, C. (2006) Chemoresponsive monolayer transistors. *Proc. Natl. Acad. Sci. USA*, **103**, 11452–11456.

102 Steinhart, M., Wendorff, J.H., Greiner, A., Wehrspohn, R.B., Nielsch, K., Schilling, J., Choi, J., and Gösele, U. (2002) Polymer nanotubes by wetting of ordered porous templates. *Science*, **296**, 1997.

103 Steinhart, M., Wehrspohn, R.B., Gösele, U., and Wendorff, J.H. (2004) Nanotubes by template wetting: a modular assembly system. *Angew. Chem., Int. Ed.*, **43**, 1334–1344.

104 Pisula, W., Kastler, M., Wasserfallen, D., Davies, R.J., Garcia-Gutierrez, M.-C., and Müllen, K. (2006) From macro-to nanoscopic templating with nanographenes. *J. Am. Chem. Soc.*, **128**, 14424–14425.

105 Steinhart, M., Murano, S., Schaper, A.K., Ogawa, T., Tsuji, M., Gösele, U., Weder, C., and Wendorff, J.H. (2005) Morphology of polymer/liquid-crystal nanotubes: influence of confinement. *Adv. Funct. Mater.*, **15**, 1656–1664.

106 Steinhart, M., Zimmermann, S., Göring, P., Schaper, A.K., Gösele, U., Weder, C., and Wendorff, J.H. (2005) Liquid crystalline nanowires in porous alumina: geometric confinement versus influence of pore walls. *Nano Lett.*, **5**, 429–434.

107 Zhi, L., Wu, J., Li, J., Kolb, U., and Müllen, K. (2005) Carbonization of disclike molecules in porous alumina membranes: toward carbon nanotubes with controlled graphene-layer orientation. *Angew. Chem., Int. Ed.*, **44**, 2120–2123.

108 Salleo, A. (2007) Charge transport in polymeric transistors. *Mater. Today*, **10**, 38–45.

109 Dimitrakopoulos, C.D. and Malenfant, P.R.L. (2002) Organic thin film transistors for large area electronics. *Adv. Mater.*, **14**, 99–117.

110 Mcculloch, I. (2005) Thin films: rolling out organic electronics. *Nat. Mater.*, **4**, 583–584.

111 Tang, C., Tracz, A., Kruk, M., Zhang, R., Smilgies, D.-M., Matyjaszewski, K., and Kowalewski, T. (2005) Long-range ordered thin films of block copolymers prepared by zone-casting and their thermal conversion into ordered nanostructured carbon. *J. Am. Chem. Soc.*, **127**, 6918–6919.

112 Duffy, C.M., Andreasen, J.W., Breiby, D.W., Nielsen, M.M., Ando, M., Minakata, T., and Sirringhaus, H. (2008) High-mobility aligned pentacene films grown by zone-casting. *Chem. Mater.*, **20**, 7252–7259.

113 Miskiewicz, P., Mas-Torrent, M., Jung, J., Kotarba, S., Glowacki, I., Gomar-Nadal, E., Amabilino, D.B., Veciana, J., Krause, B., Carbone, D., Rovira, C., and Ulanski, J. (2006) Efficient high area OFETs by solution based processing of a π-electron rich donor. *Chem. Mater.*, **18**, 4724–4729.

114 Tracz, A., Jeszka, J.K., Watson, M., Pisula, W., Müllen, K., and Pakula, T. (2003) Uniaxial alignment of the columnar super-structure of a hexa(alkyl) hexa-*peri*-hexabenzocoronene on untreated glass by simple solution processing. *J. Am. Chem. Soc.*, **125**, 1682–1683.

115 Breiby, D.W., Bunk, O., Pisula, W., Solling, T.I., Tracz, A., Pakula, T., Müllen, K., and Nielsen, M.M. (2005) Structure of zone-cast HBC-C12H25 films. *J. Am. Chem. Soc.*, **127**, 11288–11293.

116 Breiby, D.W., Hansteen, F., Pisula, W., Bunk, O., Kolb, U., Andreasen, J.W., Müllen, K., and Nielsen, M.M. (2005) *In situ* studies of phase transitions in thin discotic films. *J. Phys. Chem. B*, **109**, 22319–22325.

117 Piris, J., Pisula, W., Tracz, A., Pakula, T., Müllen, K., and Warman, J.M. (2004) Thermal switching of the optical anisotropy of a macroscopically aligned film of a discotic liquid crystal. *Liq. Cryst.*, **31**, 993–996.

118 Piris, J., Debije, M.G., Stutzmann, N., Laursen, B.W., Pisula, W., Watson, M.D., Bjørnholm, T., Müllen, K., and Warman, J.M. (2004) Aligned thin films of discotic hexabenzocoronenes: anisotropy in the optical and charge transport properties. *Adv. Funct. Mater.*, **14**, 1053–1061.

119 Pisula, W., Menon, A., Stepputat, M., Lieberwirth, I., Kolb, U., Tracz, A., Sirringhaus, H., Pakula, T., and Müllen, K. (2005) A zone-casting technique for device fabrication of field-effect transistors based on discotic hexa-*peri*-hexabenzocoronene. *Adv. Mater.*, **17**, 684–689.

120 Tsao, H.N., Räder, H.J., Pisula, W., Rouhanipour, A., and Müllen, K. (2008) Novel organic semiconductors and processing techniques for organic field-effect transistors. *Phys. Status Solidi A*, **205**, 421–429.

121 Pisula, W., Tomović, Ž., Stepputat, M., Kolb, U., Pakula, T., and Müllen, K. (2005) Uniaxial alignment of polycyclic aromatic hydrocarbons by solution processing. *Chem. Mater.*, **17**, 2641–2647.

122 Boamfa, M.I., Viertler, K., Wewerka, A., Stelzer, F., Christianen, P.C.M., and Maan, J.C. (2003) Mesogene-polymer backbone coupling in side-chain polymer liquid crystals, studied by high magnetic-field-induced alignment. *Phys. Rev. Lett.*, **90**, 025501.

123 Lee, J.-H., Choi, S.-M., Pate, B.D., Chisholm, M.H., and Han, Y.-S. (2006) Magnetic uniaxial alignment of the columnar superstructure of discotic metallomesogens over the centimetre length scale. *J. Mater. Chem.*, **16**, 2785–2791.

124 Shklyarevskiy, I.O., Jonkheijm, P., Stutzmann, N., Wasserberg, D., Wondergem, H.J., Christianen, P.C.M., Schenning, A.P.H.J., de Leeuw, D.M.,

Tomovic, Z., Wu, J., Müllen, K., and Maan, J.C. (2005) High anisotropy of the field-effect transistor mobility in magnetically aligned discotic liquid-crystalline semiconductors. *J. Am. Chem. Soc.*, **127**, 16233–16237.

125 Cristadoro, A., Lieser, G., Räder, H.J., and Müllen, K. (2007) Field-force alignment of disc-type π-systems. *ChemPhysChem*, **8**, 586–591.

126 Cristadoro, A., Ai, M., Räder, H.J., Rabe, J.P., and Müllen, K. (2008) Electrical field-induced alignment of nonpolar hexabenzocoronene molecules into columnar structures on highly oriented pyrolitic graphite investigated by STM and SFM. *J. Phys. Chem. C*, **112**, 5563–5566.

127 Damman, P., Dosiere, M., Brunel, M., and Wittmann, J.C. (1997) Nucleation and oriented growth of aromatic crystals on friction-transferred poly (tetrafluoroethylene) layers. *J. Am. Chem. Soc*, **119**, 4633–4639.

128 Wittmann, J.C. and Smith, P. (1991) Highly oriented thin films of poly (tetrafluoroethylene) as a substrate for oriented growth of materials. *Nature*, **352**, 414.

129 Zimmermann, S., Wendorff, J.H., and Weder, C. (2002) Uniaxial orientation of columnar discotic liquid crystals. *Chem. Mater.*, **14**, 2218–2223.

130 Piris, J., Debije, M.G., Stutzmann, N., van de Craats, A.M., Watson, M.D., Müllen, K., and Warman, J.M. (2003) Anisotropy in the mobility and photogeneration of charge carriers in thin films of discotic hexabenzocoronenes, columnarly self-assembled on friction-deposited poly (tetrafluoroethylene). *Adv. Mater.*, **15**, 1736–1740.

131 Bunk, O., Nielsen, M.M., Solling, T.I., van de Craats, A.M., and Stutzmann, N. (2003) Induced alignment of a solution-cast discotic hexabenzocoronene derivative for electronic devices investigated by surface X-ray diffraction. *J. Am. Chem. Soc.*, **125**, 2252–2258.

132 van de Craats, A.M., Stutzmann, N., Bunk, O., Nielsen, M.M., Watson, M., Müllen, K., Chanzy, H.D., Sirringhaus, H., and Friend, R.H. (2003) Meso-epitaxial solution-growth of self-organizing discotic liquid-crystalline semiconductors. *Adv. Mater.*, **15**, 495–499.

133 Bjørnholm, T., Hassenkam, T., and Reitzel, N. (1999) Supramolecular organization of highly conducting organic thin films by the Langmuir–Blodgett technique. *J. Mater. Chem.*, **9**, 1975–1990.

134 Josefowicz, J.Y., Maliszewskyj, N.C., Idziak, S.H.J., Heiney, P.A., McCauley, J.P., Jr, and Smith, A.B., III (1993) Structure of Langmuir–Blodgett films of disk-shaped molecules determined by atomic force microscopy. *Science*, **260**, 323–326.

135 Reitzel, N., Hassenkam, T., Balashev, K., Jensen, T.R., Howes, P.B., Kjaer, K., Fechtenkötter, A., Tchebotareva, N., Ito, S., Müllen, K., and Bjørnholm, T. (2001) Langmuir and Langmuir–Blodgett films of amphiphilic hexa-*peri*-hexabenzocoronene: new phase transitions and electronic properties controlled by pressure. *Chem. Eur. J.*, **7**, 4894–4901.

136 Laursen, B.W., Norgaard, K., Reitzel, N., Simonsen, J.B., Nielsen, C.B., Als-Nielsen, J., Bjornholm, T., Solling, T.I., Nielsen, M.M., Bunk, O., Kjaer, K., Tchebotareva, N., Watson, M.D., Müllen, K., and Piris, J. (2004) Macroscopic alignment of graphene stacks by Langmuir–Blodgett deposition of amphiphilic hexabenzocoronenes. *Langmuir*, **20**, 4139–4146.

137 Kubowicz, S., Thunemann, A.F., Geue, T.M., Pietsch, U., Watson, M.D., Tchebotareva, N., and Müllen, K. (2003) X-ray reflectivity study of an amphiphilic hexa-*peri*-hexabenzocoronene at a structured silicon wafer surface. *Langmuir*, **19**, 10997–10999.

138 Kubowicz, S., Pietsch, U., Watson, M.D., Tchebotareva, N., Müllen, K., and Thunemann, A.F. (2003) Thin layers of columns of an amphiphilic hexa-*peri*-hexabenzocoronene at silicon wafer surfaces. *Langmuir*, **19**, 5036–5041.

139 Kastler, M. (2006) Discotic Materials for Organic Electronics, Dissertation, University Mainz, Germany.

140 Wang, Z., Watson, M.D., Wu, J., and Müllen, K. (2004) Partially stripped insulated nanowires: a lightly substituted hexa-*peri*-hexabenzocoronene-based columnar liquid crystal. *Chem. Commun.*, 336–337.

141 Kastler, M., Pisula, W., Laquai, F., Kumar, A., Davis, R., Baluschev, S., Garcia-Gutierrez, M.C., Wasserfallen, D., Butt, H.-J., Riekel, C., Wegner, G., and Müllen, K. (2006) Organization of charge-carrier pathways for organic electronics. *Adv. Mater.*, **18**, 2255–2259.

142 Pisula, W., Kastler, M., Wasserfallen, D., Pakula, T., and Müllen, K. (2004) Exceptionally long-range self-assembly of hexa-*peri*-hexabenzocoronene with dove-tailed alkyl substituents. *J. Am. Chem. Soc.*, **126**, 8074–8075.

143 De Cupere, V., Tant, J., Viville, P., Lazzaroni, R., Osikowicz, W., Salaneck, W.R., and Geerts, Y.H. (2006) Effect of interfaces on the alignment of a discotic liquid-crystalline phthalocyanine. *Langmuir*, **22**, 7798–7806.

144 Grelet, E. and Bock, H. (2006) Control of the orientation of thin open supported columnar liquid crystal films by the kinetics of growth. *Europhys. Lett.*, **73**, 712–718.

145 Pisula, W., Tomovic, Z., El Hamaoui, B., Watson, M.D., Pakula, T., and Müllen, K. (2005) Control of the homeotropic order of discotic hexa-*peri*-hexabenzocoronenes. *Adv. Funct. Mater.*, **15**, 893–904.

146 Wasserfallen, D., Kastler, M., Pisula, W., Hofer, W.A., Fogel, Y., Wang, Z., and Müllen, K. (2006) Suppressing aggregation in a large polycyclic aromatic hydrocarbon. *J. Am. Chem. Soc.*, **128**, 1334–1339.

147 Liu, C.Y., Fechtenkötter, A., Watson, M.D., Müllen, K., and Bard, A.J. (2003) Room temperature discotic liquid crystalline thin films of hexa-*peri*-hexabenzocoronene: synthesis and optoelectronic properties. *Chem. Mater.*, **15**, 124–130.

148 Pisula, W., Kastler, M., El Hamaoui, B., García-Gutiérrez, M.-C., Davies, R.J., Riekel, C., and Müllen, K. (2007) Dendritic morphology in homeotropically aligned discotic films. *ChemPhysChem*, **8**, 1025–1028.

149 Brabec, C.J., Sariciftci, N.S., and Hummelen, J.C. (2001) Plastic solar cells. *Adv. Funct. Mater.*, **11**, 15–26.

150 Mayer, A.C., Scully, S.R., Hardin, B.E., Rowell, M.W., and McGehee, M.D. (2007) Polymer-based solar cells. *Mater. Today*, **10**, 28–33.

151 Schmidt-Mende, L., Fechtenkötter, A., Müllen, K., Moons, E., Friend, R.H., and MacKenzie, J.D. (2001) Self-organized discotic liquid crystals for high-efficiency organic photovoltaics. *Science*, **293**, 1119–1122.

152 Li, J., Kastler, M., Pisula, W., Robertson, J.W.F., Wasserfallen, D., Grimsdale, A.C., Wu, J., and Müllen, K. (2007) Organic bulk-heterojunction photovoltaics based on alkyl substituted discotics. *Adv. Funct. Mater.*, **17**, 2528–2533.

153 Pisula, W., Kastler, M., Wasserfallen, D., Robertson, J.W.F., Nolde, F., Kohl, C., and Müllen, K. (2006) Pronounced supramolecular order in discotic donor–acceptor mixtures. *Angew. Chem., Int. Ed.*, **45**, 819–823.

154 Feng, X., Liu, M., Pisula, W., Takase, M., Li, J., and Müllen, Kl. (2008) *Adv. Mater.*, **20**, 2684–2689.

155 Räder, H.J., Rouhanipour, A., Talarico, A.M., Palermo, V., Samorì, P., and Müllen, K. (2006) Processing of giant graphene molecules by soft-landing mass spectrometry. *Nature Mater.*, **5**, 276–280.

15
Endohedrals

Lai Feng, Takeshi Akasaka, and Shigeru Nagase

15.1
Introduction

Endohedral metallofullerenes are created by trapping metal atoms or metal clusters into a fullerene cage, which naturally combines the properties of fullerenes and metals together. This novel hybrid molecule was first indicated by mass spectrometry in as early as 1985 [1]. Six years later, the successful synthesis and isolation of La@C$_{82}$ was reported by Smalley and coworkers [2]. In the following years, great efforts have been made for the synthesis of various endohedral metallofullerenes. The encapsulated species were found to cover group III metals and most lanthanide metals, as well as their nitride clusters and carbide clusters. Their fullerene cage size also ranges from C$_{60}$ to C$_{110}$, possibly even smaller or bigger. How these metallic species combine with fullerene cage has been of primary interest in the past decade. The encapsulated atoms or clusters were widely investigated by X-ray photoemission, photoenergy loss spectroscopy, as well as the theoretical calculations [3, 4]. All these studies revealed an electron transfer interaction between the fullerene cage and entrapped metal atoms or cluster. Thus, an onion-like model could be used to describe the electronic structure of metallofullerenes, in which the interior layer composed of metal atoms or cluster is positively charged and the exterior layer composed of the fullerene cage is negatively charged. This electron transfer was expected to stabilize not only the encapsulated species, but also the fullerene cage that can sometimes be instable in the empty form. This structural feature is in remarkable contrast with that of nonmetallic endohedral fullerenes, such as N@C$_{60}$ [5], P@C$_{60}$ [6, 7], He@C$_{60}$ [8], and recently synthesized H$_2$@C$_{60}$ [9, 10], in which the nonmetal atoms have very weak interactions with the fullerene cage. In this sense, the endohedral metallofullerenes are more likely a kind of chemically hybrid molecules, whereas the nonmetallic endohedral fullerenes are physically hybrid molecules.

Due to their unique and complex three-dimensional structures, the structural determination of endohedral metallofullerenes, including the structure of fullerene cage and the position or motion of encapsulated metallic species, has become one of the biggest challenges in fullerene chemistry. Generally, it is believed that the correct

Carbon Nanotubes and Related Structures. Edited by Dirk M. Guldi and Nazario Martín
Copyright © 2010 WILEY-VCH Verlag GmbH & Co. KGaA, Weinheim
ISBN: 978-3-527-32406-4

understanding of their structures might provide important clues disclosing the formation mechanism of endohedral metallofullerenes, and can therefore be helpful in improving their production yield. In recent years, high-resolution nuclear magnetic resonance (NMR), MEM/Rietveld method [11], and X-ray crystallographic method have been employed and developed for the structural determinations of metallofullerenes. As a result, many unconventional features of metallofullerenes have been revealed and clearly demonstrated. For example, some metallofullerenes, such as $La_2@C_{72}$ [12], $Sc_2@C_{66}$ [13], $Sc_3N@C_{68}$ [14], and $Tb_3N@C_{84}$ [15], were surprisingly found to violate the isolated pentagon rule (IPR).

On the other hand, as macroscopic quantities of endohedral metallofullerenes became available in recent years, the interest in their chemical properties has been rapidly aroused, which is actually inspired by their potential applications in material science. A large number of recent studies especially focus on their different chemical reactivities induced by encapsulated metallic species. These findings are also regarded as an important aspect of the metallofullerene-science, and will be discussed in detail in this chapter.

Therefore, in the present chapter, we emphasize on the recent progresses made in the field of endohedral metallofullerenes since 2002, which involve their synthesis, isolation, structural characterization, and chemical properties, as well as applications. Importantly, we attempt to understand their structural and chemical features that are caused by the different encapsulated metallic species.

15.2
Recent Investigations in the Synthesis of Endohedral Metallofullerenes

15.2.1
The Reactive Gas Atmosphere

Endohedral metallofullerenes are usually synthesized in a modified Krätschmer–Huffman generator during the vaporization of graphite rods containing metals or metal oxides in a helium atmosphere. This way, only the fullerenes with encapsulated metals or metal carbide can be obtained. In 1999, a small amount of N_2 was introduced into the Krätschmer–Huffman generator by Dorn group, and a class of novel endohedral fullerenes with an internal trimetallic nitride (TN) cluster ($Sc_3N@C_{2n}$) was successfully synthesized in high yield [16]. Since N_2 is actually the source of the nitride atom in trimetallic nitride fullerenes, it is regarded as a "reactive" gas, in contrast to the "nonreactive" helium atmosphere. Therefore, this method is named as "trimetallic nitride template" (TNT) process. Besides N_2, NH_3 can be alternatively introduced in the synthetic process, leading to even more predominant production of TN fullerenes. By employing the reactive gas atmosphere, Dunch group claimed that the relative yield of scandium TN fullerenes is higher than 95% as compared with the empty fullerenes and conventional metallofullerenes [17–20]. Further investigations confirmed that these reactive gases suppress the empty fullerene formation and provide the soot extracts with a high purity of TN fullerenes.

Very recently, air has been tentatively introduced in the fullerene synthesis process, which has led to the discovery of an endohedral fullerene with an entrapped tetrahedral metal oxide cluster ($Sc_4(\mu_3-O)_2@I_h-C_{80}$) [21]. Although air was previously known to hamper the formation of fullerenes, herein, an increase in the air flow up to 12 Torr/min improves the production of $Sc_4(\mu_3-O)_2@I_h-C_{80}$. As air is the source of the oxygen atom in C_{80} cage, it is also considered as a reactive gas.

15.2.2
The Solid Additive

The investigations in solid additive that is packed together with the metal oxide into the graphic rod were performed with the aim to increase the yield of metallofullerenes. Up to now, the reported solid additives mainly included transition metal species and their oxides or salts. Transition metals, such as Ni [22, 23], Cu [24], and Co or CoO [25, 26], cannot be encapsulated into the fullerene cage, but can dramatically increase the yields of metallofullerenes. Typically, addition of Ni is beneficial for the production of $M@C_{2n}$ (M = La [22], Tb [23]), while addition of Cu increases the yield of $Sc_3N@C_{2n}$ by a factor of ~5 [24]. Thus, these metallic solid additives work more like catalysts in the production of metallofullerenes.

As to another class of solid additive, such as CaNCN [27], Fe_xN [28] and Cu$(NO_3)_2·5H_2O$ [29], they are found to be even more efficient than those metals or metal oxide additives in selective synthesis of NT fullerenes. Their nitride ligand actually results in the generation of nitride gas under the arc plasma condition, while the metal ion probably works as a catalyst. Particularly, the effect of Cu$(NO_3)_2·5H_2O$ on the synthesis of $Sc_3N@C_{80}$ has been investigated systematically. The proposed mechanism of this modified synthesis process explains how the added Cu$(NO_3)_2·5H_2O$ chemically adjusts the temperature, energy, and reactivity of the plasma environment. In quantifying the fullerenes obtained by vaporizing the graphite rods with increasing Cu$(NO_3)_2·5H_2O$ contents, the contributions of Cu$(NO_3)_2·5H_2O$ to the suppression of empty fullerenes and the promotion of $Sc_3N@C_{80}$ are clearly revealed. Under the optimized condition, the produced soot exclusively contains nearly $Sc_3N@C_{80}$ and other insoluble composites. This study indicated that highly selective synthesis of specific metallofullerene might be possible and practicable.

15.3
Advances in Nonchromatographic Techniques for Separation of Endohedral Metallofullerenes

15.3.1
Separation by Electrochemical Method

The bulk production of metallofullerenes usually generates complex mixtures of empty fullerenes and endohedral metallofullerenes mixed together in a carbonaceous soot matrix. Among them, the species extractable by organic solvents are in small fraction, but even this part includes a great variety of fullerenes and metallofullerenes in numerous isomeric forms. High-performance liquid chromatography (HPLC) is the most widely used method for metallofullerene separation, which is based on the different molecular weights, shapes, and polarities. However, this method usually requires multiple steps and consumes a lot of time. A more facile and faster separation method is always desirable.

Recently, alternative separation methods based on the different electrochemical or chemical properties of metallofullerenes have been developed. As early as in 1998, Alford group reported the efficient enrichment of small-bandgap fullerenes, such as C_{74} and $Gd@C_{60}$, by combined sublimation, selective reduction, and reoxidation process [16]. Recently, Akasaka and coworkers reported a more complete separation of metallofullerenes by electrochemical method [30]. $La@C_{82}$ (C_{2v} and C_s isomers) and $La_2@C_{80}$ were selectively isolated in a few steps due to their much higher first reduction potentials with respect to other fullerenes and metallofullerenes. Although further HPLC purification is necessary, the separation time is greatly reduced and macroscale separations become possible.

15.3.2
Separation by Other Chemical Methods

$Sc_3N@C_{80}$ is the most abundantly yielded metallofullerene. Note that it possesses much lower chemical reactivities relative to empty fullerenes and conventional metallofullerenes. By taking advantage of its inertness in a Diels–Alder reaction, Dorn and coworkers separated $Sc_3N@C_{80}$ directly from the soot extract by cyclopentadiene-functionalized resins [31], which react with most fullerenes and metallofullerenes other than $Sc_3N@C_{80}$. Thus, when a fullerene mixture passed through such a resin packed column, $Sc_3N@C_{80}$ was eluted out, while others were absorbed by the resins. Under the thermal condition, these absorbed fullerenes could be released again due to the reversibility of the Diels–Alder cycloaddition. As most $M_3N@C_{80}$ type metallofullerenes have nearly identical chemical properties, this method has been likewise applied to the separations of $Lu_3N@C_{80}$, $Gd_3N@C_{80}$, and $Ho_3N@C_{80}$. Also, the I_h isomer of $Sc_3N@C_{80}$ was separated from its more reactive D_{5h} isomer by a similar method.

Recently, a more convenient separation method known as "stir and filter approach" (SAFA) was developed by Stenvenson *et al.* [32]. By stirring fullerene and NT fullerene mixture with cyclopentadienyl- or amino-modified silica, most fullerenes were absorbed on the silica, while NT fullerenes, such as $Sc_3N@C_{80}$, $Sc_3N@C_{78}$, and $Sc_3N@C_{68}$, were almost free of absorption and collected by filtration. As shown in Figure 15.1, with increase in the stirring time, the NT fullerenes were gradually enriched and even pure $Sc_3N@C_{80}$ was obtained finally. However, as this absorption process is irreversible, how to recover those absorbed fullerenes becomes an open question for further studies.

In addition, by chemical method, even an isomerical purification of metallofullerene can be achieved. Echegoyen and coworkers were the first to report such a

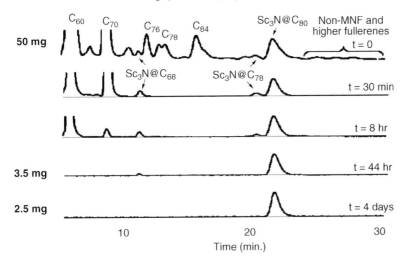

Figure 15.1 HPLC chromatograms for unreacted fullerene species in solution at various times for the reaction of diamino silica (12 mmol) with 50 mg of fullerene extract (~2.5 mmol). Chromatographic conditions are 0.8 ml/min toluene, PYE column, 360 nm UV detection, and 50 µl injection. Reproduced with permission from Ref. [32]. Copyright 2006 by The American Chemical Society.

separation for two $Sc_3N@C_{80}$ isomers ($Sc_3N@I_h\text{-}C_{80}$ and $Sc_3N@D_{5h}\text{-}C_{80}$) [33] as a prototype. As the D_{5h} isomer has 230 mV lower oxidation potential than that of the I_h isomer, an oxidant, tris(p-bromophenyl)-aminium hexachloroantimonate (TPBAH) having an oxidation potential between the first oxidations of the I_h and D_{5h} isomers, was chosen to selectively oxidize D_{5h} isomer, while keeping the I_h isomer intact. The final separation was achieved on a silica chromatographic column, where the cation of $Sc_3N@D_{5h}\text{-}C_{80}$ was absorbed, and the pure I_h isomer was collected. As the two isomers have similar retention times on HPLC columns, this redox-based method is obviously more efficient and allows macroquantity separation in comparatively shorter time.

As for conventional metallofullerenes, they usually have higher reactivities than those of empty fullerenes and NT fullerenes. It appears that an adverse strategy should be employed for their separations. It has been reported that azacrown selectively formed a complex with $La@C_{82}$ and $La_2@C_{80}$, which is insoluble in toluene [34] and can be separated from other fullerenes by filtration. $La@C_{82}$ and $La_2@C_{80}$ were finally extracted by CS_2 and collected. It is noteworthy that this separation also involves an unusual interaction between the aza compound and fullerenes, but largely weaker and reversible, as compared with the above-mentioned interaction between amino-modified silica and fullerenes. Likewise, $La@C_{82}$ and $La_2@C_{80}$ are even selectively reduced by DMF or its thermolysis derivative. By addition of $n\text{-}Bu_4NClO_4$, the stable anions of those lanthanum metallofullerenes were obtained, and then extracted by acetone/CS_2 mixture [35]. The subsequent reoxidation by a weak acid, such as dichloroacetic acid, afforded neutral metallofullerenes, which are soluble in toluene and subjected to further HPLC separation.

15.4
Structures of Endohedral Metallofullerenes Determined by X-Ray Crystallographic Method

In recent years, X-ray crystallography has been widely used in determining the three-dimensional structures of endohedral metallofullerenes, including the structure of fullerene cage, the localization of encapsulated metals or clusters, and the metal–cage bonding interaction. Comparatively, the structural determinations based on NMR spectroscopy and MEM/Rietveld method frequently appear to be incomplete and ambiguous. Thus, especially in the last few years, the structures of either newly discovered or previously reported metallofullerenes were mainly determined or updated by X-ray crystallographic method.

15.4.1
Monometallofullerenes

In monometallofullerenes ($M@C_{2n}$), the encapsulated metal ions can mostly be either trivalent or divalent. Earlier studies showed that most lanthanide metals preferably transfer three electrons to the fullerene cage, affording a class of trivalent monometallofullerenes with an open-shell structure. Their electronic structures were commonly described as $M^{3+}@C_{2n}^{3-}$. Among them, $M@C_{82}$ is the most abundantly yielded species, which sometimes possesses two isomers and has been extensively investigated by both experimental and theoretical tools. Due to the paramagnetic property, their structural determinations were based on NMR spectroscopic studies of their anions, such as $[M@C_{2v}-C_{82}]^-$ (M = La [36], Y [37], Ce [38]) and $[La@C_s-C_{82}]^-$ [39]. These studies indicated that the two isomers of $M@C_{82}$ have C_{2v} and C_s symmetries, respectively. The first definite structure of $La@C_{2v}-C_{82}$ was modeled by X-ray crystallographic method [40]. As shown in Figure 15.4, the structure of $La@C_{2v}-C_{82}$ is actually presented as its carbene derivative ($La@C_{2v}-C_{82}(Ad)$, Ad = adamantylidene). Due to the spherical shape of fullerene cage and the random motion of encapsulated La atom, the resulted multiple disorders hamper either the crystallization of pristine metallofullerene or the resolution of its X-ray structure. Nevertheless, when the $La@C_{82}$ molecule is covalently "captured" by adamantylidene, it is facile to pack its derivative in order. Thus, $La@C_{2v}-C_{82}(Ad)$ exhibits only two disorders, originating from its own chirality. As shown in Figure 15.2a, the crystallograpically modeled $C_{2v}-C_{82}$ cage for $La@C_{82}$ agrees well with the previous NMR spectroscopic studies. The encapsulated La atom is localized in an off-center position along the C_2 axis of $C_{2v}-C_{82}$ cage and close to a six-membered ring. The C1–C2 distance is 2.097 Å, indicating a broken [6,6] bond. The distances of La-C1 and La-C2 are 2.658 and 2.634 Å, respectively, suggesting the strong La-cage bonding. Comparatively, in case of carbene derivative of $Gd@C_{2v}-C_{82}$ [41] and $Y@C_{2v}-C_{82}$ [42], both their X-ray structures resemble that of $La@C_{2v}-C_{82}(Ad)$, which agrees well with the DFT calculated common structure of $M^{3+}@C_{82}^{3-}$ [43]. All these results suggest that the trivalent lanthanide ions, irrespective of whether they have f electrons or not, might possess almost the same position in $C_{2v}-C_{82}$ cage. However, it

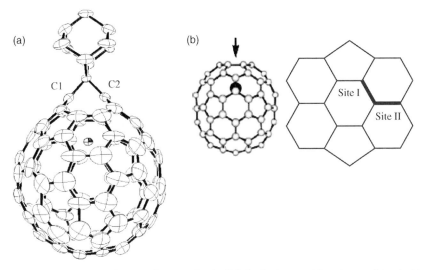

Figure 15.2 A ORTEP drawing of (a) La@C_{2v}-C_{82}(Ad) (major isomer, Ad = adamantylidene) with thermal ellipsoids shown at the 50% probability level and (b) the illustration of addition sites.

is noteworthy that, in the previously presented structure of Gd@C_{2v}-C_{82} [44] and Eu@C_{2v}-C_{82} [45] by MEM/Rietveld analysis, Gd^{3+} and Eu^{3+} are located near the C$-$C double bond on the opposite side of C_{2v}-C_{82} cage along the C_2 axis.

La@C_{72} and La@C_{74} are the so-called "missing" fullerenes because they were previously detected only by mass spectrometry and were never isolated. Recently, La@C_{72} and La@C_{74} have been isolated in the form of their derivatives by HPLC and characterized. Their X-ray structures (Figure 15.3a and b) definitely show a non-IPR C_2-C_{72} cage for La@C_{72} [46] and an IPR D_{3h}-C_{74} cage for La@C_{74} [47]. In La@C_2-C_{72}, the La^{3+} ion is localized close to the fused [5,5] junction, thus indicating their strong interaction. The distances between La and two carbons of the [5,5] junction are 2.615 and 2.606 Å, respectively. These are somewhat lower than the calculated values of 2.714 and 2.680 Å. In case of La@C_{3h}-C_{74}, the encapsulated La atom is mainly localized at a site near the dichlorophenyl group, which is a little deviated from the calculated optimal site along the C_2 axis on the I_h plane. This site shift might have been caused by the introduction of the dichlorophenyl group.

As compared with the popularity of trivalent monometallofullerenes, only a few divalent lanthanide ions, such as Tm^{2+}, Yb^{2+}, Sm^{2+}, and some divalent alkaline earth *metals, such as* Ba^{2+} *and* Ca^{2+}, *can be encapsulated into fullerene cages. Due to two-electron transfer between fullerene cage and divalent encapsulated metal,* M^{2+}@C_{2n}^{2-} *has a closed-shell structure, which enables them to possess distinct properties from those of trivalent monometallofullerenes and empty analogues. For instance,* C_{74} *and* M^{3+}@C_{74}^{3-} *are either insoluble in most organic solvents or instable under ambient conditions due to their predicated small HOMO–LUMO gaps. However, the divalent monometallofullerenes, including* Ca@C_{74}, Ba@C_{74}, *and* Yb@C_{74}, *have good solubilities and stabilities.* Ba@C_{74} *has been successfully isolated and characterized as* Ba@D_{3h}-C_{74} *via an X-ray crystallographic method* [48, 49]. *As* Ba@C_{74} *and* La@C_{74} *share the same*

(a)　　　　　　　　　　　　　　(b)

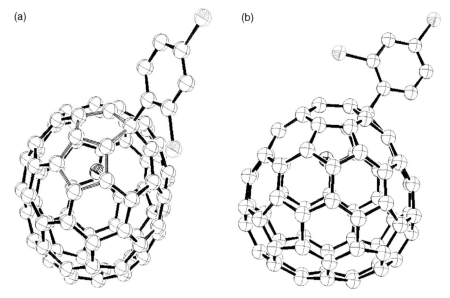

Figure 15.3 ORTEP drawings of (a) La@C_2-C_{72}($C_6H_3Cl_2$) and (b) La@D_{3h}-C_{74}($C_6H_3Cl_2$) with thermal ellipsoids shown at the 50% probability level. Solvent molecules and H atoms are omitted for clarity. The fused pentagon rings are labeled by gray color.

D_{3h}-C_{74}, it implies that both [D_{3h}-C_{74}]$^{2-}$ and [D_{3h}-C_{74}]$^{3-}$ are highly stabilized. The encapsulated Ba^{2+} ion was found to be localized in two split positions with occupancies of 0.63 and 0.37, respectively. Theoretical calculations have indicated that the most favorable endohedral site for the Ba atom is off-center under a [6,6] double bond along the C_2 axis on the σh plane of D_{3h}-C_{74} [50], resembling the optimal position of La atom in La@D_{3h}-C_{74} [51]. However, the observed two positions for Ba atom are both a little deviated from the calculated one, which is attributed to the interaction between Ba^{2+} and the cocrystallized Co(OEP) molecule.

15.4.2
Dimetallofullerenes

Most fullerene cages are large enough to encapsulate more than one metal atom. Up to now, several dimetallofullerenes have been reported [13, 52–57]. The motion of the two metals and the metal–cage interaction have been of great interest. A lot of theoretical calculations and spectroscopic studies have been carried out to reveal their possible geometrical and electronic structures. Perhaps due to their low production yields, only a few dimetallofullerenes have been studied by X-ray crystallographic method. La$_2$@C_{72} has been predicted to have a non-IPR structure as La$_2$@D_2(10611)-C_{72} by combined theoretical calculations and NMR spectroscopic studies [12]. The recent crystallographic analysis on its carbene derivative definitely confirmed the previously predicted structure [58]. As shown in Figure 15.4, there are two pairs of fused pentagons on the two poles of D_2-C_{72} cage. The two encapsulated La atoms are

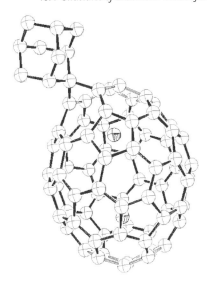

Figure 15.4 A ORTEP drawing of $La_2@D_2\text{-}C_{72}(Ad)$ with thermal ellipsoids shown at the 50% probability level. Solvent molecules and H atoms are omitted for clarity. The fused pentagon rings are labeled by gray color.

highly localized and close to the two [5,5] junctions, respectively. The shortest La–C distances (2.495 Å and 2.501 Å) also involve the carbons on the fused [5,5] junctions, indicating the strong associations between La^{3+} ions and their adjacent pentagon pairs. Likewise, $M_2@I_h\text{-}C_{80}$ (M = La, Ce) [59] and $La_2@D_{3h}\text{-}C_{78}$ [60] were also characterized in the form of their carbene derivatives. In either case, a highly localized and widely separated metallic ion pair was observed.

On the other hand, $Er_2@C_s\text{-}C_{82}$ [61] and $Er_2@C_{3v}\text{-}C_{82}$ [62] were cocrystallized with Co(OEP), respectively. The encapsulated Er_2 was presented to be in random motion. As shown in Figure 15.5, in fullerene cage of $Er_2@C_s\text{-}C_{82}$, Er_2 possesses multiple

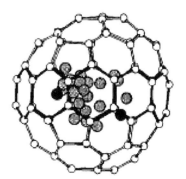 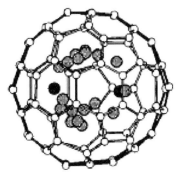

Figure 15.5 Two orthogonal views of the $Er_2@C_s\text{-}C_{82}$ molecule with all 23 erbium sites shown as hatched circles or solid circles for the two most highly occupied sites. Solid lines denote the band of 10 contiguous hexagons. Reproduced with permission from Ref. [61]. Copyright 2002 by The American Chemical Society.

positions with occupancies ranging from 0.35 to 0.01. These positions are along a band of 10 contiguous hexagons of the C_s-C_{82} cages. The Er_2 with major occupancy is close to a [6,6] junction and a [5,6] junction with Er–C distance ranging from 2.30 to 2.39 Å. Another isomer, $Er_2@C_{3v}$-C_{82} has a similar structural feature. It appears that the encapsulated Er^{3+} ions have almost equivalent associations with [6,6] junction or [5,6] junction.

Very recently, Dorn and coworkers reported a special group of dimetallofullerenes, $M_2@C_{79}N$ (M = Y, Tb), synthesized under reactive atmosphere [63]. Unlike those TN fullerenes, N atom was trapped in the fullerene cage, forming metallo-azafullerenes. $Tb_2@C_{79}N$ was cocrystallized with Ni(OEP) and studied by X-ray crystallographic method. Without taking account of the N-atom-induced dissymmetry, $C_{79}N$ cage would have I_h symmetry as that of $La_2@I_h$-C_{80}. In fact, the site of the N heteroatom definitely cannot be determined due to its serious disorders.

15.4.3
Metallic Carbide Fullerenes and Metallic Oxide Fullerenes

Since its first isolation and characterization by MEM/Rietveld analysis [64], metallic carbide fullerene ($Sc_2C_2@C_{84}$) has attracted considerable interest. By recent studies, another four metallic carbide fullerenes, including $Sc_2C_2@C_{3v}$-C_{82} [65–68], $Y_2C_2@C_{3v}$-C_{82} [68, 69], $Sc_3C_2@I_h$-C_{80} [67, 70, 71], and $Sc_2C_2@C_{2v}(6073)$-C_{68} [72], have been discovered and characterized. It is noteworthy that $Sc_3C_2@I_h$-C_{80} had been previously characterized as $Sc_3@C_{82}$ by MEM/Rietveld method [73], and has been recently revised by both X-ray crystallographic method [70] and improved MEM/Rietveld analysis [71]. The revised structure agrees well with the NMR spectroscopic studies on Sc_3C_{82} anion [70]. The structure of $Sc_2C_2@C_{3v}$-C_{82} determined by either MEM/Rietveld analysis [68] or X-ray crystallography [65] agrees well with each other. The Sc_2 pair disordered over several positions even in its carbene derivative and under low temperature (90 K) indicates its high mobility. Sc_2C_2 cluster shows a bent structure in its major form (Figure 15.6), which is consistent with its DFT-optimized structure [74]. Similarly, a bent structure of Y_2C_2 in $Y_2C_2@C_{3v}$-C_{82} was also suggested by MEM/Rietveld analysis [68].

In a very recent report, four Sc atoms were found to be trapped inside I_h-C_{80} cage together with two O atoms, forming a novel metallofullerene $Sc_4(\mu_3\text{-}O)_2@I_h$-$C_{80}$ [21]. As shown by its X-ray structure, the two O atoms are localized on two of the triangular faces of the Sc_4 tetrahedron. Like encapsulated metallic carbide, $Sc_4(\mu_3\text{-}O)_2$ was also disordered over several orientations. The Sc–O distances range from 1.964(14) to 2.11(2) Å, which are within the range of common Sc–O bond length (1.87–2.29 Å) for inorganic complex. Thus, it is believed that the two oxygen atoms are involved in organizing the four Sc atoms, whereas much longer O–C distances (2.562–2.721 Å) indicate very weak O–cage interaction.

All these studies suggest that two or more than two metals can be trapped inside a fullerene cage in form of metallic carbide or metallic oxide, yielding stable and energetically favored endohedral metallofullerenes. Since no "pure" metallic cluster fullerenes have been experimentally observed by X-ray crystallographic method,

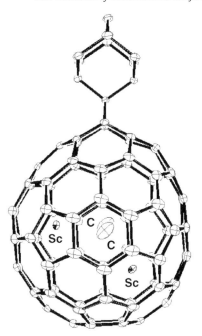

Figure 15.6 A ORTEP drawing of $Sc_2C_2@C_{3v}$-C_{82}(Ad) with thermal ellipsoids shown at the 50% probability level. Solvent molecules and H atoms are omitted for clarity.

there appears an open question about the existence of such metallofullerenes that trap more than two metal atoms in the form of metallic cluster other than the forms of metallic carbide, metallic oxide, or metallic nitride as mentioned in the following section.

15.4.4
Trimetallic Nitride Fullerenes

As the most abundantly yielded metallofullerene species, $Sc_3N@C_{80}$ was first structurally determined by the X-ray crystallographic method [16]. The encapsulated Sc_3N cluster was found to have a planar structure in an I_h-C_{80} cage. Up to now, several other TN fullerenes have been characterized likewise. The most widely studied type is $M_3N@I_h$-C_{80} (or $MM'_2N@I_h$-C_{80}), in which the encapsulated cluster is either homometallic nitride cluster or heterometallic nitride cluster such as $CeSc_2N$ [75], $ErSc_2N$ [25], Gd_2ScN [76], $GdSc_2N$ [76], $TbSc_2N$ [76], and even $ScYErN@C_{80}$ [77] with three different metals. Partial information on the X-ray structures of those homo-TN fullerenes is summarized in Table 15.1. In terms of the increasing ionic radius of M^{3+}, the structures of M_3N vary from planar Sc_3N to significantly pyramidal Gd_3N, in which the nitride ion is 0.522(8) Å out of the plane of three Gd^{3+} ions and the sum of Gd–N angle of Gd_3N as 341.6°, as shown in Figure 15.7a. Such pyramidalization of Gd_3N produces a high strain in its structure, which is energetically unfavorable, and

Table 15.1 Selected crystallographic data for M$_3$N@I$_h$-C$_{80}$, including bond length (Å) and angles (°).

	Sc$_3$N@I$_h$-C$_{80}$·5o-xylene [79]	Lu$_3$N@I$_h$-C$_{80}$·5o-xylene [79]	Tm$_3$N@I$_h$-C$_{80}$·Ni(OEP)·2benzene [80]	Y$_3$N@I$_h$-C$_{80}$(C$_4$H$_9$N)·2.5CS$_2$ [81]	Dy$_3$N@I$_h$-C$_{80}$·Ni(OEP)·2benzene [82]	Tb$_3$N@I$_h$-C$_{80}$·Ni(OEP)·2benzene [28]	Gd$_3$N@I$_h$-C$_{80}$·Ni(OEP)·1.5benzene [83]
R(M^{3+}) [84]	0.75	0.85	0.87	0.90	0.91	0.92	0.94
T (K)	90(2)	90(2)	90(2)	90(2)	100(2)	90(2)	—
M–N	1.9931(14) Sc$_1$	2.001(3) Lu$_1$	2.032(6) Tm$_1$	2.067(5) Y$_1$	2.004(8) Dy$_1$	2.056(4) Tb$_1$	2.038(8) Gd$_1$
M–N	2.0323(16) Sc$_2$	2.0819(8) Lu$_2$	2.020(6) Tm$_2$	2.051(5) Y$_2$	2.068(6) Dy$_2$	2.089(4) Tb$_2$	2.085(4) Gd$_2$
M–N	2.0526(14) Sc$_3$	2.0592(10) Lu$_3$	2.058(6) Tm$_3$	2.022(4) Y$_3$	2.044(7) Dy$_3$	2.077(4) Tb$_3$	2.117(5) Gd$_3$
M	2.205(4)–2.498(4) Sc$_1$	2.220(7)–2.577(6) Lu$_1$	2.349(10)–2.539(10) Tm$_1$	2.375(6)–2.538(6) Y$_1$	2.414(12)–2.461(10) Dy$_1$	2.423(3)–2.497(3) Tb$_1$	2.344(13)–2.600(14) Gd$_1$
M	2.145(4)–2.367(4) Sc$_2$	2.112(7)–2.496(7) Lu$_2$	2.370(9)–2.528(9) Tm$_2$	2.354(9)–2.548(6) Y$_2$	2.371(10)–2.457(9) Dy$_2$	2.434(3)–2.472(3) Tb$_2$	2.402(10)–2.482(12) Gd$_2$
M–C	2.147(4)–2.302(4) Sc$_3$	2.165(7)–2.478(7) Lu$_3$	2.357(9)–2.487(9) Tm$_3$	2.312(6)–2.451(6) Y$_3$	2.391(10)–2.424(10) Dy$_3$	2.404(3)–2.518(3) Tb$_3$	2.439(12)–2.481(18) Gd$_3$
M–N–M	120.70(7) Sc$_1$-N-Sc$_2$	Lu$_1$-N-Lu$_2$	120.6(3) Tm$_1$-N-Tm$_2$	116.9(2) Y$_1$-N-Y$_2$	122.2(3) Dy$_1$-N-Dy$_2$	116.78(18) Tb$_1$-N-Tb$_2$	111.6(2) Gd$_1$-N-Gd$_2$
M–N–M	119.21(6) Sc$_2$-N-Sc$_3$	Lu$_2$-N-Lu$_3$	117.3(3) Tm$_2$-N-Tm$_3$	122.1(2) Y$_2$-N-Y$_3$	115.5(4) Dy$_2$-N-Dy$_3$	111.63(18) Tb$_2$-N-Tb$_3$	110.3(3) Gd$_2$-N-Gd$_3$
M–N–M	118.47(7) Sc$_1$-N-Sc$_3$	121.51(11) Lu$_1$-N-Lu$_3$	122.1(3) Tm$_1$-N-Tm$_3$	119.8(2) Y$_1$-N-Y$_3$	122.0(3) Dy$_1$-N-Dy$_3$	117.65(19) Tb$_1$-N-Tb$_3$	119.7(3) Gd$_1$-N-Gd$_3$
Σ(M–N–M)	358.37	—	360.0	358.8	359.8	346.06	341.6

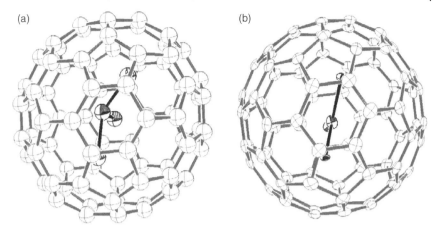

Figure 15.7 ORTEP drawings of (a) $Gd_3N@I_h$-C_{80} and (b) $Gd_2ScN@I_h$-C_{80} with thermal ellipsoids shown at the 50% probability level. Ni(OEP) and solvent molecules are omitted for clarity.

hence $Gd_3N@I_h$-C_{80} has extremely low production yield. If one of Gd^{3+} ion is replaced by Sc^{3+}, forming $Gd_2ScN@I_h$-C_{80}, the internal cluster is relaxed to a planar form (Figure 15.7b) [76]. Also, the yield of $Gd_2ScN@I_h$-C_{80} is considerably higher than that of $Gd_3N@I_h$-C_{80}. In addition, as the internal M^{3+} ions become larger, the N–M and M–C distances of $M_3N@I_h$-C_{80} are gradually extended, indicating the larger M_3N cluster induced a "punch out" effect on the fullerene cage. Note that the carbons adjacent to larger M^{3+} ions show higher pyramidalization angles (θ_p [78]), which have been proposed to be crucial for the higher chemical reactivities of $M_3N@I_h$-C_{80} with larger internal M^{3+} ions.

Recently, as higher TN fullerenes, such as $Gd_3N@C_s(39\ 663)$-C_{82} [85], $M_3N@C_s(51\ 365)$-C_{84} (M = Tm [86], Tb [15], Gd [86]), $Tb_3N@D_3$-C_{86} [28], and

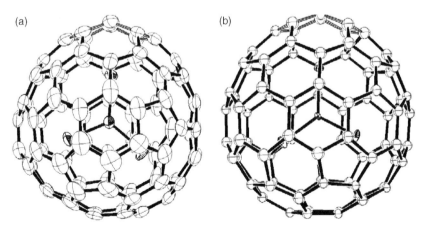

Figure 15.8 ORTEP drawings of (a) $Gd_3N@C_s(39\ 663)$-C_{82} and (b) $Tb_3N@C_s(51\ 365)$-C_{84} with thermal ellipsoids shown at the 50% probability level. Ni(OEP) and solvent molecules are omitted for clarity. The fused pentagon rings are labeled by gray color.

$Tb_3N@D_2\text{-}C_{88}$ [28], have been characterized crystallograpically, it is interesting to find that, unlike $M_3N@I_h\text{-}C_{80}$ (M = Tb, Gd), both Tb_3N and Gd_3N have flattened structures as well as extended M–N and M–C distances in these higher fullerene cages. Note that such high fullerene cages cannot efficiently entrap Sc_3N cluster, as no higher $Sc_3N@C_{2n}$ ($2n > 80$) has been reported so far. The theoretical calculations also suggest that Sc_3N cannot effectively interact with cages larger than C_{80}. On the other hand, in those smaller TN fullerenes, including $Sc_3N@D_3(6140)\text{-}C_{68}$ [14] and $Sc_3N@D_{3h}\text{-}C_{78}$ [26], Sc_3N unit retains its planar geometry, even though the shortest Sc–N bond (1.961 Å) in $Sc_3N@D_3(6140)\text{-}C_{68}$ is much shorter than the typical optimum Sc–N bond length of about 2.05 Å [87]. Up to now, the reported biggest TN cluster that can be trapped inside C_{68} cage is Lu_2Sc [88]. A slightly bigger cluster Y_3N in DFT-optimized structure of $Y_3N@D_3(6140)\text{-}C_{68}$ is forced to be even more pyromidal than Gd_3N in $Gd_3N@I_h\text{-}C_{80}$, which makes $Y_3N@D_3(6140)\text{-}C_{68}$ energetically unfavorable [87]. All these studies clearly revealed a crossing selectivity between endohedral metallic species and fullerene cages.

In addition, among the above-mentioned TN fullerenes, several species, including $Sc_3N@D_3(6140)\text{-}C_{68}$ [14], $Gd_3N@C_s(39\ 663)\text{-}C_{82}$ [85], and $M_3N@C_s(51\ 365)\text{-}C_{84}$ (M = Tm [86], Tb [15], Gd [86]), do not obey isolated pentagon rule. They either have three or one fused pentagon pair. Notably, for $Sc_3N@D_3(6140)\text{-}C_{68}$, the three fused pentagon pairs are symmetric to the D_3 axis, and the three Sc atoms of Sc_3N are adjacent to the three [5,5] junctions, respectively [14]. As for $Gd_3N@C_s(39\ 663)\text{-}C_{82}$ and $M_3N@C_s(51\ 365)\text{-}C_{84}$ (M = Tm, Tb, Gd), both of them have only one fused pentagon pair on the pole of their egg-shaped fullerene cage (Figure 15.8). Likewise, in each molecule, there is one metal atom of M_3N approaching to the only [5,5] bond. Thus, in all these non-IPR NT fullerenes, the shortest M–C separations always involve the carbon atoms at [5,5] junctions, which highlights the significant interaction between the metal ions and the fused pentagon pairs. Theoretical calculations confirmed that the charge transfer from the M_3N cluster to the fullerene cage is mainly localized in the fused pentagon pairs. Therefore, the antiaromatic fused pentagon moiety becomes more aromatic [89]. In this sense, the encapsulated metals seem to stabilize the non-IPR fullerene cage to some extent, which makes non-IPR structures possible and even relatively common in endohedral metallofullerene family.

15.5
Electrochemical Properties of Endohedral Metallofullerenes

The electrochemical properties of endohedral metallofullerenes are frequently varied with encapsulated metals or clusters. $Sc_3N@I_h\text{-}C_{80}$ and $La_2@I_h\text{-}C_{80}$ are two of the widely studied metallofullerenes. Both Sc_3N and La_2 donate six electrons to the $I_h\text{-}C_{80}$ cage. Their electronic structures are formally described as $(Sc_3N)^{6+}@(I_h\text{-}C_{80})^{6-}$ and $(La_2)^{6+}@(I_h\text{-}C_{80})^{6-}$, respectively. Nevertheless, these two metallofullerenes exhibit different electrochemical properties. The first reduction and oxidation potentials of $La_2@I_h\text{-}C_{80}$ are 960 mV higher and only 10 mV lower than those of $Sc_3N@I_h\text{-}C_{80}$

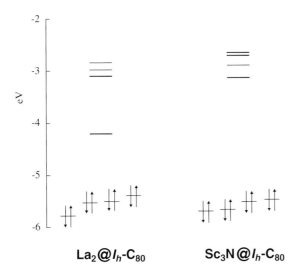

Figure 15.9 The molecular orbital diagrams of $La_2@I_h$-C_{80} and $Sc_3N@I_h$-C_{80}.

(Table 15.2). Theoretical calculations have been performed to investigate these differences. As shown in Figure 15.9, the LUMO of $La_2@I_h$-C_{80} has much lower energy level with respect to that of $Sc_3N@I_h$-C_{80}, while its HOMO possesses a comparable energy level [90]. This result suggests that $La_2@I_h$-C_{80} is much easier to be reduced relative to $Sc_3N@I_h$-C_{80}, which is consistent with its experimentally observed redox potentials. As for $M_3N@I_h$-C_{80} (M = Sc, Lu, Tm) with various internal ions, their redox behaviors are similar, which consists of two irreversible reduction steps and one reversible oxidation step in the field from 1.0 to -2.0 V (versus Fc^+/Fc). Nevertheless, their reduction potentials are slightly negatively shifted as the Pauling electronegativity of internal ion decreases from 1.36 for Sc^{3+} to 1.27 for Lu^{3+} and 1.25 for Tm^{3+} (Table 15.2). $M_2@I_h$-C_{80} (M = La, Ce) both having reversible redox steps.

In addition, on comparing $Sc_3N@I_h$-C_{80} with the D_{5h} isomer, a 230 mV difference in their first oxidation potentials was observed. Also, $La_2@I_h$-C_{80} has 340 mV higher first oxidation potential with respect to the D_{5h} isomer. Taking into account, the correlation between the first oxidation potential and the HOMO of molecule, it seems to be reasonable that the HOMO is varied with different molecular symmetries.

Recently, some investigations also focused on the correlation between the metallofullerene cage size and their electrochemical properties, which involved $Yb@C_{2n}$ ($n = 37$–42) [93] and $Gd_3N@C_{2n}$ ($n = 40$, 42, and 44) [94]. For the latter series, the first oxidation potentials of $Gd_3N@C_{2n}$ decreased with increasing cage size, going from 0.58 V for $Gd_3N@C_{80}$ to 0.32 V for $Gd_3N@C_{84}$ and 0.06 V for $Gd_3N@C_{88}$, whereas their reduction potentials were less affected. According to the authors, this correlation will make sense, if the oxidation is based on the cage and the reduction is based on the encapsulated cluster, which is the same for all the three molecules.

Table 15.2 Redox potentials (V, versus Fc$^+$/Fc).

	Sc$_3$N@I$_h$-C$_{80}$ [91]	Sc$_3$N@D$_{5h}$-C$_{80}$ [91]	Lu$_3$N@I$_h$-C$_{80}$ [91]	Lu$_3$N@D$_{5h}$-C$_{80}$ [91]	Tm$_3$N@I$_h$-C$_{80}$ [80]	Tm$_3$N@D$_{5h}$-C$_{80}$ [80]	La$_2$@I$_h$-C$_{80}$ [90]	La$_2$@D$_{5h}$-C$_{80}$ [90]	Ce$_2$@I$_h$-C$_{80}$ [90]	Ce$_2$@D$_{5h}$-C$_{80}$ [90]
$^{ox}E_1$	0.57	0.34	0.64	0.45	0.65	0.39	0.56	0.22	0.57	0.20
$^{red}E_1$	−1.27	−1.33	−1.40	−1.41	−1.43	−1.45	−0.31	−0.36	−0.39	−0.40
M	Sc		Lu		Tm		La		Ce	
$\chi^{a)}$	1.36		1.27		1.25		1.1		1.12	

a) Pauling electronegativity [92].

15.6
Chemical Reactivities of Endohedral Metallofullerenes

Studies on chemical reactivities of endohedral metallofullerenes have been hampered due to their poor availability. Early studies were usually performed on submilligram quantities of metallofullerenes, which made the understanding of their chemical properties incomplete and ambiguous. Recently, as the production of metallofullerenes increased rapidly, detailed characterizations for chemical reactions of metallofullerenes became possible. Thus, the metallofullerene chemistry has been gradually established.

15.6.1
Reductions and Oxidations

Due to the odd-number electron transfer from encapsulated metal to fullerene cage, trivalent monometallofullerenes ($M^{3+}@C_{2n}^{3-}$) have an unpaired electron on their HOMO. Losing or gaining one electron leads to their closed-shell structures. All the reported anions, including $[M@C_{2v}\text{-}C_{82}]^-$ (M = La [36], Y [37], Ce [38]) and [La@C_s-$C_{82}]^-$ [39], were synthesized by potential-controlled reduction process. They show diamagnetic properties and extraordinary stabilities even under ambient condition, which make them suitable for NMR spectroscopic studies. Especially, these anions can be facilely generated by alternative ways. The reductions even possibly occur in some solvents, such as DMF and pyridine [95]. In another report, azacrown [34] or unsaturated thiacrown [96], having a proper size, was observed to form 1: 1 complex with La@C_{2v}-C_{82}, in which La@C_{2v}-C_{82} accepted one electron and converted it to anion. It was believed that the guest and host molecular interaction induced the proximity of La@C_{82}, and those crown ethers facilitated the charge transfer process.

$M_3N@C_{2n}$ has a closed-shell structure, while their anions and cations usually have open-shell structures and lower stabilities. $[Sc_3N@C_{68}]^+$ is the first electrosynthesized cation, which was characterized by *in situ* ESR and absorption spectroscopic studies [97]. The 22 lines in its ESR spectrum originate from three equivalent Sc hyperfine splittings of $1.289 \times g$. No observable N hyperfine splitting was detected under the same experimental condition.

15.6.2
Cycloadditions

15.6.2.1 Diels–Alder Reaction
C_{60} has two types of bonds: [6,6] bond and [5,6] bond (type A and type D in Figure 15.10), but only pyracyclene-type carbons. Its [6,6] bonds are dienophilic, which enables the molecule to undergo various Diels–Alder reactions ([4 + 2] cycloadditions). However, $Sc_3N@I_h$-C_{80} consists of two types of carbon atoms: a pyrene-type carbon (intersection of three six-membered rings) and a corannulene-type carbon (intersection of a five-membered ring and two six-membered rings), thus affording [5,6] bond (type D) and another type of [6,6] bond (type B in Figure 15.10). In

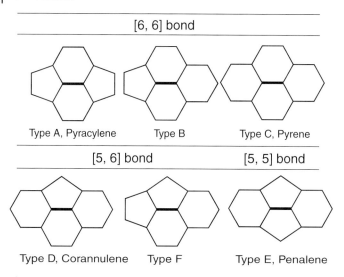

[6, 6] bond

Type A, Pyracylene   Type B   Type C, Pyrene

[5, 6] bond   **[5, 5] bond**

Type D, Corannulene   Type F   Type E, Penalene

Figure 15.10 Representation of all possible [6, 6], [5, 6], and [5, 5] bond types that may be present in a fullerene structure.

2002, Dorn and coworkers were the first to report the Diels–Alder reaction of $Sc_3N@I_h$-C_{80} [98]. The addition site was found to be a [5,6] bond instead of [6,6] bond. This distinct regioselectivity is proposed to have a relationship with the lower pentagon angle strain resulting from the addition to the corannulene-type site.

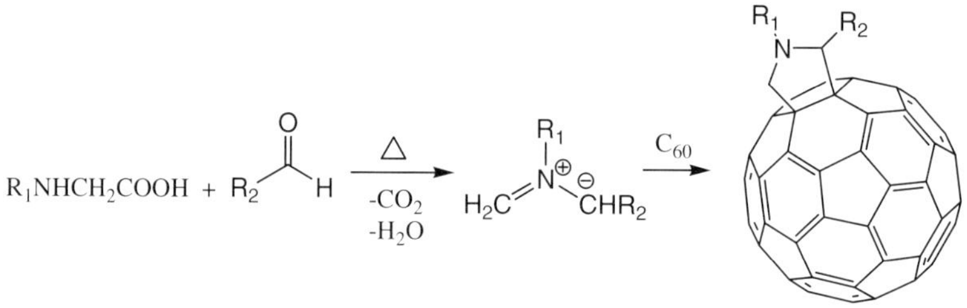

$$R_1NHCH_2COOH + R_2CHO \xrightarrow[\substack{-CO_2 \\ -H_2O}]{\triangle} \underset{H_2C}{\overset{R_1}{N^{\oplus}}}\overset{\ominus}{CHR_2} \xrightarrow{C_{60}}$$

Scheme 15.1 Prato reaction.

15.6.2.2 Prato Reactions

Prato reaction is the reaction between fullerene and azomethine yilde with 1,3-dipole character (Scheme 15.1). Azomethine ylides can be generated *in situ* from various readily accessible chemicals. The great popularity of this reaction in fullerene chemistry is due to its good selectivity on [6,6] bond and its general tolerance with a wide range of functional groups.

The reactivities of metallofullerenes toward azomethine yilde have been studied recently [99–101]. Prato reactions involving $M_3N@I_h$-C_{80} have attracted much interest. It was found that the addition of *N*-alkylazomethine occurred on the [5,6] bond of $Sc_3N@I_h$-C_{80} [102, 103], but mainly on the [6,6] bond of $Y_3N@I_h$-C_{80} [81] or $Gd_3N@I_h$-C_{80} [104, 105]. Further studies suggested that the [6,6]-

pyrrolidine adduct of $Y_3N@I_h$-C_{80} could thermally be isomerized to a [5,6]-adduct, while [5,6]-adduct of $Sc_3N@I_h$-C_{80} and [6,6]-adduct of $Gd_3N@I_h$-C_{80} remain unchanged. Further theoretical studies also confirmed their different stabilities. The DFT-calculated formation energy of the [5,6]-adduct of $Sc_3N@I_h$-C_{80} is 11.7 kcal/mol lower than that of the [6,6]-adduct, whereas, in case of $Gd_3N@I_h$-C_{80}, the formation energy of its [5,6]-adduct is 0.4 kcal/mol higher than that of its [6,6]-adduct [105]. Thus, it appears that [5,6]-adducts of $M_3N@I_h$-C_{80} (M = Sc, Y) are thermodynamically favored products and [6,6]-adducts of $Y_3N@I_h$-C_{80} are the kinetically favored ones. Whereas [6,6]-adducts of $Gd_3N@I_h$-C_{80} are both thermodynamically and kinetically favored products. Such different regioselectivities and stabilities were proposed to be attributed to their different internal metal ions, which were considered as the feature of metallofullerene chemistry.

Note that, in the reaction of $Sc_3N@I_h$-C_{80} with *N*-alkylazomethine yilde, only a thermodynamically favored [5,6]-adduct was observed. However, by employing less reactive 1,3-dipolar yilde (*N*-tritylazomethine yilde), [6,6]-pyrrolidino-adduct of $Sc_3N@I_h$-C_{80} was detected together with its [5,6]-adduct [106]. This [6,6]-adduct also can be thermally isomerized to [5,6]-adduct. Likewise, $La_2@I_h$-C_{80} reacted with *N*-tritylazomethine yilde, yielding both [5,6] and [6,6]-adduct [107]. The [6,6]-adduct can be separated from its [5,6]-isomer by crystallization process. Both the experimental and the theoretical studies suggested that the two La atoms are highly localized in the C_{80} cage of the [6,6]-adduct. Comparatively, both $Sc_3N@D_{5h}$-C_{80} and $Sc_3N@D_{3h}$-C_{78} exhibited higher reactivities than $Sc_3N@I_h$-C_{80}. Their monoadducts yielded from Prato reactions were characterized as kinetically favored [6,6]-adducts [91, 108].

15.6.2.3 Carbene Reactions

The *in situ* generated carbene by irradiation of diazocompound has high selectivity toward the [6,6] bond of C_{60}. The first reported carbene reaction of metallofullerene was conducted by irradiation of 2-adamantane-2,3-[3*H*]-diazirine and $La@C_{2v}$-C_{82} in a degassed solvent [40]. This reaction yielded only two monoadducts, indicating the high regioselectivity of carbene toward $La@C_{2v}$-C_{82}. The two monoadducts were determined by either the X-ray crystallographic method or the NMR spectroscopic studies as well as theoretical calculations. Addition sites (site I and II in Figure 15.2b) involved two [6,6]-bonds adjacent to the La^{3+} ion. Unlike the closed cyclopropane structure of a typical [6,6]-adduct of C_{60}, both the two [6,6]-adducts of $La@C_{2v}$-C_{82} have a broken [6,6]-bond due to the carbene addition, forming methanofulleroids instead of methanoadducts of $La@C_{2v}$-C_{82}. This open-cage structure was also found to be a common feature for all the reported carbene derivatives of metallofullerenes.

As this carbene reaction is quite clean and simple, it was frequently used as a probe to examine the chemical reactivities of various metallofullerenes. The reactions of adamantylidene carbene with $M_2@I_h$-C_{80} (M = La [59], Ce [59], Gd [41]), $M_2@D_{3h}$-C_{78} (M = La [60], Ce [109]), $Sc_2C_2@C_{3v}$-C_{82} [65], $Sc_3C_2@I_h$-C_{80} [70], and non-IPR $La_2@D_2$-C_{72} [58, 110] were reported recently. For $M_2@I_h$-C_{80} (M = La, Ce) and $Sc_3C_2@I_h$-C_{80}, regardless of the encapsulated metals or cluster, the additions exclusively occurred on the [6,6] bond. In contrast, the carbene additions on non-

IPR $La_2@D_2$-C_{72} [58] select either [5,6] bonds or [6,6] bond that was adjacent to the fused-pentagon pair [110]. The second addition also preferred to occur on a [5,6]-bond on the other pentagon pair. Even in longer reaction time, no multiaddition was observed, indicating the higher reactivities of the fused-pentagon regions. As to $La_2@D_{3h}$-C_{78}, the selectivity of carbene addition is comparatively lower. Both [5,6] bond and [6,6] bond are involved in the additions.

15.6.2.4 Bis-Silylation

Photoinduced addition of 1,1,2,2-tetramesityl-1,2-disilirane or a related digermirane to endohedral metallofullerenes, such as $M@C_{2v}$-C_{82} (M = La [111], Y [111], Pr [112], Ce [38], and Gd [113]), $M_2@I_h$-C_{80} (M = La [114] and Ce [115]), $Sc_2C_2@C_{3v}$-C_{82} [116], $Sc_3N@I_h$-C_{80} [117], and $Ce_2@D_{3h}$-C_{78} [109], can proceed easily. In contrast to the empty fullerenes, most of them other than $Sc_2C_2@C_{3v}$-C_{82} and $Sc_3N@I_h$-C_{80}, also react thermally with the disilirane reagent. This higher reactivity is attributed to the stronger electron donor properties of those metallofullerenes as compared with empty fullerenes.

Bis-silylation of $M_2@I_h$-C_{80} (M = La and Ce) yielded only one monoadduct. The structures of their monoadducts, especially the motion of the encapsulated metals, were elucidated by NMR spectroscopic studies and X-ray crystallographic method. The silanomethanosilano addend is bridged on the 1,4-position of a six membered-ring, thus affording two dynamically exchanged conformers. The motion of the encapsulated metals was found to depend on their individual nature. In case of disilylated $Ce_2@C_{80}$, the two Ce atoms are localized at the pole–pole plane inside the cage, which is parallel to the addition sites. However, as to disilylated $La_2@C_{80}$, the two La atoms are in a two-dimensional hopping motion between two addition sites along the equatorial plane. These two results are in clear contrast to the three-dimensional motion in pristine $Ce_2@C_{80}$ or $La_2@C_{80}$, and also their almost fixed positions in carbene- or Prato-adducts of $Ce_2@C_{80}$ or $La_2@C_{80}$.

15.6.2.5 Cycloaddition via a Zwitterion Approach

C_{60} can react with an electron-deficient acetylene and a phosphine, affording a [6,6]-methano-adduct bearing a phosphorus ylide in high yield (Scheme 15.2) [118]. Likewise, it was reported that the reaction of $Dy@C_{2v}$-C_{82} with dimethylacetylene-dicarboxylate and triphenylphosphine proceeded smoothly [119]. The merely pro-duced monoadduct is a [6,6]-methanofulleroid (Compound 1 in Figure 15.11). This high regioselective addition occurs on the [6,6]-bond that also undergoes carbene addition.

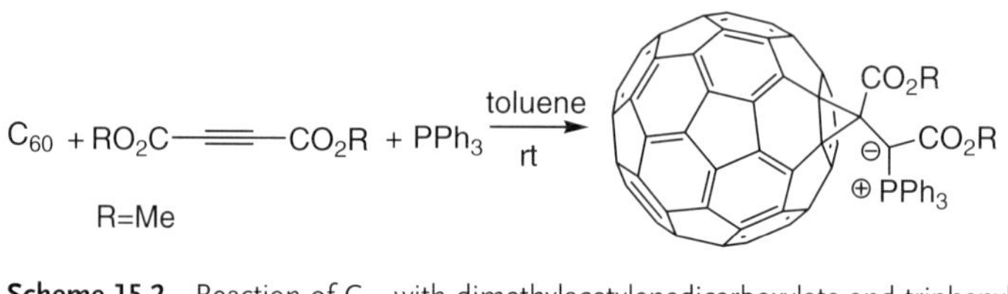

Scheme 15.2 Reaction of C_{60} with dimethylacetylenedicarboxylate and triphenylphosphine.

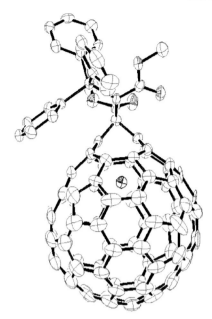

Figure 15.11 A ORTEP drawing of Compound 1 with thermal ellipsoids shown at the 50% probability level. Solvent molecules and H atoms are omitted for clarity.

15.6.3
Nucleophilic Addition

Bingel reaction is another very efficient chemical modification method in fullerene chemistry. Its mechanism involves a nucleophilic attack of a carbon anion that is *in situ* produced by deprotonation of α-halo esters or α-halo ketones. This method provides easy access to versatile fullerene derivatives as well as water-soluble fullerenes.

Bingel reaction was also performed on metallofullerenes with the aim to obtain various methano-adducts. Due to the multiple electron transfers from the encapsulated metals or clusters to the fullerene cage, metallofullerenes are not as good electron-deficient species as empty fullerenes. Nevertheless, the Bingel reaction of $M_3N@I_h$-C_{80} (M = Y [120], Gd [121]), $Sc_3N@I_h$-C_{78} [122], and $Gd_3N@C_{84}$ [123] can proceed smoothly at room temperature. The monoadduct of $Y_3N@I_h$-C_{80} was characterized as a [6,6] methanofulleroid adduct [120], different from the methano-adducts of C_{60} or C_{70}. $Sc_3N@D_{3h}$-C_{78} even readily afforded a symmetric bis-adduct with high regioselectivity [122]. Note that $Sc_3N@I_h$-C_{80} and $Gd_3N@C_{88}$ [123] do not undergo Bingel reaction under the same experimental conditions. Such inertness is attributed to the fact that smaller encapsulated cluster or larger cage size induced lower degree of pyramidalization of cage carbons.

The Bingel reaction of $La@C_{2v}$-C_{82} was reported and yielded five monoadducts [124, 125]. Four of them show diamagnetic properties, in contrast to the paramagnetic properties of $La@C_{2v}$-C_{82} and another minor product. The X-ray structure of its major diamagnetic product is shown in Figure 15.12 [124].

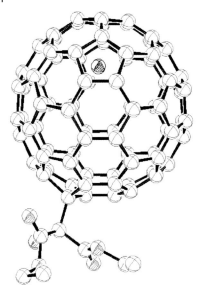

Figure 15.12 A ORTEP drawing of major monoadduct in Bingel reaction of La@C_{2v}-C_{82} with thermal ellipsoids shown at the 50% probability level. Solvent molecules and H atoms are omitted for clarity.

La@C_{2v}-C_{82} is modified by a singly bonded bromomalonate group. The addition site is far from the La^{3+} ion and has the most positive Mulliken charge density. The proposed reaction mechanism is mainly involves a nucleophilic attack of carbanion on La@C_{2v}-C_{82}. The following otherwise slowly proceeding bromo-leaving and cyclopropanation process are possibly replaced by an unidentified rapid oxidation of the intermediate. As compared with that of pristine La@C_{2v}-C_{82}, its diamagnetic monoadducts have negatively shifted the first reduction potentials and positively shifted the first oxidation potentials, suggesting their larger HOMO–LUMO gaps. On the other hand, its minor paramagnetic monoadduct was characterized as a methanofulleroid La@C_{2v}-C_{82}-C(CO$_2$C$_2$H$_5$)$_2$ by NMR spectroscopic studies on its anion [125]. La@C_{2v}-C_{82}-C(CO$_2$C$_2$H$_5$)$_2$ and above-mentioned Y$_3$N@I$_h$-C$_{80}$-C(CO$_2$C$_2$H$_5$)$_2$ were found to have common feature in their redox behaviors. Both of them exhibit high stabilities in their one-electron reductive states, which is in contrast with the previously reported retro-cycloaddition of [C$_{60}$-C(CO$_2$C$_2$H$_5$)$_2$]$^-$.

Bingel reaction is not restricted to highly acidic carbonyl compound such as bromomalonate. La@C_{2v}-C_{82} can even react with malonate in the presence of DBU at elevated temperature [126]. This reaction affords a 7,13-bismalomate derivative of La@C_{2v}-C_{82} with high regioselectivity, which dimerizes during the crystallization process, indicating its more reactive radical character.

15.6.4
Radical Reactions

Endohedral metallofullerenes behave like either a radical sponge or a radical. Various radicals can readily add to metallofullerenes and form some novel derivatives. Sc$_3$N@I$_h$-

C_{80} shows inert reactivity toward nucleophilic carbon anion in Bingel reaction, but comparable reactivity in radical reactions. Its reaction with metal generated malonate radical proceeds smoothly in refluxed chlorobenzene, yielding two methanofulleroids: $Sc_3N@I_h-C_{80}-C(CO_2C_2H_5)_2$ and $Sc_3N@I_h-C_{80}-CH(CO_2C_2H_5)$ [127]. Addition of per-fluoroalkyl groups generated from perfluoroalkyl iodides (R_fI) was carried out on a mixture of $Sc_3N@I_h-C_{80}$ and $Sc_3N@D_{5h}-C_{80}$ under dynamic vacuum and over 500 °C, yielding $Sc_3N@C_{80}(CF_3)_{2n}$ ($n \Leftarrow 6$). For their bis-CF_3 derivatives, two CF_3 groups were shown to be equivalently added on either isomer according to the ^{19}F NMR spectroscopic studies [128]. A 1,4-addition was proposed by theoretical studies, which possibly leads to the derivatives with minimum formation energies.

Radical additions of $Y@C_{82}$ with perfluoroalkyl groups generated from $AgCF_3CO_2$ were performed likewise [129]. A series of mono- and multiadducts, such as $Y@C_{82}(CF_3)$, $Y@C_{82}(CF_3)_3$, and $Y@C_{82}(CF_3)_5$, were isolated and characterized. Note that only odd numbers of CF_3 groups were added to $Y@C_{82}$, resulting in derivatives with closed-shell structures, which seem to be the thermodynamically favored products. The two isomers of $Y@C_{82}(CF_3)_5$ were proposed by the authors having an addition pattern with 1,4-additions across four contiguous six-membered rings.

Further studies on $M^{3+}@C_{2n}^{3-}$ type metallofullerenes definitely revealed their unique radical character, which was believed to arise from the unpaired electron on their HOMOs. Previously, $La@C_{82}$ was reported to undergo multiple additions of phenyl radicals [130] or fluoroalkyl radicals [131]. Also, as mentioned in the above sections, $La@C_{72}$ and $La@C_{74}$ can react with thermally generated dichlorophenyl radical in 1,2,4-trichlorobenzene, surprisingly yielding singly bonded derivatives ($La@C_{72}-C_6H_3Cl_2$ and $La@C_{74}-C_6H_3Cl_2$). In recent studies, it was found that $La@C_{82}$ even thermally reacted with toluene in the presence of 3-triphenyl-methyl-5-oxazolidinone, leading to four monoadducts commonly described as $La@C_{82}-CH_2C_6H_5$ [132]. This result indicates that $La@C_{82}$ is more reactive even toward unstable radical instead of the azomethine yilde. Alternatively, under photo-irradiation condition, $La@C_{82}$ can directly react with not only toluene, but also $\alpha,\alpha,2,4$-tetrachlorotoluene, yielding $La@C_{82}-CHClC_6H_3Cl_2$. One of isomers was fully determined. Its X-ray structure is shown in Figure 15.13. The addition site was revealed to possess the highest spin density by theoretical calculations, thus confirming the proposed radical–radical reaction.

15.7
Applications of Endohedral Metallofullerenes

15.7.1
Magnetic Resonance Imaging (MRI) Contrast Agents

In recent years, much concern has been paid to improving MRI contrast agent based on gadofullerenes. Due to their unique structures and high stabilities, the toxicity of Gd^{3+} ion can be completely shielded by fullerene cage. Also, the surface modifications of fullerenes greatly improved their water solubilities and biocompat-ibilities, affording them promising prospect as MRI contrast agent. Typically,

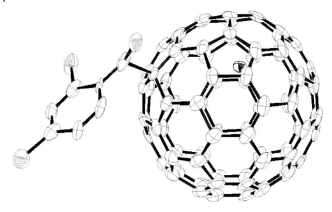

Figure 15.13 A ORTEP drawing of La@C$_{2v}$-C$_{82}$-CHClC$_6$H$_3$Cl$_2$ with thermal ellipsoids shown at the 50% probability level. Solvent molecules and H atoms are omitted for clarity.

the hydroxylated gadolinium fullerene Gd@C$_{82}$(OH)$_n$ has been shown to have much higher spin coupling with water protons than other Gd complexes. Its T1 spin-lattice relaxivity (r1) is nearly 20 times larger than the values of gadopentetate dimeglumine (Magnevist; Schering, Berlin, Germany) at 0.47 T MR imaging [133]. Another carboxylated product Gd@C$_{60}$(C(COOH)$_2$)$_n$ exhibited comparable relaxivity as those of commercial agents but decreased uptake by the reticuloendothelial system (RES) and facile excretion [134]. Recently, Gd$_3$N@C$_{80}$[DiPEG5000(OH)$_x$] has been synthesized by PEGylation of Gd$_3$N@C$_{80}$ with long-chain glycols and hydroxylation. This compound possesses both higher relaxivity and much longer residence within the tumor volume [121]. In addition, the gadofullerenes can react with some amino acid or amino glycol to yield water-soluble compounds, such as Gd@C$_{82}$O$_m$(OH)$_n$(NHC$_2$H$_4$COOH)$_l$ [135] and Gd$_3$N@C$_{80}$(N(OH)(C$_2$H$_4$O)$_n$CH$_3$)$_x$ [136], which both provided excellent MRI enhancement.

Further studies otherwise focused on the paramagnetic relaxation mechanism induced by water-soluble gadofullerenes. Generally, there are two opinions on how the encapsulated Gd^{3+} ions coupled with water protons. (1) It was believed that the unpaired f-electrons of the encapsulated Gd^{3+} may couple their spins through the electrons on the molecular orbitals to water protons that are hydrogen bonded to the cage surface. Thus, the hydrogen bonded water molecules may act like inner-sphere water and be rapidly exchanged with many surrounding water molecules. The appended groups were involved by changing the water diffusion rates [136]. (2) The Gd^{3+} may possibly exchange its spins with the surrounding water molecules through the protons on the appended groups such as hydroxyl and carboxyl groups. So the distance between protons of appended groups and fullerene cage appears to be a very crucial factor, which might explain the higher relaxivities of hydroxylated gadofullerenes relative to their carboxylated derivatives [134]. Nevertheless, it was commonly realized that hydrogen bonding induced aggregation of gadofullerenes may slow down their molecular tumbling and therefore improve their relaxivities. Since serious aggregation may either cause unexpected biotoxicity [135], the factors that affect the aggregation of gadofullerenes, such as pH values of the aqueous solution [137] and concentration of additive salts [138], have been studied.

15.7.2
Peapod and Nanorod

Since the discovery of fullerene-filled single-walled nanotubes (SWNTs), this novel nanomaterial (the so-called peapod) has attracted much interest. Filling SWNTs with endohedral metallofullerenes introduced even more interesting properties for peapod. One might expect to control the bandgap of peapod by varying the encapsulated metallofullerenes and utilize them in the engineering of field-effect transistors (FET). For instance, when $Gd@C_{82}$ was entrapped in a (11,9) SWNT, the studies based on scanning tunneling microscopy and spectroscopy suggested that the original band gap of SWNT (0.43 eV) is narrowed down to 0.17 eV where $Gd@C_{82}$ is located [139]. It was proposed that the combined contribution of elastic strain and charge transfer from $Gd@C_{82}$ to SWNT [140] may be responsible for the observed gap modulation. Interestingly, the Gd-peapod-FET exhibits ambipolar p- and n-type characteristics, which were believed to be the result of the $Gd@C_{82}$-induced band modulation. In another report, the $Dy@C_{82}$-doped semiconducting SWNT was revealed to exhibit a temperature depended p- or n-type character, indicating that the charge transfer from $Dy@C_{82}$ to nanotube varies with temperature [141]. Below about 75 K, single-electron charging phenomena dominate the transport and show irregular Coulomb blockade oscillation, implying that the insertion of $Dy@C_{82}$ splits the tube into a series of several quantum dots. Besides monometallofullerenes, dimetallofullerenes, and TN fullerenes, such as $La_2@C_{80}$ [142], $Ti_2@C_{80}$ (or $Ti_2C_2@C_{78}$) [143] and $Dy_3N@C_{80}$ [144], have been reported to form peapod with SWNT. Particularly, as to $La_2@C_{80}@SWCNT$ and $Ti_2@C_{80}@SWCNT$, the Raman studies indicate unexpectedly high intensities of the intratubular fullerene bands, which were even stronger than the SWCNT bands. This behavior was explained by a so-called "antenna effect." There are also other studies focusing on the filling rate, the structural parameters, and the charge transfer between metallofullerenes and the tube. All these studies were performed with the aim to utilize these novel hybrid materials in the architecture of new molecular devices.

Very recently, another one-dimensional nanostructures, nanorods of $La@C_{82}(Ad)$ with good crystallinity were reported by Akasaka group. Importantly, unlike the n-type character of C_{60} nanorods, $La@C_{82}(Ad)$ nanorods exhibit a p-type character [145], which is responsible for their unique prospect in the fabrication of carbon material-based FET. In addition, $La@C_{82}(Ad)$ nanorod was found to orientate perpendicularly to the magnetic field, indicating its negative magnetic anisotropy, which is in contrast to the positive magnetic anisotropy of C_{60} nanorod. Such different alignments of C_{60} and $La@C_{82}(Ad)$ nanorods are considered to be useful in the controllable construction of nanomaterials.

15.7.3
Electron Donor/Acceptor Conjugate

In 2007, the first $Sc_3N@C_{80}$-ferrocene fabricated electron donor/acceptor conjugate was reported by Echegoyen and coworkers [146]. $Sc_3N@C_{80}$ was combined with ferrocene in the form of its [5,6]-pyrrolidine adduct. By the photophysics measurements, a significant stabilization of the radical ion pair state of $Sc_3N@C_{80}$-ferrocene was observed when compared to an analogous C_{60}-ferrocene conjugate. This result might indicate a promising prospect of $Sc_3N@C_{80}$ in solar-energy conversion applications.

15.8
Concluding Remarks

According to the recent studies on the synthesis and separation techniques, the yields of metallofullerenes, especially the TN fullerenes, increase greatly, and the separation also becomes more efficient than ever before. These advances make the macro-quantities of metallofullerenes presently available in laboratory production.

On the other hand, the studies on the structures of a considerable number of metallofullerenes and their derivatives especially with the aid of X-ray crystallographic method clearly elucidated their cage structures, the motions of encapsulated metallic species, and metal–cage bonding as well as the structures of various derivatives. Due to the electron transfer from the encapsulated metallic species, the selection rules for the cage of metallofullerenes are quite different from those of empty fullerenes. It has to be mentioned that the presence of non-IPR metallofullerenes is quite frequent. Although the non-IPR fullerene cages have numerous isomers, only a few of them are experimentally observed for metallofullerenes. How the encapsulated species afford such high selectivity in the formation of non-IPR fullerene cage is one of the open questions for future work. Various encapsulated metallic species were also found to affect the reactivities of fullerene cages. Highly selective additions frequently occur on metallofullerenes, some even having low symmetries. In general, the metallofullerene chemistry is greatly featured by the significant contributions from entrapped metallic species.

Finally, although the investigation on the applications of metallofullerenes is still in its infancy, some fascinating prospects of metallofullerenes in medicine and material science have been revealed. Hopefully, as their large-scale productions become more common, the practicability of metallofullerenes will be highlighted further.

Acknowledgment

This work was supported in part by a Grant-in-Aid for Scientific Research on Innovative Areas (No. 20108001, "pi-Space"), a Grant-in-Aid for Scientific Research

(A) (No. 20245006), The Next Generation Super Computing Project (Nanoscience Project), and a Grant-in Aid for Scientific Research on Priority Area (Nos. 20036008, 20038007) from the Ministry of Education, Culture, Sports, Science, and Technology of Japan.

References

1 Heath, J.R., Zhang, Q., Liu, Y., Curl, R.F., Tittel, F.K., and Smalley, R.E. (1985) Lanthanum complexes of spheroidal carbon shells. *J. Am. Chem. Soc.*, **107**, 7779–7780.

2 Guo, T., Chai, Y., Jin, C., Haufler, R.E., Felipe Chibante, L.P., Fure, J., Michael Alford, J., Wang, L., and Smalley, R.E. (1991) Fullerenes with metals inside. *J. Phys. Chem.*, **95**, 7564–7568.

3 Akasaka, T. and Nagase, S. (eds) (2002) *Endofullerenes: A New Family of Carbon Clusters*, Kluwer Academic Publishers, Dordrecht.

4 Shinohara, H. (2000) Endohedral metallofullerenes. *Rep. Prog. Phys.*, **63**, 843–892.

5 Pietzak, B., Waiblinger, M., Murphy, T.A., Weidinger, A., Hohne, M., Dietel, E., and Hirsch, A. (1998) Properties of endohedral $N@C_{60}$. *Carbon*, **36**, 613–615.

6 Larsson, J.A., Greer, J.C., Harneit, W., and Weidinger, A. (2002) Phosphorous trapped within buckminsterfullerene. *J. Chem. Phys.*, **116**, 7849–7854.

7 Scheloske, M., Naydenov, B., Meyer, C., and Harneit, W. (2006) Synthesis and functionalization of fullerenes encapsulating atomic phosphorus. *Isra. J. Chem.*, **46**, 407–412.

8 Shabtai, E., Weitz, A., Haddon, R.C., Hoffman, R.E., Rabinovitz, M., Khong, A., Cross, R.J., Saunders, M., Cheng, P.C., and Scott, L.T. (1998) ^{3}He NMR of $He@C_{60}{}^{6-}$ and $He@C_{70}{}^{6-}$. New records for the most shielded and the most deshielded ^{3}He inside a fullerene. *J. Am. Chem. Soc.*, 6389–6393.

9 Komatsu, K., Murata, M., and Murata, Y. (2005) Encapsulation of molecular hydrogen in fullerene C_{60} by organic synthesis. *Science*, **307**, 238–240.

10 Murata, M., Murata, Y., and Komatsu, K. (2006) Synthesis and properties of endohedral C_{60} encapsulating molecular hydrogen. *J. Am. Chem. Soc.*, **128**, 8024–8033.

11 Takata, M., Umeda, B., Nishibori, E., Sakata, M., Saito, Y., Ohno, M., and Shinohara, H. (1995) Confirmation by X-ray-diffraction of the endohedral nature of the metallofullerene $Y@C_{82}$. *Nature*, **377**, 46–49.

12 Kato, H., Taninaka, A., Sugai, T., and Shinohara, H. (2003) Structure of a missing-caged metallofullerene: $La_2@C_{72}$. *J. Am. Chem. Soc.*, **125**, 7782–7783.

13 Wang, C.R., Kai, T., Tomiyama, T., Yoshida, T., Kobayashi, Y., Nishibori, E., Takata, M., Sakata, M., and Shinohara, H. (2000) Materials science – C_{66} fullerene encaging a scandium dimer. *Nature*, **408**, 426–427.

14 Olmstead, M.M., Lee, H.M., Duchamp, J.C., Stevenson, S., Marciu, D., Dorn, H.C., and Balch, A.L. (2003) $Sc_3N@C_{68}$: folded pentalene coordination in an endohedral fullerene that does not obey the isolated pentagon rule. *Angew. Chem. Int. Ed.*, **42**, 900–903.

15 Beavers, C.M., Zuo, T.M., Duchamp, J.C., Harich, K., Dorn, H.C., Olmstead, M.M., and Balch, A.L. (2006) $Tb_3N@C_{84}$: an improbable, egg-shaped endohedral fullerene that violates the isolated pentagon rule. *J. Am. Chem. Soc.*, **128**, 11352–11353.

16 Stevenson, S., Rice, G., Glass, T., Harich, K., Cromer, F., Jordan, M.R., Craft, J., Hadju, E., Bible, R., Olmstead, M.M., Maitra, K., Fisher, A.J., Balch, A.L., and Dorn, H.C. (1999) Small-bandgap endohedral metallofullerenes in high yield and purity. *Nature*, **401**, 55–57.

17 Yang, S.F. and Dunsch, L. (2005) A large family of dysprosium-based trimetallic nitride endohedral fullerenes: $Dy_3N@C_{2n}$ ($39 \Leftarrow n \Leftarrow 44$). *J. Phys. Chem. B*, **109**, 12320–12328.

18 Dunsch, L. and Yang, S. (2007) Metal nitride cluster fullerenes: their current state and future prospects. *Small*, **3**, 1298–1320.

19 Yang, S.F. and Dunsch, L. (2006) Expanding the number of stable isomeric structures of the C_{80} cage: a new fullerene $DY_3N@C_{80}$. *Chem. Eur. J.*, **12**, 413–419.

20 Yang, S.F., Kalbac, M., Popov, A., and Dunsch, L. (2006) Gadolinium-based mixed metal nitride clusterfullerenes $Gd_xSC_{3x}N@C_{80}$ ($x = 1, 2$). *ChemPhysChem*, **7**, 1990–1995.

21 Stevenson, S., Mackey, M.A., Stuart, M.A., Phillips, J.P., Easterling, M.L., Chancellor, C.J., Olmstead, M.M., and Balch, A.L. (2008) A distorted tetrahedral metal oxide cluster inside an icosahedral carbon cage. Synthesis, isolation, and structural characterization of $Sc_4(\mu_3-O)_2@I_h-C_{80}$. *J. Am. Chem. Soc.*, **130**, 11844–11845.

22 Lian, Y.F., Yang, S.F., and Yang, S.H. (2002) Revisiting the preparation of $La@C_{82}$ (I and II) and $La_2@C_{80}$: efficient production of the "Minor" isomer $La@C_{82}$ (II). *J. Phys. Chem. B*, **106**, 3112–3117.

23 Shi, Z.J., Okazaki, T., Shimada, T., Sugai, T., Suenaga, K., and Shinohara, H. (2003) Selective high-yield catalytic synthesis of terbium metallofullerenes and single-wall carbon nanotubes. *J. Phys. Chem. B*, **107**, 2485–2489.

24 Stevenson, S., Mackey, M.A., Thompson, M.C., Coumbe, H.L., Madasu, P.K., Coumbe, C.E., and Phillips, J.P. (2007) Effect of copper metal on the yield of $Sc_3N@C_{80}$ metallofullerenes. *Chem. Commun.*, 4263–4265.

25 Olmstead, M.M., de Bettencourt-Dias, A., Duchamp, J.C., Stevenson, S., Dorn, H.C., and Balch, A.L. (2000) Isolation and crystallographic characterization of $ErSc_2N@C_{80}$: an endohedral fullerene which crystallizes with remarkable internal order. *J. Am. Chem. Soc.*, **122**, 12220–12226.

26 Olmstead, M.H., de Bettencourt-Dias, A., Duchamp, J.C., Stevenson, S., Marciu, D., Dorn, H.C., and Balch, A.L. (2001) Isolation and structural characterization of the endohedral fullerene $Sc_3N@C_{78}$. *Angew. Chem. Int. Ed.*, **40**, 1223–1225.

27 Wolf, M., Muller, K.H., Skourski, Y., Eckert, D., Georgi, P., Krause, M., and Dunsch, L. (2005) Magnetic moments of the endohedral cluster fullerenes $Ho_3N@C_{80}$ and $Tb_3N@C_{80}$: the role of ligand fields. *Angew. Chem. Int. Ed.*, **44**, 3306–3309.

28 Zuo, T.M., Beavers, C.M., Duchamp, J.C., Campbell, A., Dorn, H.C., Olmstead, M.M., and Balch, A.L. (2007) Isolation and structural characterization of a family of endohedral fullerenes including the large, chiral cage fullerenes $Tb_3N@C_{88}$ and $Tb_3N@C_{86}$ as well as the I_h and D_{5h} isomers of $Tb_3N@C_{80}$. *J. Am. Chem. Soc.*, **129**, 2035–2043.

29 Stevenson, S., Thompson, M.C., Coumbe, H.L., Mackey, M.A., Coumbe, C.E., and Phillips, J.P. (2007) Chemically adjusting plasma temperature, energy, and reactivity (CAPTEAR) method using NO_x and combustion for selective synthesis of $SC_3N@C_{80}$ metallic nitride fullerenes. *J. Am. Chem. Soc.*, **129**, 16257–16262.

30 Tsuchiya, T., Wakahara, T., Shirakura, S., Maeda, Y., Akasaka, T., Kobayashi, K., Nagase, S., Kato, T., and Kadish, K.M. (2004) Reduction of endohedral metallofullerenes: a convenient method for isolation. *Chem. Mater.*, **16**, 4343–4346.

31 Ge, Z.X., Duchamp, J.C., Cai, T., Gibson, H.W., and Dorn, H.C. (2005) Purification of endohedral trimetallic nitride fullerenes in a single, facile step. *J. Am. Chem. Soc.*, **127**, 16292–16298.

32 Stevenson, S., Harich, K., Yu, H., Stephen, R.R., Heaps, D., Coumbe, C., and Phillips, J.P. (2006) Nonchromatographic "stir and filter approach" (SAFA) for isolating $Sc_3N@C_{80}$ metallofullerenes. *J. Am. Chem. Soc.*, **128**, 8829–8835.

33 Elliott, B., Yu, L., and Echegoyen, L. (2005) A simple isomeric separation of

D_{5h} and I_h Sc$_3$N@C$_{80}$ by selective chemical oxidation. *J. Am. Chem. Soc.*, **127**, 10885–10888.

34 Tsuchiya, T., Sato, K., Kurihara, H., Wakahara, T., Nakahodo, T., Maeda, Y., Akasaka, T., Ohkubo, K., Fukuzumi, S., Kato, T., Mizorogi, N., Kobayashi, K., and Nagase, S. (2006) Host–guest complexation of endohedral metallofullerene with azacrown ether and its application. *J. Am. Chem. Soc.*, **128**, 6699–6703.

35 Tsuchiya, T., Wakahara, T., Lian, Y.F., Maeda, Y., Akasaka, T., Kato, T., Mizorogi, N., and Nagase, S. (2006) Selective extraction and purification of endohedral metallofullerene from carbon soot. *J. Phys. Chem. B*, **110**, 22517–22520.

36 Akasaka, T., Wakahara, T., Nagase, S., Kobayashi, K., Waelchli, M., Yamamoto, K., Kondo, M., Shirakura, S., Okubo, S., Maeda, Y., Kato, T., Kako, M., Nakadaira, Y., Nagahata, R., Gao, X., Van Caemelbecke, E., and Kadish, K.M. (2000) La@C$_{82}$ anion. An unusually stable metallofullerene. *J. Am. Chem. Soc.*, **122**, 9316–9317.

37 Feng, L., Wakahara, T., Tsuchiya, T., Maeda, Y., Lian, Y.F., Akasaka, T., Mizorogi, N., Kobayashi, K., Nagase, S., and Kadish, K.M. (2005) Structural characterization of Y@C$_{82}$. *Chem. Phys. Lett.*, **405**, 274–277.

38 Wakahara, T., Kobayashi, J., Yamada, M., Maeda, Y., Tsuchiya, T., Okamura, M., Akasaka, T., Waelchli, M., Kobayashi, K., Nagase, S., Kato, T., Kako, M., Yamamoto, K., and Kadish, K.M. (2004) Characterization of Ce@C$_{82}$ and its anion. *J. Am. Chem. Soc.*, **126**, 4883–4887.

39 Akasaka, T., Wakahara, T., Nagase, S., Kobayashi, K., Waelchli, M., Yamamoto, K., Kondo, M., Shirakura, S., Maeda, Y., Kato, T., Kako, M., Nakadaira, Y., Gao, X., Van Caemelbecke, E., and Kadish, K.M. (2001) Structural determination of the La@C$_{82}$ isomer. *J. Phys. Chem. B*, **105**, 2971–2974.

40 Maeda, Y., Matsunaga, Y., Wakahara, T., Takahashi, S., Tsuchiya, T., Ishitsuka, M.O., Hasegawa, T., Akasaka, T., Liu,

M.T.H., Kokura, K., Horn, E., Yoza, K., Kato, T., Okubo, S., Kobayashi, K., Nagase, S., and Yamamoto, K. (2004) Isolation and characterization of a carbene derivative of La@C$_{82}$. *J. Am. Chem. Soc.*, **126**, 6858–6859.

41 Akasaka, T., Kono, T., Takematsu, Y., Nikawa, H., Nakahodo, T., Wakahara, T., Ishitsuka, M.O., Tsuchiya, T., Maeda, Y., Liu, M.T.H., Yoza, K., Kato, T., Yamamoto, K., Mizorogi, N., Slanina, Z., and Nagase, S. (2008) Does Gd@C$_{82}$ have an anomalous endohedral structure? Synthesis and single crystal X-ray structure of the carbene adduct. *J. Am. Chem. Soc.*, **130**, 12840–12841.

42 Unpublished result.

43 Mizorogi, N. and Nagase, S. (2006) Do Eu@C$_{82}$ and Gd@C$_{82}$ have an anomalous endohedral structure? *Chem. Phys. Lett.*, **431**, 110–112.

44 Nishibori, E., Iwata, K., Sakata, M., Takata, M., Tanaka, H., Kato, H., and Shinohara, H. (2004) Anomalous endohedral structure of Gd@C$_{82}$ metallofullerenes. *Phys. Rev. B*, **69**, 113412.

45 Sun, B.Y., Sugai, T., Nishibori, E., Iwata, K., Sakata, M., Takata, M., and Shinohara, H. (2005) An anomalous endohedral structure of Eu@C$_{82}$ metallofullerenes. *Angew. Chem. Int. Ed.*, **44**, 4568–4571.

46 Wakahara, T., Nikawa, H., Kikuchi, T., Nakahodo, T., Rahman, G.M.A., Tsuchiya, T., Maeda, Y., Akasaka, T., Yoza, K., Horn, E., Yamamoto, K., Mizorogi, N., Slanina, Z., and Nagase, S. (2006) La@C$_{72}$ having a non-IPR carbon cage. *J. Am. Chem. Soc.*, **128**, 14228–14229.

47 Nikawa, H., Kikuchi, T., Wakahara, T., Nakahodo, T., Tsuchiya, T., Rahman, G.M.A., Akasaka, T., Maeda, Y., Yoza, K., Horn, E., Yamamoto, K., Mizorogi, N., and Nagase, S. (2005) Missing metallofullerene La@C$_{74}$. *J. Am. Chem. Soc.*, **127**, 9684–9685.

48 Friese, K., Panthofer, M., Wu, G., and Jansen, M. (2004) Strategies for the structure determination of endohedral fullerenes applied to the example of Ba@C$_{74}$ Co(octaethylporphyrin)·2C$_6$H$_6$. *Acta Crystallogr., Sect. B*, **60**, 520–527.

49 Reich, A., Panthofer, M., Modrow, H., Wedig, U., and Jansen, M. (2004) The structure of Ba@C_{74}. *J. Am. Chem. Soc.*, **126**, 14428–14434.

50 Tang, C.M., Fu, S.Y., Deng, K.M., Yuan, Y.B., Tan, W.S., Huang, D.C., and Wang, X. (2008) The density functional calculations on the structural stability, electronic properties, and static linear polarizability of the endohedral metallofullerene Ba@C_{74}. *J. Mol. Struct. (Theochem)*, **867**, 111–115.

51 Tang, C.M., Deng, K.M., Tan, W.S., Yuan, Y.B., Liu, Y.Z., Wu, H.P., Huang, D.C., Hu, F.L., Yang, J.L., and Wang, X. (2007) Influence of a dichlorophenyl group on the geometric structure, electronic properties, and static linear polarizability of La@C_{74}. *Phys. Rev. A*, **76**, 013201.

52 Cao, B.P., Wakahara, T., Tsuchiya, T., Kondo, M., Maeda, Y., Rahman, G.M.A., Akasaka, T., Kobayashi, K., Nagase, S., and Yamamoto, K. (2004) Isolation, characterization, and theoretical study of La$_2$@C_{78}. *J. Am. Chem. Soc.*, **126**, 9164–9165.

53 Yang, S.F. and Dunsch, L. (2006) Di- and tridysprosium endohedral metallofullerenes with cages from C_{94} to C_{100}. *Angew. Chem. Int. Ed.*, **45**, 1299–1302.

54 Cao, B.P., Hasegawa, M., Okada, K., Tomiyama, T., Okazaki, T., Suenaga, K., and Shinohara, H. (2001) EELS and 13C NMR characterization of pure Ti$_2$@C_{80} metallofullerene. *J. Am. Chem. Soc.*, **123**, 9679–9680.

55 Cao, B.P., Suenaga, K., Okazaki, T., and Shinohara, H. (2002) Production, isolation, and EELS characterization of Ti$_2$@C_{84} dititanium metallofullerenes. *J. Phys. Chem. B*, **106**, 9295–9298.

56 Tagmatarchis, N. and Shinohara, H. (2000) Production, separation, isolation, and spectroscopic study of dysprosium endohedral metallofullerenes. *Chem. Mater.*, **12**, 3222–3226.

57 Yamada, M., Wakahara, T., Tsuchiya, T., Maeda, Y., Akasaka, T., Mizorogi, N., and Nagase, S. (2008) Spectroscopic and theoretical study of endohedral dimetallofullerene having a Non-IPR fullerene cage: Ce$_2$@C_{72}. *J. Phys. Chem. A*, **112**, 7627–7631.

58 Lu, X., Nikawa, H., Nakahodo, T., Tsuchiya, T., Ishitsuka, M.O., Maeda, Y., Akasaka, T., Toki, M., Sawa, H., Slanina, Z., Mizorogi, N., and Nagase, S. (2008) Chemical understanding of a non-IPR metallofullerene: stabilization of encaged metals on fused-pentagon bonds in La$_2$@C_{72}. *J. Am. Chem. Soc.*, **130**, 9129–9136.

59 Yamada, M., Someya, C., Wakahara, T., Tsuchiya, T., Maeda, Y., Akasaka, T., Yoza, K., Horn, E., Liu, M.T.H., Mizorogi, N., and Nagase, S. (2008) Metal atoms collinear with the spiro carbon of 6,6-open adducts, M$_2$@C_{80}(Ad) (M = La and Ce, Ad = adamantylidene). *J. Am. Chem. Soc.*, **130**, 1171–1176.

60 Cao, B., Nikawa, H., Nakahodo, T., Tsuchiya, T., Maeda, Y., Akasaka, T., Sawa, H., Slanina, Z., Mizorogi, N., and Nagase, S. (2008) Addition of adamantylidene to La$_2$@C_{78}: isolation and single-crystal X-ray structural determination of the monoadducts. *J. Am. Chem. Soc.*, **130**, 983–989.

61 Olmstead, M.M., de Bettencourt-Dias, A., Stevenson, S., Dorn, H.C., and Balch, A.L. (2002) Crystallographic characterization of the structure of the endohedral fullerene {Er$_2$ @ C_{82} Isomer I} with C_s cage symmetry and multiple sites for erbium along a band of ten contiguous hexagons. *J. Am. Chem. Soc.*, **124**, 4172–4173.

62 Olmstead, M.M., Lee, H.M., Stevenson, S., Dorn, H.C., and Balch, A.L. (2002) Crystallographic characterization of Isomer 2 of Er$_2$@C_{82} and comparison with Isomer 1 of Er$_2$@C_{82}. *Chem. Commun.*, 2688–2689.

63 Zuo, T.M., Xu, L.S., Beavers, C.M., Olmstead, M.M., Fu, W.J., Crawford, D., Balch, A.L., and Dorn, H.C. (2008) M$_2$@C_{79}N (M = Y, Tb): isolation and characterization of stable endohedral metallofullerenes exhibiting M-M bonding interactions inside aza[80]fullerene cages. *J. Am. Chem. Soc.*, **130**, 12992–12997.

64 Wang, C.R., Kai, T., Tomiyama, T., Yoshida, T., Kobayashi, Y., Nishibori, E., Takata, M., Sakata, M., and Shinohara, H. (2001) A scandium carbide endohedral metallofullerene: $(Sc_2C_2)@C_{84}$. *Angew. Chem. Int. Ed.*, **40**, 397–399.

65 Iiduka, Y., Wakahara, T., Nakajima, K., Nakahodo, T., Tsuchiya, T., Maeda, Y., Akasaka, T., Yoza, K., Liu, M.T.H., Mizorogi, N., and Nagase, S. (2007) Experimental and theoretical studies of the scandium carbide endohedral metallofullerene $Sc_2C_2@C_{82}$ and its carbene derivative. *Angew. Chem. Int. Ed.*, **46**, 5562–5564.

66 Iiduka, Y., Wakahara, T., Nakajima, K., Tsuchiya, T., Nakahodo, T., Maeda, Y., Akasaka, T., Mizorogi, N., and Nagase, S. (2006) ^{13}C NMR spectroscopic study of scandium dimetallofullerene, $Sc_2@C_{84}$ versus $(Sc_2C_2)@ C_{82}$. *Chem. Commun.*, 2057–2059.

67 Yamazaki, Y., Nakajima, K., Wakahara, T., Tsuchiya, T., Ishitsuka, M.O., Maeda, Y., Akasaka, T., Waelchli, M., Mizorogi, N., and Nagase, S. (2008) Observation of ^{13}C NMR chemical shifts of metal carbides encapsulated in fullerenes: $Sc_2C_2@C_{82}$, $Sc_2C_2@C_{84}$, and $Sc_3C_2@C_{80}$. *Angew. Chem. Int. Ed.*, **47**, 7905–7908.

68 Nishibori, E., Ishihara, M., Takata, M., Sakata, M., Ito, Y., Inoue, T., and Shinohara, H. (2006) Bent $(metal)_2C_2$ clusters encapsulated in (Sc_2C_2) $@C_{82}(III)$ and $(Y_2C_2)@C_{82}(III)$ metallofullerenes. *Chem. Phys. Lett.*, **433**, 120–124.

69 Nishibori, E., Narioka, S., Takata, M., Sakata, M., Inoue, T., and Shinohara, H. (2006) A C_2 molecule entrapped in the pentagonal–dodecahedral Y_2 cage in $Y_2C_2@C_{82}(III)$. *ChemPhysChem*, **7**, 345–348.

70 Iiduka, Y., Wakahara, T., Nakahodo, T., Tsuchiya, T., Sakuraba, A., Maeda, Y., Akasaka, T., Yoza, K., Horn, E., Kato, T., Liu, M.T.H., Mizorogi, N., Kobayashi, K., and Nagase, S. (2005) Structural determination of metallofuIlerene SC_3C_{82} revisited: a surprising finding. *J. Am. Chem. Soc.*, **127**, 12500–12501.

71 Nishibori, E., Terauchi, I., Sakata, M., Takata, M., Ito, Y., Sugai, T., and Shinohara, H. (2006) High-resolution analysis of $(Sc_3C_2)@C_{80}$ metallofullerene by third generation synchrotron radiation X-ray powder diffraction. *J. Phys. Chem. B*, **110**, 19215–19219.

72 Shi, Z.Q., Wu, X., Wang, C.R., Lu, X., and Shinohara, H. (2006) Isolation and characterization of $Sc_2C_2@C_{68}$: a metal-carbide endofullerene with a non-IPR carbon cage. *Angew. Chem. Int. Ed.*, **45**, 2107–2111.

73 Takata, M., Nishibori, E., Sakata, M., Inakuma, M., Yamamoto, E., and Shinohara, H. (1999) Triangle scandium cluster imprisoned in a fullerene cage. *Phys. Rev. Lett.*, **83**, 2214–2217.

74 Valencia, R., Rodriguez-Fortea, A., and Poblet, J.M. (2008) Understanding the stabilization of metal carbide endohedral fullerenes $M_2C_2@C_{82}$ and related systems. *J. Phys. Chem. A*, **112**, 4550–4555.

75 Wang, X.L., Zuo, T.M., Olmstead, M.M., Duchamp, J.C., Glass, T.E., Cromer, F., Balch, A.L., and Dorn, H.C. (2006) Preparation and structure of $CeSc_2N@C_{80}$: an icosahedral carbon cage enclosing an acentric $CeSc_2N$ unit with buried f electron spin. *J. Am. Chem. Soc.*, **128**, 8884–8889.

76 Stevenson, S., Chancellor, C.J., Lee, H.M., Olmstead, M.M., and Balch, A.L. (2008) Internal and external factors in the structural organization in cocrystals of the mixed-metal endohedrals $(GdSc_2N@I_h-C_{80}, Gd_2ScN@I_h-C_{80}$, and $TbSc_2N@I_h-C_{80})$ and nickel(II) octaethylporphyrin. *Inorg. Chem.*, **47**, 1420–1427.

77 Chen, N., Zhang, E.Y., and Wang, C.R. (2006) C-80 encaging four different atoms: the synthesis, isolation, and characterizations of $ScYErN@C_{80}$. *J. Phys. Chem. B*, **110**, 13322–13325.

78 Haddon, R.C. and Raghavachari, K. (eds) (1993) Chapter 7, in *In Buckminsterfullerenes*, VCH, New York.

79 Stevenson, S., Lee, H.M., Olmstead, M.M., Kozikowski, C., Stevenson, P., and Balch, A.L. (2002) Preparation and crystallographic characterization of a new endohedral, $Lu_3N@C_{80}·5(o$-xylene), and

comparison with $Sc_3N@C_{80}\cdot5$(o-xylene). *Chem. Eur. J.*, **8**, 4528–4535.

80 Zuo, T.M., Olmstead, M.M., Beavers, C.M., Balch, A.L., Wang, G.B., Yee, G.T., Shu, C.Y., Xu, L.S., Elliott, B., Echegoyen, L., Duchamp, J.C., and Dorn, H.C. (2008) Preparation and structural characterization of the I_h and the D_{5h} isomers of the endohedral fullerenes $Tm_3N@C_{80}$: icosahedral C_{80} cage encapsulation of a trimetallic nitride magnetic cluster with three uncoupled Tm^{3+} ions. *Inorg. Chem.*, **47**, 5234–5244.

81 Echegoyen, L., Chancellor, C.J., Cardona, C.M., Elliott, B., Rivera, J., Olmstead, M.M., and Balch, A.L. (2006) X-Ray crystallographic and EPR spectroscopic characterization of a pyrrolidine adduct of $Y_3N@C_{80}$. *Chem. Commun.*, 2653–2655.

82 Yang, S.F., Troyanov, S.I., Popov, A.A., Krause, M., and Dunsch, L. (2006) Deviation from the planarity – a large Dy_3N cluster encapsulated in an I_h-C_{80} cage: an X-ray crystallographic and vibrational spectroscopic study. *J. Am. Chem. Soc.*, **128**, 16733–16739.

83 Stevenson, S., Phillips, J.P., Reid, J.E., Olmstead, M.M., Rath, S.P., and Balch, A.L. (2004) Pyramidalization of Gd_3N inside a C_{80} cage. The synthesis and structure of $Gd_3N@C_{80}$. *Chem. Commun.*, 2814–2815.

84 Greenwood, N.N. and Earnshaw, A. (1984) *Chemistry of the Elements*, Pergamon, Oxford, UK.

85 Mercado, B.Q., Beavers, C.M., Olmstead, M.M., Chaur, M.N., Walker, K., Holloway, B.C., Echegoyen, L., and Balch, A.L. (2008) Is the isolated pentagon rule merely a suggestion for endohedral fullerenes? The structure of a second egg-shaped endohedral fullerene-$Gd_3N@C_s$(39663)-C_{82}. *J. Am. Chem. Soc.*, **130**, 7854–7855.

86 Zuo, T., Walker, K., Olmstead, M.M., Melin, F., Holloway, B.C., Echegoyen, L., Dorn, H.C., Chaur, M.N., Chancellor, C.J., Beavers, C.M., Balch, A.L., and Athans, A.J. (2008) New egg-shaped fullerenes: non-isolated pentagon structures of $Tm_3N@C_s$(51365)-C_{84} and

$Gd_3N@C_s$(51365)-C_{84}. *Chem. Commun.*, 1067–1069.

87 Popov, A.A. and Dunsch, L. (2007) Structure, stability, and cluster-cage interactions in nitride clusterfullerenes $M_3N@C_{2n}$ (M = Sc, Y; 2n = 68–98): a density functional theory study. *J. Am. Chem. Soc.*, **129**, 11835–11849.

88 Yang, S., Popov, A.A., and Dunsch, L. (2008) Large mixed metal nitride clusters encapsulated in a small cage: the confinement of the C_{68}-based clusterfullerenes. *Chem. Commun.*, 2885–2887.

89 Slanina, Z., Chen, Z.F., Schleyer, P.V., Uhlik, F., Lu, X., and Nagase, S. (2006) $La_2@C_{72}$ and $Sc_2@C_{72}$: computational characterizations. *J. Phys. Chem. A*, **110**, 2231–2234.

90 Iiduka, Y., Ikenaga, O., Sakuraba, A., Wakahara, T., Tsuchiya, T., Maeda, Y., Nakahodo, T., Akasaka, T., Kako, M., Mizorogi, N., and Nagase, S. (2005) Chemical reactivity of $Sc_3N@C_{80}$ and $La_2@C_{80}$. *J. Am. Chem. Soc.*, **127**, 9956–9957.

91 Cai, T., Xu, L.S., Anderson, M.R., Ge, Z.X., Zuo, T.M., Wang, X.L., Olmstead, M.M., Balch, A.L., Gibson, H.W., and Dorn, H.C. (2006) Structure and enhanced reactivity rates of the D_{5h} $Sc_3N@C_{80}$ and $Lu_3N@C_{80}$ metallofullerene isomers: the importance of the pyracylene motif. *J. Am. Chem. Soc.*, **128**, 8581–8589.

92 Lide, D.R. (ed.) (2000) *Handbook of Chemistry and Physics*, 81st edn, CRC, Boca Raton, FL.

93 Xu, J.X., Li, M.X., Shi, Z.J., and Gu, Z.N. (2006) Electrochemical survey: the effect of the cage size and structure on the electronic structures of a series of ytterbium metallofullerenes. *Chem. Eur. J.*, **12**, 562–567.

94 Chaur, M.N., Melin, F., Elliott, B., Athans, A.J., Walker, K., Holloway, B.C., and Echegoyen, L. (2007) $Gd_3N@C_{2n}$ (n = 40, 42, and 44): remarkably low HOMO–LUMO gap and unusual electrochemical reversibility of $Gd_3N@C_{88}$. *J. Am. Chem. Soc.*, **129**, 14826–14829.

95 Solodovnikov, S.P. and Lebedkin, S.F. (2003) ESR spectra of endohedral metallofullerene Ce@C_{82} radical anions in dimethylformamide and pyridine. *Russ. Chem. B*, **52**, 1111–1113.

96 Tsuchiya, T., Kurihara, H., Sato, K., Wakahara, T., Akasaka, T., Shimizu, T., Kamigata, N., Mizorogi, N., and Nagase, S. (2006) Supramolecular complexes of La@C_{82} with unsaturated thiacrown ethers. *Chem. Commun.*, 3585–3587.

97 Yang, S.F., Rapta, P., and Dunsch, L. (2007) The spin state of a charged non-IPR fullerene: the stable radical cation of Sc_3N@C_{68}. *Chem. Commun.*, 189–191.

98 Iezzi, E.B., Duchamp, J.C., Harich, K., Glass, T.E., Lee, H.M., Olmstead, M.M., Balch, A.L., and Dorn, H.C. (2002) A symmetric derivative of the trimetallic nitride endohedral metallofullerene, Sc_3N@C_{80}. *J. Am. Chem. Soc.*, **124**, 524–525.

99 Lu, X., He, X.R., Feng, L., Shi, Z.J., and Gu, Z.N. (2004) Synthesis of pyrrolidine ring-fused metallofullerene derivatives. *Tetrahedron*, **60**, 3713–3716.

100 Cao, B.P., Wakahara, T., Maeda, Y., Han, A.H., Akasaka, T., Kato, T., Kobayashi, K., and Nagase, S. (2004) Lanthanum endohedral metallofulleropyrrolidines: synthesis, isolation, and EPR characterization. *Chem. Eur. J.*, **10**, 716–720.

101 Feng, L., Lu, X., He, X.R., Shi, Z.J., and Gu, Z.N. (2004) Reactions of endohedral metallofullerenes with azomethine ylides: an efficient route toward metallofullerene-pyrrolidines. *Inorg. Chem. Commun.*, **7**, 1010–1013.

102 Cardona, C.M., Kitaygorodskiy, A., Ortiz, A., Herranz, M.A., and Echegoyen, L. (2005) The first fulleropyrrolidine derivative of Sc_3N@C_{80}: pronounced chemical shift differences of the geminal protons on the pyrrolidine ring. *J. Org. Chem.*, **70**, 5092–5097.

103 Cai, T., Ge, Z.X., Iezzi, E.B., Glass, T.E., Harich, K., Gibson, H.W., and Dorn, H.C. (2005) Synthesis and characterization of the first trimetallic nitride templated pyrrolidino endohedral metallofullerenes. *Chem. Commun.*, 3594–3596.

104 Cardona, C.M., Kitaygorodskiy, A., and Echegoyen, L. (2005) Trimetallic nitride endohedral metallofullerenes: reactivity dictated by the encapsulated metal cluster. *J. Am. Chem. Soc.*, **127**, 10448–10453.

105 Chen, N., Zhang, E.Y., Tan, K., Wang, C.R., and Lu, X. (2007) Size effect of encaged clusters on the exohedral chemistry of endohedral fullerenes: a case study on the pyrrolidino reaction of $Sc_xGd_{3-x}N$@C_{80} ($x = 0$–3). *Org. Lett.*, **9**, 2011–2013.

106 Cai, T., Slebodnick, C., Xu, L., Harich, K., Glass, T.E., Chancellor, C., Fettinger, J.C., Olmstead, M.M., Balch, A.L., Gibson, H.W., and Dorn, H.C. (2006) A pirouette on a metallofullerene sphere: interconversion of isomers of N-tritylpyrrolidino I_h Sc_3N@C_{80}. *J. Am. Chem. Soc.*, **128**, 6486–6492.

107 Yamada, M., Wakahara, T., Nakahodo, T., Tsuchiya, T., Maeda, Y., Akasaka, T., Yoza, K., Horn, E., Mizorogi, N., and Nagase, S. (2006) Synthesis and structural characterization of endohedral pyrrolidinodimetallofullerene: La_2@$C_{80}(CH_2)_2NTrt$. *J. Am. Chem. Soc.*, **128**, 1402–1403.

108 Cai, T., Xu, L., Gibson, H.W., Dorn, H.C., Chancellor, C.J., Olmstead, M.M., and Balch, A.L. (2007) SC_3N@C_{78}: encapsulated cluster regiocontrol of adduct docking on an ellipsoidal metallofullerene sphere. *J. Am. Chem. Soc.*, **129**, 10795–10800.

109 Yamada, M., Wakahara, T., Tsuchiya, T., Maeda, Y., Kako, M., Akasaka, T., Yoza, K., Horn, E., Mizorogi, N., and Nagase, S. (2008) Location of the metal atoms in Ce_2@C_{78} and its bis-silylated derivative. *Chem. Commun.*, 558–560.

110 Lu, X., Nikawa, H., Tsuchiya, T., Maeda, Y., Ishitsuka, M.O., Akasaka, T., Toki, M., Sawa, H., Slanina, Z., Mizorogi, N., and Nagase, S. (2008) Bis-carbene adducts of non-IPR La_2@C_{72}: localization of high reactivity around fused pentagons and electrochemical properties. *Angew. Chem. Int. Ed.*, **47**, 8642–8645.

111 Yamada, M., Feng, L., Wakahara, T., Tsuchiya, T., Maeda, Y., Lian, Y.F., Kako, M., Akasaka, T., Kato, T., Kobayashi, K., and Nagase, S. (2005) Synthesis and characterization of exohedrally silylated $M@C_{82}$ (M = Y and La). *J. Phys. Chem. B*, **109**, 6049–6051.

112 Akasaka, T., Okubo, S., Kondo, M., Maeda, Y., Wakahara, T., Kato, T., Suzuki, T., Yamamoto, K., Kobayashi, K., and Nagase, S. (2000) Isolation and characterization of two $Pr@C_{82}$ isomers. *Chem. Phys. Lett.*, **319**, 153–156.

113 Akasaka, T., Nagase, S., Kobayashi, K., Suzuki, T., Kato, T., Yamamoto, K., Funasaka, H., and Takahashi, T. (1995) Exohedral derivatization of an endohedral metallofullerene $Gd@C_{82}$. *J. Chem. Soc., Chem. Commun.*, 1343–1344.

114 Wakahara, T., Yamada, M., Takahashi, S., Nakahodo, T., Tsuchiya, T., Maeda, Y., Akasaka, T., Kako, M., Yoza, K., Horn, E., Mizorogi, N., and Nagase, S. (2007) Two-dimensional hopping motion of encapsulated La atoms in silylated $La_2@C_{80}$. *Chem. Commun.*, 2680–2682.

115 Yamada, M., Nakahodo, T., Wakahara, T., Tsuchiya, T., Maeda, Y., Akasaka, T., Kako, M., Yoza, K., Horn, E., Mizorogi, N., Kobayashi, K., and Nagase, S. (2005) Positional control of encapsulated atoms inside a fullerene cage by exohedral addition. *J. Am. Chem. Soc.*, **127**, 14570–14571.

116 Akasaka, T., Nagase, S., Kobayashi, K., Suzuki, T., Kato, T., Kikuchi, K., Achiba, Y., Yamamoto, K., Funasaka, H., and Takahashi, T. (1995) Synthesis of the first adducts of the dimetallofullerenes $La_2@C_{80}$ and $SC_2@C_{84}$ by addition of a disilirane. *Angew. Chem. Int. Ed.*, **34**, 2139–2141.

117 Wakahara, T., Iiduka, Y., Ikenaga, O., Nakahodo, T., Sakuraba, A., Tsuchiya, T., Maeda, Y., Kako, M., Akasaka, T., Yoza, K., Horn, E., Mizorogi, N., and Nagase, S. (2006) Characterization of the bis-silylated endofullerene $Sc_3N@C_{80}$. *J. Am. Chem. Soc.*, **128**, 9919–9925.

118 Chuang, S.C., Santhosh, K.C., Lin, C.H., Wang, S.L., and Cheng, C.H. (1999) Synthesis and chemistry of fullerene derivatives bearing phosphorus

substituents. Unusual reaction of phosphines with electron-deficient acetylenes and C_{60}. *J. Org. Chem.*, **64**, 6664–6669.

119 Li, X.F., Fan, L.Z., Liu, D.F., Sung, H.H.Y., Williams, I.D., Yang, S., Tan, K., and Lu, X. (2007) Synthesis of a $Dy@C_{82}$ derivative bearing a single phosphorus substituent via a zwitterion approach. *J. Am. Chem. Soc.*, **129**, 10636–10637.

120 Lukoyanova, O., Cardona, C.M., Rivera, J., Lugo-Morales, L.Z., Chancellor, C.J., Olmstead, M.M., Rodriguez-Fortea, A., Poblet, J.M., Balch, A.L., and Echegoyen, L. (2007) "Open rather than closed" malonate methano-fullerene derivatives. The formation of methanofulleroid adducts of $Y_3N@C_{80}$. *J. Am. Chem. Soc.*, **129**, 10423–10430.

121 Fatouros, P.P., Corwin, F.D., Chen, Z.J., Broaddus, W.C., Tatum, J.L., Kettenmann, B., Ge, Z., Gibson, H.W., Russ, J.L., Leonard, A.P., Duchamp, J.C., and Dorn, H.C. (2006) *In vitro* and *in vivo* imaging studies of a new endohedral metallofullerene nanoparticle. *Radiology*, **240**, 756–764.

122 Cai, T., Xu, L., Shu, C., Champion, H.A., Reid, J.E., Anklin, C., Anderson, M.R., Gibson, H.W., and Dorn, H.C. (2008) Selective formation of a symmetric $Sc_3N@C_{78}$ bisadduct: adduct docking controlled by an internal trimetallic nitride cluster. *J. Am. Chem. Soc.*, **130**, 2136–2137.

123 Chaur, M.N., Melin, F., Athans, A.J., Elliott, B., Walker, K., Holloway, B.C., and Echegoyen, L. (2008) The influence of cage size on the reactivity of trimetallic nitride metallofullerenes: a mono- and bis-methanoadduct of $Gd_3N@C_{80}$ and a monoadduct of $Gd_3N@C_{84}$. *Chem. Commun.*, 2665–2667.

124 Feng, L., Nakahodo, T., Wakahara, T., Tsuchiya, T., Maeda, Y., Akasaka, T., Kato, T., Horn, E., Yoza, K., Mizorogi, N., and Nagase, S. (2005) A singly bonded derivative of endohedral metallofullerene: $La@C_{82}CBr$ $(COOC_2H_5)_2$. *J. Am. Chem. Soc.*, **127**, 17136–17137.

125 Feng, L., Wakahara, T., Nakahodo, T., Tsuchiya, T., Piao, Q., Maeda, Y., Lian, Y.,

Akasaka, T., Horn, E., Yoza, K., Kato, T., Mizorogi, N., and Nagase, S. (2006) The bingel monoadducts of La@C$_{82}$: synthesis, characterization, and electrochemistry. *Chem. Eur. J.*, **12**, 5578–5586.

126 Feng, L., Tsuchiya, T., Wakahara, T., Nakahodo, T., Piao, Q., Maeda, Y., Akasaka, T., Kato, T., Yoza, K., Horn, E., Mizorogi, N., and Nagase, S. (2006) Synthesis and characterization of a bisadduct of La@C$_{82}$. *J. Am. Chem. Soc.*, **128**, 5990–5991.

127 Shu, C.Y., Cai, T., Xu, L.S., Zuo, T.M., Reid, J., Harich, K., Dorn, H.C., and Gibson, H.W. (2007) Manganese(III)-catalyzed free radical reactions on trimetallic nitride endohedral metallofullerenes. *J. Am. Chem. Soc.*, **129**, 15710–15717.

128 Shustova, N.B., Popov, A.A., Mackey, M.A., Coumbe, C.E., Phillips, J.P., Stevenson, S., Strauss, S.H., and Boltalina, O.V. (2007) Radical trifluoromethylation of Sc$_3$N@C$_{80}$. *J. Am. Chem. Soc.*, **129**, 11676–11677.

129 Kareev, I.E., Lebedkin, S.F., Bubnov, V.P., Yagubskii, E.B., Ioffe, I.N., Khavrel, P.A., Kuvychko, I.V., Strauss, S.H., and Boltalina, O.V. (2005) Trifluoromethylated endohedral metallofullerenes: synthesis and characterization of Y@C$_{82}$(CF$_3$)$_5$. *Angew. Chem. Int. Ed.*, **44**, 1846–1849.

130 Tumanskii, B.L. and Kalina, O.G. (eds) (2001) *Endofullerenes: Radical Reaction of Fullerene Derivatives*, Kluwer Academic Publishers, Dordrecht, The Netherlands.

131 Tagmatarchis, N., Taninaka, A., and Shinohara, H. (2002) Production and EPR characterization of exohedrally perfluoroalkylated paramagnetic lanthanum metallofullerenes: (La@C$_{82}$)-(C$_8$F$_{17}$)$_2$. *Chem. Phys. Lett.*, **355**, 226–232.

132 Takano, Y., Yomogida, A., Nikawa, H., Yamada, M., Wakahara, T., Tsuchiya, T., Ishitsuka, M.O., Maeda, Y., Akasaka, T., Kato, T., Slanina, Z., Mizorogi, N., and Nagase, S. (2008) Radical coupling reaction of paramagnetic endohedral metallofullerene La@C$_{82}$. *J. Am. Chem. Soc.*, **130**, 16224–16230.

133 Mikawa, M., Kato, H., Okumura, M., Narazaki, M., Kanazawa, Y., Miwa, N., and Shinohara, H. (2001) Paramagnetic water-soluble metallofullerenes having the highest relaxivity for MRI contrast agents. *Bioconjug. Chem.*, **12**, 510–514.

134 Bolskar, R.D., Benedetto, A.F., Husebo, L.O., Price, R.E., Jackson, E.F., Wallace, S., Wilson, L.J., and Alford, J.M. (2003) First soluble M@C$_{60}$ derivatives provide enhanced access to metallofullerenes and permit *in vivo* evaluation of Gd@C$_{60}$[C (COOH)$_2$]$_{10}$ as a MRI contrast agent. *J. Am. Chem. Soc.*, **125**, 5471–5478.

135 Shu, C.Y., Zhang, E.Y., Xiang, J.F., Zhu, C.F., Wang, C.R., Pei, X.L., and Han, H.B. (2006) Aggregation studies of the water-soluble gadofullerene magnetic resonance imaging contrast agent: [Gd@C$_{82}$O$_6$(OH)$_{16}$(NHCH$_2$CH$_2$COOH)$_8$]$_x$. *J. Phys. Chem. B*, **110**, 15597–15601.

136 MacFarland, D.K., Walker, K.L., Lenk, R.P., Wilson, S.R., Kumar, K., Kepley, C.L., and Garbow, J.R. (2008) Hydrochalarones: a novel endohedral metallofullerene platform for enhancing magnetic resonance imaging contrast. *J. Med. Chem.*, **51**, 3681–3683.

137 Toth, E., Bolskar, R.D., Borel, A., Gonzalez, G., Helm, L., Merbach, A.E., Sitharaman, B., and Wilson, L.J. (2005) Water-soluble gadofullerenes: toward high-relaxivity, pH-responsive MRI contrast agents. *J. Am. Chem. Soc.*, **127**, 799–805.

138 Laus, S., Sitharaman, B., Toth, V., Bolskar, R.D., Helm, L., Asokan, S., Wong, M.S., Wilson, L.J., and Merbach, A.E. (2005) Destroying gadofullerene aggregates by salt addition in aqueous solution of Gd@C$_{60}$(OH)$_x$ and Gd@C$_{60}$[C (COOH)$_2$]$_{10}$. *J. Am. Chem. Soc.*, **127**, 9368–9369.

139 Okazaki, T., Shimada, T., Suenaga, K., Ohno, Y., Mizutani, T., Lee, J., Kuk, Y., and Shinohara, H. (2003) Electronic properties of Gd@C$_{82}$ metallofullerene peapods: (Gd @ C$_{82}$)$_n$@SWNTs. *Appl. Phys. A*, **76**, 475–478.

140 Hirahara, K., Suenaga, K., Bandow, S., Kato, H., Okazaki, T., Shinohara, H., and Iijima, S. (2000) One-dimensional

metallofullerene crystal generated inside single-walled carbon nanotubes. *Phys. Rev. Lett.*, **85**, 5384–5387.

141 Chiu, P.W., Gu, G., Kim, G.T., Philipp, G., Roth, S., Yang, S.F., and Yang, S. (2001) Temperature-induced change from p to n conduction in metallofullerene nanotube peapods. *Appl. Phys. Lett.*, **79**, 3845–3847.

142 Debarre, A., Jaffiol, R., Julien, C., Nutarelli, D., Richard, A., and Tchenio, P. (2003) Specific Raman signatures of a dimetallofullerene peapod. *Phys. Rev. Lett.*, **91**, 085501–085504.

143 Debarre, A., Jaffiol, R., Julien, C., Richard, A., Nutarelli, D., and Tchenio, P. (2003) Antenna effect in dimetallofullerene peapods. *Chem. Phys. Lett.*, **380**, 6–11.

144 Kalbac, M., Kavan, L., Zukalova, M., Yang, S.F., Cech, J., Roth, S., and Dunsch, L. (2007) The change of the state of an endohedral fullerene by encapsulation into SWCNT: a Raman spectroelectrochemical study of $Dy_3N@C_{80}$ peapods. *Chem. Eur. J.*, **13**, 8811–8817.

145 Tsuchiya, T., Kumashiro, R., Tanigaki, K., Matsunaga, Y., Ishitsuka, M.O., Wakahara, T., Maeda, Y., Takano, Y., Aoyagi, M., Akasaka, T., Liu, M.T.H., Kato, T., Suenaga, K., Jeong, J.S., Iijima, S., Kimura, F., Kimura, T., and Nagase, S. (2008) Nanorods of endohedral metallofullerene derivative. *J. Am. Chem. Soc.*, **130**, 450.

146 Pinzon, J.R., Cardona, C.M., Athans, A.J., Gayathri, S.S., Guldi, D.M., Herranz, M.A., Martin, N., Torres, T., and Echegoyen, L. (2007) $Sc_3N@C_{80}$-ferrocene electron-donor/acceptor conjugates as promising materials for photovoltaic applications. *J. Am. Chem. Soc.*, **120**, 4241–4244.

16
Carbon Nanostructures: Calculations of Their Energetics, Thermodynamics, and Stability

Zdeněk Slanina, Filip Uhlík, Shyi-Long Lee, Takeshi Akasaka, and Shigeru Nagase

16.1
Introduction

Fullerenes or cage compounds built exclusively from carbon atoms and their metal-containing forms known as metallofullerenes were for the first time observed in the gas phase by Kroto *et al.* [1, 2] in 1985, while fullerenes in crystalline form were prepared in 1990 by Krätschmer *et al.* [3]. An enormous volume of the observed and calculated data has been accumulated (see, e.g., [4–10]). In addition to spheroidal fullerene cages, other objects like elongated cylindrical bodies known as nanotubes, prepared by Iijima [11] soon after the fullerene synthesis [3], nanocones [12], or peapods [13] have been studied extensively, partially since they could be applied, in quantum computing and molecular electronics [14]. Fullerenes and metallofullerenes have also been treated by computational and other theoretical techniques [15–17]. As their experimental characterization is frequently based [18–20] on ^{13}C NMR spectroscopy, computational support is essential for the structure elucidations as well as for a deeper insight into various measured quantities and phenomena. Fullerenes themselves are polyhedral cages containing only carbon atoms arranged into five- and six-membered rings whereas *quasi*-fullerenes contain [21] other types of cycles.

Historically, fullerenes can be related [22–26] to mass spectrometric observations of carbon clusters up to C_{15} by Hahn and coworkers [27], later on expanded [28–30] up to C_{33}. In 1960s, the first simple computation appeared [31–34] and qualitative estimations for larger carbon cages including C_{60} were presented [35–39], especially in Schultz's treatment [35] of fully hydrogenated polyhedral carbon cages (which actually agrees reasonably well even with the present advanced calculations [40]). When C_{60} was discovered [1], computations could readily supply useful supporting data [16, 17, 41], including its IR spectrum [42–48]. Fullerene research [49, 50] has developed in a close theory-experiment cooperation, mostly based on large-scale computations [51, 52]. Dozens of surveys are available [53–80]; however, the computations themselves are still reviewed quite rarely [16, 17, 41, 81–90].

Carbon Nanotubes and Related Structures. Edited by Dirk M. Guldi and Nazario Martín
Copyright © 2010 WILEY-VCH Verlag GmbH & Co. KGaA, Weinheim
ISBN: 978-3-527-32406-4

Studies of C_{60} and higher fullerenes [19, 91–93] have been widely based on the so-called isolated pentagon rule (IPR) [94, 95] – especially stable fullerenes should have all their pentagons surrounded only by hexagons. Such IPR cages can be quite numerous [96–99] and their equilibrium mixtures have been computed in agreement with experiments for C_{76}–C_{96} [100–142]. A similar isomeric interplay has been described for smaller fullerene systems like [143–146] C_{32} or C_{36} (though the IPR pattern does not exist below C_{60}). Metallofullerenes can also coexist [41] in several isomeric forms as well, for example [147–171] $Ca@C_{72}$, $Mg@C_{72}$, $Ca@C_{74}$, $Ca@C_{82}$, $La@C_{82}$, $Tm@C_{82}$, $Ti_2@C_{80}$, $Sc_2@C_{84}$, $Ti_2@C_{84}$, or $Sc_2@C_{76}$ (though violations of the IPR pattern are more likely). Isomeric sets are also produced [172–176] by derivatives of fullerenes or nanotubes. Although the interisomeric separation energies in such systems are important, they alone cannot predict the relative stabilities of the isomers. Owing to high temperatures, entropy contributions can in fact even overcompensate the enthalpy terms. Hence, the enthalpy–entropy interplay represents an essential issue in fullerene science.

16.2
Energetics and Thermodynamics of Clusters

The present quantum-chemical calculations of fullerenes start with the optimized geometries [177–181] obtained at semiempirical like [182–188] MNDO, AM1, PM3 or SAM1, *ab initio* Hartree–Fock self-consistent field (HF SCF), or density functional theory (DFT) levels. Various *ab initio* computational procedures are mostly applied using Gaussian [189, 190] and Spartan [191] program packages. Stability of the SCF wave function [192, 193] can be an issue for some fullerenic structures. The geometry optimizations are frequently followed by harmonic vibrational analysis (with an optional frequency scaling [194]) in order to check the types of the localized stationary points and also to generate IR or Raman spectra.

The relative populations or concentrations of m isomers can be expressed as their mole fractions, w_i, using the isomeric partition functions q_i. In terms of q_i and the ground state energy changes $\Delta H_{0,i}^0$, the mole fractions are given [195–197]

$$w_i = \frac{q_i \exp[-\Delta H_{0,i}^0/(RT)]}{\sum_{j=1}^m q_j \exp[-\Delta H_{0,j}^0/(RT)]}, \tag{16.1}$$

where R stands for the gas constant and T for the absolute temperature. Equation (16.1) is – under the conditions of the interisomeric thermodynamic equilibrium – a rigorous formula [198]. However, the partition functions are to be practically constructed within the rigid-rotor and harmonic-oscillator (RRHO) approximation. The partition functions q_i reflect the rotational, vibrational, electronic, symmetry, and chirality [199] contributions. The metallofullerene relative-stability treatment according to Eq. (16.1) can be further improved by an alternative approach for describing the encapsulate motions (i.e., in place of the RRHO approach). One can expect that if the encapsulate is relatively free to move within the cage, then, at sufficiently high temperatures, its behavior in different cages will bring about the same contribution to the partition functions. The contributions would then cancel out in Eq. (16.1). This simplification can be called free, fluctuating, or floating encapsulate model (FEM). In the FEM model, in addition to the removal of the three lowest vibrational frequencies, the symmetries of the cages should be treated as the highest possible one, considering the averaging effect of the large amplitude motions of the encapsulate.

Let us mention for the completeness that if the partition functions are neglected and the vibrational zero-point energy is extracted from $\Delta H^0_{0,i}$, the relative potential energies $\Delta E_{r,i}$ remain, giving the simple Boltzmann factors

$$w'_i = \frac{\exp[-\Delta E^0_{r,i}/(RT)]}{\sum_{j=1}^{m} \exp[-\Delta E^0_{r,j}/(RT)]}, \tag{16.2}$$

with no reference to entropy contributions. Clearly enough, the simple Boltzmann factors – in contrast to the properly evaluated mole fractions w_i from Eq. (16.1) – can never cross with a temperature change and thus are not really useful.

Fullerenes are cages (mostly) built from three-coordinated (sp^2) carbon atoms arranged in two types of rings, pentagons and hexagons. However, any polyhedra has to obey Euler's polyhedral theorem [200, 201] (more precisely, convex polyhedra)

$$V + F = E + 2, \tag{16.3}$$

where V denotes the number of vertexes (atoms), F is the number of faces (rings), and E is the number of edges (bonds). If only pentagons and hexagons are allowed for fullerenes, their numbers n_5 and n_6 give the total count of faces

$$F = n_5 + n_6. \tag{16.4}$$

All carbon atoms in fullerenes are three-coordinated; it must therefore hold for the number of edges

$$E = \frac{3V}{2} \tag{16.5}$$

as each bond is accounted twice. One can also count the edges through rings

$$E = \frac{5n_5 + 6n_6}{2}. \tag{16.6}$$

Hence

$$V = \frac{5n_5 + 6n_6}{3}. \tag{16.7}$$

Euler's theorem gives

$$V + F = \frac{3V}{2} + 2. \tag{16.8}$$

One can thus get for the ring counts

$$n_5 + n_6 = \frac{5n_5 + 6n_6}{6} + 2, \tag{16.9}$$

which can immediately be reduced to

$$n_5 = 12. \tag{16.10}$$

In other words, in any conventional fullerene C_n the number of five-membered rings must be equal to 12. The number of six-membered rings is variable; however, Eq. (16.7) gives a relationship to the number of carbon atoms:

$$n_6 = \frac{n-20}{2}. \tag{16.11}$$

According to Eq. (16.11), the smallest possible fullerene has the stoichiometry C_{20}. As there are always 12 pentagons, the smallest IPR cage should have 12×5 carbon atoms – known as truncated icosahedron or buckminsterfullerene C_{60}. Findings (16.10) and (16.11) are actually quite general – they are valid for any polyhedral nanocarbon form as long as it is built from just five- and six-membered rings. If for example four- and seven-membered rings are also allowed, Euler's network closure requirement takes [143] the following form:

$$2n_4 + n_5 - n_7 = 12. \tag{16.12}$$

In particular, Eq. (16.12) says that the removal of two pentagons can be compensated by an increase of four-membered rings by one. In some convenient cases (obviously, connected pentagons), this step can result in a decrease of energy as shown [143] for C_{32} cages.

Let us now move from the fullerene topology to the energetics. There is a smooth, decreasing dependency [34, 51, 86, 202–205] (cf. also G.E. Scuseria, 1993, unpublished results) of the relative heats of formation $\Delta H_{f,298}^0/n$ on the number of atoms n in carbon clusters. This finding can readily be rationalized. Let us consider only the IPR fullerenes. We deal with two types of bonds – 5/6 (between pentagons and hexagons) and 6/6 (shared by two hexagons). If these two types of bonds can be represented [206] by some uniform dissociation energies, $H_{5/6}$ and $H_{6/6}$, we can readily write for the atomization heat

$$\Delta H_{at} = 60H_{5/6} + \left(\frac{3n}{2}-60\right)H_{6/6} \tag{16.13}$$

as an IPR fullerene C_n always has sixty 5/6 bonds, while the number of the 6/6 bonds is $\frac{3n}{2}-60$. The atomization and formation heats for carbon aggregates are related by the heat of vaporization of carbon, ΔH_{vap}:

$$\frac{\Delta H_f}{n} = -\frac{\Delta H_{at}}{n} + \Delta H_{vap}. \tag{16.14}$$

Thus,

$$\frac{\Delta H_f^0}{n} = -\frac{3}{2}H_{6/6} + \Delta H_{vap}^0 + \frac{60}{n}(H_{6/6}-H_{5/6}). \tag{16.15}$$

This functional dependency can be expressed as

$$\frac{\Delta H_f^0}{n} = A + \frac{B}{n},$$ (16.16)

where B is a positive constant, A is a relatively small number, and hence Eq. (16.16) indeed gives a smoothly decreasing curve.

From such a curve, however, one cannot yet see a particular stability of C_{60} or C_{70}. Thus, a still more complex description of the fullerene synthesis is needed. Even if we treat the problem from the thermodynamic point of view, pressure should be included [207–212]. For example, the equilibrium constant $K_{60/70}$ for an interconversion equilibrium between the two clusters expressed in the partial pressures p_i gives [207–211]

$$K_{60/70} = \frac{p_{70}^{6/7}}{p_{60}} = \frac{(1-x_{60})^{6/7}}{x_{60}} P^{-1/7},$$ (16.17)

where P stands for the total pressure of the two clusters and x_{60} is the mole fraction of buckminsterfullerene. At higher pressures, C_{70} is more populated than C_{60}, but at the conditions of a saturated carbon vapor, the stability order is reversed in favor of C_{60} so that an agreement with the experiment is obtained. Although the isomeric stability problem is much better understood [198, 213–215] at present than the relative stabilities of nonisomeric carbon clusters, there are some interesting results available for the nonisomeric situation [207–211]. This is especially true for a temperature increase in the clustering degree under the saturation conditions [216]. While the equilibrium constants for cluster formation decrease with temperature, the saturated pressure increases. It is just the competition between these two terms that decides the final temperature behavior [217–220].

16.3
Stabilities of Empty Fullerenes

The relative stabilities of the small carbon clusters have frequently been studied by both theoretical [221–239] and experimental [240–263] techniques. Ion-chromatography observations by von Helden *et al.* [260, 262, 263] showed that starting from about $n = 7$, cyclic rings always exist in addition to the linear forms. Cyclic structures were for example computed [235, 236, 239, 264–270] for $C_7 - C_{13}$. C_{11} should be the first species for which the cyclic structure becomes dominant. Similar transition from cyclic to polyhedral species was expected [271] around $n = 45$. The linear and cyclic (rhombic, bicyclic) C_4 isomers were especially frequently computed. Newer estimates [272–276] suggest only a small separation between the two forms. Then, the entropy contributions lead to the linear species at higher temperatures [244, 277–279] in agreement with ion chromatography [260]. A similar effect was also computed [213, 221, 239, 280–282] for C_6.

Exhaustive topological generations of fullerene cages represent a crucial step for computational studies, and they were systematically treated by Fowler *et al.* [96, 97, 283–301] and other groups [95, 98, 99, 302–330], employing topological concepts like ring spiral [96, 97, 307], Goldberg polyhedra [287], leapfrog transformation [295], topological duals [305], or Stone–Wales transformations [303, 331–334]. It was found [307–312] that, for example, the number of C_{60} isomers is 1812. Odd-numbered cages are usually not considered in the enumerations though they are known [313, 335, 336]. There are also applications of Pólya's theorem [198, 337] to various substituted fullerenes [320, 323, 329, 330].

As von Helden *et al.* [260, 271] observed gas phase fullerene-like structures around C_{30}, smaller fullerenes have also been treated. Even the smallest fullerene C_{20} is available [338–341], and thus computed [215, 342–345]. C_{36} fullerene was also isolated [346], computed [144, 145, 347, 348], and linked to narrow nano-tubes [349–352]. Another computed smaller fullerene is [143] C_{32} though not yet isolated.

However, presently the relative populations of the isomeric IPR cages are primarily studied, from C_{76} till C_{98} [100–142]. In fact, C_{72} and C_{74} are the only species [19, 92] among C_{60}–C_{96} not isolated yet (some access to C_{74} is possible [353, 354]). A low solubility [355] in conventional solvents may be one reason for the difficulties. Hence, as C_{72} could only be recorded in the gas phase [356, 357], its structure is unknown. There is just one IPR satisfying structure for C_{72}, namely with D_{6d} symmetry [97]. However, a non-IPR (i.e., IPR-violating) structure with one pentagon–pentagon junction [148] is by a few kcal/mol lower in energy than the IPR cage. Moreover, it was demonstrated [358] in the Si_{60} case that the IPR/non-IPR stability order can be reversed by the entropy factors. Hence, the C_{72} system was investigated at semiempirical [359] and DFT level [360]. At the DFT level, the geometry optimizations were carried out [360] using the B3LYP/3-21G treatment. In the optimized B3LYP/3-21G geometries the harmonic vibrational analysis was carried out and the B3LYP/6-31G* separation energies were evaluated. The electronic excitation energies were evaluated by means of the ZINDO method [361]. The following structures were considered [360]: the IPR cage (a), two non-IPR cages [148] with one pentagon–pentagon junction (b) and (c), a structure [148] with one heptagon (d), a cage [149] with two heptagons (e), and two structures [355] each with two pentagon–pentagon junctions (f), (g). The (c) structure of C_{2v} symmetry with just one pentagon/pentagon fusion represents the lowest energy isomer. Figure 16.1 presents the DFT-computed temperature development of the relative concentrations of the seven C_{72} isomers in a high temperature region. The lowest energy structure (c) is the most populated species at any temperature. On the other hand, the IPR structure (a) is always negligible. The C_{74} case is, nevertheless, different. Recently, Shinohara and coworkers [354] recorded electronic spectrum of C_{74} anion and suggested that the cage could have D_{3h} symmetry (i.e., the only one available IPR structure [97]). This finding prompted DFT calculations [362] that treated a set of six isomers, five of them being non-IPR species. However, the computations show that the IPR structure prevails at any relevant temperature in agreement with the experiment [354].

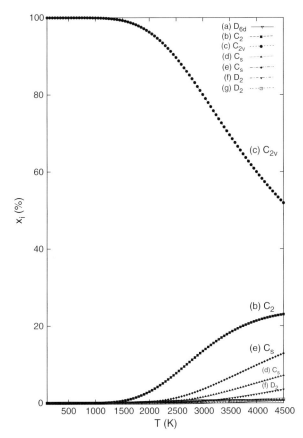

Figure 16.1 Relative concentrations [360] of the C_{72} isomers based on the B3LYP/6-31G* energetics and the B3LYP/3-21G and ZINDO entropy.

16.4
Stabilities of Metallofullerenes

The relative populations of the individual empty cages do not directly imply a stability order for the related metallofullerenes. The reason is rooted in the charge transfer from the encapsulated metal atom to the cage – the transferred charge roughly corresponds to the encapsulated atom valence (quantum-chemical calculations basically give the amount of transferred charge as nonintegral numbers). Hence, for metallofullerenes the populations of the charged cages are obviously more relevant now than those of the neutral empty species. This interplay can, for example, be illustrated [41] on $La_2@C_{80}$. For C_{80}, there are [117–119] seven IPR cages, the highest in energy being an icosahedral structure (I_h symmetry). The icosahedral cage actually undergoes a Jahn–Teller distortion as it has a fourfold degenerate HOMO, partially filled with just two electrons (owing to the Jahn–Teller effect, the cage symmetry is in fact reduced to D_2). However, upon encapsulation of two La atoms, the cage gets those six electrons in order to completely fill its HOMO. Moreover, $La_2@C_{80}$ derived from the icosahedral C_{80} cage is actually the lowest-energy isomer and it was indeed isolated [363] as the most populated $La_2@C_{80}$ species.

The stability evaluations according to Eq. (16.1) have also been applied to a few metallofullerenes like $Ca@C_{72}$, $Ca@C_{74}$, $Ca@C_{82}$, $La@C_{82}$, and other endohedrals of current interest (cf. [364–376]). $Ca@C_{72}$ [147], and recently even its second isomer [364] were isolated, however, their structures are not yet known. It follows from the very first $Ca@C_{72}$ computations [148, 149] that there are four isomers especially low in potential energy. The endohedral $Ca@C_{72}$ species created by putting Ca inside the sole IPR cage has been labeled [148] by (a). The other three $Ca@C_{72}$ isomers considered in Ref. 148 are related to two non-IPR C_{72} cages (b) and (c), and to a C_{72} structure with one heptagon (d). There is also an interesting species [149] with two heptagons (e) (see Figure 16.2). The computations [150] were performed in the 3-21G basis set for C atoms and a dz basis [365] and the effective core potential (ECP) on Ca combined with the B3LYP functional (B3LYP/3-21G~dz). The electronic excitation energies were evaluated by means of the time-dependent DFT response theory [366] at the B3LYP/3-21G~dz level. Figure 16.3 presents [150] the RRHO and FEM temperature developments of the relative concentrations of the five $Ca@C_{72}$ isomers in a high temperature region. At a low temperature the relative concentrations of the (c) and (b) structures are interchanged and beyond this point the (b) structure is always somewhat more populated.

In contrast to $Ca@C_{72}$, $Ca@C_{74}$ was not only isolated [147] but even its structure was determined [153]. According to the ^{13}C NMR spectra recorded by Achiba and coworkers [153], $Ca@C_{74}$ exhibits D_{3h} symmetry of its cage. There is only one IPR structure possible [97] for C_{74} and the sole C_{74} IPR cage has D_{3h} symmetry. A set of altogether six $Ca@C_{74}$ isomers was subjected to the stability computations [154]. It turns out, in agreement with the experiment, that the encapsulate with the IPR cage has not only favorable enthalpy but also entropy term and thus, all the remaining isomers can act as minor species at best (Figure 16.4). A similar IPR-cage dominance also operates in the $Ba@C_{74}$ system [376] as shown in Figure 16.5.

The third illustrative system, $Ca@C_{82}$, exhibits the richest isomerism among the Ca endohedrals [147, 155, 157, 366–368]. Shinohara and coworkers [155] isolated four isomers of $Ca@C_{82}$ and labeled the isomers by (I), (II), (III), and (IV). Dennis and Shinohara concluded [158, 370] from the ^{13}C NMR spectra of $Ca@C_{82}$(III) its symmetry as C_2. The ultraviolet photoelectron spectra [159] support the finding and a similarity with $Tm@C_{82}$(II) was also noted [167]. Achiba and coworkers [160] measured the ^{13}C NMR spectra and assigned the symmetry of isomers (I), (II), (III), and (IV) as C_s, C_{3v}, C_2, and C_{2v}, respectively. The $Ca@C_{82}$ species were also computed [156] and the C_{2v} structure was found as the lowest-energy isomer. There were still three other low energy species – C_s, C_2, and C_{3v} (there are some symmetry reductions in the calculations: $C_{3v}(b) \rightarrow C_s$, $C_{2v} \rightarrow C_s$, $C_2(a) \rightarrow C_1$) $C_2(b) \rightarrow C_1$, $C_s(b) \rightarrow C_1$. Figure 16.6 presents the temperature development [161] of the relative concentrations of the nine $Ca@C_{82}$ isomers in the FEM treatment (the enthalpy part from the B3LYP/6-31G*//B3LYP/3-21G~dz calculations, the entropy part evaluated at the B3LYP/3-21G~dz level). In fact, the FEM approach is in a better agreement with

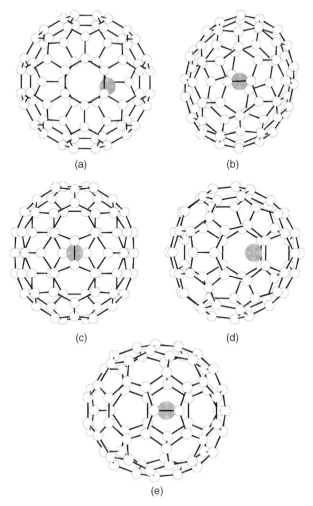

Figure 16.2 B3LYP/3-21G~dz optimized structures of Ca@C$_{72}$ isomers: (a) IPR, (b) 5/5 pair, (c) 5/5 pair, (d) seven-membered ring, (e) 2 seven-membered rings.

the available observed concentration information than the conventional RRHO approach. The observed yields [160] of the isomers were nearly equal except for the considerably less-produced C$_{3v}$ species (though the HPLC chromatograms [155] could indicate somewhat larger production differences). The FEM treatment basically reproduces the relative populations of the four isomers.

Another illustration can be served with La@C$_{82}$, that is, an electronic open-shell system. The La@C$_{82}$ metallofullerene is one of the very first endohedrals that was macroscopically produced [371]. Recently, structures of two of its isomers were clarified [163, 164] using ^{13}C NMR spectra of their monoanions generated electrochemically. The major isomer [163] was thus assigned C$_{2v}$ symmetry and the minor species [164] C$_s$. The C$_{2v}$ structure was moreover confirmed by an X-ray powder

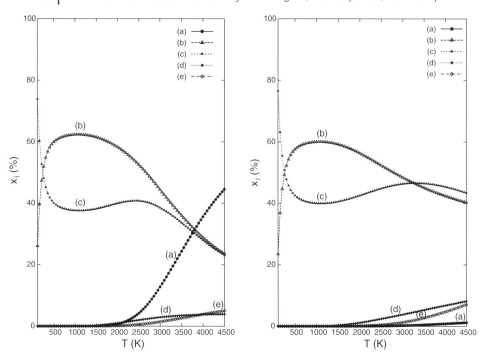

Figure 16.3 Relative concentrations of the $Ca@C_{72}$ isomers based on the B3LYP/6-31G* energetics and the B3LYP/3-21G~dz entropy, using the RRHO [150] (left) and FEM (right) treatment.

diffraction study [372]. The findings stand in a contrast to $Ca@C_{82}$ with four known isomers. Computations at *ab initio* HF and DFT levels pointed out [121, 162, 373] just three IPR cages with a sufficiently low energy after La atom encapsulation: C_{2v}, C_{3v}(b), and C_s(c). The fourth lowest La endohedral species, C_2(a), is actually already too high in energy to be significant in the experiment. A partial agreement with experiment can be achieved [374] for temperatures roughly from 1000 to 1300 K using the RRHO treatment, however, a really good agreement with the observed facts [163, 164] is reached with the FEM approach (Figure 16.7). There is still one important aspect to be considered. The fullerene and metallofullerene production is not always close to the interisomeric equilibrium. This factor may be pertinent to the $La@C_{82}$ case. Lian *et al.* [375] reported a Ni-catalyzed production of $La@C_{82}$ with a considerably variable isomeric ratio.

There is a general stability problem related to fullerenes and metallofullerenes – relative stabilities of clusters with different stoichiometries. Let us consider a series of metallofullerene formations with one common cage $X@C_n$ and variable encapsulated metals X:

$$X(g) + C_n(g) = X@C_n(g). \tag{16.18}$$

The encapsulation process is thermodynamically characterized by the standard changes of, for example, enthalpy $\Delta H^0_{X@C_n}$ or the Gibbs energy $\Delta G^0_{X@C_n}$. An

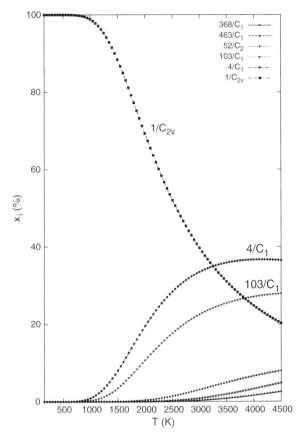

Figure 16.4 Relative concentrations of the Ca@C$_{74}$ isomers based on the B3LYP/6-31G* energetics and the B3LYP/3-21G~dz entropy, using the FEM treatment [154].

illustration will be given on a series Ca@C$_{74}$, Sr@C$_{74}$, and Ba@C$_{74}$ with the potential-energy changes computed at the B3LYP/6-31G*~dz level and the entropy part at the B3LYP/3-21G~dz level. The equilibrium composition of the reaction mixture is controlled by the encapsulation equilibrium constants $K_{X@C_n,p}$

$$K_{X@C_n,p} = \frac{p_{X@C_n}}{p_X p_{C_n}} \tag{16.19}$$

expressed in the terms of partial pressures of the components. The encapsulation equilibrium constant is interrelated with the standard encapsulation Gibbs energy change

$$\Delta G^0_{X@C_n} = -RT \ln K_{X@C_n,p}. \tag{16.20}$$

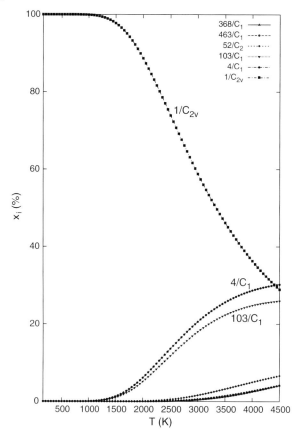

Figure 16.5 Relative concentrations of the Ba@C_{74} isomers based on the B3LYP/6-31G*~dz energetics and the B3LYP/3-21G~dz entropy, using the FEM treatment.

Temperature dependency of the encapsulation equilibrium constant $K_{X@C_n,p}$ is then described by the van't Hoff equation

$$\frac{d \ln K_{X@C_n,p}}{dT} = \frac{\Delta H^0_{X@C_n}}{RT^2}, \qquad (16.21)$$

where the $\Delta H^0_{X@C_n}$ term is typically negative so that the encapsulation equilibrium constants decrease with increase in temperature. Let us further suppose that the metal pressure p_X is actually close to the respective saturated pressure $p_{X,\mathrm{sat}}$ so that we can deal with a special case of clustering under saturation conditions [377]. While the saturated pressures $p_{X,\mathrm{sat}}$ for various metals are known from observations [378], the partial pressure of C_n is less clear (though, p_{C_n} should exhibit a temperature maximum and then vanish). As mentioned already, the computed equilibrium constants $K_{X@C_n,p}$ have to show a temperature decrease with respect to the van't Hoff equation(16.21), which, however, does not necessarily mean a decrease in yield with increase in the temperature. To clarify this aspect, let us consider the combined $p_{X,\mathrm{sat}} K_{X@C_n,p}$ term

$$p_{X@C_n} \sim p_{X,\mathrm{sat}} K_{X@C_n,p} \qquad (16.22)$$

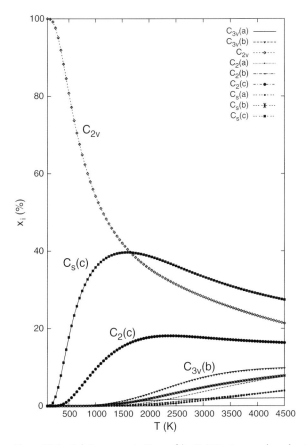

Figure 16.6 Relative concentrations of the Ca@C_{82} isomers based on the B3LYP/6-31G* energetics and the B3LYP/3-21G~dz entropy, using the FEM treatment [161].

that directly controls the partial pressures of various X@C_n encapsulates in an endohedral series (based on one common C_n fullerene). Then, we get a different picture. The considered $\Xi = p_{X,\text{sat}} K_{X@C_n,p}$ term can, indeed, frequently (though not necessarily) increase with temperature:

$$\frac{d\Xi}{dT} = \frac{d(p_{X,\text{sat}} K_{X@C_n,p})}{dT} > 0. \tag{16.23}$$

The saturated pressure is related to the heat of vaporization

$$\frac{d \ln p_{X,\text{sat}}}{dT} = \frac{\Delta H_{X,\text{vap}}}{RT^2} \tag{16.24}$$

and consequently;

$$\frac{d\Xi}{dT} = \frac{p_{X,\text{sat}} K_{X@C_n,p}}{RT^2} \left[\Delta H_{X,\text{vap}} + \Delta H^0_{X@C_n} \right]. \tag{16.25}$$

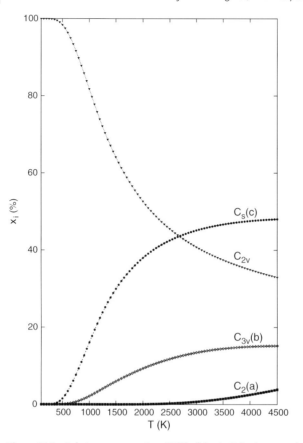

Figure 16.7 Relative concentrations [374] of the La@C_{82} isomers derived within the FEM approach.

If $\Delta H_{X,\mathrm{vap}} > |\Delta H^0_{X@C_n}|$, the $p_{X,\mathrm{sat}} K_{X@C_n,p}$ quotient increases with temperature. In fact, while for Ca@C_{74}, the $p_{X,\mathrm{sat}} K_{X@C_{74},p}$ quotient increases with temperature, it is about constant for Sr@C_{74}, and it decreases with temperature for Ba@C_{74}. It is more convenient to deal with the relative quotient $\frac{p_{X,\mathrm{sat}} K_{X@C_{74},p}}{p_{Ba,\mathrm{sat}} K_{Ba@C_{74},p}}$. The computed values [379] of the relative quotient in the series Ca@C_{74}, Sr@C_{74}, and Ba@C_{74} at a temperature of 1500 K are 4.3×10^{-5}, 5.3×10^{-3}, 1.0, and at 2000 K they are 9.9×10^{-4}, 0.030, 1.0. Incidentally, the computed stability proportions do correlate with qualitative abundances known from observations. For Ba@C_{74}, even microcrystals could be prepared [380] so that a diffraction study was possible, while for Sr@C_{74} at least various spectra could be recorded [381] in solution, and Ca@C_{74} was studied [153] only by NMR spectroscopy (while Mg@C_{74} was never observed yet). Generally speaking, under some experimental arrangements, under-saturated or perhaps even super-saturated metal vapors are also possible (for example in the ion-bombardment production technique [382, 383]).

The series of metallofullerene formations with one common cage X@C$_n$ introduced in Eq. (16.18) allows another interesting stability conclusion. Three formal reaction steps can be considered for our illustrative series Mg@C$_{74}$, Ca@C$_{74}$, Sr@C$_{74}$, and Ba@C$_{74}$: (i) double-ionization of the free metal, (ii) double charging of the empty cage, and (iii) placing the metal dication into the dianionic cage. The (ii) energy is identical for all the members in the series, and the (iii) terms should be similar as they are controlled by electrostatics. Interestingly enough, the feature that the stabilization of metallofullerenes is mostly electrostatic can be documented [384] by using the topological concept of "atoms in molecules" (AIM) [385, 386], which indeed shows that the metal–cage interactions form ionic (and not covalent) bonds. Hence, the free-metal ionization potentials should actually represent a critical yield-controlling factor – the computed relative potential-energy changes upon encapsulation $\delta_{rel} \Delta E$ and the relative observed ionization potentials of the free atoms δ_{rel}IP should be correlated according to the above three-step analysis:

$$\delta_{rel} \Delta E \sim \delta_{rel} \text{ IP}. \tag{16.26}$$

This interesting conclusion is documented in Figure 16.8 that uses both the observed second and the first ionization potentials [387] (though the second ionization potentials are more relevant for the considered series). In fact, this correlation should operate for such a homologous reaction series of metal encapsulations into any type of carbon nanostructures. Moreover, this type of reasoning should step by step explain the fullerene-encapsulation stability islands known throughout the periodic system (though the underlying calculations are quite demanding).

The above-mentioned list of examples of the comprehensive stability calculations is not exhaustive. Among the members of the family of fullerene and metallofullerene cages computed within the framework, those with the already available X-ray crystallographic analysis have a special and important position as they offer reliable ultimate tests for computations [389]. A masterpiece of such well structurally characterized species is, for example, served [388] by La@C$_{74}$ (though based on a solvent-related derivative, namely La@C$_{74}$-C$_6$H$_3$Cl$_2$). Similar solvent-produced derivative (La@C$_{72}$-C$_6$H$_3$Cl$_2$) was also used [390] for the X-ray crystallographic analysis of La@C$_{72}$, revealing its non-IPR carbon cage. In fact, a metallofullerene reaction with solvent can selectively enhance the extraction of an unexpected cage [391]. The stability evaluations for metallofullrenes with more encapsulated metal atoms are also available, for example, for Sc$_3$N@C$_{80}$ [392–394]. However, the relative-stability calculations of, for example, metal-carbide matallofullerenes with encapsulates like Sc$_2$C$_2$ [395] or Sc$_3$C$_2$ [396] (and for many other [397, 398] interesting species), are still to be carried out.

16.5
Stabilities of Nonmetal Endohedrals

Recently, the encapsulation inside the fullerene cages has been extended from metal atoms to nonmetals and even to their small molecules. N$_2$@C$_{60}$ and N$_2$@C$_{70}$ were

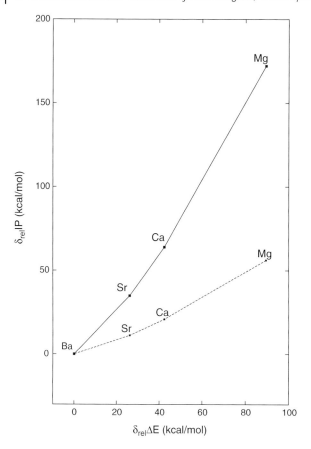

Figure 16.8 The computed relative potential-energy changes upon encapsulation $\delta_{rel}\,\Delta E$ and the observed [387] relative ionization potentials of the free atoms δ_{rel} IP for the series Mg@C$_{74}$, Ca@C$_{74}$, Sr@C$_{74}$, and Ba@C$_{74}$ (solid line – the second IPs, dashed line – the first IPs).

first prepared by Peres *et al.* [399] using heating under high pressure. Out of two thousand C$_{60}$ molecules, about one was observed to incorporate N$_2$. The nitrogen molecule containing endohedrals were present even after several hours of heating at 500 K. N$_2$@C$_{60}$ was also reported [400, 401] in the chromatographic separation after the nitrogen-ion implantation into C$_{60}$. This ion bombardment is in fact primarily used for the N@C$_{60}$ production [402, 403] though with very low yields. Still, N@C$_{60}$ and its derivatives have been studied vigorously [176, 404–407]. Other nonmetallic endohedrals are represented by complexes of fullerenes with rare gas atoms, in particular with He [20, 408–411]. Very recently, molecular hydrogen [412] and even a water molecule [413] were placed inside open-cage fullerenes. The cage with H$_2$ was subsequently closed [414]. Such nonmetallic-endohedrals systems have also been computed [415, 416]. In both the B3LYP/3-21G and the PW91/3-21G treatments, the lowest-energy N$_2$@C$_{60}$ structure found has the N$_2$ unit oriented toward a pair of parallel pentagons such that the complex exhibits D$_{5d}$ symmetry (5 : 5 structure, see

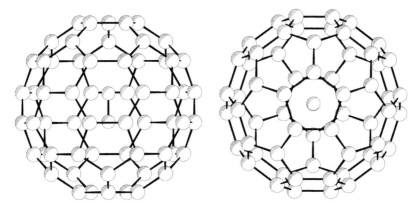

Figure 16.9 Two views of the PW91/3-21G optimized structure [416] of $N_2@C_{60}$.

Figure 16.9). Such a minimum energy structure was also computed for $NH_3@C_{60}$. The subsequent harmonic vibrational analysis confirms that the 5 : 5 structures are indeed local energy minima. For $N_2@C_{60}$, the MP2 = FC/6-31G* encapsulation energies before and after the BSSE correction are −17.5 and −9.28 kcal/mol, respectively. The BSSE-corrected MP2 = FC/6-31G* value for $NH_3@C_{60}$ is −5.23 kcal/mol. Once the corresponding entropy change ΔS_T^0 is evaluated, one can deal with the thermodynamics controlling Gibbs-energy term ΔG_T^0. Using the partition functions from the DFT calculations, the $T\,\Delta S_T^0$ term at room temperature for $N_2@C_{60}$ and $NH_3@C_{60}$ comes as −5.95 and −5.46 kcal/mol, respectively. If the entropy values are combined with the enthalpy terms derived from the BSSE corrected MP2 = FC/6-31G* stabilization energy, the ΔG_T^0 standard changes for reactions $N_2@C_{60}$ and $NH_3@C_{60}$ at room temperature read −2.64 and 1.53 kcal/mol, respectively. The values correspond to those, for example, of water dimerization in the gas phase [217].

16.6
Kinetic Control

In addition to the pristine fullerenes and metallofullerenes, their derivatives have also been studied vigorously as high expectations have been placed on them for applications ranging from nonlinear optics to medicine. Consequently, the computed properties of various available or proposed fullerene derivatives have been of immense interest. Historically, fullerene oxides were the first observed fullerene compounds and they have naturally attracted attention of both experiment and theory – see, for example, [417–427]. The oxides $C_{60}O_n$ have been observed up to $n = 4$ while computations have so far studied only mono- and dioxides.

With the monooxide, two structures have in particular been considered, originated in the bridging over a 5/6 or 6/6 bond (i.e., the bond shared by a pentagon–hexagon pair and by two hexagons, respectively) in the pristine C_{60} – 5/6 or 6/6 isomers. There had been a long standing issue that $C_{60}O$ had consistently been computed [417, 421] in disagreement with the observations. The ground state or the lowest-energy isomer in the system is calculated, almost regardless of the applied method, as the 5/6 species, while in the experiment only the 6/6 structure could originally be observed [418]. The theory-experiment disagreement cannot be explained through temperature (i.e., entropy) factors [421]. The problem becomes even more serious with $C_{60}O_2$. In the experiment, only 6/6 structures were reported [419]. However, in computations [421] the higher thermodynamic stability of the 5/6 monooxide is, somehow, conserved even with dioxides. Consequently, the thermodynamic-stability computations are unable to identify the computed most-populated isomer with the observed [419, 420] 6/6 species as among the computed bridged structures a 5/6 isomer is the lowest in energy. The disturbing problem was successfully explained as a possible kinetic control of the relative stabilities. The relatively detailed kinetic considerations [422] could indeed put the computations and the observations in agreement with each other. More recently, in experimental studies [422, 423] the yet missing 5/6 monooxide was found. However, as the 5/6 isomer was generated by photolysis, we deal with a photochemical and not with the thermochemical production which had been otherwise treated by the computations [421] (the photochemical generation would require computations of excited electronic states).

Moreover, the kinetic control of fullerene-derivative productions is obviously a more general feature, as documented for example also by calculations [426] of oxygen additions to narrow nanotubes. Yet another system where such kinetic computations were already carried out [427] is $C_{84}O$, actually as a support for photochemical studies [424]. In particular, monooxides of the two most abundant isomers of C_{84} (D_2 and D_{2d}) were computed [427]. Their full geometry optimizations and vibrational harmonic analyses were performed at semiempirical PM3 quantum-chemical level, not only for local energy minima but also for related activated complexes.

The D_{2d} C_{84} structure is the lowest-energy isomer but it is located only about 0.5 kcal/mol below the D_2 species. In each of the two cages, the shortest and the longest C−C bonds were selected. The selected bonds were then bridged by an O atom and the produced $C_{84}O$ structure was reoptimized. Hence, four species were treated: $C_{84}O$ D_2/long (6/6), $C_{84}O$ D_{2d}/short (6/6), $C_{84}O$ D_{2d}/long (5/6), and $C_{84}O$ D_2/short (6/6). In the D_2/long and D_{2d}/long structures the critical C−C bonds are actually broken (Figure 16.10). The D_2/long structure is the lowest-energy species. The related activated complexes are rather of keto-forms (Figure 16.10). Their vibrational eigenvector associated with the imaginary frequency has a clear physical meaning. In one direction it represents a motion toward dissociation of the C−O bond. In the opposite direction it leads to a formation of the second C−O bond and thus to the bridging (the computations deal only with singlet oxygen for simplicity). Though the PM3 computed activation enthalpies $\Delta H^{\neq}_{298}$ (evaluated just at room temperature) allow discussion of the kinetic control, we should also compute activation entropy $\Delta S^{\neq}_T$ (at some relevant representative temperature T) so that we can directly deal with the rate constant k, given by thermodynamic formulation of the

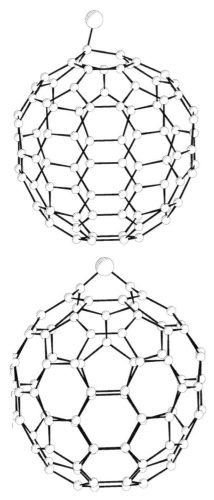

Figure 16.10 $C_{84}O$ – addition to the longest $C-C$ bond of D_{2d} C_{84}: PM3 optimized [427] minimum-energy structure and activated complex (top).

transition-state theory:

$$k = \frac{k_B T}{h} \exp\left[\Delta S_T^{\neq}/R\right] \exp\left[-\Delta H_T^{\neq}/(RT)\right]. \tag{16.27}$$

However, as long as we deal with moderate temperatures, the differences in the entropy terms should not be critical. The computed kinetic results [427] show that the kinetic order (within the thermochemical kinetics) is just reversed in comparison to the thermodynamic order. Hence, for relatively short reaction times the $C_{84}O$ D_2/short isomer should primarily be formed (in spite of the fact that it should be thermodynamically the least stable in the studied set). This finding corresponds to the conclusions for the $C_{60}O$ case [421] or for oxygen additions to narrow nanotubes [428]. The computational findings suggest that processes of fullerene derivatization could frequently be controlled by kinetics rather than by thermodynamics. The conclusion represents an important prerequisite for computational modeling of formation and properties of fullerene-derivative based materials. However, such kinetic computations are considerably more demanding and thus still quite rare compared to the vastly prevailing thermodynamic-stability calculations at present.

The suitability of density functional theory for evaluations of long-range molecular interactions has been an important computational issue [428–437] with a relevancy also for calculations of carbon nanostructures. There has been a growing evidence that the otherwise reliable B3LYP functional does not work sufficiently well at long distances, a feature otherwise important for a good description of weak complexes and kinetic transition states. Recently, Zhao and Truhlar [438–443] have performed a series of test DFT calculations with a conclusion [443] that the MPWB1K functional (the modified Perdew and Wang exchange functional MPW and Becke's meta correlation functional optimized against a kinetics database) is the best combination for evaluations of nonbonded interactions with a relative averaged mean unsigned error of only 11%. Fullerene encapsulations of nonmetal atoms and small molecules represent an interesting case for this type of computations as there are also some related observations. The MPWB1K functional was indeed tested [444] on $H_2@C_{60}$, $Ne@C_{60}$, and $N_2@C_{60}$ as a tool for evaluations of stabilization energies upon encapsulation of nonmetallic species into fullerenes. It was found that the MPWB1K values can be within a few kcal/mol from the MP2 or SCS-MP2 (spin-component scaled MP2) values so that further applications of the functional are clearly encouraged. The best estimates of the encapsulation-energy gains found for $H_2@C_{60}$, $Ne@C_{60}$, and $N_2@C_{60}$ are at least 4 kcal/mol, slightly less than 4 kcal/mol, and about 9 kcal/mol, respectively. The DFT calculations with this new family of functionals are likely to improve in the future description of some special structural and bonding situations in various carbon nanostructures.

The comprehensive stability calculations are a part of the steadily growing knowledge of the family of fullerene and metallofullerene cages, and are constantly attracting interest of more applied branches of nanoscience and nanotechnology [445] like quantum computing [446], superconductivity [447], and molecular medicine [448], while concepts for first nanocarbon-based nonotechnology devices are emerging [449–454].

Acknowledgments

The reported research has been supported by a Grant-in-aid for Scientific Research of Innovation Areas ("pi-Space"), the Twenty First Century COE Program, Nanotechnology Support Project, the Next Generation Super Computing Project (Nanoscience Project), and Scientific Research on Priority Area from the Ministry of Education, Culture, Sports, Science, and Technology of Japan, by the National Science Council, Taiwan-ROC, and by the Ministry of Education of the Czech Republic (MSM0021620857). The authors also wish to thank the following organizations for kindly permitting the reprinting of copyrighted material: American Institute of Physics; Elsevier Scientific Publishing Company.

References

1 Kroto, H.W., Heath, J.R., O'Brien, S.C., Curl, R.F., and Smalley, R.E. (1985) *Nature*, **318**, 162.

2 Heath, J.R., O'Brien, S.C., Zhang, Q., Liu, Y., Curl, R.F., Kroto, H.W., Tittel, F.K., and Smalley, R.E. (1985) *J. Am. Chem. Soc.*, **107**, 7779.

3 Krätschmer, W., Lamb, L.D., Fostiropoulos, K., and Huffman, D.R. (1990) *Nature*, **347**, 354.

4 Yoshimura, S. and Chang, R.P.H. (eds) (1998) *Supercarbon: Synthesis, Properties and Applications*, Springer-Verlag, Berlin.

5 Hirsch, A. (ed.) (1999) *Fullerenes and Related Structures*, Springer-Verlag, Berlin.

6 Dresselhaus, M.S., Dresselhaus, G., and Avouris, Ph. (eds) (2001) *Carbon Nanotubes: Synthesis, Structure, Properties and Applications*, Springer-Verlag, Berlin.

7 Benedek, G., Milani, P., and Ralchenko, V.G. (eds) (2001) *Nanostructured Carbon for Advanced Applications*, Kluwer Academic Publishers, Dordrecht.

8 Buhl, M. and Hirsch, A. (2001) *Chem. Rev.*, **101**, 1153.

9 Nagase, S., Kobayashi, K., and Akasaka, T. (1996) *Bull. Chem. Soc. Jpn.*, **69**, 2131.

10 Nagase, S., Kobayashi, K., Akasaka, T., and Wakahara, T. (2000) *Fullerenes: Chemistry, Physics, and Technology* (eds K.M. Kadish and R.S. Ruoff), John Wiley & Sons, Inc., New York, p. 395.

11 Iijima, S. (1991) *Nature*, **354**, 56.

12 Ge, M.H. and Sattler, K. (1994) *Chem. Phys. Lett.*, **220**, 192.

13 Smith, B.W., Monthioux, M., and Luzzi, D.E. (1998) *Nature*, **396**, 323.

14 Tans, S.J., Verschueren, A.R.M., and Dekker, C. (1998) *Nature*, **393**, 49.

15 Dresselhaus, M.S., Dresselhaus, G., and Eklund, P.C. (1996) *Science of Fullerenes and Carbon Nanotubes*, Academic Press, San Diego.

16 Cioslowski, J. (1995) *Electronic Structure Calculations on Fullerenes and Their Derivatives*, Oxford University Press, Oxford.

17 Scuseria, G.E. (1995) *Modern Electronic Structure Theory, Part I* (ed. D.R. Yarkony), World Scientific Publishers, Singapore, p. 279.

18 Jinno, K. and Saito, Y. (1996) *Adv. Chromatogr.*, **36**, 65.

19 Achiba, Y., Kikuchi, K., Aihara, Y., Wakabayashi, T., Miyake, Y., and Kainosho, M. (1996) *The Chemical Physics of Fullerenes 10 (and 5) Years Later* (ed. W. Andreoni), Kluwer Academic Publishers, Dordrecht, p. 139.

20 Saunders, M., Cross, R.J., Jiménez-Vázquez, H.A., Shimshi, R., and Khong, A. (1996) *Science*, **271**, 1693.

21 Taylor, R. (1995) in *The Chemistry of Fullerenes* (ed. R. Taylor), World Scientific Publishers, Singapore, p. 1.

22 Heinzinger, K. and Slanina, Z. (1993) MPG-Spiegel, No. 2, 11.

23 Baggott, J. (1994) *Perfect Symmetry. The Accidental Discovery of Buckminsterfullerene*, Oxford University Press, Oxford.

24 Aldersey-Williams, H. (1995) *The Most Beautiful Molecule. An Adventure in Chemistry*, Aurum Press, London.

25 Slanina, Z. (1998) *Chem. Intell.*, **4** (2), 52.

26 Slanina, Z. (2001) *Int. J. Hist. Eth. Natur. Sci. Technol. Med. NTM*, **9**, 41.

27 Mattauch, J., Ewald, H., Hahn, O., and Strassmann, F. (1943) *Z. Phys.*, **20**, 598.

28 Honig, R.E. (1954) *J. Chem. Phys.*, **22**, 126.

29 Drowart, J., Burns, R.P., DeMaria, G., and Inghram, M.G. (1959) *J. Chem. Phys.*, **31**, 1131.

30 Hintenberger, H., Franzen, J., and Schuy, K.D. (1963) *Z. Naturforsch.*, **18a**, 1236.

31 Pitzer, K.S. and Clementi, E. (1959) *J. Am. Chem. Soc.*, **81**, 4477.

32 Hoffmann, R. (1966) *Tetrahedron*, **22**, 521.

33 Slanina, Z. (1975) *Radiochem. Radioanal. Lett.*, **22**, 291.

34 Slanina, Z. and Zahradník, R. (1977) *J. Phys. Chem.*, **81**, 2252.

35 Schultz, H.P. (1965) *J. Org. Chem.*, **30**, 1361.

36 Jones, D.E.H. (1966) *New Scientist*, **32**, 245.

37 (a) Ōsawa, E. (1970) *Kagaku*, **25**, 854; (b) (1971) *Chem. Abstr.*, **74**, 75698v.

38 Bochvar, D.A. and Gal'pern, E.G. (1973) *Dokl. Akad. Nauk SSSR*, **209**, 610.

39 Davidson, R.A. (1981) *Theor. Chim. Acta*, **58**, 193.

40 Slanina, Z., Zhao, X., Uhlk, F., Lee, S.-L., and Adamowicz, L. (2004) *Int. J. Quantum Chem.*, **99**, 640.

41 Kobayashi, K. and Nagase, S. (2002) *Endofullerenes – A New Family of Carbon Clusters* (eds T. Akasaka and S. Nagase), Kluwer Academic Publishers, Dordrecht, The Netherlands, p. 99.

42 Wu, Z.C., Jelski, D.A., and George, T.F. (1987) *Chem. Phys. Lett.*, **137**, 291.

43 Elser, V. and Haddon, R.C. (1987) *Nature*, **325**, 792.

44 Stanton, R.E. and Newton, M.D. (1988) *J. Phys. Chem.*, **92**, 2141.

45 Weeks, D.E. and Harter, W.G. (1988) *Chem. Phys. Lett.*, **144**, 366.

46 Weeks, D.E. and Harter, W.G. (1989) *J. Chem. Phys.*, **90**, 4744.

47 Slanina, Z., Rudziński, J.M., Togasi, M., and Ōsawa, E. (1989) *J. Mol. Struct. (Theochem)*, **61**, 169.

48 Huffman, D.R. and Krätschmer, W. (1991) *Mater. Res. Soc. Proc.*, **206**, 601.

49 Braun, T. (1992) *Angew. Chem., Int. Ed. Engl.*, **31**, 588.

50 Braun, T., Schubert, A., Maczelka, H., and Vasvári, L. (1995) *Fullerene Research 1985–1993*, World Scientific Publishers, Singapore.

51 Bakowies, D. and Thiel, W. (1991) *J. Am. Chem. Soc.*, **113**, 3704.

52 Häser, M., Almlöf, J., and Scuseria, G.E. (1991) *Chem. Phys. Lett.*, **181**, 497.

53 Weltner, W., Jr. and van Zee, R.J. (1989) *Chem. Rev.*, **89**, 1713.

54 Curl, R.F. and Smalley, R.E. (1988) *Science*, **242**, 1017.

55 Kroto, H.W. (1988) *Science*, **242**, 1139.

56 Haddon, R.C. (1988) *Acc. Chem. Res.*, **21**, 243.

57 Jelski, D.A. and George, T.F. (1988) *J. Chem. Educ.*, **65**, 879.

58 Klein, D.J. and Schmalz, T.G. (1990) *Quasicrystals, Networks, and Molecules of Fivefold Symmetry* (ed. I. Hargittai), VCH Publishers, New York.

59 Ōsawa, E. (1990) *Kagaku*, **45**, 552.

60 Huffman, D.R. (1991) *Phys. Today*, **44**, 22.

61 Kroto, H.W., Allaf, A.W., and Balm, S.P. (1991) *Chem. Rev.*, **91**, 1213.

62 Kroto, H.W. (1992) *Angew. Chem., Int. Ed. Engl.*, **31**, 111.

63 Heath, J.R. (1992) *Fullerenes. Synthesis, Properties, and Chemistry of Large Carbon Clusters*, ACS Symposium Series No. 481 (eds G.S. Hammond and V.J. Kuck), ACS, Washington.

64 Fischer, J.E., Heiney, P.A., Luzzi, D.E., and Cox, D.E. (1992) *Fullerenes. Synthesis, Properties, and Chemistry of Large Carbon Clusters*, ACS Symposium Series No. 481 (eds G.S. Hammond and V.J. Kuck), ACS, Washington.

65 Slanina, Z. (1992) *Chem. Listy*, **86**, 327.

66 Smalley, R.E. (1992) *Acc. Chem. Res.*, **25**, 98.

67 Hare, J.P. and Kroto, H.W. (1992) *Acc. Chem. Res.*, **25**, 106.

68 Fischer, J.E., Heiney, P.A., and Smith, A.B., III (1992) *Acc. Chem. Res.*, **25**, 112.

69 Diederich, F. and Whetten, R.L. (1992) *Acc. Chem. Res.*, **25**, 119.

70 Haddon, R.C. (1992) *Acc. Chem. Res.*, **25**, 127.

71 Fagan, P.J., Calabrese, J.C., and Malone, B. (1992) *Acc. Chem. Res.*, **25**, 134.

72 Weaver, J.H. (1992) *Acc. Chem. Res.*, **25**, 143.

73 Hawkins, J.M. (1992) *Acc. Chem. Res.*, **25**, 150.

74 Wudl, F. (1992) *Acc. Chem. Res.*, **25**, 157.

75 McElvany, S.W., Ross, M.M., and Callahan, J.H. (1992) *Acc. Chem. Res.*, **25**, 162.

76 Johnson, R.D., Bethune, D.S., and Yannoni, C.S. (1992) *Acc. Chem. Res.*, **25**, 169.

77 Koruga, D., Hameroff, S., Withers, J., Loutfy, R., and Sundareshan, M. (1993) *Fullerene C60 – History, Physics, Nanobiology, Nanotechnology*, Elsevier, Amsterdam.

78 Cohen, M.L. and Crespi, V.H. (1993) *Buckminsterfullerenes* (eds W.E. Billups and M.A. Ciufolini), VCH Publishers, New York.

79 Erwin, S.C. (1993) *Buckminsterfullerenes* (eds W.E. Billups and M.A. Ciufolini), VCH Publishers, New York.

80 Dresselhaus, M.S., Dresselhaus, G., and Eklund, P.C. (1993) *J. Mater. Res.*, **8**, 2054.

81 Schmalz, T.G. and Klein, D.J. (1993) *Buckminsterfullerenes* (eds W.E. Billups

and M.A. Ciufolini), VCH Publishers, New York.

82 Scuseria, G.E. (1993) *Buckminsterfullerenes* (eds W.E. Billups and M.A. Ciufolini), VCH Publishers, New York.

83 White, C.T., Mintmire, J.W., Mowrey, R.C., Brenner, D.W., Robertson, D.H., Harrison, J.A., and Dunlap, B.I. (1993) *Buckminsterfullerenes* (eds W.E. Billups and M.A. Ciufolini), VCH Publishers, New York.

84 Haddon, R.C. and Raghavachari, K. (1993) *Buckminsterfullerenes* (eds W.E. Billups and M.A. Ciufolini), VCH Publishers, New York.

85 Cioslowski, J. (1993) *Rev. Comput. Chem.*, **4**, 1.

86 Wang, C.Z., Zhang, B.L., Ho, K.M., and Wang, X.Q. (1993) *Int. J. Mod. Phys. B*, **7**, 4305.

87 González, J., Guinea, F.F., and Vozmediano, M.A.H. (1993) *Int. J. Mod. Phys. B*, **7**, 4331.

88 Andreoni, W. (1993) *Electronic Properties of Fullerenes* (eds H. Kuzmany, J. Fink, M. Mehring, and S. Roth), Springer-Verlag, Berlin.

89 Slanina, Z., Lee, S.-L., and Yu, C.-H. (1996) *Rev. Comput. Chem.*, **8**, 1.

90 Cioslowski, J., Rao, N., and Moncrieff, D. (2000) *J. Am. Chem. Soc.*, **122**, 8265.

91 Achiba, Y., Kikuchi, K., Aihara, Y., Wakabayashi, T., Miyake, Y., and Kainosho, M. (1995) *Science and Technology of Fullerene Materials* (eds P. Bernier, D.S. Bethune, L.Y. Chiang, T.W. Ebbesen, R.M. Metzger, and J.W. Mintmire), Materials Research Society, Pittsburgh.

92 Achiba, Y. (1997) *Kagaku*, **52** (5), 15.

93 Mitsumoto, R., Oji, H., Yamamoto, Y., Asato, K., Ouchi, Y., Shinohara, H., Seki, K., Umishita, K., Hino, S., Nagase, S., Kikuchi, K., and Achiba, Y. (1997) *J. Phys. IV*, **7**, C2–525.

94 Kroto, H.W. (1987) *Nature*, **329**, 529.

95 Schmalz, T.G., Seitz, W.A., Klein, D.J., and Hite, G.E. (1988) *J. Am. Chem. Soc.*, **110**, 1113.

96 Manolopoulos, D.E. and Fowler, P.W. (1992) *J. Chem. Phys.*, **96**, 7603.

97 Fowler, P.W. and Manolopoulos, D.E. (1995) *An Atlas of Fullerenes*, Clarendon Press, Oxford.

98 Yoshida, M. and Ōsawa, E. (1995) *Bull. Chem. Soc. Jpn.*, **68**, 2073.

99 Yoshida, M. and Ōsawa, E. (1995) *Bull. Chem. Soc. Jpn.*, **68**, 2083.

100 Slanina, Z., Sun, M.-L., Lee, S.-L., and Adamowicz, L. (1995) *Recent Advances in the Chemistry and Physics of Fullerenes and Related Materials*, Vol. 2 (eds K.M. Kadish and R.S. Ruoff), The Electrochemical Society, Pennington, NJ, p. 1138.

101 Ettl, R., Chao, I., Diederich, F.N., and Whetten, R.L. (1991) *Nature*, **353**, 149.

102 Manolopoulos, D.E. (1991) *J. Chem. Soc., Faraday Trans.*, **87**, 2861.

103 Cheng, H.P. and Whetten, R.L. (1992) *Chem. Phys. Lett.*, **197**, 44.

104 Li, Q., Wudl, F., Thilgen, C., Whetten, R.L., and Diederich, F. (1992) *J. Am. Chem. Soc.*, **114**, 3994.

105 Hino, S., Matsumoto, K., Hasegawa, S., Inokuchi, H., Morikawa, T., Takahashi, T., Seki, K., Kikuchi, K., Suzuki, S., Ikemoto, I., and Achiba, Y. (1992) *Chem. Phys. Lett.*, **197**, 38.

106 Orlandi, G., Zerbetto, F., Fowler, P.W., and Manolopoulos, D.E. (1993) *Chem. Phys. Lett.*, **208**, 441.

107 Hawkins, J.M. and Meyer, A. (1993) *Science*, **260**, 1918.

108 Austin, S.J., Fowler, P.W., Orlandi, G., Manolopoulos, D.E., and Zerbetto, F. (1994) *Chem. Phys. Lett.*, **226**, 219.

109 Michel, R.H., Schreiber, H., Gierden, R., Hennrich, F., Rockenberger, J., Beck, R.D., Kappes, M.M., Lehner, C., Adelmann, P., and Armbruster, J.F. (1994) *Ber. Bunsenges. Phys. Chem.*, **98**, 975.

110 Michel, R.H., Kappes, M.M., Adelmann, P., and Roth, G. (1994) *Angew. Chem., Int. Ed. Engl.*, **33**, 1651.

111 Diederich, I., Whetten, R.L., Thilgen, C., Ettl, R., Chao, I., and Alvarez, M.M. (1991) *Science*, **254**, 1768.

112 Kikuchi, K., Nakahara, N., Wakabayashi, T., Suzuki, S., Shiromaru, H., Miyake, Y., Saito, K., Ikemoto, I., Kainosho, M., and Achiba, Y. (1992) *Nature*, **357**, 142.

113 Bakowies, D., Gelelßus, A., and Thiel, W. (1992) *Chem. Phys. Lett.*, **197**, 324.

114 Slanina, Z., François, J.-P., Bakowies, D., and Thiel, W. (1993) *J. Mol. Struct. (Theochem)*, **279**, 213.

115 Woo, S.J., Kim, E., and Lee, Y.H. (1993) *Phys. Rev. B*, **47**, 6721.

116 Nakao, K., Kurita, N., and Fujita, M. (1994) *Phys. Rev. B*, **49**, 11415.

117 Sun, M.-L., Slanina, Z., Lee, S.-L., Uhl k, F., and Adamowicz, L. (1995) *Chem. Phys. Lett.*, **246**, 66.

118 Hennrich, F.H., Michel, R.H., Fischer, A., Richard-Schneider, S., Gilb, S., Kappes, M.M., Fuchs, D., Bürk, M., Kobayashi, K., and Nagase, S. (1996) *Angew. Chem., Int. Ed. Engl.*, **35**, 1732.

119 Wang, C.-R., Sugai, T., Kai, T., Tomiyama, T., and Shinohara, H. (2000) *J. Chem. Soc., Chem. Commun.*, 557.

120 Kikuchi, K., Nakahara, N., Wakabayashi, T., Honda, M., Matsumiya, H., Moriwaki, T., Suzuki, S., Shiromaru, H., Saito, K., Yamauchi, K., Ikemoto, I., and Achiba, Y. (1992) *Chem. Phys. Lett.*, **188**, 177.

121 Nagase, S., Kobayashi, K., Kato, T., and Achiba, Y. (1993) *Chem. Phys. Lett.*, **201**, 475.

122 Orlandi, G., Zerbetto, F., and Fowler, P.W. (1993) *J. Phys. Chem.*, **97**, 13575.

123 Nagase, S. and Kobayashi, K. (1993) *Chem. Phys. Lett.*, **214**, 57.

124 Slanina, Z., Lee, S.-L., Kobayashi, K., and Nagase, S. (1995) *J. Mol. Struct. (Theochem)*, **339**, 89.

125 Saunders, M., Jiménez-Vázquez, H.A., Cross, R.J., Billups, W.E., Gesenberg, C., Gonzalez, A., Luo, W., Haddon, R.C., Diederich, F., and Herrmann, A. (1995) *J. Am. Chem. Soc.*, **117**, 9305.

126 Diederich, F., Ettl, R., Rubin, Y., Whetten, R.L., Beck, R., Alvarez, M., Anz, S., Sensharma, D., Wudl, F., Khemani, K.C., and Koch, A. (1991) *Science*, **252**, 548.

127 Beck, R.D., John, P.S., Alvarez, M.M., Diederich, F., and Whetten, R.L. (1991) *J. Phys. Chem.*, **95**, 8402.

128 Kikuchi, K., Nakahara, N., Honda, M., Suzuki, S., Saito, K., Shiromaru, H., Yamauchi, K., Ikemoto, I., Kuramochi, T., Hino, S., and Achiba, Y. (1991) *Chem. Lett.*, 1607.

129 Raghavachari, K. and Rohlfing, C.M. (1991) *J. Phys. Chem.*, **95**, 5768.

130 Slanina, Z., François, J.-P., Kolb, M., Bakowies, D., and Thiel, W. (1993) *Fullerene Sci. Technol.*, **1**, 221.

131 Dennis, T.J.S., Kai, T., Tomiyama, T., and Shinohara, H. (1998) *Chem. Commun.*, 619.

132 Ōsawa, E., Ueno, H., Yoshida, M., Slanina, Z., Zhao, X., Nishiyama, M., and Saito, H. (1998) *J. Chem. Soc., Perkin Trans.*, **2**, 943.

133 Slanina, Z., Lee, S.-L., Yoshida, M., and Ōsawa, E. (1996) *Chem. Phys.*, **209**, 13.

134 Slanina, Z., Lee, S.-L., Yoshida, M., and Ōsawa, E. (1996) *Physics and Chemistry of Fullerenes and Their Derivatives* (eds H. Kuzmany, J. Fink, M. Mehring, and S. Roth), World Scientific Publishers, Singapore, p. 389.

135 Slanina, Z., Lee, S.-L., Yoshida, M., and Ōsawa, E. (1996) *Recent Advances in the Chemistry and Physics of Fullerenes and Related Materials*, Vol. 3 (eds K.M. Kadish, and R.S. Ruoff), The Electrochemical Society, Pennington, NJ, p. 967.

136 Slanina, Z., Lee, S.-L., and Adamowicz, L. (1997) *Int. J. Quantum Chem.*, **63**, 529.

137 Slanina, Z., Zhao, X., Lee, S.-L., and Ōsawa, E. (1997) *Recent Advances in the Chemistry and Physics of Fullerenes and Related Materials*, Vol. 4 (eds K.M. Kadish and R.S. Ruoff), The Electrochemical Society, Pennington, NJ, p. 680.

138 Slanina, Z., Zhao, X., Lee, S.-L., and Ōsawa, E. (1997) *Chem. Phys.*, **219**, 193.

139 Slanina, Z., Zhao, X., Deota, P., and Ōsawa, E. (2000) *J. Mol. Model.*, **6**, 312.

140 Zhao, X., Slanina, Z., Goto, H., and Ōsawa, E. (2003) *J. Chem. Phys.*, **118**, 10534.

141 Zhao, X. and Slanina, Z. (2002) Fullerenes, *The Exciting World of Nanocages and Nanotubes*, Vol. 12 (eds P. Kamat, D. Guldi, and K. Kadish), The Electrochemical Society, Pennington, NJ, p. 679.

142 Zhao, X. and Slanina, Z. (2003) *J. Mol. Struct. (Theochem)*, **636**, 195.

143 Zhao, X., Slanina, Z., Ozawa, M., Ōsawa, E., Deota, P., and Tanabe, K. (2000) *Fullerene Sci. Technol.*, **8**, 595.

144 Slanina, Z., Zhao, X., and Ōsawa, E. (1998) *Chem. Phys. Lett.*, **290**, 311.

145 Slanina, Z., Uhlk, F., Zhao, X., and Ōsawa, E. (2000) *J. Chem. Phys.*, **113**, 4933.

146 Varganov, S.A., Avramov, P.V., Ovchinnikov, S.G., and Gordon, M.S. (2002) *Chem. Phys. Lett.*, **362**, 380.

147 Wan, T.S.M., Zhang, H.W., Nakane, T., Xu, Z.D., Inakuma, M., Shinohara, H., Kobayashi, K., and Nagase, S. (1998) *J. Am. Chem. Soc.*, **120**, 6806.

148 Kobayashi, K., Nagase, S., Yoshida, M., and Ōsawa, E. (1997) *J. Am. Chem. Soc.*, **119**, 12693.

149 Nagase, S., Kobayashi, K., and Akasaka, T. (1999) *J. Mol. Struct. (Theochem)*, **462**, 97.

150 Slanina, Z., Kobayashi, K., and Nagase, S. (2003) *Chem. Phys. Lett.*, **372**, 810.

151 Slanina, Z., Zhao, X., Grabuleda, X., Ozawa, M., Uhlk, F., Ivanov, P.M., Kobayashi, K., and Nagase, S. (2001) *J. Mol. Graphics Mod.*, **19**, 252.

152 Slanina, Z., Uhlk, F., Adamowicz, L., Kobayashi, K., and Nagase, S. (2004) *Int. J. Quantum Chem.*, **100**, 610.

153 Kodama, T., Fujii, R., Miyake, Y., Suzuki, S., Nishikawa, H., Ikemoto, I., Kikuchi, K., and Achiba, Y. (2003) Fullerenes, in *Fullerenes and Nanotubes: The Building Blocks of Next Generation Nanodevices*, Vol. **13** (eds D.M. Guldi, P.V. Kamat, and F. D'Souza), The Electrochemical Society, Pennington, NJ, p. 548.

154 Slanina, Z., Kobayashi, K., and Nagase, S. (2004) *Chem. Phys.*, **301**, 153.

155 Xu, Z.D., Nakane, T., and Shinohara, H. (1996) *J. Am. Chem. Soc.*, **118**, 11309.

156 Kobayashi, K. and Nagase, S. (1997) *Chem. Phys. Lett.*, **274**, 226.

157 Kimura, T., Sugai, T., and Shinohara, H. (1999) *Int. J. Mass Spectrom.*, **188**, 225.

158 Dennis, T.J.S. and Shinohara, H. (1998) *Appl. Phys. A*, **66**, 243.

159 Hino, S., Umishita, K., Iwasaki, K., Aoki, M., Kobayashi, K., Nagase, S., Dennis, T.J.S., Nakane, T., and Shinohara, H. (2001) *Chem. Phys. Lett.*, **337**, 65.

160 Kodama, T., Fujii, R., Miyake, Y., Sakaguchi, K., Nishikawa, H., Ikemoto, I., Kikuchi, K., and Achiba, Y. (2003) *Chem. Phys. Lett.*, **377**, 197.

161 Slanina, Z., Kobayashi, K., and Nagase, S. (2004) *J. Chem. Phys.*, **120**, 3397.

162 Kobayashi, K. and Nagase, S. (1998) *Chem. Phys. Lett.*, **282**, 325.

163 Akasaka, T., Wakahara, T., Nagase, S., Kobayashi, K., Waelchli, M., Yamamoto, K., Kondo, M., Shirakura, S., Okubo, S., Maeda, Y., Kato, T., Kako, M., Nakadaira, Y., Nagahata, R., Gao, X., van Caemelbecke, E., and Kadish, K.M. (2000) *J. Am. Chem. Soc.*, **122**, 9316.

164 Akasaka, T., Wakahara, T., Nagase, S., Kobayashi, K., Waelchli, M., Yamamoto, K., Kondo, M., Shirakura, S., Maeda, Y., Kato, T., Kako, M., Nakadaira, Y., Gao, X., van Caemelbecke, E., and Kadish, K.M. (2001) *J. Phys. Chem. B*, **105**, 2971.

165 Wakahara, T., Okubo, S., Kondo, M., Maeda, Y., Akasaka, T., Waelchli, M., Kako, M., Kobayashi, K., Nagase, S., Kato, T., Yamamoto, K., Gao, X., van Caemelbecke, E., and Kadish, K.M. (2002) *Chem. Phys. Lett.*, **360**, 235.

166 Kobayashi, K. and Nagase, S. (2003) *Mol. Phys.*, **101**, 249.

167 Kodama, T., Ozawa, N., Miyake, Y., Sakaguchi, K., Nishikawa, H., Ikemoto, I., Kikuchi, K., and Achiba, Y. (2002) *J. Am. Chem. Soc.*, **124**, 1452.

168 Cao, B., Hasegawa, M., Okada, K., Tomiyama, T., Okazaki, T., Suenaga, K., and Shinohara, H. (2001) *J. Am. Chem. Soc.*, **123**, 9679.

169 Nagase, S. and Kobayashi, K. (1997) *Chem. Phys. Lett.*, **276**, 55.

170 Cao, B., Suenaga, K., Okazaki, T., and Shinohara, H. (2002) *J. Phys. Chem. B*, **106**, 9295.

171 Wang, C.R., Georgi, P., Dunsch, L., Kai, T., Tomiyama, T., and Shinohara, H. (2002) *Curr. Appl. Phys.*, **2**, 141.

172 Kwon, Y.K., Tomanek, D., and Iijima, S. (1999) *Phys. Rev. Lett.*, **82**, 1470.

173 Slanina, Z., Kobayashi, K., and Nagase, S. (2003) *Chem. Phys. Lett.*, **382**, 211.

174 Slanina, Z. and Uhlk, F. (2003) *Chem. Phys. Lett.*, **374**, 100.

175 Slanina, Z., Uhlk, F., and Adamowicz, L. (2003) *Fullerene Nanotub. Carb. Nanostruct.*, **11**, 219.

176 Dietel, E., Hirsch, A., Pietzak, B., Waiblinger, M., Lips, K., Weidinger, A., Gruss, A., and Dinse, K.P. (1999) *J. Am. Chem. Soc.*, **121**, 2432.

177 Clark, T. (1985) *A Handbook of Computational Chemistry. A Practical Guide to Chemical Structure and Energy Calculations*, John Wiley & Sons, Inc., New York.

178 Boyd, D.B. (1990) *Rev. Comput. Chem.*, **1**, 321.

179 Stewart, J.J.P. (1990) *Rev. Comput. Chem.*, **1**, 45.

180 Zerner, M.C. (1991) *Rev. Comput. Chem.*, **2**, 313.

181 Hehre, W.J., Radom, L., von Schleyer, P.R., and Pople, J.A. (1986) *Ab Initio Molecular Orbital Theory*, John Wiley & Sons, Inc., New York.

182 Dewar, M.J.S. and Thiel, W. (1997) *J. Am. Chem. Soc.*, **99**, 4899.

183 Dewar, M.J.S., Zoebisch, E.G., Healy, E.F., and Stewart, J.J.P. (1985) *J. Am. Chem. Soc.*, **107**, 3902.

184 Stewart, J.J.P. (1989) *J. Comput. Chem.*, **10**, 209.

185 Dewar, M.J.S., Jie, C., and Yu, J. (1993) *Tetrahedron*, **49**, 5003.

186 Stewart, J.J.P. (1990) *MOPAC 5.0, QCPE 455*, Indiana University.

187 Stewart, J.J.P. (1999) *MOPAC 2002*, Fujitsu Ltd., Tokyo.

188 AMPAC (1997) *AMPAC, 6.0*, Semichem, Shavnee.

189 Frisch, M.J., Trucks, G.W., Schlegel, H.B., Scuseria, G.E., Robb, M.A., Cheeseman, J.R., Zakrzewski, V.G., Montgomery, J.A., Jr., Stratmann, R.E., Burant, J.C., Dapprich, S., Millam, J.M., Daniels, A.D., Kudin, K.N., Strain, M.C., Farkas, O., Tomasi, J., Barone, V., Cossi, M., Cammi, R., Mennucci, B., Pomelli, C., Adamo, C., Clifford, S., Ochterski, J., Petersson, G.A., Ayala, P.Y., Cui, Q., Morokuma, K., Malick, D.K., Rabuck, A.D., Raghavachari, K., Foresman, J.B., Cioslowski, J., Ortiz, J.V., Stefanov, B.B., Liu, G., Liashenko, A., Piskorz, P., Komaromi, I., Gomperts, R., Martin, R.L., Fox, D.J., Keith, T., Al-Laham, M.A., Peng, C.Y., Nanayakkara, A., Gonzalez, C., Challacombe, M., Gill, P.M.W., Johnson, B., Chen, W., Wong, M.W., Andres, J.L., Gonzalez, C., Head-Gordon, M., Replogle, E.S., and Pople, J.A. (1998) *Gaussian 98, Revision A. 11.1*, Gaussian, Inc., Pittsburgh.

190 Frisch, M.J., Trucks, G.W., Schlegel, H.B., Scuseria, G.E., Robb, M.A., Cheeseman, J.R., Montgomery, J.A., Jr., Vreven, T., Kudin, K.N., Burant, J.C., Millam, J.M., Iyengar, S.S., Tomasi, J., Barone, V., Mennucci, B., Cossi, M., Scalmani, G., Rega, N., Petersson, G.A., Nakatsuji, H., Hada, M., Ehara, M., Toyota, K., Fukuda, R., Hasegawa, J., Ishida, M., Nakajima, T., Honda, Y., Kitao, O., Nakai, H., Klene, M., Li, X., Knox, J.E., Hratchian, H.P., Cross, J.B., Adamo, C., Jaramillo, J., Gomperts, R., Stratmann, R.E., Yazyev, O., Austin, A.J., Cammi, R., Pomelli, C., Ochterski, J.W., Ayala, P.Y., Morokuma, K., Voth, G.A., Salvador, P., Dannenberg, J.J., Zakrzewski, V.G., Dapprich, S., Daniels, A.D., Strain, M.C., Farkas, O., Malick, D.K., Rabuck, A.D., Raghavachari, K., Foresman, J.B., Ortiz, J.V., Cui, Q., Baboul, A.G., Clifford, S., Cioslowski, J., Stefanov, B.B., Liu, G., Liashenko, A., Piskorz, P., Komaromi, I., Martin, R.L., Fox, D.J., Keith, T., Al-Laham, M.A., Peng, C.Y., Nanayakkara, A., Challacombe, M., Gill, P.M.W., Johnson, B., Chen, W., Wong, M.W. Gonzalez, C., and Pople, J.A. (2004) *Gaussian 03, Revision C. 01*, Gaussian, Inc., Wallingford.

191 Hehre, W.J., Burke, L.D., and Schusterman, A.J. (1993) *Spartan*, Wavefunction, Inc., Irvine.

192 Schlegel, H.B. and McDouall, J.J.W. (1991) *Computational Advances in Organic Chemistry* (eds C. Ögretir and I.G. Csizmadia), Kluwer Academic Publishers, Dordrecht.

193 Jensen, F. (1999) *Introduction to Computational Chemistry*, John Wiley & Sons, Ltd, Chichester.

194 Slanina, Z., Uhlk, F., and Zerner, M.C. (1991) *Rev. Roum. Chim.*, **36**, 965.

195 Slanina, Z. (1987) *Int. Rev. Phys. Chem.*, **6**, 251.

196 Slanina, Z. (1989) *Comput. Chem.*, **13**, 305.

197 Slanina, Z. (1992) *Theor. Chim. Acta*, **83**, 257.

198 Slanina, Z. (1986) *Contemporary Theory of Chemical Isomerism*, Kluwer, Dordrecht.

199 Slanina, Z., Uhlk, F., François, J.-P., and Ōsawa, E. (2000) *Croat. Chem. Acta*, **73**, 1047.

200 Coxeter, H.S.M. (1963) *Regular Polytopes*, Macmillan, New York.

201 Coxeter, H.S.M. (1969) *Introduction to Geometry*, John Wiley & Sons, Inc., New York.

202 Slanina, Z. (1998) *Chem. Intell.*, **4** (2), 52.

203 Curl, R.F. (1993) *Phil. Trans. Roy. Soc. London A*, **343**, 19.

204 Sun, M.-L., Slanina, Z., and Lee, S.-L. (1995) *Chem. Phys. Lett.*, **233**, 279.

205 Xu, C.H. and Scuseria, G.E. (1996) *Chem. Phys. Lett.*, **262**, 219.

206 Slanina, Z. and Ōsawa, E. (1997) *Fullerene Sci. Technol.*, **5**, 167.

207 Rudziński, J.M., Slanina, Z., Togasi, M., Ōsawa, E., and Iizuka, T. (1988) *Thermochim. Acta*, **125**, 155.

208 Slanina, Z., Rudziński, J.M., and Ōsawa, E. (1987) *Carbon*, **25**, 747.

209 Slanina, Z., Rudziński, J.M., and Ōsawa, E. (1987) *Collect. Czech. Chem. Commun.*, **52**, 2381.

210 Slanina, Z., Rudziński, J.M., Togasi, M., and Ōsawa, E. (1989) *Thermochim. Acta*, **140**, 87.

211 Slanina, Z., Zhao, X., Kurita, N., Gotoh, H., Uhlk, F., Rudziński, J.M., Lee, K.H., and Adamowicz, L. (2001) *J. Mol. Graphics Mod.*, **19**, 216.

212 Kiyobayashi, T. and Sakiyama, M. (1993) *Fullerene Sci. Technol.*, **1**, 269.

213 Slanina, Z. (1987) *Chem. Phys. Lett.*, **145**, 512.

214 Slanina, Z., Adamowicz, L., Bakowies, D., and Thiel, W. (1992) *Thermochim. Acta*, **202**, 249.

215 Slanina, Z. and Adamowicz, L. (1992) *Thermochim. Acta*, **205**, 299.

216 Slanina, Z. (2003) *Z. Phys. Chem.*, **217**, 1119.

217 Slanina, Z. (1991) *Chem. Phys.*, **150**, 321.

218 Dardi, P.S. and Dahler, J.S. (1990) *J. Chem. Phys.*, **93**, 3562.

219 Slanina, Z. (1992) *Thermochim. Acta*, **207**, 9.

220 Slanina, Z. (1992) *Thermochim. Acta*, **209**, 1.

221 Raghavachari, K., Whiteside, R.A., and Pople, J.A. (1986) *J. Chem. Phys.*, **85**, 6623.

222 Raghavachari, K. and Binkley, J.S. (1987) *J. Chem. Phys.*, **87**, 2191.

223 Martin, J.M.L., François, J.P., and Gijbels, R. (1989) *J. Chem. Phys.*, **90**, 3403.

224 Parasuk, V. and Almlöf, J. (1989) *J. Chem. Phys.*, **91**, 1137.

225 Liang, C. and Schaefer, H.F., III (1990) *J. Chem. Phys.*, **93**, 8844.

226 Martin, J.M.L., François, J.P., and Gijbels, R. (1990) *J. Chem. Phys.*, **93**, 8850.

227 Kurtz, J. and Adamowicz, L. (1991) *Astrophys. J.*, **370**, 784.

228 Martin, J.M.L., François, J.P., and Gijbels, R. (1991) *J. Chem. Phys.*, **94**, 3753.

229 Martin, J.M.L., François, J.P., and Gijbels, R. (1991) *J. Comput. Chem.*, **12**, 52.

230 Ewing, D.W. (1991) *Z. Phys. D*, **19**, 419.

231 Slanina, Z., Rudziński, J.M., and Ōsawa, E. (1991) *Z. Phys. D*, **19**, 431.

232 Martin, J.M.L., François, J.P., Gijbels, R., and Almlöf, J. (1991) *Chem. Phys. Lett.*, **187**, 367.

233 Watts, J.D. and Bartlett, R.J. (1992) *Chem. Phys. Lett.*, **190**, 19.

234 Parasuk, V. and Almlöf, J. (1992) *Theor. Chem. Acta*, **83**, 227.

235 Slanina, Z., Kurtz, J., and Adamowicz, L. (1992) *Chem. Phys. Lett.*, **196**, 208.

236 Slanina, Z., Kurtz, J., and Adamowicz, L. (1992) *Mol. Phys.*, **76**, 387.

237 Curtiss, L.A. and Raghavachari, K. (1995) *Quantum Mechanical Electronic Structure Calculations with Chemical Accuracy* (ed. S.R. Langhoff), Kluwer, Dordrecht.

238 Raghavachari, K. and Curtiss, L.A. (1995) *Quantum Mechanical Electronic Structure Calculations with Chemical Accuracy* (eds S.R. Langhoff), Kluwer, Dordrecht.

239 Martin, J.M.L. and Taylor, P.R. (1996) *J. Phys. Chem.*, **100**, 6047.

240 Krätschmer, W., Sorg, N., and Huffman, D.R. (1985) *Surf. Sci.*, **156**, 814.

241 McElvany, S.W., Creasy, W.R., and O'Keefe, A. (1986) *J. Chem. Phys.*, **85**, 632.

242 McElvany, S.W., Dunlap, B.I., and O'Keefe, A. (1987) *J. Chem. Phys.*, **86**, 715.

243 Krätschmer, W. and Nachtigall, K. (1987) *Polycyclic Aromatic Hydrocarbons and Astrophysics* (eds A. Léger, L. d'Hendecourt, and N. Boccara), D. Reidel, Dordrecht.

244 van Zee, R.J., Ferrante, R.F., Zerinque, K.J., Weltner, W., Jr., and Ewing, D.W. (1988) *J. Chem. Phys.*, **88**, 3465.

245 McElvany, S.W. (1988) *J. Chem. Phys.*, **89**, 2063.

246 Heath, J.R., Cooksy, A.L., Gruebele, H.W., Schmuttenmaer, C.A., and Saykally, R.J. (1989) *Science*, **244**, 564.

247 Vala, M., Chandrasekhar, T.M., Szczepanski, J., van Zee, R., and Weltner, W., Jr. (1989) *J. Chem. Phys.*, **90**, 595.

248 Vala, M., Chandrasekhar, T.M., Szczepanski, J., and Pellow, R. (1989) *High Temp. Sci.*, **27**, 19.

249 Shen, L.N. and Graham, W.R.M. (1989) *J. Chem. Phys.*, **91**, 5115.

250 Cheung, H.M. and Graham, W.R.M. (1989) *J. Chem. Phys.*, **91**, 6664.

251 Kurtz, J. and Huffman, D.R. (1990) *J. Chem. Phys.*, **92**, 30.

252 Heath, J.R., Sheeks, R.A., Cooksy, A.L., and Saykally, R.J. (1990) *Science*, **249**, 895.

253 Szczepanski, J. and Vala, M. (1991) *J. Phys. Chem.*, **95**, 2792.

254 Shen, L.N., Withey, P.A., and Graham, W.R.M. (1991) *J. Chem. Phys.*, **94**, 2395.

255 Heath, J.R., van Orden, A., Kuo, E., and Saykally, R.J. (1991) *Chem. Phys. Lett.*, **182**, 17.

256 Heath, J.R. and Saykally, R.J. (1991) *J. Chem. Phys.*, **94**, 1724.

257 Heath, J.R. and Saykally, R.J. (1991) *J. Chem. Phys.*, **94**, 3271.

258 Withey, P.A., Shen, L.N., and Graham, W.R.M. (1991) *J. Chem. Phys.*, **95**, 820.

259 Jiang, Q. and Graham, W.R.M. (1991) *J. Chem. Phys.*, **95**, 3129.

260 von Helden, G., Hsu, M.-T., Kemper, P.R., and Bowers, B.T. (1991) *J. Chem. Phys.*, **95**, 3835.

261 Zajfman, D., Feldman, H., Heber, O., Kella, D., Majer, D., Vager, Z., and Naaman, R. (1992) *Science*, **258**, 1129.

262 von Helden, G., Kemper, P.R., Gotts, N.G., and Bowers, M.T. (1993) *Science*, **259**, 1300.

263 Gotts, N.G., von Helden, G., and Bowers, M.T. (1995) *Int. J. Mass Spectr. Ion Process.*, **149/150**, 217.

264 Slanina, Z., Uhlk, F., and Adamowicz, L. (1997) *J. Radioanal. Nucl. Chem.*, **219**, 69.

265 Slanina, Z., Lee, S.-L., François, J.-P., Kurtz, J., Adamowicz, L., and Smigel, M. (1994) *Mol. Phys.*, **81**, 1489.

266 Slanina, Z., Lee, S.-L., François, J.-P., Kurtz, J., and Adamowicz, L. (1994) *Chem. Phys. Lett.*, **223**, 397.

267 Slanina, Z., Lee, S.-L., Smigel, M., Kurtz, J., and Adamowicz, L. (1995) *Science and Technology of Fullerene Materials* (eds P. Bernier, D.S. Bethune, L.Y. Chiang, T.W. Ebbesen, R.M. Metzger, and J.W. Mintmire), Materials Research Society, Pittsburgh.

268 Slanina, Z., Zhao, X., Ōsawa, E., and Adamowicz, L. (2000) *Fullerene Sci. Technol.*, **8**, 369.

269 Giesen, T.F., van Orden, A., Hwang, H.J., Fellers, R.S., Provençal, R.A., and Saykally, R.J. (1994) *Science*, **265**, 756.

270 Plattner, D.A. and Houk, K.N. (1995) *J. Am. Chem. Soc.*, **117**, 4405.

271 von Helden, G., Gotts, N.G., and Bowers, M.T. (1993) *Nature*, **363**, 60.

272 Slanina, Z. (1990) *Chem. Phys. Lett.*, **173**, 164.

273 Chase, M.W., Jr., Davies, C.A., Downery, J.R., Jr., Frurip, D.J., McDonald, R.A., and Syverud, A.N. (1985) *J. Phys. Chem. Ref. Data*, **14** (Supplement 1), JANAF Thermochemical Tables, Third Edition, Vols. 1,2.

274 Nygren, M.A. and Pettersson, L.G.M. (1992) *Chem. Phys. Lett.*, **191**, 473.

275 Johnson, B.G. (1995) *Modern Density Functional Theory: A Tool for Chemistry* (eds J.M. Seminario and P. Politzer), Elsevier, Amsterdam.

276 Watts, J.D., Gauss, J., Stanton, J.F., and Bartlett, R.J. (1992) *J. Chem. Phys.*, **97**, 8372.

277 Slanina, Z. (1989) *Chem. Phys. Lett.*, **161**, 265.

278 Slanina, Z. (1990) *Chem. Phys. Lett.*, **166**, 317.

279 Michalska, D., Chojnacki, H., Hess, B.A., Jr., and Schaad, L.J. (1987) *Chem. Phys. Lett.*, **141**, 376.

280 Slanina, Z. (1988) *Thermochim. Acta*, **127**, 237.

281 van Orden, A. and Saykally, R.J. (1998) *Chem. Rev.*, **98**, 2313.

282 Slanina, Z. (2000) *Skept. Inq.*, **24** (4), 18.

283 Fowler, P.W. and Woolrich, J. (1986) *Chem. Phys. Lett.*, **127**, 78.

284 Fowler, P.W. (1986) *Chem. Phys. Lett.*, **131**, 444.

285 Fowler, P.W. and Steer, J.I. (1987) *J. Chem. Soc., Chem. Commun.*, 1403.

286 Fowler, P.W., Cremona, J.E., and Steer, J.I. (1988) *Theor. Chim. Acta*, **73**, 1.

287 Ceulemans, A. and Fowler, P.W. (1989) *Phys. Rev. A*, **39**, 481.

288 Ceulemans, A. and Fowler, P.W. (1990) *J. Chem. Phys.*, **93**, 1221.

289 Fowler, P.W. (1990) *J. Chem. Soc., Faraday Trans.*, **86**, 2073.

290 Fowler, P. (1991) *Nature*, **350**, 20.

291 Fowler, P.W. (1991) *J. Chem. Soc., Faraday Trans.*, **87**, 1945.

292 Fowler, P.W., Batten, R.C., and Manolopoulos, D.E. (1991) *J. Chem. Soc., Faraday Trans.*, **87**, 3103.

293 Ceulemans, A. and Fowler, P.W. (1991) *Nature*, **353**, 52.

294 Fowler, P.W. and Manolopoulos, D.E. (1992) *Nature*, **355**, 428.

295 Fowler, P.W., Manolopoulos, D.E., and Ryan, R.P. (1992) *J. Chem. Soc., Chem. Commun.*, 408.

296 Fowler, P.W. (1992) *J. Chem. Soc., Perkin Trans. II*, 145.

297 Fowler, P.W., Manolopoulos, D.E., and Ryan, R.P. (1992) *Carbon*, **30**, 1235.

298 Manolopoulos, D.E., Woodall, D.R., and Fowler, P.W. (1992) *J. Chem. Soc., Faraday Trans.*, **88**, 2427.

299 Fowler, P.W. and Morvan, V. (1992) *J. Chem. Soc., Faraday Trans.*, **88**, 2631.

300 Fowler, P.W., Manolopoulos, D.E., Redmond, D.B., and Ryan, R.P. (1993) *Chem. Phys. Lett.*, **202**, 371.

301 Manolopoulos, D.E. and Fowler, P.W. (1993) *Chem. Phys. Lett.*, **204**, 1.

302 Lee, S.-L. (1992) *Theor. Chim. Acta*, **81**, 185.

303 Stone, A.J. and Wales, D.J. (1986) *Chem. Phys. Lett.*, **128**, 501.

304 Coulombeau, C. and Rassat, A. (1991) *J. Chim. Phys.*, **88**, 173.

305 Lipscomb, W.N. and Massa, L. (1992) *Inorg. Chem.*, **31**, 2297.

306 Tang, A.C., Li, Q.S., Liu, C.W., and Li, J. (1993) *Chem. Phys. Lett.*, **201**, 465.

307 Manolopoulos, D.E., May, J.C., and Down, S.E. (1991) *Chem. Phys. Lett.*, **181**, 105.

308 Liu, X., Klein, D.J., Schmalz, T.G., and Seitz, W.A. (1991) *J. Comput. Chem.*, **12**, 1252.

309 Liu, X., Klein, D.J., Seitz, W.A., and Schmalz, T.G. (1991) *J. Comput. Chem.*, **12**, 1265.

310 Liu, X., Schmalz, T.G., and Klein, D.J. (1992) *Chem. Phys. Lett.*, **188**, 550.

311 Manolopoulos, D.E. (1992) *Chem. Phys. Lett.*, **192**, 330.

312 Liu, X., Schmalz, T.G., and Klein, D.J. (1992) *Chem. Phys. Lett.*, **192**, 331.

313 Sun, M.-L., Slanina, Z., and Lee, S.-L. (1995) *Fullerene Sci. Technol.*, **3**, 627.

314 Fowler, P.W., Quinn, C.M., and Redmond, D.B. (1991) *J. Chem. Phys.*, **95**, 7678.

315 Wei, S., Shi, Z., and Castleman, A.W., Jr. (1991) *J. Chem. Phys.*, **94**, 3268.

316 Klein, D.J., Seitz, W.A., and Schmalz, T.G. (1993) *J. Phys. Chem.*, **97**, 1231.

317 Klein, D.J. (1994) *Chem. Phys. Lett.*, **217**, 261.

318 Balaban, A.T., Klein, D.J., and Folden, C.A. (1994) *Chem. Phys. Lett.*, **217**, 266.

319 Dias, J.R. (1993) *Chem. Phys. Lett.*, **209**, 439.

320 Balasubramanian, K. (1991) *Chem. Phys. Lett.*, **182**, 257.

321 Balasubramanian, K. (1991) *Chem. Phys. Lett.*, **183**, 292.

322 Balasubramanian, K. (1992) *J. Chem. Inf. Comput. Sci.*, **32**, 47.

323 Balasubramanian, K. (1992) *Chem. Phys. Lett.*, **198**, 577.

324 Balasubramanian, K. (1992) *Chem. Phys. Lett.*, **197**, 55.

325 Balasubramanian, K. (1992) *J. Mol. Spectr.*, **157**, 254.

326 Balasubramanian, K. (1993) *Chem. Phys. Lett.*, **201**, 306.

327 Balasubramanian, K. (1993) *Chem. Phys. Lett.*, **206**, 210.

328 Balasubramanian, K. (1993) *Chem. Phys. Lett.*, **202**, 399.

329 Balasubramanian, K. (1993) *J. Phys. Chem.*, **97**, 6990.

330 Balasubramanian, K. (1993) *J. Phys. Chem.*, **97**, 4647.

331 Babi,ć, D., Bassoli, S., Casartelli, M., Cataldo, F., Graovac, A., Ori, O., and York, B. (1995) *J. Mol. Simulat.*, **14**, 395.

332 Eggen, B.R., Heggie, M.I., Jungnickel, G., Latham, C.D., Jones, R., and Briddon, P.R. (1996) *Science*, **272**, 87.

333 Eggen, B.R., Heggie, M.I., Jungnickel, G., Latham, C.D., Jones, R., and Briddon, P.R. (1997) *Fullerene Sci. Technol.*, **5**, 727.

334 Slanina, Z., Zhao, X., Uhlk, F., Ozawa, M., and Ōsawa, E. (2000) *J. Organomet. Chem.*, **599**, 57.

335 McElvany, S.W., Callahan, J.H., Ross, M.M., Lamb, L.D., and Huffman, D.R. (1993) *Science*, **260**, 1632.

336 Deng, J.-P., Ju, D.-D., Her, G.-R., Mou, C.-Y., Chen, C.-J., Lin, Y.-Y., and Han, C.-C. (1993) *J. Phys. Chem.*, **97**, 11575.

337 Fujita, S. (1991) *Symmetry and Combinatorial Enumeration in Chemistry*, Springer-Verlag, Berlin.

338 Wahl, F., Wörth, J., and Prinzbach, H. (1993) *Angew. Chem., Int. Ed. Engl.*, **32**, 1722.

339 Prinzbach, H., Weller, A., Landenberger, P., Wahl, F., Worth, J., Scott, L.T., Gelmont, M., Olevano, D., and von Issendorff, B. (2000) *Nature*, **407**, 60.

340 Wang, Z., Ke, X., Zhu, Z., Zhu, F., Ruan, M., Chen, H., Huang, R., and Zheng, L. (2001) *Phys. Let. A*, **280**, 351.

341 Iqbal, Z., Zhang, Y., Grebel, H., Vijayalakshmi, S., Lahamer, A., Benedek, G., Bernasconi, M., Cariboni, J., Spagnolatti, I., Sharma, R., Owens, F.J., Kozlov, M.E., Rao, K.V., and Muhammed, M. (2003) *Eur. Phys. J. B*, **31**, 509.

342 Slanina, Z. and Adamowicz, L. (1993) *Fullerene Sci. Technol.*, **1**, 1.

343 Slanina, Z. and Adamowicz, L. (1993) *J. Mol. Struct. (Theochem)*, **281**, 33.

344 Lu, J., Re, S., Choe, Y., Nagase, S., Zhou, Y.S., Han, R.S., Peng, L.M., Zhang, X.W., and Zhao, X.G. (2003) *Phys. Rev. B*, **67**, 125415.

345 Zhang, Z.X., Pan, Z.Y., Wang, Y.X., Li, Z.J., and Wei, Q. (2003) *Mod. Phys. Lett.*, **17**, 877.

346 Piskoti, C., Yarger, J., and Zettl, A. (1998) *Nature*, **393**, 771.

347 Murry, R.L., Colt, J.R., and Scuseria, G.E. (1993) *J. Phys. Chem.*, **97**, 4954.

348 Louie, S.G. (1996) *Nature*, **384**, 612.

349 Sun, L.F., Xie, S.S., Liu, W., Zhou, W.Y., Liu, Z.Q., Tang, D.S., Wang, G., and Qian, L.X. (2000) *Nature*, **403**, 384.

350 Qin, L.C., Zhao, X.L., Hirahara, K., Miyamoto, Y., Ando, Y., and Iijima, S. (2000) *Nature*, **408**, 50.

351 Wang, N., Tang, Z.K., Li, G.D., and Chen, J.S. (2000) *Nature*, **408**, 50.

352 Slanina, Z., Uhlk, F., and Adamowicz, L. (2003) *J. Mol. Graphics Mod.*, **21**, 517.

353 Diener, M.D. and Alford, J.M. (1998) *Nature*, **393**, 668.

354 Moribe, H., Inoue, T., Kato, H., Taninaka, A., Ito, Y., Okazaki, T., Sugai, T., Bolskar, R., Alford, J.M., and Shinohara, H. (2003) The 25th Fullerene-Nanotubes Symposium, Awaji, Japan, 1P-1.

355 Kato, H., Taninaka, A., Sugai, T., and Shinohara, H. (2003) *J. Am. Chem. Soc.*, **125**, 7782.

356 Boltalina, O.V., Dashkova, E.V., and Sidorov, L.N. (1996) *Chem. Phys. Lett.*, **256**, 253.

357 Boltalina, O.V., Ioffe, I.N., Sidorov, L.N., Seifert, G., and Vietze, K. (2000) *J. Am. Chem. Soc.*, **122**, 9745.

358 Slanina, Z., Lee, S.-L., Kobayashi, K., and Nagase, S. (1994) *J. Mol. Struct. (Theochem)*, **312**, 175.

359 Slanina, Z., Zhao, X., Uhlk, F., and Ōsawa, E. (1999) *Electronic Properties of Novel Materials – Science and Technology of Molecular Nanostructures* (eds H. Kuzmany, J. Fink, M. Mehring, and S. Roth), AIP, Melville, p. 179.

360 Slanina, Z., Ishimura, K., Kobayashi, K., and Nagase, S. (2004) *Chem. Phys. Lett.*, **384**, 114.

361 Kanis, D.R., Ratner, M.A., Marks, T.J., and Zerner, M.C. (1991) *Chem. Mater.*, **3**, 19.

362 Slanina, Z., Uhlk, F., Zhao, X., Adamowicz, L., and Nagase, S. (2007) *Fulleren. Nanotub. Carb. Nanostruct.*, **15**, 195.

363 Akasaka, T., Nagase, S., Kobayashi, K., Wälchli, M., Yamamoto, K., Funasaka, H., Kako, M., Hoshino, T., and Erata, T. (1997) *Angew. Chem., Int. Ed. Engl.*, **36**, 1643.

364 Ichikawa, T., Kodama, T., Suzuki, S., Fujii, R., Nishikawa, H., Ikemoto, I., Kikuchi, K., and Achiba, Y. (2004) *Chem. Lett.*, **33**, 1008.

365 Hay, P.J. and Wadt, W.R. (1985) *J. Chem. Phys.*, **82**, 299.

366 Casida, M.E., Jamorski, C., Casida, K.C., and Salahub, D.R. (1998) *J. Chem. Phys.*, **108**, 4439.

367 Wang, L.S., Alford, J.M., Chai, Y., Diener, M., Zhang, J., McClure, S.M., Guo, T., Scuseria, G.E., and Smalley, R.E. (1993) *Chem. Phys. Lett.*, **207**, 354.

368 Kubozono, Y., Ohta, T., Hayashibara, T., Maeda, H., Ishida, H., Kashino, S., Oshima, K., Yamazaki, H., Ukita, S., and Sogabe, T. (1995) *Chem. Lett.*, 457.

369 Hopwood, F.G., Fisher, K.J., Greenhill, P., Willett, G.D., and Zhang, R. (1997) *J. Phys. Chem. B*, **101**, 10704.

370 Shinohara, H. (1997) *Fullerenes: Recent Advances in the Chemistry and Physics of*

Fullerenes and Related Materials, Vol. 4
(eds K.M. Kadish and R.S. Ruoff), The
Electrochemical Society, Pennington, NJ,
p. 467.

371 Chai, Y., Guo, T., Jin, C., Haufler, R.E.,
Chibante, L.P.F., Fure, J., Wang, L.,
Alford, J.M., and Smalley, R.E. (1991)
J. Phys. Chem., **95**, 7564.

372 Nishibori, E., Takata, M., Sakata, M.,
Tanaka, H., Hasegawa, M., and
Shinohara, H. (2000) *Chem. Phys. Lett.*,
330, 497.

373 Nagase, S. and Kobayashi, K. (1994)
Chem. Phys. Lett., **228**, 106.

374 Slanina, Z., Kobayashi, K., and Nagase, S.
(2004) *Chem. Phys. Lett.*, **388**, 74.

375 Lian, Y.F., Yang, S.F., and Yang, S.H.
(2002) *J. Phys. Chem. B*, **106**, 3112.

376 Slanina, Z. and Nagase, S. (2006) *Chem.
Phys. Lett.*, **422**, 133.

377 Slanina, Z. (2004) *J. Cluster Sci.*, **15**, 3.

378 Alcock, C.B., Itkin, V.P., and Horrigan,
M.K. (1984) *Can. Metallurg. Quart.*,
23, 309.

379 Slanina, Z., Uhlk, F., and Nagase, S.
(2007) *Chem. Phys. Lett.*, **440**, 259.

380 Reich, A., Panthofer, M., Modrow, H.,
Wedig, U., and Jansen, M. (2004) *J. Am.
Chem. Soc.*, **126**, 14428.

381 Haufe, O., Hecht, M., Grupp, A.,
Mehring, M., and Jansen, M. (2005) *Z.
Anorg. Allgem. Chem.*, **631**, 126.

382 Gromov, A., Krawez, N., Lassesson, A.,
Ostrovskii, D.I., and Campbell, E.E.B.
(2002) *Curr. App. Phys.*, **2**, 51.

383 Campbell, E.E.B. (2003) *Fullerene Collision
Reactions*, Kluwer Academic Publishers,
Dordrecht.

384 Kobayashi, K. and Nagase, S. (1999)
Chem. Phys. Lett., **302**, 312.

385 Bader, R.F.W. (1991) *Chem. Rev.*,
91, 893.

386 Bader, R.F.W. (1998) *J. Phys. Chem. A*,
102, 7314.

387 Lide, D.R. (ed.) (2004) *CRC Handbook of
Chemistry and Physics*, 85th edn, CRC
Press, Boca Raton, FL.

388 Nikawa, H., Kikuchi, T., Wakahara, T.,
Nakahodo, T., Tsuchiya, T., Rahman,
G.M.A., Akasaka, T., Maeda, Y., Yoza, K.,
Horn, E., Yamamoto, K., Mizorogi, N.,
and Nagase, S. (2005) *J. Am. Chem. Soc.*,
127, 9684.

389 Slanina, Z., Uhlk, F., Lee, S.-L.,
Adamowicz, L., and Nagase, S. (2008) *Int.
J. Quantum Chem.*, **108**, 2636.

390 Wakahara, T., Nikawa, H., Kikuchi, T.,
Nakahodo, T., Rahman, G.M.A.,
Tsuchiya, T., Maeda, Y., Akasaka, T., Yoza,
K., Horn, E., Yamamoto, K., Mizorogi, N.,
Slanina, Z., and Nagase, S. (2006) *J. Am.
Chem. Soc.*, **128**, 14228.

391 Nikawa, H., Yamada, T., Cao, B.,
Mizorogi, N., Slanina, Z., Tsuchiya, T.,
Akasaka, T., Yoza, K., and Nagase, S.
(2009) *J. Am. Chem. Soc.*, **131**, 10950.

392 Duchamp, J.C., Demortier, A., Fletcher,
K.R., Dorn, D., Iezzi, E.B., Glass, T., and
Dorn, H.C. (2003) *Chem. Phys. Lett.*,
375, 655.

393 Krause, M. and Dunsch, L. (2004)
ChemPhysChem., **5**, 1445.

394 Slanina, Z. and Nagase, S. (2005)
ChemPhysChem., **6**, 2060.

395 Wang, C.R., Kai, T., Tomiyama, T.,
Yoshida, T., Kobayashi, Y., Nishibori,
E., Takata, M., Sakata, M., and Shinohara,
H. (2001) *Angew. Chem., Int. Ed.*,
40, 397.

396 Iiduka, Y., Wakahara, T., Nakahodo, T.,
Tsuchiya, T., Sakuraba, A., Maeda, Y.,
Akasaka, T., Yoza, K., Horn, E., Kato, T.,
Liu, M.T.H., Mizorogi, N., Kobayashi, K.,
and Nagase, S. (2005) *J. Am. Chem. Soc.*,
127, 12500.

397 Nagase, S., Kobayashi, K., and Akasaka, T.
(1997) *J. Mol. Struct. (Theochem)*, **398/
399**, 221.

398 Wakahara, T., Akasaka, T., Kobayashi, K.,
and Nagase, S. (2002) *Endofullerenes – A
New Family of Carbon Clusters* (eds T.
Akasaka and S. Nagase), Kluwer
Academic Publishers, Dordrecht, The
Netherlands, p. 231.

399 Peres, T., Cao, B.P., Cui, W.D., Khong,
A., Cross, R.J., Saunders, M., and Lifshitz,
C. (2001) *Int. J. Mass Spectr.*,
210, 241.

400 Suetsuna, T., Dragoe, N., Harneit, W.,
Weidinger, A., Shimotani, H., Ito, S.,
Takagi, H., and Kitazawa, K. (2002) *Chem.
Eur. J.*, **8**, 5079.

401 Suetsuna, T., Dragoe, N., Harneit, W.,
Weidinger, A., Shimotani, H., Ito, S.,
Takagi, H., and Kitazawa, K. (2003) *Chem.
Eur. J.*, **9**, 598.

402 Murphy, T.A., Pawlik, Th., Weidinger, A., Höhne, M., Alcala, R., and Spaeth, J.-M. (1996) *Phys. Rev. Lett.*, **77**, 1075.

403 Knapp, C., Dinse, K.-P., Pietzak, B., Waiblinger, M., and Weidinger, A. (1997) *Chem. Phys. Lett.*, **272**, 433.

404 Pietzak, B., Waiblinger, M., Murphy, T.A., Weidinger, A., Höhne, M., Dietel, E., and Hirsch, A. (1997) *Chem. Phys. Lett.*, **279**, 259.

405 Cao, B.P., Peres, T., Cross, R.J., Saunders, M., and Lifshitz, C. (2001) *J. Phys. Chem. A*, **105**, 2142.

406 Kobayashi, K., Nagase, S., and Dinse, K.-P. (2003) *Chem. Phys. Lett.*, **377**, 93.

407 Wakahara, T., Matsunaga, Y., Katayama, A., Maeda, Y., Kako, M., Akasaka, T., Okamura, M., Kato, T., Choe, Y.K., Kobayashi, K., Nagase, S., Huang, H.J., and Atae, M. (2003) *Chem. Commun.*, 2940.

408 Saunders, M., Jiménez-Vázquez, H.A., Cross, R.J., and Poreda, R.J. (1993) *Science*, **259**, 1428.

409 Saunders, M., Jiménez-Vázquez, H.A., Cross, R.J., Mroczkowski, S., Freedberg, D.I., and Anet, F.A.L. (1994) *Nature*, **367**, 256.

410 Cross, R.J., Saunders, M., and Prinzbach, H. (1999) *Org. Lett.*, **1**, 1479.

411 Rubin, Y., Jarrosson, T., Wang, G.-W., Bartberger, M.D., Houk, K.N., Schick, G., Saunders, M., and Cross, R.J. (2001) *Angew. Chem., Int. Ed.*, **40**, 1543.

412 Carravetta, M., Murata, Y., Murata, M., Heinmaa, I., Stern, R., Tontcheva, A., Samoson, A., Rubin, Y., Komatsu, K., and Levitt, M.H. (2004) *J. Am. Chem. Soc.*, **126**, 4092.

413 Iwamatsu, S.-I., Uozaki, T., Kobayashi, K., Re, S., Nagase, S., and Murata, S. (2004) *J. Am. Chem. Soc.*, **126**, 2668.

414 Komatsu, K., Murata, M., and Murata, Y. (2005) *Science*, **307**, 238.

415 Slanina, Z., Uhlk, F., Adamowicz, L., and Nagase, S. (2005) *Mol. Simul.*, **31**, 801.

416 Slanina, Z. and Nagase, S. (2006) *Mol. Phys.*, **104**, 3167.

417 Raghavachari, K. (1992) *Chem. Phys. Lett.*, **195**, 221.

418 Creegan, K.M., Robbins, J.L., Robbins, W.K., Millar, J.M., Sherwood, R.D., Tindall, P.J., Cox, D.M., McCauley, J.M., Jr., Jones, D.R., Gallagher, R., and Smith, A.B., III (1992) *J. Am. Chem. Soc.*, **114**, 1103.

419 Balch, A.L., Costa, D.A., Noll, B.C., and Olmstead, M.M. (1995) *J. Am. Chem. Soc.*, **117**, 8926.

420 Winkler, K., Costa, D.A., Balch, A.L., and Fawcett, W.R. (1995) *J. Phys. Chem.*, **99**, 17431.

421 Slanina, Z., Lee, S.-L., Uhlk, F., and Adamowicz, L. (1996) *Recent Advances in the Chemistry and Physics of Fullerenes and Related Materials*, Vol. 3 (eds K.M. Kadish and R.S. Ruoff), The Electrochemical Society, Pennington, NJ, p. 911.

422 Heymann, D., Bachilo, S.M., Weisman, R.B., Cataldo, F., Fokkens, R.H., Nibbering, N.M.M., Vis, R.D., and Chibante, L.P.F. (2000) *J. Am. Chem. Soc.*, **122**, 11473.

423 Weisman, R.B., Heymann, D., and Bachilo, S.M. (2001) *J. Am. Chem. Soc.*, **123**, 9720.

424 Juha, L., Ehrenberg, B., Couris, S., Koudoumas, E., Leach, S., Hamplová, V., Pokorná, Z., Müllerová, A., and Kubát, P. (2001) *Chem. Phys. Lett.*, **335**, 539.

425 Heymann, D., Bachilo, S.M., and Weisman, R.B. (2002) *J. Am. Chem. Soc.*, **124**, 6317.

426 Slanina, Z., Stobinski, L., Tomasik, P., Lin, H.-M., and Adamowicz, L. (2003) *J. Nanosci. Nanotechnol.*, **3**, 193.

427 Slanina, Z., Uhlk, F., Juha, L., Tanabe, K., Adamowicz, L., and Ōsawa, E. (2004) *J. Mol. Struct. (Theochem)*, **684**, 129.

428 Lackes, D.J. and Gordon, R.G. (1993) *Phys. Rev. A*, **47**, 4681.

429 Kristyan, S. and Pulay, P. (1994) *Chem. Phys. Lett.*, **229**, 175.

430 Perez-Jorda, J.M. and Becke, A.D. (1995) *Chem. Phys. Lett.*, **233**, 134.

431 Meijer, E.J. and Sprik, M. (1996) *J. Chem. Phys.*, **105**, 8684.

432 Tsuzuki, S., Uchimaru, T., and Tanabe, K. (1998) *Chem. Phys. Lett.*, **287**, 202.

433 Zhang, Y.K., Pan, W., and Yang, W.T. (1997) *J. Chem. Phys.*, **107**, 7921.

434 Wesolowski, T.A., Parisel, O., Ellinger, Y., and Weber, J. (1997) *Phys. Chem. A*, **101**, 7818.

435 Patton, D.C. and Pederson, M.R. (1997) *Phys. Rev. A*, **56**, R2495.

436 Perez-Jorda, J.M., San-Fabian, E., and Perez-Jimenez, A.J. (1999) *J. Chem. Phys.*, **110**, 1916.

437 Tsuzuki, S. and Lüthi, H.P. (2001) *J. Chem. Phys.*, **114**, 3949.

438 Zhao, Y. and Truhlar, D.G. (2004) *J. Phys. Chem. A*, **108**, 6908.

439 Zhao, Y., Lynch, B.J., and Truhlar, D.G. (2005) *ChemPhysChem.*, **7**, 43.

440 Zhao, Y. and Truhlar, D.G. (2005) *J. Phys. Chem. A*, **109**, 4209.

441 Zhao, Y. and Truhlar, D.G. (2005) *Phys. Chem. Chem. Phys.*, **7**, 2701.

442 Zhao, Y. and Truhlar, D.G. (2005) *J. Phys. Chem. A*, **109**, 5656.

443 Zhao, Y. and Truhlar, D.G. (2005) *J. Chem. Theory Comput.*, **1**, 415.

444 Slanina, Z., Pulay, P., and Nagase, S. (2006) *J. Chem. Theory Comput.*, **2**, 782.

445 Guldi, D.M., and Martin, N. (eds) (2002) *Fullerenes: From Synthesis to Optoelectronic Properties*, Kluwer Academic Publishers, Dordrecht.

446 Ardavan, A., Austwick, M., Benjamin, S.C., Briggs, G.A.D., Dennis, T.J.S., Ferguson, A., Hasko, D.G., Kanai, M., Khlobystov, A.N., Lovett, B.W., Morley, G.W., Oliver, R.A., Pettifor, D.G., Porfyrakis, K., Reina, J.H., Rice, J.H., Smith, J.D., Taylor, R.A., Williams, D.A., Adelmann, C., Mariette, H., and Hamers, R.J. (2003) *Phil. Trans. Roy. Soc. London A*, **361**, 1473.

447 Chancey, C.C. and O'Brien, M.C.M. (1997) *The Jahn–Teller Effect in C_{60} and Other Icosahedral Complexes*, Princeton University Press, Princeton.

448 Bolskar, R.D., Benedetto, A.F., Husebo, L.O., Price, R.E., Jackson, E.F., Wallace, S., Wilson, L.J., and Alford, J.M. (2003) *J. Am. Chem. Soc.*, **125**, 5471.

449 Gimzewski, J.K. (1996) *The Chemical Physics of Fullerenes 10 (and 5) Years Later* (eds W. Andreoni), Kluwer Academic Publishers, Dordrecht, p. 117.

450 Harneit, W., Waiblinger, M., Meyer, C., Lips, K., and Weidinger, A. (2001) Recent advances in the chemistry and physics of fullerenes and related materials, in *Fullerenes for the New Millennium*, Vol. 11 (eds K.M. Kadish, P.V. Kamat and D. Guldi), Electrochemical Society, Pennington, NJ, p. 358.

451 Kobayashi, K., Nagase, S., Maeda, Y., Wakahara, T., and Akasaka, T. (2003) *Chem. Phys. Lett.*, **374**, 562.

452 Hiroshiba, N., Tanigaki, K., Kumashiro, R., Ohashi, H., Wakahara, T., and Akasaka, T. (2004) *Chem. Phys. Lett.*, **400**, 235.

453 Yasutake, Y., Shi, Z., Okazaki, T., Shinohara, H., and Majima, Y. (2005) *Nano Lett.*, **5**, 1057.

454 Maeda, Y., Hashimoto, M., Kaneko, S., Kanda, M., Hasegawa, T., Tsuchiya, T., Akasaka, T., Naitoh, Y., Shimizu, T., Tokumoto, H., Lu, J., and Nagase, S. (2008) *J. Mater. Chem.*, **18**, 4189.

Index

Carbon Nanotubes and Related Structures. Edited by Dirk M. Guldi and Nazario Martín
Copyright © 2010 WILEY-VCH Verlag GmbH & Co. KGaA, Weinheim
ISBN: 978-3-527-32406-4